Publisher's Note:

This facsimile edition relies on the best currently available scan of a rare book. Each page has been hand-cropped to maximize readable content, centered for readability, and sent through OCR to align the image on the page. It is understood that even after this process, a few pages contain text that is blurry or difficult to read. They are included for completeness.

VACUUM OILS as supplied to the Principal NAVIES & STEAMSHIP COMPANIES throughout the World, are stocked at the following Stations:—

HOME PORTS

- LONDON ... Vacuum Oil Co., Ltd., [illegible]
- SOUTHAMPTON [illegible]
- PLYMOUTH ... [illegible]
- *CARDIFF ... Vacuum Oil Co., Ltd., [illegible]
- *LIVERPOOL ... Vacuum Oil Co., Ltd., [illegible]
- *MANCHESTER ... Vacuum Oil Co., Ltd., [illegible]
- *HULL ... Vacuum Oil Co., Ltd., [illegible]
- IMMINGHAM ... [illegible]
- MIDDLESBROUGH Vacuum Oil Co., Ltd., [illegible]
- *WEST HARTLEPOOL Vacuum Oil Co., Ltd., [illegible]
- *NEWCASTLE ... Vacuum Oil Co., Ltd., [illegible]
- SUNDERLAND, TYNE DOCK [illegible]
- BLYTH ... [illegible]
- GLASGOW ... Vacuum Oil Co., Ltd., [illegible]
- GREENOCK ... Vacuum Oil Co., Ltd., [illegible]
- LEITH ... Vacuum Oil Co., Ltd., [illegible]
- BURNTISLAND ... [illegible]
- METHIL ... [illegible]
- *DUNDEE ... Vacuum Oil Co., Ltd., [illegible]
- *ABERDEEN ... Vacuum Oil Co., Ltd., [illegible]
- *BELFAST ... Vacuum Oil Co., Ltd., [illegible]
- *DUBLIN ... Vacuum Oil Co., Ltd., [illegible]

BALTIC, &c.

- REYKJAVIK ... [illegible]
- *TRONDHJEM ... Norsk Vacuum Oil Co. [illegible]
- [illegible]

CONTINENTAL, &c.

- [illegible] ... Vacuum Oil Co., S.A.F., [illegible]
- [illegible] ... Warehouse: Vacuum Oil Co., S.A.F., [illegible]
- [illegible] ... Vacuum Oil Co., S.A.F., [illegible]
- [illegible] ... Vacuum Oil Co., S.A.F., [illegible]
- NANTES ... Agents: Douard & Eidem [illegible]
- [illegible] ... Vacuum Oil Co., S.A.F., [illegible]
- BILBAO ... Vacuum Oil Co., S.A.E. [illegible]
- *CADIZ ... Vacuum Oil Co., S.A.E., [illegible]
- *OPORTO ... Vacuum Oil Co., [illegible]
- *LISBON ... Vacuum Oil Co. [illegible]
- *SEVILLE ... Vacuum Oil Co., S.A.E., [illegible]
- GIBRALTAR ... Gibraltar [illegible]
- *BARCELONA ... Vacuum Oil Co. S.A.E. [illegible]
- PORT ST. LOUIS DU RHONE [illegible]
- *MARSEILLES ... Vacuum Oil Co., S.A.F., [illegible]
- SAVONA ... [illegible]
- *GENOA ... Vacuum Oil Co., S.A.I. [illegible]
- LEGHORN ... [illegible]
- *NAPLES ... Vacuum Oil Co., S.A.I. [illegible]
- *VENICE ... Vacuum Oil Co., S.A.I. [illegible]
- [illegible]
- *PIRAEUS ... Vacuum Oil Co., A.G. [illegible]
- *ODESSA ... Russ. V. O. Co., Ltd., [illegible]

EASTERN

- *ALEXANDRIA ... Vacuum Oil Co., [illegible]
- ADEN ... [illegible]
- *PORT SAID ... Vacuum Oil Co., [illegible]
- *KARACHI ... [illegible]
- *BOMBAY ... Vacuum Oil Co., [illegible]
- *COLOMBO ... Vacuum Oil Co., [illegible]
- *MADRAS ... Vacuum Oil Co., [illegible]
- *CALCUTTA ... Vacuum Oil Co., [illegible]
- *RANGOON ... Vacuum Oil Co., [illegible]
- PENANG ... [illegible]
- *SINGAPORE ... Vacuum Oil Co., [illegible]
- [illegible]
- *SHANGHAI ... Vacuum Oil Co., [illegible]
- *HANKOW ... Vacuum Oil Co., [illegible]
- VLADIVOSTOCK ... [illegible]
- *KOBE ... Vacuum Oil Co., [illegible]
- *YOKOHAMA ... Vacuum Oil Co., [illegible]
- *NAGASAKI ... Vacuum Oil Co., [illegible]

COLONIAL, S. & W. Africa

- *MELBOURNE ... Vacuum Oil Co. Proprietary, Ltd., [illegible]
- NEWCASTLE, N.S.W. Vacuum Oil Co. Pty., Ltd., [illegible]
- *SYDNEY ... Vacuum Oil Co. Pty., Ltd., [illegible]
- ADELAIDE Main Office V. O. Co. Pty., Ltd., [illegible]
- PORT ADELAIDE Depot V. O. Co. Pty., Ltd., [illegible]
- BRISBANE ... V. O. Co. Pty., Ltd., [illegible]
- FREMANTLE ... Vacuum Oil Co. Pty., Ltd., [illegible]
- *WELLINGTON, N.Z. V. O. Co. Pty. Ltd., [illegible]
- *AUCKLAND, N.Z. V. O. Co. Pty., Ltd., [illegible]
- CHRISTCHURCH, N.Z. V. O. Co. Pty., Ltd., [illegible]
- DELAGOA BAY ... V. O. Co. of S. Africa, Ltd., [illegible]
- *DURBAN ... V. O. Co. of S. Africa, Ltd., [illegible]
- *EAST LONDON ... V. O. Co. of S. Africa, Ltd., [illegible]
- *PORT ELIZABETH V. O. Co. of S. Africa, Ltd. [illegible]
- *CAPE TOWN ... V. O. Co. of S. Africa, Ltd., [illegible]
- AZORES ... Vacuum Oil Co., [illegible]
- FUNCHAL ... Vacuum Oil Co.
- *LAS PALMAS ... Vacuum Oil Co.
- ST. VINCENT, C.V., [illegible]
- *SANTA CRUZ DE TENERIFFE. Vacuum Oil Co.

NORTH AMERICAN

- MONTREAL ... Imperial Oil Co., Ltd., [illegible]
- QUEBEC ... Imperial Oil Co., Ltd., [illegible]
- CHARLOTTETOWN P.E.I., Imperial Oil Co., Ltd., [illegible]
- ST. JOHN'S, Newfoundland, [illegible]
- NORTH SYDNEY, N.S. [illegible]
- HALIFAX, N.S. ... [illegible]
- ST. JOHN, N.B. ... [illegible]
- BANGOR, Me. ... Vacuum Oil Co., [illegible]
- PORTLAND, Me. ... Vacuum Oil Co., [illegible]
- *BOSTON ... Vacuum Oil Co., [illegible]
- *NEW YORK ... Vacuum Oil Co., [illegible]
- PHILADELPHIA ... Vacuum Oil Co., [illegible]
- BALTIMORE ... Standard Oil Co., [illegible]
- SPARROWS POINT Standard Oil Co., [illegible]
- CURTIS BAY, Md. ... Standard Oil Co., [illegible]
- RICHMOND, Va. ... Standard Oil Co., [illegible]
- NEWPORT NEWS, Va. ... Standard Oil Co., [illegible]
- NORFOLK, Va. ... Standard Oil Co., [illegible]
- SEWELLS POINT. ... Standard Oil Co., [illegible]
- WILMINGTON, N.C. ... Standard Oil Co., [illegible]
- CHARLESTON, S.C. ... Standard Oil Co., [illegible]
- SAVANNAH ... Standard Oil Co.
- BRUNSWICK, Ga. ... Standard Oil Co.
- JACKSONVILLE, Fla. ... Standard Oil Co.
- KEY WEST, Fla. ... Standard Oil Co.
- PORT TAMPA, Fla. ... [illegible]
- TAMPA ... [illegible]
- PENSACOLA ... Standard Oil Co.
- MOBILE, Ala. ... Standard Oil Co.
- GULFPORT, Miss. ... Standard Oil Co.
- NEW ORLEANS, La. ... Standard Oil Co. of Louisiana, [illegible]
- PORT ARTHUR, Tex. ... [illegible]
- TEXAS CITY ... [illegible]
- GALVESTON ... [illegible]
- SAN DIEGO ... Standard Oil Co., [illegible]
- SAN PEDRO, REDONDO ... Standard Oil Co., [illegible] Los Angeles.
- SAN FRANCISCO ... Standard Oil Co., [illegible]
- PORTLAND, Ore. ... Standard Oil Co., [illegible]
- ABERDEEN, Wash. ... Standard Oil Co., [illegible]
- TACOMA ... Standard Oil Co., [illegible]
- SEATTLE ... Standard Oil Co., [illegible]
- VANCOUVER ... Imperial Oil Co., Ltd., [illegible]
- VICTORIA, B.C. ... Imperial Oil Co., Ltd., [illegible]
- HONOLULU ... Standard Oil Co.

WEST INDIES & CENTRAL AMERICA

- HAVANA, Cuba ... West India Oil Co., [illegible]
- KINGSTON ... W. India Oil Co., [illegible]
- SANTO DOMINGO, S.D. West India Oil Co.
- SAN JUAN, Porto Rico, West India Oil Co.
- ST. THOMAS, D.W.I. [illegible]
- BARBADOS, B.W.I. ... West India Oil Co.
- DEMERARA, B.G. ... West India Oil Co.
- TRINIDAD, B.W.I. ... West India Oil Co., Port of Spain.
- CURACOA, D.W.I., ... [illegible]
- COLON, Panama ... [illegible]

SOUTH AMERICAN

- BAHIA ... [illegible]
- RIO JANEIRO ... [illegible]
- SANTOS ... [illegible]
- MONTEVIDEO ... [illegible]
- *BUENOS AYRES ... Vacuum Oil Co., [illegible]
- LA PLATA ... [illegible]
- ROSARIO ... [illegible]
- TALCAHUANO ... [illegible]
- ANTOFAGASTA ... [illegible]
- IQUIQUE ... [illegible]
- CALLAO ... [illegible]
- VALPARAISO ... [illegible]

Telegraphic Address "VACUUM" at all Ports marked thus*

Arrangements made for supplies to be taken on Ship's Receipt, and the Account to be rendered at home. Chief Offices for the [illegible]

Best available scan

ALPHABETICAL LIST OF ADVERTISERS.

ALPHABETICAL LIST OF ADVERTISERS.—*Continued.*

CLASSIFIED LIST OF ADVERTISERS.

PAGE

ACCUMULATORS.

General Electric Co. (*U.S.A.*) .. 92 adv.
Stewart, Duncan & Co., Ltd. .. 106
Vickers, Ltd. .. .. 19 & 20

ADVERTISEMENTS.

"Fighting Ships," 1921 Edition .. 3

AEROPLANES AND AIRSHIPS.

Armstrong, Sir W. G., Whitworth & Co., Ltd. 21
Beardmore, William & Co., Ltd. .. 24
Coventry Ordnance Works, Ltd. .. 54
De Bolotoff, Ltd. 75
Roe, A. V. & Co., Ltd. 74
Vickers, Ltd. .. .. 19 & 20
White, J. Samuel & Co., Ltd. .. 27

AEROPLANE CONTROL.

Sperry Gyroscope Co., Ltd. .. 15

AEROPLANE CORDS.

Bullivant & Co., Ltd. 139

AEROPLANE FABRICS.

North British Rubber Co., Ltd. .. 97

AIR COMPRESSORS.

Broom & Wade, Ltd. 68
De Laval Steam Turbine Co. (*U.S.A.*) 42
Escher Wyss & Co., S.A. (*Switzerland*) 43
General Electric Co. (*U.S.A.*) .. 92
Ingersoll-Rand Co. 82
Parsons, C. A. & Co., Ltd. .. 17
Reavell & Co., Ltd... 49
Seagers, Ltd. 69
Siebe, Gorman & Co., Ltd. .. 55

ALLOYS.

(see Metals, Copper, Brass, Alloys, Ingots, etc.)

ALTERNATORS.

Parsons, C. A. & Co., Ltd. .. 17

AMMETERS.

Elliott Brothers (*London*), Ltd. .. 66
Nalder Bros. & Thompson, Ltd. .. 65

PAGE

AMMUNITION (War).

Armstrong, Sir W. G., Whitworth & Co., Ltd. .. .. .. 21 adv.
Bethlehem Steel Co. (*U.S.A.*) .. 40
Bofors, A.B. (*Sweden*) 53
Brown, John & Co., Ltd. 23
Cammell Laird & Co., Ltd. .. 22
Coventry Ordnance Works, Ltd. .. 54
Rudge-Whitworth, Ltd. 139
Schneider & Cie (*France*) 36
Vickers, Ltd. .. .. 19 & 20

ANCHORS.

Bofors, A. B. (*Sweden*) 53
Fellows Bros., Ltd. 83
Hall's Patent Anchor Co., Ltd. .. 72
Hingley, N. & Sons, Ltd. 72
Mitchells, Ltd. 121
Willetts, David, Ltd. 72

ANTI-CORROSIVE COMPOSITIONS.

Briggs, W. & Sons, Ltd. 125
Crown Compositions Co. 125
Red Hand Compositions, Ltd. .. 125
Wailes Dove Bitumastic, Ltd. .. 126

ANTI-FOULING COMPOSITIONS.

Briggs, W. & Sons, Ltd. 125
Crown Compositions Co. 125
Red Hand Compositions, Ltd. .. 125
Wailes Dove Bitumastic, Ltd. .. 126

ANTI-FRICTION METALS.

Delta Metal Co., Ltd. 116
Hoyt Metal Co., Ltd. 136

ARMOUR PLATES.

Armstrong, Sir W. G., Whitworth & Co., Ltd. 21
Beardmore, William & Co., Ltd. .. 24
Bethlehem Steel Co. (*U.S.A.*) .. 40
Bofors, A.B. (*Sweden*) 53
Brown, John & Co., Ltd. 23
Cammell Laird & Co., Ltd. .. 22
Schneider & Cie (*France*) 36
Vickers, Ltd. .. .. 19 & 20

BALLOON FABRICS.

North British Rubber Co., Ltd. .. 97

BEARINGS (Ball).

Ransome & Marles Bearing Co., Ltd. 87
Rudge-Whitworth, Ltd. 139

PAGE

BEARINGS (Roller).

Ransome & Marles Bearing Co., Ltd. 87 adv.

BEARINGS (Thrust).

Broom & Wade, Ltd 68
Ransome & Marles Bearing Co., Ltd. 87

BELTING.

Dermatine Co., Ltd 108
India Rubber, Gutta Percha & Telegraph Works Co., Ltd. .. 98
Reddaway, F. & Co., Ltd. .. 140

BICYCLES.

Rudge-Whitworth, Ltd. 139

BINOCULARS.

Lawrence & Mayo, Ltd. 63

BLOWERS.

Escher Wyss & Co., S.A. (*Switzerland*) 43
Parsons, C. A. & Co., Ltd. .. 17

BOAT HOISTS.

Brown Bros. & Co., Ltd. 57

BOAT-RELEASING GEAR.

Bullivant & Co., Ltd. 139

BOILERS.

Babcock & Wilcox, Ltd. 46
Baltimore Dry Docks & Ship Building Co. (*U.S.A.*) 41
Beardmore, William & Co., Ltd. .. 24
Bethlehem Steel Co. (*U.S.A.*) .. 40
Bretagne, Ateliers et Chantiers (*France*) 37
Cammell Laird & Co., Ltd. .. 22
Delaunay Belleville (*France*) .. 37
Doxford, William & Sons, Ltd. .. 28
Forth Shipbuilding & Eng. Co., Ltd. 25
Merryweather & Sons, Ltd. .. 32
Mitchells, Ltd. 121
Pattison, C. & T.T. (*Italy*) .. 38
Schneider & Cie (*France*) 36
Smulders, A. F. Firma (Werf Gusto) (*Holland*) 14
St. Nazaire, Chantier et Ateliers (*France*) 35
Stirling Boiler Co., Ltd. 45
Thornycroft, John I. & Co., Ltd. 26

PAGE

Wallsend Slipway & Eng. Co., Ltd. 31 adv.
Werf Gusto (Firma A. F. Smulders) (*Holland*) 14
White, J. Samuel & Co., Ltd. .. 27
Yarrow & Co., Ltd... 32

BOILER COMPOSITIONS.

Red Hand Compositions Co., Ltd. 125

BOILER & ENGINE FITTINGS.

Brown, John & Co., Ltd 23
Cockburns, Ltd. 110
Eversted & Vignoles, Ltd. (*inside front cover*)
Hulburd Engineering Co., Ltd. .. 110
Metallic Valve Co. 110
Noakes, Thomas & Sons 111

BOILER SCALING TOOLS.

Gilman, Frank 71

BOILER TUBES.

(see Tubes, Boiler.)

BOLTS & NUTS.

Mitchells, Ltd. 121
Willetts, David, Ltd. 72

BOLTS & NUTS (Brass).

United Brassfounders & Engineers, Ltd. 109

BOLTS & NUTS (Bright Steel).

Talbot-Stead Tube Co., Ltd. .. 113

BORING MACHINES.

Drummond Bros., Ltd. 80
Hemp, Joshua & Co., Ltd. 77
Herbert, Alfred, Ltd. 78
Hulse & Co., Ltd. 81
Muir, Wm. & Co., Ltd. 81

BORING MILLS.

Niles-Bement-Pond Co. (*U.S.A.*) .. 79

BRASSFOUNDERS.

Auld, David & Sons, Ltd. 110
McGeoch, Wm. & Co., Ltd. .. 91
Ritchie & Whiteman, Ltd .. 121
Robinson, A. & Co., Ltd. 50
United Brassfounders & Engineers, Ltd. 109
Wallsend Slipway & Engineering Co., Ltd. 31

CLASSIFIED LIST OF ADVERTISERS.—*Continued.*

CLASSIFIED LIST OF ADVERTISERS.—*Continued.*

CLASSIFIED LIST OF ADVERTISERS.—*Continued.*

CLASSIFIED LIST OF ADVERTISERS.—*Continued.*

CLASSIFIED LIST OF ADVERTISERS.—*Continued.*

CLASSIFIED LIST OF ADVERTISERS.—*Continued.*

CLASSIFIED LIST OF ADVERTISERS.—*Continued.*

PAGE
D.
113 adv.
79 ..
112 ..
110 ..
119 ..
118 ..
44 ..
14 ..
24 ..
22 ..
112 ..
& 20 ..
21 ..
18 ..
& 20 ..
21 ..
24 ..
40 ..
53 ..
23 ..
22 ..
18 ..
115 ..
96 ..
22 ..
& 20 ..
22 ..
& 20 ..

RODS, &c.

24 ..
40 ..
116 ..

Searchlight Reflectors
STEAM TURBINES,
REDUCTION GEARING,
ALTERNATORS
Blowers, Compressors,
Condensing Plant, Etc.
Parsons & Co., Ltd
HEATON-WORKS, NEWCASTLE ON TYNE
LONDON OFFICE: 8 VICTORIA ST. S.W.1

Type C.

Type K.

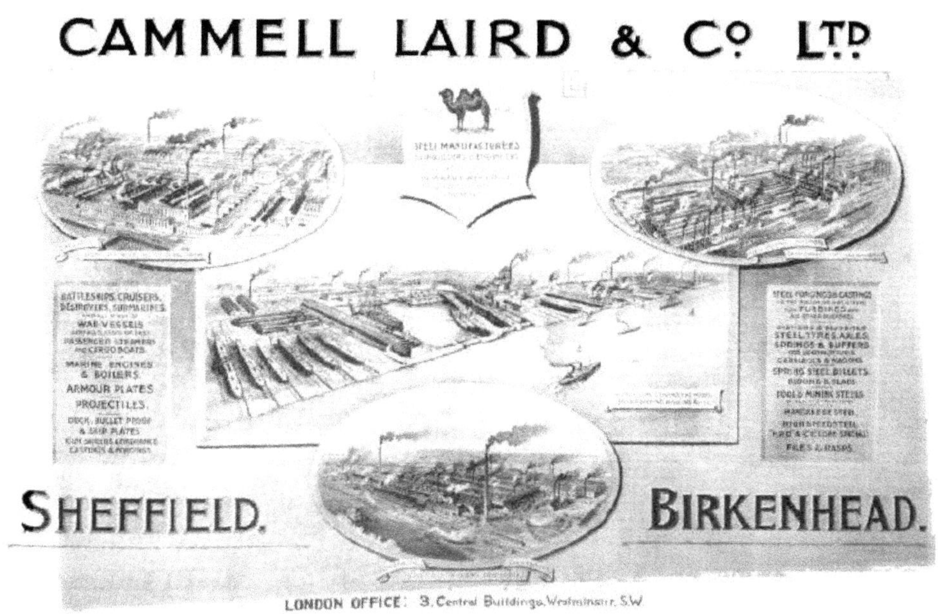
CAMMELL LAIRD & Co. LTD.
BATTLESHIPS, CRUISERS, DESTROYERS, SUBMARINES
WAR VESSELS
PASSENGER STEAMERS
CARGO BOATS
MARINE ENGINES & BOILERS
ARMOUR PLATES
PROJECTILES.
DECK, BULLET PROOF & SHIP PLATES
STEEL FORGINGS & CASTINGS
TURBINES
STEEL TYRES, AXLES
SPRINGS & BUFFERS
CARRIAGES & WAGONS
SPRING STEEL BILLETS
TOOL & MINING STEELS
HIGH SPEED STEEL
FILES & RASPS
SHEFFIELD.
BIRKENHEAD.
LONDON OFFICE: 3, Central Buildings, Westminster, S.W.

DESIGNERS & BUILDERS of WARSHIPS, CARGO & MAIL BOATS
Steam, Turbo, & Semi-Diesel
Beardmore-Tosi Diesel Engines
H.M.S. RAMILLIES
BUILT AT DALMUIR
SHEETS, SHIP & BOILER PLATES & SECTIONS, STEEL CASTINGS
STEEL SHAFTING FORGINGS & CASTINGS. LOCOMOTIVES.
WILLIAM
BEARDMORE
AND COMPANY, LIMITED.

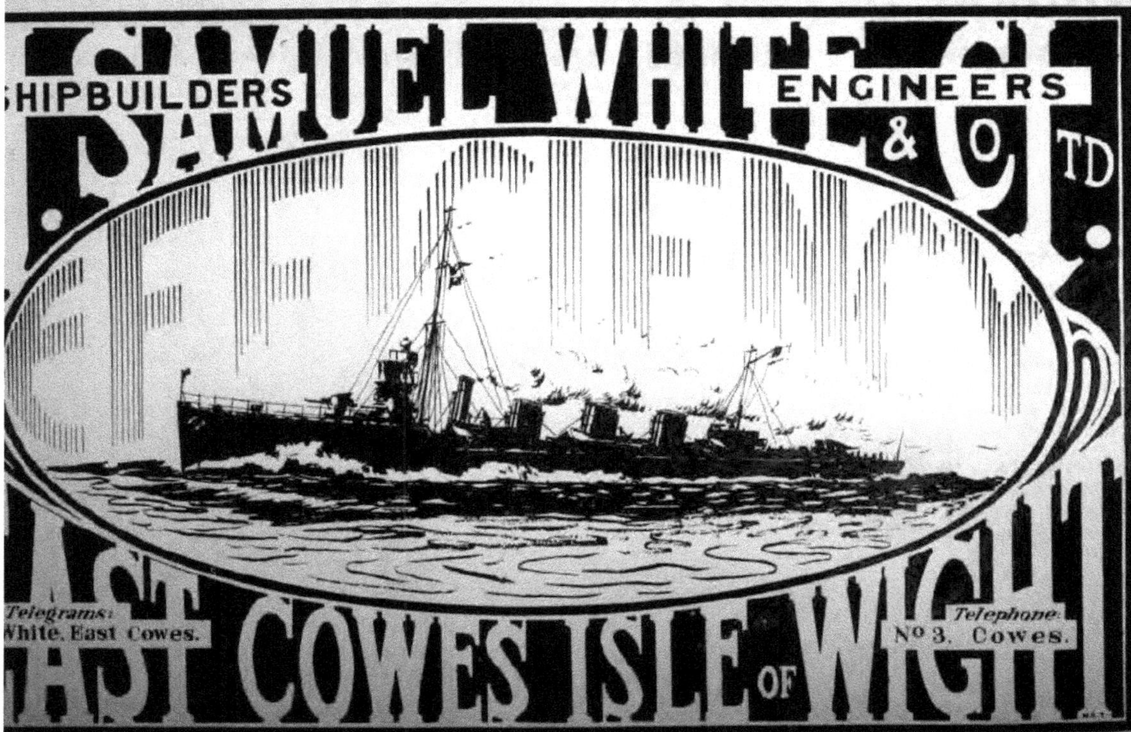
HIPBUILDERS
SAMUEL WHITE
ENGINEERS
& CO LTD
Telegrams:
White, East Cowes.
AST COWES ISLE OF WIGHT
Telephone:
No 3. Cowes.

William Doxford & Sons Ltd. Sunderland,

Telephone Nos. 412, 413, 414, 1141.

SHIPBUILDERS & ENGINEERS

Telegrams: "DOXFORD, SUNDERLAND."

BUILDERS
OF
LARGE CARGO VESSELS
AND TANKERS.

HIGH SPEED
WAR VESSELS

OIL ENGINES
OPPOSED PISTON TYPE
OF
PROVED HIGH EFFICIENCY
FOR
PROPULSION & AUXILIARIES.

OIL BURNING INSTALLATIONS
"SUPERHEAT SYSTEM."

Floating Oil Storage
on Patent Balanced Pressure
System.

Patent Electrical Oil Level
Indicator.

Demonstration 2-Stroke Marine Oil Engine
Opposed Piston Type.

FOUR-CYLINDER INSTALLATIONS

BROOKE
MARINE·MOTORS
MOTOR·BOATS
SUPER-EXCELLENCE IN CONSTRUCTION.
WRITE FOR CATALOGUE TO
J. W. BROOKE & Co., LTD.,
ADRIAN WORKS,
LOWESTOFT.
KEY-NOTE:
"RELIABILITY."

iers de St. Nazaire (Penhoët.)

CAPITAL DE 12 MILLIONS SIÈGE SOCIAL: 6 BIS RUE AUBER PARIS.

ÉT a Saint-Nazaire
.ndie a Grand Quevilly pres Rouen | France

NAVIRES DE COMMERCE.

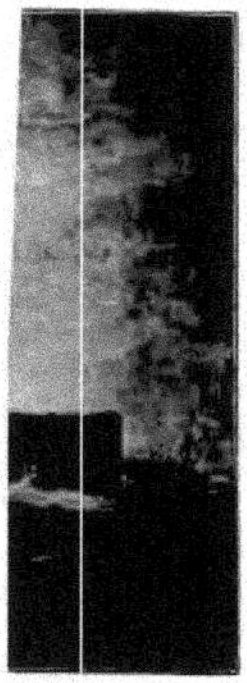

Paquebot " France "
de la Cie Générale Transatlantique.

ASSÉS
LEURS

PAQUEBOTS DE LUXE ET MIXTES,
YACHTS, CARGO-BOATS, PÉTROLIERS
REMORQUEURS.

T CHAUDIERES MARINES DE TOUS SYSTÈMES.

BETHLEHEM STEEL COMPANY
Head Office & Ordnance Works
SOUTH BETHLEHEM Pa. U.S.A.
Shipyards on Atlantic & Pacific Coasts
FIELD, COAST DEFENCE & NAVAL
ORDNANCE & AMMUNITION
from 1pdr. to 18"
BATTLESHIPS, CRUISERS,
DESTROYERS & SUBMARINES.
ARMOUR PLATE
European Office –
25 Victoria St
LONDON, S.W.
BETHLEHEM STEEL Co

DREDGING PLANT.
COALING VESSELS, COALING ELEVATORS (capacity 1000-tons per hour).
PASSENGER STEAMERS, CARGO VESSELS, TANK STEAMERS, TUGS, etc.
BOILERS, Land and Marine.
BRIDGE, ROOF & CONSTRUCTIONAL STEEL WORK of every description.
3 Positions of a 150 TON Floating Derricking and Revolving CRANE
WERF GUSTO
Firma A.F. Smulders
SHIPBUILDERS & ENGINEERS,
SCHIEDAM, HOLLAND.
Anderson Rodger, 38, Victoria Street, London, S.W.1.

The
"Robinson"
Marine Superheater
ADOPTED BY
THE PRINCIPAL
STEAMSHIP COMPANIES
ORDERS IN HAND for SUPERHEATERS for
GEARED TURBINE INSTALLATIONS of an
AGGREGATE EXCEEDING
150,000. S.H.P.
WRITE FOR FULL
PARTICULARS
THE SUPERHEATER CORPORATION LTD.
14 New Burlington Street, London, W.1.
Telephone Gerrard 3822

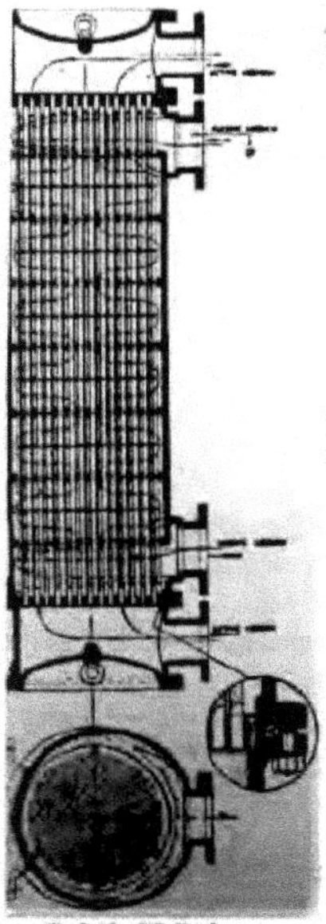

S. & K. Oil Cooler.

Section Through "Koerting" Burner and Air Register.

Koerting Duplex Oil Strainer.

SCHUTTE SHIP MANŒUVRING VALVE.

This Type of Valve is used for Marine Purposes by all the Leading Turbine Builders of the United States.

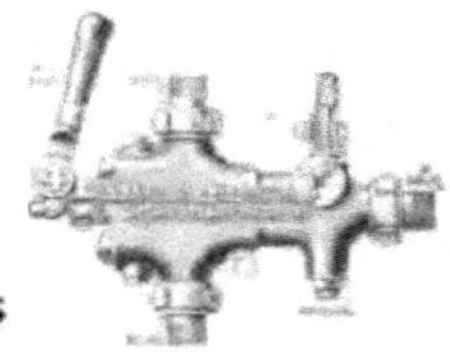

KOERTING INJECTOR.

S. & K. EVAPORATOR.

AIR COMPRESSORS FOR

DIESEL ENGINE SHIPS.

"V" Class Auxiliary Compressor, driven by Semi-Diesel Engine.

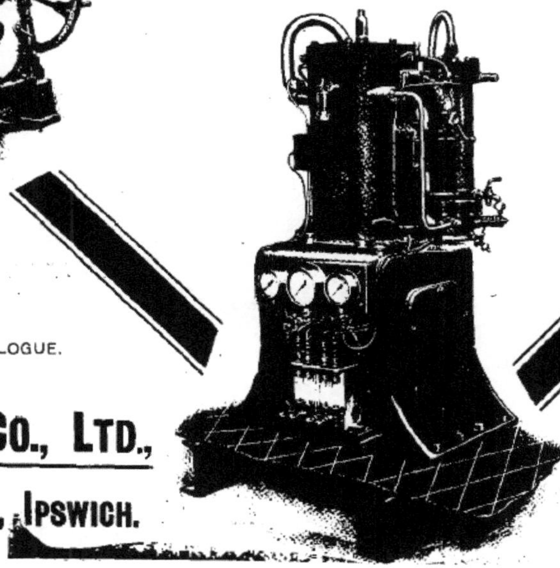

3-Stage Vertical Compressor as used on Diesel Submarine Engine.

Motor Driven Auxiliary C
3-Stage Quadruplex

WRITE TO
DEPT. "G."
FOR DESCRIPTIVE CATALOGUE.

REAVELL & CO., LTD.,

RANELAGH WORKS, IPSWICH.

Telephone:—324 and 325.
Telegrams:—"Reavell, Ipswich."

MAKERS OF MA
AND AUXILIAR
COMPRESSORS FC
DIESEL ENGINE
AND ALL OTHER PURPOS

40 adv.

EN, SONS & Co. LTD., BEDFORD

Set.

Steam Turbine Driven Generating Set.

Steam Engine Driven Boiler Room Fan.

Steam Engine Driven Generating Set.

p.

MAKERS OF

AUXILIARY MACHINERY,

AS SUPPLIED TO

BRITISH ADMIRALTY & FOREIGN NAVIES.

CRUISER, 27,000 TONS.
FITTED WITH MARCONI 25 K.W. INSTALLATION.
1½ KILOWATT CABINET.
WIRELESS TELEGRAPH Co. Ltd.
GRAPH & TELEPHONE INSTALLATIONS,
CTION FINDERS ETC. FOR USE AT SEA,
ON LAND OR IN THE AIR.
HOUSE, STRAND, LONDON, W.C.2. Telephone: CITY 8710
RAND, LONDON. Cablegrams: EXPANSE, LONDON.

ATCH" Nautical Binocular.

(designed for Marine use at night.)

inches. Magnifying power, 3 diameters.

-, 2 inches. Field of View at 1,000 yds., 180 yds.

The "Dog-Watch" Binocular has been specially designed with a view to making it as efficient as possible for use at sea under *unfavourable* conditions of *light* and *atmosphere*. Having the moderately low magnifying power of 3 diameters it will be found of great utility when the conditions are not favourable and will clearly define objects when glasses of higher power will fail to do so. The object lenses are extra large, thus admitting the greatest possible amount of light, and giving an extraordinarily brilliant image.

The "Dog-Watch," therefore, is pre-eminently suitable for night use and in unfavourable weather, and in use under these conditions, it has been pronounced by Pilots and Naval Men generally to be unsurpassed.

The body of the "Dog-Watch" is constructed of hard metal specially designed to withstand rough usage. The eye-pieces are curved to the shape of the eye, to obscure the side light, and fixed sun and spray shades are fitted to the object lenses. Loops are provided at the sides for sling or lanyard.

Captain W. Jolliffe, S.Y. "Queen Mab," writing for a Binocular suitable for night use said, "I have got several glasses on my Ship but not one of them is the slightest use at night."

A "Dog-Watch" Binocular having been sent to him we received the following letter: "I enclose cheque for 'Dog-Watch.' I consider it a very good glass for night work."

in Solid Leather Case with Shoulder Straps, £4 15s.

Code Word for Telegrams—"Vibrator"

HLIGHT" Naval Telescope.

LARGELY USED IN THE BRITISH

VY AND MERCANTILE MARINE.

The **"Flashlight"** Telescope is eminently suitable for naval purposes, the body and tube being constructed of hard drawn German Silver and, although light in weight, is exceedingly strong and is able to withstand a great amount of hard wear and rough handling.

The lenses are of the finest quality, giving a remarkably brilliant and well-defined image and a wide field of view.

For Yachting, general Marine and look-out purposes, the "Flashlight" Telescope can be strongly recommended.

and fitted with a Strap and solid leather protecting cap, as illustrated, to protect object lens when not ... Magnification 18 diameters, length when closed 17½ inches; when extended to full length 22½ inches.

Code Word for Telegrams—"Vulcanise."

NCE & MAYO, Ltd.,

)pticians, Admiralty Chart Agents for Eastern Ports,

19, NEW OXFORD STREET, LONDON.

LEXANDRIA, DELHI, BOMBAY, CALCUTTA, MADRAS, LUCKNOW, SIMLA and RANGOON.

Code Address for Telegrams—"Stockade, London."

'A1' AT KEW.

Highest Award Class "A" Kew Certificate.

Highest Award Class "A" Kew Certificate.

The "White Star" Sextant

(Latest Model)

Specification.—8 in. Special design, extra rigid club pattern, bridge handle Sextant with Spring tangent, divided on Silver to 10 seconds, best large fixed Horizon with Capps adjustment **and improved horizontal adjustment by means of a key, thus avoiding all strain on the glass.** Seven neutral shades, Blank, Prospect, and Inverting Telescope with extra power and fitted with outside object cells, also Star Binocular, two dark ... refractor to Vernier, fitted in polished Mahogany case with flush handle, lock and key, and fastening hooks. Kew "A" Certificate. Highest Class Workmanship throughout.

Price, **£27 10s. 0d.** Code Word for Telegrams—"Star ..."

The Pneumercator Tank Gauge
This Instrument is invaluable for measuring or weighing
OIL FUEL IN SHIPS' TANKS.
It has been supplied to all the principal oil burning vessels including the
"AQUITANIA."
WRITE FOR PARTICULARS.
MEASURES AND WEIGHS ALL LIQUIDS.
"KELVITE" Compasses
AND SOUNDING MACHINES.
THE "K.B.B." NAVAL COMPASS.
Supplied to all the latest War Vessels.
PRINCIPAL FEATURES.
Horizontal pointer instead of lubber line.
Coarse Magnifying prism.
Special Corrector magnets in adjustable carriers.
Gimbals supported on rollers.
THE "KELVITE" MARK IV MOTOR DRIVEN SOUNDING MACHINE.
For Sounding from the Bridge. Length of line run out indicated on top of machine. Speed and depth tables, prepared for the purpose, show the equivalent vertical depth according to speed of ship. Depth is known immediately the lead touches bottom.
The "K.B.B." NAVAL COMPASS.
"KELVITE" MARK IV MOTOR-DRIVEN SOUNDING-MACHINE.
KELVIN BOTTOMLEY & BAIRD LTD
MANUFACTURERS OF THE LATE LORD KELVIN'S INVENTIONS AND OTHER INSTRUMENTS
KBB
18 CAMBRIDGE ST GLASGOW

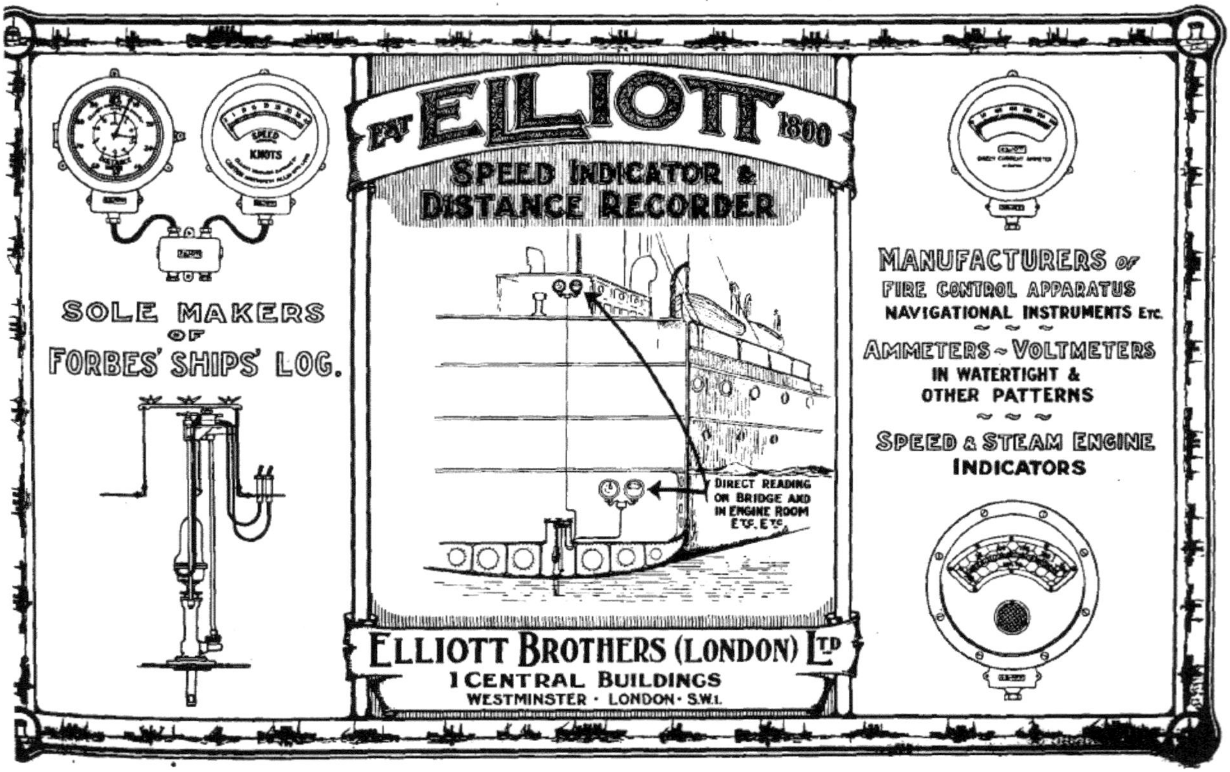
EST ELLIOTT 1800
SPEED INDICATOR & DISTANCE RECORDER
DIRECT READING ON BRIDGE AND IN ENGINE ROOM ETC. ETC.
ELLIOTT BROTHERS (LONDON) LTD
1 CENTRAL BUILDINGS
WESTMINSTER · LONDON · S.W.1.
SOLE MAKERS OF FORBES' SHIPS' LOG.
SPEED
KNOTS
MANUFACTURERS OF
FIRE CONTROL APPARATUS
NAVIGATIONAL INSTRUMENTS ETC.
AMMETERS ~ VOLTMETERS
IN WATERTIGHT & OTHER PATTERNS
SPEED & STEAM ENGINE INDICATORS

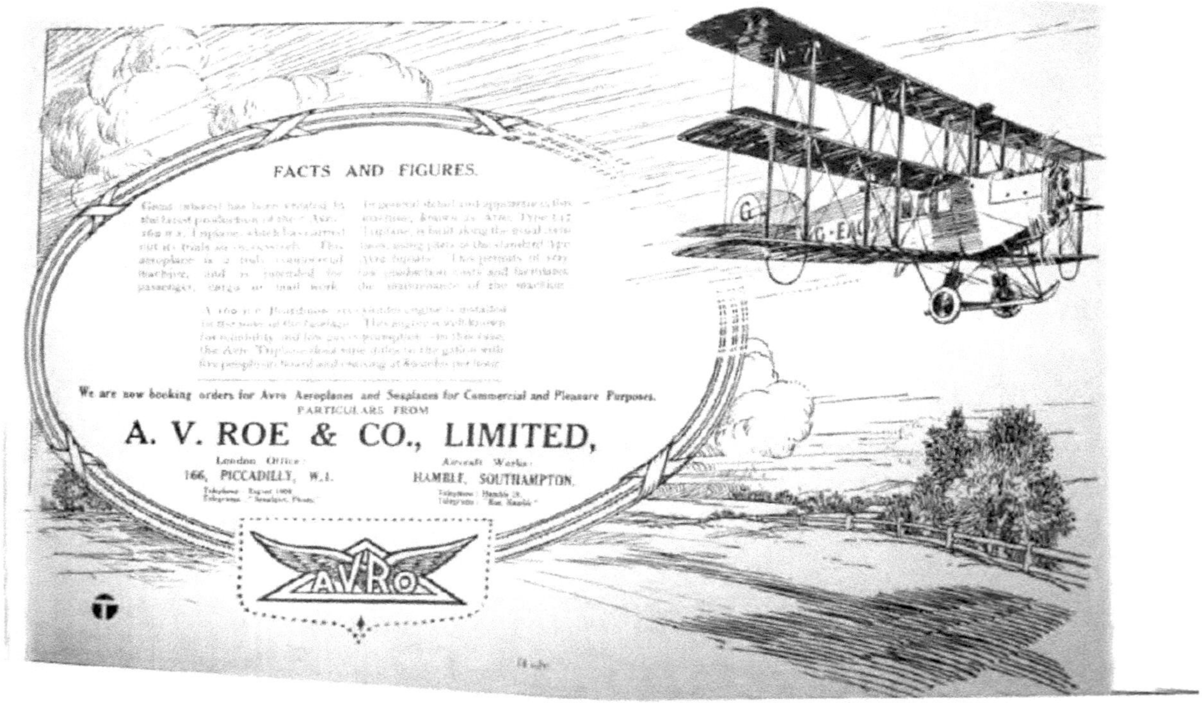
FACTS AND FIGURES.
We are now booking orders for Avro Aeroplanes and Seaplanes for Commercial and Pleasure Purposes.
PARTICULARS FROM
A. V. ROE & CO., LIMITED,
London Office:
166, PICCADILLY, W.1.
Aircraft Works:
HAMBLE, SOUTHAMPTON.
AVRO

The Globe Pneumatic Engineering Co. Ltd
"GLOBE"
CHIPPING, CAULKING and RIVETING HAMMERS.
1. Victoria St, Westminster London, S.W.1.

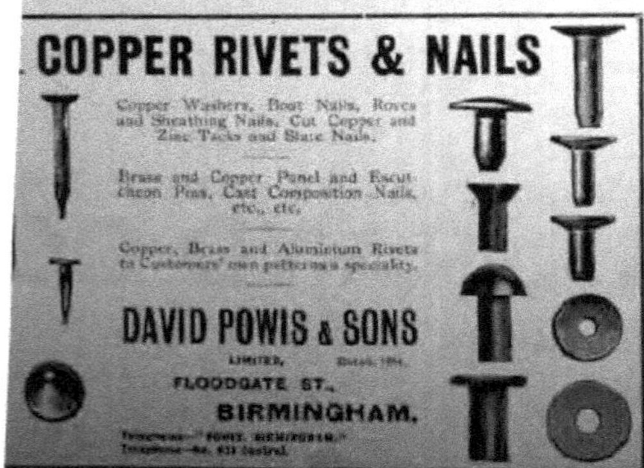

COPPER RIVETS & NAILS
Copper Washers, Boat Nails, Roves and Sheathing Nails, Cut Copper and Zinc Tacks and Slate Nails.
Brass and Copper Panel and Escutcheon Pins, Cast Composition Nails, etc., etc.
Copper, Brass and Aluminium Rivets to Customers' own patterns a specialty.
DAVID POWIS & SONS
LIMITED,
FLOODGATE ST.,
BIRMINGHAM.

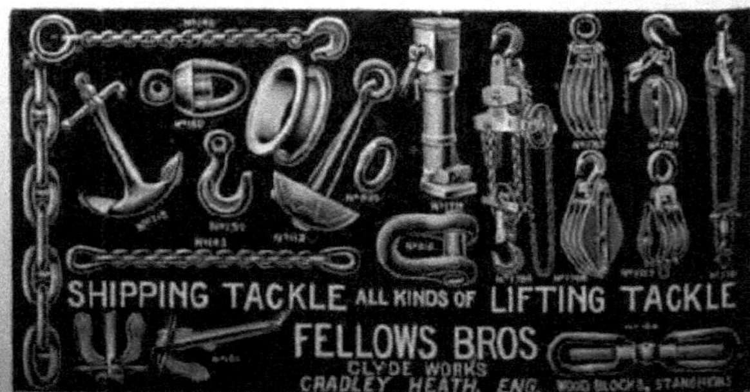

SHIPPING TACKLE ALL KINDS OF LIFTING TACKLE
FELLOWS BROS
CLYDE WORKS
CRADLEY HEATH, ENG.
WOOD BLOCKS, STANCHIONS

The Electric Ship!
GENERAL OFFICE: SCHENECTADY, N.Y.
GENERAL ELECTRIC COMPANY.
SALES OFFICES IN

The Hall Mark for Quality.

The Hall Mark for Quality.

150 K.W. Curtis Geared Turbo-Generator Set.

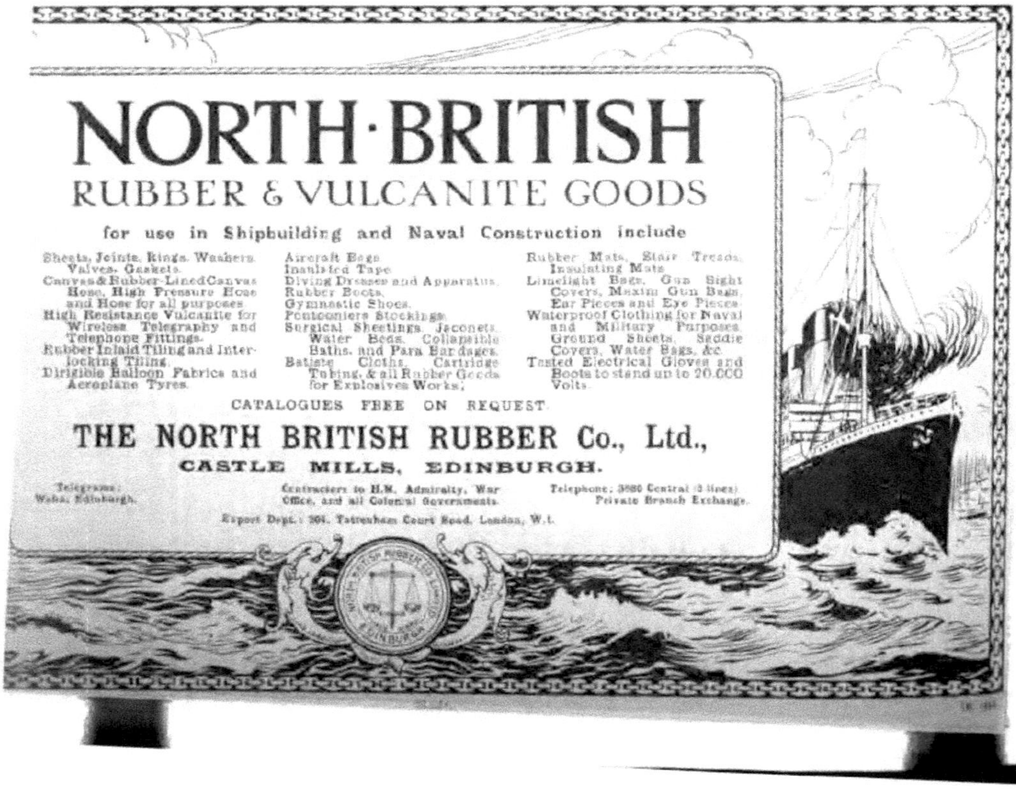

NORTH·BRITISH
RUBBER & VULCANITE GOODS
for use in Shipbuilding and Naval Construction include
Sheets, Joints, Rings, Washers, Valves, Gaskets.
Canvas & Rubber-Lined Canvas Hose, High Pressure Hose and Hose for all purposes.
High Resistance Vulcanite for Wireless Telegraphy and Telephone Fittings.
Rubber Inlaid Tiling and Interlocking Tiling.
Dirigible Balloon Fabrics and Aeroplane Tyres.
Aircraft Bags.
Insulated Tape.
Diving Dresses and Apparatus.
Rubber Boots.
Gymnastic Shoes.
Pontooniers Stockings.
Surgical Sheetings, Jaconets, Water Beds, Collapsible Baths, and Para Bandages.
Batiste Cloths, Cartridge Tubing, & all Rubber Goods for Explosives Works.
Rubber Mats, Stair Treads, Insulating Mats.
Limelight Bags, Gun Sight Covers, Maxim Gun Bags, Ear Pieces and Eye Pieces.
Waterproof Clothing for Naval and Military Purposes, Ground Sheets, Saddle Covers, Water Bags, &c.
Tested Electrical Gloves and Boots to stand up to 20,000 Volts.
CATALOGUES FREE ON REQUEST.
THE NORTH BRITISH RUBBER Co., Ltd.,
CASTLE MILLS, EDINBURGH.
Telegrams: Waba, Edinburgh.
Contractors to H.M. Admiralty, War Office, and all Colonial Governments.
Telephone: 3880 Central (3 lines) Private Branch Exchange.
Export Dept.: 204, Tottenham Court Road, London, W.1.

Rubber Tiling
An artistic and noiseless floor covering of extreme durability, introduced by The Silvertown Company 30 years ago, and laid by them on over 200 Liners, Battleships and Yachts.
Ebonite Cells
The Silvertown Company are prepared to undertake the manufacture of Accumulator and Battery Boxes of any size and shape.
Silvertown Ebonite was used by the National Physical Laboratory in conducting the exhaustive tests on which the Admiralty specification is based.
Cables
made at Silvertown include Standard Rubber Insulated Cables and Wires for Lighting and Power on Ships of all classes, and also Cables with various outside mechanical protections for Shipyard Work.
50 years' manufacturing experience.
Please write for List No. 40.
Ship Lighting Sets
Prices on application for Steam, Paraffin or Petrol Engines and Dynamos.
Enquiries invited also for Motors, Balancers, Boosters, Switchboards, and other Electrical products.
BRANCHES
HOME.
Silvertown
The India Rubber, Gutta Percha and Telegraph Works Co., Ltd.
(The Silvertown Company).
HEAD OFFICE: 106, Cannon St., London, E.C.4.
TELEGRAMS: INLAND: "Silvergray, Cannon, London." FOREIGN: "Silvergray, London."
WORKS: Silvertown, London, E.16.
INLAND: "Graysilver, Silver, London." FOREIGN: "Graysilver, London."
Codes used: Marconi International, A1 and A.B.C. 6th Edition.
BRANCHES
ABROAD.
Melbourne—21/23, Elizabeth Street.
Sydney—279, George Street.
France—Office:
31, Rue la Boetie, PARIS.
Works:
PERSAN
(Seine et Oise).

HINX-
ENGLAND'S MOST DURABLE PLUGS.
SPHINX
Admiralty Pattern.
Adopted & standardised by the Admiralty, also by the Ministry of Munitions for use on all Transport vehicles
Telegrams.
"Sphinx Birmingham"
Telephone Mid. 1742

THOS. & WM. SMITH LTD.,

Newcastle-on-Tyne.

ESTABLISHED 1782.

MAKERS OF
STEEL WIRE ROPES
FOR ALL PURPOSES
IN SHIPYARDS
AND ON BOARD SHIPS
OF ALL CLASSES.

LAUNCHING CHECKS
OF HIGH
TENSILE STRESS AND
FLEXIBILITY
A SPECIALITY.

TELEGRAPHIC AND CABLE ADDRESS:
"SMITHS, Newcastle-on-Tyne."

250-TON STEAM FLOATING CRANE.

Crane designed and constructed by Cowans, Sheldon & Co. Ld.
Pontoon designed and constructed by Sir W. G. Armstrong, Whitworth & Co. Ld.
Fitted with 5,865 feet of Thos. & Wm. Smith's High-class Crane Rope varying from 2½ in. to 5½ in. circ.

1½" × 8' Plate Shears.

Y & SCOTT, LIMITED,

to His Majesty's Army, Navy, Indian & Colonial Forces.

UTOMATIC PISTOLS, and all kinds of HIGH-CLASS SPORTING GUNS AND RIFLES.

PISTOLS.

Calibre.

Revolver.

"IMPERIAL PREMIER."

The above is an engraving of W. & C. SCOTT & SONS' highest grade Gun, which is the finest weapon in the world

Webley & Scott Signal Pistol.

As supplied to His Majesty's Navy and Army, French Russian and other Governments for Signalling Purposes

The chief features of the WEBLEY & SCOTT MILITARY AUTOMATIC PISTOLS are:—

SIMPLICITY OF CONSTRUCTION. STRENGTH OF MECHANISM

LIGHTNESS and PORTABILITY

"Proprietary" Hammerless Ejector Gun

Specially designed for Colonial use.

The above Gun has been specially designed to meet the requirements of those sportsmen, who, whilst desiring a thoroughly reliable up-to-date weapon, cannot pay the high prices hitherto commanded by the English Guns of this quality. Being made on the interchangeable principle, it will be found most suitable for use in the Colonies and up-country districts, where, in case of a break-down there are no facilities for repairs, as duplicate limbs can always be supplied when desired.

·455 High Velocity AUTOMATIC PISTOL

(As Adopted by the Admiralty).

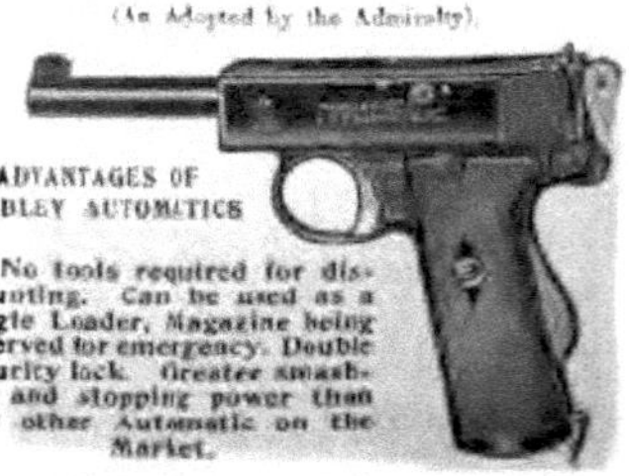

ADVANTAGES OF WEBLEY AUTOMATICS

No tools required for dismounting. Can be used as a Single Loader, Magazine being reserved for emergency. Double security lock. Greater smashing and stopping power than any other Automatic on the Market.

WORKS: 81 to 91, Weaman St. & 8 to 22, Slaney St.; OFFICES & SHOWROOMS: 86, Weaman St., BIRMINGHAM.

Telegrams: "WEBLEY, BIRMINGHAM." LONDON DEPOT: 78, Shaftesbury Avenue, W.C.

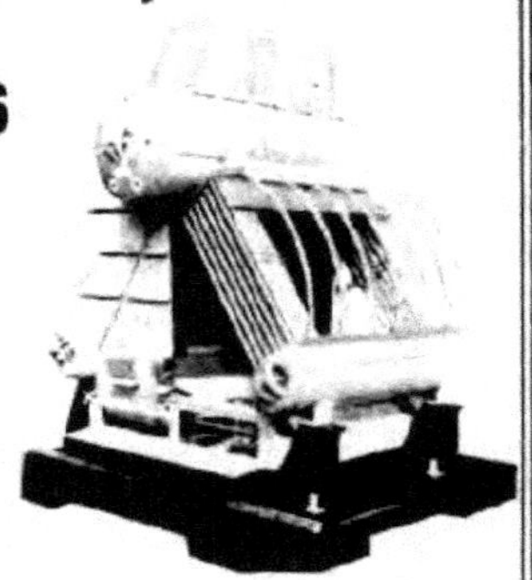

o., Ltd., Manchester.
id Compressed Hydraulic Forged
AND BRONZE
BES
nd other purposes.
Electro-Coppering and Alloys.

.RK GATE IRON & STEEL CO.

LIMITED,

ROTHERHAM, ENGLAND.

ıfacturers of High-class Steel by Siemens' Process only.

STEEL PLATES,

ARINE, LANCASHIRE, LOCOMOTIVE, LAND, AND OTHER BOILERS.

TANK PLATES, SHEETS, & CHEQUERED PLATES to Admiralty Pattern

IED TO THE BRITISH ADMIRALTY FOR MARINE AND OTHER BOILERS.

H TENSILE STEEL (in Plates, Sheets & Bars),

FOR BATTLESHIPS, TORPEDO BOATS, AND DESTROYERS.

S in rounds, squares, flats, angles, tees & sectional bars of all descriptions.

LDING AND CASEHARDENING STEEL.

---

to British Admiralty, War Office, Board of Trade, Lloyds, English & Foreign Railways.

Quasi-Arc Electric Welding.

The Quasi-Arc Process of Electric Welding is exclusively used in all the Royal Dockyards, is officially approved by Lloyd's Register of Shipping and The British Corporation for the Survey and Registry of Shipping to be used by Shipbuilders, and The Quasi-Arc electrode is the only electrode used in the entire construction of ships built under Admiralty Survey, and to the classification and Survey of Lloyd's Register of Shipping.

Rivetless Sea-going Lighter.

This Sea-going Lighter A.C. 1320, built under Admiralty Survey and launched in July, 1918, was welded throughout by The Quasi-Arc Process of Electric Welding.

Length, B.P.	126' 0"
Breadth, Mld.	16' 4"
Depth, Mld.	7' 6"
Load Displacement on 5' 0" Draft,	310 tons.
Deadweight	240 tons.
Block coefficient	·92.

Watertight bulkheads are fitted forward and aft, and a non-watertight bulkhead amidships.

Fractured Cast Steel Stern-post of the Super-Dreadnought Ramillies repaired by The Quasi-Arc Process of Electric Welding.

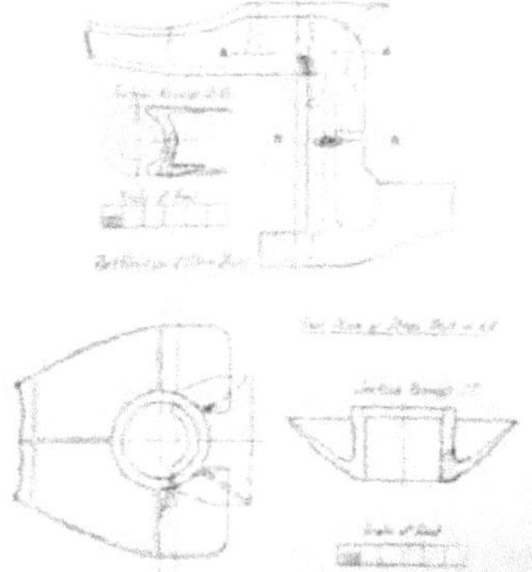

A description of the Repair to H.M.S. "Ramillies" Stern-post and other important repairs by "The Quasi-Arc Process," is given in the paper "Some Experiences with Electric Welding in Warships," by W. H. Gard, Esq., C.B., M.V.O., Assistant Director of Naval Construction to the British Admiralty, read at the Institution of Naval Architects, April, 1918.

Rivetless Coasting Vessel.

The Quasi-Arc welded coaster "Fullagar."

This ship, built under Lloyds Survey by Messrs. Cammell Laird & Co., Ltd., Birkenhead and launched Feb. 5th, 1920, was welded throughout by The Quasi-Arc Process of Electric Welding.

Length B.P. ...	150' 0"	Deadweight	500 tons.
Breadth Mld.	23' 9"	Hold Capacity	26000 C.F.
Depth Mld.	11' 6"	Gross Tonnage	425.
" to Qr. Dk.	15' 6"	Speed	10 Knots.

THE QUASI-ARC COMPANY, Ltd., 3, Laurence Pountney Hill, Cannon Street, LONDON, England.

Branches & Agencies. Glasgow, Birmingham, Newcastle, Paris, Marseilles, Genoa, Lisbon, Madrid, Rotterdam, Brussels, New York, Montreal, Johannesburg, Melbourne, Sydney, Adelaide, Copenhagen, Bombay, Calcutta, Rangoon, Osaka.

WORKS : MILE END, LONDON, E.3.

Premier
THE WORLD'S MOST EFFICIENT WELDING PLANT
FOR Ships' Construction AND REPAIRS TO Boilers, Forgings, Steel Castings Machinery &c. &c.
British Patent No 113,293.
Full particulars from THE PREMIER ELECTRIC WELDING CO. LTD, Bank Buildings, St James's Str! LONDON, S.W.1.
ALSO AT Belfast Edinburgh Glasgow Liverpool Newcastle Swansea
SIMPLE - FOOLPROOF - RELIABLE
ARC WELDER

No. 28 Contesse.

DICK'S MARINE OILS
Dick's
Marine Engine Oils
Compound Marine Engine Oils,
Marine Cylinder Oils,
Superheated Steam Oils,
Turbine Oils, Diesel Oils.
SUPPLIED TO BRITISH AND FOREIGN ADMIRALTIES AND LEADING STEAMSHIP COMPANIES IN THE WORLD.
W. B. DICK & CO., LTD.,
90, Fenchurch St., London, E.C.3.

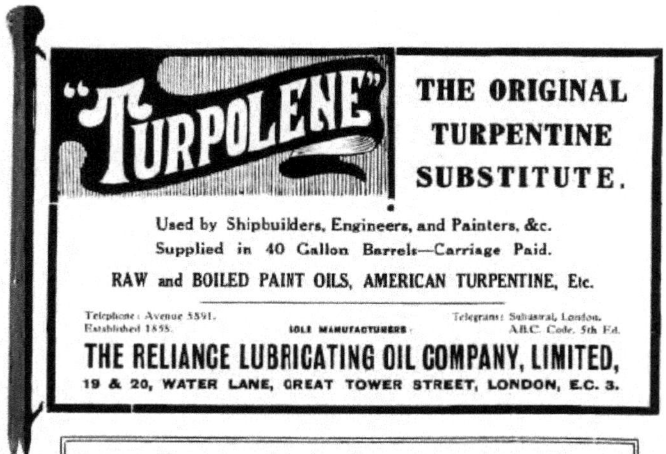
"TURPOLENE"
THE ORIGINAL TURPENTINE SUBSTITUTE.
Used by Shipbuilders, Engineers, and Painters, &c.
Supplied in 40 Gallon Barrels—Carriage Paid.
RAW and BOILED PAINT OILS, AMERICAN TURPENTINE, Etc.
Telephone: Avenue 5591.
Established 1858.
SOLE MANUFACTURERS:
Telegrams: Subastral, London.
A.B.C. Code, 5th Ed.
THE RELIANCE LUBRICATING OIL COMPANY, LIMITED,
19 & 20, WATER LANE, GREAT TOWER STREET, LONDON, E.C. 3.

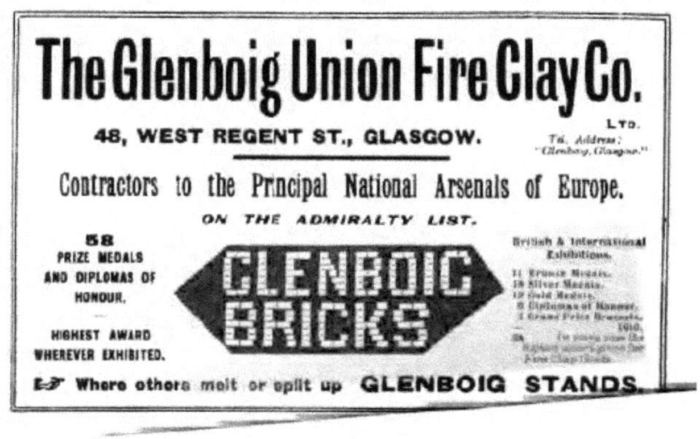
The Glenboig Union Fire Clay Co.
LTD.
48, WEST REGENT ST., GLASGOW.
Tel. Address: "Glenboig, Glasgow."
Contractors to the Principal National Arsenals of Europe.
ON THE ADMIRALTY LIST.
58 PRIZE MEDALS AND DIPLOMAS OF HONOUR.
HIGHEST AWARD WHEREVER EXHIBITED.
GLENBOIG BRICKS
British & International Exhibitions.
Where others melt or split up GLENBOIG STANDS.

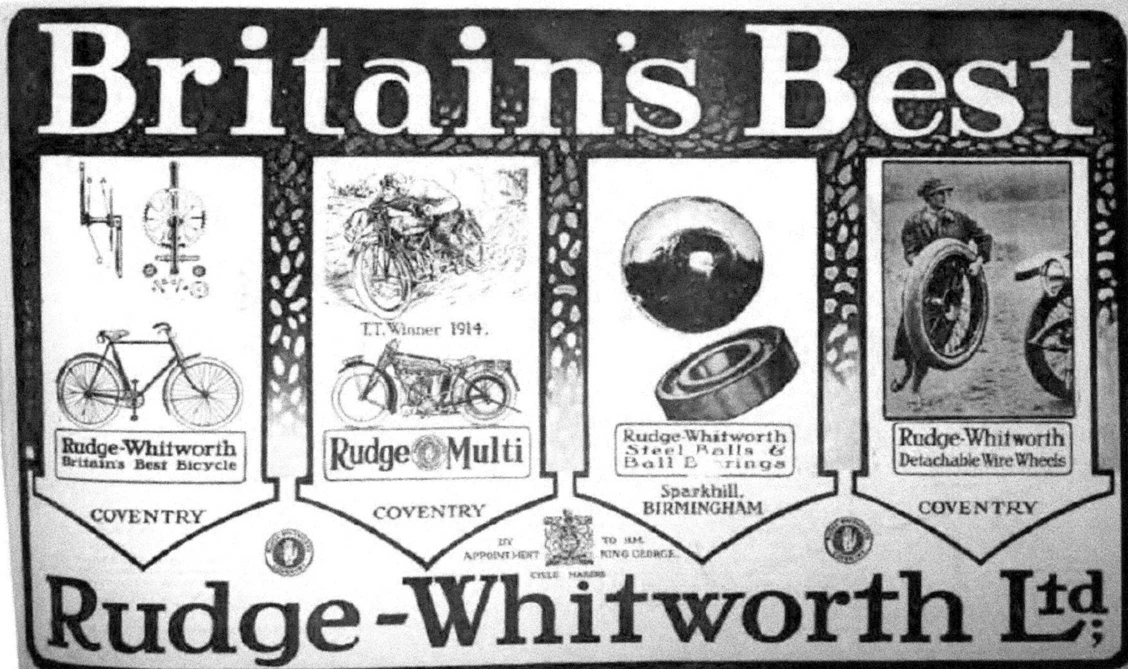
Britain's Best
T.T. Winner 1914.
Rudge-Whitworth
Britain's Best Bicycle
COVENTRY
Rudge Multi
COVENTRY
Rudge-Whitworth
Steel Balls &
Sparkhill,
BIRMINGHAM
Rudge-Whitworth
Detachable Wire Wheels
COVENTRY
By Appointment to H.M. King George
Cycle Makers
Rudge-Whitworth Ltd;

Contractors to His Majesty's Naval, Military and Civil Departments.

The United Flexible Metallic Tubing Co., Limited.

Patentees and Manufacturers of

Flexible Metallic Tubing

and Connections; Brass and Ironfounders.

Ponders End, Middlesex.

Registered Offices and Showrooms:

112, Queen Victoria Street, LONDON, E.C. 4.

Telegrams: "TUYAUX," CENTRAL, LONDON.
Telephone: BANK 773.

Sight-Setter's Headpiece with Flexible Metallic Tubing

[illegible] of carrying Oil to a Cruiser in Port by means of Flexible Metallic Tubing.
Oil Fuel is also carried out to vessels at sea in a similar way.

Flexible Metallic Tubing has become a necessity, and it is now employed in thousands of cases where rubber, which becomes charred and perishes at a comparatively low steam temperature, is absolutely useless. Oils of all description have no deterrent effect on Flexible Metallic Tubing.

In tropical climates **Flexible Metallic Tubing** has become indispensable, the action of the burning sun's rays being no detriment whatever to the tubing: moreover the dry scorching heat so deadly to rubber hose has no effect on this manufacture, whether "in stock" or in use.

The Company are Contractors to the Admiralty, and supply large quantities of various sizes of Tubing up to 12 inch interior diameter for innumerable marine purposes. Amongst others, Fuelling Warships at sea and in port, Voice Tubing (Fire Control), Charging Torpedoes, Boiler Sweeping, etc., etc., and specially for highest steam pressures. Indestructible by action of sea water.

ARSHIP TYPE.

From a drawing specially prepared for "Fighting Ships" by Oscar Parkes, Esq., M.B.

JANE'S FIGHTING SHIPS
1920

FOUNDED IN **1896** BY FRED T. JANE.

XXIII YEAR OF ISSUE.

JOINT EDITORS:
OSCAR PARKES, M.B. AND
MAURICE PRENDERGAST.

SAMPSON LOW, MARSTON
& CO., LTD: LONDON AND
EDINBURGH, MCMXX.

PREFACE.

IT was hoped that the 1919 Edition of "Fighting Ships" would be the last issue to suffer from the disabilities arising from the late War. This expectation has, unfortunately, not been realised, for the preparation of the present volume has been seriously impeded by four important problems, created by the Treaties of Peace. The four questions, of which solutions were required before the editorial work could be completed, can be summed up thus:—

(a) The surrender of ex-Enemy Warships and their partition among the Allied and United States Navies;

(b) the selection of various ex-German and ex-Austrian Warships for incorporation in the French and Royal Italian Navies;

(c) the nomination of Warships chosen for the German Post-War Fleet, in accordance with the Naval Clauses of the Treaty of Versailles; and

(d) the disposal of unfinished German naval vessels.

It was not until September and October, 1920, that definite and satisfactory replies were received to the questions concerning the Austrian and German Navies, raised under the above four headings. Inquiries were pursued in every possible quarter, in order to define the future status of the Austro-Hungarian Danube Flotilla and Turkish Navy. Our investigations met with little success, and it was decided that, lacking accurate information, the pages describing the Danube craft and Ottoman Navy should be repeated again at the end of this volume. However, at the last moment before going to Press, the desired information was communicated, and has been incorporated in a Late Addenda page. In the next Edition, the Austro-Hungarian Danube Flotilla will, of course, be distributed among the appropriate Navy Sections, while the Turkish Navy will be inserted again among the Minor European Navies

To the British, French and Italian Navy Sections, full-page Tables have been added, detailing the ex-Enemy Warships received by each country for "Propaganda and Experimental purposes." Similar paragraphs have been added to the United States and Japanese Navy Sections. The ships so distributed are not to be added to the Allied and United States Navies, but are to be broken up or destroyed under experimental tests. These vessels do not constitute an acquisition of strength to the Navies of the recipient Powers; they are for all practical purposes condemned vessels. Accordingly, we have deleted all illustrations and descriptions of such ex-Enemy units. The information tabulated, however, furnishes our readers with an explanation of the disappearance of many important ex-Enemy Warships from our pages; it also accounts for the total abolition of the Austrian Navy Section. As recorded now in our pages, the Navy of the late Dual Monarchy has been almost completely partitioned among the Fleets of it's recent opponents. In our treatment of ex-Enemy Warships, we have made one exception in the case of German Submarines allotted to Japan. These craft are fully described and illustrated because we consider they will have a marked influence on future Japanese Submarine types.

Space forbids that we should call attention in detail to all the improvements effected in this Edition, but we should like to indicate two innovations. The first of these relates to the dates attached to photographic illustrations—a feature added at the request of our correspondents. We find that, on examining the Photograph of a Warship, the question often arises: "At what date was that Photograph taken?" To answer this question, it is necessary to search through Editions of "Fighting Ships" for four, five or six years past. In the second place, so many new Photographs have been added during the past five years, and so many obsolete views replaced by more modern illustrations, we think the time has now come when we can place a date under our more modern Warship views. It should be noted carefully that the dates attached to Photographs are not, in all cases, those at which the views were taken, for it is often impossible to secure such information. The figures indicate, in the majority of cases, the Edition of this Annual in which the views first appeared. Where no date is attached to Photographs, it can be taken that such illustrations are of a "pre-War" standard, and we should be pleased to receive from our correspondents new Photographs, to replace these undated illustrations, which are becoming obsolete.

Correspondents have also called attention to the fact that our plans of Torpedo Craft have fallen into arrears. Extreme pressure of editorial work has prevented due attention being paid to this matter during the past two years. In the present Edition, the matter of Torpedo Craft plans has been thoroughly investigated, and diagrams have been added of all modern types of Destroyers and Torpedo Boats, wherever it has been possible to add such drawings.

Although our 1919 Edition reached a very high standard of accuracy, the text on every page of it has been carefully examined, corrected and improved, when preparing this volume. To nearly every Navy Section, both great and small, new textual information, illustrations and silhouettes have been added. We take particular pride in our illustrative matter: we need only mention that between 500 and 600 new views have been added or renewed in the present issue, to show that no efforts have been spared to present our readers with the finest and most up-to-date illustrations of Warship types that can be obtained, from all quarters of the world.

Turning to the various Navy Sections in detail, we may point out to our readers a few of the many improvements and changes effected this year.

Great Britain.

In this Section, the outstanding feature is a new and experimental re-arrangement of the larger Warship pages, designed to secure larger illustrative spaces. During the recent War, Battle Cruiser types were developed to such enormous lengths, the usual "Fighting Ships" photographic illustration of five inches width became too small to depict these ships distinctly. This point is well exemplified by our smaller and full length illustration of H.M. Ships *Repulse* and *Renown*. Here, H.M.S. *Repulse* is so distant, it is impossible to discern many important details in her design, such as the position of the triple 4-inch gun mountings. Accordingly, we were compelled, last year, to give full-page illustrations for Battle Cruisers over 700 feet in length. These large Photographs proved so popular with our readers, we have decided this year to give rather larger views of the more modern British Dreadnought and Battle Cruiser types. We should be glad to hear from our correspondents, if they consider that this innovation, in the way of larger Photographic reproductions, should be extended hereafter to other large Navy Sections, such as those describing the Fleets of France and the United States.

United States.

The establishment of a powerful Fleet on the Pacific Coast has brought into prominence the question of Base accommodation on that sea frontier of the United States. To this subject we have paid due attention. European naval periodicals have ignored entirely the subject of Naval Bases required for the new U.S. Pacific Fleet. Two Naval Commissions have examined the subject most carefully, and consider that no less a sum than fifty million pounds will be required to provide adequate Base, Fuel and Repair Accommodation for the U.S. Warships now stationed in those waters. The schemes outlined by the Helm and McKean Commissions have not been sanctioned officially, but their magnitude shows this matter is one which cannot be overlooked by " [illegible]

matter has been amplified. Photographs have been added of those most interesting vessels, the Italian "Mobile Batteries." It may be urged by some of our readers that these craft are "only just Warships" and should not appear in our pages. May we point out that the value of a Warship type should not be measured by comparison with such maritime *colossi* as H.M.S. *Hood* and the U.S.S. *Indiana*? Rather should the worth of a Warship be measured in relation to geographical conditions and possible strategical and tactical situations. The fact that the "Mobile Batteries" were called into existence during the late War, and played an important part therein, demonstrates that this species of fighting ship has a distinct rôle in the Italian Navy, and therefore should be described. We may add that the Royal Italian Navy is in a transitory stage of development. A new scheme of Naval Defence has been planned, in which Torpedo, Submarine and Aircraft types play an important part. This project has not yet been put into execution entirely; it involves the abolition of many of the larger and older Warship classes. It is therefore possible that ships such as those comprising the *Vittorio Emanuele* Class (which appears again in our pages) may have disappeared before the next Edition of "Fighting Ships" is issued.

Minor Navies.

Beyond the careful revision of descriptive matter, and the addition of new Photographs, wherever possible, we have few comments to offer on these Sections. Naval Correspondents have enabled us to add many new views to the Royal Danish and Royal Spanish Navy Sections. It will be observed that Germany has been treated as the Navy of a Minor European Power, and is now included between Danish and Netherlands Sections. This is in accordance with the new German Fleet Establishment. None the less, Germany occupies 21 pages, or twice the space usually occupied by Fleets of the lesser European countries. Two reasons underly the unwonted length of the German Navy Section; firstly, this nation still retains vessels, surplus to the permitted Establishment, for Mine Sweeping and other subsidiary duties; secondly, we have not received satisfactory information regarding the disposal of various older German Warship types and Fleet Auxiliaries. Until we are assured that these vessels have definitely disappeared, we are compelled to include them again.

Russian Navy.

It has, of course, been impossible to secure any trustworthy information regarding the Russian Navy for many years past. Internecine strife has also reduced it's efficiency to a very low ebb. As this Section cannot be raised, at present, to the exacting standard of accuracy demanded in the other Navy Sections, we still relegate the Russian Navy Section to the final pages of our work.

Conclusion.

To the various Ministries of Marine and Defence, and to the foreign Navy Departments, may we express here our deep appreciation of the assistance rendered in the compilation of this volume? And, whilst expressing this tribute of our gratitude, we must not omit the Naval Attachés of the various Embassies and Legations in London, who have been most courteous in attending to the editorial inquiries. Nelson said he had the happiness to command a band of brothers; and of our unofficial Correspondents throughout the world, we may say that, in contributing views and news of the World's Navies, they have been almost fraternal in their generosity.

Our Editors have been again Dr. Parkes and Mr. Maurice Prendergast. By their diligence and inexhaustible patience, they have surmounted all the intricate problems entailed in the preparation of this volume. Their labours, we feel sure, will raise the international reputation of "Jane," as a classic work of naval reference, to a still higher point of esteem. It has been said that every country gets the Press it deserves. We will not go so far as to say that nations can only produce naval annuals consonant with their naval strength, for many smaller countries with efficient Fleets produce no reference books for naval purposes. Still, we may opine with all respect, that a country, defended by so great and so famous a Fleet as the British Navy, merits the production within its shores of a Naval Annual of the very highest standard of excellence and accuracy. And it has been with that high standard in view that this volume has been prepared and is now presented to our readers.

THE PUBLISHERS.

Overy House,
100, Southwark Street,
London, S.E/1, England.

Special Notice.

CONTENTS.

S—*continued.*

LATE ADDENDA NOTES.

JAPANESE NAVY.

1920-28 NAVAL PROGRAMME. Number of Submarines unofficially reported to be 24 Ocean-going and 40 Coastal Boats, total 64. Designs reported to have been prepared for a 1700-ton Mine-laying type, similar to *O 1* (ex German *U 125*).

NAGATO CLASS :—Unofficially reported that "horizontal armour protection over machinery spaces can be taken as equivalent to 7", allowance being made in this figure for oblique angle of impact."

DESTROYERS :—*Akikaze* Class reported to be as follows :—*About* 1950 tons. Dimensions: *about* 350 × 34 × feet. Guns: 4 or 5—5.5 inch. Tubes: 6—21 inch, in two triple deck mountings. Designed S.H.P. 50,000 = 38 kts. *Max.* oil capacity: 600 tons = 3500 miles at cruising speed. *Mowo* Class :—Some (or all) boats now have 6 tubes mounted in two triples on centre line. Believed that the latest types of First and Second Class Destroyers are fitted to carry and lay mines ; but (*a*) the classes so equipped, and (*b*) number of mines carried by each boat, cannot be ascertained.

General Note.—All above information is not from any authentic source, and cannot be implicitly relied on.

(Continued on next page).

FRENCH NAVY.

The Naval Budget for 1921, presented to the Chamber, amounts to 1,236,000,000 francs. Credits are asked for the commencement of 6 Light Cruisers, 12 Destroyers, and 12 Submarines. Excepting that the number of Destroyers is 12 instead of 6, all these projected vessels are included in the French Navy Section of this Edition. Sums are also asked for the re-fitting of the ex-Enemy Ships incorporated in the Navy ; and for conversion of *Bearn* into an Aircraft Carrier.

All Dreadnoughts now in service have been (or will be) improved in the following way : Improvements in turrets, range-finders, fire-control and director appliances ; new projectiles and torpedoes ; addition of anti-aircraft and anti-submarine weapons ; flash-tight doors to ammunition hoists and passages ; improved defences to magazines against underwater attack ; enlarged W/T. range ; to be fitted for carrying seaplanes, kite-balloons or Sacconey kites. Such modifications have already been carried out in *Bretagne, France, Paris* and *Jean Bart*. *Lorraine, Provence* and *Courbet* are to be taken in hand. *Condorcet, Diderot, Vergniaud* and *Voltaire* are to be re-fitted. The Armoured Cruisers *Edgar Quinet, Waldeck-Rousseau* and *Ernest Renan* will undergo minor refits at Toulon, to keep their speeds up to 22 kts. *Jules Michelet, Jules Ferry* and *Victor Hugo* will also be overhauled at Bizerta, to keep their speeds up to 19-20 kts.

(Continued on next page).

LATE ADDENDA NOTES: GENERAL

(Continued from Page 6.)

New Warship Names listed on this Page do not appear in General Index of Ship Names. Also Ships whose names appear in General Index have no reference to the number of this page.

UNITED STATES.

The Secretary of the Navy, in his Annual Report for 1920, recommends that, in the event of the United States not entering the League of Nations, a further "Three Years' Programme" should be authorized, comprising 3 Battleships, 1 Battle Cruiser, 30 Cruisers, 8 Gunboats, 18 Destroyers, 18 Submarines, 11 Fleet Auxiliaries. Doubtful if this Programme will be sanctioned or adopted *in toto* by the new Administration.

Completion dates of new Dreadnoughts and Battle Cruisers reported thus:—

DREADNOUGHTS:—

INDIANA	1923	*MONTANA*	1924
MASSACHUSETTS	1923	*NORTH CAROLINA*	1924
IOWA	1924	*SOUTH DAKOTA*	1924

BATTLE CRUISERS:—

CONSTITUTION	1923	*LEXINGTON*	1924
UNITED STATES	1923	*RANGER*	1924
CONSTELLATION	1924	*SARATOGA*	1924

New Second Class Cruisers named thus:—**OMAHA** (*Number 4*, launched 14th Dec., 1920), *MILWAUKEE* (5), *CINCINNATI* (6), *RALEIGH* (7), *DETROIT* (8), *RICHMOND* (9), *CONCORD* (10), *TRENTON* (11), *MARBLEHEAD* (12), *MEMPHIS* (13).

JAPAN.

I.—NEW CONSTRUCTION, ADDENDA LIST.

2 BATTLESHIPS:—*OWARI* (Yokosuka D.Y.), *KII* (Kure D.Y.), both 1920-28 Programme.

5 LIGHT CRUISERS:—*ABUKUMA* (Uraga Dock Co.), *KINU* (Kawasaki Co.), both probably authorized under 1920-28 Programme; *AYASE* (Sasebo D.Y.), *OTONASE* (Sasebo D.Y.), *MINASE* (Uraga Dock Co.), all probably authorized under 1918-24 Programme.

DESTROYERS (all probably authorized under 1918-24 Programme; new construction under 1920-28 Programme not reported). *First Class*:—*Okaze, Mokaze, Tanaajikaze* (all by Maidzuru D.Y.); *Soyokaze, Suzukaze* (both by Mitsubishi Co.). *Second Class*:—*Basho, Kaido, Kakitsubata, Tsutsuji* (all by Ishikawajima Co.); *Karukaya, Kikyo, Omodaka, Yuri* (all by Kawasaki Co.); *Ayame, Botan* (both by Uraga Dock Co.); *Ajisai, Nadeshiko, Shion* (all three by Fujinagata Co.).

SUBMARINES, AUXILIARIES:—Not reported.

General Note.—All above to be commenced. Building Yard named after each ship in brackets.

II.—NEW CONSTRUCTION:—SHIPS ALREADY LISTED.

Reported that Armour Plate for four Capital Ships has been ordered in Great Britain from Messrs. Vickers, Ltd.

BATTLE CRUISERS:—*ATAGO* to be begun by Kawasaki Co., and *TAKAO* by Mitsubishi Co.

NAGATO CLASS:—Belt unofficially given as 12".

LIGHT CRUISERS:—*YURA* to be built by Sasebo D.Y.

FIRST CLASS DESTROYERS:—*Namikaze, Numakaze*, to be built by Mitsubishi Co., Nagasaki, and *Nokaze* by Maidzuru D.Y. ***Shiokaze*** launched October 22nd, 1920, and ***Hadakaze*** June 26th, 1920.

SECOND CLASS DESTROYERS:—*Ashi* and *Tsuta* by Kawasaki Co., *Hasu* and *Hishi* by Uraga Dock Co., *Sumire* and *Yamogi* by Ishikawajima Co., *Tade* and *Warabi* by Fujinagata Co. Details of this Class reported to be:—835 tons. Length, 275.6 feet. Speed: 31.5 kts. Armament: 3—4.7 inch and 2 M.G., 6 tubes in two triples. (Details open to question and require verification.) *Aoi* Class:—***Kiku*** launched October 13th, 1920.

SUBMARINES:—No information reported beyond order given to Messrs. Vickers, Ltd., for 2 sets of 12-cylinder Diesel engines, each 1200 B.H.P., similar to type of engine in British "L" class Submarines.

FRANCE.

Cancel condemnation of Destroyer ***Cavalier*** on French Addenda Page 284a. Now disrated and to be attached to Engineering School. Only 15 Destroyers of the Older ("Hurricane Deck") types now left, and only 3 T.B. of ***Mistral*** Class.

Submarine Chaser ***C 101*** sunk.

Submarine ***Coulomb*** condemned. In Submarine section, read throughout ***Carissan*** instead of ***Corissan***.

ITALY.

VULCANO reported condemned.

MINOR EUROPEAN NAVIES.

POLAND.

The four "River Gunboats" built at the Schichau Yard, Danzig, are in service. One named **MARSHAL PILSUDSKI,** 500 tons, complement 50. Two Patrol Boats purchased from Finland, both built to design by Admiral Barowski, late Chief of Naval Construction, Russian Baltic Fleet. ("Morning Post," December 14th, 1920).

FINLAND.

"Karjala Class" is reported to consist of **KARJALA** (*ex-Filin*) and **TURUNAA** (*ex-Orlan*).

Ex-Russian Torpedo Boats ***Nos. 215, 216, 218, 222*** reported no longer in existence. (This latter note also applies to Russian Baltic Torpedo Boats where these boats are referred to again).

DENMARK.

Torpedo Boat ***Svaerdfisken*** now repaired: no alteration in appearance. Submarine ***Rota*** is commissioned. Patrol Boats ***P 2, P 3*** have been condemned.

GERMANY.

Credits voted for commencement of new Light Cruiser, 1921.

NETHERLANDS.

Mine Layer **VAN MEERLANT** launched Nov. 23rd, 1920.

PORTUGAL.

ADAMASTOR has no mainmast and only 2—4.1 inch guns. ***Tamega*** is ready to launch.

RUSSIA (BLACK SEA).

GENERAL ALEXEIEFF, GENERAL KORNILOFF, two Destroyers and about 20 small craft interned at Bizerta, under French supervision. Reported that only Destroyers now left are ***Cerigo, Korfu, Derski, Pulki, Pospyeshni, Kapitan-Saken, Zavidni, Zavetni, Zvonki, Zorki.*** Also **TERETZ** said to be only surviving ship of her class. Submarine ***Buryevestnik*** reported to have been scuttled.

This Edition closed, 16th Dec., 1920.

GENERAL INDEX. (Named Ships only.)

ABBREVIATIONS: (A) signifies Austrian, (Ar) Argentine, (Brz) Brazilian, (C) Chinese, (Ch) Chilian, (D) Danish, (Du) Dutch, (Est) Esthonian, (F) French, (G) German, (Grk) Greek, (H) Hungarian (late), (I) I (J) Japanese, (N) Norwegian, (P) Portuguese, (R) Russian, (Rum) Rumanian, (S) Swedish, (Sp) Spanish, (T) Turkish (late), (U.S.) United States. No indication of nationality signifies British Empire.

A

B

D—*continued.*

E

F

G

H

R—*continued.*

S

GLOSSARY OF TECHNICAL TERMS.

ENGLISH.	GERMAN.	FRENCH.	ITALIAN.	RUSSIAN.	SPANISH.	SWEDISH	DUTCH.
1 Abaft	1 Achtern (adv) hinter (prep); achterlicher als quer ab (abaft the beam)	1 Sur l'arrière	1 A poppa	1 Korma	1 A popa	1 Akterligt	1 Achterlijker dan dw.
2 Abeam	2 dwars ab, quer ab	2 Par le travers	2 Per traverso	2 Poperek	2 Por el través	2 Tvärs ut /direction/	2 Dwars
3 Above water	3 Über wasser	3 Au dessus de l'eau	3 Sopracqua	3 Nadvodnie	3 Sobre el agua	3 Öfvervattens	3 Boven water
4 Accommodation hulk	4 Wohnschiff	4 Caserne flottante 4 Ponton	4 Pontone d'accomodamento		4 Ponton de comodidad	4 Logementsfartyg	4 Accomodatie legger
5 Ahead	5 Vorwärts	5 Avant	5 A prora (davanti)	5 Vpered	5 A proa	5 Förut /direction/ Framåt /movement	5 Vooruit
6 Air lock	6 Luftverschluss (?)	6 Ecluse de chauffe/sas	6 Serratura d'aria		6 Cerradura de aire	6 Luftslass	
7 Amidships	7 Mittschiffs	7 Au milieu	7 Al centro	7 Vseredinu	7 Al centro	7 Midskepps	7 Midscheeps
8 Ammunition ship	8 Munitionsschiff	8 Porte-munitions	8 Nave di munizioni	8 Soodno s ammunieziei; boevimi pripasami	8 Buque de municiónes	8 Ammunitionsfartyg	8 Ammunitie schip
9 Anti-aircraft (gun)	9 Fliegerabwehrgeschutz kanone (abbrev. Flak)	9 Canon contre-aeriens	9 Cannone anti-aeroplani	9 Protivo-aeroplanniya	9 Cañon antiaeroplanos	9 Luftförsvarskanon (generally "Luftkanon")	9 Anti-lucht-schip (kan
10 A.P. (projectile)	10 Panzersprenggranate	10 Projectile de rupture	10 Proiettile	10 Probivanshii-bronu	10 Proyectil	10 Halvpansargranat	
11 Armament	11 Armierung	11 Armement	11 Armamento	11 Vo-oroojenie	11 Armamento	11 Bestyckning	11 Bewapening
12 Armed liner, or Auxiliary cruiser	12 Hilfskreuzer	12 Croiseur auxiliaire	12 Nave armata / incrociatore ausilare	12 Vo-oroojeunii passajirskii-parokhod; Vspomogatelnii krieser	12 Buque armado/o crucero auxiliar	12 Hjälpkryssare	12 Gewapende lyner of hulp kruiser
13 Armour	13 Panzer	13 Cuirasse	13 Corazzato	13 Bronia	13 Coraza Blindage	13 Pansar	13 Pantser
14 Armoured cruiser	14 Panzerkreuzer	14 Croiseur cuirassé	14 Incrociatore corazzati	14 Bronenosnii-kreiser	14 Crucero acorazado	14 Pansarkryssare	14 Gepantserde Kruiser
15 Astern	15 Rückwärts	15 Arrière	15 A poppa	15 Nazad	15 A popa	15 Back	15 Achteruit
16 Athwartship	16 dwarsschiffs	16 Par le travers (d'un navire)	16 Atraverso il bastimento	16 Soodno-na-traverze	16 Por el través del buque	16 Tvärskepps	
17 Awash	17 im überfluten Zustand	17 A fleur d'eau	17 A fior d'acqua	17 Vroven-s-vodoi	17 A flor de agua	17 I övervattensläge	
18 Barbettes	18 Barbetten	18 Barbettes	18 Barbette	18 Barbeti	18 Torres-barbetas	18 Barbettorn	18 Barbette
19 Bases (of barbettes)	19 Trunk (Schacht)	19 Au dessous pied de la tourelle	19 Piattaforme	19 Vnizu	19 Plataformas	19 Fasta torn	19 Torenwal
20 Battery	20 Batterie	20 Batterie	20 Batteria	20 Battereya	20 Bateria	20 Batteri	20 Battery
21 Battle-cruiser	21 Panzerkreuzer	21 Cuirassé rapide. 21 Croiseur de bataille.	21 Incrociatore di battaglia	21 Boevoi kreiser	21 Crucero de batalla	21 Slagkryssare	21 Slag kruiser
22 Battleship	22 Linienschiff	22 Cuirassé d'escadre	22 Nave da battaglia	22 Bronenosets	22 Acorazado	22 Slagskepp	22 Slagschip
23 Beam	23 Breite	23 Largeur	23 Larghezza	23 Shirina	23 Manga	23 Bredd	23 Breedte
24 Before	24 vorn; vor (prep.) vorlicher als quer ab (before the beam)	24 Avant.	24 Avanti	24 Pered; pred	24 A proa	24 Förut	24 Voor
25 Belt	25 Gürtel	25 Ceinture	25 Cintura	25 Poyass	25 Faja	25 Gördelpansar	25 Gordel
26 Bilge keel	26 Schlingerkiel	26 Quille de roulis	26 Chiglia di rollio	26 Bokovoi keel	26 Quilla de pantoque	26 Slingerköl	26 Kimkiel
27 Binnacle	27 Kompassgehäuse	27 Boussole/Habitacle	27 Abitacolo		27 Bitácora	27 Nakterhus	27 Nachthuis

JAPANESE

(pronounce as French).

English / Japanese	No.	Characters
Fighting tops / Sento-Shorou	79	戰鬪檣樓
Ends / Hashi	77	両端
Draught / Kissui	70	吃水
Displacement / Hai-sui-Tonsu	65	排水噸數
Destroyer / Kuchiku-tei	62	驅逐艦
Deck / Kanpan	51	甲板
Cruiser / Jungo-Kan	50	巡洋艦
C. T. (Conning tower) / Shireito	46	司令塔
Complement / Teiin	43	定員
Coal / Sekitan	41	石炭
c/m (Centimetre)		センチメーター
Casemates / Kaku-Hodai	38	穹窖砲臺
Capacity / Yoryo	36	容量
Bunkers / Sekitanko	34	石炭庫
Bulkheads / Kakuheki	33	隔壁
Bridge / Kankio	32	艦橋
Boilers / Kikan	20	汽罐
Bow / Kanshu	31	艦首
Belt / Kotai	25	甲帯
Beam / Haba	23	巾
Battleship / Sento-Kan	22	戰鬪艦
Battery / Hodai	20	砲臺
Bases (of Barbettes, etc.) / Hoto-no-Kibu	19	砲塔の基部
Barbettes / Rohoto	18	露砲塔
Astern / Tomo	16	艦後
Armoured Cruiser / Kotetsu-Junyo-Kan	14	甲鐵巡洋艦
Armour / Kotetsu	13	甲鐵
Amidships / Chuo	7	中央
Ahead / Omote	5	艦前
Above water	3	水上

GLOSSARY OF TECHNICAL TERMS.—*(Continued.)*

ENGLISH.	GERMAN.	FRENCH.	ITALIAN.	RUSSIAN.	SPANISH	SWEDISH.	I
28 B.L. (gun)	28 Hinterlader	28 Canon culasse	28 Cannone a retrocarica	28 Zaryajansheesyas kazennoi ohasti	28 Cañón de retrocarga	28 Bakladdningskanon	
29 Boilers	29 Kessel	29 Generateurs/Chaudieres	29 Caldaie	29 Kotli	29 Calderas	29 Ångpannor	2 Ketels
30 Boom (port defence)	30 Hafensperre	30 Estacade	30 Asta da bome (per difesa portuale)		30 Cadena de troncos (para defensa de puerto)	30 Bomstängsel	30 Sluitboo
31 Bow	31 Bug	31 Avant	31 Prora	31 Noss	31 Proa	31 Bog	31 Boeg
32 Bridge	32 Kommando-Brücke	32 Pont	32 Ponte di comando	32 Most	32 Puente	32 Brygga	32 Brug
33 Bulkheads	33 Schötten	33 Traverses/Cloisons	33 Paratie	33 Pereborki	33 Mamparos	33 Skott	33 Schot
34 Bunker (coal)	34 Kohlenbunker	34 Soute à charbon	34 Carbone per carboniere	34 Oogolnaya yama. To bunker, groozitsya ooglem	34 Carbón de las carboneras	34 Kolboxar	34 Bunker (
35 Camouflage	35 Unsicintbarmalen	35 Camouflage	35 Camuffamento or "Camouflage"	35 Iskoostvennoe prikpitie	35 "Camouflage"	35 Maskering	35 Identitei
36 Capacity	36 Capacität	36 Exposant/Capacité de charge	36 Capacita	36 Poinyestchenie	36 Capacidad	36 Kapacitet	36 Vermoge
37 Capstan	37 Ankerspill, Ankerwinde	37 Cabestan	37 Argano		37 Cabrestante	37 Ankarspel, Spel	37 Spil
38 Casemates	38 Casematten	38 Casemates	38 Casamette	38 Kazemat	38 Casamatas	38 Kasematter	38 Kazemat
39 Centre-, (keel-) line	39 Kiellinie	39 Quille centrale/mediano	39 Paramezzale centrale	39 Srednyaya liniya: diametralnaya ploskost	39 Sobrequilla central	39 Midskeppslinje	39 Midden (
40 Chart house	40 Kartenhaus	40 Cabine de cartes	40 Sala nautica	40 Shtoormanskaya roobka	40 Caseta de derrota	40 Styrhytt	40 Kaarthu
41 Coal	41 Kohlenvorrat	41 Charbon	41 Carbone	41 Ugol	41 Carbon	41 Kol	41 Steenkol
42 Coast defence ship	42 Küstenpanzerschiff	42 Garde-côtes.	42 Nave per difesa della costa	42 Soodno beregovoi oboroni	42 Buque para defensa de costas	42 Kustförsvarsfartyg	42 Kust ver
43 Coastguard vessel	43 Küstenschutzschiff	43 Navire garde-côtes	43 Nave guarda costas	43 Soodno beregovoi okhrani	43 Buque guarda costa	43 Kustbevakningsfartyg	43 Kust bes
44 Compass	44 Kompass	44 Compas; Boussole	44 Bussola		44 Brújula	44 Kompass	44 Kompas
45 Complement	45 Besatzung	45 Equipage	45 Equipaggio	45 Kompleki	45 Dotacion	45 Besättning	45 Bemanni
46 C.T. (conning tower)	46 Kommando Turm	46 Block-haus (C.T.)	46 Torre di comando	46 Roolevaia bashnia	46 Torre de mando	46 Stridstorn	46 Comman
47 Crane	47 Kran	47 Grue	47 Grua	47 Kran	47 Grua	47 Kran	47 Krann
48 "Creep" (anti-submarine)	48 U – drache (?)	48 Gerbe sousmarine	48 "Serpeggiare" (anti-sommergibile	48 Tral	48 Arrastramiento (anti-submarino)		
49 Crow's nest	49 Krähennest	49 Nid de pie	49 Gabbia di crocetta		49 Garita para el vijia	49 Utkikskorg	49 Kraaie n
50 Cruiser	50 Kreuzer	50 Croiseur	50 Incrociatore	50 Kreiser	50 Crucero	50 Kryssare	50 Kruiser
51 Deck	51 Deck	51 Pont	51 Ponte	51 Palooba	51 Cubierta	51 Däck	51 Dek
52 Depôt ship	52 Stammschiff, Mutterschiff	52 Soutien/Convoyeur/Ravitailleur de	52 Nave di deposito/Nave appoggio	52 Soodno sklad; soodno baza	52 Buque de depósito	52 Depotfartyg	52 Depot sc
53 ,, (torpedocraft)	53 ,, fur Torpedofahrzeuge)	53 ,, (Torpilleurs).	53 ,, Siluranti		53 ,, (torpederos)	53 ,, för torpedbåtar	53 Depot (t
54 ,, (destroyers)	54 ,, (,, Zerstörer)	54 ,, (Destroyers).	54 ,, Cacciatorpediniere	54 Soodno-matka dlya kontr-minonosetzev	54 ,, (cazatorpederos)	54 ,, ,, jagare	54 ,, (t

JAPANESE.

(pronounce as French.)

No.	Term	Japanese
216	Water line / *Suishen*	吃水線
213	Upper deck / *Jokanpan*	上甲板
212	Turrets / *Yihoto*	圓砲塔
206	Trials / *Shi Unten (Kokoromi Unten)*	試運轉
194	Tons	噸
193	Torpedo tubes / *Sairai Hatsushakuan*	水雷發射管
192	Torpedo boat / *Suirai-Tei*	水雷艇
191	Torpedo / *Sairai*	水雷
182	Submerged / *Suichu*	水中
180	Submarine / *Shen-sui*	潜水
179	Stern / *Kanbi*	艦尾
178	Starboard (side) / *Ugen*	右舷
176	Speed / *Sokurioku*	速力
173	Shields / *Tate*	楯
167	Screens / *Kakusho*	スクルー
155	Ram / *Shoto*	撞頭
150	Protection to vitals / *Yobu-Bogio*	要部防禦
147	Port (side) / *Sagen*	左舷
135	Normal displacement / *Kitei*	規定
	m/m (Millimetre)	ミリメーター
	M. (Metre)	メーター
123	Masts / *Hobashira*	檣
121	Main (–deck) / *Chukan-pan*	中甲板
119	Machinery	機械
118	Lower (–deck) / *Gekan-pan*	下甲板
117	Liquid fuel / *Yekitai-Nenrio*	液体燃料
113	L. (Length) / *Nagasa*	長
112	Kts. (Knots)	節
101	Hoods / *Hogai*	砲蓋
9	(Guns	砲

GLOSSARY OF TECHNICAL TERMS.—*(Continued).*

No.	ENGLISH.	GERMAN.	FRENCH.	ITALIAN.	RUSSIAN.	SPANISH.	SWEDISH.	DUTCH.
55	Depôt ship (T.B.)	Stammschiff (für Torpedoboote)	Soutien de (Torpilleurs)	Nave appoggio (torpedine)		Buque de déposito (torpederos)	Depotfartyg för torpedbåtar	Depôt (T.B.)
56	„ (M.L.)	„ („ Motorboote)	„ (Vedettes).	„ (moto-scafo)		„ (lancha de motor)	„ „ motorbåtar	„ (M.L.)
57	„ (submarines)	Mutterschiff (für Unterseeboote)	„ (Sous-marins).	„ (sommergibili)	Soodno-matka dlya podvodnikh lodok	„ (submarinos)	„ förundervattensbåtar	„ (onderzeeërs)
58	Depression (of guns)	Senkung	Angle negatif de pointage	Depressione (cannone)	Oogol snijeniya	Punteria baja (de las piezas)	Dumpning	
59	Depth charge	Wasserbombe	Charge de fond/Bombes à explosives reglée en profondeur	Carica di fondo (esplosivo)	Bomba dlya vzriva pod vodoi	Carga de fondo (explosiva)	Vattenbomb	Diepte lading
60	Derrick	Ladebaum	Mât de charge	Albero da carico	Nod-emnaya strela	Pescante de carga	Bom	Laadboom
61	Despatch vessel	Aviso	Aviso	Nave per dispacci	Posilnoe soodno	Buque de aviso	Avisofartyg	Koerier schip
62	Destroyer	Zerstörer	Contre-torpilleur	Caccia-torpediniere	Kontr-minonosets	Cazatorpederos	Jagare	Torpedobootjager
63	Diesel engine	Dieselmaschine, Ölmaschine, Ölmotor	Moteur Diesel	Macchina "Diesel"	Dizel-motor	Máquina "Diesel"	Dieselmotor	Diesel machine
64	Dirigible	Lenkbares Luftschiff	Dirigeable	Dirigibile	Derijabl	Dirigible	Luftskepp	Bestuurbare
65	Displacement	Deplacement	Déplacement	Dislocamento	Vodoizmiestchenie	Desplazamiento	Deplacement	Waterverplaatsing
66	Division (of fleet)	Division	Division	Divisione (di una flotta)	Deviziya; devizion	División (de una flota)	Division	Eskader
67	Dock (floating)	Schwimmdock	Dock (flottant).	Bacino (galleggiante)	Dok (plavoochii)	Dique flotante	Flytdocka	Dok (drÿvend)
68	„ (graving)	Trockendock	Cale sèche/Bassin de radoub	Bacino (di carenaggio)	Dok (sookhoi)	Dique seco	Torrdocka	„ (droog)
69	Dockyard (D.Y.)	Werft	Arsenal	Arsenale	Shipyard, Verv.	Arsenal	Varv	„ werf
70	Draught (mean)	mittlerer Tiefgang	Tirant d'eau moyen	Immersione media		Calado medio	Djupgående, medel—	Diepang (gemiddeld)
71	„ (max.)	grösster Tiefgang	Tirant d'eau maximum	Immersione massima		Calado máximo	„ maximi—	„ (grootste)
72	Dreadnought	Grosskampschiff	Dreadnought	Dreadnought	Drednaout	Dreadnought	Dreadnought	
73	Drifter	Drifter, Dampflogger	Chalutier	Nave di pesca ("Drifter")	Maloe sooduo	Vapor de pesca ("Drifter")		Smak
74	Echelon	Staffellinie	Echelon. / 74 En quinconce.	Linea in "Echelon"	Eshelon	Linea en escalones	Flankformering	
75	"Electric drive"	elektrische Antrieb	Commande electrique	Comando elettrico		Impulsión electrica	Elektrisk drift	
76	Elevation (of guns)	Höhewichtung, Elevation	Angle de pointage positif	Elevazione (cannone)	Oogol vozvosheniya	Elevación (tiro de cañón)	Elevation	
77	Ends	Ends	Extrémités	Estremita	Krai	Extremos	Stävar	Einden
78	Examination Vessel	Untersuchungschiff	Arraisonneur	Nave d'ispezione		Buque de inspección	Undersökningsfartyg	Onderzoekings schip
79	Fighting tops	Marsen	Hunes/armés/militaires	Coffo militari	Poostchechnii mars	Cofas militares	Stridsmärsar	Gevechtsmars
80	Fire Control	Feuerleitung	Direction de tir	Controllo del tiro	Kontrol ognya	Regulator de tiro	Eldledning	Controle by het vuren
81	„ top	Artilleriestand im Topp	Hune télémétrique.	„ (coffa)		„ „ (cofa)	Eldledningsmärs	
82	Fire director	Zentralabfeuerung	Direction du feu	Direttore de tiro	Mekhanizm oopot reblyaemii pri strelbl zalpom	Director de tiro	Centralavfyrning	
83	Fisheries vessel	Fischerieschutzboot	Garde-pêche.	Nave peschiera	Maloe voennoe soodno dlya ohhrani ribnikh promislov	Buque de pesquería	Fiskeri-inspektionsfartyg	Controle schip (visschel
84	Flagship	Flaggschiff	Navire amiral	Nave ammiraglia	Flagmanskii korabl	Buque almirante	Flaggskepp	Vlagschip
85	Fleet	Flotte	Flotte.	Flotta		Flota	Flotta	Vloot
86	Flotilla	Flotille	Flotille.	Flottiglia	Flotiliya	Esquadrilla	Flottilj	Flotilje
87	„ leader	Führerschiff	Conducteur de flotille.	Nave capofilo	Soodno vedooshii flotiliu	Buque cabo de fila	Flottiljchefsfartyg	„ leider
88	Forecastle	Back	Gaillard d'avant.	Castello di prua	Bak	Castillo	Back	Logies
89	Foremast	Fockmast	Mât de misaine	Albero di trinchetto		Palo trinquete	Fockmast, Förlingmast	Fokkemast
90	Freeboard	Freibord	Franc-bord	Bordo libero	Zapas plavoochesti soodna	Obra muerta	Fribord	Uitwatering
91	Full load displ't	Gesamte Wasserverdrängung	Deplacement en plein charge	Carico completo (dislocamento)	Vodvizmeshenie v polnom groozoo	Cargamento completo (desplazamiento)	Deplacement, fullt rustad	Volle lading merk
92	Funnels	Schornsteine	Cheminées	Fumaioli	Trubi	Chimeneas	Skorstenar	Schoorsteen
93	Fuze	Zünder	Amorce	Spoletta	Vzrivatel	Espoleta	Tändrör	Grut
94	Gunboat	Kanonenboot	Cannonière	Cannoniera	Kanonerskaya lodka	Lancha cañonera	Kanonbåt	Kanonneerboot
95	Guns	Geschutze	Artillerie/Canons	Cannoni	Orudiia	Cañones	Kanoner	Kanonnen
96	Gyroscopic compass	Kreiselkompass	Compas gyroscopique	Busola giroscopica	Jiroskop	Brújula giroscópica	Gyroskopkompass	Gyroskoop kompas
97	Hawse	Klüse	Aussiere/Écubiere	"Hawse" (apertura per la catena dell 'ancora)	Kluz	Escobén	Klys	Slang
98	H.E. (burstingcharge)	Sprengladung	Haut explosif	Alto esplosivo (carica di scoppio)	Silno-vzrivchatii	Fuerte explosivo (carga estelladora)	Sprängladdning	

GLOSSARY OF TECHNICAL TERMS.—*Continued.*

ENGLISH.	GERMAN.	FRENCH.	ITALIAN.	RUSSIAN.	SPANISH.	SWEDISH.	DUTC
99 H.E. (shell)	99 Sprenggeschoss, Sprenggranate	99 Obu à grande capacité	99 Granata di alto esplosivo	99 Silno-varivchatii snaryad	99 Granada de fuerte explosivo	99 Spränggranat	
100 Hoist (ammunition)	100 Munitionsaufzug	100 Monte-charges	100 Elevatore di munizioni	100 Podem (in full, podemnaya mashina)	100 Ascensor para municiónes	100 Ammunitionshiss	100 Hysche (an
101 Hoods	101 Kuppel	101 Carapaces	101 Scudi	101 Pokrishki	101 Carapachos	101 Kåpor	101 Pantserkoe
102 "Horns" (of mine)	102 Kontakte, Fühlhörner (?)	102 Antennes	102 Corna (di una mina)		102 Conteras (de una mina)	102 Horn	102 "Hoorns"
103 Hospital ship	103 Hospitalschiff, Lazarettschiff	103 Navire hôpital	103 Nave ospedale	103 Gospitalnoe soodno	103 Buque-hospital	103 Lasarettsfartyg	103 Hospitaal s
104 Hull	104 Rumpf	104 Coque	104 Scafo	104 Korpoos; koozov soodna	104 Casco	104 Skrov	104 Hol
105 Hulk	105 Hulk	105 Ponton	105 Pontone	105 Blokshiv	105 Ponton	105 Pråm	105 Legger
106 Hydroplane (boat)	106 Gleitboot	106 Hydroplane (bateau)	106 Idroplano (battello)	106 gidroplane	106 Hidroplano (buque)	106 Hydroplan	
107 ,, (in subs.)	107 Tiefenruder	107 Gouvernails de profondeur				107 Djuproder	
108 Ice-breaker	108 Eisbrecher	108 Brise-glace.	108 Rompighiaccio	108 Ledokol	108 Rompehielos	108 Isbrytare	108 Ysbreeker
109 Jumping wire	109 Minenabweiser, Minendraht	109 Orin.	109 Salto del filo (?)		109 Paso del alambre	109 Avledareswire	
110 Kite balloon	110 Drache	110 Saucisse ("Captif")	110 Pallone a cervo-volante	110 Zmeikovii aerostat	110 Globo-cometa cautivo	110 Ballong	110 Kabel ball
111 Kite-balloon ship	111 Drachenschiff	111 Navire porte ballon captif	111 ,, ,, al (nave d'appoggio)	111 Soodno-matkadlya zmeikovikh aerostatov	111 ,, (buque portador)	111 Ballongfartyg	111 ,, ,,
112 Kts. (Knots)	112 See-meilen (or knotten)	112 Noeuds	112 Nodi	112 Uzell	112 Millas	112 Knop	112 Knoopen (t
113 Length (p.p.)	113 Länge zwischen den Perpendikeln	113 Longueur (entre perpendiculaires)	113 Lunghezza tra le perpendicolari		113 Eslora entre perpendiculares	113 Längd mellan perpendiklarne	113 Lengte (p.p
114 ,, (w.l.)	114 Länge auf der Wasserlinie (?)	114 ,, (à la flottaison)	114 Lunghezza alla linea di carico		114 Eslora en la linea de agua, maxima	114 ,, i vattenlinjen	114 ,, (w.
115 ,, (o.a.)	115 Länge uber alles	115 ,, (totale)	115 Lunghezza massima		115 Eslora de fuera a fuera de miembros	115 ,, överallt	115 ,, (gr
116 Light cruiser	116 kleiner Kreuzer	116 Croiseur léger	116 Incrociatore leggiera	116 Legdii kreiser	116 Crucero ligero	116 Lätt kryssare	
117 Liquid Fuel	117 Heizoil	117 Combustible liquide	117 Combustibile liquido	117 Jhidkoe toplivo	117 Combustibele liquido	117 Brännolja	117 Vloeibare b
118 Lower deck	118 Zwischen-deck	118 Faux pont	118 Ponte inferiore	118 Nijhniata paluba	118 Cubierta baja	118 Mellandäck, Trossdäck	118 Tusschend
119 Machinery	119 Maschinen	119 Machines	119 Machinaria	119 Mekhanizm	119 Maquinas	119 Maskineri	119 Werktuige
120 Magazine	120 Munitionskammer	120 Soute	120 Magazino-deposito munizioni	120 Sklad; magazin	120 Pañol de municiónes	120 Ammunitionsdurk	120 Magazyn
121 Main-deck	121 Batterie deck	121 Pont de la batterie	121 Ponte principale	121 Glavnaia paluba	121 Cubierta principal	121 Huvuddäck	121 Kuildek
122 Mainmast	122 Grossmast	122 Grand mât	122 Albero di maestra		122 Palo mayor	122 Stormast	122 Groote mas
123 Mast	123 Mast	123 Mât.	123 Albero		123 ,, macho	123 Mast	123 Mast
124 ,, (tripod)	124 Dreibein, Dreifussmast	124 ,, (tripode).	124 ,, (treppiedi)	124 Machta trepojnaya	124 ,, trípode	124 Tripodmast	
125 ,, (lattice)	125 Gittermast	125 ,, (en treillis)	125 ,, (graticcio)	125 Reshetchataya machta	125 ,, de celosia	125 Gallermast	
126 Microphone	126 Geräuschempfänger	126 Microphone	126 Microfono	126 Mikrofon	126 Micrófono	126 Mikrofon	126 Microphone
127 Mine	127 Mine	127 Mine	127 Mina	127 Mina	127 Mina	127 Mina	127 Mijn
128 Mine-layer	128 Minenleger	128 Mouilleur de mines 128 Pose-mines	128 Posa-mina	128 Minnoi zagraditel	128 Colocador de minas	128 Minutläggningsfartyg	128 Mijn legger
129 Mine-sweeper	129 Minensucher	129 Dragueur de mines	129 Draga-mina	129 Minnij tralshik	129 Rastreador de minas	129 Minsvepningsfartyg	129 Mijn veege
130 Mizzen-mast	130 Besanmast	130 Mât de artimon.	130 Albero di mezzana		130 Palo mesana	130 Mesanmast	130 Bezaans m
131 M.L. (motor launch)	131 Motorboot	131 Vedette	131 Lancia a motore	131 Motornoe patrolnoe soodno (literally, motor patrol vessel)	131 Lancha de motor	131 Motorbåt	131 M.L. motor
132 M.S. (motor ship)	132 Motorschiff	132 Bateau automobile	132 Nave a motore	132 Motornoe soodna	132 Buque de motor	132 Motorfartyg	132 M.S. motor
133 Monitor	133 Monitor	133 Monitor	133 Monitor	133 Monitor	133 Monitor	133 Monitor	133 Monitor
134 Navy yard	134 Marinewerft	134 Arsenal	134 Cantiere dello stato	134 Admiralteistve	134 Arsenal del estado	134 Orlogsvarv	134 Marine wer
135 Normal displ't	135 Normale Wasserdrängung	135 Deplacement normal	135 Dislocamento normale	135 Normalnoe vodoizmeshenie	135 Desplazamiento normal	135 Normal deplacement	135 Normale w
136 Net Cutter	136 Netzschere	136 Coupe-filets	136 Taglia-rete		136 Cortador de redes	136 Nätsax	136 Net breeke
137 Net layer	137 Netzleger	137 Mouilleur de filets	137 Posa-rete		137 Colocador de redes	137 Nätutläggningsfartyg	137 Net legger
138 Oil tanker	138 Oldampfer, Tankdampfer	138 Pétrolier	138 Nave petroliera a cisterna	138 Nalivnoe soodno	138 Buque de cisterna petrolero	138 Oljetankfartyg	138 Olie tanksh
139 Paddle-wheel	139 Rad	139 Roue	139 Ruote a palette	139 Grebnoe parokhodnoe koleso	139 Rueda de paletas	139 Skovelhjul	139 Wiel (rader
140 ,, steamer	140 Raddampfer	140 Navire à roues.	140 Piroscafo a ruote		140 Vapor de ruedas	140 Hjulångare	140 Raderboot

GLOSSARY OF TECHNICAL TERMS.—*Continued.*

ENGLISH.	GERMAN.	FRENCH.	ITALIAN.	RUSSIAN.	SPANISH.	SWEDISH.	DUTCH.
141 Patrol boat/vessel	141 Vorpostenboot, Fahrzeug	141 Patrouilleur	141 Battelo / nave di pattuglia	141 Patrolnoe soodno	141 Buque de patrulla/ Lancha de patrulla	141 Bevakningsbåt	141 Patrouille boot
142 Periscope	142 Sehrohr	142 Periscope.	142 Periscopio	142 Periskop	142 Periscopio	142 Periskop	142 Periskoop
143 Pilot vessel	143 Lotsenboot	143 Bateau pilote	143 Battello pilota		143 Bote de práctico	143 Lotsbåt	143 Loods boot
144 Pinnace (steam)	144 Dampfpinasse	144 Chaloupe à vapeur	144 Canotto a vapore	144 Kater	144 Pinaza de vapor	144 Ångslup	144 Stoom barkas
145 Pole mast	145 Stangenmast	145 Mât gaule.	145 Albero a palo		145 Palo enterizo ó tiple	145 Pålmast	
146 Poop	146 Poop, Hinterschiff	146 Dunette.	146 Cassero di poppa	146 Korma ; oot	146 Toldilla	146 Poop, Hytta	146 Poep
147 Port (side)	147 Backbord	147 Babord	147 Sinistra	147 Lievii	147 Babor	147 Babord	147 Bakboord
148 Powder (propellant)		148 Charge d'explosion	148 Polvere impellente	148 Porokhevoi	148 Pólvora impelente	148 Krut	148 Poeder
149 Propeller	49 Schraube, Propeller	149 Helice	149 Elica	149 Grebnoi vint	149 Hélice	149 Propeller	149 Schroef
150 Protection to vitals	150 Schutz an vitalen stellen	150 Protection des machines	150 Protezione delle parti vitali	150 Zastchita	150 Protection de las partes vitales	150 Skydd för vitala delar	150 Bescherming der vitale deelen
151 Protective deck	151 Schutzdeck	151 Pont cuirassé	151 Coperta proteggente	151 Bronevaya palooba	151 Cubierta de protección	151 Pansardäck	151 Beschermings dek
152 Q.F. (=R.F. in U.S.)	152 Schnell Feuer Geschutz	152 à tir rapide	152 Tiro rapido	152 Skorostrelnii	152 Tiro rápido	152 Snabbskjutande kanon	152 Snelvuur
153 Radius of action	153 Fahrbereich, Aktionsradius	153 Rayon d'action	153 Raggio d'azione	153 Radius deistviya	153 Radio de acción	153 Aktionsradie	153 Straal van actie
154 Rake (of masts, &c.)	154 Rak	154 Inclinaison sur l'arrière	154 Inclinazione (di un albero)	154 Ooklon	154 Inclinación (de un palo)	154 Fall (på master, etc.)	154 Hang (van masten)
155 Ram	155 Rammsporn	155 Eperon	155 Sperone	155 Tarann	155 Espolon	155 Ramm	155 Ram
156 Range yds. (metres)	156 Entfernung, Schlussweite	156 Portée	156 Portate-yarde/metri	156 Dalnost boya	156 Alcance yardas/metros	156 Avstånd	156 Afstand (yds./meters)
157 Range-finder	157 Entfernungsmesser	157 Télémètre.	157 Telemetro	157 Dalnomer	157 Telémetro	157 Avståndsmätare	157 Afstand zoeker
158 Repair ship	158 Reparaturschiff	158 Transport Navire-atelier	158 Nave per riparazione	158 Plavoochaya masterskaija	158 Buque de reparaciones	158 Verkstadsfartyg	158 Reparatie schip
159 River gunboat	159 Flusskanonenboot	159 Cannonière de rivière	159 Cannoniera da fiume	159 Rechnaya kanonerskaya lodka	159 Lancha cañonera de río	159 Flodkanonbåt	159 Rivier kanonneerboot
160 River monitor	160 Flussmonitor	160 Monitor de rivière	160 Monitor da fiume		160 Monitor de río	160 Flodmonitor	160 Rivier monitor
161 Royal yacht	161 Königliche Yacht	161 Yacht royal/impérial/ présidentiel	161 Yacht reale	161 Korolevskaya yakhta	161 Yate real	161 Kunglig yacht	161 Koninklijk yacht
162 Rudder	162 Ruder, Steuerruder, Seitenruder, Vertikalruder	162 Gouvernail	162 Timone	162 Rool	162 Timon	162 Roder	162 Roer
163 Salvage ship	163 Bergungsschiff	163 Navire relèveur	163 Nave di salvataggio	163 Spasatelnoe soodno	163 Buque de salvamento	163 Bärgningsfartyg	163 Bergings schip
164 S.S. (steamship)	164 Dampfer	164 Navire à vapeur	164 Vapore		164 Buque de vapor	164 Ångfartyg	164 S.S. (stoomschip)
165 S.V. (sailing vessel)	165 Segler	165 Voilier	165 Bastimento a vela		165 Buque de vela	165 Segelfartyg	165 Z.S. (zeilschip)
166 Scout	166 Aufklärungskreuzer	166 Eclaireur Estafette	166 Nave esploratrice	166 Razvedchik	166 Buque explorador	166 Spaningsfartyg	166 Verkenner
167 Screens	167 Splittertraversen	167 Ecrans	167 Schermi	167 Shirmi	167 Pantallas	167 Splintskärmar	167 Splinterschilden Schermen
168 Seaplane	168 Wasserflugzeug	168 Hydravion	168 Idrovolante	168 Gidroaeroplan	168 Hidrovolante	168 Hydroaeroplan	168 Water vliegtuig
169 Seaplane-carrier	169 Flugzeugmutterschiff	169 Porte-avions	169 ,, (nave appoggio)	169 Soodno-matka dlya gidroaeroplanov	169 ,, (buque portador)	169 Flygmaskins—depotfartyg	169 Watervliegtuig drager
170 Searchlight	170 Scheinwerfer	170 Projecteur.	170 Proiettore	170 Projektor	170 Proyector	170 Strålkastare	170 Zoeklicht
171 Searchlight-top	171 Scheinwerfertopp	171 Hune à projecteurs	171 ,, (coffa)		171 ,, (cofa)	171 Strålkastaremast	171 Controle top zoeklicht
172 Sheer (of hull)	172 Sprung		172 Curvatura dello scafo		172 Arrufo del casco	172 Sprang	172 Geer (van hol)
173 Shields	173 Schutzschildern	173 Boucliers.	173 Scudi	173 Stchit	173 Mantelete	173 Sköldar	173 Schilden
174 Sloop	174 Sloop, U-bootsabwehrkreuzer	174 Aviso/Sloop	174 Scialuppa	174 Shlup	174 Balandra		174 Sloep
175 Sounding machine	175 Lotmaschine	175 Appareil de sondage	175 Macchina da scandaglio		175 Aparato para sondar		175 Diepte peiling machine
176 Speed	176 Geschwindigkeit	176 Vitesse	176 Velocita	176 Skorost khoda	176 Velocidad	176 Fart	176 Vaart, Snelheid
177 Squadron (of fleet)	177 Geschwader	177 Escadre	177 Squadra (di una flotta)	177 Eskadra ; brigada	177 Escuadra de una flota	177 Eskader	177 Eskader (van vloot)
178 Starboard	178 Steuerbord	178 Tribord	178 Dritta	178 Pravaia storona	178 Estribor	178 Styrbord	178 Stuurboord
179 Stern	179 Heck	179 Arrière	179 Poppa	179 Korma	179 Popa	179 Akter	179 Hek
180 Submarine	180 Untersee-Boote	180 Sousmarin	180 Sottomarino	180 Podvodnaia lodka	180 Submarino	180 Undervattensbåt	180 Onderzeetorpedoboot
181 Submarine-chaser	181 U-bootsjäger	181 Chasseur de sousmarine	181 Cacciasommergibili		181 Caza-submarino	181 U-båtsjagare	181 Onderzeeër jager
182 Submerged	182 Unterwasser	182 Submergé	182 Subacqueo	182 Podvodnii	182 Sumergido	182 Under vattnet	182 Onderwater
183 Submersible	183 Tauchboot	183 Submersible	183 Sommergibile	183 Opooskaushiisya pod vodoo ; mogooshii opooskatsya	183 Sumergible	183 Dykbåt	183 Duikboot
184 Super-firing (guns)	184 Übereinandergestellt	184 Superposé/e/s (canons.)		184 Strelba s visshei pozitsii cherez nizshoon			

GLOSSARY OF TECHNICAL TERMS.—*Continued.*

	FRENCH.	ITALIAN.	RUSSIAN.	SPANISH.	SWEDISH.	DUTCH.
	185 Ravitailleur/Transport vivres	185 Nave di provviste		185 Buque de abastecimiento	185 Provianfartyg	185 Toever schip
ff	186 Navire pour Service Hydrographique	186 Nave idrografica	186 Opisnoe soodno	186 Buque hidrográfico	186 Sjömätningsfartyg	186 Controle schip
	187 Navire-Citerne	187 Nave-cisterna (acqua)	187 Nalivnoe soodno	187 Buque de cisterna (agua)	187 Vattenbåt	187 Tank schip (voor water)
	188 Annexe	188 Battello d'avviso	188 Tender	188 Bote-aviso	188 Hjälpfartyg	188 Lichter
	189 Mât de hune	189 Albero de freccia	189 Stenga	189 Mastelero de galope	189 Stang	189 Steng
	190 Mât de perroquet	190 Albero da velaccio	190 Bram-stenga	190 Mastelero de juanete	190 Bramstang	190 Bovenbram steng
	191 Torpille	191 Siluro	191 Mina	191 Torpedo	191 Torped	191 Torpedo
	192 Torpilleur	192 Torpediniera	192 Minonoseta	192 Torpedero	192 Torpedbåt	192 Torpedoboot
ohre	193 Tubes lance torpilla	193 Tubi lancia siluri	193 Minnie aparati	193 Tubos lanza torpedos	193 Torpedtub	193 Lanceerinrichting (Cuis of Kanoe)
	194 Tonnes	194 Tonnelate	194 Tonni	194 Toneladas	194 Ton	194 Tonnen
	195 Navire-Ecole	195 Nave scuola di marina	195 Oochebnoe soodno	195 Buque escuela de marina	195 Övningsfartyg	195 Opleidings schip
lsschiff	196 ,, mousses	196 ,, (ragazzi)		196 ,, (muchachos)	196 Skepp-gossefartyg	196 ,, (jongens)
schiff	197 ,, mécaniciens	197 ,, (macchinisti)		197 ,, (maquinistas)	197 Maskinistskolfartyg	197 ,, (machinisten)
hiff	198 ,, de cannonage	198 ,, (artiglieria)		198 ,, (artilleria)	198 Artilleriskolfartyg	198 ,, (kanonneers)
hiff	199 ,, d'application	199 ,, (guardie-marine)		199 ,, (guardia-marinas)	199 Kadettfartyg	199 ,, (adelborsten, etc.)
al-schiff	200 ,, navigation	200 ,, (navigazione)		200 ,, (navegacion)	200 Skolfartyg	200 ,, (navigatie)
ff	201 ,, de torpilles	201 ,, (mina)		201 ,, (minas)	201 Minskolfartyg	201 ,, (mijn)
ff	202 ,, de chauffe	202 ,, (fuochisti)		202 ,, (fogoneros)	202 Eldareskolfartyg	202 ,, (stokers)
hiff	203 ,, de torpilles automobils	203 ,, (siluri)		203 ,, (torpedos)	203 Torpedskolfartyg	203 ,, (torpedo)
dampfer	204 Transport	204 Trasporto	204 Transport; transportnoe soodno	204 Buque transporte	204 Transport	204 Transport
	205 Chalutier	205 "Trawler"	205 Traooler	205 "Trawler" (Buque de pesca)	205 Trawlare, Fiskefångare	205 Stoom visschersman
	206 Essais	206 Prove	206 Ispitaniia	206 Pruebas	206 Provturer	206 Proeftocht
	207 Compartement/caisse d'assiette de relevement	207 "Trimming tank" (cisterna di assettamento)	207 Oorovnitelnaya sisterna	207 "Trimming tank" (cisterna de equilibración)	207 Trimtank	207 Ballast tank
er	208 Remorqueur	208 Rimorchiatore	208 Booksir; booksirnoe soodno	208 Remolcador	208 Bogserbåt	208 Sleep boot
or	209 Turbo-moteur	209 Turbo generatore		209 Turbo-generador	209 Turbingenerator	
	210 Turbine	210 Turbina	210 Turbina	210 Turbina	210 Turbin	210 Turbine
ichtung, ehrfacher	211 Turbines à engrenages	211 Turbina ad ingranaggio		211 Turbina con engranaje	211 Utväxlad turbin	
	212 Tourelles	212 Torri	212 Bashnia	212 Torres	212 Torn	212 Torens
	213 Pont de gaillards	213 Ponte superiore	213 Verkhniaia paluba	213 Cubierta alta	213 Övre däck	213 Opperdek
Exhaustor	214 Base de cheminées	214 Cassa a fumo (nel fumaiolo)	214 Narodjnaya dimovaya trooba	214 Flus de reunión de chimenea	214 Rökupptag	
ufter	215 Ventilateur	215 Ventilatore	215 Ventilyator	215 Ventilador	215 Ventilator	215 Luchtkoker
	216 Flottaison	216 Linea d'acqua	216 Vaterliniia	216 La linea de flotacion	216 Vattenlinje	216 Waterlyn
raphie ahtlose	217 Telegraphie sans fils (T.S.F.)	217 Radiotelegrafia	217 Bezprovolornoe telegrafirovanie	217 Radio-telegrafia	217 Gnisttelegrafi, Radiotelegrafi	217 Telegraphie zonder draad
Luftdrähte	218 Antennes de T.S.F.	218 ,, (antenne)	218 Anteni	218 ,, (antenas)	218 Antenner, Luftnät	
(?)	219 Deflecteurs de T.S.F.	219 ,, (Stendadori aerei)		219 (Extendedores aéreos)	219 Spröt för gnistnät	
te Tür	220 Porte de cloison etanche	220 Porta stagna	220 Vodonepronicaemaya dver	220 Puerta estanca	220 Vattentät dörr	220 Waterdichte deur
hte Abteilung	221 Compartement etanche	221 Compartimento stagno	221 Vodonepronicaemoe otdelenie	221 Compartimiento estanco	221 Vattentät avdelning	221 Waterdichte afdeeling
	221a Cloison étanche					
	222 Yacht (à vapeur)	222 Yacht a vapore	222 Yakhta (parovaya)	222 Yate de vapor	222 Ångyacht	222 Yacht (stoom)

(ROYAL) BRITISH NAVY.

LIMITS OF BRITISH NAVAL STATIONS, 1912.

Note.—Since this Chart was prepared, limits of Australia and New Zealand Stations have been slightly varied. New Caledonia and the New Hebrides come within the Australia Station and Fiji within New Zealand Station.

From Chart furnished by courtesy of the Hydrographic Department, Admiralty.

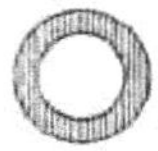

Note.—M (International Code) at masthead indicates that Submarines are in company and/or exercising. When accompanied by group from Numerical Table (Int. Code), number of Submarines are indicated by group hoist.

ROSYTH

to	miles
Aberdeen	95
Dover	473
Helder	360
Hull	260
Kronstadt	1315
Newcastle	130
Portsmouth	570
Sheerness	458
Wilhelmshaven	445

SHEERNESS

to	miles.
Dover	45
Dunkirk	70
Helder	190
Hull	200
Kronstadt	1200
Newcastle	328
Portsmouth	150
Wilhelmshaven	270

PORTSMOUTH

to	miles.
Barbados	3600
Berehaven	375
Brest	226
Cherbourg	72
Devonport	155
Gibraltar	1200
Lorient	331
New York	3080
Pernambuco	3906
Portland	55
Wilhelmshaven	620

DEVONPORT

to	miles.
Berehaven	375
Brest	140
Cape of Good Hope	5890
Cherbourg	110
Gibraltar	1050
Lorient	233
New York	2905
Portland	100
Queenstown	292
Rochefort	370
Toulon	1761

GIBRALTAR

to	miles.
Ajaccio	1000
Algiers	420
Berehaven	1170
Bizerta	875
Brest	953
Devonport	1050
Malta	980
Oran	225
Toulon	713

MALTA

to	miles.
Aden	2320
Ajaccio	450
Bizerta	240
Constantinople	702
Port Said	920
Sevastopol	1000
Spezia	580
Suda Bay	575
Suez	1010
Toulon	612

ADEN

to	miles.
Batavia	3950
Beira	2824
Bombay	1652
Cape Town	4450
Colombo	2100
Delagoa Bay	3340
Devonport	4350
Durban	3650
Karachi	1470
Mauritius	2600
Suez	1310
Tamatave	2200
Zanzibar	1770

COLOMBO

to	miles.
Bombay	900
Calcutta	1240
Devonport	6450
Fremantle (W.A.)	3105
Karachi	1400
Penang	1280
Rangoon	1207
Singapore	1600
Sydney (N.S.W.)	5450

SINGAPORE

to	miles.
Bangkok	820
Batavia	500
Devonport	8026
Hong Kong	1430
Manila	1320
Saigon	648

HONG-KONG

to	miles.
Devonport	9450
Kiao-chau	1270
Manila	650
Nagasaki	1067
Port Arthur	1415
Saigon	930
Shanghai	870
Ta-kau (Formosa)	340
Vladivostock	1927
Wei-hai-wei	1470
Yokosuka	1620

CAPE OF GOOD HOPE

to	miles.
Ascension	2395
Durban	812
Falkland Isles	4800
Fremantle (W.A.)	4850
Hobart	5527
Madagascar	1800
Port Elizabeth	428
Portsmouth	5960
St. Helena	1710
Teneriffe	4470

ESQUIMAULT

to	miles.
Honolulu	2410
San Francisco	750
Seattle	76
Yokosuka	4300

SYDNEY (N.S.W.)

to	miles.
Adelaide	1076
Auckland	1280
Batavia	4476
Brisbane	510
Cape of Good Hope	6157
Colombo	5450
Devonport (via Cape)	12,047
Hobart	630
Melbourne	570
Singapore	4980
Wellington, (N.Z.)	1234

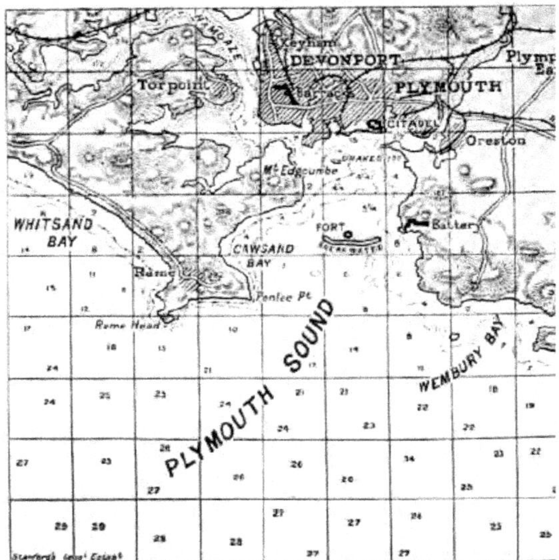

(In 2000 yard squares. Soundings in fathoms. Heights in feet.)

DEVONPORT and KEYHAM. At *Devonport* (South Yard)—two big slips, three [illegible] Dry docks:—No. 3, size 430 × 93 × 34½ feet, and three others, of which [illegible] Dock) can take a second class cruiser; other two suitable for small craft only. [illegible] (North Yard)—No slips. New basin 35½ acres, depth, 32½ft. Tidal basin [illegible] depth, 32ft.

Dry Docks in North Yard:—
Entrance lock, 730* × 95 × 44 feet.
No. 9 (double), 745 × 95 × 32 „
Keyham (North Yard) No. 10 (double), 741 × 95 × 44 feet.
„ 8 659 × 95 × 32 „
* Can be lengthened 86 feet.

In the old part of the yard there are three docks, of which one (No. 7) can [illegible] class cruiser; the others small craft only. Total employees for the two yards, 14,000 [illegible]

Admiralty Chart No. 1267. Rise of Spring Tide 15½ feet (Dockyard)

BRITISH DOCKYARD—HOME WATERS.

(Divided into 2000 yard squares. Soundings in fathoms. Heights in feet.)

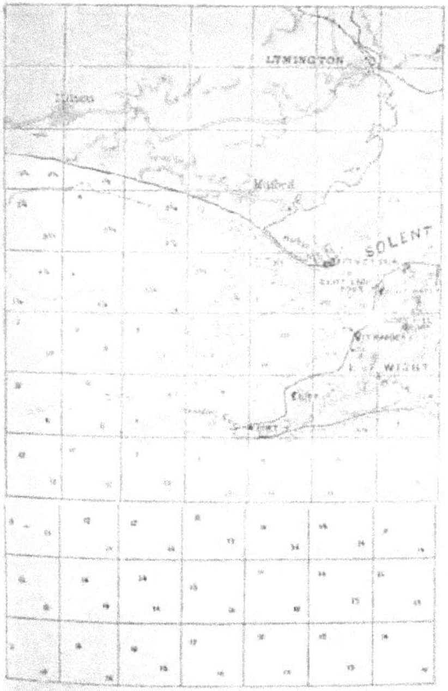

WESTERN ENTRANCE.

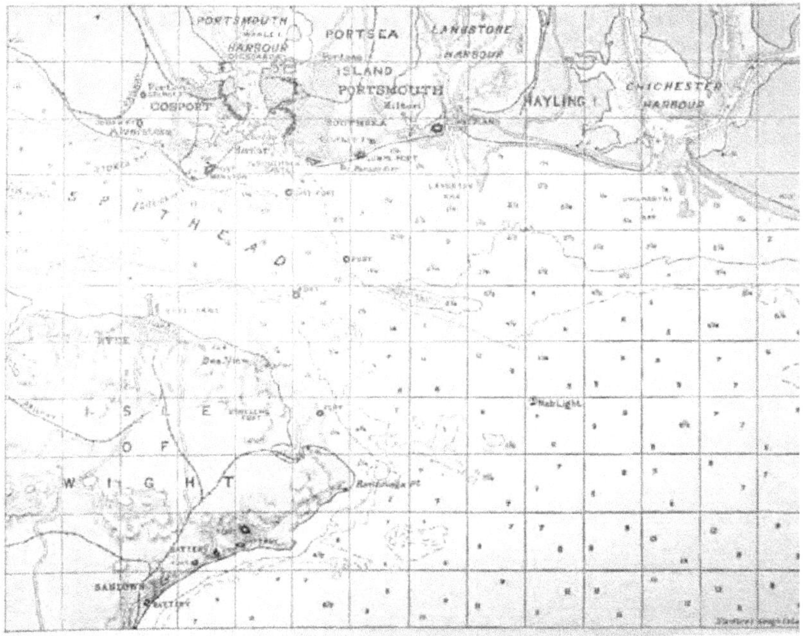

EASTERN ENTRANCE

PORTSMOUTH. One slip for building battleships or cruisers up to 750 feet.
Three large and one small basins.

Dry docks :—No. 15, 613 × 94 × 33¼ feet.
,, 14, 723 × 100 × 33½ ,,
,, 13, 563½ × 82 × 33¼ ,,
,, 12, 500½ × 80 × 32½ ,,
} Each can be lengthened by 14½ feet if desired.

Lock A, 467½ × 80 × 32¾ feet.
,, B, 467½ × 81¾ × 33¾ ,,
} Can be lengthened 17¼ ft.
,, C, 850 × 110 × 39½ ,,
,, D, 850 × 110 × 39½ ,,
} ,, ,, 71¾ ft.

and eleven smaller docks. Floating Docks 680 × 113 × 30 feet (32,000 tons) at Fountain Lake, and one small for Submarines at Haslar. Oil Fuel and Coal Storage, Mine Depot, Gunwharf, &c., &c. Total employees, about 8000 normally.

Admiralty Chart No. 2631-2045. Rise of Spring Tide, 13½ feet (Dockyard)

Note.—Breakwater from Lumps Fort due south about 2 miles.

BRITISH NAVAL PORTS AND HARBOURS—HOME WATERS.

(All maps divided into 2000 yard squares. Soundings in fathoms. Heights in feet.)

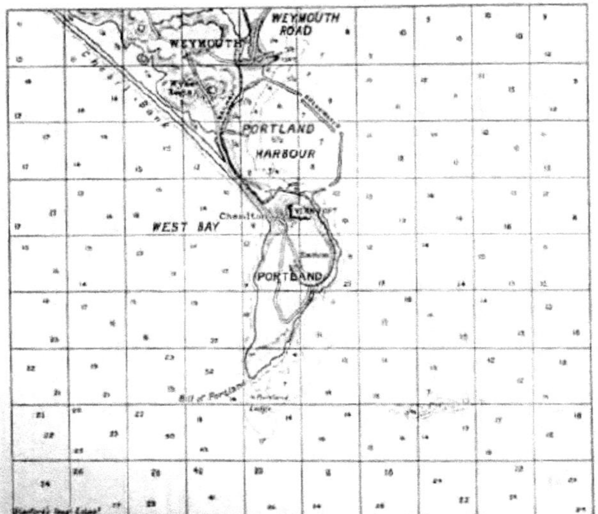

PORTLAND. 1500 acres of enclosed harbour; minimum depth, 30 feet. Floating dock suitable for destroyers and a large German Floating Dock, surrendered Sept., 1920. Coaling station and Oil Fuel Storage. N. and E. entrances 700 feet wide. Strongly fortified. Admiralty Chart No. 2255. Rise of Spring Tide, 9 feet (Bill of Portland); 6½ feet (Breakwater).

Other Naval Harbours and Stations in Home Waters.

BEREHAVEN. Good anchorage. **SCILLY.** The anchorage is moderately good. The entrance is very narrow, difficult and dangerous. **LOUGH SWILLY & BUNCRANA.** Very moderately fortified. Good anchorage. **KINGSTOWN** (DUBLIN). **TORBAY, FALMOUTH, CROMARTY FIRTH, MORAY FIRTH,** (SCOTLAND, E. COAST). **SOAPA FLOW, LOCH EWE, LAMLASH,** &c. **CAMPBELTOWN.** Submarine Training Station. **SHANDON,** GARELOCH. Anti-Submarine Station, for Experiments and Training. Oil Fuel Storage at Berehaven and Rathmullen.

DOVER. New harbour, 610 acres; depth of 30 feet at low water over half the area. E. entrance, 1 cable; W. entrance, 800 feet. Coaling station. Well fortified. Floating dock for Destroyers.

Moorings for 16 battleships, 5 large cruisers, 7 *Counties*, 4 small cruisers, and for destroyers.

(The tide at entrance is very strong, making ingress difficult.)

Admiralty Chart No. 1698-1828.

Rise of Spring Tide, 18¾ feet.

West Coast (Wales).

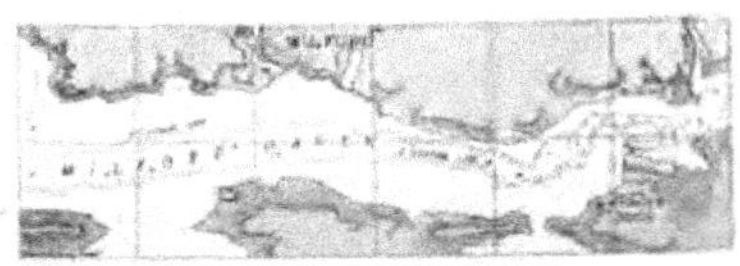

PEMBROKE. Building yard. Two large slips. Dry docks:—One, 445¾* × 75 × 25½ feet (high water). About 2500 employees.

Admiralty Chart No. 915-2393.

Rise of Spring Tide, 22½ feet (Dockyard).

* Can be lengthened by 14 feet.

BRITISH DOCKYARDS AND NAVAL HARBOURS. (HOME STATIONS).

(All maps divided into 2000 yard squares. Soundings in fathoms. Heights in feet.)

HAULBOWLINE. Dockyard. No slips. Dry docks:—
(a) Entrance, size 720 × 94 × 32½ feet.
(b) No. 1 „ 608 × 94 × 32½ „ (lengthened 1911). Employees 1000 (now more).
Admiralty Chart No. 1765. Rise of Spring Tide 11½ feet. Oil Fuel Storage.

HARWICH. Base for Torpedo Craft, Submarines, with two Floating Docks, &c. Admiralty Chart No. 2693.

IMMINGHAM. Headquarters, Humber Area. For Repairs. King's Dock (private), 733 × 56 × 22 feet.

OSEA ISLAND. River Blackwater. C.M.B. Base.

At JARROW-ON-TYNE, Admiralty Floating Dock, 680 × 113 × 36 feet (32,000 tons.)

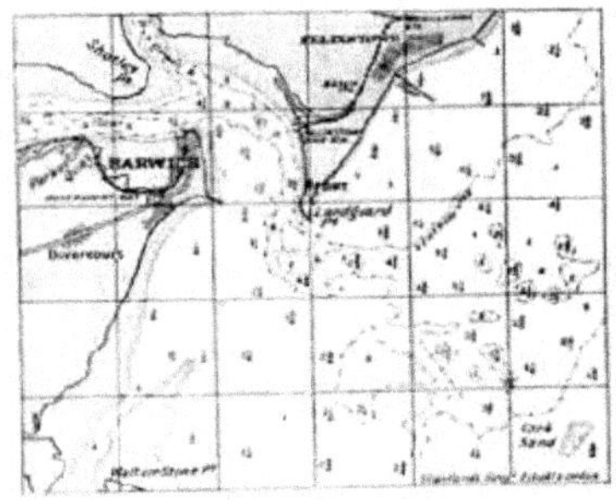

Harwich.

ROSYTH (and) **PORT EDGAR.** Fortified. Three docks, Nos. 1, 2, 3, each 854 × 110 × 40½ feet. Entrance lock 850 × 110 × 40¼ feet. Main basin 56 acres. Emergency entrance 125 feet wide × 38½ feet depth. Tidal basin 900 × 470 feet. Oil fuel. 250,000 tons concrete reservoir and 37 steel tanks for 185,000 tons. Also oil fuel storage at Port Edgar.
Admiralty Chart No. 114 B. Rise of Spring Tide, 16¼ feet.

BRITISH DOCKYARDS—**THE NORE** (Home Station).

(In squares of 2000 yards. Soundings in fathoms. Heights in feet.)

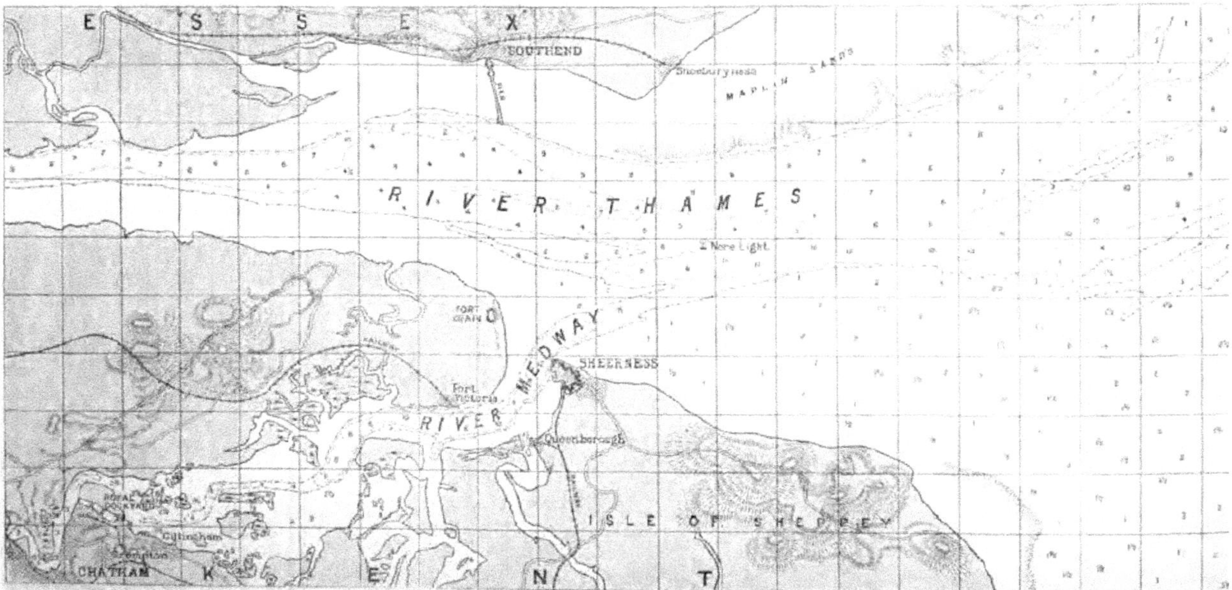

CHATHAM. Three building slips. Three large closed basins. Dry docks :—

No. 9,	656½ × 84 × 33¼ feet.	
,, 8,	457½ × 82 × 33 ,,	Each can be lengthened 17½ feet.
,, 7,	457½ × 82 × 33 ,,	
,, 6,	457½ × 80 × 32½ ,,	
,, 5,	491½ × 80 × 32½ ,,	

There are three other docks suitable only for small craft (Nos. 2, 3, 4). Also floating docks, including one German, 720 × 170 × — feet (Kiel, 40,000 tons) delivered Sept., 1920, with smaller dock inside. Total employees, 12,000. Admiralty Chart No. 1834-1607-1185. Rise of Spring Tide, 18½ feet.

SHEERNESS. No slips. One small basin. Dry docks :—Five, all small, and a small craft only. Destroyer base. Flying school and aircraft base at Fort Grain. dock, 680 × 113 × 36 feet (33,000 tons). Employees, 2,[illegible]00.

Admiralty Chart No. 1833. Rise of Spring Tide 16 feet.

BRITISH HARBOURS—**EAST COAST** (Home).

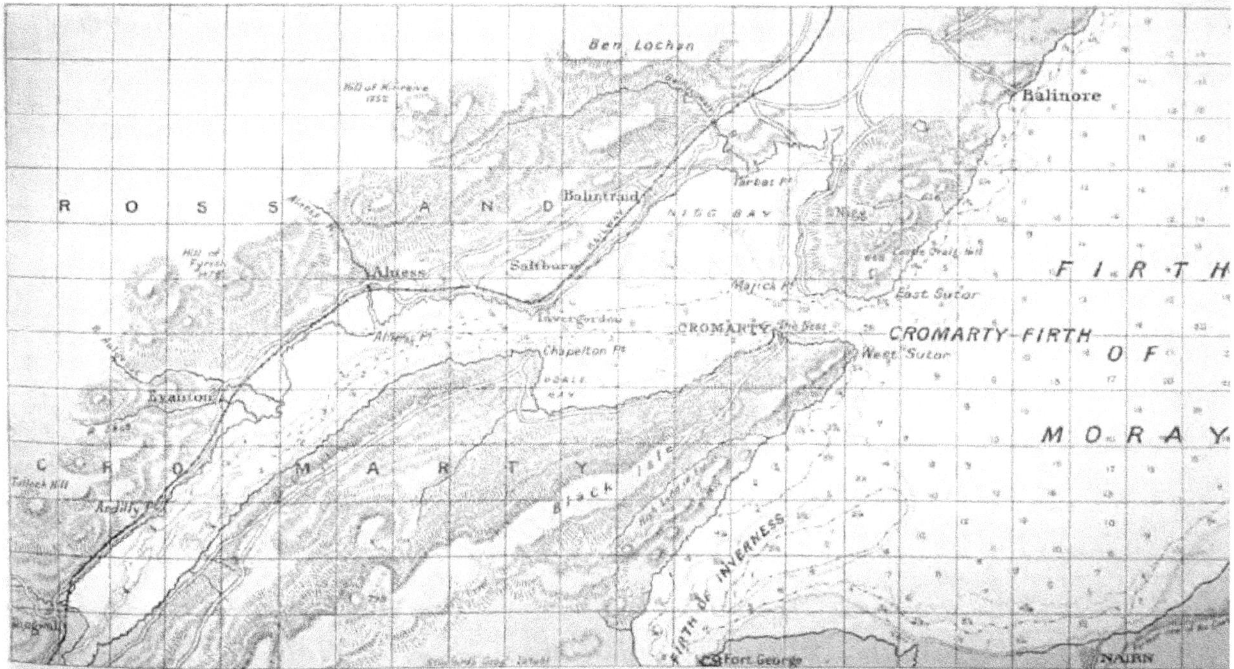

Errata: East Sutor now *North* Sutor.
West Sutor now *South* Sutor.

Cromarty Firth. Fleet Anchorage. **INVERGORDON.** Dockyard and Headquarters, N. Scotland Area.

ng Dock for T.B.D. Destroyer Anchorage may be used again. **Admiralty Chart Nos. 2180, 2581, 3729. At LYNESS, Orkneys, Fuel, Accommo**

(To uniform scale. Divided into 2000 yard squares. Soundings in *fathoms*.)

Note.

A Map of Salonika is published with Greek Navy.

GIBRALTAR. *West Harbour*—No slips. Dry docks:—

No. 1 (Prince of Wales), size, 851¼ × 90 × 38¾ feet.		Each can be lengthened by 40 ft., but actually only one caisson available for this purpose for these three Docks.
„ 2 (Queen Alexandra) „ 552 × 90 × 38¾ „		
„ 3 (King Edward) „ 450¾ × 90 × 38¾ „		

There is another dock (No. 4) for torpedo craft.

Area of harbour, 450 acres, depth 30 feet at low water. Well fortified—guns mounted 1000 feet above water. Oil Fuel Depot.

Admiralty Chart No. 144. Rise of Spring Tide, 3¼ feet (New Mole)

Private Docks.

PORT SAID. Floating Dock: 295 × 61 × 18 feet.

ALEXANDRIA. Naval and Military base during War:—
Dry Dock: 533 × 64 × 23 feet.

SUEZ CANAL. 90 miles long, 31 feet deep × 108 feet wide at bottom.

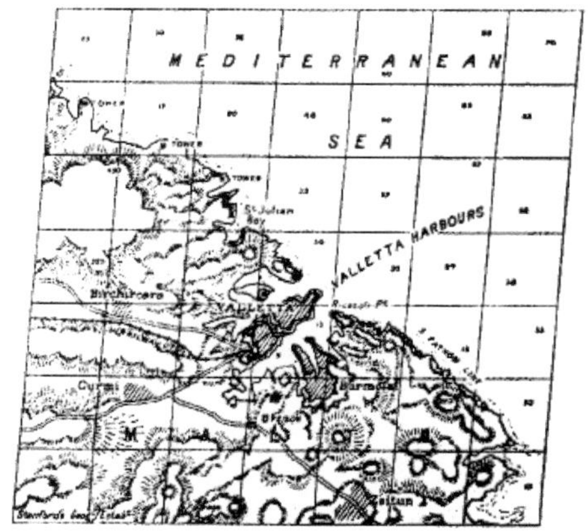

MEDITERRANEAN.

MALTA. One small building slip. Dry docks:—

No. 1 (double), 536½ × 73 × 23¾ feet.	Also one floating dock.
„ 2 (Hamilton), 520 × 94 × 33¼ „	
„ 3 (Somerset), 468 × 80 × 32¾ „	
„ 4 790* × 95 × 34½ „	
„ 5 (single), 550 × 95 × 34½ „	

* Can be lengthened by 40 feet.

Employees, 5000.

Area of war harbour, about 100 acres. New works completed 1908. Well fortified. Good anchorage. Base for Mediterranean Fleet. Large Aircraft Station being built here 1918—19.

Admiralty Chart No. 194. Rise of Spring Tide 2 feet.

(Divided into 2000 yard squares. Uniform scale. Soundings in fathoms. Heights in feet.)

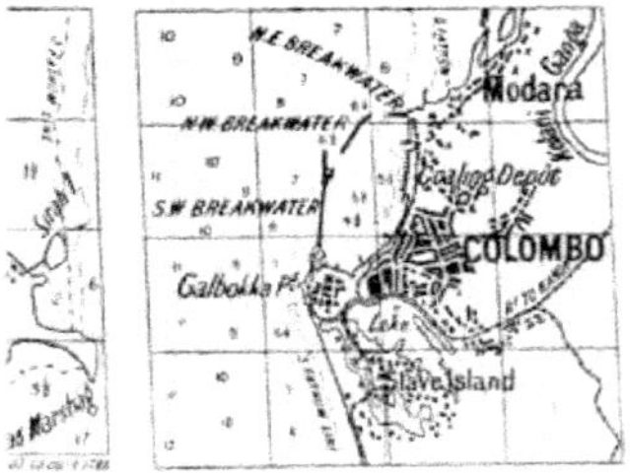

INDIES.

miles. Admiralty Chart No. 7

Excellent and deep harbour, sheltered by breakwaters. Admiralty Chart No. 3686

awkward entrance. Coaling station. Fortified. One dler.

bour. Coaling. Dry dock: 345 × 46 × 15 ft.

nd roads. Average anchorage, 10 fathoms. Tanjong (2) Albert, 471 × 59½ × 21. Keppel Harbour: (1) 380 3 × 100 × 34 (*Dreadnought* and to take any warship).

ee Royal Indian Marine Section.

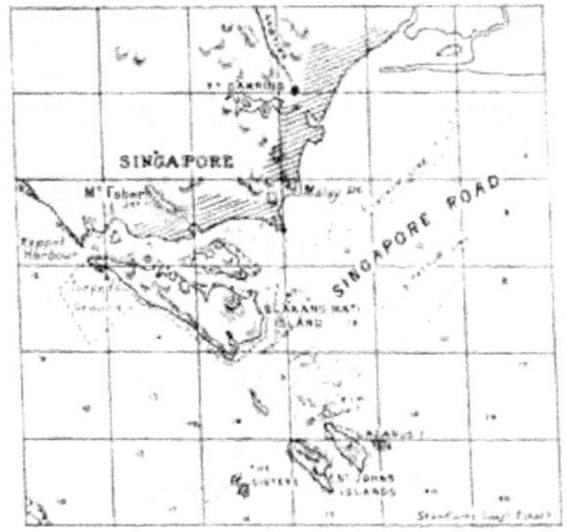

SINGAPORE.

STATION.

Dry docks:—Admiralty No. 1, 574½ × 95 × 39½ feet (Butterfield & Swire), 750 × 88 × 34½ feet (*Dreadnought*) 30 feet (*Dreadnought*). Hope Dock, 432½ × 84 × 24 feet. three smaller, able to take torpedo craft, etc. Oil Fuel

are the property of the Hong Kong & Whampoa Dock Average depth of harbour, 40 feet. Admiralty Chart fuel.

ations begun but never finished. Coaling station.

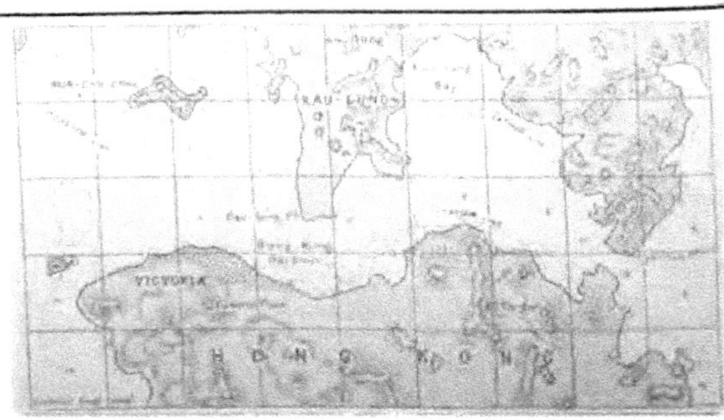

HONG-KONG. (China Station).

(Maps to Uniform Scale. Divided in 2000 yard squares. Soundings in fathoms.)

BERMUDA.

N. America and West Indies.

BERMUDA. Naval Dockyard. Fortified. Large floating dock, 545 × 92¾ × 33 feet (*Dreadnought*, 17,500 tons). Shallow harbour. Admiralty Chart No. 360. Oil Fuel Depot.

PORT ROYAL (Jamaica). Good Anchorage and Oil Fuel Depot.

[illegible]TOWN (Barbados). Coaling. Open roadstead. There is a small dock here, able to take vessels drawing up to 14 feet.

ST. JOHN'S (Newfoundland). There is a private dock here, 569¼ × 85 × 24 feet. (*Dreadnought, if lightened.*)

[illegible]INIDAD. 4,000 tons floating dock, 305 × 65 × 18 feet.

[illegible] ESQUIMAULT and Docks in the Dominion of Canada transferred to Royal Canadian Navy Section.

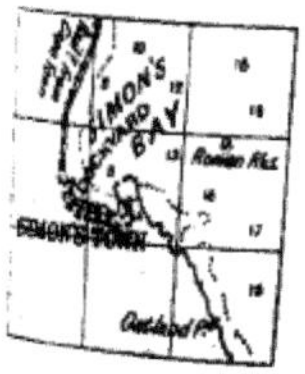

Africa Station.

CAPE COAST CASTLE (British Gold Coast, W. Africa). Fortified coaling station. Anchorage average 28 feet.

ST. HELENA. Fortified coaling station. 10-fathom harbour.

SIERRA LEONE (W. Africa). Fortified coaling station. Anchorage, 6-16 fathoms.

SIMON'S BAY (Cape of Good Hope). Naval dockyard. Selborne Dock, 740* × 95 × 36½ feet. (*Dreadnought.*) Tidal basin, 28 acres, 30 feet deep. Fortified moderately. Admiralty Chart No. 1849. There is a dock at Capetown, 500 × 66 × 24½ feet, and a floating one at Durban, 425 × 70 × 23 feet. Congella Dock, Durban (begun Dec., 1919), 1140 × 110 × 41 feet. (*Dreadnought.*)

* Can be lengthened 23½ feet.

South America.

[illegible]T STANLEY (Falkland Is.) Good deep harbour. Fortified coaling station.

BRITISH COMMERCIAL HARBOURS AND IMPORTANT SHIPPING PORTS.

British Private Docks, at Home, able to take Dreadnoughts.

There are 235 dry docks of considerable size in U.K. outside Royal Yards.

HEBBURN (Graving)	six, 700 × 90 × 28½ feet.
BRISTOL (Avonmouth)	„ 855 × 100 × 34 „
SOUTHAMPTON (No. 5)	„ 745 × 91 × 28½ „
„ (No. 6)	„ 782 × 100 × 31 „
LIVERPOOL (Canada Graving)	„ 925½ × 94 × 31 „
„ (Brocklebank Graving) ...	„ 799½ × 93¾ × 31 „
„ (Gladstone)	„ 1050 × 120 × 43 „
BIRKENHEAD (No. 3)	„ 750 × 85½ × 26 „
„ (No. 6)	„ 708 × 82 × 30½ „
„ (No. 7)	„ 861 × 92 × 33½ „
GLASGOW (No. 4 not complete)	„ 1020 × 110 × 40 „
BELFAST (Thompson)	„ 850 × 96 × 35½ „
„ (Alexandra)	„ 787¾ × 80 × 25½ „

Principal Commercial Harbours (1914).

(In order of importance, with approximate tons of shipping clearing annually.)

London (9 million); Liverpool (7 million); Cardiff (6¾ million); Newcastle and district (4½ million); Hull (2½ million); Glasgow (2 million); Southampton (1½ million); and Newport, Blyth, Swansea, Sunderland, Leith, Dover, Kirkcaldy, Grimsby, Middlesbrough, Grangemouth (all doing about 1 million); Harwich, Manchester, Bristol, Goole, Hartlepool, Belfast, and Folkestone (all doing from ¾ to ½ million).

Mercantile Marine.

From "Bureau Veritas" 1919—20 figures. Ships under 100 tons *gross* excluded.

8418 Steam and Motor Ships 18,531,742 tons *gross*.
101 Vessels (with Auxiliary Motors) of 42,508 tons *gross*.
4562 Sailing Vessels of 687,985 tons *gross*.

During 1919 British Shipyards launched 612 Vessels of 1,620,442 tons *gross*; some of these for foreign account.

BRITISH WARSHIP BUILDERS.

Note.—The headings give the abbreviated titles by which builders are mentioned on later Ship Pages. With a few exceptions, all details given below were kindly approved or furnished in 1919 by the firms mentioned.

Armstrong.

Note.—Hitherto it has been the custom to refer to this firm as "Elswick." The Elswick Shipyard has ceased naval construction, but was re-opened 1920 for Mercantile Work only. From now onwards, the firm will be mentioned as "Armstrongs" or "Armstrong Whitworth."

SIR W. G. ARMSTRONG, WHITWORTH & CO., Ltd. (Newcastle-on-Tyne). *Armstrong Naval Yard.* 81 acres. Frontage about a mile. Nine slips, 1000 to 500 ft. long. *Walker Shipyard.* Area 30 acres. Frontage 317 yards. Six slips from 650 to 450 ft. long. Total employees, including ordnance, armour plate, marine engine and steel works, about 31,000.

Beardmore.

WM. BEARDMORE & CO., Ltd. (Naval Construction Works, Dalmuir, nr. Glasgow). All classes of ships, naval and mercantile, up to largest size and highest speed. Engines of all types and sizes. Six big slips, four smaller and four 500 feet being laid out 1920. Employees circa 10,000. This yard is designed for vessels up to 900 ft. long. [illegible] basin of 8 acres, 28 ft. at low water [illegible]

rd.

RE). Area of yard, 108 acres. Six slips
cks, five small and No. 6, 708 × 80 × 12
acres; inner basin, 2¾ acres. Annual
itput. Establishment consists of North
assenger steamers of cross-channel type
w) Yard, warships and mercantile vessels
light cruisers, flotilla leaders (speciality),
r Diesel Engines. Equipment of yard
000.

.

Works, Clydebank, (Glasgow). Area,
berths: Five of 950 to 600 ft. in length,
hs are commanded by derricks, tower and
d basin: 5¼ acres in area, 30 ft. depth,
t. width respectively. Basin commanded
f 5 to 30 tons capacity. Builders of war
st dimensions, inclusive of machinery and
, geared-cruising and all-geared types.
of H.M.S. *Hood*), 313,600 H.P. (exclusive
shipyard and engine works, 8500 to 10,000.

RTON, N.B.). Area: 60 acres. Building
s, one 475 ft., one 910 ft. Numerous cranes
Destroyers, torpedo boats, submarines and
ry, &c.

.

Sunderland). Builders of vessels up to
rbine and Diesel marine engines up to any
unkering craft. Output over 100,000 tons

d.

ERING CO. (Govan, Glasgow). Area:
build ships up to 900 ft. in length. Dock:
Wet basin: 900 ft. long and ten docks up
antile ships, engines, boilers, &c., of all types.

Wolff.

arge slips in main yard (two 860 × 99 ft.).
ion in new yard. Two large docks and three
s: 20,000. Also at Govan, Glasgow, 7 large
acilities at Bootle, Liverpool, and at South-
.

Leslie.

ebburn Yard, Hebburn-on-Tyne). Twelve
× 21 ft. on blocks; 26 ft. draught at quay
I.H.P. per annum. Designers and builders
Engine every type of war vessel. Builders
s. Employees: 6-7000.

rs.

(Hebburn-on-Tyne). All classes of war and
rating engines, water-tube and Scotch boilers.
and machinery. Fourteen slips. Employees:
29 ft.; Jarrow, 440 × 70 × 18 ft.; also
. Employees: 9000.

Scott's.

SCOTT'S SHIPBUILDING & ENGINEERING CO., Ltd. (Greenock, N.B.). Slips for six large and four small vessels; fitting-out basin; graving dock. Employees: 6000. Makers of heavy oil engines and licensees for Laurenti type of submarines.

Stephen.

ALEX. STEPHEN & CO., Ltd. (Linthouse, Govan, near Glasgow). Build destroyers and torpedo craft; also mail, passenger and cargo steamers. Machinery: all types, 40,000 H.P. output per annum. Boilers: Scotch, cylindrical and all water-tube types. Seven building berths for building ships up to 700 ft. long. Water front: 1500 ft. Area of yard: 52 acres. Repairs of all classes to hulls and machinery at firm's Govan graving dock. Works adjacent to Glasgow Docks. Employees: 3500.

Swan Hunter.

SWAN, HUNTER & WIGHAM RICHARDSON, Ltd. (Wallsend-on-Tyne). Twenty-one building berths, fifteen of which are served by overhead electric cranes. Four berths covered in. Employees: about 8000. Annual *gross* shipbuilding capacity (1918): 150,000 tons. Engine works: 100,000 H.P. output per year.

The dry docks dept. includes a large repairing yard with two graving docks and two floating docks. Engine works have developed the Neptune and Polar marine oil engines. Total area of works: 78 acres. Water frontage: 4000 ft. Shipyard also at Southwick-on-Wear, with three building berths. Allied firms are the Wallsend Slipway & Engineering Co., Ltd., Wallsend; Barclay, Curle & Co., Ltd., of Whiteinch, Govan, Elderslie and Glasgow.

Thornycroft.

JOHN I. THORNYCROFT & CO., Ltd. (Woolston, Southampton). Builders of torpedo boats, destroyers, light cruisers, merchant vessels up to 400 ft. Thirteen building berths, including three covered in. One hauling-up slip for ships up to 170 ft. long. Machinery output for 1918 was 211,843 H.P. Employees: about 5,000. Water frontage: 2000 ft. Opposite Southampton Docks. Specialities: Turbines, water-tube boilers, oil fuel gear. Workshops in Southampton Docks, adjacent to fitting-out dock for hull and machinery repairs. Total floor area: 25,000 sq. ft.

Vickers.

VICKERS, Ltd. (Barrow-in-Furness). Area of shipbuilding department only: about 50 acres. Thirteen building berths, viz., one, 520 ft. long; two (covered-in), 420 and 330 ft. long; three, 450 to 350 ft. long; five (covered-in), 330 ft. long; all berths served by electric or steam travelling cranes, lattice work derricks and electric winches. Fitting-out basin equipped with two 150-ton Giant Hammer-head cranes, 30-ton electric jib crane, two 10-ton steam derrick cranes, one 18-ton steam travelling crane, and numerous other cranes, winches and capstans; also 20-ton floating crane. One floating dock, 2000 tons lift; new floating dock, 420 × 69½ (clear) ft., to lift 5200 tons. One graving dock, 500 × 60 × 22 ft. Shops for building boilers, engines, electric equipment, shells, &c. Employees (engineering and shipbuilding): about 30,000. (Oversea Branch: Canadian Vickers Co., Montreal.)

White.

J. SAMUEL WHITE & CO., Ltd. (Cowes). The oldest shipyard on the Admiralty List. Light cruisers, gunboats, flotilla leaders, destroyers, submarines, minelayers, patrol boats, turbine pinnaces, and small naval craft of every kind. Seven building berths. Engine works for reciprocating and turbine engines of highest powers, "White-Forster" water-tube boilers, "White" Diesel engines, and "White" patent oil fuel installations. Hammer-head 80-ton crane. Employees: 4600.

Yarrow.

YARROW & CO., Ltd. (Scotstoun, Glasgow). Area of yard: 12 acres. Water frontage: 750 ft. Nine building berths, for ships up to 400 ft. long. Wet basin for fitting-out, 350 × 85 ft., served by 50-ton crane. Specialities are destroyers, fast yachts, vessels for shallow river navigation, both of stern-wheel type and of type propelled by screws working in tunnels. Yarrow water-tube boilers and Yarrow superheater, both for land and marine use. Also vessels propelled by internal-combustion motors. Employees: about 2000. (Overseas Branch: Yarrows, Ltd., Victoria, B.C.—see Canadian Navy Section.)

Other Private Firms in the British Isles.

majority build engines, &c. Those whose names are in heavy type built Monitors, Sloops, Minesweepers, Patrol Gunboats, Patrol Boats, &c. during the War.

Shipbuilders.

ABDELA, Isaac J. & Mitchell, Ltd., Queen's Ferry, Chester.
AILSA Shipbuilding Co., Ltd., Troon.
ALLEY & McLellan, Ltd., Polmadie, Glasgow.
ARDROSSAN Dry Dock and Shipbuilding Co., Ltd., Ardrossan.
AUSTIN, S. P. & Sons, Ltd., Sunderland
AYRSHIRE Dockyard Co., Irvine, N.B.
BAIRD Bros., North Shields.
BARCLAY CURLE & Co., Ltd., Whiteinch, Elderslie, Stobcross, Kelvinhaugh, Govan, all near Glasgow.
BARTRAM & Sons, Ltd., Sunderland.
BEACHING Bros., Ltd., Great Yarmouth.
BLUMER, John & Co., Sunderland.
BLYTH Shipbuilding and Dry Docks Co., Ltd., Blyth.
BLYTHSWOOD Shipbuilding Co., Scotstoun, Glasgow.
BOW, McLachlan & Co., Paisley.
BROOK & Co., Lowestoft.
BROWN, George & Co., Greenock.
BURNTISLAND Shipbuilding Co., Burntisland (near Rosyth, N.B.)
CAIRD & Co., Ltd., Greenock, N.B.
CALEDON Shipbuilding and Engineering Co., Ltd., Dundee & Stannergate.
CAMPBELTOWN Shipbuilding Co., Campbeltown.
CAMPER & NICHOLSON, Ltd., Gosport.
CHAMBERS, J., Lowestoft.
CLYDE Shipbuilding and Engineering Co., Ltd., Port Glasgow.
COASTER CONSTRUCTION Co., Rossie Island, Montrose.
COCHRANE & Sons, Ltd., Selby.
COLBY Bros., Lowestoft.
CONNELL, Charles & Co., Ltd., Scotstoun.
COOK, Welton & Gemmell, Ltd., Thornaby Yard, Beverley.
COX & Co., Engineers, Ltd., Falmouth.
CRABTREE & Co., Ltd., Great Yarmouth.
CRAIG, Taylor & Co., Ltd., Stockton-on-Tees.
CRAN & SOMERVILLE, Leith.
CRICHTON & Co., South Saltney, Chester.
CROWN, John & Sons, Ltd., Sunderland.
CUMMING, D. M., Ltd., Blackhill Yard, Glasgow.
DAY, Summers & Co., Southampton.
DIBBLES (1918), Ltd.
DIXON, Sir Raylton & Co., Ltd., Cleveland Yard, Middlesbrough-on-Tees.
DOBSON, William & Co., Newcastle-on-Tyne.
DUBLIN Dockyard Co., Ltd. Dublin.
DUNCAN, R. & Co., Ltd., Port Glasgow.
DUNDEE Shipbuilding Co., Ltd., Dundee (also *v.* Goole Shipbuilding Co.)
DUNLOP, Bremner & Co., Ltd., Port Glasgow.
DUTHIE Torry Shipbuilding Co., Aberdeen.
EARLE'S Shipbuilding and Engineering Co., Hull.
EDWARDS & Co., Ltd., Milwall.
ELTRINGHAMS, Ltd., Wallington Quay, South Shields.
FAY, J. G. & Co., Ltd., Southampton.
FELLOWS & Co., Ltd., South Town Dry Docks, Great Yarmouth.
FERGUSON Bros., Port Glasgow.
FIFE, W. & Sons, Fairlie.
FINCH, E. & Co., Ltd., Chepstow.
FLEMING & Ferguson, Ltd., Paisley.
FORBES, J. & G., Sandhaven, Fraserburgh.
FORTH SHIPBUILDING Co., Alloa.
FULLERTON, J. & Co., Paisley.
FURNESS Shipbuilding Co., Haverton Hill-on-Tees.
GARDINER Shipbuilding and Engineering Co., Hamworthy, Poole.
GOOLE Shipbuilding & Repairing Co., Ltd., Hull, and Associated Firm, The Dundee Shipbuilding Co., Ltd., Dundee.
GRANGEMOUTH Dockyard Co., Grangemouth.
GRAY, W. & Co., Ltd., West Hartlepool and Seaton, Teesmouth.
GREENOCK Dockyard Co., Ltd., Greenock.
HALL, A. & Co., Ltd., Aberdeen.
HALL, Russell & Co., Ltd., Aberdeen.
HAMILTON, William & Co., Ltd., Port Glasgow.
HARKESS, W. & Sons, Ltd., Middlesbrough.
HAWTHORNS & Co., Ltd., Leith.
HENDERSON, D. & W. & Co., Ltd., Partick, Glasgow.
HEPPLES (1919), Ltd., South Shields.
HILL & Sons, Chas., Bristol.
HILL, Richard & Co., Poole, Dorset.
INGLIS, A. & J., Glasgow.
IRVINE'S Shipbuilding & Dry Docks Co., Ltd., Middleton Yd., Hartlepool.
LAING, Sir James & Sons, Ltd., Sunderland.
LARNE Shipbuilding Co., Larne Harbour.
LEWIS, J. & SONS, Aberdeen.
LITHGOW'S Ltd. (late Russell & Co.), Port Glasgow.
LIVINGSTONE & COOPER, Hessle, Hull.
LLOYD ROYAL BELGE (late Brys & Gylsen), Whiteinch, Glasgow.
LOBNITZ & Co., Ltd., Renfrew.
LYTHAM Shipbuilding & Engineering Co., Lytham, Lancs.
MACKIE & Thomson, Ltd., Govan.
MANCHESTER Dry Dock Co., Ltd., Manchester.
McALLISTER, R. & Sons, Dumbarton.
MacCOLL & Thomson, Ltd., Govan.
MacCOLL, Ltd., Abercorn Basin, Belfast.
McGREGOR & Sons, Kirkintilloch.
McMILLAN, A. & Sons, Ltd., Dumbarton.
MEDWAY Shipbuilding Co.,
MENZIES & Co., Leith.
MERSEY Engineering Works Co., Ltd., Liverpool.
MONMOUTHSHIRE Shipbuilding Co., Chepstow.
MONTROSE Shipbuilding Co., Ltd., Montrose.
MORRIS, J. & D., Pelaw Main (Tyne District).
MUNRO, A., Ardrishaig.
MURDOCH & Murray, Ltd., Port Glasgow.
NAPIER & Miller, Ltd., Old Kilpatrick.
NEWCASTLE Shipbuilding Co., Hebburn-on-Tyne.
NORTH OF IRELAND Shipbuilding Co., Londonderry.
NORTHUMBERLAND Shipbuilding Co., Ltd., Howdon-on-Tyne.
OLDERFLEET Shipbuilding Co., Larne.
OSBOURNE Graham & Co., Ltd., Sunderland.
OUSE Shipbuilding Co., Hook, near Goole.
PHILIP & Son, Ltd., Dartmouth.
PICKERSGILL, Wm. & Sons, Ltd., Sunderland.
PIMBLOTT, I. & Sons, Ltd., Northwich.
PITCHERS, Ltd., Great Yarmouth.
PRIESTMAN, J. & Co., Sunderland.
RAMAGE & Ferguson, Leith.
READHEAD, J. & Sons, Ltd., South Shields.
RENNIE RITCHIE & NEWPORT Shipbuilding Co., Ltd., Rutherglen (Glasgow), Whiteinch (Glasgow) and Wivenhoe, Essex.
RENNOLDSON, Chas. & Co., South Shields.
RENNOLDSON, J. P. & Sons, South Shields.
RICHARDSON, DUCK & Co., Ltd., Stockton-on-Tees.
ROBERTSON, A. & Co., Sandbank, Argyllshire.
ROBERTSON, A. W. & Co., Canning Town.
RODGERS, A. & Co., Port Glasgow.
ROPNER & Sons, Ltd., Stockton-on-Tees.
ROSE Street Foundry and Engineering Co., Inverness.
SCARR, Henry, Hessle.
SCARR, T. H., Howden.
SCOTT & Sons, Bowling, nr. Glasgow.
SHORT Bros., Ltd., Pallion, Sunderland.
SHRUBSALL, H., E., Greenwich.
SIMONS, Wm. & Co., Ltd., Renfrew.
SIMPSON, Strickland & Co., Ltd., Dartmouth.
SMITH'S Dock Co., Ltd., Middlesbrough-on-Tees.
SUMMERS & PAYNE, Ltd., Southampton.
STOTHERT & Co., Ltd., Bristol.
SUNDERLAND Shipbuilding Co., Ltd., Sunderland.
THOMAS, W. & Sons, Amlwich, N. Wales.
THOMPSON, J. L. & Sons, Ltd., Sunderland.
THOMPSON, R. & Sons, Ltd., Sunderland.
TYNE Iron Shipbuilding Co., Ltd., Willington-Quay-on-Tyne.
WALKER, C. H. & Co., Ltd., Sudbrook, nr. Chepstow, Mon.
WALKER, R. C., Grimsby.
WARREN, W. H., New Holland, Lincs.
WATSON, J. S., Gainsborough.
WHITE Bros., Itchen, Southampton.
WHITE, Wm. & Sons, Cowes, I.W.
WHITEHALL Shipbuilding Co., Whitby.
WILLIAMSON, R. & Sons, Workington.
WILLOUGHBY Bros., Ltd., Plymouth.
WOOD, SKINNER & Co., Ltd., Newcastle-on-Tyne.
WORKMAN, CLARK & Co., Ltd., Belfast.
YARWOOD, W. J. & Sons, Castle Dock, Northwich, Cheshire.

Marine Engine Builders (*Exclusive of Diesel Engine builders*).

ABERNETHY, J. & Co., Aberdeen.
AITCHISON, Blair & Co., Clydebank.
ALLEN, W. H., Sons & Co.
AMOS & Smith, Ltd., Hull.
BELL, HEMSLEY, Ltd., Southampton.
BLAIR & Co., Ltd., Stockton-on-Tees.
CAMPBELL & Caldermoore, Paisley.
CENTRAL Marine Engine Works, West Hartlepool.
CLARK, G., Ltd., Sunderland.
COOPER & Greig, Dundee.
DICKINSON, J. & Sons, Ltd., Sunderland.
DUNSMUIR & Jackson, Ltd., Govan.
ELLIOTT & Garrood, Ltd., Beccles.
FISHERS, Ltd., Paisley.
GAUDLIE, Gillespie & Co., Glasgow.
GREY, G. T., South Shields.
HOLMES, C. D. & Co., Ltd., Hull.
HUTSONS, Ltd., Kelvinhaugh, Glasgow.
HOSKING, T. J., Ltd., London.
KINCAID, J. G. & Co., Ltd., Greenock.
LEWIS, J. & SONS, Aberdeen.
LIDGERWOOD, W. V. V., Speedwell Ironworks, Coatbridge.
McCOLL & Pollock, Ltd., Sunderland and Belfast.
McKIE & Baxter, Glasgow.
METROPOLITAN-VICKERS, Ltd., Trafford Park, Manchester, Rateau Turbines.
MILLER & Maclie, Ltd., Glasgow.
MUIR & Houston, Ltd., Glasgow.
MUMFORD & Co., Colchester.
NORTH-EASTERN Marine Engineering Co., Ltd., Wallsend-on-Tyne, and Sunderland.
PARSON'S Marine Steam Turbine Co., Ltd., Wallsend.
PLENTY & Son, Ltd., Newbury.
RANKIN & Blackmore, Greenock.
RICHARDSONS, WESTGARTH & Co., Ltd., Hartlepool.
RITCHIE, J., Partick.
ROWAN, David & Co., Glasgow.
ROSS & Duncan, Govan, Glasgow.
RUSSEL & Co., Glasgow.
SHIELDS Engineering and Dry Dock Co., Ltd., North Shields.
SISSON, W. & Co., Ltd., Gloucester.
VAUXHALL and West Hydraulic Engineering Co., Ltd., Luton, Bedford.
WALLSEND Slipway and Engineering Co., Ltd., Wallsend (Dock 540 ft. long).
WILLANS & ROBINSON, & Co., Rugby.

Diesel Engine Builders.

(From "The Motorship," New York.)

	Type of Engine.
Cammell Laird & Co., Birkenhead	Fullagar (2-cycle)†
North British Diesel Engine Co., Glasgow	Nlseco (4-cycle)
Clyde Shipbuilding & Engineering Co., Govan, Glasgow	Carels (2-cycle)
Harland & Wolff Diesel Dept., Glasgow and Belfast	Burmeister & Wain (4-cycle)
Wm. Doxford & Sons (Northumberland Shipbuilding Co., Ltd.), Sunderland	Junkers (2-cycle, Solid-injection)
Wallsend Slipway, Wallsend-on-Tyne	——?——?——†
North Eastern Marine Engineering Co., Wallsend-on-Tyne	Werkspoor (4-cycle)
John I. Thornycroft & Co., London	Carels (2-cycle)†
Wm. Denny Bros. & Co., Ltd., Dumbarton	Sulzer (2-cycle)
J. Samuel White & Co., East Cowes	White M.A.N. (2-cycle)
*Vickers Limited, Barrow-in-Furness	Vickers (4-cycle)
Mirrlees, Bickerton & Day, Stockton-on-Tees	Mirrlees (4-cycle)†
Willans, Robinson & Co., Rugby	Willans (4-cycle)†
Yarrows Limited, Scotstoun	M .A .N. (2-cycle)†
Fairfield Shipbuilding & Engineering Co., Port Glasgow	M. A. N. (2-cycle)†
Scotts Shipbuilding & Engineering Co., Greenock	Ansaldo-Fiat (2-cycle)
Alex. Stephen & Sons, Linthouse (Glasgow)	——?——————?——
Sir W. G. Armstrong, Whitworth & Co., Newcastle-on-Tyne	Armstrong-Whitworth, Schneider-Carels, Sulzer† (2 & 4-cycle)
Richardsons Westgarth, Middlesborough	Carels (2-cycle)
Wm. Beardmore & Co., Ltd., Dalmuir	Beardmore & Tosi (4-cycle)†
Swan, Hunter & Wigham Richardsons Ltd., Newcastle-on-Tyne	Neptuno & Polar (2-cycle)†
Palmers Shipbuilding & Engineering Co., Jarrow-on-Tyne	——?——————?——
Barclay, Curle & Co., Glasgow	Burmeister & Wain (4-cycle)
Norris, Henty & Gardners Ltd., Patricroft	Gardner (4 cycle)
Union Ship Engineering Co. Ltd., Montrose	——?—— (4-cycle)
Ruston & Hornsby, Lincoln	Nlseco (4-cycle)†
J. Brown & Co., Clydebank	——————†
Bellis & Morcom, Birmingham	Nlseco †
Crossley Bros., Openshaw	——————†
G. & J. Weir, Cathcart, Glasgow	——————†

* Also building engines in association with Messrs. Petters, Yeovil.
† With various other firms, built engines to Vickers, Ltd. plans during war.

Colour of Ships, 1920.

Warships in Home Waters and Mediterranean, all grey. Light Cruisers on China Station, Sloops on China and New Zealand Stations, white hulls, grey upperworks. Sloops in East Indies and Red Sea, and Light Cruisers in East Indies, white with primrose-yellow masts, funnels and yards. Sloops for North America and West Indies and African Stations, grey all over.

Fishery Protection Trawlers (Home Waters), black hulls and upperworks; funnels may be varied between black, white and grey, and may have rings for identification purposes. Sloops on Fisheries Protection and Patrol Duties painted as fully commissioned men-of-war.

Twin Screw Mine Sweepers in commission with three-fifths crews, painted as Reserve Ships.

Surveying Ships, believed to have white hulls with grey or yellow funnels.

Destroyers of 1st, 2nd, 3rd, 4th, 6th and Reserve Flotillas bear numbers on bows; also bands on funnels, as explained on a later page. Submarines have their Class Letter and Numbers painted on sides of Conning Tower.

Uniforms.

Admiral of the fleet one stripe more than a full Admiral.
In relative ranks, Engineers have the same *with the curl*, and with purple between the stripes (1915 (1918
" " Surgeons " " " " red " "
Paymasters " " " " white " "
Naval Instructors " " " " blue " "

Note.—On the right sleeve only, inverted chevrons are worn for every year of War Service since August 5t The first and lowest chevron in silver is for War Service up to Dec. 31st, 1914. The upper gold chevr for each year's service from Jan. 1st, 1915. Also worn by the Royal Marines, R.A.N., R.C.N., R.N.R., R.N.V.R., R.N. Sick Berth Reserve and officers of the Mercantile Marine serving in H.M. Ships and Aux

* Ring separated from top stripe.

Navy Estimates and Personnel, 1920-21.

£84,372,300. Personnel: *maximum* estimated, 136,000 Officers and Men, decreasing to by end of financial year. The reduced Fleet Establishment will be based on a pers 127,500 Officers and Men.

Board of Admiralty.

First Lord: The Right Hon. Walter Hume Long, M.P.

First Sea Lord: Admiral of the Fleet Earl Beatty, G.C.B., O.M., D.S.O.

Second Sea Lord:

Third Sea Lord:

Fourth Sea Lord:

Civil Lord:

Director of Naval Construction: Sir Eustace H. Tennyson D'Eyncourt, K.C.B.

BRITISH NAVAL ORDNANCE, HEAVY, B.L. (Capt. R. V. WALLING, R.G.A., retired.)

NOTE.—All details are unofficial, but believed to be approximately correct. Particulars given, for reference purposes, of some older marks of guns, now dismounted from H.M. Ships.

Designation, Calibre, ins.	Mark.	Length.	Weight of piece without B.M.	Weight of Projectile.	M.V.	M.E.	Weight of Charge.	Max. R.P.M.	REMARKS.
		cals.	tons cwts. qrs.	lbs.	f. s.	f. t.	lbs.		
18	I	35 ?	149 0 0 ?	3,333 ?	?	?	?	?	Monitors and "Furious" originally mounted those pieces, but they have been removed.
15	"B"	45	96 0 0	1950	2500	84,500	450	1.2	Particulars doubtful. Mk. I also reported to be 42 cals. long.
	I	45	96 0 0	1920	2655	84,070	428	1.2	
13.5	V	45	74 18 1	1250 / 1400	2700	63,190	296	1.5	With tapered inner "A" tube, without forward step, weighs 1 cwt. more. Obsolete Marks: I; II; III; IIIa, b. c, d, e, and f; IV.
12	XII	50	65 13 1	850	3010	53,400	307	2	Dreadnoughts and Battle Cruisers.
	XI } XI*	50	65 13 1	850	3010	53,046	307	2	
	X	45	56 16 1 }	850	2800	47,800	258	2	
	X*	45	56 12 1						
	IX	40	50 0 0	850	2650	39,200	246	2	Mk. IX guns in "Commonwealth" and "M 1" only :—Wt. of Charge, 254 lbs. Marks: VI, IV, V, Vw, VIII, VIIIe, VIIIv, IX, IXe, IXv, IXw, X, X*, XI, XI* and XII have been introduced into the Naval Service.
10	VII	45	30 4 2	500	2850	26,945	146¾	3	Only mounted in "Triumph," now sunk.
	VI } VI*	45	36 10 0	500	2800	27,181	146¾	3	Only mounted in "Swiftsure." Mk. VI*, land guns. Wt. given is that without "Ring, connecting hydraulic buffer." A Mk. V pattern was sealed, but the gun was never made.
9.2	XI	50	28 0 0	380	3000	23,700	128¼	4	
	XI*	50	27 15 2	380	3000	22,930	128¼	4	
	Xv	46	27 19 0	380	2800	20,685	120	4	Only two Mk. Xv guns in the Service.
	X	46	27 19 0	380	2800	20,685	120	4	There is one Mk. $\frac{X}{IX}$ gun for Naval Service.
	VIII	40	24 6 3	380	2347	14,520	66	4	"Terrible" only; two guns from "Powerful" mounted in R.G.A. Battery at Scapa Flow. Twenty different Marks have been made and fourteen introduced in Naval Service.
MEDIUM, B.L.									
7.5	III } III*	50	15 9 3	200	3000	12,481	54¼	5	Monitors, removed from "Swiftsure," Mks. IV and IV* were for "Triumph," sunk. New Mark in "Hawkins" and "Raleigh," of which no details available.
	II** } V	50	14 16 2	200	2800	12,500	61	5	
	I	45	13 9 2	200	2600	9340	61	5	"Devonshire" class only. Other Marks in Naval Service: I* and I** of 45 cal./L; II and II* of 50 cal./L.
6	XVIII	50	?	100	?	?	?	7	New ships.
	XIV	50	?	100	3100	6665	?	7	"Cleopatra" class.
	XI* } XI	50	8 8 2	100	3000	6240	32½	7	
	VIII } VII	45	7 7 2	100	2750	5250	23	7	Mk. VIII left hand guns for "T" mounting in "County" class. Many new Marks of 6 in. introduced during the War, including a short Mk. XIX for L.S. (Travelling Carriage.)
5.5	I	50	6 4 0	82	2950	4520	23½	12	Woolwich Arsenal. Details doubtful. Mounted in "Furious" and "Hood"
4.7	I	50	3 14 0	45	3000	2800	14	12	Woolwich Arsenal. Modified 4.7 inch Q.F. V (L.S.). New T.B.D's.

Also :—
Q.F. 6 in., many marks } quite obsolete, but were mounted in D.A.M.S. and S.D.V.
Q.F.C. 6 in., " " } during the War.
Q.F. 4.7 in. Mk. V, Land Service only.

Mk. IV, three-motion B.M., Mk. IV "B," same, converted to single motion.
Mk. III, " Mk. III "B," " "
Mk. II, " Mk. II "B."
Mk. I,

'ISH NAVAL ORDNANCE, TORPEDOES, MINES, &c.

LIGHT GUNS.—(See Note on previous page).

Weight of Projectile.	M.V.	M.E.	Weight of Charge.	Max. R.P.M.	REMARKS.
lbs. 30	f.s. 3000	f.t. 1631	lbs. 11¼	?	B.L.
31	3000	?	5¼	12	B.L.
31	3000	?	9½	12	B.L.
31	3000	1934	7½	20	Semi-automatic; extensively used on AA. mountings.
31	2300	1137	[illegible]	20	Semi-automatic; lighter model of Mk. V. Obsolete Marks: I, II and III Q.F. and four marks of Q.F.C.
14½	3050	992	5½	30	Semi-automatic; for AA. mountings.
12½	?	?	[illegible]	?	Q.F.; for submarines.
12½	[illegible]	[illegible]	[illegible]	25	Q.F.; for A.A.C. mountings.

loes.

ks of 21 inch, 18 inch and 14 inch torpedoes ... the third without. 21 inch Mk. IV* in ... ght 14 inch Mk. XI for use by "Cuckoo"

es.

various Marks of mines produced during ... Spherical types with swinging-bar contact ... " type. Also spherical types with horn ... mines; (5) Special types for laying by

Bombs.

nces to these types: Experimental 3300 lb. ... of 1100 lbs. T.N.T.), 530 lb., 230 lb. (with ... to be impact, delay action and special types ... various British types adopted).

harges.

... 330 and 1120 lb. types.

lights.

n latest ships). About 8—36 inch and 2— ... tle Cruisers; 4—36 inch in Light Cruisers; ... h mountings were used in *Iron Duke*, *King* ... *New Zealand* Classes, but these are said to ... d SL. Latest SL. is the 36 inch "Harrison" ... ns in a spirit flame. Beam has a purplish ... arget much better than older types. Auto-

s (PVs.)

but there are about twenty marks, roughly ... "M" type said to be "otter gear" of ... 18 kts. speed, keeping depth by ordinary ... to be for Battleships and older Cruisers up ... sers, &c., up to 25 kts. speed. Depth kept ... any oscillations," the hydrostatic valve by ... "porpoising." Length about 12 feet, ... Destroyers reported to use special types ... [illegible] in large ships by chains from ship ... [illegible] for sliding bar chains at stern. ... their PVs. from stern.

Naval Ordnance.—*continued.*

Also introduced (for Anti-Submarine purposes) during War:—

11 inch howitzer mounted in some ships of "Devonshire" and "County" Classes.
200 lb. stick bomb howitzer in some Special Service Vessels.

Thornycroft Depth Charge 9.2 inch Thrower in Destroyers, P-boats, some Sloops and Special Service Vessels.

Also:—

Q.F., 12 pr. 12 cwt. Mk. I.
" " " Mk. II (war model); mostly for merchantmen.
" 12 pr. 18 cwt. Mk. I, for "G," "H" and "I" classes of T.B.D.
" 12 pr. 8 cwt. Mk. I (boat gun).
" 14 pr. Mk. I (H.M.S. "Swiftsure"—now removed).
" 6 pr.
" 3 pr. } have been used as AA. guns.
" 3 pr. (semi-automatic, Vickers design) } Saluting guns in big ships.
" 2½ pdr. (automatic) modified "pom-pom" for AA. work.
" 77 mm.—captured German field guns, adapted for use in some submarines.

Machine Guns:—

.303 inch Vickers.
.303 inch Maxim.
.303 inch Lewis.

Rifles:—

.303 inch Lee-Enfield (Long).

Pistols:—

.45 inch Webley automatic (H.V.).

EXPLANATORY NOTES TO BRITISH NAVY SECTION. *(Introduced 1919, Revised 1920.)*

DISPLACEMENTS.

THE figures given are, in all cases, "Navy List Displacement" at normal load. Where figures are available, full sea-going displacements at deep load (exclusive of oil fuel) are added. It must be pointed out that these statistics may under-estimate the present displacements of British warships. Additions made during the War have greatly increased displacements. After the Battle of Jutland, about 2000-3000 tons of extra weights were added to British capital ships (*a*) by the re-inforcement of barbette crowns and strengthening of decks round magazines and shell rooms; (*b*) by the provision of flash-tight doors to ammunition trunks and magazines. Dreadnoughts and Battle Cruisers completed before the War were also burdened with director fire controls and enlarged searchlight equipment. Capital Ships and Light Cruisers also received paravanes, aeroplanes (with small flying-off platforms), kite-balloon winches, &c.

In the case of Light Cruisers, these additions resulted in displacement being increased by 500-800 tons. Pre-War types of Destroyers had as much as 100 tons added, in the shape of depth charges and depth charge throwers, paravanes, extra oil fuel and anti-aircraft guns. Such increases of displacement naturally resulted in deeper draughts and loss of *maximum speed*. In the case of armoured ships, the thickest section of waterline armouring is sometimes *submerged* at deep load draught.

COMPLEMENTS.

Figures given are those now ruling, but during the War complements were greatly increased. The *Renown*, for example, is said to have had a complement of 1500 towards the end of the War.

GUNNERY NOTES.

Throughout all classes, the War has resulted in increased and improved systems of gunnery controls (which, in recent forms, are allied to wireless telegraphy), by the installation or addition of director controls, large base range-finders, &c. Range-finders for anti-aircraft guns are also mounted. Rather noticeable additions are the provision of range-clocks at prominent points on the fore or after ends of large ships; the painting of deflection scales on barbette shields and "Sceptre Gear" on masts. The "Sceptre Gear" may be added originated in H.M.S. *Sceptre* when that boat was attached to the Harwich Force, and her experimental "gear" was improvised from empty cigarette tins.

Before the War, the tendency of British gunnery was towards flat trajectories with big danger zones and moderately efficient range-taking. The present trend is towards high-angle fire and better abnormal ranges. High-angle fire gives decreased danger zones, but this drawback can be surmounted by high-efficiency range-finders and controls. A very large number of lighter guns were removed from British Warships, to provide armaments for Defensively Armed Merchant Ships, Submarine Decoy Vessels, Fleet Auxiliaries, Mine Sweepers, Patrols, Small Monitors, &c., &c. Many of the guns so surrendered have not yet been remounted in the Warships from which they were removed, and it is doubtful if they will ever be replaced now.

The only new heavy calibres introduced were the 18-inch and 14-inch guns (both no longer mounted in 1920), but numerous marks of guns from 7.5-inch calibre downwards have been added, including new types of 5.5, 4.7 and 4-inch guns. Anti-aircraft guns are, of course, distributed among all classes. It is projected that all Capital Ships should have a uniform armament of 2—4 inch AA. guns. Examination of armaments will show that the mono-calibre system has become more or less standard, guns of 15 to 12 inch calibre being mounted in Capital Ships, 7.5 to 5.5 inch marks in Light Cruisers, and 4.7 to 4 inch guns in Destroyers. In the earlier "C" Class Cruisers, the mixed 6 inch and 4 inch armament was gradually altered to a uniform 6 inch on the centre-line, the single bigger calibre gun amidships on the flying deck replacing 6—4 inch in the waist.

TORPEDO NOTES.

The increase of tubes in Light Cruiser types has been remarkable. In the case of the "D" type Light Cruisers, as many as 12 tubes are mounted in triple revolving deck mountings. The new British type of triple tube, it may be remarked, weighs less than the preceding type of twin tube. The triple tube is said to have been introduced in Destroyers, not so much with a view to increasing torpedo fire, as for the sake of securing light and handy stowage for extra torpedoes. The motives underlying the addition of 18-inch secondary tubes in some T.B.D. are dealt with at length in the General Notes for the "Admiralty R" Class boats. These 18-inch tubes fired "cold" torpedoes; discharge being effected from the navigating position by means of hydro-pneumatic firing gear, a wandering lead and a firing pistol, the last named usually being stowed in the Captain's pocket. The same type of firing gear was also used for the 21-inch tubes and "heater" type torpedoes. Submerged stern tubes were removed from all Dreadnoughts and Battle-Cruisers during the War.

The *submerged* tubes in the *Renown* Class of Battle-Cruisers, *Courageous* type of Cruisers, and "C" Light Cruisers, have proved very unsatisfactory, as the bars bend when attempting to launch torpedoes at high speeds. These *submerged* tubes will probably be removed and replaced by *above water* positions.

SEARCHLIGHTS.

Controls have been devised for these, by which searchlights are controlled on target in conjunction with the guns. Battleships and Light Cruisers usually mount 8 to 4—36 inch searchlights, with a pair of 24-inch searchlights for signalling purposes, both by day and by night. The cramped bridges of Dreadnoughts and Battle-Cruisers proved very inconvenient to the signalling staff; there was also difficulty in bringing signal halyards inboard to the narrow bridges. Accordingly, searchlights were adapted for day signalling, and their use is very prevalent now. Notes on the Harrison Lamp type of SL. are given on the preceding page. Shutters, similar to the iris diaphragm of cameras, were introduced for Searchlights. These allowed SL. to be ready for instant use. With the arc burning, and the shutter closed, all light was cut off; on opening out the shutter, the light was brought into use at once, while the width of the beam could be increased or narrowed by adjustments of the shutter.

Curiosity may be aroused by the peculiar searchlight towers, added to masts and funnels of British ships, usually called "coffee boxes." The old system of SL. control consisted in having an observer stationed behind and above or to one side of the SL., the observer giving—or rather shouting—corrections in bearing, elevation, etc., to the operators. This system of control was found to be too crude and inefficient for war purposes. Accordingly, the "coffee box" tower was devised. In this, the searchlight parties are located *within* the tower, and *below* the searchlights, vision being given by sighting slits, while the searchlights themselves are worked by small motors.

Flotilla Leaders have usually 1 or 2—24 inch searchlights, and Destroyers 1—20 inch searchlight, both types being controlled in conjunction with guns. In the "Modified R" and "S" Destroyers, the searchlight is mounted on and revolves with the after pair of tubes—a detail which has been added to some of the "M" Classes of T.B.D.

AIRCRAFT NOTES.

One of the remarkable features of the War was the way in which all classes of ships (excepting Destroyers) were equipped to carry and fly off small aeroplanes from barbette crowns in big ships, and from small platforms in Light Cruisers. In the *Carlisle* and "D" types of Light Cruisers, a high "Arc de Triomphe" hangar was combined with the fore bridges, the fore portion of which can be closed by wind screens or roller shutters. These hangars have been stripped and lightened as much as possible; aircraft are rarely stowed in them now. The turret-roof platforms in Dreadnoughts and Battle Cruisers have also been dismantled in many ships. In the *Delhi* group of the "D" Light Cruisers and *Caledon*, revolving aeroplane platforms have made their appearance; they appear to be simply the turret-roof flying-off platform minus the supporting turret.

Kite-balloon winches were added to practically all types of ships from Dreadnoughts to Trawlers, but, of course, excepting Submarines. Experiments have, however, been carried out by a "K" Class Submarine, diving with a Kite-Balloon in tow. In the last months of the War, it was the custom for the Grand Fleet to go to sea with about 50 aeroplanes and 30 kite-balloons.

MINE LAYING.

Before the War, this duty was considered the *métier* of specially-constructed ships. At the end of hostilities, practically all types of ships from Cruisers down to Trawlers and C.M.B's. were engaged in sowing of "eggs." Mine laying by Destroyers proved extremely successful, and the mine-laying Submarine has fully justified its adoption. During the War, Destroyers of the "V" and "W" types were often converted for Mine-laying duties within twenty-four hours. Such boats had, of course, laying chutes and mine rails permanently built into their hulls. For conversion, it was only necessary to remove the after set of tubes, and sometimes the quarter-deck 4-inch guns, ship the mines and rig screens. On returning from a laying expedition, the boats were converted back again for normal Flotilla duties within a day.

Mine-laying Destroyers are not included in the 1920 Establishment of Destroyer Flotillas, but it should be recollected that boats which retain chutes and deck rails can be converted for this duty as rapidly as during the War. If asked to indicate two of the most remarkable developments of the War, we would be inclined to single out (*a*) the conjoint use of aircraft and warships, and (*b*) the general adoption of mine laying for all types of war vessels.

MINE SWEEPING.

The invention of the paravane resulted in all classes of ships becoming mine-sweepers of a very efficient type. The pre-war type of sweep, worked by two vessels, was extremely slow in operation and subject to continual breakdowns. The money, time and thought expended on the development of the paravane has been fully justified. In fact, this device is one of the British inventive triumphs of the War, and was adopted by the Navies of all our Allies. Compared with the pre-war form of sweep, it has been said that the "PV" can sweep three times the area in a third of the time required under the old system. Destroyers working PV's. at high speed can sweep great areas in a remarkably short space of time.

Of the types of paravanes used, space forbids description. Roughly, there are two forms, (*a*) high speed for use by Warships, and (*b*) "otter gear," for low speed ships, particularly of the mercantile class. The High Speed Submarine Sweep (H.S.S.S.) is an explosive type of PV., towed horizontally from the stern of a Destroyer. It is fitted with a firing pistol and is equivalent to a towed 14-inch torpedo. Explosive Paravanes can be fired in three ways: (*a*) on impact, (*b*) by tension of towing wire when target is fouled, (*c*) by electric current controlled from towing ship. The hydrostatic valve "cuts out" firing gear if the PV. rises above 20-ft. depth. H.S.M.S. stands for High Speed Mine Sweep; this type being fitted with extremely powerful cutter jaws for severing moorings of mines.

As regards methods of towing: Capital Ships and Light Cruisers usually tow PVs. from a skeg or sliding bar affixed to stem. Destroyers, by reason of their light draught, are less liable to hit mines forward, and therefore their PVs. are towed from their sterns. It may be remarked that the streaming of PVs. at high speeds may result in *maximum* speed of the towing ship being reduced by about one or two knots.

ARMOUR PROTECTION.

No notable increases have been made in vertical and external protection. But, in the case of horizontal armour, thickening of protection has taken place. Already there is a tendency to decrease the sections of vertical and external armour and to increase the strength of decks against high-angle fire and aero bombs. Alterations so effected to British Dreadnoughts have already been noted under the heading of Displacements. Torpedo and mine attack has produced bulge protection—a device which appears likely to become a standard feature in large British Warships.

The use of a *complete* internal citadel, with or without bulge protection, will probably become a standard feature in all battle-line units.

War experience has proved the practical inutility of the conning tower. The cramped internal space, poor vision, and the difficulty of dispersing cordite fumes, have led to the abandonment of the conning tower. In many British Dreadnoughts, it merely serves now to house a large gyroscopic compass. In the Falklands Battle, Admiral Sir Doveton Sturdee declined to use the conning tower, and directed the action from the control top on the fore tripod of H.M.S. *Invincible*—a position which has found much favour as an action station for the chief executive staff, on account of the great visibility given.

MACHINERY NOTES.

In this section, we have few remarks to make, beyond noting the fact that the geared turbine and "all oil" fuel system have secured a distinct success. At present, designed speeds must be taken with much reserve, on account of the large weights added to British ships during the War. On deep load, the units of the *Royal Sovereign* Class fitted with bulge protection cannot get much above 20 kts. The "Town" Classes of Light Cruisers, owing to added weights, are said to be good only for 22-23 kts., Destroyers, after six months' service and with foul hulls, may have their speeds reduced to 24 kts. In the North Sea, large ships were able to economise weight by reducing their fuel capacities.

On the other hand, Destroyers had to have their fuel capacities enlarged, on account of the necessity for these craft to act as an anti-submarine screen for the Battle Fleet. Until the fuel capacities of the Destroyers were enlarged, the endurance of the large ships was limited, not to the fuel carried in their own bunkers, but to the fuel carried by the tanks and bunkers of the Destroyer anti-submarine screen. If the Destroyers had left the Fleet at sea to return to their bases for fuel, the principal Battle Squadrons would have been naked to submarine attack.

It may be asked why the fuel capacities stated for Destroyers are apparently reversed, the larger capacity being given first, instead of the smaller. The explanation is, Destroyers have "peace" and "war" capacities, the "peace" fuel being carried above the water line in tanks. In war, these above water tanks cannot be used, on account of the danger from fire. However, Destroyers did, during the War, use the "peace tanks," it being arranged that they should first burn off the fuel in these compartments, so that they would be empty by the time Destroyers came into the "danger areas" of the North Sea.

APPEARANCE.

Perhaps the best example to the changes caused by the War may be realised by comparing the 1913-14 views of H.M.S. *Neptune* with views of the same ship as she is to-day. In pre-war types of Capital Ships, extra tops have been added to masts for the main director tower, control tops have been enlarged, bridges re-built to leave conning towers unhampered, and searchlight towers added to funnels and masts—all changes resulting in a great alteration to appearance.

One of the most hideous excresences is the ad "clinker screen" to the fore funnel—a screen devised clinkers and gases from the uptakes fouling the uppe a following wind. In Destroyers, the "woodbine" f greater elevation than the other stacks serves the Topmasts were generally reduced to stumps on the o war, but now that ships are reverting to oceanic c topmasts are in favour for securing large W/T. range ships, the fore topmast has been completely jettisone aerials rigged to a yard attached to the "starfish" control top, or directly to the "starfish" itself. T rig has a particularly unartistic look. The alt necessary, when a Range-Finder was added over the top; topmasts had to be dismantled in order to secu for the R.F. on all points of the horizon.

Dazzle-painting was used in practically all types has been abandoned now, but illustrations are give painted British Warships, so as to place this form of record in our pages. From the air, "dazzled" ships difficult to identify than those painted a uniform co this reason alone, variegated colouring will probab war-paint feature. The illustrations of H.M.S. *St* show the peculiar canvas triangles rigged out from between funnels in the case of other ships not illustr view to breaking up the ship's outline and makin more difficult. The U.S. Navy made experiments wi range finder" system of painting—a specialised fo painting which gave ships an extraordinary aspect.

Camouflage painting—that is the use of colours object merge into it's surroundings or background— British Warships, when they were serving in the A Gallipoli Campaign. In those cases, ships inshore v be attacked by Submarines on their seaward beam, were made to make ships indistinguishable from the hi of Gallipoli and the Adriatic Coast.

Another form of camouflage we illustrate is th dummy guns and torpedo tubes on the screens of Destroyers. The sea and cloud effects on the pa were very good, but the shadows and lighting of the s tubes, etc., were, of course, at variance with the othe details of the boats. This discord, however, could onl at close range.

FLEET ESTABLISHMENT.

In the pages describing H.M. Navy, all Ships on List, 1920, have been deleted. These vessels are not practically all will pass on to the Sale List at the first as they are not required for the Fleet Establishmen grounds we have expunged all such ships from our p may be brought back into service, to meet special In this event, details of vessels restored to the Activ Lists will probably be found on reference to our 191

BRITISH WARSHIPS: RECOGNITION SILHOUETTES.

By E. L. King.

Scale: 125 feet to 1 inch.

"Fly" class.
(River G.B.).

Killerig.

Sir Bevis.

Francol class.
(Oilers).

Sir Hugo.

Later "Hunt" class.
Alresford, Leamington, &c.
(Minesweepers and Tenders, &c.)

Generally same as later "Hunt" class.

Beaufort class
(Surveying Ships).

Racer.
(Salvage Vessel).

Dwarf class.

Espiegle.
(Sloop).

TWO FUNNELS ABREAST.

Aphis class.
(River Gunboats. Two funnels abeam wide apart).

Scale: 250 feet to 1 inch.

Argus.
(Aircraft Carrier. No funnels. Masts, derricks, chart house, jackstaff and wind screen over dotted area all disappear).

M22, M29, M31, M33.
(Minelayers. Guns now replaced by derricks).

Eagle.
(Aircraft Carrier. Funnel and island superstructure on starboard beam).

Marshal Soult.

Furious.
(Aircraft Carrier).

Erebus.
Terror.

Courageous,
Glorious.

Gorgon.
(Sale to Roumania pending).

Royal Sovereign class.

(One Funnel types continued on next page.)

BRITISH WARSHIPS.—RECOGNITION SILHOUETTES.

(E. L. KING.)

Note.—Majority of Ships on this page of Yacht and/or Mercantile Build.

Scale: 125 feet to 1 inch.

HUSSAR.
(*Despatch Vessel.*)

SNAPDRAGON.
(*Target Towing Ship.*)

CORNFLOWER
GODETIA

Other Ships.

"FLOWER" class (*Acacia, Arabis, Azalea* types).
Are Fleet Sweeping Vessels Sloops).

CHELMSFORD, CHELTENHAM
(*First type built.*)

LEWES.
(*Second type built.*)

"RACECOURSE" class
(*Paddle Minesweepers.*)

[illegible]

Scale: 250 feet to 1 inch.

ADAMANT.
ALECTO.
(*For S/M.*).

MAIDSTONE.
(*S/M Depot Ship.*)

ARK ROY
(*Aircraft Ca*

TITANIA.

LUCIA.

AMBROSE.

(*Submarine Depot Ships.*)

WOOLWICH.

GREENWICH.

SANDHURST.

(*T.B.D. Depot Ships.*)

CYCLOPS.

ASSISTANCE.

(*Repair Ships.*)

PEGAS
(*Aircraft C*

DILIGENCE.

BLENHEIM.

(*T.B.D. Depot Ships.*)

VULCAN.
(*Submarine Dep*

BRITISH WARSHIPS.—RECOGNITION SILHOUETTES. **(Revised 1920).**

(E. L. KING).

TWO FUNNELS. **Light Cruiser and Battleship Types.** TWO FUNNELS.

Scale of all: 250 feet to 1 inch.

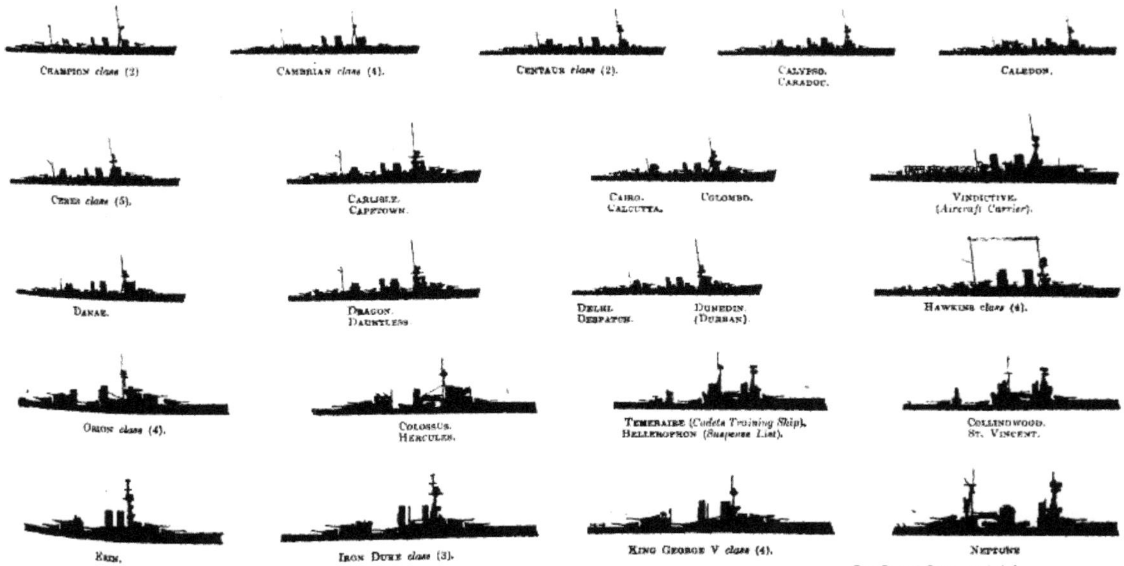

(Two Funnel Types

BRITISH WARSHIPS.—RECOGNITION SILHOUETTES.

(E. L. KING.)

(**Revised and re-arra**

TWO, THREE & FOUR FUNNELS.

TWO, THREE & FOUR FU

Scale of all: 250 feet to 1 inch.

COMMONWEALTH.

RENOWN.
REPULSE.

QUEEN ELIZABETH *class* (5).

HOOD.

THREE FUNNELS.

ARETHUSA *class* (6).

PRINCESS MAR
(*Minelaye*

CAROLINE *class*
(other 5 ships).

CLEOPATRA.
(6th Unit, "Carolin

NEW ZEALAND.

LION, PRINCESS ROYAL.

TIGER.

FOUR FUNNELS.

ANTRIM.
CARNARVON.

WEYMOUTH *class*.

BIRMINGHAM.
LOWESTOFT.

CHATHAM *cla*

BRITISH DESTROYERS.—RECOGNITION SILHOUETTES.

TWO, THREE & FOUR FUNNELS. (E. L. KING.) (Revised and re-arranged 192[illegible]) TWO, THREE & FOUR FUNNE[illegible]

Scale of all: 128 feet to 1 inch.

Yarrow "M"

As "Yarrow M" but no tubes.

RAPID.

Thornycroft "S."

Spencer, Shakespeare, Wallace. (*Leaders.*)

Admiralty "Modified R."

Yarrow "S."

Admiralty "S."

Bruce, Douglas, &c.
"Large Admiralty type." (*Leaders.*)

"V" *class.* (*Leaders.*)
(*High foremast. Not all fitted with 6 tubes yet*).

Thornycroft "V" and "W" *classes.*
(6 tubes).

Admiralty "R's" (A).

Admiralty "R's" (B).
(With Poulsen W/T.)

Admiralty "V" and "W" *classes* (6 tubes).
(Admiralty "Later W" (*Verity, Wren, &c.*) almost same.

Admiralty "Later W" *class.*
(*Whitehed, Veteran, &c.*).
Thornycroft *Wishart, Witch,* similar.

Thornycroft "R" *class.*
(*Mainmast may be between tubes in a few boats*).

Thornycroft "M" *class.*

Admiralty "V" *class* (5 tubes).

Similar to 5 and 6 tube boats.

Admiralty "V" *class* (4 tubes).

Grenville, Seymour, Saumarez.
(*Leaders.*)

Nimrod, Abdiel* (*Leaders*).
(*No tubes; funnels all same height)

BRITISH C. M. B., M. L., PATROL BOATS AND SUBMARINES.—RECOGNITION SILHOUETTES.

(E. L. KING.)

Scale: 62·5 ft. to 1 inch (or 1/750th).

55-ft. C.M.B., at rest.*

55-feet C.M.B. at full speed showing typical stern wave*

Motor Launches.

* In some boats, the short mast shown at conning position is fitted instead of the two larger masts.

ONE FUNNEL. **(Scale 125 feet to 1 inch).**

P 31—59.
Patrol Boats.

R 1—4, 7, 8, 11

R 9 10.

R

Scale: 85 feet to 1 inch.

"H 21—32 & Later" *class.*

"E 7—20 & Later" *class.*

"G" *class.*

A. { L 1—8 / L 9—33 } *classes.*

K 2, 3, 6, 8—11, 14, 15, 22.

B. { L 1—8 / L 9—33 } *classes.**

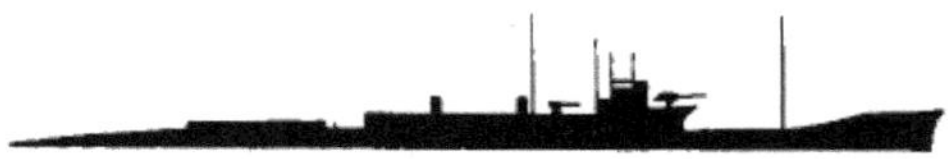

K 5, 12, 16.

C. { L 1—8 / L 9—33 } *classes.**

"M" *class.*

L 50 *class.**

* Vary a little in details. Deck before C.T. m hull may have straight run down to stern. ing and shape of gun screen also vary in tv

1913-14 BRITISH DREADNOUGHTS.

(Royal Sovereign Class.)

ROYAL SOVEREIGN (20th April, 1915), **ROYAL OAK** (17th Nov., 1914), **RESOLUTION** (14th Jan., 1916), **RAMILLIES** (12th Sept., 1916), **REVENGE** (29th May, 1915).

Normal displacement, 25,750 tons (about 31,250-33,500 tons *full load*). Complement, 937-997.

Length (*over all*), 624½ feet. Beam, 88½ feet.* *Mean* draught, 27 feet. Length (*p.p.*), 580 feet.

*For *R.S.* and *R.O.* With bulge protection, *Revenge* and *Resolution* 101' 5", *Ramillies* 102½ feet.

Guns:
- 8—15 inch, 42 cal. } **Dir. Con.**
- 14—6 inch, 50 cal. }
- 2—3 inch (*anti-aircraft*)
- 4—3 pdr.
- 5 M.G.

Torpedo tubes (21 inch):
- 4 *submerged*

Armour (H.T.):
- 1" Fo'xle over Battery
- 1½"—1¼" Upper
- 2", 1½", 1" Main
- 2½", 1" (forw'd) } Lower
- 4", 3", 2½" (aft) }
- Special Protection:
- 1½"—1" Internal screened between end barbettes

Armour (K.C.):
- 13" Belt
- 6"—4" Belt (ends)
- 1" Belt (bow)
- 6", 4" Bulkheads (f. & a.)
- 6" Battery
- 10"—7" Barbettes
- 13"—5" Gunhouses
- 1½" Funnel uptakes
- 9"—3" C.T. Base
- 11" C.T. (6"—3" Hood)
- 6" Fore Com. Tube
- 6" Torp. Con. Tower
- 4" Tube (T.C. Tower)

Royal Sovereign.

1919 Photo, Cribb, Southsea.

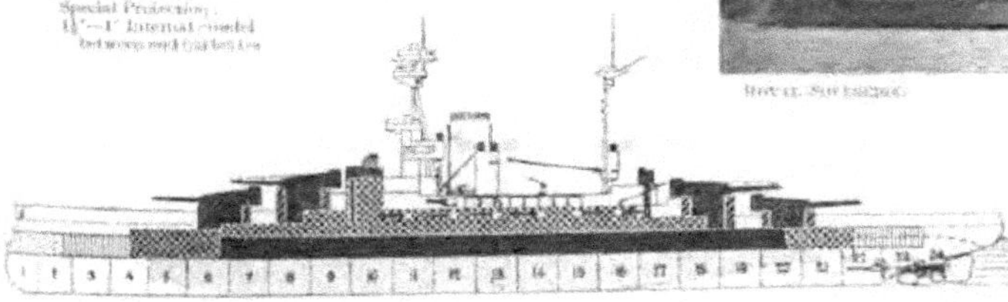

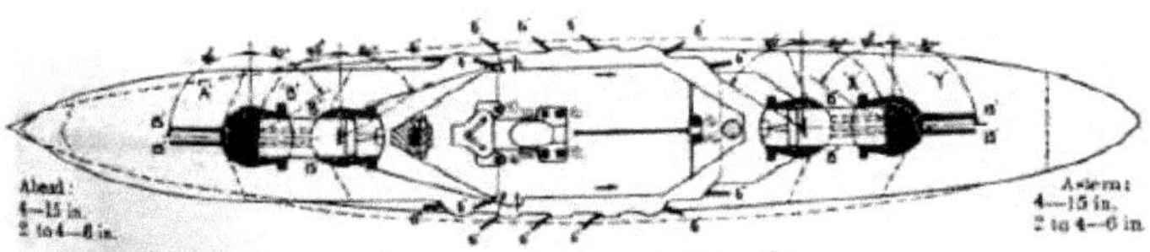

Ramillies. Aircraft View.

1919 Official R.A.F. Ph

Broadside: 8—15 in., 7—6 in., 2—21 in. tubes.

Machinery: Turbine, Parsons. Boilers: (see *Notes*). 4 screws. Designed H.P. 40,000 = 23 kts. *without* bulges, about 21½-22 *with* bulges. Fuel: Oil only, *normal*, 900 tons; *maximum*, about 3400 tons. Coal: 140 tons (for culinary purposes and boats).

Gunnery Notes.—Much as *Queen Elizabeth* class. Battery differently disposed in these ships. 6 inch batteries are wet in head seas, but dwarf walls in battery retain water and it is rapidly drained away.

Armour Notes.—Thicknesses much as *Queen Elizabeth* class, but armour differently distributed. Barbettes 6"—4" as they descend behind belt. Gunhouses, 13" face, 11" sides and rear; crowns 5". In these ships 2" protective deck has a high 2" slope behind belt, so that flat part of protection can be put on main deck and at top of belt, instead of a deck lower. Internal protection is very good, and with protective bulge, defence against underwater attack is very strong. *Royal Sovereign* and *Royal Oak* have yet to be fitted with bulges.

Engineering Notes.—Designed to burn coal, but while building "all oil fuel" was adopted, and it was hoped that 23 kts. would be secured without increase of H.P. Addition of bulges has brought speed down again to about 22 kts.

Name	Builder	Machinery	Laid down	Completed	Trials:—	Boilers
Royal Sovereign	Portsmouth Y.	Parsons	Jan. '14	May, '16	41,115 = 21.6	18 Babcock
Royal Oak	Devonport Y.	Hawthorn	Jan. '14	May, '16	40,360 =	18 Yarrow
Resolution	Palmer	Palmer	Nov. '13	Dec. '16	41,406 =	18 Yarrow
Ramillies	Beardmore*	Beardmore	Nov. '13	Sept. '17	42,356 = 21.5	18 Babcock
Revenge	Vickers	Vickers	Dec. '13	Mar. '16	42,962 = 21.9	18 Babcock

*Towed to Liverpool and completed by Cammell, Laird & Co.

General Notes.—Begun under 1914-15 Estimates. *Revenge* first named *Renown*. *Ramillies* injured herself a and was delayed in completion. They are fine ships, but suffer rather from reduced freeboard. They ai ships, those with bulge protection being splendid gun platforms. With bulges, normal displacement is rai 29,350 tons (33,500 *full load*).

1912-13 BRITISH DREADNOUGHTS. ("Fast Division.")

(Queen Elizabeth Class.)

QUEEN ELIZABETH (Oct., 1913), **WARSPITE** (Nov., 1913), **VALIANT** (4th Nov., 1914),
BARHAM (31st Dec., 1914), **MALAYA** (18th Mar., 1915).

Normal displacement, 27,500 tons (about 31,600-33,000 *full load*). Complement, 955-1016.

Length (*p.p.*), 600 feet. Beam, 90½ feet. { *Mean* draught, 30⅔ feet. / *Max.* ,, 33½ ,, } Length (*over all*) { *Q.E.* 643¾ ft. / Rest 639¾ ,,

Guns:
8—15 inch 42 cal. } Dir. Con.
12—6 inch, 50 cal. }
2—3 inch (anti-aircraft).
4—3 pdr.
5 M.G.
Torpedo tubes (21 inch):
4 *submerged* (broadside).

Armour (H.T.):
Deck:
1" Fo'xle (over battery).
2"—1½" Upper ....
1½" Main ....
1" Middle ....
3" (ends) } Lower.
1" (amidships) }
Special Protec.:
2"—1" Internal citadel between end barbettes.

Armour (K.C.):
Vertical:
13" Lower belt ....
6"—4" Upper belt ....
6"—4" Belt (ends) ....
6", 4" Bulkheads (f. & a.)
6" Battery ....
10"—7" Barbettes .....
11" Gunhouses ....
1½" Funnel uptakes ...
6"—3" C.T. base .....
11" C.T. (6"—2" Hood)..
4" Fore com. tube ....
6" Torpedo C.T. ......
4" Tube (T.C. tower) ..

Barham.

1920 Photo, Sword, Weym

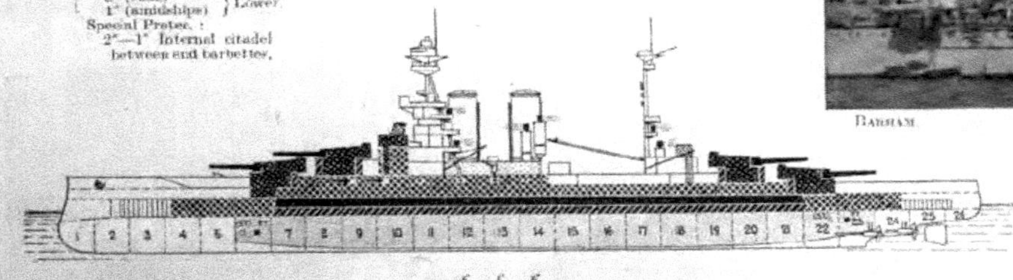

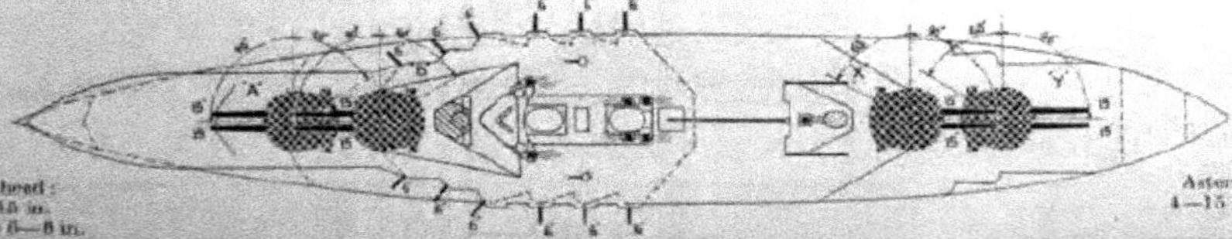

Ahead:
4—15 in.
to 6—6 in.

Astern:
4—15 in.

Broadside: 8—15 in., 6—6 in., 2—21 in. tubes.

Machinery: Parsons turbine, but *Barham* has Brown-Curtis. Geared cruising turbines in all ships. Boilers: 24 large tube (see *Notes*). Designed H.P. 75,000 = 25 kts. Fuel: Only oil, 650 tons *normal*, 3400 tons *max*. Coal: 100 tons for ordinary purposes and boats.

Notes.—First four begun under 1912 Estimates. *Malaya*, extra sh Federated Malay States. Estimated cost (average), £2,336,75 Annual upkeep = £170,416 per ship. In appearance and gene these five ships are the finest in the British Navy. Their decks a ably clear, and internal arrangements are very spacious. Taken they present the most successful type of capital ship yet designed.

Gunnery Notes.—Range of 15 inch only limited by *max.* visibility. E these guns has been enlarged. 6 inch controlled in two groups fr towers on middle bridges. *Queen Elizabeth* completed with four e at stern; these were found useless, guns removed and ports plated came into service with 2—6 inch in shields on fo'xle deck amidshi too, have been removed, but in a head sea they were said to have to whole upper deck battery.

Armour Notes.—Belt is 13" at w.l. only, 6" on upper edge, 8" an applied in vertical strakes. Barbettes, 6" and 4" within belt. 1½ to battery, but no rear screens—only dwarf walls to retain and water admitted to battery. Rear bulkhead to battery is 6" diag where it crosses centre line. Internal protection of these ships b

Engineering Notes.—"All oil" installation very successful. These splendidly, and can maintain a high average speed for long peri turbines on wing shafts, with cruising turbines geared at forward turbines on inner shafts.

Name	*Builder*	*Machinery*	*Laid down*	*Completed*	*Trials*	*Boilers*
Queen Elizabeth	Portsmouth	Wallsend	Oct. '12	Jan. '15	57,130=	24 Babcock
Warspite	Devonport	Hawthorn	Oct. '12	Mar. '15	77,510=	24 Yarrow
Valiant	Fairfield	Fairfield	Jan. '13	Feb. '16	71,112=	24 Babcock
Barham	Clydebank	Clydebank	Feb. '13	Oct. '15	76,575=	24 Yarrow
Malaya	Elswick	Wallsend	Oct. '13	Feb. '16	76,074=	24 Babcock

1912-13 BRITISH DREADNOUGHTS. ("Fast Division.")

For Plan and Description, v. opposite page.

MALAYA. *Photo, Cribb, Southsea, 192*

QUEEN ELIZABETH.* *Photo, Graphic Union.* (20th Nov., 1918).

*Other ships without fore topmast, *Barham* only, has St. top half-way up foremast.

MALAYA (Aircraft view). 1919 *Official R.A.F. Photo.*

1912 BRITISH DREADNOUGHTS.

(Iron Duke Class.)

BENBOW (Nov., 1913), **EMPEROR OF INDIA** (ex-*Delhi*, Nov., 1913),
IRON DUKE (Oct., 1912), **MARLBOROUGH** (Nov., 1912).

Normal displacement, 25,000 tons. *Full load*, 28,800 tons. Complement, 995 to 1022.

Length (*o.a.*), 622¾ feet. Beam, 89½ feet. { *Mean* draught, 28½ feet. / *Max.* ,, 32¾ ,, } Length (*p.p.*), 580 feet.

-5 in. (Mk. V.) } **Dir. Con.**
inch, 45 cal. }
inch (anti-aircraft)
pdr.
3.
nding)
o tubes (21 inch):
nmerged (broadside)

Armour (K.C.):
Vertical
12"-8" Lower belt ......
9" Middle belt ....
8" Upper belt ....
6"—4" Belt (ends) ....
8", 6", 4" Bulkheads (f. & a.) ....
6" Battery* ....
10"—7" Barbettes .....
11"— " Gunhouses....
11" Funnel uptakes ...

1911 BRITISH DREADNOUGHT. (Purchased 1914).

ERIN (*ex Reshadieh*), laid down Feb., 1911, launched Sept., 1913, purchased and completed Aug., 1914.

Normal displacement, 23,000 tons; 25,250 *full load*. Complement, 1130.

Length (*o.a.*,) 559½ feet. Beam, 91 ft. 7¼ in. { *Max* draught, 30 ft. 11 in. / *Mean* " 28 ft. 5 in. } Length (*p.p.*), 525 feet.

Guns (Vickers) :
- 10—13.5 inch, 45 cal. } **Dir.**
- 16—6 inch, 50 cal. } **Con.**
- 2—3 inch (anti-aircraft)

Torpedo tubes (21 inch) :
- 4 *submerged*

Armour (H.T.) :
Decks
- 1½" Fo'xle (over battery)
- 1½" Upper (beyond battery)
- 1½" Main ....
- 3" (ends)
- 1" (amidships) } middle

Special Protection :
- 1½" internal citadel ...

Armour (K.C.) :
Vertical
- 12"—9" Lower belt ....
- 8" Upper belt ....
- 8"—5", 4" B'lkh'ds (f. & a.)
- 5" Battery ....
- 1" Battery traverses ..
- 10"—8" Barbettes ......
- 11" Gunhouses ....
- 1" Funnel uptakes ....
- 12" C.T. (6" tube) ....
- 4" Torpedo C.T. (3" tube)

1919 Photo, Topic

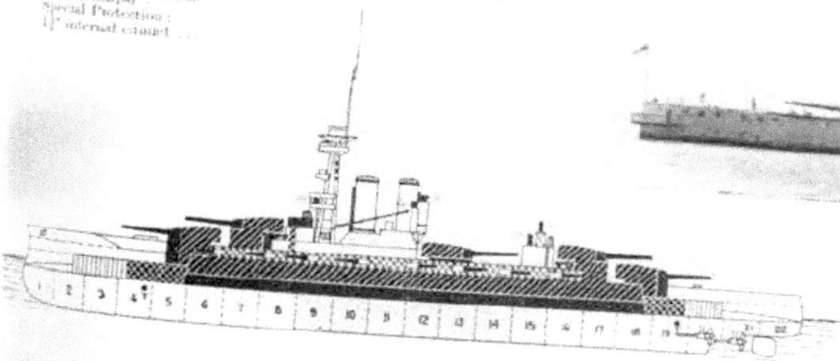

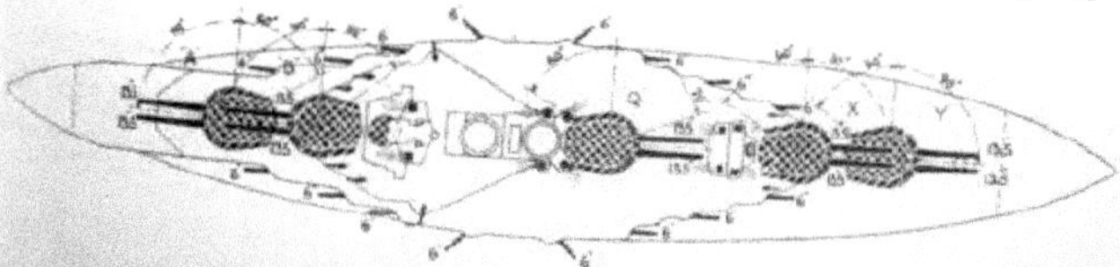

Broadside : 10—13.5 in., 8—6 in., 2—21 in. tubes.

Machinery (by Vickers) : Parsons turbines. 4 screws. Designed H.P. 26,500 = 21 kts. Boilers : 15 Babcock. Coal : *normal*, 900 tons ; *maximum*, 2120 tons + 710 tons oil. Built by Vickers. Refit 1920.

Torpedo Note.—Elswick side-loading tubes. Stern submerged tube removed during War.

Gunnery Notes.—Six 6 inch guns in place no longer mounted.

 BRITISH DREADNOUGHTS.

(KING GEORGE CLASS.)

KING GEORGE V* (Oct., 1911), CENTURION (Nov., 1911), AJAX (March, 1912).

Normal displacement, 23,000 tons (about 25,500 *full load*). Complement, 812 to 840.

Length (*over all*), 597½ feet. Beam, 89 feet. { *Mean* draught, 27½ ft. / *Max.* „ 30 ft. 10 in. } Length (p.p.), 555 feet.

* For Foreign Readers: in English "V" is pronounced "The Fifth."

Guns:
10—13.5 inch (M.V.) } **Dir.**
12—4 inch } **Con.**
2—4 inch (anti-aircraft *Cent.*)
2—3 inch (anti-aircraft *Ajax, K.G.V.*)
4—3 pdr.
5 M.G.
(1 landing)
Torpedo tubes (21 inch):
2 *submerged* (broadside).

Armour (H.T.):
Deck
1¾", 1½" Upper ....
1½" Main ....
1" Middle ....
2½"—1" (bow).. } lower
3" (stern)...... } lower
4" (over rudder) } lower
Special Protn. ....
—" Internal screens to magazines, &c. ......

Armour (K.C.):
Vertical
12" Lower belt ....
9" Middle belt ....
8" Upper belt ....
4", 6", 8" B'lkh'ds (fore)
2" 10" 8" B'lkh'ds (aft)
6"—4" Belt (bow) ....
2½" Belt (stern) ......
10"—7" Barbettes .....
11"— " Gunhouses.....
1½" Funnel uptakes ...
3½"—1" Battery (4 inch guns) ....
1" C.T. base (5" tube within) ....
11" C.T. (4" hood) .....
6" Torpedo C.T. (4" tube)

KING GEORGE V. 1919 P

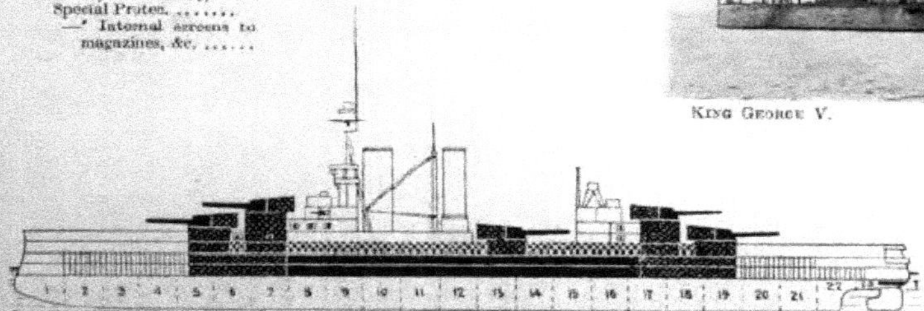

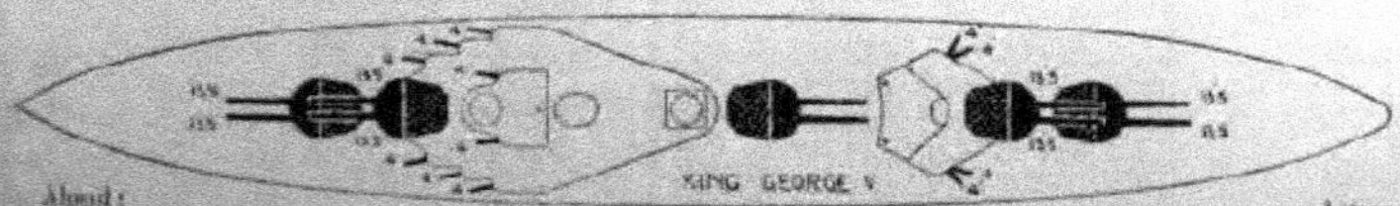

Ahead: 4—13·5 in.

Astern: 4—13·5 in.

Broadside: 10—13.5 in., 1—21 in. tube.

Machinery: Turbine (Parsons). Boilers: 18 Babcock or Yarrow. Designed H.P. 27,000 = 21 kts. Coal: *normal*, 900 tons; *maximum* (*A. & C.*) 3150 tons, (*K. G. V.*) 2870 tons + 850 tons oil in all three.

Name.	*Builder.*	*Machinery.*	*Laid down.*	*Completed.*	*Trials:* 30 hours.	*Trials:* Full power.	*Bo*
King George V	Portsmouth Y.	Parsons	Jan. '11	Nov. '12	19,808 =	28,005 = 21·4	Bab
Centurion	Devonport Y.	Hawthorn	Jan. '11	May '13	—	28,026 = 21·88	Yar
Ajax	Scotts	Scotts	Feb. '11	Oct. '13	19,820 =	28,092 = 21·06	Bab

Armour Notes.—As *Orion* class; internal protection is more extensive in these ships, but intern continuous between end barbettes. Battery for 4-inch guns on forecastle, 2" face, 3½" ports, 1" r ships. Deck armour more extensive in these ships.

Torpedo Notes.—Stern submerged tube removed during War.

General Notes.—Begun under 1910 Estimates. Are generally similar to *Orion* class. Average c about £85 per ton. On first commission, *Centurion* had experimental form of fire-control for searchlights, centrally controlled. *King George V*, also an experimental form of coaling gea during the War. *King George V* has anti-rolling tanks. Annual Cost £151,395.

Class Differences.—*K. G. V* has "Concentration and Torpedo Control Station" in extra top low on legs support main control top. *Centurion* and *Ajax* have short struts reaching halfway up mas

To Distinguish.—From *Iron Duke* class, by large oval forward funnel and absence of upper deck 6"

(ORION CLASS.)

ORION (Aug., 1910), **THUNDERER** (Jan., 1911), **MONARCH** (March, 1911), **CONQUEROR** (May, 1911).

Normal displacement, 22,500 tons (about 25,000 *full load*.) Complement, 806 to 828.

Length (*p.p.*), 545 feet. Beam, 88½ feet { *Mean* draught, 26⅔ feet. / *Max* „ 30¾ feet. } Length (*over all*), 581 feet.

Guns:
10—13.5 inch (M.V.) } **Dir.**
12—4 inch } **Con.**
1—4 inch (anti-aircraft)
1—3 inch (anti-aircraft)
4—3 pdr.
5 M.G.
(1 landing)
Torpedo tubes (21 inch):
2 *submerged* (broadside).

Armour (Nickel):
Decks { 1¾" Upper ....
1½" Main ....
1" Middle ....
2½" Lower ....
4" over rudder ......
Special protection:
—" Internal screens to magazines, &c. .....

Armour (K.C.):
Vertical { 12" Lower belt ....
9" Middle belt ....
8" Upper belt ....
4", 6", 8" B'lkh'ds (fore)
6"—4" Belt (bow) ....
2½" Belt (stern) .....
2½", 8", 10" B'lkh'ds(aft)
10"—7" Barbettes .....
11"—" Gunhouses....
1½" Funnel uptakes ...
11" C.T. (6", 3" hoods)..
5" Fore com. tube ....
3" Torpedo C.T. & tube

Topical Photo, add

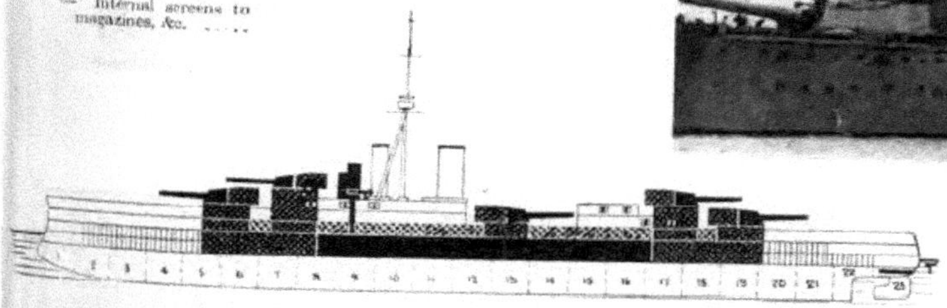

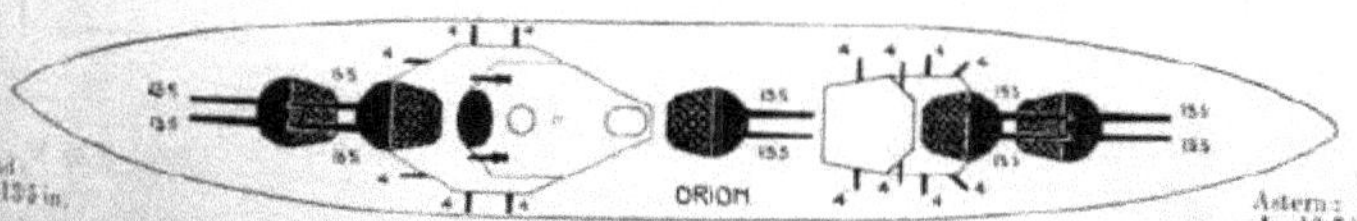

Ahead: 4—13·5 in.

Astern: 4—13·5 in.

Broadside: 10—13.5 in., 1—21 in. tube

ORION.

1910 Pho

Machinery: Parsons turbine. Designed H.P. 27,000 = 21 kts. 4 screws. Boilers: 18 (see *Notes*). Coal: *normal*, 900 tons; *maximum*, 3300 tons coal + 800 tons oil.

Armour Notes.—12" belt very narrow and submerged at deep load. Barbette bases, 6"—3" within belt. Internal vertical screens to magazines, &c., not continuous between end barbettes.

Torpedo Notes.—Stern submerged tube removed during War.

General Notes.—*Orion* begun under 1909 Estimates; the others are ships which were known as "Conditional *Dreadnoughts*," laid down under the same Estimates. Average cost, £1,887,870. In these ships, protection is given to the boats for the first time. On her first cruise the *Orion* in the Bay of Biscay rolled 21°. This rolling was attributed to her small bilge keels, 6 feet only. During 1912, improved bilge keels were fitted and the ships roll far less. On trials, *Conqueror* made a *maximum* of over 23 kts. *Monarch* was originally named *King George V*.

Name	Builder	Machinery	Laid down	Completed	Trials: 30 hrs.	Trials: 8 hrs. full power (mean)	Boilers
Orion	Portsmouth Y.	Wallsend	Nov.'09	Jan.'12	18,966 = 19.5	22,165 = 21.02	Babcock
Conqueror	Beardmore	Beardmore	Apl.'10	Nov.'12	19,100 = 19.5	28,400 = 22.13	Babcock
Thunderer	Thames I.W.	Thames I.W.	Apl.'10	June'12	18,321 = 19.5	27,416 = 20.8	Babcock
Monarch	Elswick	Hawthorn	Apl.'10	Apl.'12	19,236 =	28,155 = 21.88	Yarrow

(Colossus Class).

COLOSSUS (April, 1910) & HERCULES (May, 1910).

Normal displacement, 20,000 tons (about 22,250 *full load*). Complement, 631 and 845.

Length (*waterline*), 540 feet. Beam, 85 feet. Draught { (*normal*), 27 feet. (*max.*) 31 feet } Length *over all*, 546 feet.
Length, (*p.p.*) 510 feet.

Guns:
- 10—12 in. Mk. XI, 50 cal. } **Dir.**
- 12—4 inch } **Con.**
- 1—4 inch (anti-aircraft) (*Col.*)
- 1—3 inch (anti-aircraft) (*Herc.*)
- 4—3 pdr.
- 5 M.G.
- (1 landing)

Torpedo tubes (21 inch):
- 2 *submerged*

Armour (K.C.):

Vertical.
- 11" Lower belt ....
- 8" Upper belt ....
- 1½", 2", 5" & 8" Bulkheads ....
- 7"—2½" Belt (bow) ....
- 2½" Belt (stern) ......
- 10"—7" Barbettes .....
- 11"—" Gunhouses.....
- 1" Funnel uptakes ....
- 11" C.T. (6" & 3" hoods)
- 5" Fore com. tube ....
- 3" Torpedo C.T. ......
- 3" Tube, T.C.T. ......

Armour (Nickel):

Decks.
- 1½" Main ....
- 1¾" Middle ....
- 2½"—1¾" (bow) } *Lower*
- 3" (stern)
- 4" (over rudder) }

Special Protection: ......
- —" Screens to magazines &c. ....

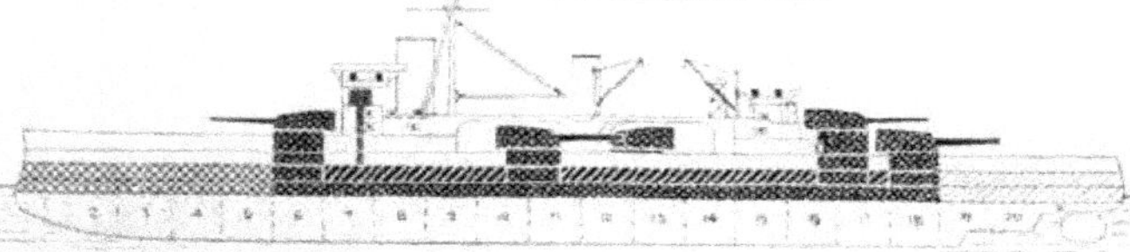

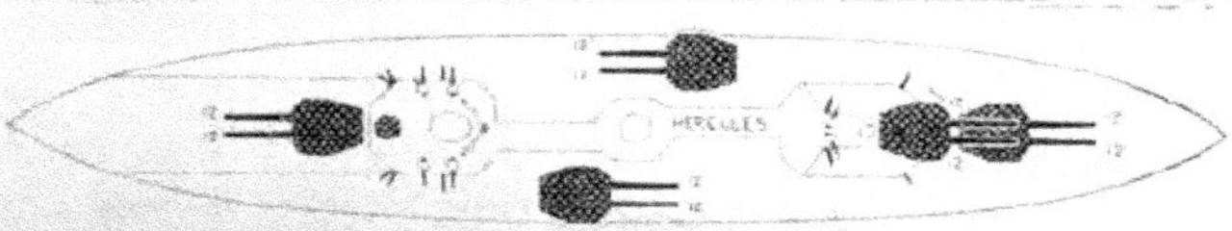

Ahead: 4 to 6—12 in.

Broadside: 8 to 10—12 in., 1—21 in. tube.

Astern: 6 to 8—12 in.

Machinery: Parsons turbine. Designed H.P. 25,000 = 21 kts. **Boilers**: Yarrow or Babcock. **Coal**: *normal*, 900 tons; *maximum*, 2900 tons + 800 tons oil.

Gunnery Notes.—Broadside of 10—12 inch not often used, as decks are badly strained by firing over them, but are fired in action. *Hercules* fired full broadsides at Jutland, but deck over Ward Room afterwards sagged badly.

Armour Notes.—Armouring different to *Neptune*. These ships are "soft enders." On plan, belt forward should stop at Section 2, and belt at stern at Section 20. Beam barbettes, 11" on exposed faces, 6" within belt; other barbettes, 4" inside belt. *Neptune's* internal protection not used in these ships, internal screens not being continuous between end barbettes. Weights of armour: 4560 tons vertical + 2010 tons decks = 6570 tons.

Torpedo Notes.—Stern tube removed during War. 21 inch tubes introduced in these ships, substituted for 18 inch tubes, and torpedo data changed in consequence.

Engineering Notes.—Much as *Neptune*.

Name	Builder	Machinery	Laid down	Completed	Trials: 30 hours	Trials: 4 hours f.p.	Boilers	Best recent speed
Colossus	Scott S. & E. Co.	Scott	July, '09	July '11	18,000 = 19.6	27,334 = 21.3	18 Babcock	19
Hercules	Palmer	Palmer	July, '09	Aug. '11	18,000 = 19.9	26,555 = 21.5	Yarrow	

Hercules. *Photo, Topical.*

Colossus. 1919 *Photo.*

General Notes.—These ships are similar to the *Neptune*, differing mainly in disposition of anti-torpedo armament, of tubes, armouring and internal protection. Begun under the 1909 Estimates, but they were laid down later than is usually the case. Average cost, £1,700,000 = about £87 per ton. Placed in Reserve, 1919.

Note to Plans.—After half of flying boat deck removed 1914-15. "Coffee-box" S.L. tower round 2nd funnel. 4 inch guns decked in. *Hercules* has splinter shields fitted around 4 inch; open ports in *Colossus*.

NEPTUNE (Sept., 1909).

Normal displacement, 19,900 tons (about 22,000 *full load*). Complement, 813.

Length (*waterline*), 540 feet. Beam, 85 feet. {Draught (*normal*), 27 feet / ,, (*max.*) 30 ,,} Length *over all*, 546 feet (p.p. 510 feet).

Guns:
- 10—12 inch, 50 cal. } Dir.
- 12—4 inch } Con.
- 2—3 inch (anti-aircraft)
- 4—3 pdr.
- 5 M.G.
- (1 landing)

Torpedo tubes (18 inch):
2 submerged

Armour (K.C.):

Vertical
- 10" Lower belt ....
- 8" Upper belt ....
- 5" & 8" Bulkheads ....
- 7"—2½" Belt (bow) ...
- 2½" Belt (stern) ......
- 9" Barbettes ....
- 11"— " Gunhouses.....
- 11" C.T. (5" tube) .....
- 1½" Torpedo C.T......
- 1" Funnel uptakes ....

Armour (Nickel):

Decks
- 1¾" Main ....
- 1¾" Middle ....
- 1½" (bow) } lower
- 3" (stern) }

Special protection:
- "— " Internal screens over magazines and machinery spaces ...
- " Bulkheads to keel....

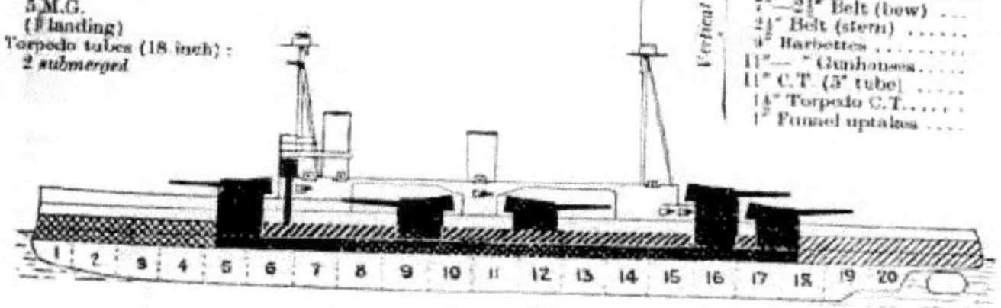

Ahead:
4 to 6—12 in.

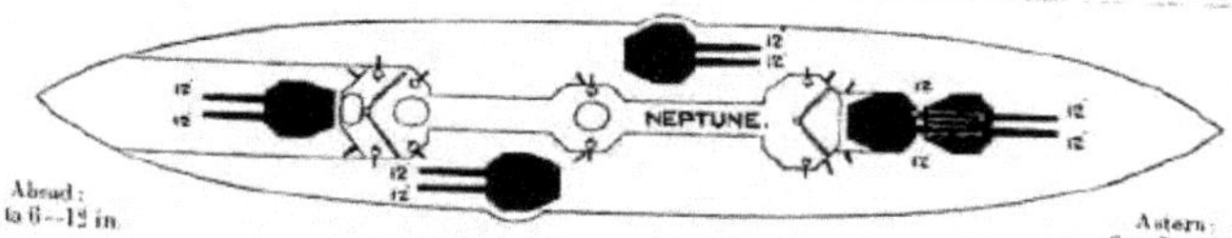

Astern:
6 to 8—12 in.

Broadside: 8—12 in., 1—18 in. tube.

Machinery: Parsons turbine. Designed H.P. 25,000 = 21 kts. Boilers: Yarrow. Coal: *normal* 900 tons; *maximum* 2710 tons + 790 tons oil fuel.

Armour Notes.—Armouring generally as *St. Vincent* class, but further internal sub-division. Internal screens are continuous between end barbettes and well inboard. Beam barbettes 10" on exposed faces. On plan, Sections 1-3, 2½" belt should be shown rising nearly to upper deck.

Gunnery Notes.—Firing of beam barbettes over decks (to get 10—12 inch broadside) not often resorted to, as decks get badly stressed by blast.

Torpedo Notes.—*Submerged* stern tube removed during War.

Engineering Notes.—Consumption on trial was 1.46 lbs. per h.p. at full power. Turbines: H.P. rotor diameter, 82 inch; L.P. rotor diameter, 109 inch. R.P.M. 332. Boiler pressure: 235 lbs. per square inch. Heating surface: 63,630 square feet. Total weight of machinery with auxiliaries: 1109 tons + water to working level, 2036¼ tons. Engine room complement, 231.

Name	*Builder*	*Machinery*	*Laid down*	*Completed*	*First Trials:* 30 hrs.	8 hrs f.p.	*Boilers*	*Best recent speed*
Neptune	Portsmouth	Clydebank	Jan. '09	Jan. '11	18,375 = 19	27,121 = 21	Yarrow	

NEPTUNE.

1919 *Photo, Sadler & R*

General Notes.—Begun under 1908 Estimates. Cost to build, £86.8 per ton. Cost of machinery, £258,000. heightened 1912. Has the best internal protection of all the 12 inch Dreadnoughts—in fact her interm was not equalled in later Battleships till the *Queen Elizabeth* and *Royal Sovereign* classes came into serv in Reserve, 1919.

Note to Plans.—"Coffee-box" S.L. towers added to 2nd funnel and mainmast, director top to foremast as screen" to fore funnel. All 4 inch guns decked in. Forward half of flying deck removed 1914-15.

(ST. VINCENT CLASS.)

ST. VINCENT (Sept., 1908), & COLLINGWOOD (Nov., 1908).

Normal displacement, 19,250 tons. *Full load*, about 22,900 tons. Complement, 813 and 823.

Length (*waterline*), 530 feet. Beam, 84 feet. { *Mean* draught, 27 feet / *Max.* „ 31½ feet. } Length (*over all*), 536 feet (*p.p.* 500 feet).

Guns:
- 10—12 inch, — cal. } **Dir. Con.**
- 12—4 inch }
- 2—3 inch (anti-aircraft)
- 4—3 pdr.
- 5 M.G.
- (1 landing)

Torpedo tubes (18 inch):
- 2 *submerged* (broadside).

Armour (K.C.):

Vertical
- 10" Lower belt ....
- 8" Upper belt ....
- 7" & 2" Belt (bow) ....
- 2" Belt (stern) ....
- 8"—5" Bulkheads ....
- 9" Barbettes ....
- 11"— " Gunhouses.....
- 11" Fore C.T. (5" tube)
- 8" After C.T. (4" tube).

Armour (K.N.C.):

Decks
- 1¾", 3" Main ....
- 1¾" Middle ....
- 1½", 3" Lower ....
- Special Protec. ....*
- —" Internal screens to magazines, boilers and engine rooms ....
- —" Bulkheads to keel ..

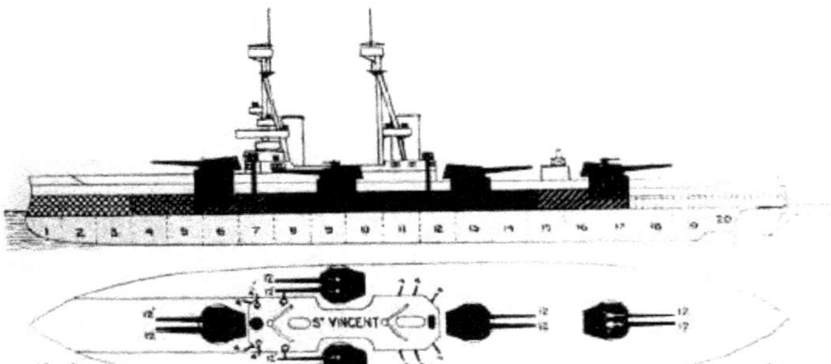

Ahead: 4 to 6 12 in.

Broadside 8—12 inch, 1—18" tube.

Astern: 6 to 8 12 in.

Machinery: Parsons turbine (as *Dreadnought*). 4 screws. Boilers: Yarrow or Babcock. Designed H.P. 24,500 = 21 kts. Coal: *normal*, 900 tons; *maximum*, 2,800 tons + 940 tons oil + 190 tons patent fuel.

Torpedo Notes.—Submerged stern tube removed during War.

Armour Notes.—Armour generally as *Bellerophon* class, but thinner belt at bow and stern. On plans, bow belt in Sections 1-3 should be shown rising almost to upper deck. Internal screens are continuous between end barbettes and set well inboard. Barbette bases, 5" within belt. Exposed faces of beam barbettes, 10".

Engineering Notes.—In this class (and all later Dreadnoughts up to *Iron Duke* class), no cruising turbines. An extra stage at forward end of H.P. turbine is used when running at cruising speed. For full speed, bye-pass valves used. H.P. rotor diameter, 68 inch. L.P. rotor diameter, 92 inch. 324 R.P.M. Pressure, 235 lbs. Heating surface: 63,414 square feet. Weight of main and auxiliary engines: 1072¼ tons + water to working level = 1083¼ tons. Average consumption on trials: 1.8 lbs. per H.P.

ST. VINCENT (1). * 1919

ST. VINCENT (2). * 1919

* After Control Top removed since above photographs were taken.

Name.	*Builder.*	*Machinery and Boilers by.*	*Laid down.*	*Completed.*	*Trials: 8 hrs. full power.*	*Bo*
St. Vincent	Portsmouth	Scott Eng. & S. Co.	Dec. '07	May, '10	25878 = 21·2	Babc
Collingwood	Devonport	Hawthorn L.	Feb. '08	April '10	26319 = 20·4	Yarro

General Notes.—Begun under 1907 Estimates. These ships are simply enlarged *Bellerophons* and seco on original *Dreadnought* design. *Collingwood* did not reach designed speed on first trials. Inte very complete. Underwent large refit in 1912-13. *Vanguard* lost by explosion during the War. *Collingwood* in Reserve or employed as Gunnery Training Ships.

(Bellerophon Class, 1 Ship).

TÉMÉRAIRE (Begun January, 1907. Launched August, 1907. Completed May, 1909).

Normal displacement, 18,600 tons (about 22,000 full load). Complement, 733 to 797.

Length (p.p.) 490 feet. (waterline), 520 feet. (over all), 526 feet. | Beam, 82½ feet | Max. draught, 30' 11" ; Mean draught, 27¼ feet.

Guns: **Dir. Con.**
10—12 inch, 45 cal.
10—4 inch
4—3 pdr.
5 M.G.
1 (landing)
Torpedo tubes (18 inch, M. '04):
2 *submerged* (broadside)

Armour (K.N.C.):
1¾"—2" Main
3"—1¾" Middle
3"—1½" Lower
Special protec.
—" Screens to magazines, boiler and engine rooms

Armour (K.C.):
10"—9" Lower belt
8" Upper belt
7"—6" Lower & upper belts (bow)
5" Belt (stern)
8" After bulkhead
10"—9" Barbettes
11"—" Gunhouses
11" Fore C.T. (4" tube)
8" After C.T. (4" tube)

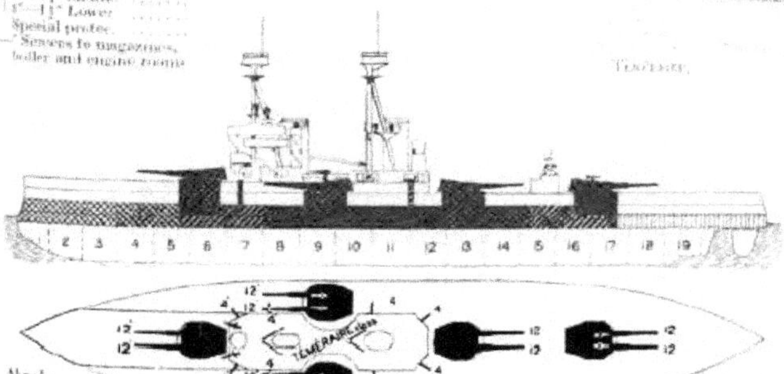

Ahead: 4 to 6—12 in.

Astern: 4 to 6—12 in.

Broadside: 8—12 in., 2—18 in. tubes.

Machinery (by Hawthorn Leslie): Parsons turbine. 4 screws. Details as *Dreadnought.* Boilers: 18 Yarrow. Designed H.P. 23,000 = 20.75 kts. Coal: *normal*, 900 tons; *maximum*, 2648 tons + 842 tons oil + 170 tons patent fuel. Trials: 8 hrs. F.P., 24619 S.H.P. = 21 kts. Best recent speed, 18 kts.

Armour Notes.—Armouring is well planned. Internal screens runs between end turrets and are set well inboard. Exposed faces of beam barbettes, 10". AA. guns removed 1919-20.

Engineering Notes.—H.P. rotor diameter, 68 inch. L.P. rotor diameter, 92 inch. 324 R.P.M. Pressure: 235 lbs. Heating surface: 55,530 square feet. Weight of main and auxiliary engines: 1614½ tons. Water to working level, total weight = about 1936 tons. Trial consumption averaged 1.55 lbs.

Torpedo Notes.—Stern submerged tube removed during War.

General Notes.—Begun under 1906 Estimates at Devonport D.Y. Now serves as Cadets' Training Ship.

Temeraire.

1920 Photo, Abrahams, De

Note.

Sister ship **BELLEROPHON** on Disposal List at end of 1920, for possible trans Dominion Navy or sale to Foreign Navy.

1919 Official R.A.F. Photo

1916 BRITISH BATTLE CRUISER. (HOOD—*continued*.)

For Plan and Description, v. opposite page.

Hood. (Amidships detail view). 1920 *Photo. Abrahams, Devonport.*

Hood (on trials). 1920 *Photo. Hopkins, Southsea.*

Hood (from Starboard Quarter). 1920 *Photo. Abrahams, D*

1916 BRITISH BATTLE CRUISER.

(Anson Type.—1 Ship).

HOOD (Begun 1st Sept., 1916, Launched 22nd August, 1918, left Clyde for Trials, 9th January, 1920).

Normal displacement, 41,200 tons (14,600 tons *full load*). Complement 1475.

Length { p.p. 810 ft. / o.a. 860 ft. } Beam { w.l. / *outside bulges* 104 ft. } Draught { *mean* 28½ ft. / *max.* 31½ ft.

Guns:
- 8—15 inch, 42 cal. } **Dir. Con.**
- 12—5·5 inch, 50 cal. } **Dir. Con.**
- 4—4 inch (anti-aircraft)
- 4—3 pdr.
- 5 M.G.
- (1 landing)

Torpedo tubes (21 inch):
- 2 *submerged* (P. & S.)
- 4 *above water* in pairs

Armour (K.C.):

Vertical.
- 3" Side (submerged) ..
- 12", 7", 5" Side (amidships) ....
- 6"—5" Side (forward)..
- 6" Side (aft)..........
- 5" 4" Bulkh'ds (f. & a.)
- 12"— " Barbettes ....
- 15" Face } Turrets
- 12"—11" Sides } Turrets
- 12" & 9" C.T. ....
- 6" C.T. Base ....
- 6" Director Tower ....
- 4"—3" Torp. Control Tower ....
- —" P. & S. R.F. Towers (5.5") ....
- —" Funnel Uptakes ...
- 1" (H.T.) Shields to 5.5" guns ....

Armour (H.T.):

Decks.
- 2" Forecastle ...
- 1" U.D. amidships
- 1½"—3" Main Deck
- 3" M.D. over Maga
- 1½"—1" L.D. forw.
- 3"—1" L.D. aft ..

Crowns.
- 3" Director Tower
- 5" C.T. ....
- 3" Torpedo Contr.
- 5" Turrets ....

Special protection (H.T.)
1½" Vertical screen Magazines, Boiler and Rooms, Bulges and ancy Spaces.

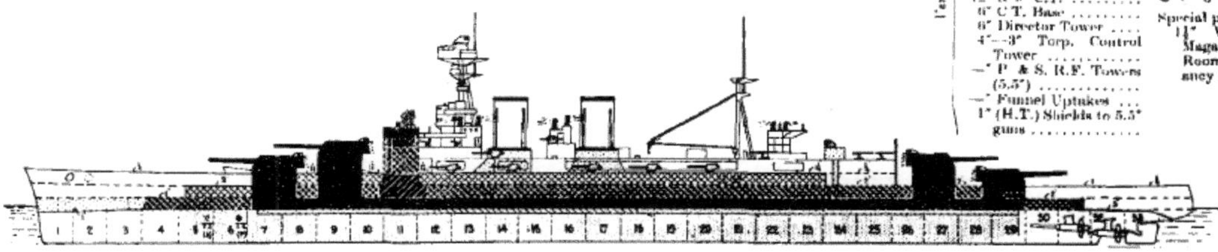

Ahead:
4—15 inch
6—5·5 inch

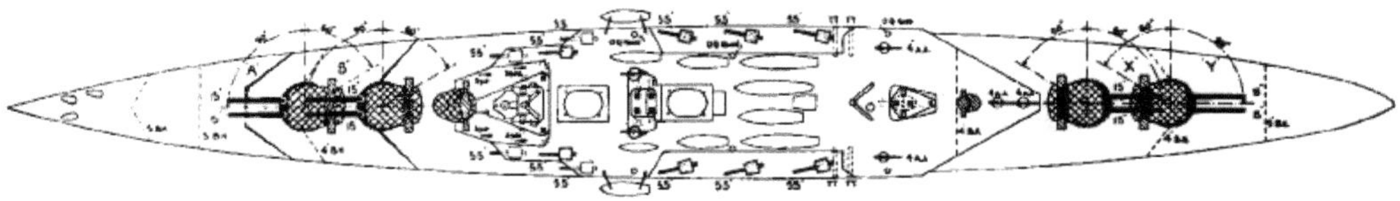

Astern:
4—15 inch

Broadside: 8—15 inch, 6—5·5 inch, 3—21 inch tubes

Machinery (by Builders): Brown-Curtis (geared) turbines. 4 screws. Boilers: 24 Yarrow (small tube). Designed S.H.P. 144,000 = 31 kts. Fuel: Oil only. 1200 tons *normal*, 4,000 tons *maximum*.

Gunnery Notes.—Barbette heights over l.w.l.: A, 32 ft.; B, 42 ft.; X, 31½ ft.; Y 21½ ft. All turrets bear 150° each beam. Designed to mount 16—5.5 inch, but the four after guns were removed before completion. 8—30 inch controlled, and 4—24 inch signalling S.L. Stem attachment for PVs. of new design.

Armour Notes.—Vertical side armour is backed by strong 2"—1" H.T. plating, not included in thicknesses given. Area of 12" armour at w.l., 562 ft. long by 9½ ft. deep. Gun houses, new type with flat crowns, small square sighting ports cut low in face for laying over open sights. On roofs, armoured cases slightly wider than F.R. to allow R.F. to be traversed for fine adjustments. Barbettes, 6"—5" as they descend through decks. C.T. is an enormous, elaborate, most expensive and ponderous structure; in upper stages, it consists of two shells, 12" outer, 9" inner, with air space between. The slope inboard of hull side detracts from effects of plunging fire. A perpendicular, dropped from top sides, just meets outer edges of bulges, which are of the improved "D'Eyncourt-Hopkinson" type.

Trials (unofficial figures).—At 42,200 tons, 15—,000 S.H.P. = 32.07 kts. (run in bad weather, wind force 6 B.S.) on Arran Mile. At 44,800 tons, 31.00 kts. were attained. On ⅔ths power, 25 kts. easily secured.

Notes on Original Design.—The original 1915 Design embodied same length and beam, but draughts were 25½ feet *mean*, 36,300 tons, and 28 feet deep. Speed: 32 kts. Belt 8", barbettes 9", much thinner deck armour, and only 2—21 inch *submerged* T. T. Four ships ordered to this design April, 1916. In the design produced after Jutland (not approved till 1917), 5,000 tons extra protection was worked in. By use of small-tube boilers, 24,000 S.H.P. gained on *same* machinery weights as for *Renown* class.

General Notes.—Begun under Emergency War Programme. Originally, there were four ships in this class, *Anson*, *Hood*, *Howe*, *Rodney*. They were begun in the autumn of 1916, to meet the German Battle Cruisers, *Graf Spee*, *Ersatz Freya* and *Ersatz* "A," which were laid down in 1916. Contractors were: *Anson* (Armstrong), *Howe* (Cammell Laird), *Rodney* (Fairfield), *Hood* (as Table). The enemy having ceased work on all his large ships, in 1917, *Anson*, *Howe* and *Rodney* were stopped in March, 1917, and dismantled to clear slips after the Armistice, but not before £1,034,000 had been expended on them. These ships were redesigned to meet the lessons of Jutland. In *Hood* the outstanding feature is the huge areas covered by heavy armour, strong framing, &c.—in fact, the general scheme of protection is on the lines of German Battle Cruisers. Lower deck quarters and messes cramped, very indifferently lit and ventilated. Cost, *about* £6,025,000; annual upkeep, £339,000.

1915 BRITISH BATTLE CRUISERS

(RENOWN CLASS)

REPULSE (8th January, 1916), **RENOWN** (4th March, 1916).

Normal displacement, *Renown*, 26,500 tons* (about 32,000-32,700 *full load*). Complement, *Renown*, 999 ; *Repulse* 1016.

Length (*p.p.*), 750 feet. { o.a. 794 feet *Repulse* / o.a. 794 feet 1½in. *Renown* } Beam { *Repulse* — feet* / *Renown* 90 feet 2 in. } Draught { *Repulse* —* feet (*mean*), —* feet (*max.*) / *Renown* 26½ feet (*mean*), 30 feet (*max.*) }

*Alterations effected in *Repulse*, 1920, not known. *Renown* (outside bulges).

Guns:
- 6—15 inch, 42 cal. } **Dir.**
- 17—4 inch, 50 cal. } **Con.**
- 2—3 inch (anti-aircraft)
- 4—3 pdr.
- 5 M.G.
- (1 landing)

Torpedo tubes (21 inch) :*
- 2 *submerged* (*Renown*)
- 8 *above water* in pairs (*Repulse*)

*See Torpedo Notes.

Armour (K.C.):
Vertical
- 6" (amidships) }
- 4" (within bow) } Lower belt
- 3" (stern) }
- 4" Fore b'lkhead }
- 3" After b'lkh'd }
- 1½" Upper belt ....
- 7"—4" Barbettes ....
- 11"—7" Gunhouses........
- 1½" Funnel uptakes ....
- 2" C.T. base (3" tube within) ....
- 10" C.T. ....
- 6" Sighting hood over C.T. ....
- 3" Torpedo C.T. ....

Armour (H.T.):
Crowns Decks
- 1½"—½" Fo'xle ..
- 1½"—1¼" Upper ..
- 3"—¾" Main (2"
- 2½" Bow
- 3½"—3" Stern ..
- —" Barbettes ..
- 3" C.T. and hoo
- 1½" Torpedo C.T

Special protection:
Modified bulges
20 ft. deep, fill
oil ....

Repulse
9" (amidships)
6" Upper belt
Other detail
but bulge

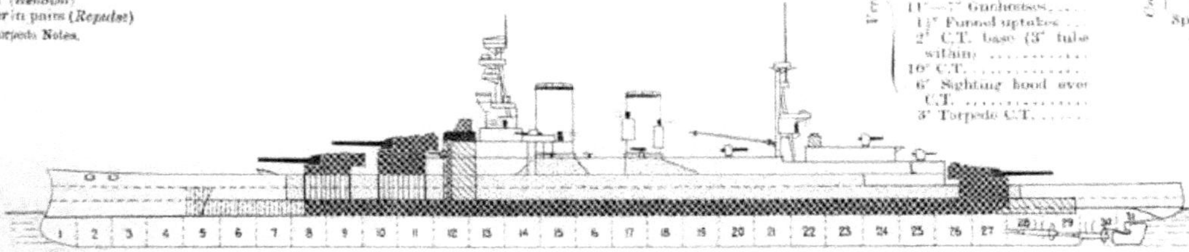

Ahead:
4—15 in.
8—4 in.

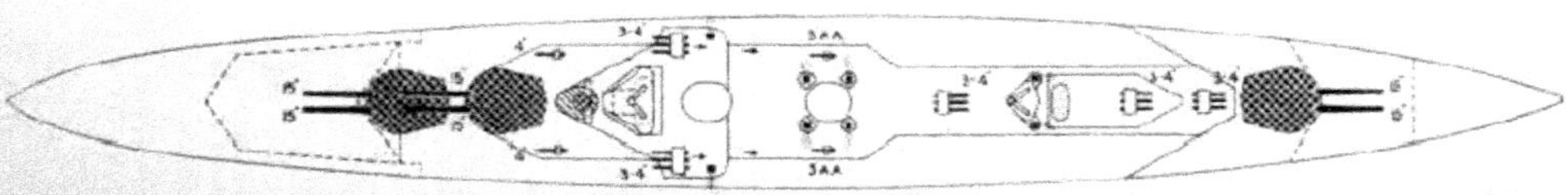

Astern:
2—15 in.
12—4 in.

Broadside: 6—15 in., 13—4 in., 1—21 in. tube.

Machinery: Brown-Curtis (direct drive) turbines. 4 screws. Boilers: 42 Babcock & Wilcox. Designed H.P. not exactly specified, but expected to be 110,000 to 120,000 S.H.P. for 30 kts. In service, S.H.P. 112,000 = about 31.5 kts. Fuel (oil only): 1000 tons *normal*; *Repulse* 4243 tons *maximum*; *Renown* 4289 tons *maximum*.

Gunnery Notes.—15-inch have range only limited by maximum visibility. Director tower under control tower on foremast. 4-inch triples have 2 director towers, and all guns can be worked from either tower or half the 4-inch from one tower. If towers are destroyed, 4-inch can work independently. 4-inch triples are clumsy and not liked. They are not mounted in one sleeve; have separate breech mechanism; gun crew of 23 to each triple. First salvo fired by forward 15-inch of *Renown* did considerable damage forward, and she had to be docked for repairs.

Torpedo Notes.—Submerged tubes a failure. *Renown*, on 1919 re-fit, had submerged tubes removed and replaced by 4 above water tubes in 4 twin mountings on main or upper deck.

Armour Notes.—Armouring adapted from *Invincible* and *Indefatigable* classes. On re-fit 1919-20, *Repulse* re-armoured on w.l. with 9" K.C. and 6" K.C. between main and upper decks, extending over sections 8-27 on plans. Belt, about 9 ft. deep. *Special Protection.*—Bulges in *Renown* [illegible], 1920.

Engineering Notes.—Turbines similar to *Tiger*. For full description, v. "Engineering," April 11th, 1919. Boilers: [illegible] lbs. per sq. in. Heating surface: [illegible] sq. ft. Consumption at full speed: about 1400 tons oil fuel per day.

Name	*Builder*	*Machinery*	*Begun**	*Completed*	*Trials:* *H.P.* *Kts.*	B[illegible]
Repulse	Clydebank	Clydebank	Jan.25,'15	Aug.,1916	119,025 = 31.7	Skelmo[illegible]
Renown	Fairfield	Fairfield	Jan.25,'15	Sept.,1916	126,300 = 32.68	Arran [illegible]

*To Battle Cruiser design.

General Notes.—Provided for by 1914-15 Navy Estimates; first designed as slightly modified *Royal* [illegible] tracted for on that basis, and begun 1914, but building was not pushed on actively after the outbreak [illegible] the Falklands Battle it was decided that these two ships should be re-designed as Battle Cruisers. [illegible] was prepared in ten days, and builders received sufficient details by January 21st, 1915, to begin b[illegible] designs were not finished and approved till April, 1915. Intended that they should be completely [illegible] months, but this time was somewhat exceeded. Both ships have turned out remarkably well a[illegible] credit on their designers and builders. Internally, they are most spacious, but they are lightly [illegible] guns "shake them up" considerably. Costs: *Repulse*, £2,627,401; *Renown*, £2,662,578; but [illegible] not inclusive of all charges, and are liable to revision. Refits: *Renown* (1919-20), £169,738; *Re*[illegible] £860,684; which is equivalent to the cost of a new *Carlisle* type Light Cruiser.

(For Plan and Description, v. opposite page.)

RENOWN. *Photo, Cribb, Southsea*, 1919.

REPULSE. RENOWN.

1919 *Photo, Sub-Lieut. Vickers, R.N.*

To distinguish: Navigating bridge in *Repulse* extends aft almost to forefunnel.

RENOWN (starboard quarter). 1920 *Photo, Cribb, Southsea*

1912 BRITISH BATTLE CRUISER.

TIGER (Begun June, 1912, Launched December, 1913, Completed October, 1914).

Normal displacement, 28,500 tons. *Full load, about* 35,000 tons. Complement 1185.

Length (*waterline*), 675 feet. Beam, 90½ feet { *Mean* draught, 28½ feet. / *Max.* „ 34 „ } { Length (*o.a.*) 704 feet. / Length (*p.p.*) 660 feet. }

Guns:
8—13.5 inch (M.V.) } Dir.
12—6 inch (M. XII) } Con.
2—3 inch (anti-aircraft)
4—3 pdr.
5 M.G.
(1 landing)
Torpedo tubes (21 inch):
4 *submerged* (broadside)

Armour (K.C.):
9" Lower belt . . . .

1912 BRITISH BATTLE CRUISER.

TIGER.

For Plan and description, v. opposite page.

Tiger. 1920 Illustration.

Tiger (Aircraft Recognition View). 1919 *Official R.A.F. photo.*

1909 BRITISH BATTLE CRUISERS.

LION and PRINCESS ROYAL.

For Plan and description, see next page.

Princess Royal. 1919 *Photo, Topical.*

Princess Royal (Aircraft Recognition View). 1919 *Official R.A.F. photo.*

1909 BRITISH BATTLECRUISERS.

(Lion Class.)

LION (August, 1910), **PRINCESS ROYAL** (April, 1911).

Normal displacement, 26,350 tons. *Full load*, 29,700 tons. Complement, 1085 & 1061.

Length (*w.l.*) 675 feet. Beam, 88½ feet. { *Mean* draught, 27⅔ feet. } { Length (*over all*) 700 feet.
{ *Max.* draught, 31⅔ feet. } { Length (*p.p.*) 660 feet.

Guns:
- 8—13.5 inch (M.V.) } Dir. Con.
- 14—4 inch, 50 cal. }
- 2—3 inch (anti-aircraft)
- 4—3 pdr.
- (*P.R.* 2—2 pdr. pom-pom)
- 5 M.G.
- (1 landing)

Torpedo tubes (21 inch):
- 2 *submerged* (broadside)

Armour (K.C.):

Vertical.
- 9", 6", 5" Lower belt ..
- 6", 5", 4" Upper belt ..
- 4" Bulkheads ....
- 9"—8" Barbettes .....
- 9" Gunhouses ....
- 1½" Funnel uptakes ...
- 10" C.T. (4" tube K.N.C.)
- 6"—3" Sighting hood .
- 1" Torpedo C.T. ......

Armour:

Decks
- 1" Upper ....
- 1½"—1" Amidships } Lower
- 2½" Ends }

Special protection:
- —" Magazine screens

Lion. 1920 Photo, Crib

Name	*Builder*	*Machinery*	*Laid down*	*Completed*	Trials (mean) *At ⅘ power*	*Full power*	B
Lion	Devonport Y.	Vickers	Nov.'09	Jun.'12	54,763 = 24·5	73,802 = 27	Ya
Princess Royal	Vickers	Vickers	May,'10	Nov.'12	53,315 =	76,510 = 28·52	Ya

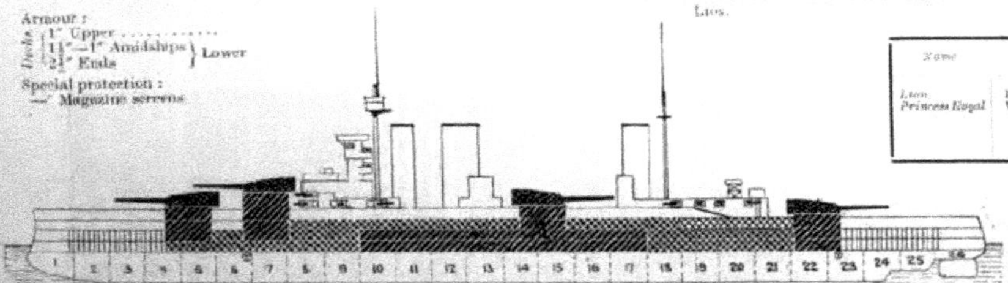

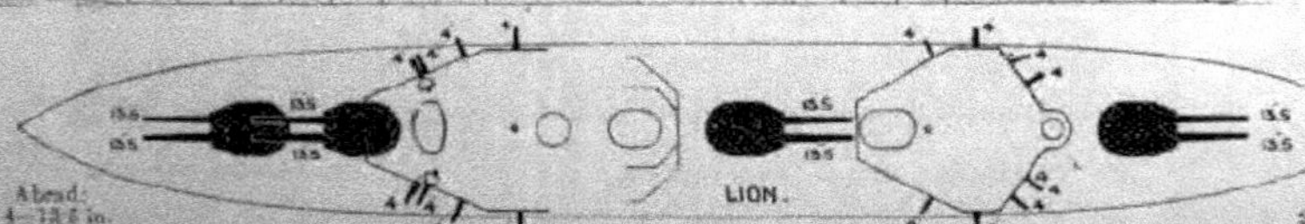

Ahead:
4—13·5 in.

Astern:
2—13·5 in.

Broadside: 8—13·5 in., 1—21 in. tube.

Machinery: Parsons turbine. 4 screws. Boilers: 42 Yarrow. Designed H.P. 70,000 = 28 kts. Coal: *normal* 1,000 tons; *maximum*, 3500 tons + 1135 tons oil.

Armour Notes.—Barbettes are 3" and 1" below upper and main deck. Magazine scree between end barbettes. 2,900 tons vertical + 1,300 tons deck armour = 5,200 tons

Engineering Notes.—*Lion* on trials used coal only. No attempt was made to force these ships up to a phenomenal speed. In service, they have attained rather more than 100,000 H.P. Coal consumption: 950 tons a day at full power. trial burned only 1·16 lbs. per h.p. per hour.

General Notes.—The *Lion* begun under the 1909 Estimates; the *Princess Roy* "contingent Dreadnoughts" of the same year's estimates. On trials, fumes fro rendered the fire control station then over fore funnel on tripod, almost uninh were consequently made. Average cost per ship, £2,087,000 = about £79¾ per to was spent on altering each. A similar but slightly larger ship of this type, Quee in the Battle of Jutland. *P. Royal* refit 1920 = £29,512. Both ships offered to American Navy—probably Chile—during the summer of 1920 but proposal not

Ship Differences.—Starfish on *Lion's* mainmast is above funnel top, in *Princess R* *Lion* has had foremost 4 inch removed from after battery, *Princess Royal* thr

Plan Notes.—Additional top to foremast, which is now a tripod. Searchlight tow screen around third funnel. A.A. gun abaft foremost battery, which now cons each beam.

(Presented to the British Navy by the Dominion of New Zealand.)

(INDEFATIGABLE TYPE.)

NEW ZEALAND (July, 1911).

Note.—For AUSTRALIA, a sister ship, see Royal Australian Navy Section.

Normal displacement, 18,800 tons; about 22,000 tons *full load*. Complement, 853.

Length (*waterline*), feet. Beam, 80 feet. { Mean draught, 26½ feet. / Max. " 30 feet. } Length (*over all*), 590 feet (p.p., 555 feet).

Guns:
8—12 inch, 50 cal. } **Dir. Con.**
14—4 inch.
1—4 inch (anti aircraft)
4—3 pdr.
2—2 pdr. pom-pom
5 M.G.
(1 landing)
Torpedo tubes (18 inch):
2 submerged (broadside)

Armour (K.C.):
Vertical
6" Belt (amidships) . . .
5"—4" Belt (forward)
4" Bulkheads . . . .
7", 4", 3" Barbettes . . .
7" Gunhouses . . . .
1½"—1" Funnel uptakes (Nickel) . . . .
10" C.T. . . . .
6"—3" Hood over C.T.
4" Com. tube . . . .
1" Torpedo Director Tower . . . .

Armour—*continued*.
Decks
1" Main . . . .
2"—1" Amidships } Lower
2½" Fore and aft } Deck
Special protection:
—" Screens to Magazines

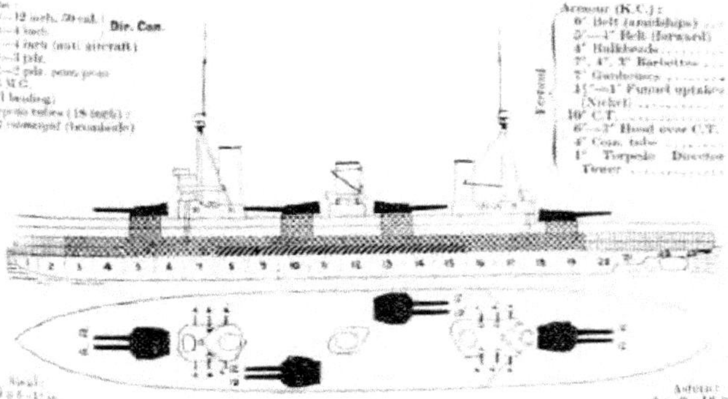

Broadside: 8—12 in., 2—18 in. tubes.

Machinery: Parsons turbine. 4 screws. Designed H.P. 44,000 = 25 kts. Boilers: Babcock. Coal: *normal*, 1000 tons; *maximum*, 3170 tons coal + 840 tons oil.

Armour Notes.—On plans, belt should not extend to stem or stern. 6" belt from lower to middle deck, Sections 7-17; 5" belt, Sections 5, 6 and 18, 19; 4" belt, Sections 3, 4 and 20-21; 2½" deck, Sections 1, 2 and 22, 23. Armour for funnel uptakes not shown. Magazine screens not continuous between end barbettes.

Torpedo Notes.—Stern tube removed during War.

Name	*Builder*	*Machinery*	*Laid down*	*Completed*	*Trials (mean):* 30 hours ¾	8 hours *full*	*Boilers*	*Best recent speed*
New Zealand	Fairfield	Fairfield	June,'10	Nov.,'12	31,291=24.77	46,894=26.3	Babcock	

General Notes.—Design is generally an enlarged *Invincible*. Special feature is that the amidship guns are farther apart in order to permit of fairly free use on either broadside. She is a better sea-boat than the *Invincibles*. This ship was originally laid down for the Dominion of New Zealand, but finally presented to the British Navy. *Indefatigable*, the original ship of this type, was sunk in the Battle of Jutland.

New Zealand.

Note to Photo.—Control tops enlarged and director towers fitted. Searchlights grouped on towers round middle funnel and after superstructure built up. 4—4 inch guns removed to provide extra cabins for Lord Jellicoe's tour, 1919.

1916 BRITISH CRUISERS.

(COURAGEOUS CLASS—2 Ships.)

COURAGEOUS (5th February, 1916), **GLORIOUS** (20th April, 1916).

Normal displacement, about 18,600 tons (*about* 22,700 tons *full load*). Complements *C*, 842; *G*, 829.

Length { p.p. 735 ft. / o.a. 786½ ft. } Beam (outside bulges), 81 ft. Draughts { *mean* 22½ ft. / *max.* 26 ft. }

…our:
- 3" Belt (amidships) ...
- 2" Belt (within bow) ..
- 2" (H.T.) Fore bulkhead
- 3"–1" (H.T.) After Bulkhead ....
- 1½" Side over belt.....
- 7"–3" Barbettes .....
- 9"–7" Gunhouses.....
- 1"–1½" Funnel uptakes
- 2" C.T. base.........
- 3" Tube ....
- 10" C.T. ....
- 6" Sighting hood ......
- 3" Torpedo C.T. (2" tube)

Armour—*continued.*

Decks (H.T.):
- 1" Forecastle ....
- 1" Upper ....
- 1¾"–¾" Main (between barbettes) ....
- 1½" Lower (stern, flat)
- 3" Lower (stern, over rudder) ....
- 2" Lower (on slopes) ..

Cupolas (?):
- 4½" Barbettes (since increased) ....
- 3" C.T. and sighting hd.
- 1¾"–¾" Torpedo C.T.

Torpedo protection (H.T.):
- Modified bulges 25 ft. deep filled with oil fuel ....
- 1½"–1" Outer internal screens between barbettes ....
- ¾" Inner screen to boiler and engine room vents

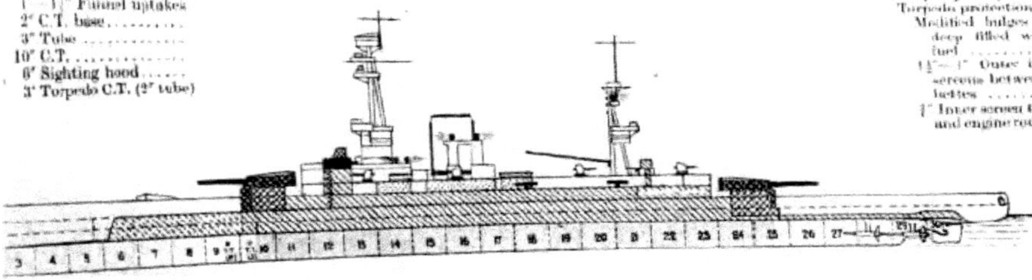

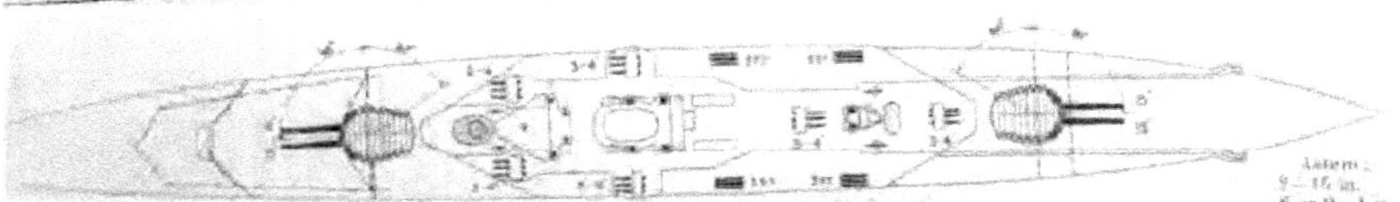

Broadside: 4—15 in. 12—4 in. 6—21 in. tubes.

Astern: 2—15 in. 6 or 9—4 in.

…(geared) turbines. 4 screws. Boilers: 18 Yarrow (small tube).

…kts. Fuel (oil only): *normal*, 750 tons; *maximum*, 3160 tons.

…mitted by maximum visibility. 15-inch director tower on foremast under control … 4-inch director tower on each mast, all 4-inch can be controlled …, 5° depression. Triple 4-inch guns are reported to be an indifferent type of … from one tower. … tory, is not good, as loaders are liable to obstruct one another. Arcs for 4-inch: … 130° for forward centre-line triple, 150° after C.L. triple.

…ted. Positions of above water tubes as noted above.

…ouring on Light Cruiser lines, the 3" belt being built up of 2" plating on 1" shell … thickened during completion. Gunhouses: 13" face, 7" sides, 11" rear, 4½" roof … 7"–6" above upper deck, 6"–4" above main deck, 3" below main deck.

…ent of machinery as in Light Cruiser *Champion*. 4-shaft geared turbines and … 30·83 kts. in service. A rather singular point is that these ships are 1½ kts … when in normal trim.

…on barbette crowns. 2 aeroplanes carried.

Name	*Builder*	*Machinery*	*Begun*	*Completed*	*Trials*
Courageous	Armstrong	Parsons	May, 1915	Dec., 1916	93,780 = 31·58
Glorious	Harland & Wolff, Belfast	Harland & Wolff	Mar., 1915	Nov., 1916	91,165 = 31·6

General Notes.—Emergency War Programme ships. Design formulated by Lord Fisher in 1915, with a view to Baltic operations—hence the shallow draught. The armament was planned to utilise available guns and mountings, and the ships were to have been built in 12 months. The lines are remarkably fine, the beam at the fore barbette being only 71 feet. On *trials*, *Courageous* met heavy weather and was driven into a head sea, straining her hull just forward of the fore barbette. Doubling plates were added here, and subsequent trials showed the defect had been overcome. *Glorious* strengthened in the same way a year later as a precautionary measure. No figures are available as to cost, but they are said to have run to three millions apiece. *Furious*, of slightly modified design, was converted while building into an Aircraft Carrier.

GLORIOUS.

1919 Photo, Cribb, Southsea.

GLORIOUS. Aircraft Recognition View.

1919 Official R.A.F. Photo.

GLORIOUS.

1920 Photo, Abrahams, Devonport.

COMMONWEALTH, as re-built 1918. 1919 *Illustration.*

(KING EDWARD CLASS.)

COMMONWEALTH (May, 1903). (Re-built 1918).

Normal displacement, 16350 tons. *Full load,* 17,050 tons. Complement, 753.

Length { *centerline,* 439 feet. / *over all,* 457 feet. } Beam { at *w.l.,* 78 feet. / outside bulges, feet. } { *Mean* draught, 26¾ feet. / *Max.* .. 28 feet. }

Guns:
- 4—12 in., IX, 40 cal. { **Dir.**
- 4—9·2 in., IX, 45 cal. { **Con.**
- 4—6 inch, VII
- 8—12 pdr.
- 2—3 inch (anti-aircraft)
- 6—3 pdr.
- 2—3 pdr. (anti-aircraft)
- 5 M.G.
- (1 landing)

Torpedo tubes (18 inch):
- 4 *submerged* (broadside).

Armour (Krupp):
- 9" Belt (amidships) ...
- 7"—3" Belt (forward) ..
- 2½" Belt (aft) ....
- 2"—1" Main, middle, lower decks ....
- 12"—6" Barbettes (N.C.)
- 12"—8" Gunhouses (K.C.)
- 8" Lower deck side ....
- 7" Battery (2" traverses)
- 4" Secondary turrets ..
- 12" C.T. (6" tube) .....

Special protection:
- Bulges, — ft. deep ...

Machinery: 2 sets 4-cylinder triple expansion. 2 screws. Boilers: 16 Babcock. Designed H.P. 17,500 (*n.d.*), 18,000 (*f.d.*) = 18·5 kts. (12 since bulge protection was added). Coal: *normal,* 950 tons; *maximum,* 2194 tons + 380 tons oil fuel.

Name.	*Builder.*	*Machinery*	*Laid down*	*Completed.*	*30 hours ¾ power.*	*8 hours full power.*	*Boilers.*	*Best recent speed.*
Commonwealth	Vickers	Fairfield	June '02	1905	12,769 = 17·9	18,238 = 19·01	16 Babcock	

General Notes.—Built under 1901-2 Estimates. Rebuilt 1918, when bulge protection, director controls, extra search-lights and fore tripod mast were added. Note that all 6-inch guns have been removed from battery; 4—6 inch are mounted on upper deck in shields. Serves as sea-going Gunnery Ship for qualifying gunlayers.

CARNARVON. 1919 *Illust.*

(DEVONSHIRE CLASS.)

CARNARVON (Oct., 1903). **ANTRIM** (Oct., 1903).

Displacement, 10,850 tons. Complement, 653.

Length (*over all*), 473½ feet. Beam, 68½ feet. *Maximum* draught, 25½ feet.

Guns: 4—7·5 inch I, 6—6 inch VII, 8—3 pdr., 5 machine, (1 landing). Torpedo tubes: 2 *submerged.* Armour (Krupp): 6"—4½" Belt (amidships), 2¼" Belt (forward), 2" Deck, Bulkheads (aft), 6" Barbettes, 7½"—4½" Hoods to these, 6" Casemates, 10" Coaning to Machinery: 2 sets vertical triple expansion. 2 screws. Boilers, various: 17-15 water-cylindrical. Designed H.P. 20,500 (*n.d.*), 21,000 (*f.d.*) = 22·25 kts. Coal: *normal* *maximum* 1,750 tons (*Carnarvon* + 250 tons oil).

Name.	*Built at*	*Laid down.*	*Completed*	*Trials—30 hours at ¾*	*Trials—8 hrs. full power.*	*Boilers*
Antrim	Clydebank	March, '02	July, '05	11,628 = 21·21	21,604 = 23·20	17 Yarrow
Carnarvon	Beardmore	Oct., '02	May, '05	15,232 = 21·45	21,460 = 23·1	17 Niclausse

General Notes.—Cost, per ship, about £850,000. Built under 1901 Estimates. *Antrim* for Signals and W.T. *Carnarvon,* sea-going Training Ship for Cadets.

("E" CLASS—2 SHIPS.)

EMERALD (18th May, 1920), **ENTERPRISE** (23rd Dec., 1919).

Displacement, 7600 tons. **Complement**,

Length { *p.p.* 565 feet } { *o.a.* feet } Beam, $54\frac{1}{2}$ feet. Draught { *mean* feet. } { *max.* feet. }

Guns:
7—6 inch, 50 cal. (**Dir. Con.**)
2—4 inch A.A.
4—3 pdr.
2—2 pdr. pom-pom.
1 M.G.
Torpedo tubes (*21 inch*):
12, in four triple deck mountings, *above water*.

Armour (H.T.):
3" Side (amidships)
$2\frac{1}{4}$"–$1\frac{1}{2}$" Side (bow)
2" Side (stern)
1" Upper deck (amidships)
1" Deck (over rudder)

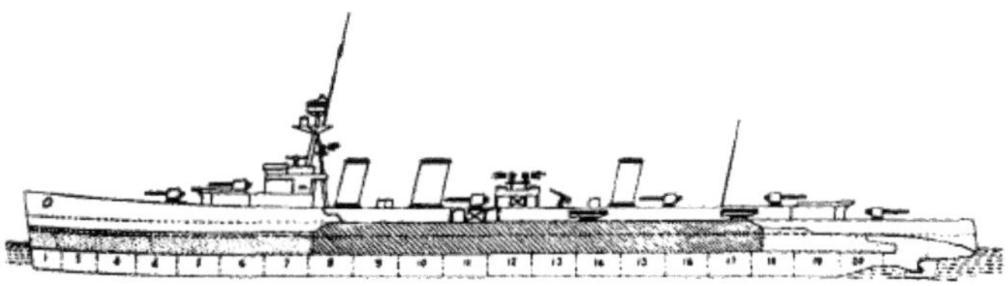

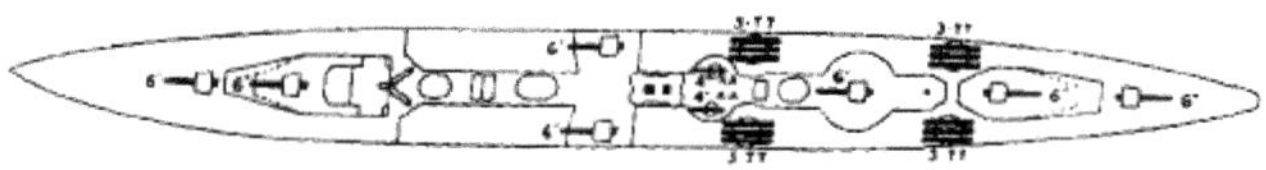

Ahead: 3 to 4—6 inch.

Broadside: 6—6 inch, 6—21 inch T.T.

Astern: 3 to 4—6 inch.

Completing.

Machinery: Turbines, 4 sets Brown-Curtis (geared), 4 screws. Designed S.H.P. 80,000 = 33 kts. (light), 32 kts. (full load). Boilers: 8 Yarrow. Fuel (oil only): *normal*, tons; *maximum*, about 1000 tons.

Name	*Builder*	*Machinery*	*Begun*	*Completed*	*Trials* H.P. kts.	*Turbines*
Emerald	*Armstrong	Wallsend	Sept., 1918			Brown-Curtis
Enterprise	†Clydebank	Clydebank	June 28/18			Brown-Curtis

General Notes.—Begun under Emergency War Programme. A third ship, *Euphrates*, ordered from Fairfield Co., cancelled.

Towed to *Chatham D.Y., †Devonport D.Y., for completion.

LASS—5 SHIPS.)

A (—, 1919). **CARLISLE** (9th July, 1918),

E, **COLOMBO** (18th Dec., 1918).

Complement, *about* 330-350.

, 43½ feet. Draught { *mean* 14 feet / *max.* 16½ „

Armour (H.T.):
3" Side (amidships) ...
2¼"—1½" Side (bow) ..
2" Side (stern) ....
1" Upper deck (amids.)
1" Deck over rudder ..
(No C.T. or tube)

CARLISLE. (*Capetown same.*) 1919 *Photo.*

Costs.—*Cairo* £801,800, *Calcutta* £866,681, *Colombo* £749,144.

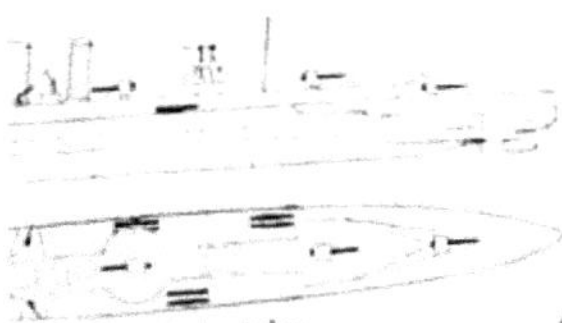

—6 in., 4—21 in. tubes.

Astern: 2—6 inch.

rown-Curtis or Parsons types. Designed S.H.P. rrow. Fuel (oil only): *normal*, 300 tons; *maximum*,

y	*Begun*	*Completed*	*Trials: H.P. kts.*	*Turbines*
erd	Nov. 28, '17 / Oct., 1917 / 1917	Sep. 18, '19 / 1920 / Nov. 11, '18	40,930 = 28·45	B-Curtis
ard	Feb. 28, '18 / Dec., 1917	June, '19		B-Curtis

Pembroke D.Y. for completion.

ordered June-July, 1917, that is, *after* the first three units of "D . In these ships, "trawler bows" were added, to remedy defects g tower also abolished, and hangar for aeroplanes combined with November, 1918, hangars have been stripped and lightened as much here now. Unofficially referred to as "Repeat C Class."

6-inch gun between foremast and first funnel. From *Ceres*

CAIRO (and Calcutta, Colombo). 1920 *Photo by courtesy of Messrs. Cammell Laird & Co. (builders).*

("D CLASS," THIRD GROUP— SHIPS.)

DESPATCH (24th Sept., 1919), **DIOMEDE** (29th April, 1919).

Length { p.p. 445 feet / o.a. „ } Beam, 46½ feet. Draught { mean 14½ feet / max. „ } = 4765 tons.

("D CLASS," SECOND GROUP—3 SHIPS.)

DELHI (23rd Aug., 1918), **DUNEDIN** (19th Nov., 1918), **DURBAN** (29th May, 1919).

Length { p.p. 445 feet / o.a. 472½ „ } Beam, 46½ feet. Draught { mean 14 feet / max. 16¼ „ } = 4650 tons.

("D CLASS," FIRST GROUP—3 SHIPS.)

DANAE (26th Jan., 1918), **DAUNTLESS** (10th April, 1918), **DRAGON** (29th Dec., 1917).

Length { p.p. 445 feet / o.a. 471 „ } Beam, 46 feet. Draught { mean 14½ feet / max. 16½ „ } = 4650 tons.

Complements of all, *about* 350.

Guns:
- 6—6 inch, 50 cal. (**Dir. Con.**)
- { 2—4 inch AA. *Diomede, Despatch* }
- 2—3 inch AA. (*in others*)
- 4—3 pdr.
- 2—2 pdr. pom-pom.
- 1 M.G.

Torpedo tubes (21 inch):
- 12 in 4 triple deck mountings.

Armour (H.T.):
- 3" Side (amidships) ...
- 2", 1¾", 1½" Side (bow and stern) ....
- 1" Upper deck (amids.)
- 1" Deck over rudder ..

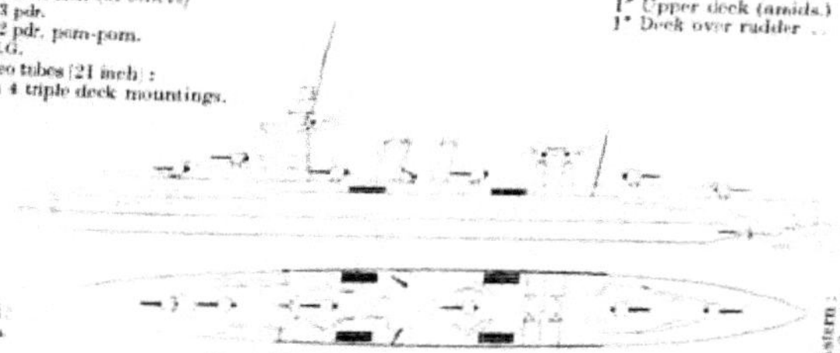

Ahead: 2—6 in.

Astern: 2—6 in.

Broadside: 6—6 in., 6—21 in. tubes.

Machinery: Turbines (all geared), Brown-Curtis or Parsons types. Designed S.H.P. 40,000 = 29 kts., 2 screws. Boilers: 6 Yarrow (small tube). Oil fuel only: *normal*, 300 tons; *maximum*, 1050 tons.

Name	*Builder*	*Machinery*	*Ordered*	*Begun*	*Completed*	*Trials H.P. Kts.*	*Turbines*
Despatch *Diomede*	Fairfield* Vickers*	Fairfield Vickers	Mar., 1918 Mar., 1918	July, 1918 June, 1918			Brown-Curtis
Delhi *Dunedin* *Durban*	Armstrong Armstrong Scotts*	Wallsend H. Leslie Scotts	Sept., 1917 Sept., 1917 Sept., 1917	29 Oct., '17 Nov., 1917 Jan., 1918	June, 1919 Sept., 1919	41,381 = 28·54 41,268 = 29·19	Brown-Curtis Brown-Curtis
Danae *Dauntless* *Dragon*	Armstrong Palmer Scott	Wallsend Scott	Sept., 1916 Sept., 1916 Sept., 1916	Dec., 1916 Jan., 1917	July, 1918 Dec., 1918 Aug., 1918	40,463 = 40,035 =	 Brown-Curtis

*Towed to following Dockyards for completion: *Diomede* to Portsmouth, *Despatch* to Chatham, *Durban* to Devonport.
† With PVs. out, and on deep draught.

DANAE.

1919 *Photo, Seward, Weymouth.*

DRAGON. (DAUNTLESS has had longer cutwater and 3 extra 21 inch TT. mounted in echelon).

1920 *Photo, Cribb.*

DELHI (DESPATCH, DUNEDIN). Revolving aeroplane platform.

1920 *Photo, by Builders.*

General Notes.—Emergency War Programme ships. Note that first three were ordered *before Carlisle* class. Design generally as *Ceres* class, but lengthened about 20 feet, to add a sixth 6-inch between foremost and first funnel ... triple tubes. Costs: *Delhi* £840,182, *Dunedin* £872,304.

Cancelled Ships (Third Group).—*Daedalus* (Armstrong) ... 1918, with *Diomede* ...

(Ceres Class—5 Ships.)

CARDIFF (ex *Caprice*, 12th May, 1917), **CERES** (24th March, 1917), **COVENTRY** (ex *Corsair*, 6th July, 1917), **CURACOA** (5th May, 1917), **CURLEW** (5th July, 1917).

Displacement, 4190 tons. Complement, 327.

Length { *p.p.* 425 feet / *o.a.* 450 ,, } Beam, 43½ feet. Draught { *mean* 14 feet. / *max.* 16¼ ,,

Guns:
- **5—6 inch, 50 cal. (Dir. Con.)**
- **2—3 inch AA.**
- **4—3 pdr.**
- **1 M.G.**

Torpedo tubes (21 inch):
- **8 *above water*, in 4 *double* mountings.**

Armour (H.T.):
- 3" Side (amidships) . . .
- 2¼"—1½" Side (bow) . .
- 2" Side (stern) . . . .
- 1" Upper deck (amids.)
- 1" Deck over rudder . .
- (Harvey or Hadfield)
- 3" C.T. . . . .
- — Tube . . . .

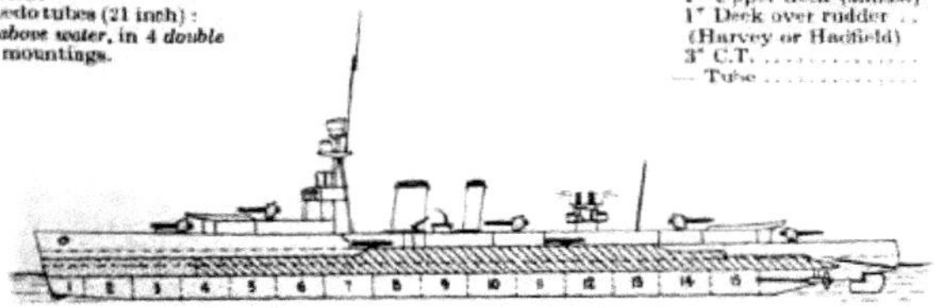

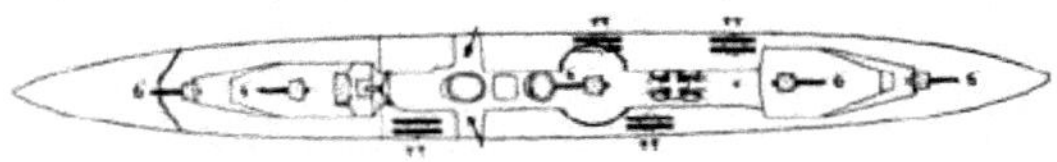

Ahead: 2—6 in.

Broadside: 5—6 in., 4—21 in. tubes.

Astern: 2—6 in.

Machinery: Turbines (all-geared), see Table. 2 screws in all. Designed S.H.P. 40,000 = 29 kts. Boilers: Yarrow. Fuel (oil only): 300 tons, *normal*. *maximum*, 950 tons.

Name	*Builder*	*Machinery*	*Begun*	*Completed*	*Trials* H.P. kts.	*Turbines*
Cardiff	Fairfield	Fairfield	July, 1916	July, 1917	41,450 = 28·96	B.-Curtis A.G.
Ceres	Clydebank	Clydebank	Apl 26,'16	June, 1917	39,425 = 29·1	B.-Curtis A.G.
Coventry	Swan Hunter	Wallsend	Aug., 1916	Feb., 1918	39,967 =	B.-Curtis A.G.
Curacoa	Pembroke D.Y.	Harland & Wolff	July, 1916	Feb., 1918	40,428 =	B.-Curtis A.G.
Curlew	Vickers	Vickers	Aug., 1916	Dec., 1917	40,240 = 28·07	Parsons A.G.

***General Notes.*—All Emergency War Programme ships, ordered April 1916. Very wet forward, to remedy which defect the later *Carlisle* and *Dragon* classes were given "trawler" bows.**

To distinguish: From *Dragon* and "D" Cruisers.—No 6 inch between foremast and first funnel. The same point separates this class from the *Caledon* class. *Coventry* flies Broad Pennant, Commodore (*D*).

Curlew. 1919 *Photo, Seward, Weymouth*

Costs.—£751,330 (liable to revision later). *Curacoa* refitted 1919-20, £72,598.

Coventry. 1919 *Photo, Messrs. Swan Hunter & Wigham Richardson.*

(IMPROVED BIRMINGHAM CLASS—4 SHIPS.)

EFFINGHAM (building), **FROBISHER** (20th March, 1920), **HAWKINS** (Oct., 1917), **RALEIGH** (29th Aug., 1919).

Normal Displacement, 9750 tons (as completed over 10000). Complement, 490.

Length { p.p. 565 feet / o.a. 605 " } Beam { 65* / outside bulges 58* feet / 65 " } Draught { mean 17½ feet / max. 20¼ " }

*Approximate.

Guns:
7—7.5 inch, 50 cal. (**Dir. Con.**)
6—12 pdr.
4—3 inch AA.
4—3 pdr.
2—2 pdr. pom-pom
— M.G.
Torpedo tubes (21 inch):
4 above water
2 submerged.

Armour (H.T. or Nickel):
3"—2" Side (amidships)
2½"—1½" Side (bow)
2½"—2" Side (stern)
1" Upper deck (amids.)
1½"—1" Deck over rudder (Hawkins)
? C.T.
Anti-Torpedo Protection:
Bulges, 5 ft. deep
Longitudinal Bulkheads, below lower deck.

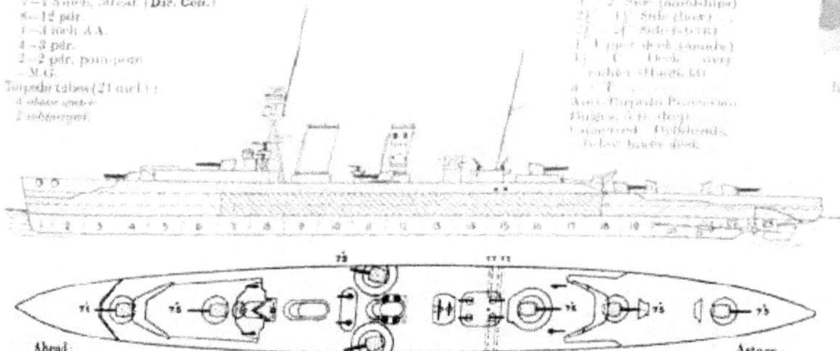

Ahead:
2 to 4—7.5 in.

Astern:
2 to 4—7.5 in.

Broadside: 6—7.5 in., 3—21 in. tubes.

Machinery: Turbines, Brown-Curtis or Parsons (geared cruising). Designed S.H.P. in *Hawkins*, 60,000 = 30 kts. (other 3 ships, with oil fuel only, expected to develop 70,000 S.H.P. and about 31 kts.). 4 screws. Boilers: 12 Yarrow (small tube). Fuel: Coal *normal*, 800 tons; *maximum*, 1000 tons + 1500 tons oil. *Hawkins* has these proportions of fuel, but other three ships may be modified to take about 3000 tons oil only, *max.* capacity.

Gunnery Notes.—7.5's are a new Vickers semi-auto. B.L. mark on centre-pivot mountings and with H.A. elevation. Very handy guns, but at H.A. (30°) elevation breeches close to deck. Hoists very noisy and interfere with telephonic communication with control stations. Unofficially stated that projectile is light and muzzle velocity high, so that guns have practically a range only limited by *maximum* visibility. Armaments originally projected for this class were (a) mixed battery of 9.2 inch and 6 inch, (b) 14—6 inch. On plans, 2—12 pdr. (in lidded ports in shelter deck) between forward 7.5's not shown.

Torpedo Notes.—*Above water* tubes are single (side-loading ?), mounted athwartships on upper deck below mainmast. *Submerged* tubes abeam of C.T. In *Repulse* class, *Glorious* class and *Centaur* class, *submerged* tubes have proved very unsatisfactory, as bars bend when attempting to fire torpedoes over 20 kts. Possible that *submerged* tubes have not been mounted in *Hawkins*.

Armour and Protection Notes.—Armour on Light Cruiser lines. Estimated that bulges and sub-division will keep ship afloat, even if all three boiler-rooms and both engine-rooms are flooded. Bulges modified *Furious* type filled with oil fuel.

Name	Builder	Machinery	Begun	Completed	Trials H.P. kts.	Turbines
Effingham	Portsmouth D.Y.	H. & Wolff	April, '17	No.		Brown-Curtis
Frobisher	Devonport D.Y.	Wallsend	Aug., '16	No.		Brown-Curtis
Hawkins	Chatham D.Y.	Parsons	June, '16	July 25, '19		Parsons
Raleigh	Beardmore*	Beardmore	Dec. 9, '15	1920	71350 = 31	Brown-Curtis

*Completed by Devonport D.Y.

HAWKINS. 1920 Photo, Robertson, Gourock, N.

General Notes.—Begun under Emergency War Programme, summer of 1915 and ordered December, 1915. They were designed to work in German Light Cruiser types, and also for hunting enemy commerce raiders in Atlantic—hence their big fuel capacities. Since oil fuel might not be obtainable at distant naval stations and ports, they were given some coal and 4 small coal-burning boilers arranged to provide sufficient steam for cruising speeds. In the post-war era, they will serve as Flagships to Light Cruiser Squadrons or Foreign Stations. Internally, design is very spacious. *Hawkins* has proved a very good sea-boat. One unit of this class, *Cavendish* (Harland & Wolff), was accelerated in building, and completed as the Aircraft Carrier *Vindictive* (q.v.). Average cost for *Hawkins* and *Raleigh*, £1,512,670.

HAWKINS.

(...lass—3 Ships.)

(24th Jan., 1917), **CARADOC** (23rd Dec., 1916).

tons. Complement, 344.

42½ feet. Draught { *mean* 14¼ feet. / *max.* 16¼ " }

Armour (H.T.):
3" Side (amidships) ...
2¼"—1½" Side (bow) ..
2½"—2" Side (stern) ..
1" Upper deck (amids.)
1" Deck over rudder ..
(Harvey or Hadfield)
6" C.T. ....
4" Tube..............

3 in., 4—21 in. tubes. Astern: 2—6 in.

...ns 2 screws. Designed S.H.P. 40,000 = 29 kts.
..., 300 tons; *maximum*, 935 tons.

...ess and mounted in later Light Cruisers has about 10° elevation.

	Begun	*Completed*	*Trials H.P.*	*Trials kts.*
...d	Mar. 17, '16	Mar., '17	47,887	=
...ke	Feb. 7, '16	June 21, '17	43,312	=
	Feb., '16	June, '17	41,196	=

...wn as "Baby Dreadnoughts." When new, they could make ...being almost unfightable in a head sea. *Cassandra*, of this class, ...ine, soon after Armistice was signed.

CALYPSO. 1919 *Photo, by courtesy of Messrs. Hawthorn Leslie & Co. (Builders).*

CARADOC. 1920 *Photo.*

To distinguish

Caledon class from *Centaur* class

Caledon class:

1. *Three* yards on foremast.
2. AA guns abeam *between* funnels.
3. *Deck* tubes.
4. *Raking* stem.

Note.—Searchlight bridge before forefunnel not fitted in *Caledon*. Revolving aeroplane platform before SL. Tower added to *Caledon* 1919. Not certain that it has been added to other two ships.

Centaur class:

1. *Two* yards on foremast.
2. AA guns on centre line *before and abaft* S.L. tower.
3. Tubes *submerged*.
4. "*Yacht*" stems.

Note.—S.L. bridge before forefunnel not fitted in *Centaur*.

CENTAUR. 1920 Photo.

(CENTAUR CLASS—2 SHIPS.)

CENTAUR (6th Jan., 1916), **CONCORD** (1st April, 1916).

Displacement, 3750 tons. Complement, 336-329.

Length, (*p.p.*) 420, (*o.a.*) 446 feet. Beam, 42 feet. Draught { *mean* 14½ feet, *max.* 16¼ „ }

Guns:
- 5—6 inch, 50 cal. (**Dir. Con.**)
- 2—3 inch AA.
- 2—2 pdr. pom-pom
- 1 M.G.

Torpedo tubes (21 inch):
- *2 submerged*

Plans:—Generally same as *Caledon* class.

Armour (H.T.):
- 3" Side (amidships) ..
- 2¼"—1½" Side (bow) ..
- 2½"—2" Side (stern) ..
- 1" Deck (amidships) ..
- 1" Deck over rudder ..

Ahead: 1—6 in.

Astern: 2—6 in.

Broadside: 5—6 in., 1—21 in. tube.

Machinery: Parsons I. R. turbines. 4 screws. Boilers: 8 Yarrow. Designed S.H.P. 40,000 = 28·5 kts. Fuel (oil only): *normal*, 306 tons; *full load*, 824 tons.

Name	*Builder*	*Machinery*	*Begun*	*Completed*	*Trials: H.P. kts.*
Centaur	Armstrong	Vickers	Jan., 1915	Aug., 1916	31,679 = 26·8
Concord	Armstrong	Vickers	Feb., 1915	Dec., 1916	40,908 =

General Notes.—Emergency War Programme ships. When new, could make 30-32 kts. Specially well sub-divided. Submerged tubes unsatisfactory; may be replaced by twin or triple deck tubes in the future. *Centaur* had her bow and stern blown off by mines during War. Were known as "Tyrwhitt's Dreadnoughts" during the War. Roll a lot and very wet forward.

Very like *Champion* and *Calliope*, v. illustration of *Calliope* on a later page.

(CAMBRIAN CLASS—LATER 4 SHIPS.)

CAMBRIAN (3rd March, 1916), **CANTERBURY** (21st Dec., 1915), **CASTOR** (28th July, 1915), **CONSTANCE** (12th Sept., 1915).

Displacement, 3750 tons. Complement, 323.

Length, (*p.p.*) 420, (*o.a.*) 446 feet. Beam, 41½ feet. Draught, 16¼ feet *max*, 14¾ *mean*.

Guns:
- 4—6 inch, 50 cal. (**Dir. Con.**)
- 1—4 inch AA. (*Canterbury*)
- 2—3 inch AA. (other ships)
- 2—2 pdr. pom-poms

Torpedo tubes (21 inch):
- 6 *above water* (*Canterbury*)
- 4 *above water*, 2 *submerged* } (other ships)

Armour (H.T. and Harvey):
- 3" side (amidships) ...
- 2¼"—1½" Side (bow) ..
- 2½"—2" Side (aft) ....
- 1" Deck (amidships)...
- 1" Deck over rudder ..
- 6" C.T. } Removed
- 4" Tube } from some

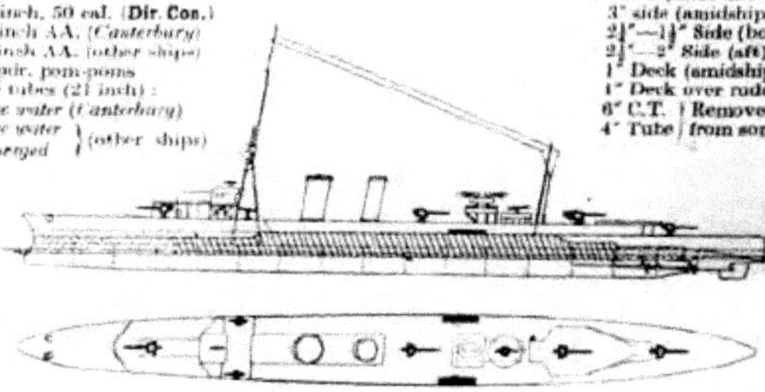

Ahead: 1—6 in.

Broadside: 4—6 in. (2 or 3—21 in T.T. ?).

Astern: 2—6 in.

Machinery: Turbines, as *Notes*. 4 screws. Boilers: 8 Yarrow in all. Designed S.H.P. 40,000 = 28.5 kts. Fuel (oil only): *normal*, 420 tons; *maximum*, 841 tons.

Name	*Builder*	*Machinery*	*Begun*	*Completed*	*Trials: H.P. kts.*	*Turbines*
Cambrian	Pembroke	Cammell Laird	Dec., 1914	May, 1916		Parsons I.R.
Canterbury	Clydebank	Clydebank	Oct. 14, '14	May, 1916	40,100=	Brown-Curtis
Castor	Cammell Laird	Cammell Laird	Oct. 28, '14	Nov., 1915	42,337=	Parsons I.R.
Constance	Cammell Laird	Cammell Laird	Jan. 25, '15	Jan., 1916	41,892=	Parsons I.R.

General Notes.—Begun under 1914-15 Estimates. Design generally same as *Champion* and *Calliope*.

1914 BRITISH LIGHT CRUISERS.

(Cambrian Class—First Two Ships.)

CALLIOPE (Dec. 17th, 1914), CHAMPION (May 29th, 1915).

Displacement, 3750 tons. Complement, 324.

Length, 420 (p.p.), 446 (o.a.) feet. Beam, 41½ feet. Draught { mean 14½ feet, max. 16 „

Calliope { Guns:
4—6 inch (**Dir Con.**)
2—3 inch A.A.
1 M.G.
Torpedo tubes (21 inch):
2 *submerged**
4 *above water*.

* Probably removed.

Champion { 4—6 inch (**Dir Con.**)
1—4 inch A.A.
1—3 pdr. A.A.
Torpedo tubes (6—21 inch):
above water

Armour (H.T.)
4" Side (amidships) . . .
2¼"—1½" Forward . . . .
2½"—2" Aft . . . .
1" Deck (amidships) . . .
1" Deck over rudder hd. (Hadfield)
6" C.T. } removed ? . .
2" Tube }

1919 *Photo.*

Calliope (before 1919-20 re-fit).

Plans generally same as *Cambrian* class on a preceding page.

Note.

Calliope very badly damaged by oil fuel fire at sea, November, 1919. Re-fitted at a cost of £85,000.

Ahead:
1—6 inch.

Astern:
2—6 inch

Broadside: 4—6 in., —21 in. tubes.

Machinery: Turbines (all-geared), Parsons. Screws: 4 in *Calliope*, 2 in *Champion*. Designed S.H.P. in *Calliope*, 37,500; in *Champion*, 40,000 = 28·5 kts. in both ships. Boilers: 8 Yarrow. Fuel (oil only): 405 tons, *normal*; *maximum*, 772 tons.

Name	*Builder*	*Machinery*	*Laid down*	*Completed*	*First trials:*	*Turbines*	*Boilers*	*Best recent speed*
Calliope	Chatham	Parsons	Jan. '14	June '15	39,317 = 28	Parsons	Yarrow	
Champion	Hawthorn	Hawthorn	Mar. '14	Dec. 20, '15	{ 30,299 = / 41,000 = 29	A.G.		

General Notes.—Begun under 1913-14 Estimates, with 6 *Caroline* class, on next page. Same hull, &c., but all-geared turbines adopted. Also carry less oil fuel, have thicker side plating, and—of course—one funnel less. Originally had 3—6 inch and 6—4 inch guns.

Photo Wanted.

Champion.

(CAROLINE CLASS—6 SHIPS.)

CAROLINE (Sept. 29th, 1914), **CARYSFORT** (Nov. 14th, 1914), **CLEOPATRA** (Jan. 14th, 1915), **COMUS** (16th Dec., 1914), *also* **CONQUEST** (20th Jan., 1915), **CORDELIA** (Feb. 23rd, 1914).

Displacement, 3750 tons (about 4780 tons *full load*). Complement, 325.

Length (*p.p.*), 420 feet. Beam, 41½ feet. Draught (*max.*) 16, (*mean*) 14½ feet. Length (*over all*), 446 feet.

Caroline, Carysfort, Comus—Guns:
4—6 inch (**Dir. Con.**)
2—3 inch AA.
1 M.G.
Tubes:
4—21 inch *above water* in 2 pairs.

Armour (H.T.):
3" Side (amidships), 2¼"—1½" forward, 2½"—2" aft, 1" Deck (amidships), 1" Deck over rudder head. C.T. removed from all.

Conquest—Guns:
4—6 inch (**Dir. Con.**)
1—4 inch AA.
2—2 pdr. pom-poms
1 M.G.
Tubes:
As *Cleopatra*

Cleopatra:
4—6 inch (**Dir. Con.**)
2—4 inch AA.
2—2 pdr. pom-poms
1 M.G.
Tubes:
8—21 inch *above water* in 4 pairs.

Cordelia:
4—6 inch (**Dir. Con.**)
1—4 inch AA.
1 M.G.
Tubes:
As *Cleopatra*

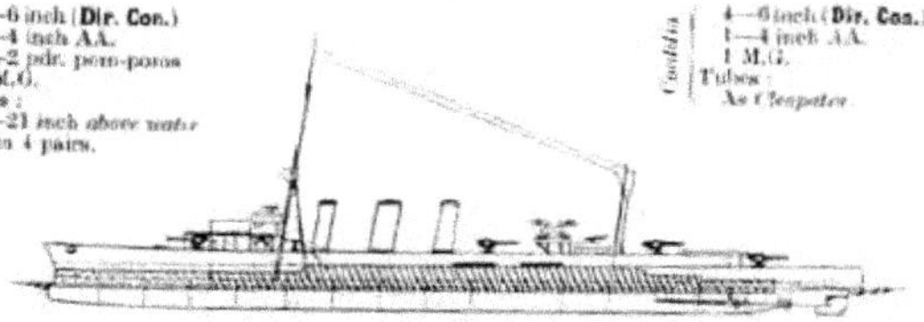

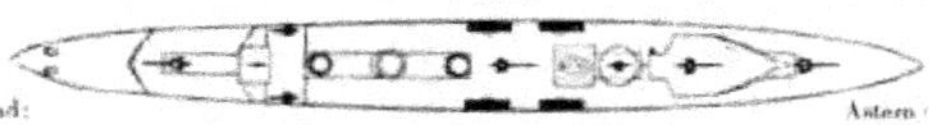

Ahead: 1—6 in.

Broadside: 4—6 in., 2 or 4—21 in. tubes.

Astern: 2—6 in.

Machinery: Turbine: Parsons or Brown-Curtis. Geared cruising Parsons in *Conquest*. **Boilers**: 8 Yarrow. Designed H.P. 40,000 = 28·5 kts. Fuel (oil only): *normal*, 482 tons; *maximum*, 917 tons.

Name	*Builder*	*Machinery*	*Laid down*	*Completed*	*Trials*	*Turbines*	*Boilers*	*Best recent speed*
Caroline	Cammell-Laird	Cammell-Laird	Jan. '14	Dec. '14	30,155 = 27	Parsons I.R.	Yarrow in all.	
Carysfort	Pembroke	Hawthorn	Feb. '14	June, '15	29,[illegible] = 28·[illegible]	Brown-Curtis		
Cleopatra	Devonport	Cammell-Laird	Feb. '14	June, '15	31,[illegible]90 =	Parsons		
Comus	Swan, Hunter	Wallsend	Nov. '13	May, '15	22,136 =	Parsons I.R.		
Conquest	Chatham	Parsons	Mar. '14	June, '15	31,025 = *	Parsons I.R.		
Cordelia	Pembroke	Hawthorn	July, '13	Jan. '15	30,716 =	Parsons I.R.		

General Notes.—*Caroline* of this class was built within twelve months. Belong to the 1913-14 Estimates. Originally had 2—6 inch and 8—4 inch: in 1916-17, 3—6 inch and 6—4 inch and tripod mast added: in 1918, 4—6 inch and ?—4 inch. Much overloaded and roll badly, but very strongly built. Nearly all mined during War, but kept afloat, and none were lost 1914-18.

*Originally by Scotts, but re-engined 1919-20. S.H.P. is for 1915 trials.

CLEOPATRA. 1919 Photo, Topical.
Only ship with 8 tubes [illegible] and 2—4 inch A.A. abaft [illegible] tripod mast.

CAROLINE (and *Carysfort, Comus*).

CONQUEST.

CORDELIA. 1920 Photo.

(Arethusa Class—6 Ships)

GALATEA (May 14th, 1914), **INCONSTANT** (July 6th, 1914),
ROYALIST (Jan. 14th, 1915), **PENELOPE** (Aug. 25th, 1915), **PHAETON** (Oct. 21st, 1914),
UNDAUNTED (April 28th, 1914).

Displacement, 3500 tons. Complement, 318.

Length (*p.p.*), 410 feet. Beam, 39 feet. { *Mean* Draught, 13½ feet. / *Max.* „ 15¾ „ } Length (*over all*), 436 feet.

Guns :
Galatea, Inconstant, Phaeton, Royalist :—
3—6 inch } **Dir.**
4—4 inch } **Con.**
2—3 inch AA.
Penelope as above, but 1—4 inch AA. in place of 2—3 inch AA.

Guns :
Undaunted :—
2—6 inch } **Dir.**
6—4 inch } **Con.**
1—4 inch AA. } *Undaunted*
2—2 pdr. AA. }
2—3 inch AA.—*Aurora*

and—in all—
Torpedo tubes : 8—21 inch *above water*
Some fitted as Mine Layers during war.

Armour (H.T.) :
3" Amidships ....
2¼—1½" Forward ....
2¼—2" Aft ....
1" Deck (amidships)...
1" Deck over rudder hd. (Hadfield)
C.T. removed ....

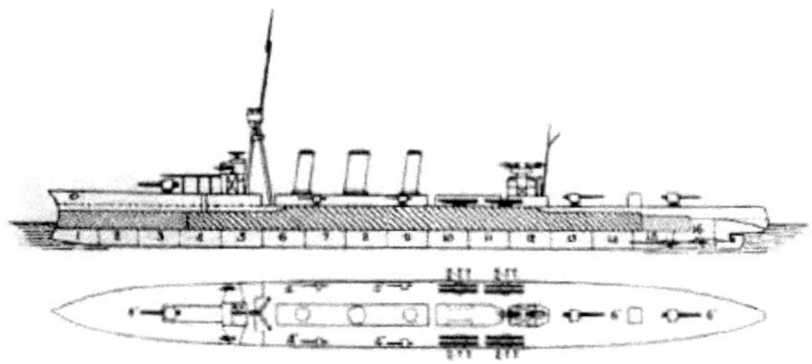

Ahead :
1—6 inch.

Broadside : 2 or 3—6 in., 3 or 2—4 in., 4—21 in. tubes.

Astern :
1—6 inch.
2—4 inch.

Machinery : Turbine (see *Notes*). Boilers : 8 Yarrow in all. Designed H.P. 40,000 = 28.5 kts. Fuel : Oil only, 482 tons *normal* ; 810 tons *maximum*.

Armour Notes.—C.T. removed and replaced by revolving launching platform for a seaplane or aeroplane. Side armour serves dual purpose of providing protection and giving skin plating.

Gunnery Notes.—Beam 4-inch forward very wet and cannot be fought in a head sea. Accordingly, 2—4 inch right forward have been taken out of five ships and a third 6-inch mounted on centre-line between mainmast and after 6 inch gun.

Undaunted. Mast abaft SL. tower. 1919 *Photo, Topic*

Royalist. *Galatea, Inconstant, Penelope, Phaeton,* 1920 *Photo, Seward, Weymo*

Name	*Builder*	*Machinery*	*Laid down*	*Completed*	*Trials*	*Turbines*	*Boilers*
Galatea	Beardmore	Beardmore	Jan.'13	Dec.'14	30,796 =	Parsons I. R.	Yarrow small tube (8 in all)
Inconstant	Beardmore	Beardmore	April '13	Jan.'15	30,357 =	Parsons I. R.	
Royalist	Beardmore	Beardmore	June '13	Mar.'15	31,054 =	Parsons I. R.	
Penelope	Vickers	Vickers	Feb.'13	Dec.'14	32,032 =	Parsons I. R.	
Phaeton	Vickers	Vickers	Mar.'13	Feb.'15	34,595 =	Parsons I. R.	
Undaunted	Fairfield	Fairfield	Dec.'12	Sep.'14	32,161 = 27·5	Brown-Curtis	

General Notes.—Built under 1912 Estimates. *Arethusa* of this class lost in the War. All have tripod masts cramped internally and poor sea boats. *Inconstant* flies broad pennant Commodore (S). *Aurora* prese Canada, 1920.

(Chatham Class—Later Ships)

BIRMINGHAM (May, 1913), **LOWESTOFT** (April, 1913).

Displacement, 5440 tons. Complement, 433.

Length (*p.p.*), 430 feet. Beam, $49\frac{7}{8}$ feet. { *Mean* draught, $15\frac{3}{4}$ feet. / *Max.* „ $17\frac{1}{4}$ „ } Length (*over all*), 457 feet.

Guns:
- 9—6 inch (**Dir. Con.**)
- 1—3 inch anti-aircraft
- 4—3 pdr.
- 2 machine

Torpedo tubes (21 inch):
- 2 *submerged*

Armour (Nickel and H.T.):
- 3" Side (amidships) ....
- $1\frac{1}{2}$" Side (forward) ....
- $1\frac{3}{4}$" Side (aft) ....

(C.T. removed)

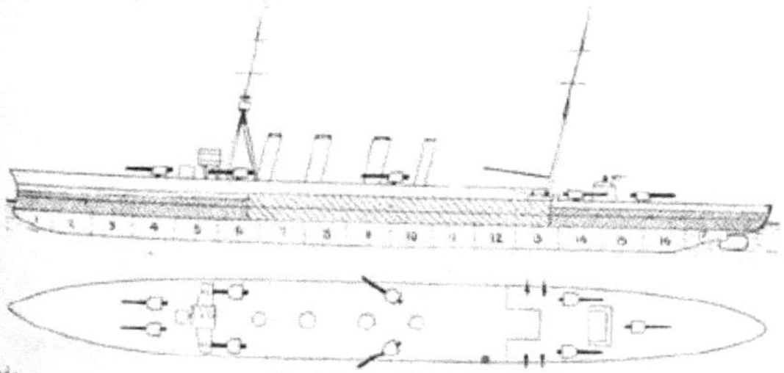

Ahead:
2 to 4—6 in.

Broadside, 5—6 in., 1—21 in. tube.

Astern:
3 to 5—6 in.

Machinery: Parsons turbine. Boilers: 12 Yarrow. Designed H.P. 25,000 = 25·5 kts. Coal: *normal*, tons; *maximum*, 1165 tons + 235 tons oil = 4680 miles at 10 kts.

Name	*Builder*	*Machinery*	*Laid down*	*Completed*	*Trials: Full power.*	*Boilers*	*Best recent speed*
Birmingham	Elswick	Hawthorn	June '12	Jan. '14	26,519 = 25·5	Yarrow	23
Lowestoft	Chatham Y.	Fairfield	July, '12	Apl. '14	26,350 = 25·6	Yarrow	24

Belong to 1911 Estimates. Average cost of above two ships £356,768. *Nottingham* of this class sunk in the War.

Birmingham. 1919 *Photo, Dr. B. H. Pidcock.*

Lowestoft. 1920 *Photo.*

Following Notes also for *Southampton*, *Dublin* and *Chatham* (next page):—

Armour Notes.—Vertical side plating from 2 ft. 7 in. below w.l. to (*a*) upper deck amidships, (*b*) 1 ft. below upper deck fore and aft. Armour is really 2" nickel amidships, 1" forward and aft, but side plating is added for total thicknesses given.

Machinery Notes.—Impulse reaction turbines, viz., on each shaft, one ahead impulse turbine plus an ahead reaction turbine, an impulse astern turbine and an astern reaction turbine.

Large Refits 1919-20: *Birmingham, Lowestoft, Dublin, Weymouth.*

(CHATHAM CLASS—EARLIER SHIPS.) 1919 *Photo.*

SOUTHAMPTON (May, 1912), **DUBLIN** (April, 1912).

Displacement, 5400 tons. Complement, 429-440.

Length (*p.p.*), 430 feet. Beam, 49½ feet. { *Mean* draught, 15¾ feet. / *Max.* „ 17' 8" } Length (*over all*), 457 feet.

Guns (**Dir. Con.**):
8—6 inch (M. XI)
1—3 inch anti-aircraft
4—3 pdr.
4 or 2 machine
1 landing.
Torpedo tubes (21 inch):
2 *submerged*.

Armour:
2" Deck on slopes ......
3" Side (amidships)........

C.T. may have been replaced by aeroplane platform.

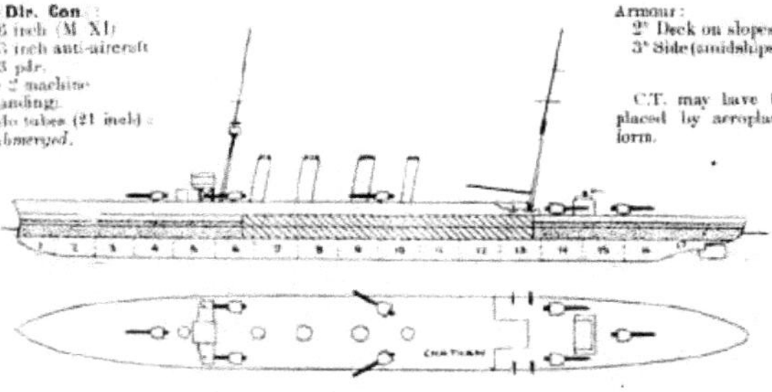

Ahead: 3—6 in.

Broadside: 5—6 inch, 1—21 inch tube.

Astern: 3—6 in.

Machinery: Curtis turbine in *Southampton*, 2 screws; others Parsons, 4 screws. Boilers: Yarrow. Designed H.P. 25,000 = 25.5 kts. Coal: *normal* 750 tons; *maximum* 1240 tons + 260 tons oil.

Name	Builder	Machinery	Laid down	Completed	Trials ⅘ power	Trials Full power	Boilers	Best recent speed
Southampton	Clydebank	Clydebank	Apr. '11	Feb. '12	23,607 =	26,096 = 25.3	Yarrow in all	24
Dublin	Beardmore	Beardmore	Apl. '11	Mar. '13	22,096 =	26,041 = 25.7		

Belong to 1910 Estimates. Average cost £[illegible]. Tripod masts fitted to all. Australian *Melbourne*, *Sydney*, *Brisbane* and [illegible] of similar class. Chatham presented to New Zealand 1920.

WEYMOUTH. (WEYMOUTH CLASS.) 1919 Ph…

WEYMOUTH (Nov., 1910), **YARMOUTH** (April, 1911), **DARTMOUTH** (Feb., 1…

Displacement, 5250 tons. Complement, 433.

Length (p.p.), 430 feet. Beam, 48½ feet. { *Mean* draught, 15 feet. / *Max.* „ 17½ „ } Length (*over all*), 453…

Guns (**Dir. Con.**):
8—6 inch (M. XI)
1—3 inch anti-aircraft
4—3 pdr.
4 machine
(1 landing).
Torpedo tubes (21 inch):
2 *submerged*.

Armour: (Nickel).
2" Deck Amidships
1"—¾ Deck (ends)
(C.T. removed)......

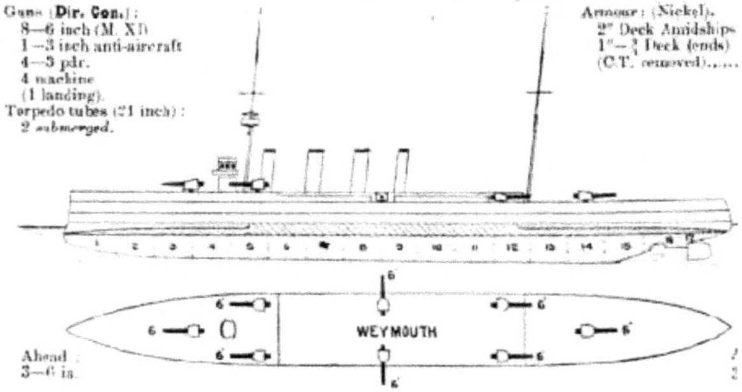

Ahead: 3—6 in.

Broadside: 5—6 inch, 1—21 inch tube.

Machinery: Parsons turbine (compound re-action), 4 screws, except *Yarmouth*, Curtis, 2… Boilers: 12 Yarrow. Designed H.P. 22,000 = 25 kts. Coal: *normal* 750 tons; *maximum* … tons + 260 tons oil = 5600 miles at 10 kts.

Name	Builder	Machinery	Laid down	Completed	Trials 30 hrs. at ¾	Trials 8 hrs. full power	Boilers
Dartmouth	Vickers	Vickers	Feb. '10	Oct. '11	[illegible]	[illegible]	Yarrow
Weymouth	Elswick	Parsons Tur. Co.	Jan. '10	Oct. '11	[illegible] =	[illegible]	
Yarmouth	London & Glas. Co.	L. & G. Co.	Jan. '10	Apl. '12	—	[illegible]	

General Notes.—These three belong to the 1909 Estimates. Average cost £[illegible]. [illegible]

Old Light Cruisers.

1920 *Photo, Abrahams, Devonport.*

HIGHFLYER (June, 1898). Large re-fit, 1919. 5600 tons. Complement, 450. Dimensions: 350 (*p.p.*) × 54 × 22 feet (*max.* draught). Guns: 11—6 inch wire (M. VII, VIII),* 4—3 pdr., 4 Maxims, 1 landing. Torpedo tubes (18 inch): 2 *submerged*. Armour: 3" Armour deck, Protection to vitals, 5" Engine hatches, 6" Conning tower. Machinery: 2 sets 4 cylinder triple expansion. 2 screws. Boilers: Belleville. Designed H.P. 10,000 = 19·5 kts. Coal: *normal*, 500 tons; *maximum*, 1120 tons. Complement, 481. Large refit, 1919.

*3—6 inch are B.L. and other 8 are Q.F.

HYACINTH. As *Highflyer* above. On Disposal List, 1920, but sale to India under consideration.

Note.

Old Cruisers *Bellona* and *Diadem*, Stokers' Training Ships and non-effective.

All Light Cruisers of *Bristol*, *Boadicea*, *Topaze*, *Talbot*, *Astræa* and *Apollo* Classes, on Disposal List, for sale and sold; except old craft like *Naiad*, *Hermione*, &c., used as Harbour Depot, Parent, Accommodation Ships, and entirely non-effective for war purposes.

Coast Defence Vessel.

1919 *Photo, Cribb, Southsea.*

GORGON (ex *Nidaros*, 9th June, 1914).

Displacement, 5700 tons. Complement, 303.

Length (*o.a.*), 310 feet. Beam, 73⅝ feet. *Max.* Draught, 16½ feet.

Guns (Elswick):
- 2—9·2 inch } Dir. Con. and
- 4—6 inch } H.A. Elevation.
- 2—3 inch anti-aircraft.
- 4—2 pdr. pom-poms.

Armour:
- *7" Belt ....
- *4" Ends ....
- 2" Deck ....
- 4" Citadel ....
- 8" Turrets ....
- 6" Secondary guns ....
- 8" Fore C.T. ....
- 6" Aft C.T. ....
- * Not fitted ?

Sale to Rumania under consideration.

Machinery: Triple expansion. 2 screws. Boilers: Yarrow. Designed H.P. 4000 = 13 kts
Coal: *maximum* 364 tons + 170 tons oil.

Name	*Builder*	*Machinery*	*Begun*	*Completed*	*Trials.*
Gorgon	Armstrong	H. Leslie	June, 1913	June, 1918	

Notes.—There were originally two ships in this class, begun at Elswick in May-June, 1913, for the Norwegian Nav as the Coast Service Battleships *Bjoergvin* and *Nidaros*. Construction was stopped when war began, and they la in the Tyne in an incomplete condition until 1915, when they were taken over for the British Navy and name *Glatton* (ex *Bjoergvin*) and *Gorgon* (ex *Nidaros*). *Glatton* blew up at Dover during September, 1918. As original designed, this ship had a beam of 55½ feet and a speed of 15 kts. Beam has been increased and speed reduced by t addition of "blisters" to the hull. The belt may never have been fitted.

1915 Photo

TERROR (18th May, 1916).

mplements, 226 and 223.

, 88 feet. *Mean* Draught, 11 feet.

F. & A. | 4" Box Citadel (over magazines) ......
.... | 6" C.T. ....
ses .... | 1" Fo'xle & Upper D'ks
4" Main Deck (Slopes)
2" Main Deck........
$1\frac{1}{2}$"-$\frac{3}{4}$" Lower Deck ..
Anti-torpedo Pro. Bulges.

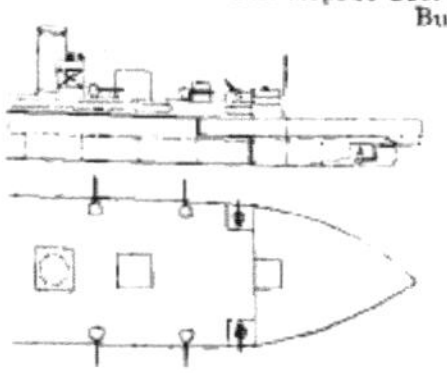

ers: Babcock. Designed H.P. 6000 = 12 kts.
y.

10,000 yards. Smoke screen apparatus fitted. *Erebus*

into 50 w.t.c. *Erebus* hit full amidships by distance-
a fortnight and rails were fitted to "blisters" of both
rs. *Terror* hit by three torpedoes in succession—2 right
The third torpedo hit the bulge and did no harm at all.

Begun	Completed	Trials
ct., 1915	Sept., 1916	7244 H.P. = 14·1 kts.
ct., 1915	Aug., 1916	6235 H.P. = 13·1 kts.

d as an improved "Abercrombie" type to outrange the
Coast. Their speed, considering their great beam, is
. Dover Patrol.
Hamilton.

MARSHAL SOULT — 1920 *Photo, Abrahams, Devonport.*

MARSHAL SOULT (17th June, 1915).

Displacement, 6670 tons. Complement, 228.

Length, 340 (*p.p.*), $355\frac{3}{4}$ (*o.a.*) feet. Beam, $90\frac{1}{4}$ feet. Draught, $10\frac{1}{2}$ feet.

Guns: 2—15 inch (Dm. Cox., H. A. elevation), 8—4 inch, 2—12 pdr., 2—3 inch AA., 2—2 pdr. AA. Armour: 8" Barbette, 13"—$4\frac{1}{4}$" Gunhouse; in both, 4" Bulkheads fore and aft, 6" C.T., 4"—1" Box Citadel over Magazine, 1" Fo'xle Deck, 2"—$1\frac{1}{2}$" Upper Deck, Lower Deck 3" at bow, $1\frac{1}{2}$" at stern, 1" Navig. Position. Deep Bulge Protection.

Machinery: 2 sets Diesel. 4 screws. Designed H.P. 1500 = 6.7 kts. Fuel: 235 tons, *maximum* oil only.

Name	*Builder*	*Machinery*	*Begun*	*Completed*	*Trials*
M. Soult	Palmer	Vickers	Jan., 1915	Nov., 1915	1898 H.P. = 6·6 kts.

General Notes.—Emergency War Programme. Has Diesel engines originally built for R.F.A. Oiler *Trefoil*. Sister ship *M. Ney*, now Hulk attached to Fort Blockhouse, Portsmouth.

Large China Gunboats.

(INSECT CLASS—12 BOATS.) 1919 Photo.

APHIS (1915), **BEE** (1916). Both by Ailsa Co.

CICALA (December, 1915), **COCKCHAFER** (December, 1915), **CRICKET** (December, 1915), **GLOWWORM** (February, 1916). All by Barclay, Curle.

GNAT (1915), **LADYBIRD** (1915). Both by Messrs. Lobnitz.

MANTIS (1915), **MOTH** (1915). Both by Sunderland S.B. Co.

SCARAB (1915), **TARANTULA** (1915). Both by Wood, Skinner & Co.

645 tons. Complement, 53. Dimensions: 230 (p.p.), 237½ (o.a.) × 36 × 4 feet. Guns: 2—6 inch, 2—12 pdr., 6 M.G.; also 2—3 inch AA. in a few. Machinery: Triple expansion. Twin screws in tunnels fitted with Messrs. Yarrow's patent balanced flap. Boilers: Yarrow. Designed H.P. 2000 = 14 kts. Fuel (coal or oil): 34 tons ed.

Note.—Emergency War Programme. Ordered February, 1915. Completed November, 1915–April, 1916. Built to Yarrow design under Messrs. Yarrow's supervision. Originally intended that these boats should proceed to [illegible] ... transported in sections overland and be re-erected and re-floated on a tributary of the Danube, to fight the Austro-Hungarian Danube Flotilla. [illegible] "China Gunboats." [illegible] Gunboats for the Chinese Rivers—hence their name of "China Gunboats." [illegible] the great "drive" [illegible] Tigris and Euphrates. [illegible] May 7th, 1919; all repaired.

British Danube Flotilla.

1919 Photo.

(FLY CLASS—16 BOATS.)

Small China Gunboats.

BLACKFLY, BUTTERFLY, CADDISFLY, CRANEFLY, DRAGONFLY, FIREFLY, GADFLY, GRAYFLY, GREENFLY, HOVERFLY, MAYFLY, SAWFLY, SEDGEFLY, SNAKEFLY, STONEFLY, WATERFLY. Built by Messrs. Yarrow, Ltd., in sections, shipped out to Mesopotamia and erected at Abadan, 1915-6—See Notes below.

98 tons. Complement, 22. Dimensions: 120 (p.p.), 126 (o.a.) × 20 × 2 feet. Guns: 1—4 inch, 1—12 pdr., 1—6 pdr. (not in all), 1—3 pdr. AA., 1—2 pdr. pom-pom, 5 or 4 M.G. Machinery: 1 set triple expansion engines. Single screw in tunnel fitted with Messrs. Yarrow's patent balanced flap. Boilers: 1 Yarrow, burning coal or oil. Designed H.P. 175 = 9.5 kts. Fuel: 5 tons coal + 10 tons oil.

Notes.—Twelve of these craft are said to have been ordered by the Admiralty in February, 1915, from Messrs. Yarrow, Ltd. They were intended to police the Tigris against Arab guerillas. After capture of *Firefly* by Turks, the number was increased to 16. *Firefly* was re-captured in the British advance from Kut-el-Amara to Bagdad. They were built in sections and shipped out, the port and starboard sections being numbered and painted red and green, to facilitate erection by native labour at Abadan. *Firefly* began to be erected August, 1915, and entered service in November of the same year. The last of the original twelve were shipped out in August, 1915, and were in service by March, 1916, the extra four (*Blackfly*, *Caddisfly*, *Hoverfly*, *Sedgefly*) following them later. All Emergency War Programme. Eight of these transferred to War Office. *Greenfly* captured, looted and burnt by Arab rebels, 1920, and recaptured by British troops, October, 1920.

	Begun.	Sent out.
Blackfly	12/15	4/16
Butterfly	2/15	7/15
Caddisfly	12/15	4/16
Cranefly		7/15
Dragonfly		7/15
Firefly	2/15	7/15
Gadfly		8/15
Grayfly		8/15

	Begun.	Sent out.
Greenfly		
Hoverfly	2/15	9/15
Mayfly	11/15	9/15
Sawfly	2/15	4/16
Sedgefly	2/15	6/15
Snakefly	11/15	9/15
Stonefly		9/16
Waterfly	2/15	

Gunboats: 1st Class.

1919 Photo.

HUSSAR (1894). ex-Torpedo Gunboat. Serves as Yacht and Despatch Vessel, C.-in-C., Mediterranean. 1070 tons. Guns: 1—4·7 inch. I.H.P. 2500 = 17·5 kts. (12 best speed 1919). Coal: 120 tons. Complement, 120.

1919 Photo.

DWARF, THISTLE (1898-99). 710 tons. Complement, 90. Dimensions: 180 x 33 x 9¾ feet. Guns: 2—4 inch, 4—12 pdr., 4 or 6 machine. Designed H.P. 1300 = 13·9 kts. Yarrow boilers. Coal: 145 tons.

Old Sloops.

Note.—Fleet Sweeping Vessels (Sloops) are listed after the Submarine Section on later pages. *Merlin*, *Mutine*, *Fantome* listed with Surveying Ships.

Photo, G. M. Carter, Esq.

ESPIEGLE (1900-02). 1070 tons. Complement, 121-112. 4—4 inch, 2 or 3 M.G. I.H.P. 1400 = 13¼ kts. Boilers: Niclausse. Coal: 105—225 tons.

Patrol Gunboat.

(KIL CLASS—1 BOAT.)

Photo, Abrahams, Devonport, 1920.

KILLERIG (July, 1918). Built by Smiths Dock Co. 890 tons. Complements, 39. Dimensions: 170 (*p.p.*), 182 (*o.a.*) × 30·30¼ × 11¼ feet. Guns: 1—4 inch. Designed H.P. 1400 = 13 kts. Machinery: 1 set inverted triple expansion. 1 screw. Boilers: Cylindrical return tube. Coal: 330 tons.

General Notes.—Emergency War Programme. 85 built or ordered during War. All sold or construction cancelled, 1920, except above vessel (attached to Dartmouth R.N. College) and *Kilmun* (listed on later page) converted to a Cable Repair Ship.

Flotilla Leaders and Destroyers.

Notes.

(*v. following pages.*)

Total over Table on next page is composed thus:—22 Leaders (+ 2 *completing*), 170 Destr (+ 4 *completing*) = 198. In the following Table, dimensions and H.P. in some classes been simplified to an average for the class. Dates refer to the time the first boat was down and the last boat completed. All building dates are given in Tables on subseq pages.

In Table, first and higher speed is that designed and to be attained on *trials* on light load drau the second and lower speed is for deep load draught. In fuel column, first figure is "pea stowage; second figure is "war" stowage where "peace" tanks above water are not u and so reduce capacity.

The following Notes are for Flotilla Leaders and Destroyers from *M* up to *V* and *S* Classes Searchlights, 1—20 or 24 inch, and 2—10 inch signalling. In the War Leaders, *V* and *S* boa 20—24 inch searchlights are controlled on target in conjunction with guns. Also direct for tubes. R.F. in all leaders and latest *V*—*W* types.

Many boats carry 2 depth charge throwers, depth charges and PV's of H.S. types. *Abdiel*, a various other boats, have been used as Minelayers, conversions being carried out with 24 hours in the case of the *V* and *W* boats.

Features which help in picking out War Design boats are thus:—Admiralty boats have rou and slightly hollowed stems (except *Warwick*, mined and re-built with straight stern, ar Admiralty "Modified W" boats, with straight and slightly sloping sterns). Thornycro boats have straight stern. Yarrow boats (from Later Yarrow M's) have strongly slopir sterns. White boats sometimes have strongly ribbed funnels. Hawthorn Leslie boats cage tops to funnels *without* caps. Thornycroft boats have usually one big flat-sided funnel, an *all* funnels have caged caps. Thornycroft boats stand high out of water and look big. Yarro boats have very sharp bows. When towing Kite Balloons, after mast is removed and W/T aerials are spread to two poles abeam and between tubes.

Funnel Markings, 1920.

One *narrow black* band on *after* funnel for boats of 1st Flotilla.
Two " " bands " " " 2nd "
Three " " bands " " " 3rd "
One *broad white* band " " " 4th "
One *narrow white* band on *fore* " " Nore Flotilla.
One " " band on *after* " " Devonport Flotilla.
One " " band on *each* " " Portsmouth "
Leaders, Captains (*D*), broad black band at tops of funnels.
Black or white disc on fore funnel for Division (or "Half") Leaders.
Boats change their marks on transfer from one Flotilla to another.
System for Mediterranean T.B.D. (6th Flotilla) not reported.

	First begun.	Last completed.	Displacements, tons.	Dimensions. Length × Beam × Dr'ft (*o.a.*) (*Max.*)	Designed S.H.P. = kts. speed	Fuel tons.	Guns. (*aa* = Anti-aircraft.)	Torpedo Tubes. No. and size.	Complement.
Design"	1917	1919	1801	332½ × 31¾ × —	40,000 = 36·5/31	505/480 *oil*	All 5—4·7 inch 1—3 inch *aa*	All 6—21 inch	All 164
er)	1916	1921	1740	329 × 31¾ × —	40,000 = 36/31	500/400 *oil*			
	1916	1917	1316—1339	312 × 29½ × 11¾	27,000 = 34/31	370/320 *oil*	4—4 inch 1—3 inch *aa*	6—21 inch	115
	1915	1916	1666—1673	325 × 31¾ × 11¾	36,000 = 34/31	510/413 *oil*	4—4 inch 2—2 pdr.	(*Abdiel* : none) 4—21 inch	116
	1914	1916	1604—1687						104
7 inch)	1918	1920	1345	312 × 30⅝ × —	30,000 = 35/32	370/320 oil	4—4·7 inch 2—2 pdr.	6—21 inch	127
inch)	1918	1921	1325	312 × 29½ × —	27,000 = 34/31	350/320 *oil*			
	1917	1919	930	269½ × 25¾ × —	23,000 = 36/32	255/215 *oil*	All 3—4 inch 1—2 pdr.	All 4—21 inch	All 90
	1917	1919	1087	275¼ × 27½ —	29,000 = 36/32	300/250 *oil*			
	1917	1919	1075	276 × 26⅝ —	27,000 = 36/31	300/250 *oil*			
	1917	1918	1315	312 × 29½	30,000 = 35/32	370/320 *oil*	All 4—4 inch 1—3 inch *aa*	6—21 inch	127
	1917	1918	1280—1316	312 × 29 × 11¼	27,000 = 34/31	370/320 *oil*		6—21 inch	127
	1916	1918	1325	312 × 30⅝ × 11¾	30,000 = 35/32	375/325 *oil*		6—21 inch	110
	1916	1918	1272—1339	312 × 29½ × 11¾	27,000 = 34/31	370/320 *oil*		6,5 or 4—21 inch.	110
ed R	1916	1917	1076—1091	276 × 26¾ × 11¾	27,000 = 36/31	300/250 *oil*	All 3—4 inch 1—2 pdr.	All 4—21 inch	All 82
	1915	1917	897—923	271½ × 25¾ × 10½	23,000 = 36/31				
	1915	1917	1035	274 × 27½ × 11	29,000 = 35/31	285/220 *oil*			
	1915	1917	1036—1096	275 × 26½ × 11¾	27,000 = 36/31	290/240 *oil*			
	1913	1916	1033	273 × 27½ × 11	27,500 = 35/31	275/210 *oil*			

7 Admiralty Large Design.

MACKAY. 1920 Photo, [illegible] Ltd.

5 *Cammell Laird*: **Bruce, Campbell, Douglas, Mackay** (ex *Claverhouse*) **Malcolm.**
2 *Hawthorn Leslie*: **Montrose, Stuart.**

Displacement: 1801 tons. Dimensions: 320 (p.p.), 332½ (o.a.) × 31¾ × 10½ feet (mean) draught. Guns: 5—4.7 inch (Dir. Con.), 1—3 inch AA. (*Stuart*, 2—2 pdr. AA.; *Malcolm*, 1—[illegible] pdr. AA.). Tubes: 6—21 inch in two triple mountings. Machinery: Parsons (all-geared) turbines. Designed S.H.P. 40,000 = 36.5 kts.* Boilers: Yarrow. Oil: about 500/450 tons. Complement, 164.

*On trials, light load draught; on deep load, 31 kts. with same S.H.P.

General Notes.—Emergency War Programme boats. *War Losses*: *Scott* (Cammell Laird). *Cancelled* 1918: *Barrington, Hughes* (both Cammell Laird).

To Distinguish.—By 4.7 inch guns in 5 positions. Deck chart-house and bridge; thin funnels, equal in height. AA. gun platform well abaft 2nd funnel. Variations are 2 or 3 yards on topmast and small topmast to mainmast in one or two boats.

	Begun.	Launch.	Comp.
Bruce ..	12/5/17	26/2/18	30/5/18
Campbell ..	10/11/17	21/9/18	21/12/18
Douglas	30/6/17	8/6/18	2/9/18
Mackay ..	5/3/18	21/12/18	?6/19
Malcolm ..	27/3/18	29/5/18	29/5/19
Montrose ..	4/10/17	10/6/18	14/9/18
Stuart ..	18/10/17	22/8/18	21/12/18

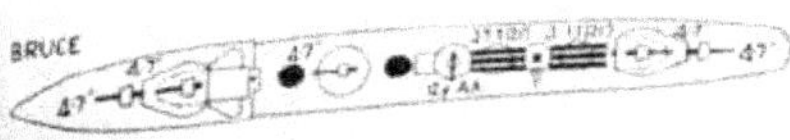

3 + 2 Thornycroft Type.

SPENSER. 1920 Photo, Seward, Weymouth.

5 *Thornycroft*: **Keppel,* Rooke,* Shakespeare, Spenser, Wallace.** 1740 tons. Dimensions: 31[illegible] (p.p.), 329 (o.a.) × 31¾ × 10½ feet (mean) draught. Guns: 5—4.7 inch (Dir. Con.), 1—3 inch AA., 2 or 1—[illegible] pdr. pom-pom. Tubes: 6—21 inch in 2 triple deck mountings.

Machinery: Brown-Curtis all-geared turbines. Designed S.H.P. 40,000 = 36 kts.† 2 screws. Boilers: Yarrow. O[illegible] about 550/460 tons.

†Light load on trials; on deep load, same S.H.P. = 31 kts.

General Notes.—Built under War Emergency Programme. Appearance almost exactly same as *Bruce, Campbell*, &[illegible] but these boats have the usual big, flat-sided Thornycroft funnels with caged caps. Complement, 164. No W[illegible] Losses. *Cancelled* 1918: *Saunders, Spragge.*

	Begun.	Launch.	Comp.	Trials.
*Keppel** ..	10/18	4/20	....	....
*Rooke** ..	11/18	16/7/20	....	....
Shakespeare ..	10/16	7/7/17	10/17	38.74
Spenser ..	10/16	22/9/17	12/17	37.76
Wallace ..	8/17	26/10/18	2/19	37.72

* *Keppel* completed by Portsmouth D.Y. *Rooke* to be completed by one of H.M. Yards.

(Plans as for *Bruce* opposite.)

5 Admiralty "V" Leaders.

VAMPIRE (topmast to mainmast). 1920 Photo, [illegible]

2 *Cammell Laird*: **Valentine,* Valhalla:** 1339 tons (1523 deep load).
2 *Denny*: **Valkyrie, Valorous*:** 1325 tons (1510 deep load).
1 *White*: **Vampire:** 1316 tons (1500 deep load).

Dimensions: 300 (*p.p.*), 312 (*o.a.*) × 29½ × 9 feet (*mean*), 11¼ (*max.*) draught. Guns: 4—4 inch (Mk. V. Dir. Con.), 1—3 inch AA. Tubes: 6—21 inch in two triples.† Machinery: Turbines, all geared-types. *Valentine* and *Valhalla* Parsons, others Brown-Curtis. Designed S.H.P. 27,000 = 34 kts. (31 deep load). 2 screws. Boilers: 3 Yarrow, except *Vampire* with White-Forster. Oil: about 370/320 tons. Complement, 115.

General Notes.—Emergency War Programme boats. Are exactly the same as the Admiralty V Destroyers, but specially fitted as Leaders. No War Losses.

To distinguish.—From all Admiralty Thornycroft V's and W's with six tubes: *high* foremast, bridges rather larger, more boats carried abeam of 2nd funnel, searchlight platform between tubes extended to take standard compass. After mast stands *against* after superstructure. From Admiralty Modified W's (*Whitshed, &c.*), by funnels and 3 inch AA. gun: (*Whitehall, Venomous, &c.*), by 3 inch AA. gun only; from Thornycroft Modified W's, by funnels and 3 inch AA. gun. *Valentine* has two stump masts for W/T. aerials on each side of bandstand.

		Begun.	Launch.	Comp.
Valentine	..	7/8/16	24/3/17	27/6/17
Valhalla	..	8/8/16	22/5/17	31/7/17
Valkyrie	..	25/5/16	13/3/17	16/6/17
Valorous	..	25/5/16	[illegible]/5/17	21/8/17
Vampire	..	10/10/16	21/5/17	22/9/17

*Have served as Minelayers.

†*Vampire* has always had 6 tubes; added to *Valhalla* 1920, *Valkyrie* and *Valentine* are to be so fitted, but still had 4 T.T. at end of 1920.

(Plans as "6-tube V's," *Vanessa, Vanity, &c.*, on later page.)

3 Marksman (later) Type.

SEYMOUR. 1919 Photo, C

3 *Cammell Laird*: **Grenville, Saumarez, Seymour.*** First 1,666 tons, last two 1673 tons. Dimensio (p.p.), 325 (o.a.) × 31¾ × 10 feet (mean), 12½ (max.) draught. Guns: 4—4 inch (Dir. Con.), 2—2 pdr. p 1 M.G. Tubes: 4—21 inch in pairs. Machinery: Parsons turbines. 3 screws. Designed S.H.P. 36,0 kts. Boilers: Yarrow. Oil: about [illegible] tons. Complement, 116.

General Notes.—All Emergency War Programme. Same hull design as *Nimrod*, but have only three funnels ar bring 4 inch gun forward. The Admiralty V type Destroyers were derived from the above design in t One boiler room and one funnel deleted, reducing length by 15 feet p.p. and S.H.P. by 9,000; amidshi gun transposed to after superstructure.

War Losses.—*Hoste* (Cammell Laird). *Removals.*—*Anzac* (Denny) presented to Australia, 1919. *Parker* on list, 1920.

		Begun.	Launch.	Comp.
Grenville	..	19/6/15	17/6/16	11/10/16
Saumarez	..	2/3/16	14/10/16	21/12/16
Seymour	..	23/11/15	31/8/16	30/11/16

*Has been used as a Minelayer.

SEYMOUR

2 Marksman Type.

1915 Photo.

[illegible] no longer carried.

1 *Cammell Laird*: **Abdiel.** 1687 tons. Parsons turbines. Armament: 4—4 inch, 2—2 pdr. [illegible] This boat served as a Minelayer during War, carrying about 60-70 mines [illegible] Parsons turbines. Complement, 116. Dimensions, H.P., speed, etc., as *Nimrod*, below.

1919 Photo.

1 *Denny*: **Nimrod,** 1608 tons. Dimensions: 325 × 31¾ × 11¾ feet. Armament: 4—4 inch (Dir. Con.), 1—3 inch AA., 2—2 pdr. AA., 1—2 pdr. pom-pom. 4—21 inch tubes in pairs. Machinery: Turbines, Parsons (geared cruising). 3 screws. Boilers: Yarrow. Designed H.P. 36,000 = 34 kts. Fuel: 510/413 tons oil. Complement, 106.

Note.—1914-15 Programme.

	Begun.	Launch.	Comp.
Nimrod ..	9/10/14	12/4/15	25/8/15

Deleted 1920.—5 boats on Disposal List. *Nimrod* will also go on Disposal List at an early date. (Also 3 *Faulknor* class sold to Chile and 1 Special Boat, *Swift*, on Disposal List.)

2 Thornycroft "Improved W" Destroyers.

1920 Photo, Seward, Weymouth.

WISHART.

2 *Thornycroft*: **Wishart, Witch.** [illegible] tons normal [illegible]. Dimensions: [illegible]. Guns: 4—4.7 inch [illegible], 2—2 pdr. pom-poms. Torpedo tubes: 6—21 inch in two triples. Machinery: [illegible] turbines (all-geared types). 2 screws. Boilers: 3 Yarrow. Designed S.H.P. [illegible] = [illegible] kts. [illegible] Oil fuel: about [illegible] tons. Complement, [illegible].

General Notes.—Begun under Emergency War Programme. Differ from Admiralty "Modified W's" in dimensions, H.P., speed and a few other details.

To distinguish.—Big, flat-sided fore funnel, [illegible] both funnels nearly equal in height. High mainmast. These T.B.D. stand high out of water.

	Begun.	Launch.	Comp.
Wishart ..	6/18	....	7/20*
Witch ..	6/18	....	....

*Both completed by Devonport D.Y.

(Plans generally as for *Wolverine* on next page.)

14 "Admiralty Modified W."

WHITSHED. 1919 *Photo, Messrs. Swan Hunter (Builders).*

VERITY. 1920 *Photo, Seward, Weymouth.*

7 BOATS, APPEARANCE AS ABOVE PHOTO:—

1 *Clydebank*: **Veteran.**
2 *Swan-Hunter*: **Whitshed, Wild Swan.**
4 *White*: **Witherington, Wivern, Wolverine, Worcester.***

Completed by *Portsmouth D.Y.

To distinguish.—Proportions of funnels reversed compared with other V's and W's. These boats have *thick* fore funnel and *thin* after funnel. No AA. guns abaft 2nd funnel. 2 pdr. pompoms in echelon between funnels. *Wivern* has plain SL. tower without compass platform on fore side.

WOLVERINE

7 BOATS, APPEARANCE AS ABOVE PHOTO:—

1 *Beardmore*: **Vansittart.**
2 *Clydebank*: **Venomous** (ex *Venom*), **Verity.**
1 *Denny*: **Volunteer.**
1 *Fairfield*: **Wanderer.**
1 *Swan Hunter*: **Whitehall.***
1 *Yarrow*: **Wren.†**

Completed by *Chatham D.Y. †Pembroke D.Y.

To distinguish.—Difficult to distinguish from other W's and six tube V's. Only distinctive feature is the 2—2 pdr. guns abaft after funnel, in echelon. *Vansittart*, *Wanderer*, have plain SL. towers without compass platform on fore side.

WANDERER

General Notes for both Types.

Displacement: 1325 tons *average* (1500 *full load*). Dimensions: 300 (*p.p.*), 312 (*o.a.*) × 29½ × 9 feet (*mean*) draught. Guns: 4—4.7 inch (Dir. Con.), 2—2 pdr. pom-poms. Tubes: 6—21 inch, in two triple mountings. Machinery: Turbines (all-geared type)—all above probably Brown-Curtis, but *Whitehall* (and perhaps one or two others) Parsons. Designed S.H.P. 27,000 = 34 kts.* 2 screws. Boilers: Yarrow, except *White* boats with White-Forster. Oil: *about* 305-367/320 tons. Complement, 127-130.

*Light load draught on *trials*; on deep load 31 kts. with same S.H.P.

Notes.—Begun under War Emergency Programme, but cost of completion of all comes under post-war Estimates. Differ from preceding V's in armament. Sometimes—but unofficially—referred to as "Repeat W Class." No War Losses, none of these boats being finished till 1919-21.

Boats cancelled 1918-19.—*Vashon, Vigo*, (ex-*Vantage*) *Vengeful* (Beardmore). *Vigo, Virulent, Volage, Volcano, Wistful* (ex-*Vigorous*) (Clydebank). *Votary, Wager, Wake, Waldegrave, Walton, Whitaker* (Denny). *Wave, Weasel, White Bear* (Fairfield). *Welcome, Welfare, Wellesley* (Hawthorn Leslie). *Wheeler, Whip, Whippet* (Scott). *Whitehead, Willoughby, Winter* (Swan-Hunter). *Werewolf, Westphal, Westward Ho!, Wrangler* (White). *Wye, Yeoman, Zealous, Zebra, Zodiac* (Yarrow). *Warren* (Chatham D.Y.). *Watson* (Devonport D.Y.). *Whelp* (Pembroke D.Y.).

	Begun.	Launch.	Comp.
Veteran	30/8/18	26/4/19	1919
Whitshed	6/18	1/19	7/19
Wild Swan	7/18	10/5/18	11/19
Witherington	27/9/18	16/1/19	10/19
Wivern	19/8/18	15/4/19	1919
Wolverine	8/10/18	17/7/19	7/20
*Worcester**	20/12/18	1919	......

Completed by *Portsmouth D.Y.

	Begun.	Launch.
Vansittart	1/7/18	17/4/19
Venomous	31/5/18	21/12/18
Verity	17/5/18	19/3/19
Volunteer	16/4/18	17/4/19
Wanderer	1918	1/5/19
Whitehall	6/18	11/9/19
Wren†	1/18	11/11/19

Completed by *Chatham D.Y. †Pembroke D.Y.

6 "Yarrow S."

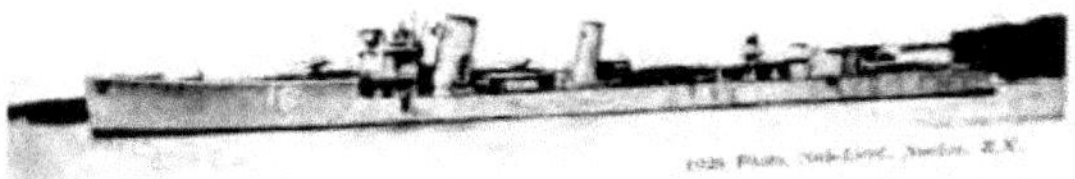

1920 Photo, [illegible], R.N.

TORCH.

6 Yarrow: **Tomahawk** (ex [illegible]), **Torch** (ex [illegible]), **Tumult, Turquoise, Tuscan, Tyrian.** Displacement, 930 tons. Dimensions: 260 [illegible] × 25 [illegible] × 9 [illegible] feet (mean) draught. Guns: 3—4 inch (Mk. IV Dir. Con. with 30° elevation), 1—2 pdr. pom-pom. Tubes: 4—21 inch in pairs. Machinery: Brown-Curtis (direct drive) turbines. Designed S.H.P. 27,000 = 36 kts.* Boilers: Yarrow. Oil: about 255/215 tons. Complement, 90.

*On light load draught; on deep load, 32 kts. with same S.H.P.

General Notes.—Emergency War Programme boats. Otherwise as "General Notes" to Admiralty S boats. *Tryphon* wrecked on Mudros, 1919, salved and towed to Malta; placed on Sale List, 1920.

To distinguish.—From Admiralty and Thornycroft S: three [illegible] funnel, sloping Yarrow stern. From other types: see Distinction Notes for Admiralty S boats. No War Losses.

	Begun.	Launch.	Comp.	Trials.*
Tomahawk ..	4/17	16/5/18	7/18	36.59
Torch ..	4/17	16/3/18	5/18	39.10
Tumult ..	4/17	17/9/18	12/18	35.70
Turquoise ..	6/17	9/11/18	5/19	39.6
Tuscan ..	6/17	1/3/19	6/19	36.90
Tyrian ..	4/17	2/1919	2/2/19	39.72

(Plan as "Admiralty S" type.)

*Trials run on 1065 tons load displacement.

5 "Thornycroft S."

1919 Photo, Messrs. Thornycroft (Builders).

SPEEDY.

5 Thornycroft: **Speedy, Tobago, Torbay, Toreador, Tourmaline.** 1087 tons. Length: 266 [illegible] (p.p.), 275 (o.a.) × 27 [illegible] × 10 [illegible] feet (mean) draught. Guns: 3—4 inch (Mk. IV Dir. Con. with 30° elevation), 1—2 pdr. pom-pom. Tubes: 4—21 inch in pairs. Machinery: Brown-Curtis (all geared) turbines. 2 screws. Designed S.H.P. 29,000 = 36 kts.* Boilers: Yarrow. Oil: about 300/250 tons. Complement, 90.

*On light load draught; on deep load, 32 kts. with same S.H.P.

General Notes.—Emergency War Programme boats; otherwise as "General Notes" to Admiralty S Class. Designed with 2—18 inch tubes on racks athwartships under bridges, for purpose explained in notes to "Admiralty S" boats; these 18 inch tubes now removed. No War Losses.

To distinguish.—From Admiralty S and Yarrow S: note that fo'xle 4 inch is raised*; funnels of equal height; stand higher out of water. From other classes, distinction notes as for Admiralty S boats.

*Not in *Tourmaline* ?

	Begun.	Launch.	Comp.	Trials.
Speedy ..	3/17	1/6/18	8/18	38.51
Tobago ..	7/17	15/7/18	10/18	38.21
Torbay ..	11/17	6/3/19	7/19	Yes
Toreador ..	11/17	7/12/18	4/19	36.40
Tourmaline	3/18	4/19	1919	Yes

(Plan as "Admiralty S" type.)

50 Admiralty "S."

2 *Beardmore*: **Tactician, Tara.**

9 *Clydebank*: **Scimitar, Scotsman, Scout, Scythe, Seabear, Seafire, Searcher, Seawolf, Simoom.**

6 *Denny*: **Senator, Sepoy, Seraph, Serapis, Serene, Sesame.**

2 *Doxford*: **Shamrock, Shikari.***

5 *Fairfield*: **Sikh, Sirdar, Somme, Spear, Spindrift.**

4 *Hawthorn Leslie*: **Tenedos, Thanet, Thracian,† Turbulent.**

4 *Palmer*: **Steadfast, Sterling, Stonehenge,* Stormcloud.**

4 *Scott*: **Strenuous, Stronghold, Sturdy, Swallow.**

3 *Stephen*: **Sabre, Saladin, Sardonyx.**

6 *Swan Hunter*: **Shark, Sparrowhawk, Splendid, Sportive, Tilbury, Tintagel.**

5 *White*: **Tribune, Trinidad, Trojan, Truant, Trusty.**

*Completed by Chatham D.Y. and †by Sheerness D.Y.

1075 tons. Dimensions: 265 (p.p.), 275-277 (o.a.) × 26⅔ × 11 feet (mean) draught. Guns: 3—4 inch (Mk. IV *Dir. Con.* with 30° elevation), 1—2 pdr. pom-pom. Tubes: 4—21 inch in pairs. (*See Notes*). Machinery: Turbines (all-geared type). Brown-Curtis (A.G.) in all except following:—Palmer boats, Parsons (A.G.); *Tilbury, Tintagel*, Parsons (A.G.). Designed S.H.P. 27,000 = 36 kts. 2 screws. 3 Yarrow boilers in all, except White boats with White-Forster. Oil: about 250/300 tons. Complement, 90.

General Notes.—Emergency War Programme boats, but cost of completion of about 45 boats included in post-war Estimates. Design derived from "Admiralty Modified R" boats. During the War, the 21 inch Marks of Torpedoes were found unsuitable for close-range night actions with enemy T.B.D., as these torpedoes, on being fired, dive and run deep for a considerable distance, so under-running hostile Destroyers. Accordingly these boats were designed with a single 18 inch tube on each beam at break of fo'xle, making six tubes in all. It was intended these 18 inch tubes should be fixed, but only Thornycrofts carried this out with their boats. The use of training tubes resulted in a lot of flared hull being cut away to the detriment of weatherly qualities. These 18 inch have now been removed from all boats except *Tintagel* and a few others. Reported to be not quite so successful as the Admiralty M's and R's. They are wet with sea on beam or bow; they ride well in head seas, but throw spray over bridges. Bows not strengthened for submarine attack by ramming. No War Losses. *Saturn, Sycamore* (both Stephen) cancelled. *Stalwart, Success, Swordsman, Tattoo, Tasmania*, presented to Royal Australian Navy, 1918.

To distinguish.—Long fo'xle, sheered and slightly turtle-backed. Funnels about equal in height, the extra height of fore funnel not being very prominent. Wedge-shaped bridges built off fo'xle. Features in these boats which also appear in the R and Modified R types are:—Mounting of searchlight on after pair of tubes; pom-pom on platform just before mainmast; after 4 inch in bandstand; boats abeam of 2nd funnel. As the openings under forebridges can be screened in, they are not a reliable identification feature.

SIRDAR

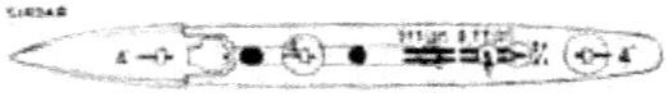

TILBURY. 1920 Photo, J. Sparshall, Esq., Smy.

		Begun.	Launch.	Comp.
2 B'more	*Tactician*	21/11/17	7/8/18	23/10/18
	Tara	21/11/17	12/10/18	9/12/18
9 Clydebank	*Scimitar*	30/5/17	27/2/18	4/18
	Scotsman	10/12/17	30/3/18	6/18
	Scout	25/10/17	27/4/18	6/18
	Scythe	14/1/18	25/5/18	7/18
	Seabear	13/12/17	6/7/18	9/18
	Seafire	27/2/18	10/8/18	11/18
	Searcher	20/3/18	11/9/18	11/18
	Seawolf	30/4/18	2/11/18	1/19
	Simoom	30/5/17	26/1/18	3/18
6 Denny	*Senator*	10/7/17	2/4/18	7/6/18
	Sepoy	6/8/17	22/5/18	6/8/18
	Seraph	4/10/17	8/7/18	25/12/18
	Serapis	4/12/17	17/9/18	21/3/19
	Serene	2/2/18	30/11/18	30/4/19
	Sesame	13/3/18	30/12/18	28/3/19
2 Doxf'd	*Shamrock*	11/17	8/18	1919
	Shikari	1/18	14/7/19	1921*
5 Fairf'd	*Sikh*	8/17	7/5/18	29/6/18
	Sirdar	8/17	6/7/18	6/9/18
	Somme	11/17	10/9/18	4/11/18
	Spear	3/18	9/11/18	17/12/18
	Spindrift	4/18	30/12/18	2/4/19

*Completing by Chatham D.Y.

		Begun.	Launch.
4 H. L.	*Tenedos*	6/12/17	21/10/18
	Thanet	13/12/17	5/11/18
	Thracian	17/1/18	5/3/20
	Turbulent	14/11/17	1919
4 Palmer	*Steadfast*	9/17	8/8/18
	Sterling	10/17	8/10/18
	Stonehenge	3/18	1919
	Stormcloud	5/18	30/5/19
4 Scott	*Strenuous*	3/18	9/11/18
	Stronghold	3/18	5/18
	Sturdy	4/18	......
	Swallow	9/17	1/8/18
3 Stephen	*Sabre*	10/9/17	23/9/18
	Saladin	10/9/17	17/2/19
	Sardonyx	25/3/18	27/5/19
6 S. H.	*Shark*	9/17	9/4/18
	Sparrowhawk	9/17	14/5/18
	Splendid	9/17	10/7/18
	Sportive	2/18	19/9/18
	Tilbury	11/17	13/6/18
	Tintagel	12/17	9/8/18
5 White	*Tribune*	21/8/17	28/3/18
	Trinidad	15/9/17	8/5/18
	Trojan	3/1/18	20/7/18
	Truant	14/2/18	18/9/18
	Trusty	11/4/18	6/11/18

†Completing by Sheerness D.Y.

**Stonehenge* wrecked near Kara Burun, Gulf of Smyrna, end of 1920. Salvage considered unlikely.

19 Admiralty "W"

WHIRLWIND. 1920 Photo, Messrs. Swan-Hunter (Builders).

2 Beardmore: **Wakeful, Watchman.***
2 Beatty: **Walker,* Westcott.**
2 Doxford: **Walpole, Whitley.**
2 Fairfield: **Walrus,* Wolfhound.**
2 Hawthorn Leslie: **Warwick,* Wessex.**
2 Palmer: **Waterhen, Wryneck.**
2 Scott: **Westminster, Windsor.**
1 Stephen: **Voyager.†**
2 Swan Hunter: **Whirlwind,* Wrestler.**
2 White: **Winchester, Winchelsea.**

* Served as Mine Layers during War and still have mine rails and chutes at stern.
† *Voyager* has always had 6 tubes and is not an "Admiralty V" re-armed.

Displacement: 1288-1316 tons (1300 average). Dimensions: 300 (p.p.), 312 (o.a.) × 29½ × 9 feet (mean), 11¼ (max.) draught. Guns: 4—4 inch (Mk V Dir. Con.), 1—3 inch AA. Torpedo tubes: 6—21 inch in two triple deck mountings. Machinery: "All-Geared" turbines, Brown-Curtis in all except Palmer boats with Parsons. 2 screws. Designed S.H.P. 27,000=34 kts.* Boilers: 3 Yarrow, except White boats with White-Forster. Oil: about 370/320 tons. Complement, 127

* Light load draught on *trials*; on deep load at 1480 tons, 31 kts. with same S.H.P.

General Notes.—All Emergency War Programme. The Notes to the Admiralty V's also apply to these boats, the only difference being in the increased number of tubes in above. The triple tubes are actually lighter than the "Admiralty R" Mark of twin tubes mounted in all preceding classes back to the M's. No War Losses.

To distinguish.—From Improved W's: 3 inch AA. gun in bandstand abaft 2nd funnel: from V's with 6 tubes: by higher mainmast: from *Vampire*: by shorter foremast, no extension to S.L. platform for standard compass, after mast stands away from after superstructure.

	Begun	Launch.	Comp.		Begun.	Launch.	Comp.
Wakeful ..	17/1/17	6/10/17	11/17	*Waterhen* ..	7/17	26/3/18	17/7/18
Watchman ..	17/1/17	2/12/17	1/18	*Wryneck* ..	7/17	13/5/18	11/18
Walker ..	26/3/17	29/11/17	12/2/18	*Westminster* ..	4/17	24/2/18	18/4/18
Westcott ..	30/3/17	14/2/18	12/4/18	*Windsor* ..	4/17	21/6/18	28/8/18
Walpole ..	5/17	12/2/18	7/8/18	*Voyager* ..	17/5/17	24/6/18	24/6/18
Whitley ..	6/17	13/4/18	10/18	*Whirlwind* ..	5/17	21/6/18	28/8/18
Walrus ..	2/17	27/12/17	8/3/18	*Wrestler* ..	7/17	15/12/17	3/18
Wolfhound ..	4/17	11/3/18	27/3/18	*Winchester* ..	21/5/17	15/12/17	15/3/18
Warwick ..	10/3/17	28/12/17	18/3/18	*Winchelsea* ..	12/6/17	1/2/18	29/4/18
Wessex ..	23/5/17	12/3/18	11/5/18				

(Plans as for "6 tube V's" *Vanessa, Vanity, &c.*, on next page.)

2 Thornycroft "W"

WOOLSTON. 1919 Photo, Messrs. Thornycroft (Builders).

2 Thornycroft: **Wolsey, Woolston.** 1315 tons. Dimensions: 300 (p.p.), 312 (o.a.) × 29½ × 9 feet (mean draught). Guns: 4—4 inch (Mk. V, Dir. Con.), 1—3 inch AA. Tubes: 6—21 inch in two triple deck mountings. Machinery: Brown Curtis turbines (all-geared type). 2 screws. Designed S.H.P. 30,000 = 35 kts.* Yarrow boilers. Oil: about 370/320 tons. Complement, 127.

* Light load draught on *trials*; on deep load at 1512 tons, 31 kts. with same S.H.P.

General Notes.—Emergency War Programme. Differ from other Admiralty W's in H.P., speed, and a few other particulars. No War Losses. Plans as "6-tube V's" on next page.

To distinguish.—Big flat-sided after funnel. Fore funnel does not look so prominently raised, owing to extra height of after funnel. Hulls stand high out of water.

	Begun.	Launch.	Comp.	Trials.		Begun.	Launch.	Comp.	Trials.
Wolsey	3/17	16/3/18	14/5/18	36.64	*Woolston*	4/17	27/14/18	28/6/18	37.11

2 Thornycroft "V."

Same appearance as *Woolston* above,
but main mast shorter.

2 Thornycroft: **Viceroy, Viscount.** 1325 tons. Dimensions: 300 (p.p.), 312 (o.a.) × 30' 7" × 9¼ feet (mean) 11¼ feet (max.) draught. Guns: 4—4 inch (Mk. V Dir. Con.), 1—3 inch AA. Tubes: 6—21 inch in 2 triple deck mountings. Machinery: Brown Curtis (all-geared) turbines. 2 screws. Designed S.H.P. 30,000 = 35 kts.* Boilers: 3 Yarrow. Oil fuel: *about* 375/320 tons. Complement, 110.

* Light load draught on *trials*; on deep load at 1512 tons, 31 kts. with same S.H.P.

General Notes.—Emergency War Programme boats. Differ from Admiralty V design in dimensions, H.P. and speed. No War Losses.

To distinguish.—As Notes to *Wolsey* and *Woolston*, except in height of mainmast.

	Begun.	Launch.	Comp.	Trials.
Viscount	12/16	29/12/17	3/18	37.66
Viceroy	12/16	17/11/17	4/18	36.5

(Plans as "6 tube V's" on next page.)

20 Admiralty "V."

VECTIS. Only boat with "H" mainmast. 1920 *Photo, Seward, Weymouth.*

2 *Beardmore*: **Vanessa, Vanity.**
1 *Doxford*: **Vega.**
1 *Fairfield*: **Vendetta.**
1 *Hawthorn Leslie*: **Verdun.**
1 *Stephen*: **Vidette.**
2 *Swan Hunter*: **Violent, Vimiera.**
1 *White*: **Vectis.**
1 *Yarrow*: **Vivien.**

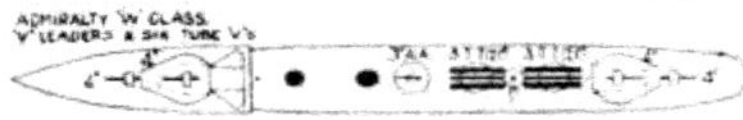

All above boats have, or will have, 6—21 inch tubes in 2 triple deck mountings (1920 alteration). Otherwise as General Notes.

To distinguish.—Identical now with "Admiralty W" boats, but have short mainmasts.

VIVACIOUS. 1920 *Photo, Seward, Weymouth.*

1 *Beardmore*: **Vancouver.**
1 *Doxford*: **Velox.**
1 *Fairfield*: **Venetia.**
1 *Hawthorn Leslie*: **Versatile.**
1 *Stephen*: **Vesper.**
1 *White*: **Vortigern.**
1 *Yarrow*: **Vivacious.**

All above have (or will have) five 21 inch tubes in one triple (forward) and one double (after) deck mountings (1920 alteration). All served as Mine Layers during War and still retain mine rails and chutes at stern. Otherwise as General Notes.

To distinguish.—The mixed T.T. mountings are special to these boats, and render them distinctive, but all boats have not yet received the forward triple tubes.

VENTUROUS, as Minelayer. (*Vanoc, Vanquisher,* same. All three boats have *high* mainmast like "W" cla 1919 *Ph*

2 *Clydebank*: **Vanoc, Vanquisher.**
1 *Denny*: **Venturous.**

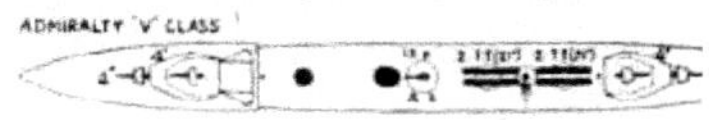

All above have four 21 inch tubes in two twin deck mountings. Served as Minelayers during War, and still ret rails and chutes at stern. Otherwise as General Notes.

To distinguish.—Only V—W boats now with two twin T.T.

General Notes.

Displacement: 1272-1339 tons (1480 *deep load*). Dimensions: 300 (*p.p.*), 312 (*o.a.*) × 29½ × 9 feet (*mean*), 11 draught. Guns: 4—4 inch (Mk. V. Dir. Con.), 1—3 inch AA., 1—2 pdr. pom-pom (not in all). Torped As noted above. Machinery: "All-Geared" turbines: Brown-Curtis in all, except Doxford and Swan boats with Parsons. 2 screws. Designed S.H.P. 27,000 = 34 kts. (light load draught on *trials*) = 31 k load draught). Boilers: 3 Yarrow in all, except White boats with White-Forster. Weight of machine tons. Oil: *about* 370/320 tons. Complement, 110. *Trials*: *Vivacious* 35.01, *Vivien* 36.79.

All Emergency War Programme. These boats are of remarkable size and power for Destroyers; in fact, five with slight modifications, have been converted to Flotilla Leaders.

	Begun.	Launch.	Comp.
Vancouver ..	15/3/17	28/12/17	9/3/18
Vanessa ..	16/5/17	14/3/18	27/4/18
Vanity ..	28/7/18	3/5/18	21/6/18
Vanoc ..	20/9/16	14/6/17	8/17
Vanquisher ..	27/9/16	18/8/17	9/17
Venturous ..	9/10/16	21/9/17	29/11/17
Vega ..	....	1/9/17	12/17
Velox ..	1/17	17/11/17	1/4/18

	Begun.	Launch.
Vendetta ..	....	3/9/17
Venetia ..	....	29/10/17
Verdun ..	13/1/17	21/8/17
Versatile ..	31/1/17	31/10/17
Vesper ..	27/12/16	15/12/17
Vidette ..	1/2/17	28/2/18
Violent ..	11/16	1/9/17
Vimiera ..	10/16	22/6/17
Vectis ..	7/12/17	4/9/17
Vortigern ..	17/1/16	15/10/17
Vivacious ..	7/16	3/11/17
Vivien ..	7/16	16/2/18

War Losses.—*Vehement* (Denny). *Verulam* (Hawthorn Leslie) and *Vittoria* (Swan Hunter) lost 1919 in Baltic o

6 Yarrow "M."

1920 Photo, [illegible] Ltd.

TRUCULENT.

1920 Photo, Gieves, Ltd.

RIVAL (3 [illegible] T.T.)

3 Yarrow: **Relentless, Rival, Sabrina, Sybille, Truculent, Tyrant.** Displacements: [illegible] tons. Dimensions: [illegible] (p.p.), [illegible] × 25¾ × 9½ feet (mean), [illegible] draught. Guns: 3—4 inch [illegible], 1—2 pdr. pom-pom,* 1 M.G. Torpedo tubes: 4—21 inch [illegible]. Machinery: Brown-Curtis turbines. 2 screws. Designed S.H.P. 23,000 = 36 kts. Boilers: 3 Yarrow. [illegible] Complement, 82.

*Not in Rival now, used for Instructional Duties.

General Notes.—All Emergency War Programme. Note that these boats are geared turbines, ... Accordingly, there is no Yarrow "R" design.

War Losses.—Strongbow, Surprise, Ullswater (all Yarrow).

	Begun.	Launch.	Comp.	Trials.*
Sabrina	8/15	24/7/16	9/16	36.9
Sybille	8/15	5/2/17	2/17	39.15
Truculent	3/16	24/3/17	5/17	[illegible]
Tyrant	2/16	19/5/17	5/17	[illegible]

*Displacements on trial were 1010–1014 tons for Relentless and Rival; Sabrina, [illegible] tons; Sybille, [illegible] tons; Tyrant and Truculent, [illegible] tons.

YARROW "M."

4 "Thornycroft R."

1920 Photo.

TEAZER.

1920 Photo, Gieves, Ltd.

ROSALIND.
(Mainmast of Taurus stepped abaft of pom-pom platform.)

4 Thornycroft: **Retriever, Rosalind.** 1035 tons. Dimensions: 274 (o.a.) × 27⅓ × 11 feet (mean) draught. Also **Taurus, Teazer**, 1065 tons. Dimensions: 274 (o.a.) × 27 × 11 feet (mean) draught. Following details for all six boats: Length (p.p.), 265 feet. Mean draught, [illegible] feet. Guns: 3—4 inch (Mk. IV), 1—2 pdr. pom-pom, 1 M.G. Tubes: 4—21 inch in pairs. Machinery: Brown-Curtis (all-geared) turbines. 2 screws. Designed S.H.P. 29,000 = 35 kts. Boilers: 3 Yarrow. Oil fuel: about [illegible] tons. Complement, 82.

General Notes.—All War Emergency Programme. Are separated from the "Thornycroft M's" in the same way as the "Admiralty R's" are divided from the "Admiralty M's," viz., by after 4 inch guns being in bandstand and machinery being "all geared" turbines. No War Losses. Radiant of this class now Siamese Phra Ruang.

	Begun.	Launch.	Comp.	Trials.
Radiant	12/15	5/11/16	2/17	39.67
Retriever	3/16	15/1/17	3/17	36.9
Rosalind	10/15	14/10/16	12/16	37.60
Taurus	3/16	10/3/17	5/17	39.27
Teazer	3/16	21/4/17	7/17	40.22

THORNYCROFT "R."

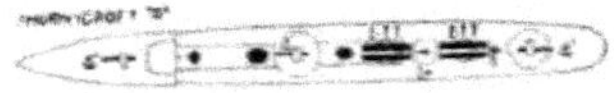

10 Admiralty "Modified R."

1 *Beardmore*: **Ulster.** 1086 tons. Brown-Curtis A.G. turbines.
1 *Doxford*: **Umpire.** 1091 tons. Brown-Curtis A.G. turbines.
1 *Fairfield*: **Undine.** 1090 tons. Brown-Curtis A.G. turbines.
2 *Palmer*: **Urchin, Ursa.** 1085 tons. Parsons A.G. turbines.
2 *Scott*: **Tirade, Ursula.** 1076 tons. Brown-Curtis A.G. turbines.
1 *Swan Hunter*: **Tower.** 1087 tons. Brown-Curtis A.G. turbines.
2 *White*: **Trenchant, Tristram.** 1085 tons. Brown-Curtis A.G. turbines.

Dimensions: 265 (*p.p.*), 275½-276 (*o.a.*) × *about* 26¾ × 10½ feet (*mean*), 11¾ (*max.*) draught. Guns: 3—4 inch (Mk. IV with 30° elevation in *Ulster* and *Ursa* DIR. CON. in all), 1—2 pdr. pom-pom. Tubes: 4—21 inch in pairs. Machinery: "All-geared" (A.G.) turbines as noted above. 2 screws. Designed S.H.P. 27,000 = 36 kts. Boilers: 3 Yarrow in all except *Trenchant* and *Tristram* by White with 3 White Forster. Oil: *about* 300/250 tons. Complement, 82. Weight of machinery: 410 tons.

General Notes.—All Emergency War Programme. These boats are said to be an attempt by the Admiralty to combine the "Admiralty R" design with that of the Yarrow "Later M's." They are said to be not quite so satisfactory as the Admiralty R's, but good sea-boats.

To distinguish.—Very long and high fo'xle, which is not sheered or turtle-backed as in "S" boats. Charthouse built on fo'xle. Fore funnel *very close* to foremast. Boats in davits abeam of after funnel. Quarter-deck abaft of mainmast is shorter than in "Yarrow M" boats, and has 4 inch in bandstand.

War Losses.—*Ulysses* (Doxford).

	Begun.	Launch.	Comp.
Ulster	19/9/16	10/10/17	21/11/17
Umpire	....	9/6/17	8/17
Undine	....	22/3/17	26/5/17
Urchin	....	7/6/17	8/17
Ursa	....	23/7/17	10/17
Tirade	....	21/4/17	6/17
Ursula	....	8/17	9/17
Tower	9/16	5/4/17	8/17
Trenchant	17/7/16	23/12/16	30/4/17
Tristram	23/9/16	24/2/17	30/6/17

ADMIRALTY MODIFIED R

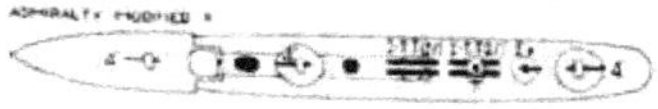

TOWER. 1919 *Photo, by courtesy of Messrs. Swan Hunter* (Buil...

1919 *Photo*, T...

URCHIN. Some boats have platform on after pair of tubes, as above.

THISBE. 1919 Photo, by courtesy of Messrs. Hawthorn Leslie (Builders).

STARFISH. *Note.*—*Satyr, Skate, Sharpshooter, Starfish, Stork* only, have mainmast as above illustration for Poulsen W/T. 1919 *Photo.*

3 *Beardmore*: **Satyr, Sharpshooter,** Parsons A.G. turbines. **Tancred,** Brown-Curtis A.G. turbines.
7 *Clydebank*: **Restless, Rigorous, Romola, Rowena, Skate, Tarpon,* Telemachus.***
3 *Denny*: **Rob Roy, Rocket, Redgauntlet.**
1 *Doxford*: **Redoubt.**
1 *Fairfield*: **Tempest.**
6 *Harland & Wolff (Govan)*: **Salmon, Skilful, Springbok, Sylph, Tenacious, Tetrarch.**
5 *Hawthorn Leslie*: **Sarpedon, Stork, Starfish, Thisbe, Thruster.**
3 *Stephen*: **Sceptre, Sturgeon, Tormentor.**
4 *Swan Hunter*: **Radstock, Raider, Sorceress, Torrid.**
1 *White*: **Sable.**

**Tarpon, Telemachus,* Minelayers.

Displacements vary from 1094 to 1036 tons (1065 *average*). Length (*p.p.*), 265 feet (*o.a.* varies from 274 to 276 feet). Beam, 26¾ feet. *Mean* draught, 9 feet. *Max.* draughts, 11¼ to 15 feet. Guns: 3—4 inch (Dir. Con.), 1—2 pdr. pom-pom (1 M.G., but not in all). Torpedo tubes: 4—21 inch in pairs. Machinery: Turbines—"all-geared" types. Clydebank, Denny, Fairfield, Harland & Wolff, Stephen, Swan Hunter and White boats, Brown-Curtis A.G. turbines; Doxford and Hawthorn Leslie boats, Parsons A.G. turbines. Beardmore boats as separately noted. Designed S.H.P. 27,000=36 kts. 2 screws in all. Boilers: 3 Yarrow or Modified Yarrow, except *Sable* by White, with White-Forster boilers. Oil fuel: 296-285-250-213 tons. Complement, 82.

		Begun.	Launch.	Comp.
3 Beardmore	*Satyr*	15/1/16	27/12/16	2/2/17
	Sharpshooter	23/5/16	27/2/17	2/4/17
	Tancred	6/7/16	30/6/17	1/9/17
7 Clydebank	*Restless*	22/9/15	12/8/16	10/16
	Rigorous	22/9/15	30/9/16	11/16
	Romola	25/8/15	14/5/16	8/16
	Rowena	25/8/15	1/7/16	9/16
	Skate	12/1/16	11/1/17	2/17
	Tarpon	19/4/16	10/3/17	4/17
	Telemachus	12/4/16	21/4/17	6/17
3 Denny	*Rob Roy*	15/10/15	29/8/16	15/12/16
	Rocket	28/9/15	2/7/16	7/10/16
	Redgauntlet	30/9/15	23/11/16	7/2/17
(Dox'd)	*Redoubt*	......	28/10/16	3/17
(F'field)	*Tempest*	......	26/1/17	26/3/17
6 H. & W.	*Salmon*	27/8/15	7/10/16	20/12/16
	Skilful	20/1/16	3/2/17	26/3/17
	Springbok	27/1/16	9/3/17	30/4/17
	Sylph	30/8/16	15/11/16	10/2/17
	Tenacious	25/7/16	21/5/17	12/8/17
	Tetrarch	26/7/16	20/4/17	2/6/17
5 H. L.	*Sarpedon*	27/9/15	1/6/16	2/9/16
	Stork	10/4/16	25/11/16	1/2/17
	Starfish	26/1/16	27/9/16	16/12/16
	Thisbe	13/6/16	8/3/17	6/6/17
	Thruster	2/[illegible]/16	10/1/17	30/3/17
3 Stn.	*Sceptre*	10/11/15	18/4/17	26/5/17
	Sturgeon	10/11/15	11/1/17	26/2/17
	Tormentor	1/5/16	22/5/17	22/8/17
4 S. H.	*Radstock*	9/15	3/6/16	9/16
	Raider	9/15	17/7/16	10/16
	Sorceress	11/15	29/8/16	12/16
	Torrid	7/16	10/2/17	5/17
(White)	*Sable*	26/12/15	28/6/16	30/11/16

General Notes.—All Emergency War Programme boats. In 1920, in Reserve, Local Defence Flotillas or attached Training Establishments.

To distinguish.—Note that the after 4 inch is mounted in a bandstand.

War Losses.—*Simoom* (Clydebank), *Recruit* (Doxford), *Tornado* (Stephen), *Torrent* (Swan Hunter), *Setter* (White).

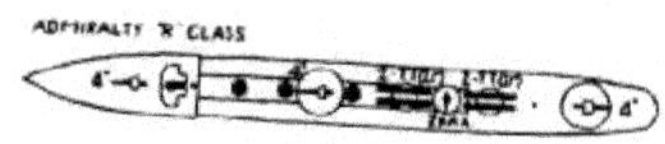

On Suspense List, end of 1920.

(For transfer to Dominion Navies or Sale to Foreign Navies.)

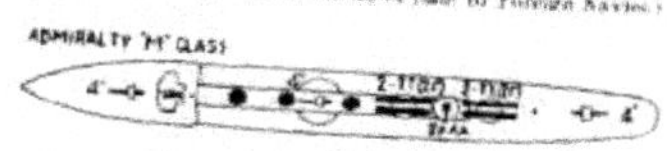

8 Admiralty "M" type: **Marne, Michael, Milbrook, Mons, Moorsom, Napean, Obedient, Ossory.** (For details, *v.* page 109, 1919 Edition.)

2 Yarrow "M" type: **Morning Star, Musketeer.** (For details, *v.* page 110, 1919 Edition.)

2 "Thornycroft M."

READY. 1919 *Photo, Messrs. Thornycroft.*

2 *Thornycroft*: **Rapid, Ready,** 1033 tons. Dimensions: 265 (*p.p.*), 274 (*o.a.*) × 27½ × 11 feet. Guns: 3—4 inch, 1—2 pdr. pom-pom, 1 M.G. Tubes: 4—21 inch in pairs. Machinery: Brown-Curtis turbines, 27,500 S.H.P. = 35 kts. 3 Yarrow boilers. Oil: *about* 285-220 tons. Complement, 82.

Note.—Emergency War Programme. Proved very successful boats. *Patriot, Patrician*, transferred to Royal Canadian Navy, 1920.

	Begun.	Launch.	Comp.	Trials
Rapid ..	8/15	7/16	9/16	35.45
Ready ..	9/15	8/16	10/16	34.30

Note.

66* Admiralty "M"	
2 Thornycroft "M"	
8 Yarrow "M"	
2 Hawthorn "M"	
3 ex-Turkish	Deleted 1920.
3 ex-Greek	Sold, on Sale and Disposal Lists, not being required for future Establishment.
17 "L"	
13 "K"	
21 "I"	
17 "H"	
13 "G"	

*As these pages are prepared, the "Admiralty M" boats, *Marne, Milne, Moss, Moorsom, Napier, Penn*, and *Peregrine*, are for sale to Rumania. *Penn* unofficially reported to have been purchased by Rumania.

Old T.B.D. *Fame* still on China Station for Target Towing and other duties, but will probably be disposed of soon.

8 P C Boats (Converted Patrol Boats)

PC 43, 56, 60, 70, as above view. 1919 *Photo, Messrs. J. S. White & Sons (Bu*
PC 71—74, no mainmast.

1 *Barclay Curle*: **PC 56** (1917). 682 tons.
1 *Caird*: **PC 43** (1917). 682 tons.
4 *White*: **PC 71—PC 74** (1918). All 694 tons.
2 *Workman Clark*: **PC 60,** (1917). 682 tons. **PC 70** (1918). 694 tons.

Dimensions: 682 ton boats: 233 (*p.p.*), 247 (*o.a.*) × 25¼ × 8.8¼ feet. 694 ton boats: 233 (*p.p.*), 247 (*o.a.* 8 feet. 682 and 694 ton boats reported to have shallow bulge protection. Guns: 1—4 inch, 2—12 pdr tubes: removed. Carry 24-30 D.C. Machinery: Parsons or Brown Curtis geared turbines. Yar (White-Forster in *PC 71—74*). 2 screws. Designed S.H.P. 3,500 = 20 kts. Oil: 154 tons in 682 ton tons in 694 boats; all oil. These two types carry extra fuel in bulges. Complement, *about* 50-55.

General Notes.—All built under Emergency War Programme. Design as P-boats, but these craft were c modified while building, to act as Submarine Decoy Vessels or "Q-boats." The after 4 inch gun was hi various forms of dummy deck loads, *e.g.*, bales or packing cases of merchandise, trusses of hay; in a was located within a collapsible pantechnicon furniture van, or under a dummy boat built in folding was expected that, on account of shallow draught, torpedoes fired by U-boats would under-run thes while, if hit by torpedo, bulge protection and special fillings would keep them afloat long enough to de boat. Now employed in A/S. Flotillas.

Removed 1920.—11 boats for sale or disposal.

8 P-Boats (Patrol Boats.)

1919 Photo, Robertson, Gourock, N.B.

1 *Hamilton*: **P38**, (1917).

1 *Harkess*: **P46** (1917).

3 *Readhead*: **P31** (1916), **P47, P48** (1917).

3 *White* (*Cowes*): **P40, P52** (both 1916), **P59** (1917).

Displacement: 613 tons. Dimensions: 230 (*p.p.*), 244½ (*o.a.*) × 23¾ × 7 feet 7 inch. Guns: Designed to mount 2—4 inch but only *P 52* has this armament; others, 1—4 inch and (in all) 1—2 pdr. pom-pom. Tubes: Originally 2 single 14 inch, removed from old Torpedo Boats and fixed on quarters, but removed from nearly all boats and replaced by rails and 30 D.C. Machinery: Brown-Curtis or Parsons geared turbines. 2 screws. Yarrow boilers (White-Forster in White boats). Designed S.H.P. 3500 = 20 kts. Oil fuel: *normal* 50 tons, *max.* 93 tons. Complement, 50-54.

General Notes.—All built under Emergency War Programme. Were designed to relieve destroyers of patrol and escort work and submarine hunting. Outline scheme for these boats stipulated; minimum size consistent with sea-keeping qualities, simplicity of construction, and adequate speed to run down submarines; also shallow draught; low upperworks to reduce visibility; and economy of fuel. Built of mild steel, but with hard steel stem for ramming submarines. Large rudder area and hull strongly cut up aft to give rapid turning. Proved very useful boats and an excellent anti-submarine type in all weathers. Could do 23 kts. when new. Those converted to Decoy Ships or "Q-boats" listed as *PC* boats on the preceding page.

War Losses.—*P 12*, *P 26*.

Deleted 1920.—35 boats for sale or on Disposal List.

P-boat (aircraft view).

1919 Official R.A.F. photo.

3 Torpedo Boats.

1 *Palmer* boat: **35** ([illegible]). 315 tons. Dimensions: 181½ × 17½ × 7½ feet. Armament: 2—12 pdr., 3—1 inch tubes. H.P. 4000 = 26 kts. Parsons turbines. Oil: 40/41 tons.

1 *Hawthorn* boat: **33** ([illegible]). 328 tons. Dimensions: 180¼ × 18½ × 7 feet. Armament: 2—12 pdr., 3—18 inch tubes. H.P. 4000 = 26 kts. 3 screws. Parsons turbine. Oil fuel: 48/41 tons.

1 *White* boat: **116** (1903). 197 tons. Dimensions: 166 × 17½ × 7 feet. Armament: *Nil.* H.P. 2950 = 25 kts. 44 tons.

C. M. B. (Coastal Motor Boats).

Length	Number and/or series letter.	Lewis Guns	Torpedoes (18 in.).	Depth Charges	Motors	B.H.P. = kts.
70 ft.	**100, 122, 123**	4	2‡	—	Ty. Y 12†	375 = 28†
55-ft.	**e**	4	1	4	Ty. Y 12	375 = 40
	80c	4	2	—	S'nb'm.	275 = 42
	bd	4	2	(2)‡	Green	275 = 38
	b	4	1	4	Green	275 = 38-40
	a	4	1	4	Ty. Y 12	250 = 33-35
40-ft.§	Numbered without series letter	2-4	1	—	Ty. / Fiat / Green	350 / ... / 375 = over 30*
		2				

† To have 900 B.H.P. motors; may make 47 kts. when re-engined. ‡ Torpedo Tubes; others carry torpedoes in troughs. ‡ [illegible] only * [illegible] cedar hull, Thornycroft V 12 engine and Y 12 carburetter; has made 43 kts.

In "Motors" column: Ty = Thornycroft; S'nb'm = Sunbeam.

Letters *a, b, bd,* &c., refer to type of motors and whether boats have single or double torpedo troughs.

63 boats, numbered between *CMB 1* and *CMB* [illegible] given in August, 1920 Navy List as in commission, viz.:—26 unlettered boats, *1—12*[illegible]; 17 boats, *14a—7*[illegible]; 7 boats, *21b—*[illegible]; 9 boats, [illegible]*bd—*[illegible]*bd*; 2 boats, [illegible]; 2 boats, [illegible].

Photo wanted.

Third type: **70-ft.** Some of these used as Minelayers. Details as Table.

C. M. B. 55-FT. TYPE. 1919 *Photo, Cribb, Sou*

Second type: 55-ft. 10 tons. Dimensions: 55 × 11 × 3 feet. Armament: 4 Lewis guns in pairs, wit inch torpedoes in troughs at stern. Some modified to take only 1—18 inch torpedo and 4 D.C. Othe when used for low speed work could carry 4 torpedoes in dropping gears over beams. Have done 41 k over in service. W/T. fitted in these. Complement, 3. Other details as Table.

No photo available.

C. M. B. 40-FT. TYPE.

First type: 40-ft. Dimensions: 40 × 8½ × 3 feet. 1 torpedo carried in central trough over stern.

General Notes.—Above types designed by Messrs. Thornycrofts Ltd. Full descriptions have appeared in technic It need only be remarked that they discharge their torpedoes tail first from troughs over stern, then swerve course and allow torpedoes to run past to target. All stepped hydroplane type hulls with chine to dam bow wave. Considering their small size, they have excellent nautical qualities. First boats were secre in 1916, and ran trials by night. 13 lost during War and 4 or more in Baltic operations. Boats commission can be ascertained from current Navy List. It is little use stating totals and numbers these craft can be laid up and placed out of commission at very short notice. Other details a For full description v. "Engineer," April 18th, 1919, and Messrs. Thornycroft's own descriptive broc

M. L. (Motor Launches).

1919 *Photo.*

M.L. 381. Some boats have mainmast right at stern for spreading W/T. aerials.

Third type: 9 boats: **M.L. 562–580.** Built 1918. No details known, but similar to first and second types below.

Second type: 52 boats: **M.L. 54–542.** Built 1915. 37 tons. Dimensions: 80-88 × 12½ feet × 3' 10". Guns: 1—3 inch AA. or lesser calibre of AA. and other types. B.H.P. 440-450 = 19 kts. Machinery: 2 sets of Standard petrol motors. Petrol: *about* 1850 gallons. Complement, 9.

First type: 4 boats: **M.L. 4, 6, 8, 54.** Built 1915. 34 tons. Dimensions: 75 × 12 × 3½ feet. Guns: 1—3 pdr. or 2 pdr. AA. and/or 1 or 2 M.G. or Lewis guns. B.H.P. 440-450 = 19 kts. Petrol: *about* 1650 gallons. Complement, 8.

General Notes.—Mostly built in U.S. and Canadian Yards and shipped across the Atlantic. During War, fitted with D.C., smoke boxes, hydrophones, &c., but special gear now removed from nearly all. Petrol consumption heavy.

War Losses: Twenty-four boats. Also 5 or 6 post-war losses.

111 Submarines (108 Completed, *4 Completing*).

No.	Type	First begun	Last completed	Displacement (tons)	H.P.	Max speed (kts.)	Fuel (tons)	Complement	T Tubes	Max. draught (ft.)
10	*R 1–12.*	1917	1918	420 / 500	240 / 1200	9.5 / 15	13	20	6	12
3	*M 1–3.*	1917	1920	1600 / 1950	...	15 / 9	...	60	4	10
6	*L 50–71.**	1917	1921	960 / 1150	2400 / 1600	17.5 / 10.5	78	...	6	13½
18	*L 9–33.*	1916	1920	890 / 1070	2400 / 1600	17.5 / 10.5	76	36	4-6	13½
8	*L 1–8.*	1916	1918							
1	*K 26.*	1918	1921	...	...	...	...	...	...	...
12	*K 2–16, K 22.*	1916	1918	1880 / 2650	10000 / 1400	24 / 9.5	200	55	8	16
30	*H 21–52.*	1917	1919	440 / 500	480 / 320	13 / 10.5	16	22	1	11¼
7	*G 3–14.*	1915	1916	700 / 975	1600 / 800	14 / 10	44	30	5	13½
16	*E 23–E 55.*	1915	1917	662 / 807	1600 / 840	15 / 10	45	30	3-5	12½

Notes to above Table.

Excepting *H 21-52* of "Holland" type, all above types are to Admiralty design. Above Table and description classes on later pages, arranged alphabetically by Class Letters. Rough division by types and building dates thus:—

55 *Ocean-going* boats: *L 1–71*, *M 1–3* (Monitors), *K 2–16* and *K 22*, *K 26*, *H 1–14*.
46 *Sea-going* boats: *H 21–52*, *E 23–55*.
10 *Coastal* boats: *R 1–12*.
— *Minelayers*: Various units of *L* and *E* types; exact total not known.

Various surrendered German Submarines are in service under the White Ensign, but only for experiments. By agr ment with all the Allies, such ex-German boats will not be added to the British Navy, but must be broken up wh trials are finished. At the end of 1920, *about* 12 German Submarines were still attached to Port Flotillas, viz.: *U 101*,* *U 133*, *U 141*, *U 126*, *U 93*, *U 91*, *U 86*, *UB 132*, *UB 121*, *UB 64*, and Minelaying Boats *U 126*, *U 124*, *UC 95*.

* On Suspense List for transfer to Dominion or Foreign Navy.

10 R Class.

Photo wanted.

R 12.

Photo wanted.

R 9, 10. (Bridge screen lower and longer than in other boats. Hull sheers up gradually to bows from C.T.)

R 3, 4. (Compare bridge screen with boats below.) 1920 *Photo, Cribb, Sou*

R 1, 2, 7, 8, 11. *Photo, Cribb, Southsea, 19*

R 1—4 by Chatham D.Y., ***R 7, R 8*** by Vickers, ***R 9, R 10*** by Armstrong Whitworth, ***R 11, R 12*** by Camm All launched and finished 1918.

Single hull type to Admiralty designs. Dimensions: 163 × 15¾ × 11½ feet. Tubes: 6—18 inch bow, surface trim, 2 tubes above water and 4 submerged. Machinery: 1 set 8-cylinder 4-cycle air injection " developing 240 B.H.P. at 375 r.p.m. and built by Ruston & Hornsby, North British Diesel Engine Co. and D.Y. Oil capacity small, but have an abnormal submerged endurance. Other details as Table.

Notes.—Built under Emergency War Programme. These boats were produced as a "submarine destroyer of sub or Contre-Submarine Type. Their outstanding feature is that—unlike nearly all other submarines in th they are faster below water than on the surface. Rudder is placed before propeller, and is small in area. is close to surface. Steering on surface is very poor. It was intended that they should submerge and chase and use their six bow tubes by salvo, specially big torpedo-compensating tanks with rapid flooding placed in bows.

Cancelled.—R 5, R 6 (Pembroke D.Y.).

3 M Class Submarine Monitors.

M 1. 1919 Photo, M. Baridson, Esq.

M 1 ex *K 18* (1918), ***M 2*** ex *K 19* (——), both by Vickers. ***M 3*** ex *K 20* (——), by Armstrong Whitworth.

Few details available. Admiralty double-hulled type. Dimensions: *about* 290 × 20 × 10 feet. Guns: 1—12 inch, 35 cal.; 1—4 inch AA. disappearing. Tubes: 4—18 inch bow. Machinery: Type not known, but reported to be similar to that of "L" class boats. Oil fuel: tons. Other details as Table.

Gunnery Notes.—12 inch Mk. XI, as mounted in *King Edward VII* class. Reported that gun is loaded and laid to high angle elevation; then boat is dived to about 12-20 feet, leaving muzzle of 12 inch gun above water, and periscopes. There is a bead sight, on gun-muzzle, so that gun can be sighted by periscope and fired when running at shallow submersion. To re-load it is necessary to return to surface.

General Notes.—Begun under War Emergency Programme. These "Submarine Monitors" are a unique design, but reports are not available of *M 1's* performance during War, so that their practical worth is a matter of conjecture.

Cancelled.—*M 4* (ex-*K 21*) by Armstrongs.

L Classes.

L 71. 1920 Photo, Glover, Ltd.

L 56. Deck before C.T. not raised. 1920 Photo, Glover, Ltd.

6 L Class. (Third "L" series: *L 51—L 71*).

(Of these, 3 delivered 3 completing.)

2 *Armstrong Whitworth*: ***L 52***, ***L 53**** (both 1919).
1 *Denny*: ***L 54***† (1919).
1 *Fairfield*: †***L 56*** (1919).
1 *Beardmore*: §***L 69***§ (1918).
1 *Scotts*: ***L 71*** (1919).

* Completed by Chatham D.Y., 1920-21, for surface running *only*.
† Completed by Devonport D.Y., 1920-21.
§ Completed by Rosyth D.Y., 1920-21.

Admiralty saddle-tank type. Dimensions: 230½ (*p.p.*), 235 (*o.a.*) × 23½ × 13¼ feet. Guns: 2—4 inch. Tubes 6—21 inch all in bows. Machinery: 2 sets 12-cylinder solid-injection, Vickers type Diesel engines by Ruston & Hornsby, Willans & Robinson, Bellis & Morcom, Armstrong, Thornycroft. Oil: 78 tons. Other details as Table.

Notes.—All begun under Emergency War Programme. Equipment as Notes to *L 9—33* on next page. Can be distinguished by the long C.T. and two guns.

Cancelled.—*L 50, 51, 60, 61* (C. Laird), *L 57, 58* (Fairfield), *L 59, L 70* (Beardmore), *L 63, 64, 72* (Scott), *L 65* (Swan Hunter), *L 66* (Pembroke D.Y.), *L 67, 68* (Armstrong), *L 73, L 74* (Denny). *L 62* not traced as completing or cancelled; hull probably sold to salvage firm.

Sunk (1919).—*L 55* in Baltic Operations.

L Classes—*Continued.*

L 33. Hull has straight slope down to stern. 1920 *Photo, by courtesy of Messrs. Swan-Hunter (Builders).*

L 20. 1919 *Photo, Cribb, Southsea.*

L 17. (*L* 11, 12, 14, same, with short screen round gun.) 1920 *Photo, Cribb, Southsea.*

L Classes—*Continued.*

L 16 (and L 15) with deck before C.T. raised. 1920 *Photo Abrahams, Devonpo*

1920 *Photo, Abrahams, Dev*

L 9. Straight slope to stern; deck before C.T. raised. Now has screen before gun, like *L 20*. Ignore app strut to foremast; it is actually part of sheer-legs in dockyard, not deleted with rest of background

18 L Class (Second "L" Series, *L 9—33*).

14 *Vickers:** ***L 11. L 12. L 14. L 17–20*** (all 1918). ***L 21–L 23*** (1919), ***L 24—27*** (1919-20).
2 *Fairfield:* ***L 15*** (1917), ***L 16*** (1918).
1 *Denny:* ***L 9*** (1918).
1 *Swan Hunter:* ***L 33*** (1919).

* *L 23* completed by Chatham D.Y.; *L 25, 27*, by Portsmouth D.Y.

Admiralty saddle-tank type. All details as *L 1—8* on next page, with these variations:—Length (*p.p.*), 229 feet feet (*o.a.*) All 6—21 inch tubes in bows; there are no beam tubes in these boats.

Notes.—All begun under Emergency War Programme. Equipment of these boats is extensive, *e.g.*, refri machinery for storage batteries, gyro compass with repeaters, 3 periscopes (one for night work), directiona phones, &c. Breastwork revolves with gun.

Cancelled (1919).—*L 28—32* (Vickers).

Not ordered.—*L 13, L 37—49.*

Not traced.—*L 34—L 36.* These boats neither appear in official List of Cancelled Boats or in List of Submarin pleting 1920. Possibly hulls sold to salvage company.

War Loss.—*L 10.*

L Classes—*Continued.*

L 8. (Deck before C.T. raised.) 1920 *Photo, Cribb, Southsea.*
L 7 believed to be same in appearance.

L 6. Now has screen to gun, which has platform sponsoned out from C.T. 1920 *Photo, Abrahams, Devonport.*

Photo wanted.

L 5

L Classes—*Continued.*

Photo wanted.

L 3.

L 2. (L 1, L 4 believed same in appearance.) 1920 *Photo, Cribb, Southsea.*

8 "L Class" (First "L" series: *L 1–8*).

4 *Vickers*: **L 1** ex *E 57*, **L 2** ex *E 58*, **L 3** (all 1917), **L 4** (1918).
1 *Swan Hunter*: **L 5** (1918).
1 *Beardmore*: **L 6** (1918).
2 *Cammell Laird*: **L 7**, **L 8** (both 1917).

Admiralty saddle-tank type. Dimensions: 222 (*p.p.*), 231 (*o.a.*) × 23½ × 13¼ feet. Guns: 1—3 inch AA, or 4 inch. Torpedo tubes: 6, viz., four bow, two beam, all 21 inch. Mine-laying boats have no beam tubes—only four bow—21 inch and carry mines in saddle tanks. Machinery: 2 sets 12 cylinder solid injection Vickers type. Oil: 76 tons. Other details as Table.

General Notes.—All built under Emergency War Programme. No War Losses.

ved K Class

ording.

ounched ——— and towed to Chatham D.Y. for completion.

ly reported that the class was designed to mount 2—5.5 inch tubes. Also unofficially reported that she will have 4 sets of .S. type) instead of steam turbines as in earlier "K Class."

outh D.Y.), K 27 (Vickers).

K Class.

1920 Photo.

a high fairwater *before* C.T., like K 5.

d in next column.)

K Class.—*Continued.*

K 10. *1919 Photo, Cribb, Southsea.*

K2, K5 (both by Portsmouth D.Y., 1916), ***K6*** (Devonport D.Y., 1916), ***K8, K9, K10*** (all by Vickers, 1916), ***K11, K12*** (both by Armstrong Whitworth, 1916-17), ***K14, K22,*** ex K 13 (both by Fairfield, 1916), ***K15*** (Scotts, 1917), ***K16*** (Beardmore, 1917).

All double-hulled type, designed by Admiralty for service as "Fleet Submarines" with Grand Fleet. Dimensions: 334 (p.p.), 337 (o.a.) × 26½ × 16 feet. Guns: 1—4 inch, 1—3 inch AA., and (in some) 1 D.C. thrower. Torpedo tubes: 8—18 inch, viz., four bow, four beam. Machinery: Parsons (V.-A.) or Brown-Curtis (A.G.) turbines, see Notes below. Oil fuel: 200 tons normal, 200 tons max. Other details as Table.

Engineering Notes.—These are not the first steam-driven boats in British Navy, having been preceded by *Swordfish (S 1)*, now scrapped. Surface machinery consists of combined steam turbines and Diesel engine. 2 sets of single reduction turbines, one H.P. and one L.P. in each set, with double helical gearing. 2 screws. 2 Yarrow small-tube type boilers with forced draught, boilers, turbines, funnels being lagged with incombustible non-conducting materials. Small electric motors are fitted for lowering funnels and closing w.t. hatches over funnel wells. Boiler room air vents closed by hydraulic power. To assist in diving quickly, or getting away after breaking surface (while steam engines are being started up), an 800 B.H.P. 8-cylinder Diesel motor is fitted, which can also be used for surface cruising. Drive from Diesel engine is through electric motors, so that these boats have three systems of transmission, (a) geared turbines for steam drive, (b) Diesel and electric transmission, (c) electric battery drive when submerged. 2 high-power compressors to charge 2500 lbs. air bottles; 2 low-power compressors to blow main ballast tanks after breaking surface. 2 electric-power bilge pumps. Hydraulic rams for raising periscopes and telescopic masts. Hydro-electric controlling gear to hydroplanes, forward hydroplanes being of housing type. Engine-room ratings wear oilskins, as water may be slipped down funnels and cause flashing-back of boilers.

Gunnery and Torpedo Notes.—Were first completed with 1 or 2 guns before and below or abaft C.T., on superstructure deck, but these were removed. For a time, these boats had only 1—3 inch, but K 17 at time of her loss mounted 1—5.5 inch. There were also 2—18 inch tubes in superstructure, above water when in surface trim, making ten tubes in all. The original three guns and superstructure tubes were removed to D.A.M.S. and S.D. Vessels (Q-boats).

General Notes.—All built under Emergency War Programme. Were first completed with flush level bows, but showed a tendency to trim by the head and dive "on their own." To remedy this, bows were raised, as shown in above illustrations. Accommodation for officers and men is remarkably spacious for submarines. Are said to be rather hot on account of steam system of propulsion, but in the North Sea this proved rather an advantage than otherwise. Considering that these boats were of a highly experimental type, they have turned out most remarkably well. Have quick roll like T.B.D., but if speed is adjusted to length of sea, they ride like cruisers. If seas are shipped over C.T., water may enter control room deck, causing earths on switchboard and men using C.T. ladder receive severe shocks.

War Losses.—K 1, 4, 17. K 13 also foundered, but was salved and re-numbered K 22.

(J Class.)

All presented in 1919 to Royal Australian Navy, q. v. for details. *War Loss.*—J 6.

30 H Class.

Photo wanted.

H 47.

H 32. (H 21—31, H 41, 42, 49, same.) 1920 Photo, Gieves, Ltd.

12 *Vickers*: **H 21—H 32** (launched 1917-18).
6 *Cammell Laird*: **H 33, H 34, H 37—H 40** (launched 1918-19).
6 *Armstrong Whitworth*: **H 41—H 46** (launched 1918-19).
4 *Beardmore*: **H 47—H 50** (launched 1918-19).
2 *Pembroke D.Y.*: **H 51** (1918), **H 52** (1919).

Single-hull "Holland" (Electric Boat Co.) type modified by Admiralty. Dimensions: 164½ (*p.p.*), 171 (*o.a.*) × 15¾ feet. Guns: Majority have 1—3 inch AA. Tubes: 4—21 inch bow. Machinery: 2 sets of Diesel engines, 8 cylinder, 4-cycle, air-injection "H" type, developing 240 B.H.P. at 375 r.p.m., and built by Vickers, Ruston & Hornsby, &c. Oil: 16 tons. Other details as Table.

Notes.—All built under War Emergency Programme. H 5[illegible], [illegible], cancelled. No War Losses.

7 G Class.

Photo, Dr. J. A. Prendergast.

All for Dominion Service, except "G 11."

G3, G4, G5 (all Chatham D.Y., 1916), **G6** (Armstrongs, 1916), **G10, G13, G14** (Vickers, 1916). Admiralty double-hull type. Dimensions: 185 (*p.p.*), 187 (*o.a.*) × 22½ × 13¼ feet. Guns: 1—3 inch AA. Torpedo tubes: 5, viz., 2—18 inch bow, 2—18 inch beam, 1—21 inch stern. Diesel engines: Vickers, Armstrong (2 sets). Oil: 44 tons. Other details as Table.

Notes.—All Emergency War Programme. Built as "Bight and North Sea Patrols," being the first genuine ocean-going boats in British Navy. Had their bows raised during War.

War Losses.—G 7, 8, 9, 11.

18 E Class.—14 Do[illegible]

E 32. 1920 Photo, Cribb, Southsea.

E 7—20 & LATER CLASS:—18 boats.

E 23† (Vickers, 1915), **E 27**† (Yarrow, 1917), **E 31** (Scott, 1915), **E 32**† (White, 1916), **E 33** (Thornycroft, 191[illegible] **E 35** (J. Brown, 1916), **E 38**† (Fairfield, 1916), **E 39, E 40** (launched by Palmers, 1916-17 and completed [illegible] Armstrongs), **E 42** (Cammell Laird, 1915), **E 44** (Swan Hunter, 1916), **E 45**,* **E 46*** (Cammell Laird, 191[illegible] **E 48**† (launched by Fairfield 1916, completed by Beardmores), **E53**, (Beardmore, 1915-16), **E 55**† (Denny, 191[illegible]

*These are Minelayers.

†To be retained for Dominion Service.

All Admiralty saddle-tank type. Built under Emergency War Programme. First British type of submarine with interna[l] sub-division by w.t. bulkheads and with beam tubes. Dimensions: 180 (*p.p.*), 181 (*o.a.*) × 22½ × 12½ feet. Tubes: 3 to 5—18 inch, viz., 2 bow, 0, 1, or 2 beam, 1 stern. Minelayers marked * have no beam tubes, and carry about 20 mines. Diesel engines: 2 sets of Vickers, Sulzer, Carels, M.A.N., or other types. Oil: 40—45 tons *normal*, 50 tons *max.* Other details as Table.

Notes.—Performed splendid work during War, and proved most satisfactory. Heavy losses sustained by this class are an index to the arduous work they performed, 1915-18.

War Losses.—E 1,* 3, 5, 6 (E 1—6 type), E 7, 8,* 9,* 10, 13, 14, 15, 16, 17, 18, 19,* 20, 22, 24, 26, 28, 30, 34, 36, 37, 47, 49, 50 (all E 7—56 and later type).

*Scuttled in Baltic.

HERMES (11th September, 1919).

Displacement, 10,400 tons *normal*. Complement { 500 R.N. / 130 R.A.F. } = 680.

Length { *p.p.* 548 feet / *o.a.* 558½ " } Beam { *w.l.* / outside bulges, 70 feet / over flying deck, 84 feet } Draught { *mean* 18 feet. / *max.* — " }

Guns (**Dir. Con.**):
- 10—6 inch, 50 cal.
- 4—4 inch, AA.
- 4—3 pdr.
- 5 M.G.
- (1 landing)

Torpedo tubes: *not known*

Armour:
Not known—probably of Light Cruiser type.
Anti-Torp. Pro. Bulges.

No plans available.

LAUNCH OF H.M.S. "HERMES." *Builders' Photo by courtesy of Sir W. G. Armstrong, Whitwor*

Aircraft Notes.—Reported to be similar to *Argus*, but with various improvements. Will have funnel placed r on starboard beam. 2 hangars, each with electric-power lift and cranes at stern.

To carry 20 sea or aeroplanes. Special ventilation system to lessen danger of fire from petrol fumes.

Machinery: Parsons (all geared) turbines. Designed S.H.P. 40,000 = 25 kts. 2 screws. Boilers: Yarrow or Babcock. Fuel (oil only): 2000 tons, *max.*

Name	Builder	Machinery	Begun	Completed	Trials
Hermes	Armstrong Whitworth Devonport D.Y.*	Parsons	Jan., 1918	1921 ?	

* Towed here for completion, January, 1920.

General Notes.—Begun under Emergency War Programme. Advanced to launching stage to clear slip for building a new Cunard liner.

EAGLE (June, 1918), late Battleship.

Displacement, 26,200 tons. Complement { — R.N. / — R.A.F. } = ——.

Length { (*p.p.*) 625 feet / (*o.a.*) 661 „ } Beam, 92 feet.* *Mean* draught, 27 feet.

* At *w.l.* about 101 feet outside bulges.

Guns:
- 6—6 inch, 50 cal.
- 2—3 inch, AA.
- 4 M.G.
- (1 landing)

Torpedo Tubes (21 inch): *not known.*

Armour:
- ..." Side (amidships) ......
- ..." Side (ends) ....
- ..." Deck ....

Special protection:
Bulges over about four-fifths of length, projecting about 6 feet from side.

EAGLE. 1920 *Photo* Cribb, Southsea.

(No plans available.)

EAGLE. 1920 *Photo, Gieves, Ltd.*
Bow view showing funnel placed on extreme starboard beam.

Machinery: Brown-Curtis (A.G.) turbines, by J. Brown & Co., Clydebank. Designed S.H.P. 55,000 = 24 kts. (No details available of trials). Boilers: . Fuel: No details available.

General Notes.—Designed and begun by Armstrong Whitworth, February, 1913, for Chile, as *Almirante Cochrane*, a Dreadnought Battleship and sister to Chilean *Almirante Latorre*. All work on this ship ceased in August, 1914, and she lay on her slip until 1917, when her purchase was negotiated with the Chilean Government. Her design was modified to an Aircraft Carrier by the Director of Naval Construction. Commissioned for ship and flying trials, April, 1920. Reported to be a very qualified success and, as an Aircraft Carrier, inferior to *Argus*. Was placed in dockyard hands, November, 1920. Cost uncertain. Bought from Chile for £1,334,358, sums subsequently expended believed to have raised total cost to nearly £3,000,000.

VINDICTIVE (17th Jan., 1918). Late Light Cruiser *Cavendish*.

Displacement, 9750 tons. Complement { R.N., about 620 / R.A.F., „ 80 } about 700

Length 565 feet (*p.p.*), 605 feet (*o.a.*). Beam, 65 feet. Draught { *Mean* $17\frac{1}{4}$ feet. / *Max.* $20\frac{1}{4}$ feet.

Guns:
4—7·5 inch, cal.
4—12 pdr.
4—3 inch AA.
4 M.G.
(1 landing)
Torpedo tubes (21 inch):
2 submerged.
4 above water.

Armour:
3"—2" Side (amidships)..
$2\frac{1}{4}$"—$1\frac{1}{2}$" Side (f. & a.) ...
1" Upper Deck (amid.) ...
$1\frac{1}{2}$" Lower Deck (aft) ...
Anti-Torp. Pro.:
Bulges 10 ft. deep.

Vindictive.

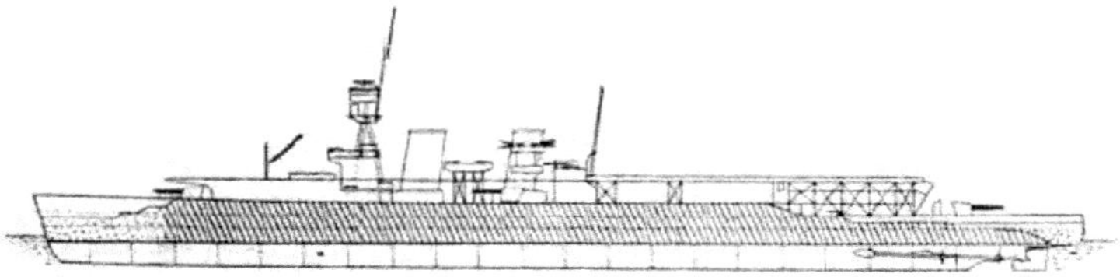

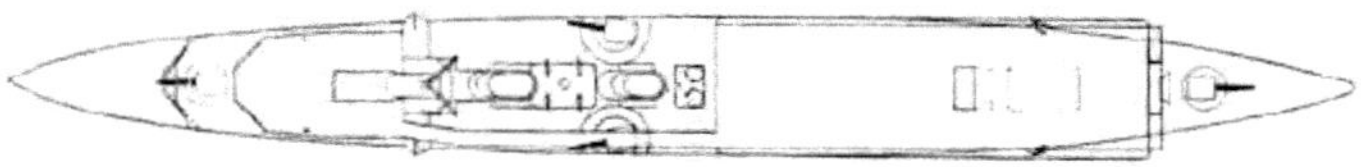

Ahead:
3—7·5 in.

Astern:
3—7·5 in.

Broadside: 3—7·5 in.; 3—21 in. tubes.

Machinery: Parsons (all geared) turbines. 4 screws. Boilers: Yarrow. Designed S.H.P. 60,000 = 29·75 (*cf.* trials). Fuel: 800 tons coal and oil *normal*, 1000 tons coal + 1420 tons oil *maximum*.

Gunnery Notes / *Armour Notes* / *Engineering Notes* } As *Hawkins*, *Raleigh*, &c., on a previous page.

Name	*Builder*	*Machinery*	*Begun*	*Completed*	*Trials: H.P. = kts.*
Vindictive	Harland & Wolff, Belfast	Harland & Wolff, Belfast	July, 1918	Oct., 1918	63,000 = 29.12

Aircraft Notes.—No reliable details available. Reported to carry six aeroplanes.

General Notes.—Built under Emergency War Programme. Commenced under name of *Cavendish*, as a unit "Improved *Birmingham*" type of Light Cruisers. Modified by builders before delivery as an Aircraft C Ran aground at 15 kts. on Middle Ground Shoal, Baltic, 6th July, 1919, and lay there for eight days. Re-fl after discharging 2212 tons. Frames and plating badly damaged. *About* £200,000 spent in repairing her at mouth D.Y., 1920-21.

ARQUS (2nd Dec., 1917). Late Liner.

Normal displacement, 15,775 tons. Complement { R.N. / R.A.F. } = 495.

Length (*p.p.*), 535 feet (*w.l.*) 560 feet (*o.a.*) 565 feet. Beam, 68 feet. Draught, 22¾ feet, *mean*.

Guns:
2—4 inch, 50 cal., Mk. V.
4—4 inch AA.
Torpedo tubes:
Nil.

Armour:
Nil.

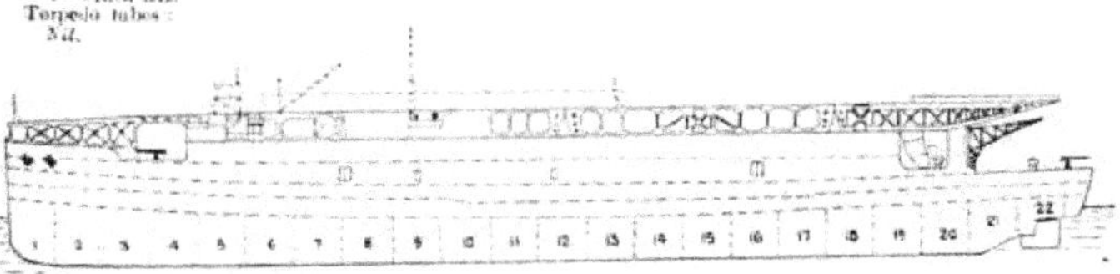

Machinery (by builders): Parsons turbines. 4 screws. Designed S.H.P. 22,000 = 20·5 kts. Can make 20·75 kts. for short periods, but 20 kts is usually best speed under ordinary conditions. Boilers: 12 cylindrical (6 D.E. and 6 S.E.), with Howdens forced draught. Fuel: 2000 tons oil.

Gunnery Notes.—Two extra 4 inch AA, now mounted, where SL. are shown in Profile over Section 9.

**Capacity, &c.*—Hangar is 350 ft. long by 68 ft. wide (*over all*) and 48 ft. clear width, 20 ft. clear height. It is divided into four sections by fire-proof screens, and can accommodate 20 seaplanes.

**Stores, &c.*—Torpedoes are carried for Sopwith "Cuckoo" type torpedo-dropping aeroplanes; aero-bombs spare parts, wings, propellers, &c. Full equipment is carried for maintenance and repair of aircraft. There are large carpenters' and engineers' workshops, for executing rapid repairs.

**Handling Gear.*—Two electrically controlled lifts for raising aircraft from hangar to flying deck. Forward lift, 30 ft. × 36 ft. After lift, 60 ft. × 18 ft. When forward lift is at flying deck level, two roller platforms slide to the sides and uncover well opening. When lift descends, the platforms are closed together and give a 20-ft. platform for flying off. When a deck load of aeroplanes is carried, wind-breaking palisades can be raised simultaneously to 14 ft. above flying deck. Two derricks with electric winches amidships on flying deck, and two electric cranes at stern on hangar deck level; all to pick up aeroplanes from the water.

**Landing.*—Wind safety mattress (or landing net) fitted, laid horizontally fore and aft. 'Planes have scissor grabs on undercarriage, which grip net on landing, check speed, and prevent 'planes from being blown sideways, and carried over ship's side by wind, after they have lost flying speed. Steam jet indicators fitted for day use and special illumination for night landings. There is a wide safety net all round flying deck.

*For all details in these notes, we are indebted to "Engineering," which published a full description of H.M.S. *Argus*, in its issue of March 28th, 1919.

Argus. 1920 *Photo*, [illegible]

Argus (starboard view). 1919 *Photo*, C.

**Engineering Notes.*—At the time designs were got out, a ¼-inch scale model was prepared for testing in the air at the National Physical Laboratory, Teddington, to solve various structural problems and to test eddy-effects of heat. It was found that the emission of hot furnace gases from the usual type of funnels creat[e] serious air disturbances, safe landings would be very difficult. Accordingly, horizontal smoke-ducts w[ith] expelling fans were fitted, to deliver all furnace gases and smoke out over stern. Designed for 18 kts., but m[odifica]tions during conversion raised speed by 2 kts.

General Notes.—Begun 1914, by Beardmore, for Italian Lloyd Sabaudo Line, as S.S. *Conte Rosso*. All work on her [stopped] in 1914. She was purchased in 1916 and converted to Aircraft Carrier. Completed September, 1918. In [appear]ance, she is quite unique, and she is sometimes called "The Floating Island." As a mobile hangar and aerodrome, she has proved very successful. Placed in dockyard hands, to be modified for secret trials, September, 1920.

FURIOUS (15 August, 1916). Late Cruiser.

Normal Displacement, *about* 19,100 tons, about 22,900 tons *full load*. Complement, { R.N. / R.A.F. } = 737.

Length, (*p.p.*) 735 feet, (*o.a.*) 786½ feet. *Beam, 88 feet. Draught, { *mean* 21½ feet. / max. 25 feet.

*Outside bulges.

Guns (**Dir. Con.**):
10—5·5 inch, -- cal.
5—3 inch AA.
4—3 pdr.
5 M.G.
(1 landing.)
Torpedo tubes (21 inch):
11 *above water*.
2 *submerged*.

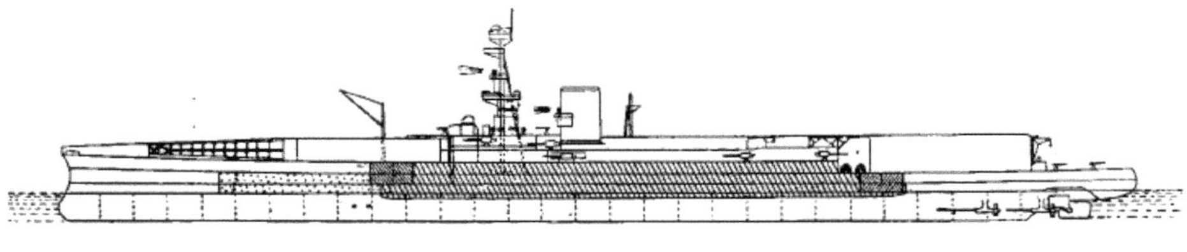

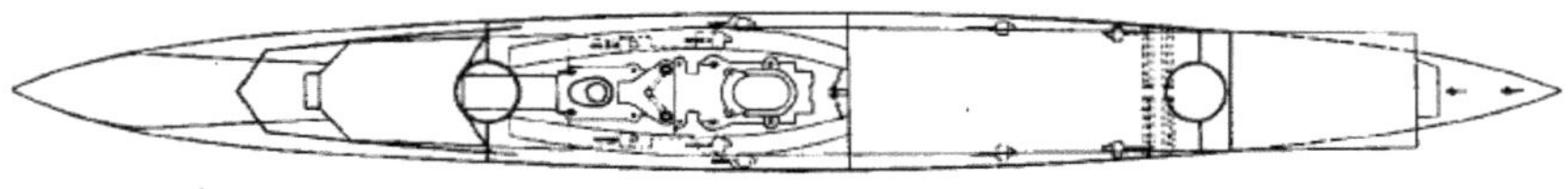

Armour:
3" Belt (amidships).....
2" Belt (bow). ....
3"-2" Bulkheads F. & A
7"-3" Barbettes..........
1" Decks (H.T. at stern
3"-1½" Decks (H.T. at
10" C.T. ....
6"-3" Sighting hood. .
3" C.T. tube ....
2" C.T. base..............
Anti-Torp. Pro.
Shallow bulges ....
1" H.T. vertical. ....
¾" Funnel and vent sc

Ahead:
6—5·5 in.

Broadside. 5—5.5 in., 9 (?)—21 inch tubes.

Astern:
2—5·5 in.

Machinery (by Wallsend Co.): Brown-Curtis (all geared) Turbines. 4 screws. Boilers: Yarrow. Designed S.H.P. 90,000 = 31 kts. *Trials*: 90,820 = kts. Fuel (oil only): 750 tons *normal*, 3,393 tons *maximum*.

Armour Notes.—3" Belt consists of 2" plating over 1" shell plating, as in Light Cruisers. Belt forward is set inboard, as plans. Barbettes are still in, though no large guns are mounted in them. Fo'xle, Upper Decks, 1" H.T. Main Deck, 1¾"—1". Lower deck at stern, 1¾"—1" flat, 2" slopes, 3" over rudder head—all H.T.

Internal Protection.—Bulge deeper than that of *Glorious* and *Courageous*, but not filled with oil fuel. There is a complete internal citadel of 1" vertical plating from barbette to barbette, well inboard and covering magazines, shell and boiler rooms. Outside and below this citadel and on each beam, are 1" screens, 250-300 ft. long, covering boiler and engine rooms. From within the citadel rise the ¾" screens protecting funnel and boiler room vents.

Gunnery Notes.—Originally designed to mount 2—18 inch and 11—5.5 inch guns, provision being made to mount 4—15 inch if the 18 inch guns proved unsatisfactory. The fore 18 inch gun was never mounted, being replaced by flying-off deck and hangar. When in service in 1917-18, she fired the after 18 inch a *few* times, and it is said "it shook her up considerably." In 1918, the after 18 inch was also taken out, flying-on deck and after hangar added. At the same time, 1—5.5 inch was removed and other 10—5.5 inch re-distributed. The after tripod mast, with gunnery controls was also removed.

Torpedo Notes.—*Submerged* tubes are not satisfactory, and may be removed. There are four sets of triple 21 inc on upper deck aft, worked on a new principle, and firing latest type torpedoes. Also 2 single tubes on up abaft funnel (not shown on plan). On re-construction, the 3" (after) Torpedo Control Tower was remo controls re-located.

General Notes.—Built under Emergency War Programme, by Armstrong Whitworth. Begun June, 1915; co July, 1917; re-built November, 1917-March, 1918. Designed as a modified *Courageous*, but altered to Carrier (see Gunnery and Aircraft Notes). Since conversion, she is said to be rather light, and is good 32-33 kts. Including cost of alterations, this ship is said to have absorbed nearly six million pounds.

Aircraft Notes.—When first completed, she carried about four Short seaplanes and six Sopwith "Pups." Th easily flown off bows, but there was difficulty in getting seaplanes shipped inboard by derricks. Experime made to use the forward flying-off deck for landing also, but it was found too dangerous. Later, the flying-on deck was added, and proved satisfactory. Forward flying-off deck is about 160 feet long, tapers has collapsible wind-breaking palisades. There is also a power-operated lift before C.T. to hoist seapla forward hangar to flying-off deck. The after flying-on deck is about 300 feet long by 100 feet wide, with net just abaft funnel. There is also another lift aft for raising seaplanes from after hangar. Narrow each side of the funnel connect the flying-off deck with flying-on deck, small trolleys on rails being used seaplanes forward or aft. No seaplanes are now carried, but only aeroplanes of "Cuckoo" (torpedo & and other types. Ten 'planes stowed in each hangar, can be raised and flown off in 3 minutes. "Blimps" be landed on flying-on deck.

FURIOUS. **(For details, v. opposite page.)** 1919 *Illustration.*

FURIOUS (present colour) 1920 *Photo, Topical.*

FURIOUS. Aircraft View. 1919 *Official R.A.F. Photo.*

Aircraft Carriers—*Continued.*

1919 *Illustration.*

PEGASUS (ex G. E. Railway Co. s.s. *Stockholm*, built by J. Brown & Co., ——, purchased 1917). 3300 tons. Dimensions: 330 (*p.p.*), 332 (*o.a.*) × 43 × 15 (*mean*), 15⅝ (*max.*) feet. Guns: 2—12 pdr. and 2—12 pdr. AA. S.H.P. 9500 = 20.25 kts. (20.8 trials). Machinery: Brown-Curtis turbines (all-geared). 2 screws. Boilers: Cylindrical. Oil fuel: 300 tons. Complement, 258 R.N. and R.A.F.

1919 *Illustration.*

ARK ROYAL (Blyth S.B. Co., 1914, purchased 1914). 7450 tons. Dimensions: 352¼ (*p.p.*), 366 (*o.a.*) × 50½ × 18 feet (*mean*). Guns: 4—12 pdr., 2 M.G. I.H.P. 3000 = 11 kts. Machinery: Vertical triple expansion. 1 screw. Boilers: Cylindrical. Oil: 500 tons. Complement, 180 R.N. and R.A.F.

Minelayers.

Note.

The 20th (Minelaying) Flotilla was abolished on February 1st, 1920, but various Destroye incorporated in the 1st, 2nd and 3rd Flotillas, Reserve and L.D.F. still retain thei rails and chutes at stern, and could take up Minelaying Duty at very short notice.

Minelaying Submarines not known exactly, but are divided between "E" and "L" c Each Minelaying "E" boat can carry *about* 20 mines stowed in saddle tanks.

1919 *Illustrati*

PRINCESS MARGARET (built 1913-14, chartered and converted during War, pur 1919). 5440 tons. Complement, 215. Dimensions: 395¼ (*o.a.*) × 54 × 16¾ feet draught). Guns: 2—4.7 inch, 2—12 pdr., 2—6 pdr. AA., 1—2 pdr. pom-pom. 15,000 = 22.5 kts (23.15 on *trials*). Oil fuel: 585 tons.

(*Continued on next pag*

1919 *Illustration.*

Note.—Photo taken *before* conversion to Minelayers. Guns in shields removed and replaced by derricks.

M 29, M 31 (Workman Clark), **M 33** (Harland & Wolff, Belfast). All launched and completed during 1915. 535 tons. Complement, 75. Dimensions: 170 × 31 × 6¾ feet. Guns: Not known, probably a few small Q.F., AA. or M.G. Designed H.P. 400 = 10 kts. Machinery: Triple expansion. 2 screws. Boilers: Yarrow. Oil fuel: 45 tons. *M 30* lost during the War.

Notes.—All originally Small Monitors built under Emergency War Programme.

1919 *Illustration.*

Note.—Photo taken before conversion to Mine Layer. Guns fore and aft replaced by derricks.

M 22 (1915). Built by Sir Raylton Dixon & Co. 540 tons. Dimensions: 170 × 31 × 6 feet. Guns: Not known, probably a few small Q.F., AA. and M.G. Machinery: Triple expansion engines and White-Forster boilers. I.H.P. 650 = 12 kts. 2 screws. Oil: 25-28 tons.

Notes.—Small Monitor built under Emergency War Programme.

Paddle Wheel Mine Sweepers.

1919 *Photo, Robertson, Gourock, N.B.*

LEWES (Mar., 1918). Both by Fleming & Ferguson. Paddle Minesweeper, design by Admiralty, adapted from Ailsa Co. design. Displacement, 820 tons. Complement, 50. Dimensions: 235 (*p.p.*), 249¾ (*o.a.*) × 29 ft. 0½ in. (58 feet outside paddle boxes) × 6½-7 feet. Guns: 1—12 pdr. (12 cwt.), 1—3 inch AA. Designed I.H.P. 1400 = 15 kts. (actually 1500 = *about* 14½ kts.). Machinery: Diagonal compound. Boilers: Cylindrical return tube. Coal: 156 tons.

Notes.—Built under Emergency War Programme.

1919 *Photo, Robertson, Gourock, N.B.*

CHELMSFORD (April, 1916), by Ailsa S.B. Co., **CHELTENHAM** (Ardrossan Co., June, 1916). Paddle Minesweepers, built to designs by Ailsa Co. Completed, April-Oct., 1916. Displacement, 810 tons. Complement, 50. Dimensions: 235 (*p.p.*), 245¾ (*o.a.*) × 29 (58 feet over paddles) × 6¾ feet (*mean* draught). Guns: 2—12 pdr. Designed H.P. 1400 = 15 kts. Machinery: Inclined compound. Boilers: Cylindrical return tube. Coal: 156 tons.

Notes.—All built under Emergency War Programme.

SIR HUGO. Mast *before* funnel. 1919 *Photo, by courtesy of Messrs. Osbourne Graham.*

SIR BEVIS. Mast *abaft* funnel. 1919 *Photo by courtesy of Blyth S.B.*

("24 CLASS"—2 SHIPS.)

SIR BEVIS (May, 1918), by Barclay, Curle. **SIR HUGO** (September, 1918), by Greenock & Grangemouth Co.

Both 1320 tons. Dimensions: 267½ (*p.p.*), 276½ (*o.a.*) × 34½ × 12 feet. Guns: 2—4 inch, as designed; but may have only 3 pdr. guns or none at all now. Machinery: Inverted 4 cylinder triple expansion. Boilers: Cylindrical return smoke tube. 1 screw. Designed H.P. 2500 = 17 kts. Coal: 260 tons. Complement, 82.

Cancelled 1919: *Galtee More* (sister to *Sir Bevis*), *Sunstar* (sister to *Sir Hugo*), both by Swan Hunter.

Deleted 1920: 20 Ships for Sale or on Disposal List.

General Notes.—Officially rated as "Fleet Sweeping Vessels (Sloops)" and built under Emergency War Pro[illegible] Although not built as Submarine Decoy Vessels (or "Q-boats"), they have a resemblance to the Standa[illegible] Ships. Originally intended there should be 24 ships, 12 as *Sir Hugo* and 12 as *Sir Bevis*. The two ships give[illegible] page not yet for sale or disposal, but will probably be disposed of within a year.

"Q" Boat—Flower Type—7* Ships.

1919 Photo, Messrs. Lobnitz & Co. (Builders).

Coreopsis. 1919 Photo, Abrahams, Devonport.

Appearance Note.—Mainmast of *Gardenia* removed during War. Not certain that it has been re-stepped.

(Anchusa Type—6 Ships.)*

COREOPSIS (Sept., 1917), **DIANTHUS** (Nov., 1917), **GARDENIA** (Dec., 1917), **HAREBELL** (May, 1918), all by Barclay, Curle. **BRYONY** (Oct., 1917), by Armstrong, Whitworth. **HIBISCUS** (Nov., 1917), by Greenock & Grangemouth Co.

All 1290 tons. Completed Aug., 1917-June, 1918. Dimensions: 250 (*p.p.*), 262¼ (*o.a.*) × 35 × 11¾-12 (*mean*), 12¼-13¾ (*max.*) feet draught. Designed to mount 2—12 pdr., 1—7.5 howitzer or 1—200 lb. stick-bomb howitzer, 4 Depth Charge Throwers, but actually armed with 2—4 inch, 1 or 2—12 pdr. and D.C. Throwers. Machinery: 4-cylinder triple expansion. Boilers: 2 cylindrical. 1 screw. Designed H.P. 2500 = 17 kts., but some only about 16 kts. on this power; about 2800 = 17 kts. actually. Coal: 200 tons. Complement, 93. Also see *General Notes*.

Deleted 1920.—5 ships on Sale List or for disposal. Also *Syringa*, sold to Egypt. *Chrysanthemum* (for Fleet Target Service) transferred to a later page.

* Not Included:—**AURICULA** (Oct., 1917), **CEANOTHUS** (June, 1917), both by Armstrong Whitworth; **GILIA** (Barclay, Curle, Feb., 1918); **LYCHNIS** (Hamilton, Aug. 1917); **SPIRAEA** (Simons, Nov., 1917); **SWEET BRIAR** (Oct., 1917), **TUBEROSE** (Nov., 1917), both by Swan Hunter. All seven on Suspense List at end of 1920 for transfer to Dominion Navies or Sale to Foreign Navies. Details as *Anchusa* type above.

(Aubrietia Type—1 Ship.)

HEATHER (Greenock & Grangemouth Co., June, 1916).

All 1250 tons. Completed Aug.-Oct., 1916. Dimensions: 255¼ (*p.p.*), 267¾ (*o.a.*) × 33½ × 11½ (*mean*), 12½ (*max.*) feet draught. Designed to mount 3—12 pdr., 2—3 pdr. AA., but have 2—4 inch, 1—3 pdr. AA. and D.C. throwers. Machinery, boilers, as "Anchusa Class," opposit 1 screw. Designed H.P. 2500 = 17.5 kts., but actually can only make 15-16½ kts. with this powe and require 3000 I.H.P. for 17½ kts. Coal: 255 tons. Complement, 80.

Deleted 1920.—3 ships for sale or on Disposal List.

General Notes to Anchusa and Aubrietia types.—Ah are Fleet Sweeping Vessels (Sloops) of the "Flower Classes," modifi in external build for service as Convoy Sloops. Disguised as mercantile vessels, these ships accompanied Convoy and acted as Submarine Decoy Vessels, or "Q-boats," hence their popular name—the "Q-boat Flowers." Ea builder was allowed to adopt a form of mercantile disguise, with the result that, by appearance, these ships a divided into many types. All were built under Emergency War Programme.

War Losses.—Owing to the nature of the service these ships were engaged on, war losses were heavy. *Anchusa, Arbut, Bergamot, Eastnor Castle, Cassiopeia* (all by Armstrong Whitworth), *Cowslip* (Barclay, Curle), *Gaillardia* (Blyth S.B. Co.), *Rhodode dron, Salvia* (both by Irvine S.B. Co.), *Tulip* (Richardson, Duck), were all lost during the War. In all, 39 vesse of *Anchusa* and *Aubrietia* types were built. Although nominally rated as "Fleet Sweeping Vessels," they ha no gallows at stern, and it is doubtful if they carry sweeping gear, unless of "otter" types.

"Flower" Classes—17 Ships.

VALERIAN.

Note.—In *Clematis, Cornflower, Cyclamen, Hollyhock, Magnolia, Wistaria,* mainmast removed during War, but orders have been issued that all Sloops retained for Post-War Fleet or Foreign Stations are to have mainmast re-stepped, K.B. platform and cork filling removed.

(ARABIS TYPE—9 SHIPS.)

CORNFLOWER (Mar., 1916), by Barclay, Curle. **GODETIA** (Jan., 1916), by Connell. **WALLFLOWER** (Nov., 1915), **WISTARIA** (Dec., 1915), both by Irvine S.B. Co. **CROCUS** (Dec., 1915), **CYCLAMEN** (Feb., 1916), both by Lobnitz. **LUPIN** (May, 1916), by Simons. **VERBENA** (Blyth S.B. Co., Nov., 1915), **VALERIAN** (C. Rennoldson, Feb., 1916).

All 1250 tons. Completed, Dec., 1915-June, 1916. Dimensions: 255¼ (*p.p.*), 267¾ (*o.a.*) × 33½ × 11 (*mean*), 11¼-11¾ *feet* (*max.* draught). *Designed* armament: 2—4.7 inch or 4 inch, 2—3 pdr. AA., but some have only 1—4.7 or 4 inch, with or without the 2—3 pdr. AA. A few are only armed with 2—3 pdr. AA. Designed I.H.P. 1400 = 17 kts., but for this speed they really have to develop about 2400 I.H.P. Machinery: 1 set 4-cylinder triple expansion. Boilers: 2 cylindrical. 1 screw. Coal: 130 tons *normal*, 260 tons *max.* = *about* 2000 miles at 15 kts. Complement, 79. Also see *General Notes.*

General Notes to "Arabis," "Azalea," and "Acacia" Types.—Single-screw Fleet Sweeping Vessels (Sloops), all built under Emergency War Programme. Popularly known as the "Flower Class," but are also referred to as the "Cabbage Class," or "Herbaceous Borders." Have triple hulls at bows to give extra protection against loss when working up mines. Very successful ships, but can only make designed speed with difficulty. Good sea-boats. Copies of this class were built in British yards for the French Navy. The last series of "Flowers" converted to Convoy Sloops (*q.v.*). Several mined during War and extensively re-built.

War Losses.—*Alyssum, Aster* (both Earle S.B. Co.), *Arabis* (Henderson), *Begonia* (Barclay, Curle), *Genista* (Napier & Miller), *Mignonette* (Dunlop & Bremner), *Lavender, Nasturtium* (both McMillan), *Primula* (Swan Hunter), all lost during War. *Gentian* (Greenock & Grangemouth Co.) and *Myrtle* (Lobnitz), sunk in Baltic, 1919, during anti-Bolshevist operations. *Geranium, Marguerite, Mallow,* presented to Australian Navy. *Peony* (McMillan), sold out. In all, 72 of these ships were built.

Deleted 1920.—12 for Sale or Disposal; *Gladiolus* sold to Portugal, *Asphodel* sold to Denmark, *Lobelia* presented to Newfoundland. *Snapdragon* (for Fleet Target Duties) transferred to a later page.

CLEMATIS. *Photo added*

(AZALEA TYPE—2 SHIPS.)

CLEMATIS (July, 1915), by Greenock & Grangemouth Co. **HELIOTROPE** (Lobni 1915).

All details as *Acacia* Type below, but these ships were designed to mount 2— or 4 inch and 2—3 pdr. AA. Nearly all so armed, but one or two have no 3 pdrs. Com 79. Also see *General Notes.*

Deleted 1920.—4 for Sale or Disposal. *Zinnia* sold to Belgium.

(ACACIA TYPE—6 SHIPS.)*

FOXGLOVE (Mar., 1915), **HOLLYHOCK** (May, 1915), both by Barclay, Curle. **BL** (July, 1915), **MAGNOLIA** (June, 1915), both by Scotts S. B. Co. **LAB** (June, 1915), by Connell & Co. **VERONICA** (Dunlop, Bremner, May, 1915).

All 1200 tons. Completed, May-Sept., 1915. Dimensions: 250 (*p.p.*), 262½ (*o.a.* 11 (*mean*), 11¼-12 feet (*max.* draught). Designed to mount 2—12 pdr. and 2—3 pdr. still armed in this way; others have (*a*) 2—4 inch, (*b*) 1—4 inch, (*c*) 1—4 inch and (*d*) 2—12 pdr., with or without the 2—3 pdr. One or two armed *only* with 1 or Designed H.P. 1400 or 1800 = 17 kts., but actually require about 2200 I.H.P. for Machinery: 1 set 4-cylinder triple expansion. Boilers: 2 cylindrical. 1 screw. tons *normal*, 250 tons *max.* = *about* 2000 miles at 15 kts. Complement, 77. Also *Notes.*

Deleted 1920.—15 ships for Sale or Disposal; also *Jonquil* sold to Portugal; *Dahlia* loaned to Board of A Fisheries.

* *Not Included*: **HONEYSUCKLE** (Lobnitz, April, 1915), on Suspense List, and transfer to Dominion Navy or sale to Foreign Navy. Details as *Acacia* Type

1919 Photo, Robertson, Gourock, N.B.

1920 Photo, Seward, Weymouth.

SHERBORNE.—Masts of equal height. There are minor variations in these ships, e.g., Mistley has small round funnel, Leamington very high foremast, and so on. Gallows, sweeping gear and/or guns now removed from many.

(TWIN SCREW "HUNT" CLASS—LATER TYPE—47 SHIPS.)

ALRESFORD* (Jan., 1919), **LEAMINGTON*** (ex *Aldeburgh*, Aug., 1918), both by Ailsa S.B. Co., **BADMINTON*** (Mar., 1918), **SWINDON** (ex *Bantry*, 1919), both by Ardrossan Co., **BURSLEM*** (ex *Blakeney*, Mar., 1918), by Ayrshire Co., **CAMBERLEY** (Dec., 1918), **CARSTAIRS*** (ex *Cawsand*, 1919), **CATERHAM** (Mar., 1919), all three by Bow, McLachlan, **FORRES***† (ex *Fowey*, Nov., 1918), **DORKING** (Sept., 1918), **DUNDALK** (Jan., 1919), **DUNOON** (Mar., 1919), all four by Clyde S.B. Co., **FERMOY** (1919), by Dundee S.B. Co., **FAVERSHAM** (July, 1918), **FORD** (ex *Fleetwood*, Oct., 1918), **RUGBY** (ex *Filey*, Sept., 1918), all three by Dunlop & Bremner, **GAINSBOROUGH** (ex *Gorleston*, Feb., 1918), **GRETNA** (April, 1919), **NORTHOLT** (June, 1918), all three by Jos. T. Eltringham, **IRVINE** (Dec., 1917), **KENDAL** (Feb., 1918), **LYDD** (ex *Lydney*, Dec., 1918), all three by Fairfield Co., **MALLAIG** (Oct., 1918), **MALVERN** (Feb., 1919), **MARAZION*** (1919), all three by Fleming & Ferguson, **MARLOW** (Aug., 1918), **MISTLEY*** (ex *Maryport*, Oct., 1918), both by W. Harkness, **NAILSEA** (ex *Newquay*, Aug., 1918), **NEWARK** (ex *Newlyn*, June, 1918), **WEYBOURNE** (Feb., 1919), all three by A. & J. Inglis, **PANGBOURNE** (ex *Padstow*, Mar., 1918), **PETERSFIELD*** (ex *Portmadoc*, Mar., 1919), **ROSS** (ex *Ramsey*, 1919), all three by Messrs. Lobnitz, **SUTTON** (ex *Salcombe*, May, 1918), by Messrs. McMillan, **SALTASH** (July, 1918), **SALTBURN** (Oct., 1918), **SELKIRK** (Dec., 1918), all three by Messrs. Murdoch & Murray, **SHREWSBURY** (Feb., 1918), **YEOVIL** (Aug., 1918), both by Messrs. Napier & Miller, **ELGIN** (ex *Troon*, Mar., 1919), **SHERBORNE*** (ex *Tarbert*, June, 1918), **TIVERTON** (Sept., 1918), **TONBRIDGE** (Nov., 1918), **TRALEE** (Dec., 1918), **TRING*** (ex *Teignmouth*, Aug., 1919), **TRURO*** (April, 1919), all seven by Messrs. Simons.

*Will probably be retained permanently: the others will be disposed of as opportunity offers.

†*Forres* contracted for by Messrs. Dunlop Bremner, but contract and materials transferred to Clyde S.B. Co.

All Emergency War Programme. First ship delivered Feb., 1918; last about March, 1920. Displacement, 800 tons. Dimensions: 220 (*p.p.*), 231 (*o.a.*) × 28 ft. 6½ in. × 7½ (*mean*), — feet (*max. draught*). Guns: A few have 1—4 inch and 1—12 pdr. AA. or 2—3 pdr., but majority have only 1—6 or 3 pdr. Many not armed now. Machinery: Vertical triple expansion. 2 screws. Designed I.H.P. 2200 = 16 kts. Boilers: Yarrow. Coal: 185 tons. Complement, 74.

General Notes.—Six have been re-named and completed as Surveying Ships (*q.v.*). Those originally named after coastal towns, watering places, fishing ports, &c., were all re-named after inland towns and villages, to obviate misunderstanding of signals and orders.

Losses (after Armistice).—*Penarth* (ex *Ronsdorf*?, built by McMillan), *Penarth* (Lobnitz), *Kinross* (Fairfield). No War Losses.

Cancelled Ships.—*Alton* (ex Arbroath), *Ashburton*, both by Ailsa S.B. Co.; *Bideford*, *Bolton* (ex *Beaumaris*), both by Ardrossan D.D. & S.B. Co.; *Bloxham*, by Ayrshire Co.; *Atherstone*, *Bala*, *Bathgate*, all three by Clyde S.B. Co.; *Battle*, *Bewdley*, *Buckley*, all three by Dundee S.B. Co.; *Northrepps*, by Messrs. Lobnitz; *Tain*, *Wembdon*, *Yealmpton*, all three by Messrs. Simons; *Clifton*, *Croston* (ex *Colyton*), *Cavan* (ex *Clonakilty*), *Cashel*, all four by Bow McLachlan; *Curragh*, *Fleet*, *Frome*, *Grays*, *Kew*, *Kingussie*, *Knole*, *Ness*, all nine by Eltringham; *Okehampton*, *Oundle*, *Radnor*, *Reading*, *Retford*, *Ringwood*, *Runcorn*, *Shifnal*, *Southwick*, all nine by Lobnitz; and 2 unnamed, by Fleming & Ferguson.

Sold, For Sale, or Disposal, 1920.—41 ships, which have been deleted from List.

Destroyer Depot Ships.

Note.—Cruiser *Essex*, Accommodation Ship for T.B.D., is non-effective.

1920 *Photo, Abrahams, Devonport.*

SANDHURST (ex S.S. *Manipur*, Harland & Wolff, purchased 1915 and converted by Workman, Clark). 11,500 tons. Dimensions: 470 (*p.p.*) × 58 × 20 feet (*max.* draught). Guns: 4—4 inch, 2—6 pdr. AA. I.H.P. 4000 = 10.5 kts. Coal: 1475 tons. Complement, 269. Cyl. boilers.

Tenders: *Avalanche*

1919 *Photo, Dr. J. A. Prendergast.*

GREENWICH (Dobson & Co., completed by Swan Hunter. Purchased 1915). 8600 tons. Dimensions: 390 (*p.p.*) × 52 × 19¾ feet (*max.* draught). Guns: 4—4 inch, 2—6 pdr. AA. I.H.P. 2500 = 11 kts. Coal: 960 tons. Complement, 241. Cyl. boilers.

Tenders: ———

Destroyer Depot Ships—(*Continued*).

WOOLWICH (Scotts S. & E. Co., 1912). 3380 tons. Dimensions: 320 × 40 × 14¼ feet. Guns: 2—4 inch. H.P. 2600 = 13.5 kts. Coal: 370 tons. Complement, 263.

Tenders: ———

(*For T.B.D. in Reserve.*) 1919 *Photo.*

DILIGENCE (ex *Tabaristan*. 1907. Purchased 1913). 7400 tons. Dimensions: 390 (*p.p.*) × 46 × — feet. Guns: 4—4 inch. H.P. 3000 = 14.2 kts. Coal: 780 tons. Complement, 258.

(*For T.B.D. in Reserve.*) 1920 *Photo, Gieves, Ltd.*

DIDO (1896). Old Cruiser of 5600 tons. Dimensions: 350 (*p.p.*) × 54 × 22 feet. Guns: Not known. Designed H.P. 9600 (*f.d.*) = 19.5 kts. Coal: 1076 tons *max.*

Destroyer Depot Ships—(*Continued*).

BLENHEIM. 1

BLENHEIM (1890). Converted 1907. 9000 to[ns. Dimen]sions: 375 (*p.p.*) × 65 × 25¾ feet (*max.* draug[ht]) ... 2—3 pdr. pom-pom. Designed H.P. 20,000 ... Coal: *normal*, 624 tons; also 650 tons carried i[n] ... (1 cwt. each) for destroyers.

Tenders: 2 Drifters.

(*For T.B.D. in Reserve.*)

HECLA (purchased 1878). 5600 tons. Dim[ensions] ... (*p.p.*) × 38¾ × 21½ feet (*mean* draught). G[uns] ... 1 M.G. Designed H.P. 2400 = 13.4 kts. ... Complement, 266.

Submarine Depot Ships.

Note.—Light Cruiser, **INCONSTANT** is attached to 1st Submarine Flotilla.

1920 Photo, T. B. Manning, Esq.

AMBROSE (Clyde S.B. Co., 1915. Purchased 1915). 6480 tons. Dimensions: 387½ (p.p.) × 47½ × 20½ feet. Guns: 2—12 pdr. I.H.P. 6350 = 14½ kts. Complement, 240.

Tenders: ———

1920 Photo, T. D. Manning, Esq.

TITANIA (Clyde S.B. Co., 1915, purchased 1915). 5200 tons. Dimensions: 335 (p.p.) × 46¼ × 18¼ feet. I.H.P. 3200 = 14.5 kts. Torpedo tubes: 2. Coal: 498 tons. Complement, 245. Cyl. boilers.

Tenders: *Maidstone, Moonshine.*

Submarine Depot Ships—*Continued.*

1920 Photo, Abrahams, Devonport.

LUCIA (Furness Withy Co., ex German Prize, *Spreewald*, converted by Clyde S.B. Co., 1916). 6005 tons. Dimensions: 366 (o.a.) × 45¼ × 19½ feet. Guns: 2—3 pdr. AA. I.H.P. 2700 = [illegible] kts. Coal: 700 tons. Cyl. boilers. Complement, 245.

Tenders: *Dreadful, Shooting Star.*

Photo wanted.

PANDORA (ex *Seti*, purchased from Russia, 1914). 1350 tons. Dimensions: 230 × 43 × 16 feet. Guns: *Nil.* H.P. 2200 = 11 kts. Coal: 580 tons. Complement, 70.

Tenders: ———

1919 Photo, Cribb.

MAIDSTONE (Scott's S. & E. Co., 1912). 3600 tons. Dimensions: 355 × 45 × 17¼ feet. Guns: *Nil.* H.P. 2800 = 14·3 kts. Coal: 465 tons. Complement, 159.

Tenders: *Alecto, Iridescence.*

Submarine Depot Ships—*Continued.*

ADAMANT and **ALECTO** (both Laird, 1911). 935 tons. Dimensions: 190 × 32½ × 11 feet (mean draught). Guns: *Adamant*, 1—4 inch. Designed H.P. 1400 = 14 kts. Coal: 180 tons. Complement, 63.

1920 Photo, G. Dell, Esq.

ROYAL ARTHUR (ex *Centaur*) (1891). 7700 tons. Dimensions: 360 (p.p.) × 60¾ × 26¾ feet. Guns: *Nil.* I.H.P. 12,000 = 20 kts., but 7000 = 15-16 kts. may be present maximum. Coal: 1230 tons. Complement, —.

Note.—Will be relieved by *Pandora* and paid off, at end of 1920.

Tenders to *Royal Arthur*: *Gleaming.*

Submarine Depot Ships—*Continued.*

1920 Photo, Seward, Weymouth.

VULCAN (1889). Rebuilt 1908. 6620 tons. Dimensions: 350 × 58 × 23¼ feet (*max.* draught). Guns: 2—4.7 inch. Armour: 5" deck (amidships), 6" conning tower, 5" engine hatches. Designed H.P. 12,000 = 20 kts. Can do now about 16 kts. Coal: *maximum* 1347 tons. Durr boilers (1909). Complement, 255.

Tender: *Vulcan II* (late *Onyx*).

Following Submarine Depot Ships are non-effective:—

Marshal Ney (Monitor), *Dolphin* (Sloop) } Fort Blockhouse, Portsmouth.

St. George, *Forth*, *Thames* } Old Cruisers, paid off 1920. Will probably be sold and broken up.

Depot Ship for Anti-Submarine Establishment and Patrols.

No photo available.

GIBRALTAR (1892). Old Cruiser of 7700 tons. Dimensions: 360 (*p.p.*) × 60 × 23¾ feet (*mean* draught). Guns: Not known. Designed H.P. 10,000 (*n.d.*) = 15 kts. (*nominal*). Coal: 1200 tons *max.*

Tenders: *v.* Establishment and Training Ships Page at end of British Navy pages.

Depot Ships for C.M.B. and M.L.

(None.)

Hospital Ship.

Photo wanted.

BERBICE. Details wanted.

Repair Ships (R.F.A.).

No photo available.

SCOTSTOUN Begun, February, 1916. Launched, Completed, July, 1916. Built by Yarrows, as Repair Ship for "Insect" and "Fly" Classes of Gunboats. 300 tons. Dimensions: 122 (p.p.), 130 (o.a.) × 31 × 3½ feet draught. I.H.P. 200 = 7.7 kts. Compound surface condensing engines. Loco. boiler. Oil: 20/36 tons.

CYCLOPS (ex *Indrabarah*, 1905). 11,300 tons. Dimensions: 460 (p.p.) × 55 × 21 feet. Guns: 6—4 inch. Machinery: Triple expansion. Designed H.P. 3500 = 11.75 kts. Coal capacity: 1595 tons. Fitted as Floating Workshop and Distilling Ship. Complement, 294.

Repair Ships—(*Continued*).

ASSISTANCE (1901). 9600 tons. Dimensions: 436 (*p.p.*) × 53 × 20 feet. Guns: 19—3 *pdr.* Machinery: Triple expansion. Boilers: Cylindrical. Fitted with Howden's forced draught. Designed H.P. (*f.d.*) 4000 = 12 kts. Coal: 2180 tons. Fitted as Floating Workshop and Distilling Ship. Complement, 283.

Tender: *Landfall.*

Distilling Vessel & Store Carrier (R.F.A.).

1920 *Photo, Abrahams, Devonport.*

BACCHUS (Hamilton & Co., 1915, purchased 1915). About 3500 tons. Dimensions: 295 × 44 × 12½ feet. Guns: Not known. I.H.P. = kts. Coal: 673 tons.

Oil Tankers (R.F.A.).

Oil Tankers *Appleleaf* (ex-*Texol*), *Brambleleaf*, *Cherryleaf*, *Orangeleaf*, *Pearleaf*, *Plumleaf* (ex-*Trinol*) are under management.

1920 *Photo, G. Bell, Esq.*

PETROBUS.

PETROBUS, PETRELLA, PETRONEL (Dunlop, Bremner, 1918). 1024 tons. Dimensions: 164 × 28 × 11½ feet. I.H.P. 500 = 9-10 kts. Own oil: 50 tons. Cylindrical boilers. Complement, 16.

Photo wanted.

1 funnel aft, 1 mast aft just before funnel

DREDGOL (Simons, 1918). About 7000 tons. Dimensions: 326 × 54½ × 18¾ feet. I.H.P. 2400 = 11 kts. Own oil: 300 tons. Cylindrical boilers.

1920 *Photo, Glenn, Ltd.*

SERBOL.

FRANCOL (Earle S.B. Co.), **MONTENOL** (W. Gray), **SERBOL** (Caledon S.B. Co.). All details as *Helpol* below, but of different appearance. All launched 1917.

Oil Tankers (R.F.A.)—*Continued.*

SLAVOL.

1920 Photo, Abrahams, Devonport.

BELGOL (Irvine D.D. Co.). **CELEROL** (Short Bros.), **FORTOL** (McMillan), **PRESTOL** (Napier & Miller), **RAPIDOL** (W. Gray), **SLAVOL**, **VITOL** (Greenock and Grangemouth Co.). All launched 1917. 4900 tons. Dimensions: 335 × 41½ × 20½ feet. I.H.P. 3375 = 14 kts. Own oil: 300 tons. Cylindrical boilers. Complement, 39.

Photo wanted.

1 mast forward.
1 funnel aft.

BIRCHOL, BOXOL (Barclay, Curle), **EBONOL** (Clyde S.B. Co.), **ELDEROL, ELMOL** (Swan Hunter), **LARCHOL, LIMOL** (Lobnitz), **HICKOROL** (———). All launched 1917. 1925 tons. Dimensions: 220 × 34 × 12 feet. Triple expansion engines and cylindrical boilers. I.H.P. 700 = 12 kts. Own oil: 40 tons. Complement, 19.

Oil Tankers (R.F.A.)—*Continued.*

Photo wanted.

1 mast forward.
1 funnel aft.

DISTOL, PHILOL, SCOTOL (Tyne Iron S. B. Co.), **KIMMEROL, VISCOL** (Craig, Taylor). All launched 1916. 1920 tons. Dimensions: 220 × 34¾ × 12½ feet. I.H.P. 700 = 9 kts. Own oil: 40 tons. Cylindrical boilers. Complement, 19.

1920 Photo, Coates, Harwich.

MIXOL (Greenock & Grangemouth Co., 1916). **THERMOL** (Caledonian S.B. Co., 1916). 4145 tons. Dimensions: 270 × 38½ × 20¼ feet. I.H.P. 1200 = 11 kts. Oil: 150 tons.

Photo wanted.

2 masts fore and aft, 1 funnel aft.

TREFOIL (Pembroke, 1913). 4060 tons. Dimensions: 280 × 39 × 18¼ feet. 2 sets 6-cylinder 2-cycle Diesel engines. H.P. 1500 = 12 kts. Deadweight capacity: 2090 tons. Own oil: 200 tons.

Oil Tankers (R.F.A.)—*Continued.*

Photos wanted.

"Servitor," "Attendant," Carol"
one mast forward, one funnel aft.

SERVITOR (Chatham, 1914). 2007 tons. Dimensi
× 34 × 13 feet. 2 sets 4-cylinder 2-cycle Diese
B.H.P. 450 = 8 kts. Deadweight capacity: 1000 to
oil fuel: 29 tons.

ATTENDANT (Chatham, 1913). **CAROL** (Devonpo
1935 tons. Dimensions: 200 × 34 × 13 feet. I
= 8 kts. Coal: 60 tons.

PERTHSHIRE (——). 5865 tons gross. I.H.P. 300

BURMA (Greenock, 1911). 3945 tons. H.P. 1200
Cargo capacity: 2500 tons oil fuel. Own fuel: 2

PETROLEUM (Wallsend, 1901). 9900 tons. Dime
× 48¾ × 24 feet. H.P. 2000 = 13 kts. Oil:

KHARKI (1899, bought 1900). 1130 tons. H.P. 77
Fuel: 90 tons.

INDUSTRY (1901). 1460 tons. H.P. 750 = 10
180 tons.

Water Carrier. *(Atlantic Fleet.)*

PETROL *About* 300 tons. I.H.P. 607.

1920 Photo.

SVIATOGOR (ex-Russian *Sviatogor*, Armstrongs, 1915-16). tons. Dimensions: 297 (*w.l.*) 323½ (*o.a.*) × 71 × feet. Guns: ? 3 sets triple expansion engines 10,000 I.H.P. = kts. 10 S.E. boilers.

Notes.—Immensely strong in framing. Has powerful pumps and can serve as Salvage and Distilling Ship.

No photo available.

ALEXANDER (ex-Russian *Aleksandr Nevski*). Built by Armstrongs. Begun June, 1916 (for Russian Government), launched December, 1916, completed December, 1917. tons. Dimensions: 273 (*w.l.*), 284 (*o.a.*) × 63 ft. 10 in. × feet. Guns: ? Machinery: 3 sets triple expansion, viz., 2 sets totalling 5000 H.P. to after screws, 1 set 2500 H.P. to bow screw = 7500 I.H.P. total. Speed: kts. 8 S.E. boilers. Coal: tons.

Notes.—As *Sviatogor* above. Also has a V-shaped recess in stern, into which following ship can insert her stem and add additional power for ice-breaking.

For Fleet Target Service.

Photo wanted.

CHRYSANTHEMUM (Nov., 1917). Built by Armstrongs. Ex Fleet Sweeping Vessel (Sloop) of *Anchusa* series (*Flower* Type). 1290 tons. Dimensions: 262½ (*o.a.*) × 35 × 11¾ feet. Guns: Not known. Machinery: 4-cylinder triple expansion. 1 screw. 2 cylindrical boilers. Designed I.H.P. 2500 = 16 to 17 kts. Coal: 260 tons *max.*

1920 Photo, *G. Bott, Esq.*

SNAPDRAGON (Napier, Dec., 1915). Ex Fleet Sweeping Vessel (Sloop) of *Arabis* type, "Flower" classes. 1250 tons. Dimensions: 267¾ (*o.a.*) × 33½ × 11 feet. Guns: Not known. Machinery and boilers as *Chrysanthemum*. I.H.P. 1400 = 17 kts. (as designed), about 15 kts. (actual). Coal: 260 tons *max.*

Also Tugs *St. Cyrus*, *St. Genny*, *St. Issey* (see below). Trawlers *May*, *Ouse* (given on a later page).

Fleet Tugs.

ST. ABBS, ST. AUBIN, ST. BEES, ST. BONIFACE, ST. BREOCK, ST. CLEARS, ST. CYRUS, ST. DAY, ST. DOGMAEL, ST. FAGAN, ST. GENNY, ST. ISSEY, ST. JUST, ST. MARTIN, ST. MONANCE, ST. OLAVES, ST. SAMPSON.

No official details available. About 450 tons. I.H.P. about 1200 = about 12 to 13 kts. May have 1—12 pdr. gun. For General Fleet or Squadron duties.

Also many other Tugs, Tanks, &c., but as these are not attached to Fleets or Squadrons, they are not listed.

Salvage Vessels.

1920 Photo.

RACER (1884, rebuilt 1916-17). 970 tons. I.H.P. 1100 = 11 kts. Re-engined with starboard machinery of old T.B. 8 submersible electric, steam centrifugal and compressed air pumps throwing 3000 tons per hour. 2—17 ton derricks.

Also:—

***REINDEER** (1884). 970 tons. Re-engined by Fairfield Co., 1918. I.H.P. 1200 = kts.

***MARINER** (1883). Ex-Sloop and [illegible] 970 tons [illegible]

Despatch Vessel.

1920 Photo, Mr. Cooke, Middlesbrough, U.K.

WATERWITCH Begun, May 6th, 1914. Launched, October 17th, 1914. Completed, June 7th, 1915. Late Turkish *Rechid Pasha*, purchased 1915. Built by Fairfield 400 tons. Dimensions: 165 × 26 × 6½ feet (*mean* draught). Designed H.P. 625 = kts. *Trials:* 615 = 13.1. Triple expansion engines and cylindrical boilers. 2 screws. Coal: 34 tons. Complement, 14.

Royal Yachts.

ALEXANDRA (1907). 2050 tons. Dimensions: 275 (*p.p.*) × 40 × 12½ feet (*mean* draught). Guns: 2—7 pdr. (bronze). Parsons turbines. H.P. 4500 = 18.25 kts. (19.1 on first [illegible]) Yarrow boilers. Coal: 270 tons. Complement, 173.

Royal Yachts—*Continued.*

VICTORIA AND ALBERT (1899). 4700 tons. Dimensions: 380 (*p.p.*) × 40 × 18 feet (*mean* draught). Guns: 2—6 pdr. (bronze). H.P. 11,800 = 20 kts. Belleville boilers. Coal: *normal*, 350 tons; *maximum*, 2000 tons. Comp. 336.

Admiralty Yacht.

1920 Photo.

ENCHANTRESS (1903). 3470 tons. Dimensions: 320 × 40 × 15 feet. Guns: 4—3 pdr. H.P. 6400 = 18 kts. Yarrow boilers. Coal: 350 tons.

Yachts.

1920 Photo, Abrahams, Devo

TRIAD Yacht, S. N. O. Persian Gulf. (Caledonian S 1909, purchased 1915). 2354 tons. Dimensions: 35 × 15¾ feet. Guns: 1—12 pdr., 4—3 pdr., H I.H.P. 2235 = 14 kts. Coal: 486 tons.

Photo received too late for inclusion in this edition.

SURPRISE (ex-Russian *Razvyet*, taken over 1920). wanted. Yacht and Despatch Vessel, C.-in-C., Medite

1920 Photo, T. D. Manning

ALACRITY (ex Russian *Mlada*, taken over 1919). wanted. Yacht and Despatch Vessel, C.-in-C., China.

Note.—Convoy Sloop *Bryony* has served as Yacht, S. N. O., Hunt Class T. S. Minesweeper *Petersfield* also serves on South American Station.

1919 Photo, Perry & Son.

CACHALOT, ICEWHALE.

Emergency War Programme. Built 1915, by Smiths Dock Co. 336 tons. Dimensions: 139¾ × 25 × 6½ feet. Guns: 1—12 pdr. H.P. 1000 = 12-13 kts. Coal: 60 tons. Complement, 26.

Also *Fair Helga*, *Fair Magda*, *Noble Nora* and *Transvaalia*.

Admiralty Trawlers.

Photo and details wanted.

Dee (*ex-Battleaxe*). **Garry** (*ex-Goldaxe*). **Kennet** (*ex-Iceaxe*). **Liffey** (*ex-Stoneaxe*).

No details known of above "Axe Class" Trawlers, but believed to have been ordered originally by the Russian Government for Patrols in White Sea.

Admiralty Trawlers—*Continued*.

(All to be retained permanently.)

Photo wanted.

"CASTLE" type.*

Photo wanted.

"MERSEY" type.*

Photo wanted.

"STRATH" type.*

*Lists also wanted showing division of following 23 ships among above three types.

Blackwater (*ex-William Inwood*).
Boyne (*ex-William Jones*).
Cherwell (*ex-James Jones*).
Caine (*ex-Isaac Chant*).
Derwent (*ex-John Brice*).
Doon (*ex-Fraser Eaves*).
Ettrick (*ex-Samuel Jamieson*).
Eden (*ex-Thomas Jones*).
Erne (*ex-John Chivers*).
Esk (*ex-Thomas Jarvis*).
Foyle (*ex-John Edmund*).
Itchen (*ex-Thomas Haggerty*).
May (*ex-Alexander Hills*).
Ness (*ex-Andrew Palmer*).
Nith (*ex-Andrew Jewer*).
Ouse (*ex-Andrew King*).
Rother (*ex-Anthony Aslett*).
Stour (*ex-Daniel Fearall*).
Test (*ex-Patrick Bowe*).
Teviot (*ex-George Ireland*).
Ure (*ex-Henry Jennings*).
Waveney (*ex-James Connor*).
Wear (*ex-John Bomkworth*).

Belong to "Castle," "Mersey," and "Strath" types. Usually as follows:—(*a*) 665 tons, 11 kts. speed, 204 tons coal; (*b*) 547 tons, 10½ kts., 164 tons coal; (*c*) 429 tons, 10 kts., 95 tons coal. Guns: Usually 1—12 pdr., but a few have 1—4 inch.

Admiralty Drifters.

Photo wanted.

Astral.
Avalanche.
Cascade.
Cirrus.
Daylight.
Eddy.
Fiery Cross.
Fog Bow.
Full Moon.
Fumarole.
Gloaming.
Halo.
Icepack.
Indian Summer.
Irridiscence.
Landfall.
Lunar Bow.
Mist.
Moonshine.
Onyx.
Overfall.
Sheen.
Shooting Star.
Silhouette.
Stormwrack.
Sunset.

Displace 199 tons. Speed: 9 kts. Gun: Usually 1—6 pdr. Coal: 34-39 tons.

Photo wanted.

C.D. 1. C.D. 7.

Displace 150 tons. Speed: 9 kts. Gun: 1—3 pdr. Coal: 31 tons.

Note.—Admitted that the above List is far from complete; it is merely inserted in the hope that it will elicit additional data and corrections from Correspondents. At the time this page was prepared, 123 Drifters were named in current Official "Navy List." Drifters named above are all Tenders to Battleships, Battle Cruisers, Leaders and Flotillas, Depôt Ships, Establishments, &c., or form part of the [illegible] Drifters at the Home Naval [illegible]

Surveying Service.

1919 Photo, Abrahams, Devonport.

BEAUFORT (ex *Ambleside*, 1919), **COLLINSON** (ex *Amersham*, 1919), by Ailsa S.B. Co. **FITZROY** (ex *Pinner*, ex *Portreath*, 1919), **FLINDERS** (ex *Radley*, 1919), by Lobnitz & Co. **KELLET** (ex *Uppingham*, 1919), **CROZIER** (ex *Verwood*, ex *Ventnor*, 1919), by Simons & Co. Converted Twin Screw Minesweepers of "Hunt Class." 800 tons. Dimensions: 231 (o.a.) × 28½ × 7½ feet. Guns: 1—3 pdr. I.H.P. 2200 = 16 kts. Machinery: Vertical triple expansion. 2 screws. Boilers: Babcock or Yarrow. Coal: 185 tons.

ENDEAVOUR. 1919 Photo.

ENDEAVOUR (Fairfield, 1912). 1280 tons. Dimensions: 200 × 34 × 19 feet. Guns: 1—3 pdr., 2 Maxims. H.P. 1100 = 13 kts. Coal: 221 tons. Complement, 44. Specially built for survey.

Surveying Service—*Continued*.

Photo wanted.

FANTOME (1901). 1070 tons. Guns: 2—4 inch, 1—3 pdr., 2 M.G. Speed: 13½ kts. Coal: 195 tons. Complement, 135.

MERLIN. 1920 Photo, R. C. L.

As *Merlin* above, but

(a) Very wide yard on foremast.

(b) Very big ensign gaff on mainmast.

(c) Boom for sounding machine on quarter deck, raked at an angle of about 45° over port beam; when seen dead astern, gives appearance of mizzen mast.

MUTINE.

MERLIN (1901). 1070 tons. **MUTINE** (1900). 980 tons. Guns: 2—3 pdr., 2 M.G. I.H.P. 1400 = 13½ kts. Belleville boilers. Coal: *about* 160-263 tons.

Cable Ship.

Photo wanted.

KILMUN (Smith's Dock Co., 1919). Ex Patrol Gun "Kil" Class. 890 tons. Dimensions: 182 (o.a.) × 30 feet. I.H.P. 760 = kts. Coal: tons.

Fishery Protection Vessels.

(*v. third column, next page.*)

Coastguard Cruiser.

Paid off.—May be sold.

JULIA (ex *Maretanza*, 1897). 316 tons. H.P. 650. 2 old 7 pdr. M.L. Coal: 53 tons.

BRITISH NAVY: SHIPS FOR TRAINING DUTIES AND EXPERIMENTS.

Nearly all described on preceding pages.

Gunnery.

Battleships ..	*Erin.* *Collingwood.* *Agamemnon.** *Swiftsure.**
Cruisers ..	*Courageous*, *Glorious* } Turret Drill.
	Commonwealth (Sea-going Gunnery Ship).
Monitors ..	*Erebus*, *Marshal Soult*, *Terror* } Director and Fire Control.
Light Cruiser ..	*Canterbury.*
Gunboat ..	*Excellent.*
Destroyers ..	*Sharpshooter.* *Stork.* *Tenacious.*
Motor Launches	255, 320, 508.
Diving Tenders ..	*Gannet, Raven, Vivid.*
Tug ..	*St. Fagan.*
Trawlers ..	*Flintshire, Nith, Stour, [illegible].*

*For Target Service.

Torpedoes: (Whitehead Experiments and Training) and Minelaying.

Light Cruiser ..	*Champion.*
Monitors ..	*M 22,* M 29,* M 31,* M 33.**
Destroyers ..	*Sable, Sylph, Tactician, Tara, Tarpon, Wakeful, Winchester.*
Patrol Boats ..	*P 46, P 47, P 48, P 52.*
Motor Launches	*335, 443.*
Tug ..	*St. Maurice.*
Trawlers ..	*Derwent, Kate Lewis, Itchen, Strathcoe*†, *Ure.*
Drifters ..	*Cirrus, Fogbow.*

*Mine Layers. †Temporary.

Submarines.

Depot Ships	*Vulcan.* *Vulcan II* (late *Onyx*).
Submarines	*H 27, 31, 32, 33, 47, 48, 52, L 24, M 2, R 1.*
Depot Ships	*M. Ney.* *Dolphin.*
Tenders (*Hunt* Class)	*Forres.* *Ross.*
M.L.	(One.)

Anti-Submarine.

Cruiser	*Gibraltar.*
Patrol Boats	*PC 43, 56, 64, 74.*
Motor Launches	*378, 388.*
Submarines	(Not known.)
Whaler	*Ieradale.*
Trawlers	*Blackwater.* *Teviot.*
Drifters	*Blizzard.* *Loll.*

Mechanical Training.

Destroyer	*Rival.*

Under Orders of Director of Compasses.

Destroyer	*Rapid.*

Patrols and Minesweeping. Training and Fisheries Protection.

Sloops	*Godetia, Harebell, Lupin.*
"Hunt" Class ..	*Badminton, Burslem, Leamington, Mistley, Sherborne, Tring, Truro.*
Trawlers ..	*Boyne, Cherwell, Colne, Dee, [illegible], Ettrick, Exe, Garry, Kennet, Liffey.*

Signals and W/T Experiments.

Cruiser	*Antrim.*
Destroyers	*Truant.* *Umpire.*
Patrol Boat	*P 59.*
Whaler	*Cachalot.*

Navigation.

Hunt Class	*Alresford.* *Carstairs.* *Caterham.*
Gunboat	*Dryad* (late *Rattler*).

Cadets.

Battleship	*Temeraire.*
Cruiser	*Carnarvon.*
Destroyer	*Sturgeon.*
Torpedo Boat	*T.B. No. 116.*
Motor Launch	487.
Patrol Gunboat	*Killerig.*

EX-ENEMY SHIPS ASSIGNED TO GREAT BRITAIN, 1920.

All are to be broken up, disposed of or destroyed for experimental purposes, and cannot be added to H.M. Navy. This list is exclusive of 10 Battleships, 5 Battle Cruisers, 5 Light Cruisers, and 33 Destroyers, scuttled at Scapa Flow, June 21st, 1919, not salved, and nominally assigned to Great Britain. Also exclusive of 124 Submarines, surrendered 1918-19, also assigned to Great Britain.

NATIONALITY.	BATTLESHIPS.	CRUISERS.	DESTROYERS.	TORPEDO BOATS.*	REMARKS.
GERMAN ..	BADEN (a)	NURNBERG (a)	S 137 (a) S 54 (a) (d) T 186 T 166	V 105 V 106 V 108	All in Firth of Forth or at Portsmouth.
	HELGOLAND	STETTIN	V 128 S 53 (a) (e) T 184 T 165	A 95 A 78 A 61 A 39	*Selected by Poland:—V 108, A 80, A 69, A 68, A 64, A 59. Selected by Brazil:—V 105, V 106, A 93, A 81, A 78, A 74. All to be retained by Poland and Brazil for Police Duties. Other boats to be broken up.
	POSEN	STUTTGART	V 125 (a) V 44 (a) T 183 T 164	A 94 A 76 A 59 A 38	
	RHEINLAND (b)	DANZIG	B 98 (e) V 28 T 182 T 163	A 93 A 75 A 55 A 37	
	WESTFALEN	MÜNCHEN	G 95 V 26 T 180 T 161	A 92 A 74 A 54 A 36	(a) Salved at Scapa Flow.
		LÜBECK	G 92 (a) S 24 T 179 T 160	A 91 A 70 A 53 A 35	(b) Being broken up in Holland.
			V 82 (a) T 197 T 178 T 159	A 90 A 69 A 52 A 34	(c) Seized at Scapa Flow, June 21st, 1920, afterwards wrecked.
			V 81 (a) (d) T 195 T 176	A 89 A 68 A 49 A 33	(d) Wrecked or foundered at sea.
			V 73 (a) T 193 T 174	A 88 A 66 A 48 A 31	(e) Was assigned to Japan. Beached at Rosyth and back broken. Assigned to Great Britain for breaking up.
			V 71 T 192 T 173	A 87 A 65 A 46 A 29	
			T 189 T 169	A 86 A 64 A 45 A 28	
				A 81 A 63 A 44 A 27	Also Submarine Depot Ship *Cyclops* to be broken up by Great Britain.
				A 80 A 62 A 41	
AUSTRIAN ..	ERZ. FERDINAND MAX	SANKT GEORG	(Nil)	73 F 50 E No. 8	Disposal of Submarines not known.
	ARPAD	K. KARL VI		72 F 51 T No. 9	Also Mine Layers *Chamaleon*, *Salamander*, to be broken up by Great Britain.
	BABENBERG	A. SPAUN		71 F 53 T No. 10	
	HABSBURG	ASPERN		70 F 55 T No. 14	
	BUDAPEST	SZIGETVAR		68 F 56 T No. 18	
	MONARCH	K. u. K. M. THERESA		67 F 57 T	
		LEOPARD		66 F 58 T	
		PANTER		65 F 59 T	
				64 F	
TURKISH ..	—	—	—	—	v. Late Addenda Page.

ROYAL INDIAN MARINE.

Officially Revised, 1920, by courtesy of Commander Scott, Officiating Director of the Royal Indian Marine, Bombay.

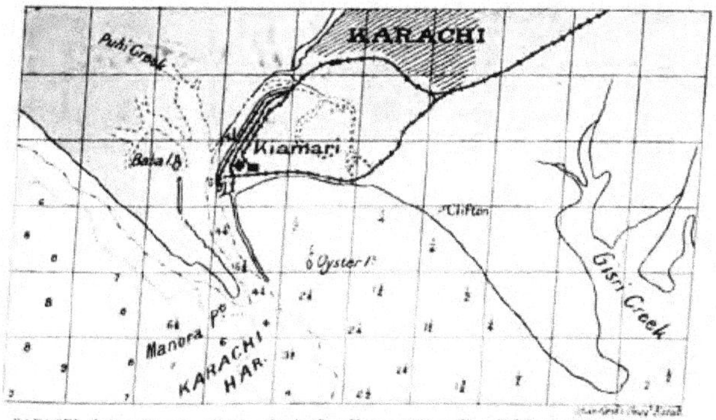

KARACHI (Indian Ocean). Graving Dock (Port Trust), 293½ × 50 × 12¾ feet. Admiralty Chart No. 40.

R.I.M. Dockyard, Bombay.

One dock, 1000 × 100 × 34 feet; and four smaller.

Private Docks.

CALCUTTA.—Eleven, 535–202 feet long, 76–41 feet at entrance, 23–8 feet over sill H.W.O.S.

BOMBAY.—Merewether, 525 × 65 × 28½ feet; Ritchie, 405 × 66 + 18 feet; Mazagon, 430 × 16 × 15½ feet.

RECOGNITION SILHOUETTES.

Scale: 1 inch = 160 feet.

NEARCHUS.

MINTO.

CANNING.

NORTHBROOK.

HARDINGE.

DUFFERIN.

Troopships.

Photo received too late for inclusion.

LAWRENCE (ex-*New Lawrence*, launched by Messrs. Beardm[ore], July 30th, 1919, completed December 27th, 1919). 1259 t[ons]. Dimensions: 225 (*p.p.*), 284½ (*o.a.*) × 34 × 8¾ feet (*mea*[*n*]), 11 feet (*max.*) draught. Complement, 97. Guns: 4—12 [pdr.]. Turbines. 2 screws. 2 Babcock boilers. S.H.P. 1900 = [...] kts. Oil: *normal*, 144 tons; *max.*, 153 tons.

Photo received too late for inclusion.

CLIVE (launched by Messrs. Beardmore, December 10th, 191[9], completed April 29th, 1920). 2100 tons. Dimensions: 24[...] (*p.p.*), 270¼ (*o.a.*) × 38½ × 10⅓ feet (*mean*), 14 feet (*max.*) draught. Complement, 111. Guns: 3—3 pdr. Turbines. 2 screws. 2 w.t. boilers (make unknown). S.H.P. 1700 = 14½ kts. Oil: 179 tons *normal*, and 199 tons *max.*

NORTHBROOK (Clydebank, 1907). 5820 tons. Dimensions: 360 × 51½ × 18½ feet. H.P. 7249 = 17 kts. Armament: 6—4.7 inch, 6—3 pdrs. Coal: 965 tons. Complement [...]

Troopships—(*continued*).

DUFFERIN (Vickers, 1904). 7457 tons. Dimensions: 437 × 53 × 20½ ft. H.P. 10,150 = 19 kts. Coal: 1200 tons. Armament: 8—4·7 inch, 8—3 pdrs. Complement, 265.

HARDINGE (Fairfield, 1900). 6520 tons. Dimensions, 407 × 51 × 19 feet. H.P. 9366 = 19 kts. Fitted with Howden's forced draught. Armament: 6—4·7 inch, 6—3 pdrs., 4 Maxims. Complement, 265.

Minto. *Photo, W. Ernest Williams, Esq., R.I.M.*

MINTO (Birkenhead, 1893). 960 tons gross. Dimensions: 203½ × 31½ × 12½ feet. H.P. 2025 = 14·5 kts. Guns: 4—3 pdr.

Troopships—(*Continued*).

No photo available.

DALHOUSIE (Greenock, 1886). 1920 tons. H.P. 2200 = 13·75 kts. Coal: 280 tons. Armament: 2—4 inch, 6—6 pdrs.

Steamers.

Nearchus. *Official Photo.*

NEARCHUS (Beardmore. Begun Oct., 1913; launched 15th Nov., 1914; completed 7th Dec., 1914). 925 tons. Dimensions: 180 (p.p.) × 29 × 11 feet. S.H.P. 850 = 11 kts. (trials 650 = 11·75). Parsons geared turbines. 1 Yarrow boiler. Oil: 120 tons.

MAYO 1125 tons gross. 240 × 34 × 13 feet.

Despatch Vessel.

(Used as Depot Ship, 1920.)

Photo by favour of N. O'Toole, Esq.

CANNING (ex old *Lawrence*, ex *Lawrence*, 1886). Paddle wheel steamer. 1154 tons. Dimensions: 212 × 32½ × 18½ feet. Guns: 4—6 pdr., 4 machine. H.P. 1200 = 13·5 kts. Coal: 270 tons.

Surveying Ships.

Photo, W. Ernest Williams, Esq.

INVESTIGATOR (Vickers, 1907). 1185 tons. 254 × 3[illegible] I.H.P. 1500 = [illegible] kts. Coal: 172 tons. No guns.

Photo wanted.

PALINURUS (Cammell Laird, 1907). 444 tons. 140 × [illegible] (max. draught). I.H.P. 186 = 10·7 kts. Coal: 64 tons.

River Craft.

BHAMO. Composite Stern-wheel Vessel. 172 tons gr[illegible] × 30 × 2 feet. Guns: 2 machine.

SLADEN. Paddle-wheel River Steamer. 270 tons gr[illegible] × 30 × 2 feet. Guns: 5 machine.

There are also about 24 yard craft, launches and small [illegible] mining vessels. Also various stern wheel and tunnel s[illegible] built during War by British Shipyards for service on [illegible] Euphrates.

Federated Malay States.

Bore expense of building the battleship **MALAYA**.

COMMONWEALTH OF AUSTRALIA.

ROYAL AUSTRALIAN NAVY.

Personnel: *Minister of the Navy*—The Rt. Hon. Sir Joseph Cook, P.C.

GEELONG. Port Philip, Melbourne. Base for Submarines.

Navy Yards.

COCKATOO ISLAND (SYDNEY), with one completed slip for building Light Cruisers and another large slip building. Employees about 1500-2000. May be closed down owing to labour troubles.

GARDEN ISLAND. Repairs only.

Also a Government yard at Walsh Island, N.S.W., devoted to mercantile construction.

Private Docks, &c.

ADELAIDE (S. AUSTRALIA). Commercial harbour, 7 miles from the town.
BRISBANE (QUEENSLAND, AUSTRALIA). 25 miles up river. Navigable for ships drawing 20 feet. Coaling station. Dry dock: 431½ × 55 × 19 feet.
FREMANTLE (W. AUSTRALIA). Dry dock, originally designed as 591 feet long, but enlarged to 694 feet and may now be 910 × 100 × 34 feet (L.W.O.S.), divisible into two sections (*Dreadnought*).
HOBART (TASMANIA). Coaling station. Anchorage average, 10 fathoms.
KING GEORGE SOUND (W. AUSTRALIA). Coaling station. Fortified.
SYDNEY (N.S.W., AUSTRALIA). Fortified coaling station and base for R.A.N. Harbour excellent. Bar 27 feet at low water. Dry docks: (Sutherland) 638 × 84 × 32 feet (*Dreadnought*), (Fitzroy) 477 × 59 × 21½ feet. *Woolwich* dock (private) 675 × 85 × 28 feet (*Dreadnought*), and four smaller.
MELBOURNE. Commercial harbour. One dry dock: 470 × 80 × 27 feet, and three smaller. Port Phillip has an area of 800 square miles.
NEWCASTLE (N.S.W., AUSTRALIA). Government floating dock, 426 × 64 × 25 feet. (9500 tons.)

T.B.D. & SUBMARINE SILHOUETTES (E. L. KING).

Scale: 125 feet to 1 inch.

Anzac (*Leader*).

Parramatta *class* t.b.d.

J 1—J 5 (*Submarines*).

J 7 (*Submarine*).

RECOGNITION SILHOUETTES.

Scale: 1 inch = 160 feet.

GAYUNDAH *class*.

PROTECTOR.

PLATYPUS.

MARGUERITE.

KURUMBA.

GERANIUM, MALLOW.

PIONEER, PSYCHE.

ENCOUNTER.

As *Chatham* class, British Navy.

BRISBANE, MELBOURNE, SYDNEY.

1910 BATTLE-CRUISER (25 kts.)

(Indefatigable Type—1 ship.)

AUSTRALIA (Oct. 25th, 1911).

Normal displacement, 18,800 tons (about 21,300 *full load*.) Complement, 820. (Flagship.)

Length (*waterline*), feet. Beam, 80 feet. { *Mean* draught, 26½ feet. / *Max.* draught, 30 feet. } Length, (*p.p.*) 555 feet. (*over all*), 590 feet.

Guns (**Dir. Con.**):
8—12 inch, 50 cal.
14—4 inch, 50 cal.
1—4 inch anti-aircraft
1—3 inch anti-aircraft
(1 landing)
4—3 pdr.
5 machine
Torpedo tubes (18 inch):
2 *submerged* (broadside).

Armour (K.C., K.N.C., Nickel):
6" Belt (amidships)...
4" Belt (ends) ....
2½-1" Decks..........
7"-3" Barbettes ......
7" Gunhouses ....
10" Conning tower......
4" C.T. tube..........
1" Torp. dir. tower
1½"-1" Funnel uptakes
(Hull + armour = 14,7?? tons.)

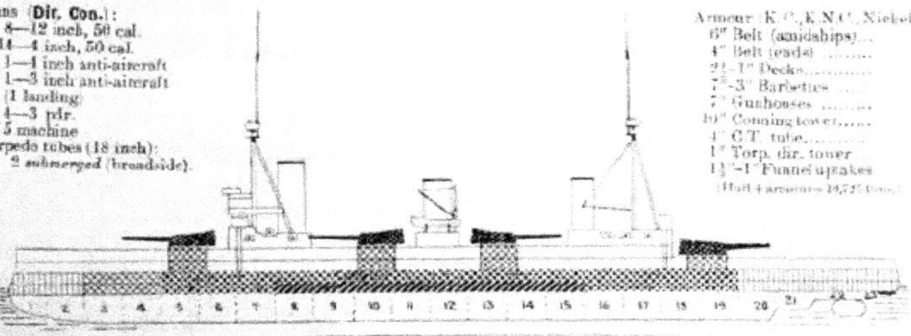

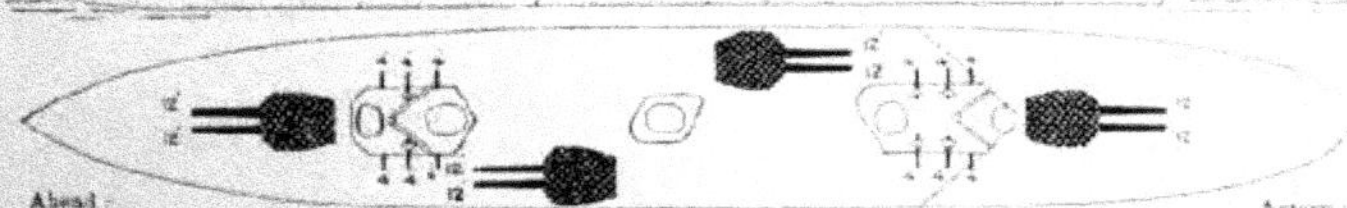

Ahead: 4 to 6—12 in.

Broadside: 8—12 in.; 1—21 in. tube

Astern: 4 to 6—12 in.

Machinery: Parsons turbines by builders. 4 screws. **Boilers**: Babcock. **Designed H.P.** 44,000 = 25 kts. **Coal**: *normal* 1000 tons, *maximum* 3176 + 840 tons oil fuel. **Endurance**: about 6,300 miles at 10 kts.

Armour Notes.—Main belt 11 feet deep, closed by 4" bulkheads and with 1" deck above. Lower deck 1" amidships, 2½" at ends outside belt. For more detailed particulars, *v. New Zealand*, British Navy Section.

Name	*Builder*	*Machinery*	*Laid down*	*Completed*	*Trials (mean):* 30 hrs. at ¾	8 hrs. full power	*Boilers*	*Best recent speed*
Australia	J. Brown, Clydebank	J. Brown, Clydebank	June '10	June '13	29,800 = 21·5	48,422 = 26·83	Babcock	

General Notes.—An improved *Indefatigable*, and a sister-ship to *New Zealand*, described on a previous page. Gunnery and Torpedo Training Ship, 1920.

AUSTRALIA. (Topmasts and topgallant masts as Silhouette).

(Chatham Class—4 Ships.)

MELBOURNE (May, 1912), **SYDNEY** (Aug., 1912), **BRISBANE** (Sept., 1915), and **ADELAIDE** (July, 1918).

Displacement, 5400 tons. Complement, 390-392.

Length (*p.p.*), 430 feet. Beam, 49½ feet. { *Mean* draught, 15¾ feet. / *Max.* „ 17¾-18¼ feet. } Length (*over all*), 457 feet.

Guns (**Dir. Con.**):
- 8—6 inch, 50 cal.
- 1—3 inch anti-aircraft
- 4—3 pdr.
- 4 machine

Torpedo tubes (21 inch):
- 2 broadside (*submerged*)

Armour (Hadfield):
- 2" Deck (on slopes)

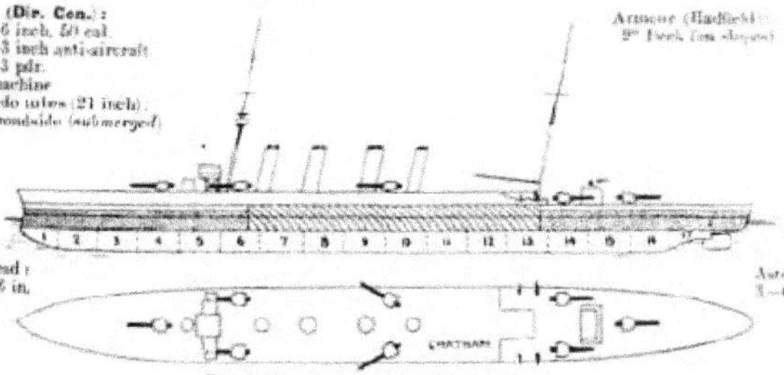

Ahead: 3—6 in.

Astern: 3—6 in.

Broadside: 5—6 in., 1—21 in. torpedo tube.

Machinery: Parsons turbine. 4 screws. Boilers: Yarrow. Designed H.P. (*n.d.*) 22,000, 25,000 *f.d.* = 25 kts. Coal: *normal* 750 tons, *maximum* about 1240 tons + about 260 tons oil fuel. *Brisbane*, 1196 coal + 260 oil.

Gunnery Notes.—Electric ammunition hoists.

Torpedo Notes.—7 torpedoes carried. 4 or more searchlights.

Armour Notes.—Internal protection by longitudinal and transverse bulkheads. Double bottom extends over magazine and machinery spaces. Conning Tower in *Sydney* (and others) replaced by revolving platform carrying a small aeroplane or seaplane.

Engineering Notes.—300 r.p.m. = full power. Boilers: 3 drum small tube type. Unifux condensers.

Name	Builder	Machinery	Laid down	Completed	Trials: 30 hrs. at ¾	Trials: 8 hrs. full power	Boilers	Best recent speed
Melbourne	Cammell-Laird	Cammell-Laird	Apl. '11	Jan. '13		25,800 = 25.7	Yarrow	
Sydney	L. & Glasgow	L. & Glasgow	Feb. '11	Jun. '13	21,466 =	25,539 = 25.7	„	
Brisbane	Sydney	Vickers	Jan. '13	Yes			„	
Adelaide	Sydney		1915	1921				

General Notes.—Practically same design as British *Chatham* class. Served during War in North Sea.

Melbourne. 1919 Photo.

Brisbane. 1920 Photo, Abrahams, Devonport.

Obsolete Light Cruisers.

(Sea-going Training Ship.)

ENCOUNTER (1902, *presented* 1919). 5880 tons. Complement, 457. Dimensions: 376 (*o.a.*)×56×20¾ feet (*max.* draught). Armament: 11—6 inch, 4—12 pdr., 4—3 pdr., 3 M.G., 1 landing. 2—18 inch tubes, *sub.* Designed H.P. 12,500=21 kts. 12 Dürr boilers. Coal: 1300 tons *max.*

Photo, Symonds & Co.

PSYCHE (1897), 2135 tons and **PIONEER** (1899), 2200 tons. Complement, 234. Armament: *Psyche*, 8—4 inch, 3 M.G. *Pioneer* has only 4—3 pdr. Tubes: 2—14 inch *above water*. Designed H.P. 7000 (*f.d.*)=20 kts. (not over 16 kts. at present). Coal: 520—540 tons.

12 Destroyers.

No.	Type	Date	Displacement	H.P.	Speed	Oil	Complement	Tubes	M dr
			tons		kts.	tons			
1	*Anzac*	'16-'17	1666	36,000 *t*	34	416/515	116	4	[illegible]
5	*Admiralty "S"*	'17-'19	1075	27,000 *t*	36	254/301	90	4	[illegible]
3	*Swan*	'15-'16	700	11,300 *t*	26	173/189/157	69	3	8½
3	*Parramatta*	'09-'12		10,600 *t*					

Flotilla Leader.

1 Marksman (later) Type.

1920 *Photo, Abrahams, Devo*

1 *Denny:* **Anzac.** 1,666 tons. Dimensions: 315 ft. (*p.p.*), 325 ft. (*o.a.*) × 31 ft. 10 in. × 12 ft. 1¼ in. (*max.* aft). Guns: 4—4 inch, 2—2 pdr. pom-poms, 1 M.G. Torpedo tubes: 4—21 inch in 2 rev. deck too Machinery: Brown-Curtis turbines. 3 screws. Boilers: Yarrow small tube. Designed S.H.P. 36,000 = Oil, 416/515 tons=about 2500 miles at 15 kts. Complement, 116. Begun January 31st, 1916. La January 11th, 1917. Completed April 24th, 1917. Begun under Emergency War Programme and pres Australia, 1919. Sister to *Seymour*, &c., of British Navy, but fo'xle of *Anzac* is 2¼ feet higher, after funnels and built in sideways, and she has a few other small differences.

Destroyers—*Continued.*

1920 *Photo, Messrs. Swan Hunter.*

5 "Admiralty S" type.

2 *Beardmore*: ***Tasmania, Tattoo.***

1 *Doxford*: ***Success.***

1 *Scott*: ***Swordsman.***

1 *Swan Hunter*: ***Stalwart.***

1075 tons. Dimensions: 265 (*p.p.*), 276 (*o.a.*, *Success* 277) × 26¾ × 9 feet (mean) draught. Armament: 3—4 inch (Mk. IV with 30° elevation *Dir. Con.*), 1—2 pdr. pom-pom, 4—21 inch tubes in two twin deck mountings, 1—24 inch searchlight controlled in unison with guns. Machinery: Brown-Curtis all-geared turbines. 2 screws. Boilers: 3 Yarrow small tube. Designed S.H.P. 27000, = 36 kts. Fuel (oil only): 254/301 tons = about 2900 miles at 15 kts. Complement, 90.

	Begun.	Launch'd	Comp.
Tasmania	18/12/17	22/11/18	29/1/19
Tattoo	21/12/17	28/12/18	7/4/19
Success	....	29/6/18	4/19

	Begun.	Launch.	Comp.
Swordsman	....	28/12/18	3/19
Stalwart	1/18	23/10/19	4/19

Note.—Built under Emergency War Programme and presented to Australian Navy, 1919. For any further notes, refer to description of "Admiralty S Class" in British Navy Section.

Destroyers—*Continued.*

WARREGO. 1919 *Photo.*

Parramatta, Yarra (1910), ***Warrego*** (1911), ***Huan*** (ex *Derwent*, 1914), ***Swan*** (Dec., 1915), ***Torrens*** (Aug. 1915). Dimensions: Length (o.a.) 250¾ *Huon, Swan, Torrens*, 252¾ *Yarra*, 246 *Parramatta, Warrego* × 24¼ × 8 feet. Armament: 1—4 inch, 3—12 pdrs., 3—18 inch tubes. Parsons turbines. Yarrow boilers. Radius 2500 kts. [illegible] fuel. *Yarra* by Denny, Dumbarton; *Parramatta* and *Warrego* by London & Glasgow Engineering Co., *Warrego* being re-erected at Cockatoo N.Y., Sydney. *Swan, Huon* and *Torrens* built at Cockatoo N.Y., Sydney. Are modified British "I" Class boats.

Torpedo Boat (*First Class.*)

No photo available.

Countess of Hopetoun. Tenders to Cerberus. No details known.

6 Submarines.

No.	Class	Begun	Comp.	Displacement (Tons.)	B.H.P. / E.H.P.	Speed (kts.)	Endurance	Tubes	Comp't.
6	J 7	1916	1917	1200 / 1760	3600 / 1400	19 / 9.5	4,000 miles at 12 kts.	6	44
	J 1—5	1915	1916	1260 / 1820					

6 Ocean-going Boats.

J 7. (C.T. set well aft. Quite different in appearance to other boats). 1920 *Illustration.*

J 5 (revolving breastwork to gun). 1920 *Photo, Cribb, Southsea.*

J 4 (and J 2, 3) revolving breastwork to gun). 1920 *Photo, Abrahams, Devonport.*

J 1 (gun platform on pillars). 1920 *Photo, Cribb, Southsea.*

"J Class"—6 Boats.

6 *Admiralty* type: **J 7** (Feb., 1917), **J 5** (Sept., 1915), both by Devonport D.Y. **J 4** (Feb., 1916), **J 3** (Dec., 1915) by Pembroke D.Y. **J 2** (Nov., 1915), **J 1** (Nov., 1915), both by Portsmouth D.Y. *J 1—5*, completed April 1916; *J 7*, Nov., 1917. Admiralty double-hull type. Dimensions: 270 (*p.p.*), 274½ (*o.a.*) × 23½ × 1 Armament: 1—4 inch gun and 6—18 inch tubes (4 bow, 2 beam). Surface machinery: 3 sets of Vickers 12 cy solid injection, direct reversing 4 cycle 1200 B.H.P. Diesel engines (14½ in. bore × 15 in. stroke, 380 r.p.m.), 80/91 tons. Built under Emergency Programme as "Bight Patrols." On completion, were the fastest engined S/M in the world, and were built on the strength of erroneous reports that German boats at sea or construction could make 18-20 kts. *J 6* lost during War. Presented to Australia, 1919. Other details as

Fleet Sweeping Vessels (Sloops).

MARGUERITE. No photo available, but generally as above view of British *Primrose* (*Azalea* class), now disposed of.

Appearance as *Mallow*.

GERANIUM (Greenock & Grangemouth Co. Begun Aug., 1915; launched 8th Nov., 1915; completed March, 1916). **MARGUERITE** (Dunlop Bremner. Begun July, 1915; launched 23rd Nov., 1915; completed Jan., 1916). Fleet Sweeping Vessels (Sloops) of *Arabis* type. 1250 tons. Dimensions: $255\frac{1}{4}$ (*p.p.*), $267\frac{3}{4}$ (*o.a.*) × $33\frac{1}{2}$ × $10\frac{3}{4}$ feet (*mean*), $11\frac{3}{4}$ (*max.* draught). Guns: 1—4·7 inch, 2—3 pdr. (AA. in *Geranium*). Designed I.H.P. 2000 = 17 kts. Trials: *Geranium*, 2312 = 17; *Marguerite*, 2309 = 16·1. Machinery: boilers, screws, as *Mallow*, opposite. Coal: 200 tons = 2050 miles at 15 kts. Complement, 79. Built under Emergency War Programme; presented to Australian Navy, 1919.

Fleet Sweeping Vessel (Sloop).

No photo available.
See illustration for *Valerian*—
British Navy Section.

MALLOW (Barclay Curle. Launched 13th July, 1915; completed Sept., 1916.) Fleet Swe Vessel (Sloop) of *Acacia* type. 1200 tons. Dimensions: 250 (p.p.), $262\frac{1}{2}$ (o.a.) × 33 × 10 (mean), $11\frac{1}{4}$ (max. load). Guns: 1—12 pdr. (12 cwt.), 2—3 pdr. AA. I.H.P. 1800 = 17 Trials: 2328 = 16·3 kts. Machinery: 1 set triple expansion inverted and 2 cylindrical bo Coal: 250 tons = 2000 miles at 15 kts. Complement, 77. Built under Emergency War Progran presented to the Australian Navy, 1919.

Photo wanted.
1 funnel, 2 pole masts, yacht bow.

UNA (ex-German Surveying Ship *Komet*, built about 1911, captured 1914). 1438 tons. Dimensi $210\frac{1}{2}$ × 31 × $15\frac{3}{4}$ feet. Guns: 3—4 inch, 2—12 pdr. H.P. 1300 = 16 kts. Coal: 270 t Complement, 114.

Note.—Sloop **FANTOME** is on the Australian Station for Surveying Duties. Believed that sh maintained by British Navy and not by the Australian Navy. She is listed with other Br Surveying Ships on an earlier page.

Obsolete Gunboats.

PROTECTOR (Elswick, 1884). 920 tons. Guns: 3—4 inch, 2—12 pdr., 4—3 pdr. H.P. 1600 = 14 kts originally. Re-boilered 1910-11. Tender to Williamstown Gunnery School. (*South Australia*).

GAYUNDAH (Elswick, 1884, reported to have been re-built at Cockatoo N.Y., Sydney, 1914). 360 tons. Guns: 1—4·7, 2—12 pdr. Speed originally: 10·5 kts.

Also *Cerberus*, now Naval Depot at Williamstown, with Cruiser *Psyche* and 1st Class T.B. *Countess of Hopetoun* and *Coogee, Gannet, Gayundah, Protector* as tenders.

Submarine Depot Ship.

PLATYPUS (J. Brown, Clydebank. Begun Sept. 2nd, 1914; launched Oct. 28th, 1916; completed Mar., 1917.) 2,460 tons. Dimensions: 310 (*w.l.*), 325 (*o.a.*) × 44 × 15¾ feet (*max.* draught). Guns: *nil* I.H.P. 3600 = 15½ kts. Two sets triple expansion reciprocating engines and 4 cylindrical return-tube boilers. 2 screws. Coal: 450 tons. Complement, 357.

Fuel Ships (R.F.A.)

ONE SHIP, to be begun at Cockatoo, N.Y., 1920-21, similar to *Biloela* below.

Photo wanted.

BILOELA. Fleet Collier, built at Cockatoo Island D.Y. Completed 1920. 10,000 tons. 369½ × × feet. Carries 4000 tons coal, 1250 tons oil; 750 tons fresh water. Own coal, 1000 tons.

Fuel Ships—*Continued.*

No photo available.

KURUMBA (Swan Hunter. Begun Sept., 1915; launched 1916; completed Jan., 1917; machinery by Wallsend Tanker. 8350 tons. Dimensions: 365 (*p.p.*), 377½ × 45½ × 23¼ feet *max.* load draught. Guns: 3 (calibre not I.H.P. 2000 = 10 kts. Triple expansion engines and S.E. cyl boilers. 2 screws. Fuel: *max.*, 688 tons coal + 257 exclusive of cargo. Complement, 65–100.

Armed Patrol Vessels.

No photo available.

COOGEE (———). 762 tons *gross.* H.P. = 16 kts. 1—4·7 inch, 2—3 pdr. Coal: 210 tons.

No photo available.

GANNET (ex *Penguin*, ———). 208 tons *gross.* H.P. kts. Guns: 1—12 pdr. Coal: 87 tons.

No photo available.

MOORILYAN (———). 1349 tons *gross.* H.P. 2350 = Guns 1—4·7 inch Q.F., 2—3 pdr. Hotchkiss. Coal: 280

Miscellaneous.

NUSA. 64 tons. Ex-German ship seized at Kawieng, 1914.

SUMATRA. No details known.

FRANKLIN (ex-*Adele*). 288 tons. Training ship for Tender to R.A.N. College, Jervis Bay.

TINGIRA. Training Ship.

DOMINION OF NEW ZEALAND.

Corrected 1920, by courtesy of Commander J. A. Williams, R.N., Naval Adviser to the New Zealand Government.

Battle Cruiser **NEW ZEALAND** was built at the expense of the Dominion.

Docks.

AUCKLAND. Coaling station. Excellent harbour; average 40 feet deep. Dry dock (Calliope): 521 × 66 × 33 feet. Also one smaller, able to take torpedo craft.

PORT CHALMERS (S. ISLAND). Otago dry dock, 500 × 70 × 21 feet, *and another smaller.*

RECOGNITION SILHOUETTES.

Scale: 160 feet = 1 inch.

PHILOMEL.

CHATHAM.

Light Cruisers.

Photo wanted.

Appearance as illustration of *Southampton* and *Dublin* (British Navy Section).

CHATHAM Nov., 1911.

Displacement: 5400 tons. Complement: 420—440 as Flagship.

Guns:
- 8—6 inch, Mk. XII
- 1—3 inch AA.
- 4—3 pdr.
- 4 or 2 M.G.
- 1 landing.

Torpedo tubes (21 inch):
- 2 *submerged*

Armour:
- 2" Deck (on slopes)
- 3" Side (amidships)

For Plan, *v.* page 80. *Southampton* and *Dublin*.

Machinery: 4—shaft Parsons turbines and Yarrow Boilers. Designed S.H.P. 25000 = 25·5 *Trials*: 25901 = 25·7. Coal: *normal*, 750 tons; *max.*, 1240 tons + 260 tons oil. Begun under Estimates by Chatham D.Y., Jan., 1911, and completed Dec., 1912. Machinery by Thames Presented to New Zealand, 1920.

To be commissioned 1921 as Training Ship.

PHILOMEL (Jan., 1890). 2575 tons. Complement, 222. Dimensions (*o.a.*) 278 × 41 × 16½ (*max.*) feet. Armament: Nil. Torpedo Tubes: Nil. Armour: 2½" Deck. Designed H.P. 7500 = 19 kts. (*f.d.*) Coal: 450 tons.

Fleet Sweeping Vessel (Sloop) **VERONICA** is on the New Zealand Station, but it is believed that she is maintained as a unit of Royal Navy and not as a charge on the New Zealand Government. She is accordingly listed in British Navy Section on an earlier page.

Gunboat.

(Training Service.)

AMOKURA (ex *Sparrow*, Greenock, 1889). Composite. 805 tons. Dimensions: 165 × 31 × Guns: Nil. H.P. 1200 = 13 kts. Boilers: Cylindrical. Coal: 105 tons.

Yacht and Lighthouse Tender.

HINEMOA (Cartsdyke, 1876, re-fitted 1907). 282 tons *gross*. Dimensions: 207¼ × 25½ I.H.P. 542 = 12 to 13 kts.

Cable Ship.

TUTANEKAI (Port Glasgow, 1896). 811 tons *gross*. Dimensions: 205 × 30 × — fee 1600 = 14 kts.

Mining Vessels.

JANIE SEDDON (1902), **LADY ROBERTS** (1901), both built at Paisley. 109 Dimensions: 95 × 18 × — feet. I.H.P. 320 = 10 kts.

DOMINION OF CANADA.

Officially Revised, 1920, by courtesy of the Deputy Minister, Department of the Naval Service, Ottawa.

Minister of Marine: Hon. Col. C. C. Ballantyne. *Personnel:* *Uniforms:* As British Navy.

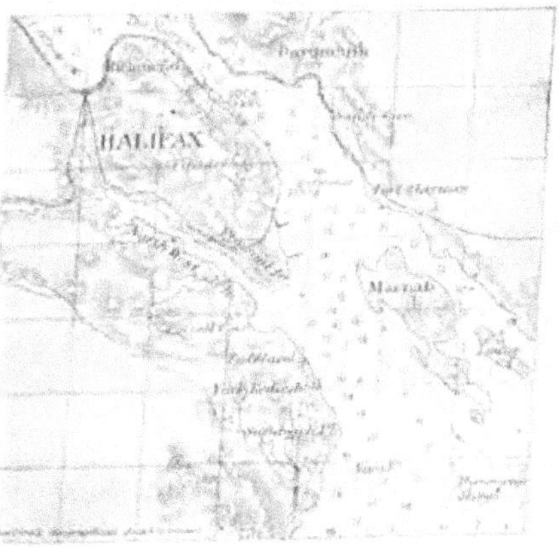

Dockyards.

Atlantic:—

HALIFAX (Nova Scotia).—Formerly Dockyard. Coaling Station. Anchorage averages 13 fathoms. Examination Base during War. Admiralty Chart No. 2410. (For Docks, see below.)

Note.—Closed temporarily, but can be reopened and carry out work as during war, 1914-18, at 14 days' notice.

Pacific:—

ESQUIMAULT.—Naval Dockyard. Admiralty Chart No. 576. (For Docks, see below.)

Note.—Remains open, but some shops closed. Can be in full working order at 7 days' notice.

Private Shipyards.

CANADIAN VICKERS CO. (Montreal).—Affiliated to Messrs. Vickers, Ltd., Barrow-in-Furness. ... details available of this establishment. "H" class Submarines built here during the War. ... large Floating Dock, see Docks below.

MESSRS. YARROW, LTD. (Lang's Cove, Esquimalt).—Branch of Messrs. Yarrow, Ltd., Scotstoun. Can build ships up to 2000 tons. Slipway 315 × 20 feet (3000 tons capacity). Fitting-out W... 500 feet long. Repair shops; one 60-ton sheer legs; one 10-ton floating derrick. Yard covers ... acres, with Government Dock adjoining.

Also many other Yards engaged on mercantile work.

Docks.

COLLINGWOOD. Collingwood Dry Dock Co., 524 × 60 × 15½ feet.

ESQUIMAULT. Government Dock 450–480 × 65 × 26½ feet. Also 2400 tons patent slip. At Skinner's Cove, New Dock, 1150 × 120 × 38 (H.W. *over sill*) feet, *projected*. To be divisible into two lengths of 650 and 550 feet.

ST. JOHN (New Brunswick). At Courtenay Bay, new dock *building*, 1150 × 133 (*at coping*) × 42 fe... (*over sills*).

HALIFAX (Nova Scotia). Dry dock: Halifax Graving Dock Co., 572 × 89 × 29 feet (*Dreadnought*).

KINGSTON (Ontario). Government dock, 280 × 55 × 18 feet.

MONTREAL. Duke of Connaught floating dock (Proprietors, Canadian Vickers Co.), 600 × 100 × 2... feet. (*Dreadnought.*) Two other private docks, 400 × 45 × 10 feet.

PORT ARTHUR. Dry dock, 700 × 74½ × 15½ feet.

PORT DALHOUSIE. Muir Bros'. dock, 260 × 46 × 10 feet.

OWEN SOUND. Dock, 300 × 60 × 12 feet.

QUEBEC: Canadian Government dry dock, 1150 × 120 × 30 feet, divisible into two docks, 650 an... 500 feet long respectively. (*Dreadnought.*) Levis dock, 600 × 62 × 20½ feet. Three floating docks 1600, 2200, and 2500 tons lift.

VANCOUVER, B.C.—New Dock *building* at Burrard Inlet by Messrs. J. Coghlan & Co., with Government subsidy, 712 × 116 × 30 feet (H.W. *over sill*).

RECOGNITION SILHOUETTES.

Scale: 1 inch = 160 feet.

Malaspina. Acadia. Cartier. Lady Grey.

Government Vessels.

Montcalm.
(Government Vessel.)

Aurora.

DESTROYERS.—**Scale: 1 inch = 160 feet.**

Patriot, Patrician (L.G.A.)

SUBMARINES.—**Scale: 1 inch = 80 feet.**

B 14, B 15, CC 1, CC 2

Light Cruiser.

Aurora. 1920 *Photo, Abrahams, Devon*

(Arethusa Class—1 Ship).

AURORA (30th September, 1913).

Displacement, 3500 tons. Complement, 318.

Length, (p.p.) 410 feet, (o.a.) 436 feet. × 39 feet × draught { 13½ feet mean, 15½ feet max.

Guns:
2—6 inch } Dir. Con.
6—4 inch }
2—3 inch A.A.
Torpedo Tubes:
8—21 inch in 4 T.D. twin mountings.

Armour (H.T.):
3" Side (amidships)
2¼"–1¼" Side (forward)
2¼"–2" Side (aft)
1" Deck (amidships)
1" Deck over rudder

Machinery: Parsons I.R. turbines and 8 Yarrow boilers. Designed S.H.P. 40,000 = 28.5 k 30,417 = 27.4. Oil: 482 tons *normal*, 810 tons *max*. Built by Devonport D.Y. Begun Octo completed October, 1914. Begun under 1912 Estimates. Presented to Canada 1920. Original *Glasgow* was offered and accepted, but was replaced by *Aurora*.

2 Destroyers.

Patriot. 1919 *Photo, Messrs. Thornycroft (Builders).*

2 *Thornycroft "M."* **Patriot, Patrician,** 1004 tons. Dimensions: 265 (p.p.) 271 (o.a.) × 27½ × 11 feet. Guns: 3—4 inch, 1—2 pdr. pom-pom, 1 M.G. Torpedo tubes: 4—21 inch in two twin deck mountings. Machinery: Brown-Curtis turbines and 3 Yarrow boilers. S.H.P. 27,500 = 35 kts. Oil: about 250 tons (peace), 400 tons (war). Complement, 82. Presented to Canada 1920.

	Begun	*Launch*	*Comp.*	*Trials*
Patriot	7/15	4/16	6/16	35.4
Patrician	6/15	6/16	7/16	35.4

4 Submarines.

Photo wanted.

2 *Holland type:* ***H 14, H 15*** (1918). Displacements: 364 tons *surface*, 434 tons *submerged*. Dimensions: 150¼ × 1 12½ feet. Torpedo tubes: 4—18 inch (bow). B.H.P. 480 = 13 kts. *on surface*; H.P. 320 = 11 kts. *submerged (for at one hour continuous running)*. Complement 20-22. Built under Emergency War Programme and presented Canadian Navy, 1919.

Photo wanted.

1 *Holland type:* ***CC 2*** (Seattle Con. & D.D. Co., 1914). **Displacements: 310 tons *on surface*, 373 tons *submerge*** Dimensions: 157 ft. 6¼ in. × 14 ft. 11½ in. × 12·1 feet. Torpedo tubes: 3—18 inch. H.P. 600 *on surface* = 15 kts 260 *submerged* = 10½ kts.

1 *Holland type:* ***CC 1*** (Seattle Con. & D.D. Co., 1914). **Displacements: 313 tons *on surface*, 373 tons *submerge*** Dimensions: 144 ft. 3½ in. × 15 ft. × 12·1 ft. Torpedo tubes: 5—18 inch. H.P. 600 *on surface* = 13 kts.; 2 *submerged* = 10½ kts.

Note.—Above two boats were begun for Chilean Navy as *Antofagasta* and *Iquique*. They were purchased by Canada jus before outbreak of war.

Depot Ships—*(Nominal)*.

Stationed at Halifax.

Guelph (1917). Drifter of 110 tons. 84 × 19·3 × 10 feet. Speed: 9 kts. Complement, 11. No guns.

Stationed at Esquimalt.

Naden (1913). Motor Vessel of 118 tons. 80 × 20 × 9 feet. No guns.

B: CANADIAN GOVERNMENT VESSELS.

(From Official List. Arranged *alphabetically*. All speeds are *maximum*.)

Notes.

All Details of Canadian Government Vessels officially revised 1920, by courtesy of the Deputy Minister, Department of the Naval Service, Ottawa.

Department of the Naval Service.

With the exception of *Malaspina*, all guns dismantled and removed to store.

Services ships were engaged on, 1920, are abbreviated thus :—

FP-A :—Fisheries Protection, Atlantic.

FP-P :—Fisheries Protection, Pacific.

HS-A :—Hydrographic Survey, Atlantic.

HS-P :—Hydrographic Survey, Pacific.

Department of the Naval Service—Continued.

ACADIA.

ACADIA (1913). Steel. 1650 tons. Dimensions : 170 × 35.5 × 12.5 feet. H.P. 1200 = 12 kts. Guns : 1—4 inch Q.F., 2—12 pdr. Q.F. 12 cwt. Coal : 260 tons. Complement, 42. (**HS-A**).

Arleux (**FP-A**). ***Givenchy*** (**FP-P**).
Armentieres (**FP-P**). ***Loos*** (**FP-A**).
Arras (**FP-P**). ***Thiepval*** (**FP-P**).

Steel Trawlers, all built 1918. 136 tons. Dimensions : 130 × 25 × 14 feet. Guns : 1—12 pdr. Q.F. 12 cwt. Speed : 10 kts. Complement, 18.

BAYFIELD (1889). Steel. 550 tons. Dimensions : 140 × 24 × 10.5 feet. Speed : 11 kts. Coal : 100 tons. Complement, 25 (**HS-A**).

BRADBURY (1915). Steel. 500 tons. Dimensions : 151 × 27½ × 8 feet. Speed : kts. Complement, 13. (*Fisheries Patrol*).

Department of the Naval Service—Continued.

CARTIER (1910). Steel. 850 tons. Dimensions : 17 × 12 feet. Guns : 3—12 pdr. 12 cwt. Q.F. H.P. kts. Coal : 150 tons. Complement, 34. (**HS-A**).

(*For Givenchy, v. Arleux*).

GULNARE (1893). Steel. 1 screw. 500 tons. Dim 137 × 20.5 × 12 feet. Speed : 10 kts. Coal : Complement, 34. (*Tidal Survey—Atlantic.*)

HOCHELAGA (1900). Steel. 930 tons. Dimension × 27.5 × 16 feet. Guns : 2—12 pdr. 12 cwt. Q.F. 12 kts. Complement, 60. (**FP-A**).

LILLOOET (1908). 760 tons. Speed : 11½ kts. tons. (**HS-A**).

(*For Loos, v. Arleux*.)

MALASPINA.

MALASPINA (1913). Steel. 1 screw. 820 tons. Di 160 × 26.5 × 11.5 feet. Guns : 1—6 pdr. Q.F. H H.P. 1350 = 14½ kts. Coal : 260 tons. Compl (**FP-A**).

(*Continued on ne*

Department of the Naval Service—Continued.

PETREL (1892). Steel. 1 screw. 400 tons. Dimensions: 116 × 22 × 9 feet. Speed: 11 kts. Coal: 65 tons. Complement, 24. (**FP-A.**)

RESTLESS (1906). Wood. 205 tons. Dimensions: 271 × 17 × 9 feet. Speed: 8½ kts. Complement, 8. (**HS-A.**)

STADACONA (1899). Iron Yacht. 682 tons. Dimensions: 168 × 33.5 × 12.5 feet. Speed: 12½ kts. Guns: 1—4 inch Q.F., 1—12 pdr. 12 cwt. Complement, 62. (**FP-P.**)

(*For Thiepval, v. Arleux.*)

Also 4 other vessels under 300 tons, 1 iron-built, 1 composite-built, and 2 wooden-built.

Department of Marine and Fisheries.

ABERDEEN (1894). 1330 tons. H.P. 1510=13 kts. Coal: 125 tons.

ARANMORE (1890). 500 tons (net registered). Speed, 13 kts. Coal: 200 tons.

BELLECHASSE (1912). 576 tons. H.P. 900=11¾ kts. Coal: 74 tons.

DOLLARD (1913). 322 tons (net registered). Speed, 11 kts. Coal: 100 tons.

DRUID (1902). 1000 tons. Speed, 13 kts. Coal: 300 tons.

ESTEVAN (1912). 2100 tons. H.P. 1500=12½ kts. Coal: 350 tons.

GRENVILLE (1914). tons. H.P. 900= kts. Coal: 100 tons.

Department of Marine and Fisheries (continued).

LADY GREY.

LADY GREY (1906). 1080 tons. H.P. 2300=14 kts. Coal: 200 tons.

LADY LAURIER (1902). 1970 tons. H.P. 1800=13 kts. Coal: 175 tons.

LAMBTON (1909). 510 tons. Speed, 11 kts. Coal: 92 tons.

MONTCALM.

MONTCALM (1904). 3270 tons. H.P. 3225=14 kts. Coal: 425 tons.

NEWINGTON (1899). 475 tons. Speed, 10 kts. Coal: 85 tons.

PRINCESS (1896). 850 tons. Speed, 11 kts. Coal: 124 tons.

QUADRA (1891). 1260 tons. Speed, 11 kts. Coal: 209 tons.

STANLEY (1888). 1890 tons. Speed, 15½ kts. Coal: 250 tons.

Department of Customs.

MARGARET (1913). 950 tons. Guns: 2—6 pdr. Designed H.P. 2000=15½ kts. max. speed. Coal: 200 tons.

Department of Militia and Defence.—

Three small vessels *Alfreda* (1901), 270 tons; *Armstrong* (1903), tons; and *Beryl* (1903), 45 tons. All 9½—8 kts. speed.

Department of Public Works.

TYRIAN (1869). 1300 tons. Speed, 10 kts. Coal: 340 tons.

Also *Speedy* (1896). 420 tons. 12 kts. speed.

Department of Railways and Canals.

DRUMMOND (1910). 4718 tons. Speed: 9¾ kts. Coal: tons.

LEONARD (1914). 3365 tons. Speed: 14 kts. Coal: 960 tons.

McKEE (1908). 4557 tons. Speed: 9¾ kts. Coal: tons.

NORTHUMBERLAND (1891). 1255 tons. Speed: 13 kts. Coal: 60 tons.

PRINCE EDWARD ISLAND (Ferry) (1915). 2793 tons. Speed: 15 kts. Coal: 320 tons.

SCOTIA (1901). 2550 tons. Speed: 12 kts. Coal: 120 tons.

SCOTIA No. 2 (1915). 1400 tons. Speed: 15 kts. Coal: 300 tons.

(Also 13 other smaller vessels.)

Department of Agriculture:—

ALICE (1907). 550 tons. Speed, 11½ kts.

And 1 smaller vessels

NEWFOUNDLAND.

SILHOUETTES.

Scale: 160 feet = 1 inch.

DAISY

WATCHFUL.

LOBELIA

1919 *Builders' photo, Messrs. Hall Russell & Co.*

WATCHFUL (Hall, Russell, 1912, purchased from Royal Navy, 1920). 612 tons. Dimensions: 154 × 25 × 11¾ feet. H.P. 800 = 11.5 kts. Guns: 2—3 pdr. Coal: 150 tons. Complement, 72.

Sloop.

As British "Flower Class" Sloops— Illustration of "Valerian."

LOBELIA (Simons, 1916, accepted from Royal Navy, 1920). Ex-Fleet Sweeping Vessel of *Arabis* type, "Flower Class." 1250 tons. Dimensions: 267¾ (*o.a.*) × 33½ × 12 feet, draught. Guns: Not known. I.H.P. 1400 = *about* 15 kts. 1 screw. Cyl. boilers. Coal: 260 tons.

No photo available.

DAISY (Aberdeen, 1911, taken over from Royal Navy, 1920). 510 tons. H.P. 500 = 10 kts. Coal: 107 tons.

PETREL. Armed Patrol Vessel. Speed about 10 kts. Guns: 1—3 pdr. Q.F.

Also old *Briton* (ex-*Calypso*) as Training Ship.

EGYPT.

ENSIGN & JACK. SULTAN'S STANDARD.

Sloop.

Photo wanted.

SYRINGA (Workman, Clark, 1917, bought 1920). 1290 tons. Dimensions: 262½ (*o.a.*) × 35 × 12 feet. Guns: Not known. Designed H.P. 2500 = 16 to 17 kts. 1 screw. Cylindrical boilers. Coal: 200 tons. Ex-Fleet Sweeping Vessel of *Anchusa* series ("Flower" Class).

Yacht.

MAHROUSSA (1865, rebuilt 1905). 3417 tons. Parsons turbine. H.P. 5500 = 17·5 kts.

Also 2 other small Yachts, and 3 Despatch Vessels.

River Gunboats.

MELIK, SHEIK, SULTAN (1897). 140 tons. Speed, 11 kts. 2 screws in tunnels. Guns: 1—12 pdr. + 1—24 pdr. howitzer, 3 machine.

FATEH, NASIR, ZAFIR (1896). 128 tons. Stern-wheel Steamers. Speed, 12 kts. Guns: 1—12 pdr. + 1—24 pdr. howitzer, 4 machine.

ABU KLEA, HAFIR, METEMMEH, TAMAI. Stern-wheel Steamers. Guns: 1—22 pdr. in first; 1—9 pdr. in others.

Also 24 other River Steamers, built 1908-1885.

Transport, &c.

GHARBIEH. 3700 tons. Also another steamer, **MUKBIR,** of 420 tons.

There are also 5 Paddle-wheel Steamers and about 15 Tugs.

Coastguard Vessels.

ABDUL MONEIM (Clydebank, 1902). 598 tons. Guns: 1—3 pdr. H.P. 1000 = 13 kts.

NOUR-EL-BAHR (1881). 450 tons. Guns: 1—3 pdr. H.P. 870 = 13.5 kts.

RESHID. About 650 tons. Guns: 2—6 pdr.

Ports and Lights Administration.

AIDA. 723 tons. 1 gun. H.P. 130 = ?

Also about 25 launches and miscellaneous craft.

(*Type unknown*).

MANAGEM. Loaned to R. N. during War, and mounted 1—12 pdr. 12 cwt. gun.

NYASALAND.

Gunboat.

GWENDOLEN (1897). 350 tons. Guns: 4—6 pdr., 4 machine. Speed 11½ kts.

Adventure, Pioneer. (1892). 30 tons. Speed, 12 kts.

Dove, Paddle Gunboat of 20 tons. Speed, 8 kts. Armed with 1 M.G. Armed German Vessel *Wentzel*, captured on Lake Nyasa during War. Present disposal unknown.

NORTHERN RHODESIA.

Two Thornycroft motor patrol boats, ***Mimi & Tou-tou,*** origi intended to act as tenders to Greek seaplanes. They requisitioned and shipped out to Lake Tanganyika in where they sank the German gunboat *Hedwig von Wissmann* captured the *Kingani*, now British ***Fi-fi.***

BRITISH EAST AFRICA.

Rose (1901). Speed, 8 kts.

M'Vita, Patrol Boat.

NATAL.

Before the war, only possessed a Tug, ***Sir John*** (1897), an Surveying Ship, ***Churchill.***

NIGERIA.

1920 *Photo, C. E. Eldred, Esq.*

HERCULES (ex-*Lady Crundall*, 1906, bought for Nigeri Marine, 1920). tons. Speed: 12 kts.

Note.—"Naval Operations—Vol. I" mentions capture of Germ Armed Vessel *Herzogin Elizabeth* and salvage of a sunk German Gunboat, re-named *Sokoto*. These ships may now in the Nigerian Marine.

BRITISH PROTECTORATE.

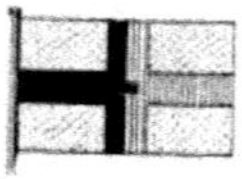

SARAWAK.

From information furnished by courtesy of the Secretary of the Sarawak Advisory Council, London, 1920.

No photo available.

ALICE LORRAINE (Kuching, 1904). 125 tons.

Zahora. 1920 *Official Photo.*

ZAHORA (1894). 225 tons. Armed with 1 small gun.

No photo available.

ADEH (1884, bought from Sarawak and Singapore S.S. Co.). Paddle Wheel Steamer of 300 tons. Has 1 small gun.

No Photo available.

L'AUBAINE. No details known.

Also 8 other small craft and launches for River Service.

UNITED STATES FLEET.

Revised from 1919-20 Edition of Official Handbook, "Ships' Data, U.S. Naval Vessels," and from information furnished by courtesy of the Navy Department, Washington, D.C.
All Photographs, unless otherwise acknowledged, by courtesy of the U.S. Navy Recruiting Bureau and Lieut.-Commr H. Reuterdahl, U.S.N.R.F.

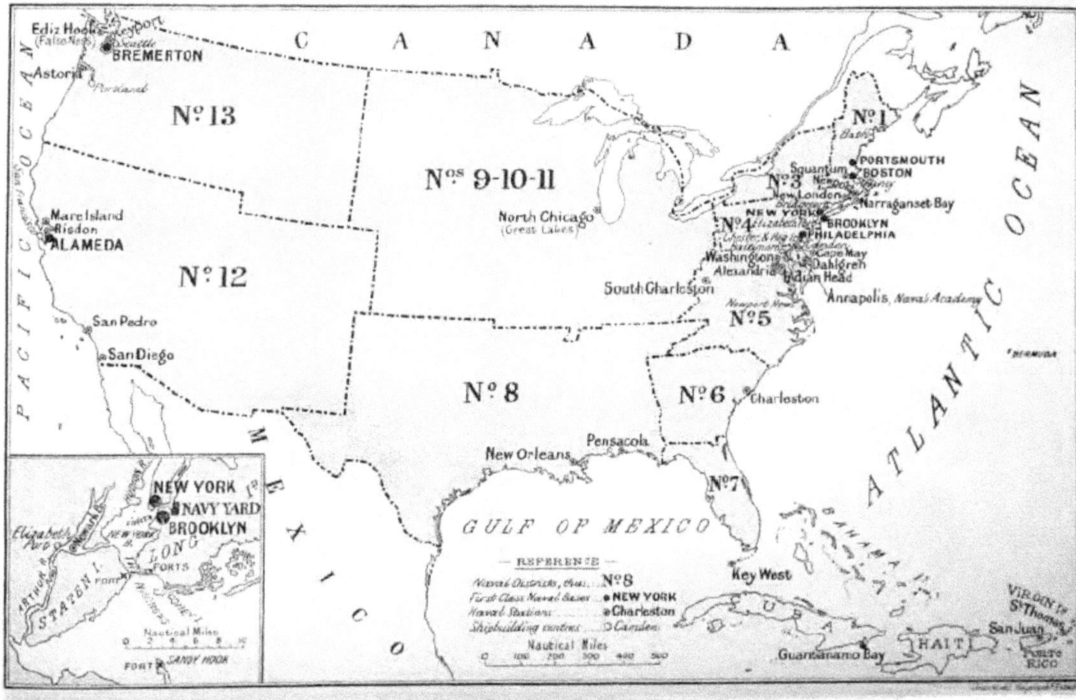

GENERAL MAP, showing Division into Naval Districts, Navy Yards, Naval Stations, &c. (Inset: Key Map, New York District.)

ATLANTIC.

NEW YORK

to	miles
Bermuda	680
Boston (Mass.)	315
Brest	2954
Cherbourg	3066
Colon	1985
Galveston	1910
Gibraltar	3370
Havana	1320
Havre	3125
Jamaica	1495
Key West	1180
La Guayra	1840
Liverpool	3062
New Orleans	1710
Norfolk Navy Yard	300
Pernambuco	3696
Philadelphia (Pa.)	240
Portsmouth Navy Yard	330
Portsmouth (England)	3060
Wilhelmshaven	3570

NEW YORK (via Panama Canal)

to	miles
Callao	3358
Colon	1981
Hong Kong	11,684
Iquique	3857
Manila	11,585
Puget Sound	6074
San Francisco	5299
Shanghai	10,885
Sydney	9611
Valparaiso	4630
Yokohama	9835

KEY WEST

to	miles
Bermuda	1100
Boston (Mass.)	1387
Galveston	730
Jamaica (Port Royal Harbour)	850
New York	1180
Pensacola	235
Philadelphia	1020
Portsmouth Navy Yard	1400

HAMPTON ROADS NAVY OPERATING BASE (measured from Norfolk Navy Yard)

to	miles
Halifax, N.S.	774
Portsmouth (N.H.)	548
Boston	290
Philadelphia	252
Sandy Hook	272
Washington (D.C.)	187
Bermuda	667
St. Thomas, V.I.	1296
San Juan (Puerto Rico)	1966
Port-au-Prince (Hayti)	1178
Vera Cruz	1755
Port Royal (Jamaica)	1280
Havana	965
Colon (Canal Zone)	1779
Ponta Delgada (Azores)	2298
Cherbourg	3275
Gibraltar	3357
Falkland Islands	6108
San Diego (Cal., via Panama Canal)	4662
San Pedro (Cal., via Panama Canal)	4732

GULF.

GUANTANAMO BAY (Cuba)

to	miles
New York	1315
Portsmouth (N.H.)	1494
Hampton Roads	1129
Philadelphia	1331
Colon (Atlantic Entrance Panama Canal)	686
Key West	587
Brest	3766
Plymouth (England)	3707
Gibraltar	3766
Buenos Ayres	5187
Rio de Janeiro	4087

PACIFIC AND ASIATIC.

SAN FRANCISCO

to	miles
Astoria (mouth of Columbia River)	562
Bremerton (Puget Sound)	800
Race Rock (Juan de Fuca Strait)	740
Esquimault	741
Guantanamo Bay (via Panama Canal)	3980
Hampton Roads (Norfolk, Va., via Panama Cnl.)	5069
Key West (via Panama Canal)	4344
Mare Island	22
Panama Canal (Pacific Entrance)	3240
San Diego	450
San Pedro	360
Tokuyama Bay (Japan)	4950
Valparaiso	5140
Hakodate (Japan)	4250
Nagasaki (Japan)	5033
Vladivostok	4554
Kiauchau (Kwantung Pen.)	5408

MARE ISLAND

to	miles
Auckland (N. Zealand)	5685
Esquimault	750
Honolulu	2095
Kiauchau	5600
Manila	7500
Panama	3421
Puget Sound	800
Yokosuka	4500

BREMERTON (Puget Sound)

to	miles
Ediz Hook	[illegible]
Astoria	[illegible]
San Diego	[illegible]
San Pedro	[illegible]
Pearl Harbour (Honolulu)	[illegible]
Pacific Entrance (Panama Canal)	[illegible]
Guantanamo Bay (via Panama Canal)	[illegible]

PEARL HARBOUR (Honolulu)

to	miles
New York (via Panama Canal)	[illegible]
Auckland (N.Z.)	[illegible]
Cavite (P.I.)	[illegible]
Esquimault	[illegible]
Guam	[illegible]
Panama Canal (Pacific Entrance)	[illegible]
San Francisco	[illegible]
Saigon	[illegible]
Singapore	[illegible]
Sydney	[illegible]
Yokosuka	[illegible]

KAVITE

to	miles
Amoy	[illegible]
Hong Kong	[illegible]
Kiau-chau	[illegible]
Nagasaki	[illegible]
Saigon	[illegible]
Singapore	[illegible]
Vladivostock	[illegible]
Yokosuka	[illegible]

FLAGS.

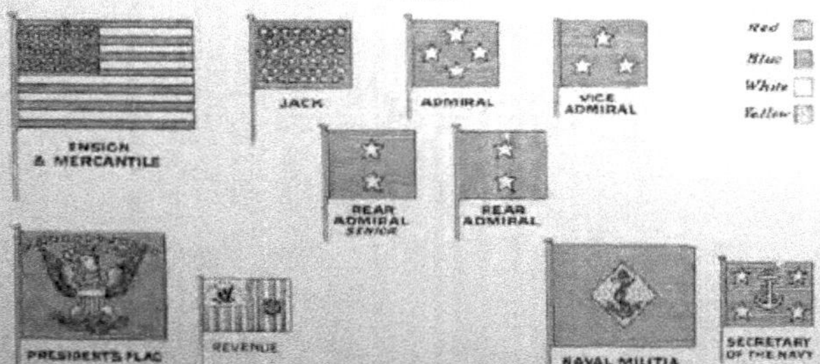

Senior officer when of or below rank of captain flies a blue triangular flag. Assistant-Secretary of Navy has a flag same as Secretary's, with colours reversed, i.e., white ground and blue anchor and stars.

Coinage.—Dollar (100 cents)—approximately 4s. 1½d. British (rate fluctuates).

Overseas Possessions.—Panama Canal Zone, Porto Rico, Virgin Islands, Alaska and Aleutian Islands, Guam, Tutuila, Wake and Johnston Islands, Philippine Islands.

Uniforms.

Admiral of the Navy. Rear-Admiral. Captain. Commander. Lieut.-Commander. Lieutenant. Ensign.

Note.—Lieutenants, junior grade, have 1½ stripes. Chief Warrant Officers one stripe broken with blue. Line Warrant Officers star without any stripe. Staff Warrant Officers under Chiefs have no sleeve mark. Engineers same as Line Officers (interchangeable). Other branches than executive wear no sleeve star, but have badge of branch over top stripe.

Mercantile Marine.

From "Bureau Veritas" 1919-20 figures. Ships under 100 tons gross excluded.

3327 steam and motor ships of 10,131,529 tons gross.

100 vessels (with auxiliary motors) of 125,335 tons gross.

1784 sailing vessels of 1,002,816 tons gross.

During 1919, U. S. shipyards launched 1651 ships of 4,075,385 tons gross; some of these for foreign account.

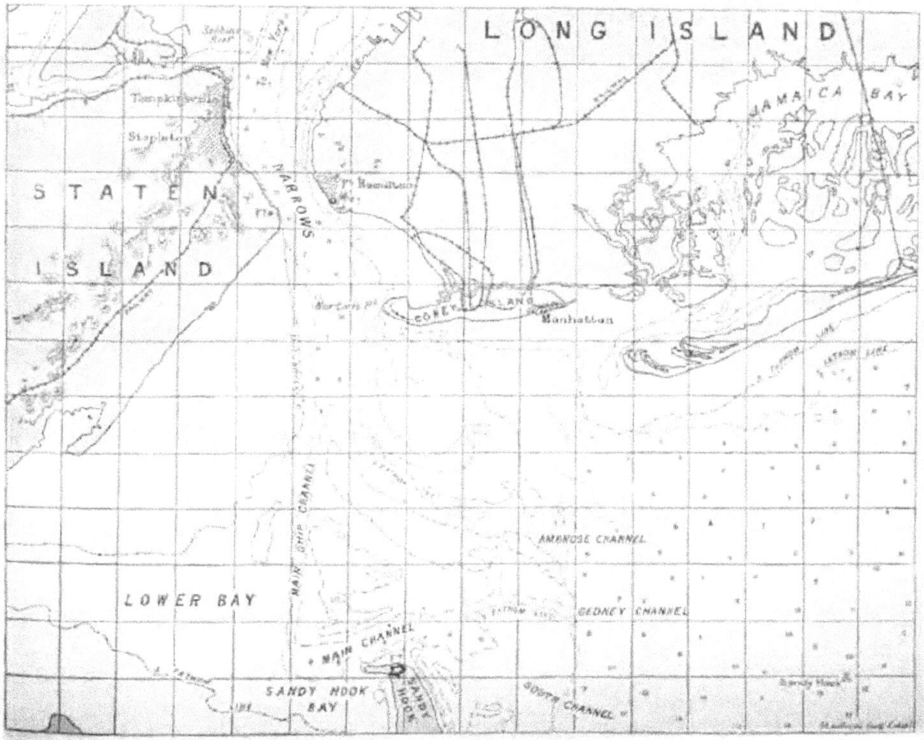

General Note to all Navy Yards Atlantic and Pacific Coasts.

In 1919 U.S. Shipping Board placed orders for 20,000 ton floating docks and eight 10,000 floating docks. When delivered, these wil assigned to the most suitable ports.

Navy Yards.

NEW YORK (& **BROOKLYN**).—3rd Na District. Depot and shipbuilding yard. No. 1 to be re-built. Slip No. 2 suitable building largest types of Dreadnoughts Battle-Cruisers built 1917-21. New pl also installed, workshops re-arranged renovated, new offices and general stores bu and magazine stores greatly enlarged, 1917-

Dry docks:

(1) granite, 330 × 66 × 25 feet.
(2) concrete, 440 × 90 × 29 feet.
(3) wood 613 × 105¼ × 29½ feet.
(4) granite and concrete, 700 × 12(35 feet.

Naval hospital here. Admiralty Chart 3204, 2491. Rise of Spring Tide, 4½ feet. Long Island: Naval Diving School, Bay Shore Rockaway Aviation Stations.

Note.—Private docks. The Robins D.D. C(Brooklyn has two timber docks: (1) 485 × × 20 feet; (2) 620 × 85 × 25 feet.

Private Floating Docks and Patent Slips.

Station	Docks No.	Docks Capacity—tons.	Slips No.	Slips Capac tons
BROOKLYN ...	12	1100—25000*	4	1000-
NEW YORK ...	4	2000—3500†		
HOBOKEN ...	5	1000—10000‡		
STATEN I. ...	4	1800—4200	2	1000-
JERSEY CITY...	3	1800—4200		
PERTH AMBOY.	3	1500—2500	1	10
WEEHAWKEN ...	1	1800		
NEWBURGH ...	...	...	1	10

* [illegible]

† [illegible]

‡ [illegible]

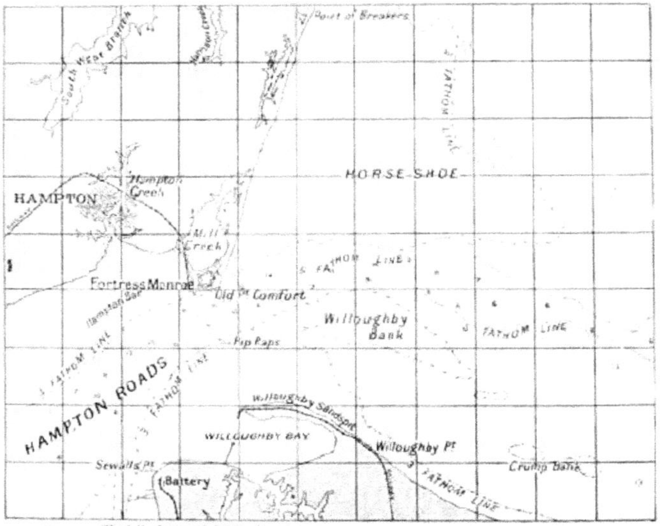

(Norfolk chart on left joins here.)

(HAMPTON ROADS AS ABOVE MAP.)

NORFOLK, Va.—5th Naval District. To be the dockyard section of the Hampton Roads Navy Operating Base. Depôt and shipbuilding yard. Naval hospital here. One or more slips for building Dreadnoughts or Battle-Cruisers begun 1917. (1) Wood dock, 460 × 85 × 25½ feet; (2) granite, 303 × 60 × 25½ feet; (3) granite dock, 550 × 112 × 34 feet (4). New concrete dock, 1022 ft. long × 110 × 42½ feet, divisible into 2 sections, about 650 and 350 feet long; has electric towing gear, 50-ton electric crane, and hydraulic lifts for rapid handling of repair materials, &c. Capable of being emptied in 90 mins.; floating pontoon crane 150 tons, and auxiliary 25 tons on hoist. Two smaller docks building 1920. In 1917 new foundry, workshops, and plant for making mines begun, new plant installed and old renovated. Three patent slips here each 1200 tons. Admiralty Chart No. 2818, 2843a.

HAMPTON ROADS NAVY OPERATING BASE (continuation to N. and N.W. of Norfolk Chart).—5th Naval District. Site on ground of Jamestown Exposition was purchased here in 1917, where it is intended that in conjunction with the Norfolk N.Y., a great Naval Base shall be established which will be developed into the principal warship port on the Atlantic coast. Plans as laid down during 1917 by the Navy Department, contemplated the following works: Submarine and Aviation Bases; Training Station for 10,000 men; Fuel Station (for coal and oil); Depots for fleet stores, mines, torpedoes and anti-submarine nets, &c.

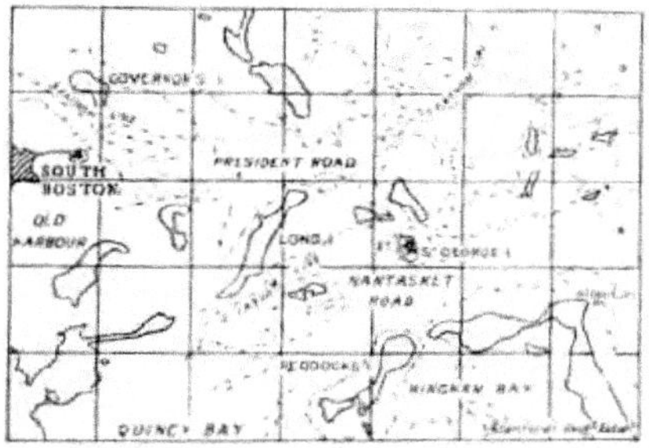

Admiralty Chart No. 1227,2482. Rise of Spring Tide, 5 feet (Charlestown Naval Yard).

League Island.
Admiralty Chart No. 2564.

Portsmouth N.H.
Admiralty Chart No. 2482,2487.
Rise of Spring Tide, 8¼ feet.

Navy Yards (Atlantic Coast).

BOSTON, Mass.—1st Naval District. Depôt. One granite dock, 389 × 46 × 26 feet; one granite and concrete dock, 729 × 101½ × 30½ feet. Commonwealth Dock (completed 1920), concrete, granite and rock, 1200 × 149 × — feet, divisible into 668 feet (inner) and 520 feet (outer) sections. Naval hospital here. Also two wooden private docks and four patent slips 1000—2500 tons.

SQUANTUM, Mass. Late Destroyer Plant operated by Bethlehem Co. Repair Base for T.B.D. and S/M. 10 building slips with electric bridge cranes and three wet slips. Maximum length of slips 310 ft. (p.p.) One 20-ton stiff leg derrick; one 10-ton guy derrick; wet slips have one 25-ton and two 10-ton bridge cranes. Yard covers 60 acres approximately. Water-front about 3600 feet.

LEAGUE ISLAND, PHILADELPHIA Pa.—4th Naval District. Depôt. Slips Nos. 2 and 3 can build Dreadnoughts or Battle Cruisers. New workshops, foundry and Marine Barracks, 1917-18. One wooden dock, 420 × 89 × 25½ feet; second dock, granite and concrete 707 × 104 × 30 feet. New dock, 1022 × 144 × 42½ feet, divisible into two sections, viz., 684 feet outer section and 338 feet inner section. Pier 1000 feet long, with 350-ton crane. Aircraft base and flying ground; also Government seaplane factory. Naval hospital here.

WASHINGTON. No docks. Yard devoted to ordnance and torpedo construction. Naval hospital here.

PORTSMOUTH, N.H.—1st Naval District. One granite and concrete, 720 × 101½ × 30½ feet. Naval hospital here.

CHARLESTON, S.C.—6th Naval District. 2nd class Navy yard. Dry dock, 503 × 113 × 34½ feet. 2800 ton patent slip finished 1920. Big new dock authorised and to be begun.

Naval Stations, &c.

ANNAPOLIS. Naval Academy.

NARRAGANSETT BAY, R.I.

NEWPORT, R.I.—2nd Naval District. Chief torpedo station. Manufactory of torpedoes, etc. Naval war college and apprentice-training station. Naval hospital and coal depôt.

CAPE MAY, N.J.—4th Naval District. Base for Submarines and Navy Airships. 310 acres bought 1919 for developing this Station.

NEW LONDON, Conn.—2nd Naval District. Submarine Training Station.

NORTH CHICAGO. Training Station for Great Lakes.

SOUTH CHARLESTON (W.Va.). U.S. naval ordnance plant, 207 acres site. Equipped for producing 20-inch shells and 20-inch armour plate. Building Mk. XII 6-inch guns, 1920.

ALEXANDRIA (Va.). U.S. Navy Torpedo Factory. Output, 8 torpedoes per day. Testing range on Potomac River near Piney Point. Employees, 1500.

Note.—Charleston, Norfolk, and Bradford, R.I., are stations for petrol and oil fuel.

PRIVATE DOCKS, &c.

(exclusive of those in New York and Brooklyn districts, previously listed).

NEW LONDON (Conn.), three patent slips, 1000–2000 tons. PHILADELPHIA, one floating dock (350 tons) and two patent slips 1000 and 2500 tons). CAMDEN (N.J.), three patent slips, 1200–150 tons. BALTIMORE (Md.), Wm. Skinner & Sons dry dock, 600 × 80 × 22½ feet. Columbian Iron works Dock (wood), 457 × 80 × 21 feet. Maryland Steel Co., Sparrow Point, wood and steel floating dock, 20,000 tons. Also one 3000 ton floating dock and two patent slips (2000 and 1200 tons). At Portland (Me.), Savannah (Ga.), Jacksonville (Fla.), small patent slips of 1200 tons (Jacksonville one 4500 ton floating dock).

U. S. PRIVATE SHIPYARDS.—ATLANTIC.

Warship builders only. Revised from details furnished by firms named, 1919.

Bath I.W.

BATH IRONWORKS (BATH, ME.). Build **battleships**, scout cruisers, destroyers and torpedo vessels, mercantile vessels of fast passenger types and large steam yachts of very fast design. Slips: one 600 ft., one 500 ft., three 350 ft. Cranes: one 100-ton stiff leg derrick (electric), one 35 ton floating sheer-legs (steam), three 15-ton electric locomotive cranes for steel runways. Yard covers 15 acres. Water-front 3000 feet. Plant up-to-date. Yard builds boilers, engines, equipment for ships, &c. Employees: 1200 (peace), 2000 (war).

Bethlehem Fore River Plant.

FORE RIVER PLANT, BETHLEHEM SHIPBUILDING CORPORATION (QUINCY, MASS.). 21 launching slips, 41 building slips, 20 building slips for submarines and four for 300 ft. destroyers. One 20-ton lift hammer-head crane, one 75-ton lift gantry crane, 16 overhead cranes over slips, capacity 5 and 10 tons and one 50-ton lift overhead crane. Yard covers 85 acres and water-front 1½ miles. Yard builds boilers, machinery and equipment. Plant thoroughly modern. Employees, 14,356.

Cramps.

MESSRS. THE WM. CRAMP & SONS SHIP & ENGINE-BUILDING CO. (PHILADELPHIA, PA.). Build dreadnoughts, battle cruisers, scout cruisers, destroyers and mercantile vessels of all types. 8 slips. Longest ship which can be built, 850 feet. One 100-ton floating derrick with 25-tons auxiliary lift; one 70-ton lift revolving hammer-head crane: numerous other cranes to lift 5–25 tons. Graving dock 433 ft. *over all*; width at entrance 69 ft. (top), 48 ft. (bottom); depth of water 20¼ ft. H.W.O.S. and 15½ ft. L.W.O.S. Patent slips: one 3000-ton and one 1000-ton railways. Area of yard 175 acres. Water-front 6300 ft. Yard builds boilers, machinery and naval equipment. Plant is up-to-date. Employees, 7000 (peace), 11,500 (war).

Bethlehem Harlan Plant.

(LATE HARLAN & HOLLINGSWORTH CO.)

HARLAN SHIPBUILDING PLANT (WILMINGTON, DELAWARE). 35 acres. Water frontage, 1800 feet. Depth at low tide, 22 feet. 5 slips, able to build up to 15,000 tons. Dry dock, 330 × 90 × 13½ feet. Equipment thoroughly up-to-date. Employees, 1800.

Note.—Above details are not up-to-date.

"Hollands."

ELECTRIC BOAT CO. (GROTON, CONN.). Submarines of the "Holland" type, which are chiefly built at the Bethlehem Fore River Plant.

Lake Co.

LAKE TORPEDO BOAT CO. (BRIDGEPORT, CONN.). Submarines of the "Lake" type.

Newport News.

NEWPORT NEWS SHIPBUILDING & DRY DOCK COMPANY (NEWPORT, VA.). 125 acres. Deep water alongside. Water-front, ¾ mile. One 140-ton revolving derrick, one 100-ton sheer-leg, one 50-ton floating derrick, one 10-ton ditto. Very large modern shops. Total employees 11,500. Eight big slips. Dry docks: (2) 804 × 80 × 30 feet; (3) 537½ × 79 × 24½ feet; (1) 593 × 50 × 24¼ feet.

New York S.B. Co.

NEW YORK SHIPBUILDING CO. (CAMDEN, N.J.). Can build dreadnoughts, cruisers, destroyers, mine-layers and sweepers, naval tenders and all classes of mercantile vessels, with boilers, machinery, and all equipment. 9 big slips, 4 smaller and 10 for destroyers. Yard can build ships up to 1000 ft. long, and covers 190.6 acres. Water-front about ⅘ths of a mile. Plant up-to-date. All machine tools and cranes electrically driven. Hydraulic and compressed air plant. Employees about 7000 (peace), 10,000 (war).

Todd, Brooklyn.

TODD SHIPYARDS CORPORATION (TEBO YACHT BASIN YARD, BROOKLYN, N.Y.). Three slips, 250 ft. long. Two 30-ton floating cranes. Two floating docks: (1) 105 × 22 ft. to lift 600 tons *d.w.* (2) 130 × 64 ft. to lift 1100 tons *d.w.* Yard covers 25 acres. Plant is modern. Employees, 1000. Associated firms:—Robin D.D. & Repair Co., Brooklyn, N.Y.; Tietjen & Lang D.D. Co., Hoboken (10,000-ton floating dock); Clinton Plant, Brooklyn; Quintard Iron Works, N.Y. (build engines, &c., for ships built by Todd Co.); White Fuel Oil Engineering Co., N.Y. Also see Pacific Coast for Tacoma and Seattle Yards.

Also a very large number of other shipyards solely engaged on mercantile construction.

U. S. A.—PANAMA CANAL ZONE.

Canal about 50 miles long. Channel 300–1000 feet wide at bottom; depth 41–85 feet. Time of passage about 10 hours.

PACIFIC ENTRANCE TO PANAMA CANAL. Heavily fortified. Graving Dock, No. 1, Balboa, 1000 × 110 × 41½ feet. Also fuel depot for 200,000 to 250,000 tons of coal at Balboa. 7—250 ton floating cranes. U.S. warships are stationed in the Canal Zone, where facilities exist for refitting ships.

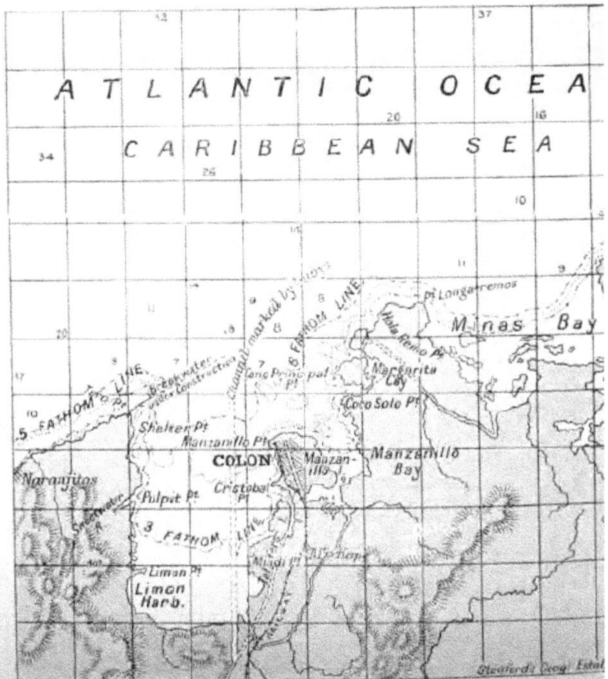

ATLANTIC ENTRANCE TO PANAMA CANAL. Heavily fortified. Dock and Fuel at Cristobal same size as Balboa (see opposite). In the Canal, Gatun, San Miguel and Mir Locks (over 1000 × 110 × 41½ feet) can be used as docks.

U.S. NAVY YARDS AND STATIONS—PACIFIC.

Charts to Uniform Scale. Divided into 2000 yards square. Soundings in fathoms.

Minor Naval Stations. [illegible]

EDIZ HOOK (or FALSE NESS). Recommended that a Naval Station be created here, to operate 18 torpedo boat destroyers and 12 submarines in peace time and double these numbers in war. Also Naval Air Station for coastal airships, K.B., flying boats, seaplanes, aeroplanes also recommended. Credits for commencing work not yet secured.

*KEYPORT (Bainbridge Island), Puget Sound). Sub-Base to Bremerton Navy Yard. Torpedo station. Recommended that a submarine station be located here, to operate 12 boats (peace), 24 (war).

ASTORIA (Columbia River, Oregon). At Tongue Pt. re-establishment of Naval Station for 12 submarines and 8 torpedo boat destroyers (double these numbers for war time) and aircraft recommended. Also anchorage recommended for 8 super-dreadnoughts. No work begun yet, but $250,000 voted to commence torpedo boat destroyer and submarine station, 1920.

* [illegible]

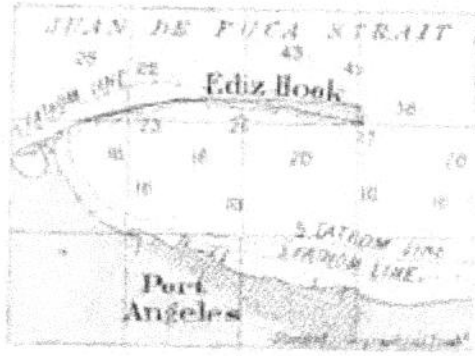

Ediz Hook. Admiralty Chart No. 1711.

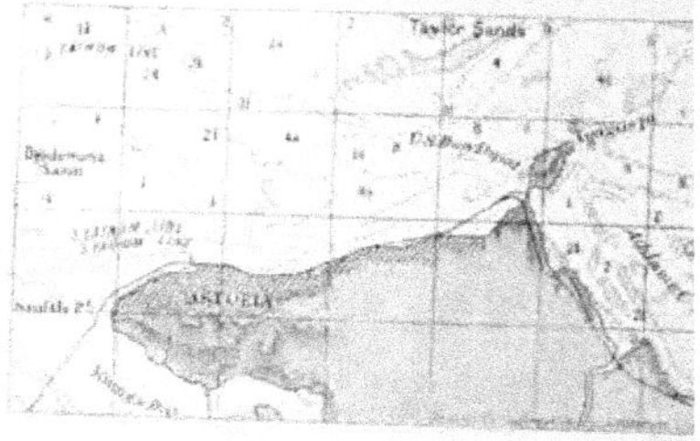

Astoria. Admiralty Chart No. 2839.

MARE ISLAND. (12th Naval District). Depot and Navy Yard (22 miles N. of San Francisco, of which small Map is given on next page). Slip to build Dreadnoughts of 600 feet length, or above. Destroyers also built.

Dry Docks: (1) Granite ...... 418 × 88 × 27½ feet
(2) Concrete ...... 687 × 102 × 31½ "

Mare Island. Admiralty Chart No. 2887.
(For Docks, see previous page.)

(HISDON), San Francisco. Temporary plant for building torpedo boat destroyers erected here during war and operated by Bethlehem Co. Believed to have reverted to control of Navy Department, but will probably be sold.

LAS ANIMAS, Cal. Naval Hospital.*

SAN DIEGO, Cal. About two million dollars voted 1920 for Marine Barracks, Naval Station, Fuel Depôt, Repair and Training Stations. Is Supply, Repair and Fuel Base for torpedo boat destroyers and submarines. North Island Air Station (1200 acres) is jointly used by U.S. Navy and Army Air Services. Very large projects, involving an outlay of 30½ million dollars have been prepared, to convert San Diego into a Fleet Operating Base, Fleet, Fuel, Supply and Repair Station, but no credits have been secured for this work.

SAN PEDRO, Cal. Station for Submarines exists here. No details known. (For Chart, v. next page.)

* A new and very large Central Naval Hospital has been projected at Balboa, near San Diego, but credits not allowed for this scheme yet.

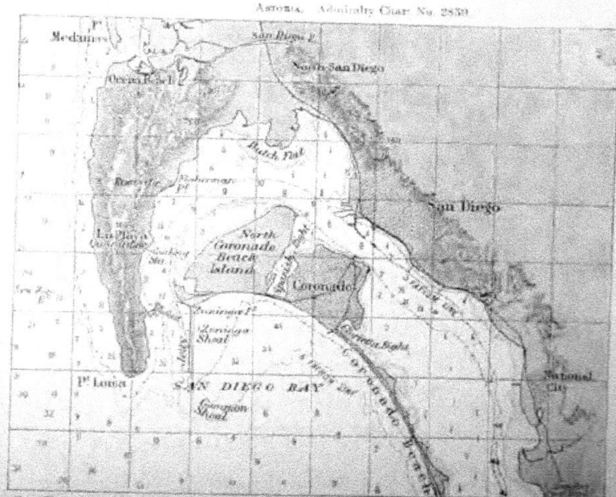

San Diego. Admiralty Chart No. 2885.

Principal Guns in the U. S. Fleet.

(Official Details of New Marks added, 1920.)

Built at Washington Gun Factory, proved at Indian Head, Dahlgren, Va., and Potomac Range.

Notation	Nominal Calibre.	Mark or Model.	Length in Calibres.	Weight of Gun.	Weight of A.P. Shot	Service Initial Velocity.	Maximum penetration firing capped A.P. direct impact against K.C. armour			Approximate Danger Space average ship, at			Muzzle Energy
							9000	6000	3000	10,000	5000	3000	
	inch.			tons.	lbs.	ft. secs.	in.	in.	in.	yards.	yards.	yards.	ft-tons.
A^8	16	I	45—50	105	2100	2800	...	...		...	...	...	98,406
A^7	14	IV	50	81	1400	2800	...	...		...	...	...	76,087
A^7	14	...	45	63½	1400	2600	18	...	...	...	...	...	...
A^6	13	I. & II	35	61¼	1130	2000	8	9½	12	110	310	550	31,333
A^6	12	VII	50	56·1	870	2950	11·0	13·9	17·5	...	...	...	52,483
A^6	12	VI	45	53·6	870	2850	10·6	13·3	16·6	...	...	...	48,984
A^5	12	V	45	52·9	870	2700	9·8	12·3	15·5	...	...	...	43,964
A^4	12	III & IV	40	52·1	870	2600	9·3	11·7	14·8	...	...	...	40,768
A^3	12	III & IV	40	52·1	870	2400	8·3	10·5	13·3	39†	126	264	34,738
A^2	12	I & II	35	45·3	870	2100	7·2	8·8	11·2	100	390	580	26,596
A^2	10	III	40	34·6	510	2700	6·9	9·0	11·9	140	400	700	25,772
...	10	I & II	30	25·1	510	2000	5·0	6·1	8·0	...	...	...	14,141
A	8	VI	45	18·7	260	2750	4·1	6·1	8·6	45†	...	...	13,630
B	8	V	40	18·1	260	2500	4·0	5·3	7·5	45	...	...	11,264
C	8	III & IV	35	13·1	200	2100	3·6	4·2	6·0	...	...	...	7,948
C	6	XII	53	10	105	3000	...	...	...	...	...	...	6,551
C	6	VIII	50	8·6	105	2800	2·3	3·2	5·2	...	...	...	5,707
D	6	VI	50	8·3	105	2600	2·2	2·9	4·7	75	250	470	4,920
D*	6	IX	45	7·0	105	2250	2·1	2·5	3·8	...	...		3,685
E*	6	IV, VII	40	6·0	105	2150	2·1	2·4	3·6	35	150	355	3,305
E*	5	VII	51	5·0	50	3150	1·4	1·8	3·4	...	...	...	3,439
	5	VI	50	4·6	50	3000	1·4	1·7	3·2	...	...	...	3,122
	5	V & VI	50	4·6	60	2700	1·6	2·0	3·5	...	...	...	3,032
F*	5	II, III, IV	40	3·1	50	2300	1·4	1·7	2·6	...	...	...	1,834
F*	4	VIII	50	2·9	33	2800	1·2	1·5	2·6	...	...	...	1,794
F*	4	VII	50	2·6	33	2500	1·2	1·4	2·2	...	...	...	1,430
F*	4	III, IV, V, VI	40	1·5	33	2000	...	1·2	1·7	...	...	..	915
F	3§	X	50	1·15	13	2700	...	...	...	...	...	...	657
F	3	V, VI, S-A	50	1·0	13	2700	...	0·8	1·2	...	..	...	658
F*	3	II, III	50	0·9	13	2700	...	0·8	1·2	...	...	...	658

* = Brass cartridge case. † = Calculated for a 20 ft. target.
Guns of 1899 and later have Vickers breech, etc. All guns use nitro-cellulose.
§ Anti-aircraft gun.

16 inch, 45 cal. in *Massachusetts* class, (6) *Constellation* class (6) *Maryland* class (4).

14 inch, 50 cal. mounted in *California* (2), and *New Mexico* (3) cl

14 inch, 45 cal. mounted in *Texas* (2), *Nevada* (2) and *Pennsylvan* classes.

12 inch, 50 cal. mounted in *Arkansas* (2) class.

12 inch, 45 cal. mounted in *Delaware* (2), *Utah* (2), *Louisiana* *Kansas* (4), and *S. Carolina* (2) classes.

12 inch, 40 cal. mounted in *Maine* (3), *New Jersey* (5), and *Oza* classes.

6 inch, 53 cal. in *Massachusetts* (6), *Constellation* (6), and Cruiser (10) classes.

For Notes on Ordnance, Torpedoes, &c., v. next page

Secretary of the Navy: Hon. J. Daniels.

Chief Constructor: Rear-Admiral D. W. Taylor.

Naval Attaché (London): Rear-Admiral A.P. Niblack, U.S.N.

Assistant Naval Attaché (London): Commander H. L. Pence, U.S

Personnel.

Estimated Establishment, 1920 .. 143,000 Officers and M

No secret is made of the fact that very serious deficiencies e in *personnel* at present. Early in 1920, the deficiency estimated at 3,765 officers and about 68,000 men. At pre new entries and enlistments fall far short of compensating retirements, discharges, desertions and officers or invalided out.

Shortage of officers has been assessed thus:—

For:—	Required:—	Available
Dreadnoughts	1500	900
Pre-Dreadnoughts	1300	440
Cruisers	1800	200
165 T.B.D.	1800	875
88 S/M.	1500	270
	8100	2685

a deficiency of 5415 officers, exclusive of those required T.B.D. and S/M. Tenders, Fleet Auxiliaries, &c. Within next three or four years 1500 officers must be available appointment to the Dreadnoughts, Battle Cruisers and Sc Cruisers now building or completing. During 1920, m ships had to be placed out of commission to provide neces ranks and ratings for the more important units of the Atla Pacific Fleets. When the Atlantic Fleet was sent to Gu tanamo Bay, 1920, for Fleet Exercises, a most exhaus overhaul of *personnel* was necessary to provide neces skilled ratings. Even so, the Fleet could only be manne a $\frac{7}{10}$ths basis and steam at 10 kts.

All Photographs by courtesy of U.S. Navy Recruiting Bureau and Lieut.-Commr. H. Reuterdahl, U.S.N.R.F., added 1920.

Tennessee class.

TENNESSEE. Completing. Details amidships, looking aft, showing part of 5 inch gun battery, AA. gun, director tower, SL. towers by after funnel and new type F.C. tops.

TENNESSEE. Completing. Detail View of new U.S. type of fore-bridges, also C.T., chart house, 5 inch director tower, forward cage mast.

New Mexico class.

NEW MEXICO. Aircraft View (broadside).

NEW MEXICO. Aircraft View (from starboard bow).

(Continued on next page.)

1919 & 1917. NEW U. S. DREADNOUGHTS (Battleships—*Single Calibre.*)

(MARYLAND CLASS—4 SHIPS.)

COLORADO*, MARYLAND** (March, 1920), ***WASHINGTON and ***WEST VIRGINIA.***

Normal displacement, 32,600 tons. *Full load*, 33,590 tons. Complement, 1410.

[illegible]

Guns (Dir. Con.):
8—16 inch, [illegible]
14—5 inch, 51 cal.
4—3 inch (A.A.) [illegible]
4—6 pdr. (saluting)
Torpedo tubes (21 inch):
2 *submerged*

Armour:
[illegible]

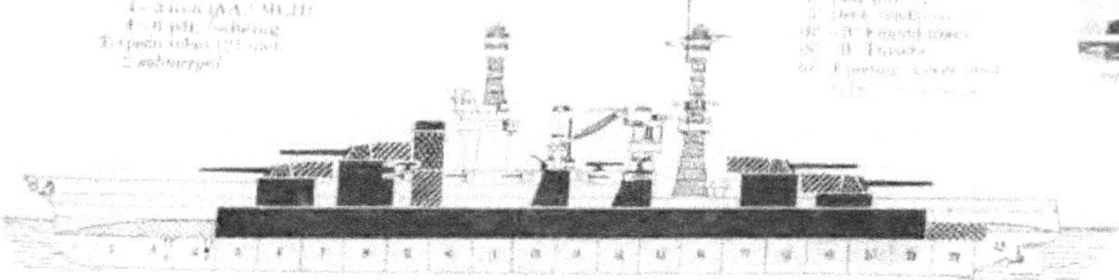

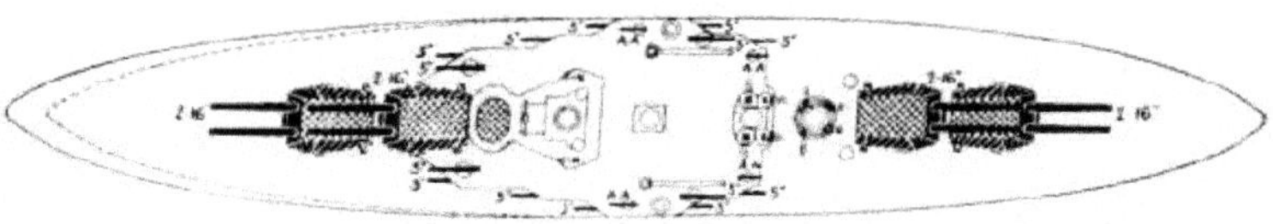

Ahead:
4—16 in.
8—5 in.

Broadside: 8—16 inch, 7—5 inch, 1—21 inch torpedo tube.

Astern:
4—16 in.
6—5 in.

Machinery: In *Colorado* and *Washington*, Westinghouse turbines and electric drive; in *Maryland* and *W. Virginia*, G.E. turbines with electrical drive. Designed S.H.P. 28,500 = 21 kts. 4 screws. Boilers: 8 Babcock & Wilcox. Fuel (oil only): *normal* tons; *maximum* tons.

Gunnery Notes.—16 inch are a new model, successfully proved at Indian Head, 1917.

Armour Notes.—Generally as those describing *California* class on next page and *New Mexico*, *Pennsylvania* and *Nevada* classes on subsequent pages.

Name	*Builder and Machinery*	*Laid down*	*Completed*	*Trials*	*Boilers*	*Best recent speed*
Colorado	New York S.B. Co.	May '19			Babcock	
Maryland	Newport News	Apl. '17			Babcock	
Washington	New York S.B. Co.	Jan., '19			Babcock	
W. Virginia	Newport News	Apl., '20			Babcock	

General Notes.—Authorised 1916 as *No.* 45 (*Colorado*), *No.* 46 (*Maryland*), *No.* 47 (*Washington*), and *No.* 48 (*W. Virginia*). *Maryland* and *W. Virginia* to be fitted with Division Flag Officers' Quarters. Except for change in primary armament, and slight increase in displacement, they are identical in nearly all respects with *California* class, described on next page.

Addenda Note.—Minor errors in areas of armour shown on plans, as notified for *Tennessee* Class on next page, also apply entirely to *Maryland* Class on this page.

Engineering Notes.—"Electric Drive" is identical with that for *California* Class (next page), but, in these ships, electi installation has been extended. Part of steam generated in boilers is diverted for running six auxiliary turb generators, supplying current to anchor gear, workshop lathes, refrigerating plant, bakeries, &c. Guns are al electrically manœuvred.

(CALIFORNIA class—2 ships).

CALIFORNIA (Nov., 1919) and **TENNESSEE** (April, 1919).

Normal displacement 32,300 tons. *Full load*, 33,190 tons. Complement, 1411 (max. 1458).

Length (*w.l.*), 600 feet. Beam, 97 feet 3½ inches. { *Mean* draught, 30 feet / *Max.* " 31 feet } Length (*over all*), 624 feet.

Guns (**Dir. Con.**):
- 12—14 inch, 50 cal., Mk. IV.
- 14—5 inch, 51 cal.
- 4—3 inch (anti-aircraft), Mk. III.
- 4—6 pdr. (saluting).
- 2—1 pdr.
- 2 M.G.
- 2 landing.

Torpedo tubes (21 inch):
- 2 *submerged*.

Armour:
- 14" Belt
- 8" Belt (aft)
- 2" Deck (ends)
- 15"—9" Funnel bases
- 18"—9" Turrets
- 16" Conning tower and tube

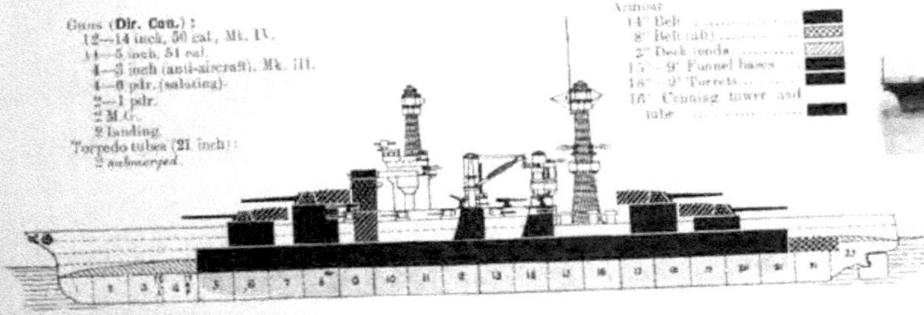

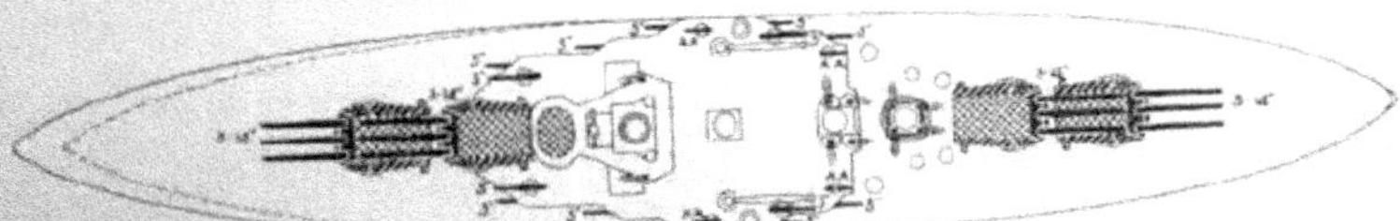

Broadside: 12—14 in., 7—5 in.; 1—21 in. torpedo tube.

Ahead:
6—14 in.
8—5 in.

Astern:
6—14 in.
6—5 in.

TENNESSEE. [illegible]

Detail Views given on earlier pages.

Machinery: *California*, G. E. turbines with electric drive. *Tennessee*, Westinghouse turbines with electric drive. 4 screws in both ships. Boilers: 8 Express Bureau type in *California*. 8 Babcock in *Tennessee*. Designed H.P. 28,500 = 21 kts. Fuel (oil only): *normal* 2200, *maximum* 3328 tons.

Gunnery Notes.—The 14 inch mountings are specially designed for large degree of elevation, up to 30°.

Armour Notes.—Generally as *New Mexico*, *Pennsylvania* and *Nevada* classes on subsequent pages.

Engineering Notes.—Generally as those relating to *New Mexico* on next page. Four alternating current motors, one in each propeller shaft, supplied by current at 3400 volts, from 2 turbo-generators. Motors are wound for 36 and 24-pole connections, operating on former as squirrel-cage machines, and on latter as wound-rotor machines. Oil for trials: 1467 tons and 187 tons feed water. Estimated weight of machinery: *California*, 1805 tons, *Tennessee*, 1983 tons. Heating surface: 50984 sq. ft. for Express Bureau type boilers in *California*; 41,768 sq. ft. + 4168 sq. ft. (superheated) for Babcock boilers in *Tennessee*.

Name	Builder	Machinery	Laid down	Completed	Refit	Trials	Boilers	Best recent speed
California Tennessee	Mare Island Y. New York Yard		Oct. '16 May '17	[illegible]			Express Babcock	

General Notes.—Authorised 1915 as Nos. 43 (Tennessee) and 44 (California). This above design is practically identical with New Mexico class on next page. The anti-torpedo protection in these ships has been most carefully studied, and is believed to be excellent. * Unofficial Notes. † Estimated date of completion.

Addenda Note: Just before going to press, blue print sketches of this Class (*Maryland* Class) were communicated by the Navy Department. Plans are accurate, with following exceptions:—

(*a*) Only fore half of C.T. base is armoured.

(*b*) Armoured uptakes to funnels do not rise above the forecastle deck level.

(*c*) The shallow 8" belt at stern extends farther aft to a little distance beyond the rudder pintle.

All above Notes also apply to *Maryland* Class on preceding page.

(PENNSYLVANIA CLASS—2 SHIPS).

PENNSYLVANIA (March, 1915) & **ARIZONA** (June, 1915).

Normal displacement, 31,400 tons. *Full load*, 32,567 tons. Complement* (*A*.) 1321, (*P*.) 1323 & 1435 as flagship.

Length (*waterline*), 600 feet. Beam, 97 feet. { *Mean* draught, $28\frac{3}{4}$ feet / *Max.* „ $29\frac{3}{4}$ „ } Length (*over all*), 608 feet (*p.p.* 596 ft.).

Guns (**Dir. Con.**):
12—14 inch, 45 cal.
14—5 inch, 51 cal.
4—3 in. (AA.), Mk. III.
4—3 pdr. (saluting)
2—1 pdr.
2 M.G.
2 landing
Torpedo tubes (21 inch):
2 *submerged*

* War:—
Penn. 1574. *Ariz.* 1620

Armour:
14" Belt (amidships)
8" Belt (aft)
3" Deck (ends)
15"—9" Funnel base
18"—9" Turrets
16" Conning tower & tube
(Total, 8072 tons.)

Ahead:
6—14 in.
2—5 in.

Astern:
6—14 in.

Broadside: 12—14 in., 7—5 in., 1—21 in. torpedo tube.

Machinery: *Pennsylvania*, Curtis (geared cruising) turbines. 4 screws. *Arizona*, Parsons, (geared cruising) turbines. 4 screws. Boilers: 12 Babcock. Designed H.P. (*Pennsylvania*), 31,500 (*Arizona*), 34,000=21 kts. Fuel: oil only, *normal* 2,322 tons (694,830 gallons).

* *Armour Notes.*—Generally as for *Nevada* class on next page. Increase of armour weight due to increased internal protection against submarine explosions and greater length of belt. Armour for each triple barbette, $226\frac{1}{2}$ tons. *Arizona* has cement backing to belt, instead of teak, and armoured fire-control tops.

* *Gunnery and Fire Control Notes.*—14 inch guns mounted in single sleeve, and can be fired as one piece. Maximum elevation enlarged to 30 in 1917. Triple positions weigh about 650 tons each (guns, mountings and armour). Hoists deliver 1 round per 40 secs., which may be improved to 2 rounds per minute, but three rounds fired per turret per minute probably represents actual rate of fire in service. Breech blocks worked by hand power. Interior of the shields to 14 inch guns very roomy and well arranged.

Each barbette has a built-in R.F. in rear of shield, ends projecting beyond sides and protected by armoured hoods.

Engineering Notes.—Oil fuel carried for trials, 1548 tons and 200 tons feed water. Heating surface, 51,322 sq. feet. Weight of Machinery: *Pennsylvania*, 2306 tons; *Arizona*, 2162 tons. 4—300 k.w. 125-volt turbo-generators in *Pennsylvania*: *Arizona* same, but 120-240 volts. In *Pennsylvania*, 240 r.p.m.=full power; in *Arizona*, 220 r.p.m. =full power.

Name	Builder	Machinery	Laid down	Completed	Trials Full Power 2 12 hrs.	Boilers	Best recent speed
Pennsylvania	Newport News	Newport News	Oct.'13	June '16	29,366 = 21·05	Babcock	
Arizona	New York Yard	New York Yard	Mar.'14	Oct. '16		Babcock	

General Notes.—*Pennsylvania* authorised 1912 as No. 38, *Arizona* 1913 as No. 39. Both ships are enlarged and improved *Nevadas*. They have proved excellent sea boats, very steady gun platforms, and have proved to be very economical ships. Their general design is marked by great simplicity and a very high standard of excellence. Living quarters are very roomy and well ventilated.

* Unofficial Notes.

ARIZONA (*Pennsylvania* same appearance).

Note polygonal tops and S.L. platforms high up masts—added 1919-20. This alteration will probabl[y] ... Dreadnoughts completed before 1920.

Aircraft View of "Pennsylvania." on earlier pages.

Note to Plans.—4—5 inch guns removed during War. On plans delete first four and last four ...

(Nevada Class—2 Ships).

OKLAHOMA (March, 1914) & NEVADA (July, 1914).

Normal displacement, 27,500 tons. *Full load*, 28,400 tons. Complement, 1350 (*N*), 1347 (*O*).

Length (*waterline*), 575 feet. Beam, 95½ feet. { *Mean* draught 28½ feet. / *Max.* „ 29⅔ „ } Length (*over all*), 583 feet.

Guns (Dir. Con.):
- 10—14 inch, 45 cal.
- 12—5 inch, 51 cal.
- 4—3 inch (AA.) Mk. III.
- 4—6 or 3 pdr. (saluting)
- 2—1 pdr.
- 2 machine.
- 1 landing.

Torpedo tubes (21 inch):
2 submerged

*War: Nevada, 1508, Oklahoma, 1628.

Armour:
- 13½" Belt (amidships)
- 8" Belt (aft)
- 13½" Bulkheads
- 13½" Funnel base
- 3" Deck
- 18"—9" Triple Turrets
- 16"—9" Double turrets
- 16" Conning tower and tube

(Total weight 7664 tons.)

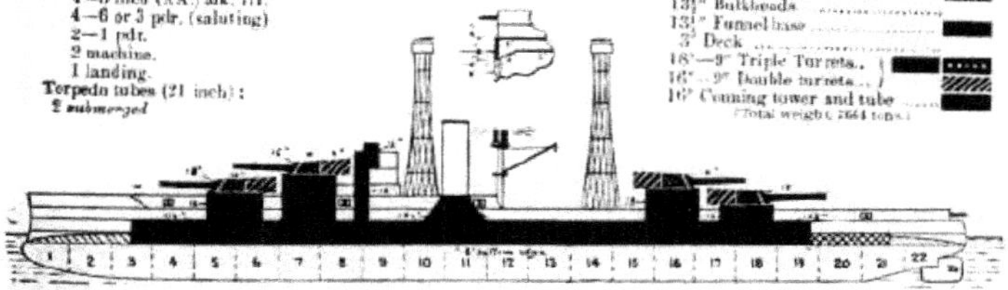

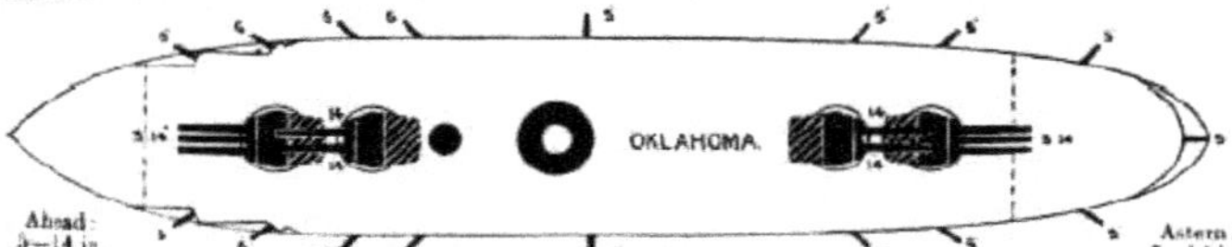

Ahead: 5—14 in. —5 in.

Astern: 5—14 in. —5 in.

Broadside: 10—14 in., 6—5 in., 1—21 in. torpedo tube.

Machinery: *Oklahoma*, Triple expansion; *Nevada*, Curtis (*geared cruising*) turbines. 2 screws in both. Boilers: *O*, 12 Babcock; *N*, 12 Yarrow. Designed H.P. 24,800 = 20·5 kts. Coal: none. Oil, 598,400 gallons (2000 tons), *maximum capacity*. Radius of action: 4000 miles at full speed, 10,000 miles at 10 kts.

*Armour *Notes.*—Main belt is 400 feet long by 17½ feet wide; 8½ feet of it being below *l.w.l.* Lower edge is 8". The ends are unarmoured; the battery also. Plates are applied in vertical strakes. Two protective decks, upper 3" flat, lower 1½" flat, 2" on slopes. Barbette bases are 13½" thick, but turrets are only 4½" where below protective deck and behind belt. Barbette shields: 18" port plate for triple positions, 16" port plate for twin positions 10" sides, 9" back, 5" roof. Sighting slits in conning tower closed by splinter-proof shutters. There is a signalling station protected by 16" armour behind conning tower. These ships mark a new era in naval construction, being the first to embody the "everything or nothing" idea in the matter of protection.

*Gunnery *Notes.*—Guns in the triple turrets in one sleeve, can be fired as one piece. Maximum elevation of 14inch guns enlarged to 30° in 1917.

Engineering Notes.—*Nevada* has 2 H.P. and 2 L.P. Curtis turbines. Cylinders of *Oklahoma* are H.P. 35", I.P. 59", L.P. (2) 78". Stroke, 48". Total heating surface, 48,000 sq. feet. Weight of machinery, *Nevada*, 1880 tons; *Oklahoma*, 1898 tons. Electric installation in both is 4 generating sets of 300 k.w., 125 volts, 2400 amp. each. Boilers are in 6 compartments and occupy less than 70 feet of length. Boilers are large tube "Express" type, those specially designed by Messrs. Babcock & Wilcox for *Oklahoma* proving most satisfactory. No superheaters. Electric-driven f.d. blowers proved too unreliable, and were replaced by steam turbine-driven blowers. All oil fuel carried in double bottom; no wing tanks.

Name	*Builder*	*Machinery*	*Laid down*	*Completed*	*Trials: Full Power—12 hrs.*	*Boilers*	*Best recent speed*
Oklahoma	N. York Shipb'g	N.Y. Shipb'ld'g	Oct. '12	May, '16	9,700 = 20·58	Babcock	22
Nevada	Fore River Co.	Fore River	Nov. '12	Mar. '16	25,312 = 20·53	Yarrow	

General Notes.—Authorised 1911 as *No.* 36 (*Nevada*), 37 (*Oklahoma*).

* Unofficial Notes.

Oklahoma. Now has Chart House as *Nevada* below. *Copyright Photo, O. W. Waterman.* (*Add*

Aircraft View on earlier pages.

Nevada. 1920 :

Note to Plans.—8—5 inch guns removed from upper and main decks during War and ports plated up. On pl
foremost and last three pair of 5 inch guns and 1—5 inch over stern.

(Texas Class—2 Ships.)

NEW YORK (Oct., 1912) & TEXAS (May, 1912).

Normal Displacement, 27,000 tons. *Full load*, 28,367 tons. Complement, 1462 (N.Y.), 1188 (T.) & 1561 as Flagship).

Length (*waterline*), 565 feet. Beam 95½ feet. { *Mean* draught, 28½ feet. / *Max.* " 29½ " } Length (*over all*), 573 feet.

Guns (**Dir. Con.**):
- 10—14 inch, 45 cal.
- 16—5 inch, 51 cal.
- 2—3 inch AA., Mk. III.
- 4—3 pdr. (saluting)
- 2—1 pdr.
- 2 M.G.
- 2 landing.

Torpedo tubes (21 inch):
- 4 *submerged.*

* War: N.Y., 1354 (Flag-ship, 1528); *Texas*, 1612

Armour (Midrale):
- 12" Belt (amidships) ....
- " Belt (bow) ....
- " Belt (aft) ....
- " Deck..................
- " Upper Belt ...
- 14"—8" Terrets
- " Barbettes..........
- " Battery ....

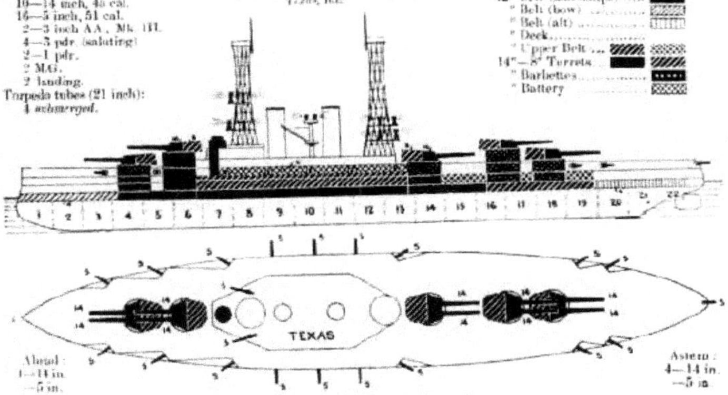

Broadside: 10—14 in., 8—5 in.; 2—21 in. torpedo tubes.

Machinery: Vertical triple expansion, 4 cylinder. 2 screws. Boilers: 14 Babcock (8 with superheaters). Designed H.P. 28,100=21 kts. Coal: *normal* tons; *maximum*, 2918 tons in *New York*; 2960 tons in *Texas*. Oil: 400 tons.

Engineering Notes.—Builders of turbine engines in the U.S. refused to adopt the standards laid down by the Navy Department. Accordingly, in these ships, reciprocating engines were reverted to, to show the turbine builders that the Navy Department was determined to have turbines built to official specification, or else the older type of engine would be taken up again. Cylinders: H.P. 39", I.P. 63", L.P. (2) 83". Stroke: 48". Grate area: 1554 sq. feet. Heating surface: 62513 + 2957 sq. feet (superheaters). Weight of machinery: 2375 tons. Electrical installation: 4 sets each of 300 k.w., 125 volts, 2400 amps., by General Electric Co.

Name	*Builder*	*Machinery*	*Laid down*	*Completed*	*Trials* * *Full Power*	*Boilers*	*Best recent speed*
Texas	Newport News	Newport News	Apl.,'11	Mar.,'14	28,373 = 21·05	Babcock	22
New York	New York Yard	New York Yard	Sept.,'11	Apl.,'14	29,687 = 21·47	Babcock	

* Both ships were light on trials—about 750 tons under normal displacement.

General Notes.—Authorised 1910 as *No. 34* (*N.Y.*) and *35* (*Texas*). Both ships fitted as flagships. First design for these ships is said to have included 15—12-inch guns in five triple turrets. Are very economical ships.

Note to Plans.—Delete (*a*) the foremost pair of 5 inch guns, (*b*) the fifth 5 inch gun from bows on each beam, (*c*) the 5 inch gun over stern, thus reducing 21—5 inch on plans to present armament of 16—5 inch.

Texas 19

Note that one ship (identity uncertain) is being refitted at Puget Sound N.Yd., 1920 with big octagon tops and S.L. platforms on masts above level of funnel caps (like *Pennsylvania* Class). This ship appears in background of *Pass. Ndro* illustration on U.S. Auxiliary Pages. Aircraft View given on earlier pages.

Texas 1

Note.—These ships are distinguishable from the *Arkansas* class owing to the funnels being smaller and nearer foremast; also only *three* turrets aft.

(Delaware Class—2 Ships).

DELAWARE (February, 1909), & NORTH DAKOTA (November, 1908).

Normal displacement, 20,000 tons. Full load, 22,060 tons. Complement, 1371 (*D*), 1476 (*N. D.*)

Length (*waterline*), 510 feet. Beam, 85¼ feet. { *Mean* draught, 27 feet. / *Max* „ 28⅓ „ } Length (*over all*), 518¾ feet.

* War: *Delaware*, 1384; *N. Dakota* 1284.

Guns (Dir. Con.):
- 10—12 inch, 45 cal.
- 2—5 inch, 51 cal.*
- 2—3 inch AA., Mk. III.
- 4—6 or 3 pdr. (saluting)
- 2 machine.
- 2 landing.

Torpedo tubes (21 inch).
- 2 *submerged*.

* *v. Gunnery Notes.*

Armour (Krupp and Midvale):
- 11" Belt (amidships)
- 3" Belt (ends)
- " Deck (slopes)
- 10" Upper belt (amidships)
- 12"-8" Turrets (N.C.)
- Turret bases (N.C.)
- 5" Battery (amidships) (N.C.)

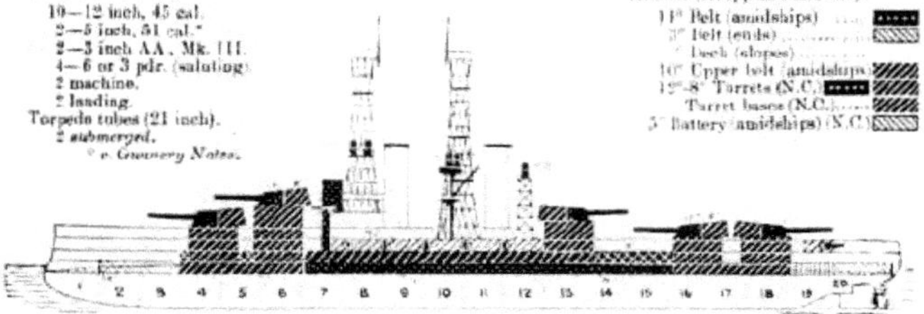

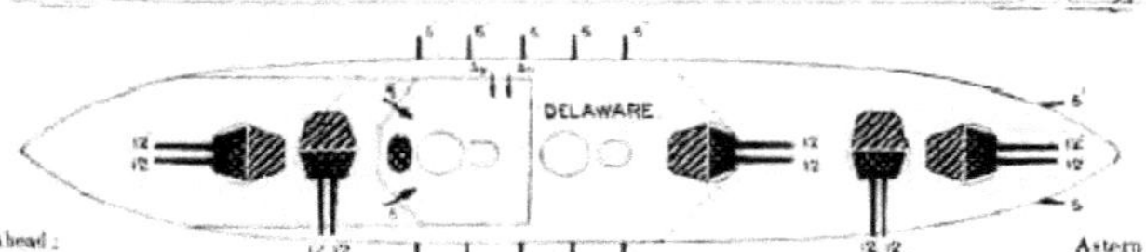

Ahead: 4—12 in. 2—5 in.

Astern: 4—12 in.

Broadside: 10—12 in., 1—5 in., 1—21 in. torpedo tube.

Machinery: *Delaware*, 2 sets triple expansion. 2 screws. *N. Dakota*, originally Curtis turbines but re-engined 1916-17 with Parsons turbines and reduction gear. 2 screws. Boilers: 14 Babcock. Designed H.P. 25,000 = 21 kts. Coal: *normal*, 1016 tons; *maximum*, 2732 tons in *Delaware*, 2740 tons in *N. Dakota*. Also 380 tons oil.

Gunnery Notes.—Delete all main deck 5-inch guns in plans (removed 1920).

Armour Notes.—Main belt 8 feet wide, 6⅔ feet of it below waterline at full load displacement, 11" thick amidships, upper belt, 7⅔ feet wide. Splinter-proof armour on uptakes and ventilators within citadel and in battery and upper deck.

Engineering Notes.—Cylinders of *Delaware* are 38½" (H.P.) 57" (I.P.) 2-76" (L.P.) × 48" stroke. Full speed, *Delaware* = 128 revs. *N. Dakota*, 263 r.p.m. for 21 kts., 229 for 19 kts., 142½ for 12 kts.* Trial consumption: 1·83 lbs. per H.P. all purposes. At 12 kts *Delaware* burned 100 tons per 24 hours, *N. Dakota* 140 tons.* Grate area: 1439 sq. feet. Heating surface: 61,943 sq. feet. Weight of machinery: *D.* 2036 tons, *N.D.* 2017 tons.* 4—300 k.w. 125-volt turbo-generators. *Delaware* was the first U.S. ship to have forced lubrication which proved so successful, it was afterwards fitted to all the older battleships with reciprocating engines.

Nominal Radii: About 6500 at 12 kts., 4600 at 18 kts., 3000 at 21·5 kts. for *N. Dakota*.* Somewhat more for *Delaware*.

Name.	*Builder.*	*Laid down*	*Completed*	*Trials Full power*	*Boilers*	*Best recent speed*
Delaware	Newport News	Nov., 1907	Feb., '10	—19·74 24,578 = 21·56*	Babcock	21·98
North Dakota	Fore River Co.	Dec., 1907	Apl., '10	— 31,300 = 21·01	Babcock	22·25

Delaware, after steaming 12000 miles, did 21·86 max. and averaged 21·32 for 24 hours.

Delaware. (Battery 5-inch guns dismounted.) 1920.

Aircraft view on earlier pages.

North Dakota (1919).

These ships have especially large boilers. The grates are 7ft. long by 14¼ft. wide = about 100 sq. ft

General Notes.—*N. Dakota* has a torpedo defence control platform under s.l. platform, abaft

Authorised 1906, as *No. 28* (*D*) and *29* (*N.D.*) Cost, without armament, £789,200. *N. D.* refit, 1916.

* These particulars relate to first trials of *North Dakota*, when originally engined with Curtis turbines. available of any trials run with new Parsons turbines and reduction gear.

(S. Carolina Class—2 Ships.)

SOUTH CAROLINA (1908) & MICHIGAN (May, 1908).

Normal displacement, 16,000 tons. *Full load* displacement, 17,617 tons. Complement, 1095* (M.) 1163 (S.)

Length (*waterline*), 450 feet. Beam, 80½ feet. { *Mean* draught, 24½ feet. / *Max.* „ 27 feet. } Length (*over all*), 452¾ feet.

Guns (**Dir. Con.**):
- 8—12 inch, 45 cal.
- 12—3 inch, 12 pdr.
- 2—3 inch AA., Mk. [illegible]
- 2—3 pdr.
- 2 M.G.
- 2 landing

Torpedo tubes (21 inch):
- 2 submerged.

(Total about [illegible] tons)

War Complements: Michigan 1151; S. Carolina 1354, flagship 1378.

Armour (S.C. Krupp; M. Midvale):
- 11"—9" Belt (amidships)
- 1½" Belt (ends)
- 3" Armour deck (slopes) (*see notes*)
- 10" Bulkheads
- 12"—8" Turrets (N.C.)
- 10"—8" Turret bases (N.C.)
- 10"—8" Lower deck (redoubt)
- 12" Conning tower (N.C.)
- 9" Tube
- (Total about [illegible] tons)

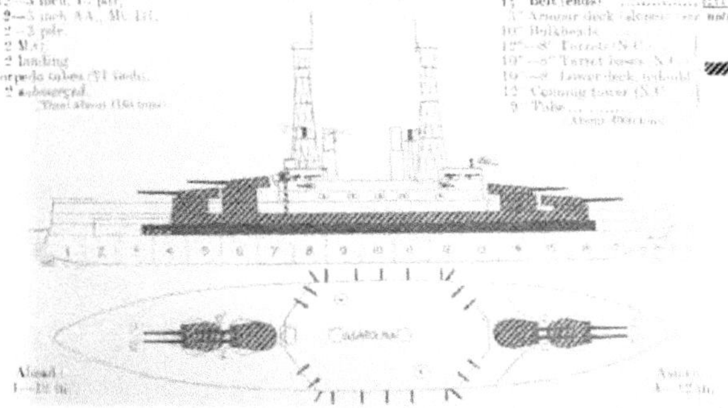

Broadside: 8—12 in., 6—3 in.; 1—21 in. torpedo tube.

Machinery: 2 sets vertical 4 cylinder triple expansion. 2 screws (outward turning). Boilers: 12 Babcock (1906 model) in 3 compartments. Designed H.P. 16,500 = 18·5 kts. Coal: *normal* 900 tons; *maximum* 2437 tons in *Michigan*, 2453 tons in *S. Carolina*.

Armour Notes.—Main belt, 8 feet wide; 6¾ feet of it below waterline; top and bottom, 9"; waterline, 11"—10½". Redoubt belt, 8" at upper edge, 10" lower. Low turrets with small bases. Fronts of turrets, 12"; sides, 8". Redoubt, 300 ft. long. Armour deck forward and amidships is 1½" on slopes, aft it is 3".

Gunnery Notes.—Hoists, electric. Big guns manœuvred electrically. Arcs of fire, 270° each turret. Fore turret guns 24 ft. above water, "after fore turret" guns 32 feet. Barbette of this rises 12 feet above the deck. Twelve hoists for 3 inch guns. *South Carolina* fitted with new ammunition hoists and latest type of fire-control system, 1917.

Engineering Notes.—Cylinders are 32" (H.P.), 52" (I.P.), 2—72" (L.P.) × 48" stroke. Pressure, 265 lbs. Revolutions full power = 121. Machinery, etc., weighs 1555 tons in *Michigan* and 1533 tons in *S. Carolina*. Grate area, 1050 sq. feet. Heating surface, 47,220 sq. feet. 4 sets 200 k.w. 125-volt turbo-generators. The first large U.S. ships to be fitted for burning oil fuel, though they do not appear to carry oil. Electric-driven f.d. blowers.

Name	*Builder*	*Laid down*	*Completed*	*Trials (mean)*	*Boilers*	*Best recent speed*
S. Carolina	Cramp	Dec., '06	Mar., '10	17,600 = 18·86	Babcock	19·68
Michigan	N.Y. Ship Bldg. Co.	Dec., '06	Jan., '10	13,253 = 17·95 16,016 = 18·79	Babcock	20·01

General Notes.—Authorised 1905 as Nos. 26 (*Michigan*) and 27 (*S. Carolina*). These ships, though laid down after, were projected *before* the British Dreadnought, and so may be considered as the first "Dreadnoughts" (*i.e.*, all big gun ships). They are good sea-boats, but roll heavily. In January, 1918, *Michigan's* lattice foremast fell during a gale. She has been re-equipped with a new lattice foremast of the latest type.

* Unofficial Notes.

Michigan.

Note to Plan.—10—3 inch, 14 pdr., either temporarily or permanently dismounted during War. They the upper deck battery.

San Juan, Puerto Rico.

Admiralty Chart No. 478,3408. Rise of Spring Tide, 1 foot.

NEW ORLEANS, La.—8th Naval District. Floating dock, 525 × 100 × 28½ feet.

PENSACOLA, Fla.—8th Naval District. 2nd class yard. 8th Naval District. Big New Aviation Station for seaplanes, dirigibles, kite-balloons, &c. Floating dock, 450 × 82 × 27 feet (12800 tons) ordered by Shipping Board to be stationed here, when ready.

KEY WEST.—7th Naval District. No docks. Submarine and Aircraft Station.

Small Naval Stations may also exist at San Juan (Porto Rico) and St. Thomas (Virgin Islands).

Private Docks.

NEW ORLEANS: Two floating docks, 2000 and 2500 tons.

MOBILE (Ala.): Pinto Dry Dock, 232 × 80 × 17 feet, and two floating docks, 1000 and 1400 tons.

GALVESTON: Dry Dock, 159 × 33.5 × feet, divisible into two sections.

Private Shipyards.

Various New Yards established 1917-18. Do not undertake Naval construction, all being devoted to mercantile contracts.

GUANTANAMO BAY (Cuba). Fleet Anchorage and Fortifying Grounds. Small repairs taken here. Fuel Depot. Admiralty Chart No. 904.

U. S. NAVY YARDS AND STATIONS—PACIFIC.

Divided into 2000 yard squares. Uniform scale. Soundings in fathoms. Heights in feet.

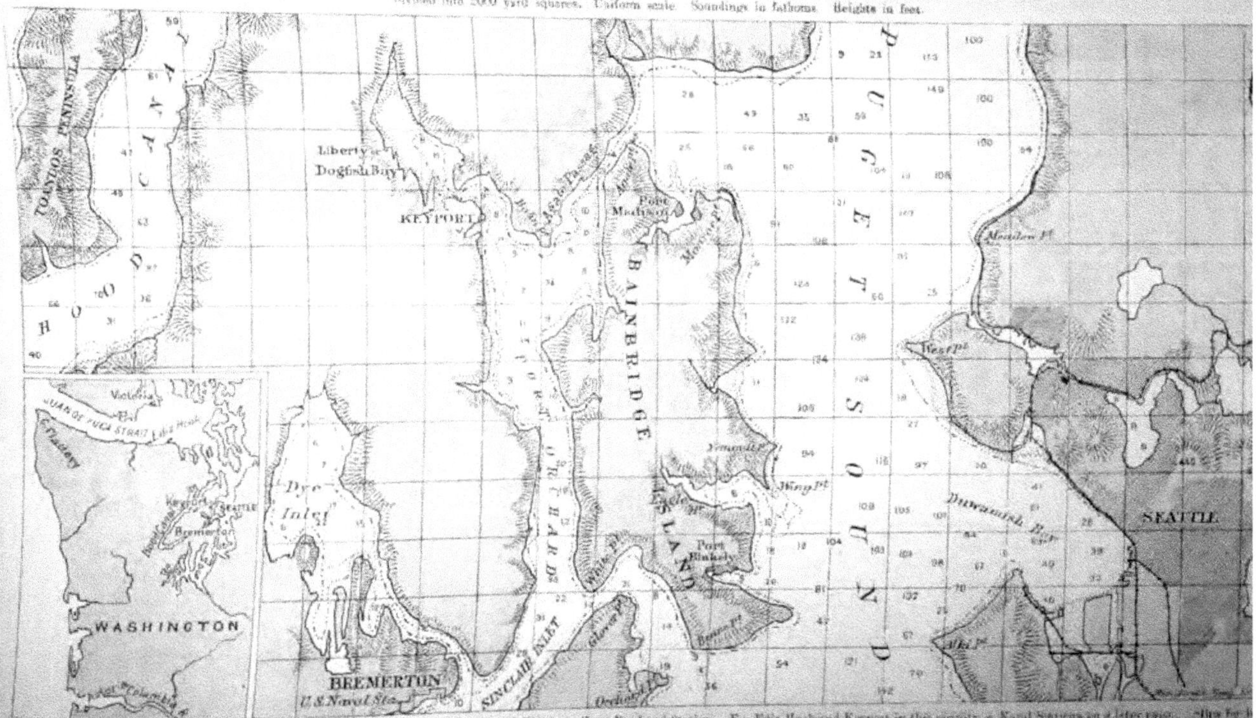

Sketch map of district.

BREMERTON, PUGET SOUND.—13th Naval District. Navy Yard and Station. For Eltz Hook and Keyport in this vicinity, v. Naval Stations on a later page. Slips for b[illegible] Scout Cruisers or Auxiliaries begun here during 1917. Dry docks: (1) Wood, 618×73×27¼ feet; (2) Granite and concrete, 801¼×113×35¾ feet; (3) Concrete, 827¼×124×2[illegible] Two slips, 440×48 feet. Recommended that the yard, &c., be developed into a 1st class Fleet Base, with two 1000 feet and two 472 feet graving docks, &c., at a cost of 44 million [illegible]

U. S. HARBOURS.—PACIFIC.

Divided into 2000 yard squares. Uniform scale. Soundings in fathoms. Heights in feet.

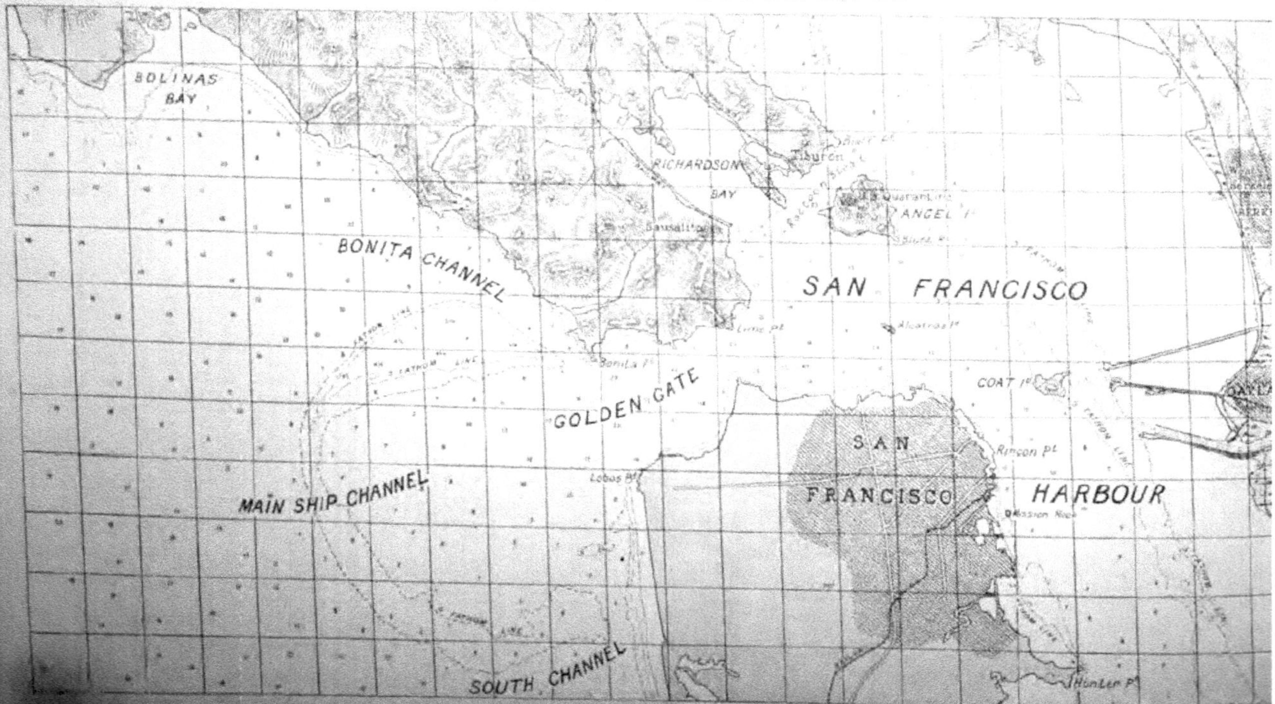

SAN FRANCISCO FLEET OPERATING BASE.—Site secured by free gift at ALAMEDA, within Bay and opposite Hunters Point. No details available, but cost estimated at fifty million dollars ... Recommended that South, Bonita and Main Ship Channels be dredged and protected by breakwaters to allow safe and rapid passage of Battle Squadrons [illegible]

Divided into 2000 yard squares. Uniform scale. Soundings in fathoms. Heights in feet.

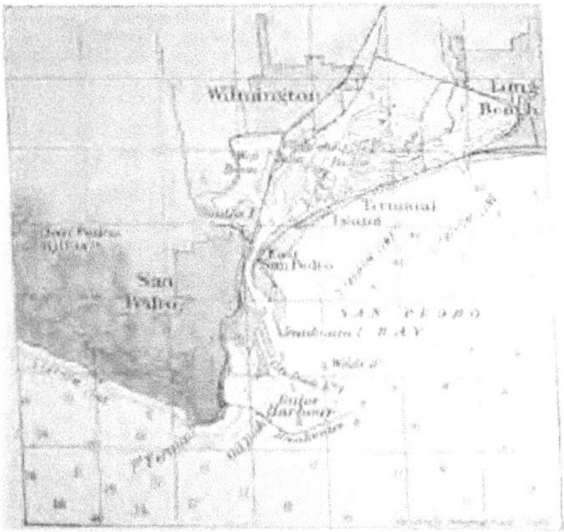

San Pedro (e. preceding page). Admiralty Chart No. 108.

Private Ship Yards.—PACIFIC COAST.

(Warship Builders only.)

Bethlehem Union Plant.

BETHLEHEM UNION PLANT (San Francisco).—38 acres. Three dry docks (Hunters Point): (1) 750 × 103 × 30 feet (28,000 tons), (2) 485 × 97 × 24 feet (10,000 tons), (3) 1110 × 110 × 41½ feet. Three small floating docks of 2700 and 1800 tons and a fourth of unknown capacity. Also a 12,000 ton floating dock bought from Ames S.B. Co., 1920. Six slips of 900 feet long, fully equipped with electric cranes, etc., and 4 smaller. Five wharves 585 × 50 feet, berthing for 15 average sized vessels. One sheer leg 100 tons, one 40 tons. 2 marine railways. All plant dates from 1910 or later. *Employees*: Average 2300. Enlarged 1917. Also yard for mercantile construction at Alameda.

California S.B. Co.

CALIFORNIA SHIPBUILDING CO. (Long Beach, Cal.). Submarine builders.

Todd Shipyards Corporation.

SEATTLE YARDS. Floating dock, (1) 325 × 100 (extreme) × 25 feet, for 6000 tons; (2) 465 × 85 × 27½ feet (12,000 tons); (3) floating dock for 2500 tons; one patent slip, 3000 tons.
TACOMA YARD. No details available. Is building Cruisers. 12,000 ton floating dock completing at end of 1918.

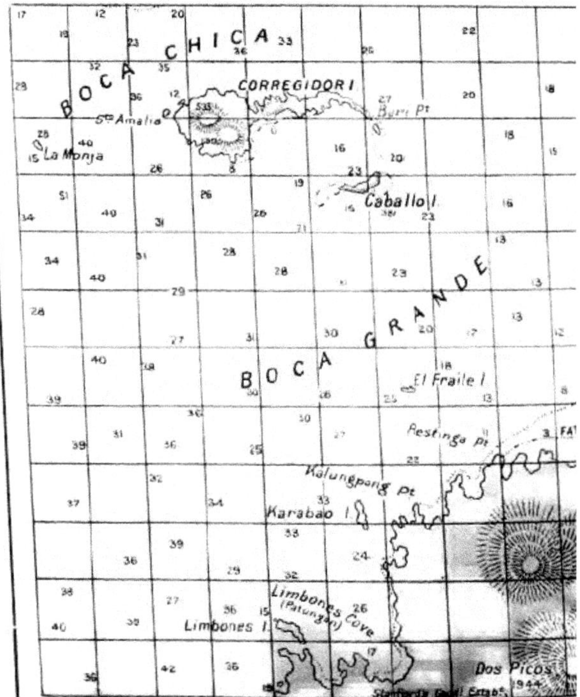

Entrance to Manila Bay. (Asiatic Station).

(Continued

Divided into 2000 yard squares. Uniform scale. Soundings in fathoms. Heights in feet.

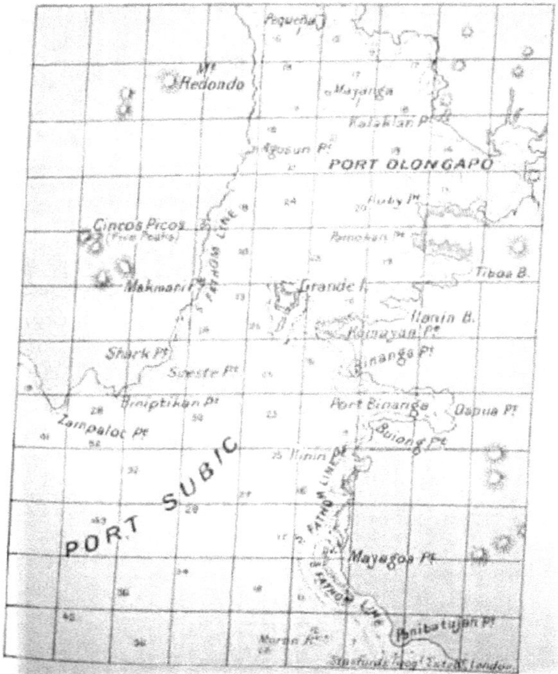

Olongapo & Port Subic.

Naval Stations.—2nd Class, Pacific.

PEARL HARBOUR (1) Honolulu. (2) Pearl Harbour, on S. side of Island of Oahu, 10-15 miles west of Honolulu. Large dock here, 1001 × 114 × 34¾ feet. Recommended that Harbour be developed into a 1st Class Fleet and Submarine Base, with extra dry docks and exten fuel and fleet stores. (Credits not secured for this work.)

TUTUILA (SAMOA).

Recommended that GUAM be developed as a Fuel and Repair Station for U. S. Naval and Mercan vessels bound for Far East. Credits not approved for this work yet.

Alaska.

Fuel and Supply Base recommended.

Asiatic Naval Stations.

CAVITE (P.I.). Small dry dock at Manila. Map of entrance on preceding page.

OLONGAPO (P.I.). Dewey Floating dock, 18,500 tons, 501' 9¾" × 100' × 37' 6".

Tutuila (Samoa).
Admiralty Chart No. 1730.

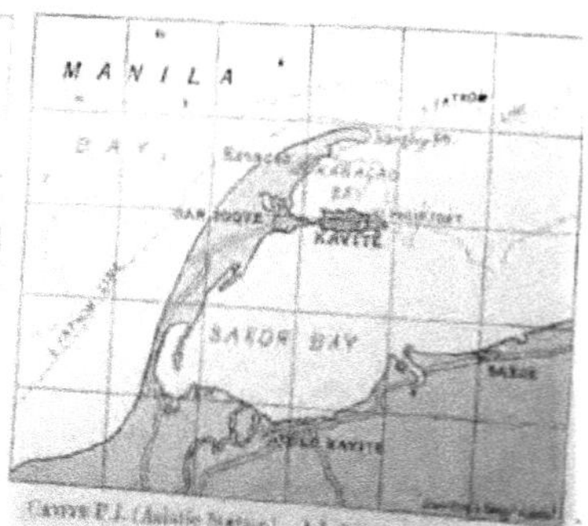

Cavite P.I. (Asiatic Station). Admiralty Chart No. 975,976.

Note.—Starred Paragraphs (•*•) are official data, furnished to "Fighting Ships" by courtesy of the Navy Department.

I.—Ordnance, &c.

DIRECTORS.—In all Battleships back to *New Jersey* Class, Battle-Cruisers, New Scout Cruisers. Type of Director not known. Vickers' Director System under trial, 1920, in T.B.D. *Hopewell* and *Thomas*. Range Clocks on masts and Deflection Scales on gunhouses, as in British Navy.

•*•*RANGE FINDERS.*—Made in various sizes from short 3-feet base range finders for navigation to 30-feet base range finders for use in turrets. Standard practice in U.S. Navy is now to mount one range finder in each turret.

•*•*HIGH EXPLOSIVES.*—T.N.T. standard high explosive for U.S. Navy. Some Amatol used. A new explosive, known as T.N.X., or Tri-Nitro-Xylol, almost identical with T.N.T. in properties, was developed during the War to eke out production of T.N.T.

•*•*NON-RICOCHET SHELLS.*—Flat nosed type; does not ricochet from surface when fired at elevation of over 2°. Made in 3 inch, 4 inch, 5 inch and 6 inch sizes for guns of equivalent bores. Delay action fuze commences on impact with water. Use of this shell converts gun into what is practically a long range depth charge thrower, the flat-nosed shell acting as depth charge.

•*•*STAR SHELLS.*—Range, three to six land miles. Made for 3 inch, 4 inch and 5 inch guns. Time fuze lights lamp and expels parachute and burner through base of shell. Normally, shell detonates at 1000 feet. Burner, 800,000 candle power, burning for 30 seconds, and illuminating sea for one mile diameter.

•*•*DAVIS NON-RECOIL GUN.*—Consists in effect of two guns, breech to breech. Recoil is absorbed by expulsion of powder charge case from rear gun. (In original mark, as experimented with, charge of buckshot set in vaseline was expelled from rear gun to neutralise recoil.) Made in 9 pdr. size, with non-richochet, A.P. and H.E. shells. Aimed by tracer bullets, a sub-calibre Lewis gun being mounted on chase of the Davis gun.

II.—Torpedoes.

•*•Torpedoes in U.S. Navy are all turbine-driven. Manufactured in 18 inch and 21 inch sizes, 21 inch for Battleships and Destroyers, and 18 inch and 21 inch for Submarines. (*No official details available of marks, ranges, speeds, charges and types of heater.*)

III. Mines.

•*•Made in various sizes and of constructions particularly adapted for work they are expected to perform. Anchor similar to British type.

Unofficial Notes.—Mk. IV type of Mine reported used for Harbour and Coastal Defence; no details known. The anti-submarine type of mine is reported to weigh 1400 lbs. and to have 300 lbs. T.N.T. charge. Body of mine, spherical, 3 feet diameter. Form of detonation uncertain; said to consist of antennae fitted with the Earle Patent Magnetic Pistol. Sinker is box-shaped and has wheels to gauge of laying ship's mine-discharging rails. Mine and sinker float together immediately after laying, while a 90 lb. plummet (attached to ¼ inch steel wire) sinks. The "plummet cord" is measured off to same depth at which mine is to float below surface. When plummet cord is spun full out, latch is unlocked and mine and sinker separate. Mine still floats on surface, while mine-mooring cable begins to spin off drum and sinker descends. Immediately the plummet touches bottom, the running-of mine-mooring cable is stopped and cable is locked. Sinker (or anchor) then comes d to bottom, drawing down the mine to the desired depth below the surface. Soluble saf plug used. Antennae said to consist of thin conducting wires and magnetic pistols, float out in a star pattern all round body of mine and giving a contact diameter of 50 feet.

IV.—Aircraft Bombs.

•*•(*a*) 18 lb. (modified 3 inch shell). Explosive charge, T.N.T.
(*b*) 20 lb. Cooper. Explosive charge, Amatol (to be replaced by T.N.T.)
(*c*) 163 lb.* Explosive charge, 117 lbs. T.N.T.
(*d*) 216 lb.* Explosive charge, 172 lbs. T.N.T.
(*e*) 230 lb. (British deisgn).* Explosive charge, T.N.T.
(*f*) 270 lb.* Explosive charge, 217 lbs. T.N.T.
(*g*) 520 lb. (British design).* Explosive charge, T.N.T.

Note.—Types marked * for Anti Submarine use. Excepting 230 lb. type, all fitted with U. Mark VI fuse. Mark VI fuse (and fuse of 230 lb. type) both armed by revolving of aeri propeller. Delay action feature consists of burning powder train.

V.—Anti-Submarine.

•*•*DEPTH CHARGES.* Mk. IV.—Charge 600 lbs. of T.N.T. Dimensions: 24 inch diamete 28 inch long. 100 feet effective radius, hydrostatic firing valve, similar to Mks. II and II Depth Charges.

•*•Depth Charge, Mks. II and III. Charge: 300 lbs. of T.N.T. Dimensions: 18 inch diameter 28 inch long. Effective within a radius of 70 feet from point of detonation. Limit o detonation adjustable between 36 and 300 feet. Safety device prevents detonation due t gun blast, etc., detonation above six fathoms, or when thrown overboard with firing devic in the off position. Detonation by hydrostatic pressure only. During War, destroyers carried from 30 to 50 depth charges and vessels on submarine patrols up to 100 depth charges.

•*•*Y-GUN OR DEPTH CHARGE PROJECTOR.*—Two barrels set at an angle of 90°, each barrel 3 feet long and 6 inch bore. 300 lb. depth charges fastened to arbors, inserted in barrels of Y and common powder charge exploded at junction of barrels; range of about 30 yards obtained.

HYDROPHONES.—K-tube fish type, 30 miles acoustic radius, but operating vessel must stop engines and auxiliaries while listening. SC- and MB-tube types, 3 miles acoustic radius; built into hull and insulated against noise; can be used without stopping ship.

(Also *v.* Non-Richochet Shells; Aircraft Bombs, Sweeping Gear.)

VI.—Sweeping Gears.

PV's as British Navy of H.S.S.S. and H.S.M.S. types, but with slight modification of inhaul gear. Methods of attachment also same as British Navy. During War, U.S. Sweepers also used the French Ronar'ch form of sweep (a combination of kites and "explosive knives"), which may have been adopted by U.S. Navy and improved.

VII.—Searchlights.

No details known, but in latest ships may be of controlled type, as in British Navy.

ONE FUNNEL.

MONTEREY.

MONADNOCK.

OZARK class (4 ships)
Variations inset—Centre "Tallahassee," Right "Tonopah.")

HELENA class (2 sh

GLOUCESTER.

SYLPH.

VIXEN.

YANKTON.

SCORPION

GOPHER.

ANNAPOLIS, VICKSBURG.

ALERT.

All Types on this page of Monitor and Yacht build.

MAYFLOWER
(Presidential Yacht).

GLACIER
(Supply Ship).

RAINBOW
(Submarine Tender).

BUSHNELL
(Submarine Depot Ship).

(Continued o

ONE FUNNEL—*continued.*

PATAPSCO *class.*

SONOMA & ONTARIO.

BIRD *class.*
(Mine Sweepers.)

WHEELING.

SACRAMENTO.

LEBANON.
(Target Repair Ship.)

GENERAL ALAVA.
(Transport.)

POMPEY.
(Supply Ship.)

LEONIDAS.
(T. B. D. Depot Ship.)

AJAX,* HANNIBAL.†
(Collier.) (Surveying Ship.)

BRUTUS.
(Collier.)

DOLPHIN.

PANTHER.
(Destroyer Depot Ship.)

BUFFALO, DIXIE & PRAIRIE.
(T. B. D. Depot Ships.)

SATURN.
(W/T. Repair Ship.)

CELTIC.
(Supply Ship.)

MELVILLE.
(T. B. D. Depot Ship.)

SAVANNAH.
(Submarine Depot Ship.)

PROMETHEUS & VESTAL.
(Repair Ships.)

FULTON.
(Submarine Depot Ship.)

BEAVER.
(Submarine Depot Ship.)

BLACK HAWK.
(T. B. D. Depot Ship.)

BRIDGEPORT.
(Repair Ship.)

(Continued on next page.)

ONE FUNNEL—*continued.*

BRIDGE.
(Supply Ship.)

HENDERSON.
(Transport.)

CULGOA.
(Supply Ship.)

VULCAN.
(Collier.)

RAPPAHANNOCK.
(Supply Ship.)

HANCOCK.
(Transport.)

BRAZOS, CUYAMA.
(Oil Tankers.)

Will be similar to
"Brazos," "Cuyama."

PYRO, NITRO
(Ammunition Ships.)

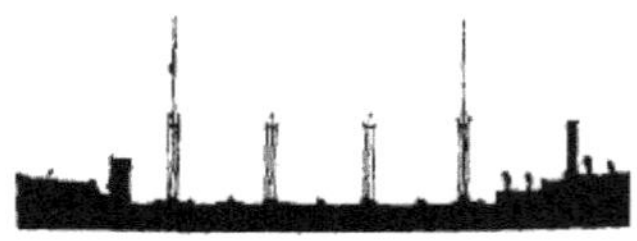

KANAWHA & MAUMEE.
(Oil Tankers.)

PROTEUS.
(Collier and Oiler.)

ONE FUNNEL—*continued.*

NEVADA, OKLAHOMA.

ARIZONA, PENNSYLVANIA.

NEW MEXICO class (3 ships).
* Variations in N.M. as inset.

TWO FUNNELS ABREAST.

WISCONSIN.

ALABAMA,* ILLINOIS.
* Variations as inset.

NEPTUNE, JASON, ORION.
(Colliers and Oilers.)

ONE "QUADRUPLE" FUNNEL.

MASSACHUSETTS *class* (6 ships).
Not yet completed.

Note: Adoption of cage or tripod masts not settled when drawing was made. Masts left blank.

TWO FUNNELS.

MONOCACY.
PALOS.
(River Gunboats)

PADUCAH *class* (2 ships).

NASHVILLE.

DENVER *class* (6 ships).

CAMDEN
(Submarine Depot Ship).

MARBLEHEAD.

ALBANY

NEW ORLEANS

BALTIMORE.
(Mine Layer).

SAN FRANCISCO.
(Mine Layer).

OLYMPIA.

CHICAGO.

MINNEAPOLIS.

AROOSTOOK.
SHAWMUT.
(Mine Layers).

COAST BATTLESHIP, No. 4.

COAST BATTLESHIPS, 1 & 2.

KENTUCKY.

(Continued on ne

TWO FUNNELS—*Continued.*

S. CAROLINA *class* (2 ships).

Half-cased funnels. Appearance distinctive,

DELAWARE *class* (2 ships).

Distinguish from "Utah" class by both cage masts being before funnels.

AA. guns in *Delaware* only, or in both ships may be mounted in derrick head tops, as inset.

UTAH *class* (2 ships).

TENNESSEE *class* (2 ships).

ARKANSAS *class* (2 ships).

TEXAS *class* (2 ships).

Can be distinguished by number and arrangement of after turrets, also by spacing of funnels and cage masts.

When completed, will be closely similar to *Tennessee* class.

MARYLAND *class* (4 ships).

CONSTELLATION *class* (6 ships).

Not yet completed.

Adoption of cage or tripod masts not decided when drawing was prepared, and masts were left blank.

THREE FUNNELS.

MAINE class (3 ships).
Low freeboard aft; big funnels.

NEW JERSEY class (5 ships).
[illegible]

CONNECTICUT.
[illegible]

BROOKLYN.
Flagship (D).

NEW HAMPSHIRE, VERMONT.
[illegible] variations in *Vermont.*

LOUISIANA, KANSAS, MINNESOTA.
[illegible]

ROCHESTER.
Flagship (D).

MONTANA, CAROLINA.
(SEATTLE is similar.)

FOUR FUNNELS.

COLUMBIA.

*CHARLESTON, ST. LOUIS.
*Variations in [illegible].

PITTSBURGH, PUEBLO.

FREDERICK, HUNTINGTON, HURON, as *Pittsburgh, Pueb*[illegible] (1) cage mast is close to first funnel and has ventilator shaft same; (2) no paired ventilat[illegible] before mainmast; (3) mainmast has only one [illegible]

Four Funnel Scout Types on next page.

FOUR FUNNELS.—(*continued*).

SALEM.

BIRMINGHAM, CHESTER*
(Variations as inset.)

New Scout Cruisers, Nos. 1—34.
Not yet completed.

Scale: 1 inch - 80 feet.

ANTI-SUBMARINE CRAFT.

SC boats.

EAGLE boats.

SUBMARINES. **Special Scale : 1 inch = 80 feet.**

D, E, F *classes* (7).

H & K *classes* (10).

L1—4, L9—11 (7).

M 1.

(Like *R21—27* but smaller, and no gun mounted).

N1—7.

O1—10.

R1—20.

R21—27.

S2, S14—17.
(In some boats, higher periscope housing, as shaded).

For Destroyer Types, v. next two pages.

ISABEL.

DOWNES.§

CUSHING, § ERICSSON.§
(Variation as inset.)

WAINWRIGHT.§

CONNER,* STOCKTON.*

BALCH.§

O'BRIEN.§

CONYNGHAM.§

DRAYTON *class* (9 boats).†

CASSIN.§

McDOUGAL.§

PORTER.§

DRAYTON *class* (8 boats).†

DUNCAN.§

NICHOLSON.§

ALLEN,§ DAVIS.§

DRAYTON *class* (4 boats).†
T.T. between funnels.
* Flush-deckers (3 funnels).
† Flivver types.

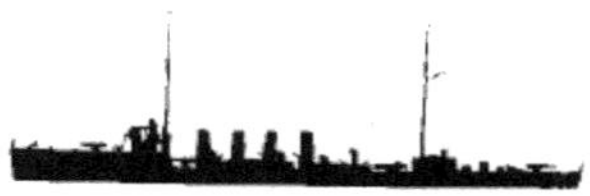

AYLWIN.§

§ "Thousand Tonner" Types.

SAMPSON,§ WILKES.§

Note.—All Silhouettes for "Four Funnel Flush Decker" Types. For "*Conner*," "*Stockton*," v. preceding page.

PREBLE.

All "Flush Deckers" will eventually be modified to above appearance.

As "*Preble*," but at present, after 4-inch gun i quarter deck and 3-inch AA. gun is on deck he *Brooks* class have 5-inch guns and twin SL. tower between boat skids.

A. P. UPSHUR, ALDEN, AULICK, BAILEY, BROOKS, B DECATUR, HULBERT types.

CALDWELL.

MANLEY.

WICKES.

NEW MEXICO *class*—(*continued*).

NEW MEXICO. Detail View.

Pennsylvania class.

PENNSYLVANIA. Aircraft View.

PENNSYLVANIA *class*—(*continued*).

PENNSYLVANIA. *Aircraft and Detail View, looking forward,* and showing new type of tops an SL. positions on cage masts (1920 alterations). Port 5 inch director tower appears at brea of deck levels; main director tower evidently abaft funnel. Note these details on No. gunhouse:—(*a*) aeroplane, (*b*) 3 inch AA. guns mounted in angles formed by wings an fuselage of aeroplane, (*c*) small-base R.F. protected by shield. In turrets Nos. 1, 2, 3, end of large-base R.F. will be seen projecting beyond and through sides of gunhouses—a R.F position normal to various U.S. Dreadnought types. It is said to be unsatisfactory, on account of "shimmer effect," created by confusing light rays being reflected off sides of gunhouses into lenses of the R.F.

Nevada class.

OKLAHOMA. Aircraft Detail View from bows to amidships. Port 5-inch director tower over No. 4 5-inch gun position. 3 inch AA. gun is after of two guns mounted on shelter deck abeam of C.T. (Part of photographing 'plane's wing cuts across upper right-hand corner of this photo.)

Texas

Aircraft V
starboard
New tops
positions
sylvania
and Det
are being
this Class

Arkansas class.

WYOMING. Aircraft View, full broadside.

(Continued on next

AIRCRAFT VIEWS OF U. S. DREADNOUGHT TYPES.

(Added 1920.)

ARKANSAS *class (continued).*

WYOMING. Aircraft View from starboard bow.

Utah class.

FLORIDA. Aircraft View from starboard bow. Note similarity of turret positions with *Delaware* Class below. Distinguish by different positions of masts and funnels. Also round tops to this Class; square tops in *Delaware* Class.

Delaware class.

NORTH DAKOTA. Aircraft View from starboard bow. See Notes to *Utah* Class above.

U.S. DREADNOUGHTS. (Battleships—*Single Calibre.*)

(MASSACHUSETTS CLASS—6 SHIPS.)

INDIANA, MASSACHUSETTS, MONTANA, NORTH CAROLINA, IOWA, SOUTH DAKOTA.

Normal displacement, 43,200 tons (*full load*, 45,000 tons). Complement, 1,415.

Length (w.l.), 660 feet. Beam, 105 feet. { *Mean* draught, 33 feet. / *Max.* } Length (o.a.), 684 feet.

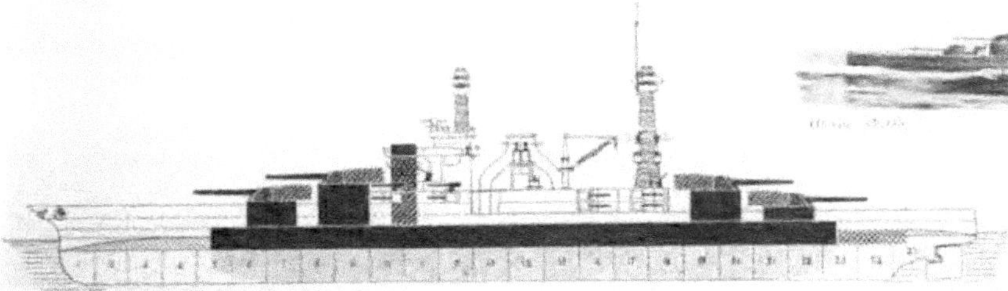

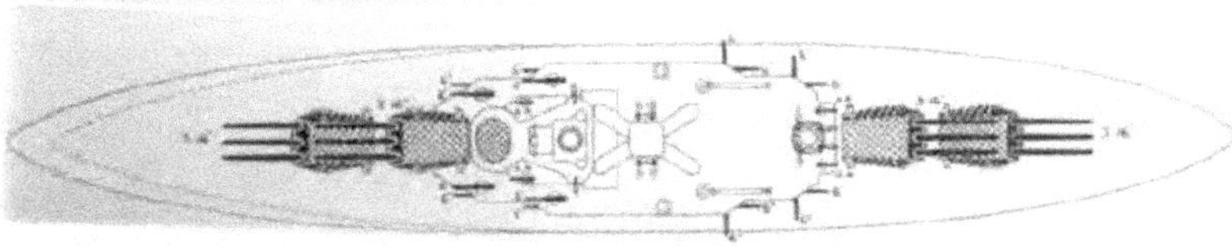

Guns (Dir. Con.):
- 12—16 inch, 50 cal. Mk. [illegible]
- 16—6 inch, 53 cal. Mk. XII.
- 4—3 inch, anti-aircraft Mk. III
- 4—6 pdr. (saluting)

Torpedo tubes (21 inch):
- 2 submerged.

Armour:
- 16" Belt
- 18"—9" Barbettes
- Anti-torpedo pro.—Bulges

Machinery: Turbines, with electric drive (as Table). 4 screws. Boilers (as Table). Designed H.P. 60,000 = 23 kts. Fuel (oil only): *normal*, tons; *maximum*, tons.

Name	*Builder*	*Machinery*	*Laid down*	*Completed*	*Last trials*	*Turbines*	*Boilers*
Indiana	New York N.Yd.		1920	Jan. '23*		Westinghouse	12 White-Forster
Massachusetts	Bethlehem, (Quincy)	Bethlehem, (Quincy)	1921?			G. E.	12
Montana	Mare Isl. N.Yd.	—	1920			Westinghouse	12 Bureau Express
N. Carolina	Norfolk, N.Yd.	—	1920			do.	12
Iowa	Newport News	Newport News	May, '20			G. E.	12 White-Forster
S. Dakota	New York N.Yd.	—	Mar., '20	Dec, '22*		Westinghouse	12

* Contract Dates.

Engineering Notes.—Westinghouse turbines are complete expansion, impulse-reaction type, developing fu[ll power at] 265 lbs. steam pressure, 50° superheat and 28½" vacuum. Electric drive consists of 2 turbo-alternat[ors of] 28,000 KVA (2-phase, 5000 volts, 1800 r.p.m.), serving one induction motor on each of the four shafts. [The] motors each develop 15,000 H.P. at 223 r.p.m. with 16 poles. For fractional speeds, motors are arranged [...] giving each 3000 H.P. at 150 r.p.m. Estimated weight of propelling machinery: 2778 tons. Heatin[g surface] (a) for White-Forster type boilers, 82,800 sq. ft. + 8940 sq. ft. superheating surface; (b) for Bureau Exp[ress] 70,040 sq. ft. + 12,192 sq. ft. superheating surface.

General Notes.—Authorized 1917: *Nos.* 49 (*S. Dakota*), 50 (*Indiana*), 51 (*Montana*). Authorized 1918: [52 (*N.*] *Carolina*), 53 (*Iowa*), 54 (*Massachusetts*). None laid down by June 1st, 1919. Estimated cost, per sh[ip, without] guns or armour, 15½ million dollars, *before* war; revised estimate on same basis *after* war, 21 million dol[lars per] ship. The hideous "quadruple" funnel was devised to keep smoke away from fire-control tops.

(New Mexico *class*—3 ships.)

NEW MEXICO (April, 1917), **IDAHO** (June, 1917) and **MISSISSIPPI** (January 1917).

Normal displacement, 32,000 tons. *Full load*, 33,000 tons. Complement (*N.M.*) 1283, (*I.*) 1260, (*M.*) 1310.

Length (*waterline*), 600 feet. Beam, 97¼ feet 4½ in. { *Mean* draught, 30 feet. *Max.* " 31 " } Length *over all*, 624 feet.

Guns (**Dir. Con.**):
12—14 inch, 50 cal., Mk. IV
14—5 inch, 51 cal.
4—3 inch (AA.), Mk. III
4—6 or 3 pdr. (saluting)
2—1 pdr.
2 machine
1 landing
Torpedo tubes (21 inch):
2 *submerged*

Armour:
14" Belt (amidships)
8" Belt (aft)
" Deck ends
15"—9" Funnel base
18"—9" Turrets
16" Conning tower & tube

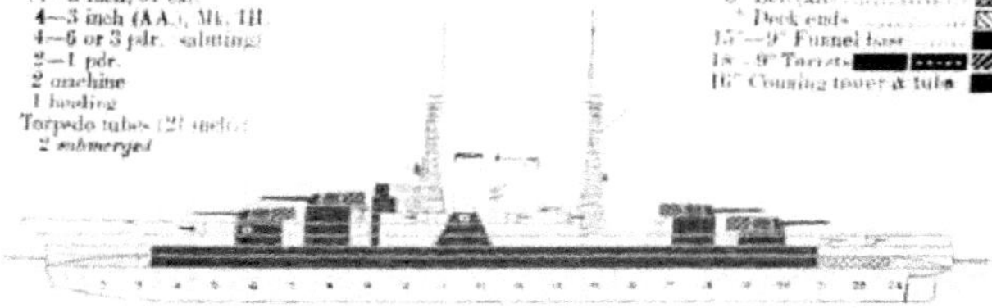

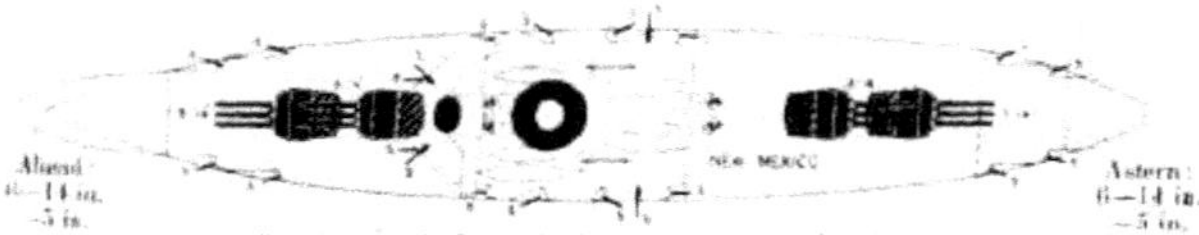

Ahead: 6—14 in. —5 in.

Astern: 6—14 in. —5 in.

Broadside: 12—14 in., 7—5 in., 1—21 in. torpedo tube.

Machinery: *New Mexico*: G.E. turbines with electric drive (see *Notes*). *Idaho*: Parsons 4-shaft (geared cruising) turbines. *Mississippi*: Curtis 4-shaft (geared cruising) turbines. 4 screws in all three ships. Boilers: 9 Babcock & Wilcox. Designed H.P., *New Mexico* 27,500, others 32,000 = 21 knots. Fuel: Oil only, 1467 tons.

*__Armour and Gunnery Notes.__—Generally as **Pennsylvania** and **Nevada** classes, but have still greater degree of internal protection. It is unofficially reported that they have *triple* bottoms and double longitudinal bulkheads, the space between outer bulkhead and inner skin being filled up with shock-absorbing materials. As designed, it was intended they should mount 45 calibre 14 inch guns, but while building, the more powerful 50 calibre model of the 14 inch gun was adopted.

Engineering Notes.—***New Mexico*** has 2 turbine-driven generating units and 4 propelling motors. Generators are bipolar alternators. For economical cruising speeds of 15—18 knots and under, one generator is used on 36-pole connection. For higher speeds, 24-pole connection will be employed. One generator can drive ship at 19 kts. in emergency. 170 propeller r.p.m.—full speed. Estimated weight of machinery: ***New Mexico*** 2251 tons, *Idaho* 2285 tons, *Mississippi* 2298 tons. Estimated heating surface, 53,158 sq. feet + 4176 feet superheaters in *New Mexico*. 280 lbs. pressure at steam chest. Electric installation, 6—300 k.w., 120—240 volt turbo-generators in ***New Mexico***, 4 in other ships. Oil on trials 1467 tons and 299 feed water.

Name	*Built at*	*Machinery*	*Laid down*	*Completed*	*Trials (hrs)*	*Boilers*	*Best recent speed*
New Mexico	New York Yard	New York Yard	Oct., '15	May, '18	31,197 = 21·08	Babcock	21
Idaho	N. York S.B. Co.	N. York S.B. Co.	Jan., '15	Mar., '19		Babcock	21·32
Mississippi	Newport News	Newport News	Apr., '15	Dec., '17		Babcock	

General Notes.—Authorised 1914 as No. 40 (*New Mexico*), 41 (*Mississippi*) and 42 (*Idaho*). ***New Mexico*** fitted with Division Flag Officer's quarters. Originally, only two ships were to have been built, but the sale of the old ***Idaho*** and ***Mississippi*** to Greece contributed two-thirds of the cost for a third unit. ***New Mexico*** was first named ***California***. The type is derived from the ***Nevada*** design through the ***Pennsylvania*** class, upon which all the above are improvements. The electrical transmission of ***New Mexico*** was adopted through the excellent results given by the Melville-Macalpine electric drive system in the Aircraft Carrier *Langley*.

*Unofficial Notes.

Idaho. 1919 Photo, New York Shipbu[illegible]

Aircraft and Detail Views on earlier pages.

Mississippi. 1919 Copyright photo, O. W. Water[illegible]

1920 *Note.*—Electric drive of *New Mexico* has proved entirely successful from all points of view. She has prov[illegible] economical at all speeds than other two ships, and has most remarkable manoeuvring power, since chan[illegible] ahead to astern can be made with the greatest facility. In 1920, 28,820 S.H.P. = 21, and 21,650 S.H.P. [illegible] kts. for 8 hours.

Note to Plans.—These are not uniform in scale with other plans in this book. Delete (*a*) the 4—5 inch guns bel[illegible] abeam of first barbette; (*b*) the 4—5 inch guns abaft fourth barbette, thus reducing the 22—5 inch sh[illegible] plans to the present armament of 14—5 inch.

(ARKANSAS CLASS—2 SHIPS.)

ARKANSAS (Jan., 1911), & WYOMING (May, 1911).

Normal displacement, 26,000 tons. *Full load*, 27,213 tons. Complements, 1481 (A) 1478 (W)*

Length (*waterline*), 554 feet. Beam, 93¼ feet. { *Mean* draught, 28½ feet. *Max.* " 29½ feet. } Length (*over all*), 562 feet.

As Flagships 1520 (A) 1517 (W).

Guns (**Dir. Con.**):
12—12 inch, 50 cal.
16—5 inch, 51 cal.
2—3 inch A.A., Mk. III
4—6 or 3 pdr. (saluting)
2—1 pdr.
2 machine
2 landing
Torpedo tubes (21 inch):
2 *submerged*.

Armour (Midvale):
11" Belt amidships
5" Belt (ends)
12"—9" Turrets
11" Turret bases
6½" Casemates (*see Notes*)
12" Conning tower

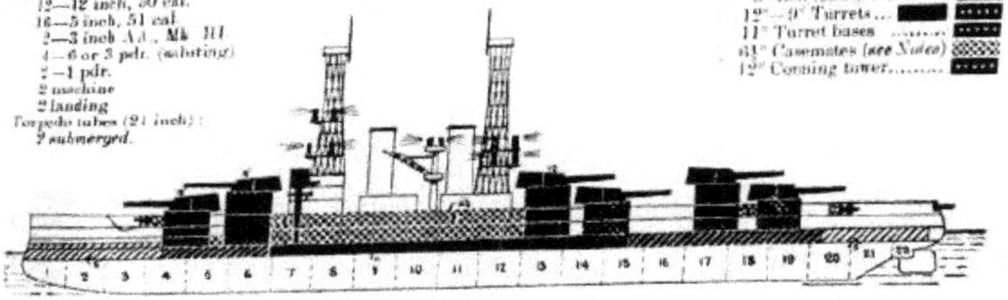

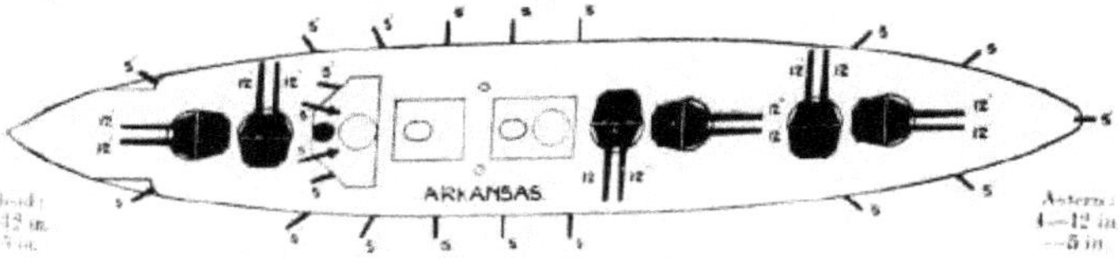

Ahead: 4—12 in. —5 in.

Astern: 4—12 in. —5 in.

Broadside: 12—12 in., 8—5 in., 1—21 in. torpedo tube.

Machinery: Parsons turbine. 4 screws. Boilers: 12 Babcock. Designed H.P. 28,000 = 20·5 kts. Coal: *normal* 1669 tons; *maximum*, 2754 tons in *Arkansas*; 2704 tons in *Wyoming*. Oil, 400 tons. Radius of action, about 8000 miles at 10 kts.

Gunnery Notes.—Height of guns above water: No. 1 turret, 28½ feet; No. 2, 36½ feet; No. 3, 33 feet; No. 4, 25 feet; No. 5, 24 feet; No. 6, 23½ feet.
Arcs of training: (1) 300°, (2) 270°, (3) 290°, (4) 260°, (5) 330°, (6) 300°.

Armour Notes.—Main belt 400 feet long, 9" on bottom edge, 11" on upper edge. Upper belt is same, but is 11" on bottom edge and 9" on upper. Internal protection is by 1½" high tensile longitudinal bulkheads. Two protective decks over all machinery and magazine spaces, [illegible] at ends. 6½" armour and 1½" splinter bulkheads to funnel uptakes, up to main deck.

Engineering Notes.—Approximately 148 r.p.m. = 10·5 kts., 215 = 15 kts., 300 = 20 kts., 320 = 22 kts. Grate area: 1428 sq. feet. Heating surface: 64,234 sq. feet. Weight of machinery: 2178 tons in *Arkansas*, 2093 tons in *Wyoming*. Electrical installation: 4 sets each 300 k.w., 125 volts, 2400 amps., by General Electric Co. Searchlights: 16. It was first intended that these ships should have combined turbines and reciprocating engines, but the plan was dropped in favour of complete turbine propulsion.

Name	*Builder*	*Machinery*	*Laid down*	*Completed*	*Trials (mean)* 24 hrs at 12	*Trials (mean)* 4 hrs. f.p.	*Boilers*	*Best recent speed*
Arkansas	N.Y. Shipbld. Co.	N.Y. Shipbld. Co.	Jan., '10	Sept., '12	—	28,533 = 21·05	Babcock	21·43
Wyoming	Cramp	Cramp	Feb., '10	Sept., '12	20784 = 19·21	31,437 = 21·22	Babcock	22·45

General Notes.—Authorised 1909 as No. 32 (*Wyoming*), 33 (*Arkansas*). Contracts awarded Sept. & Oct., 1909. Both fitted as flagships. Freeboard: Forward, 25'; amidships, 19' 2"; at No. 6 turret, 18'; at stern, 16' 3".

*Unofficial Notes.

ARKANSAS. 1910 *Copyright photo, O. W.*

WYOMING. 1919 *rg.*

Aircraft views given on earlier pages.

Note to Plans. 5—5 inch removed from upper and main decks during War and ports plated ... with not known, except that stern 5 inch shown on plans is removed ... positions.

(Utah Class—2 Ships.)

UTAH (Dec. 1909) & FLORIDA (May, 1910)

Normal displacement 21,825 tons. *Full load*, 23,033 tons. Complement, 1 [illegible] (F.), 1271 (U.) [illegible]

Length (*waterline*), 510 feet. Beam, 88½ feet. (*Mean* draught, 28½ feet; *Max.* [illegible] 30 [illegible]) Length (*over all*), 521½ feet.

* As Flagships, [illegible]

Guns (**Dir. Con.**):
- 10—12 inch, 45 cal.
- 16—5 inch, 51 cal.*
- 2—3 inch anti-aircraft Mk. III
- 4—6 pdr. (saluting)
- 4 machine
- 2 landing

Torpedo tubes (21 inch):
- 2 submerged

* *v.* Gunnery Notes.

Armour (Midvale):
- 11" Belt amidships
- 5" Belt (ends)
- Deck (slopes)
- 10" Upper belt (amidships)
- 12"—8" Turrets (N.C.)
- Turret bases
- 6½" Battery amidships
- 6½" Casemates
- Conning tower

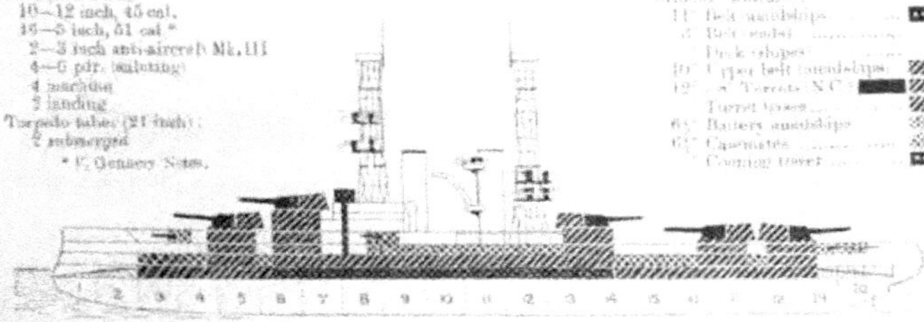

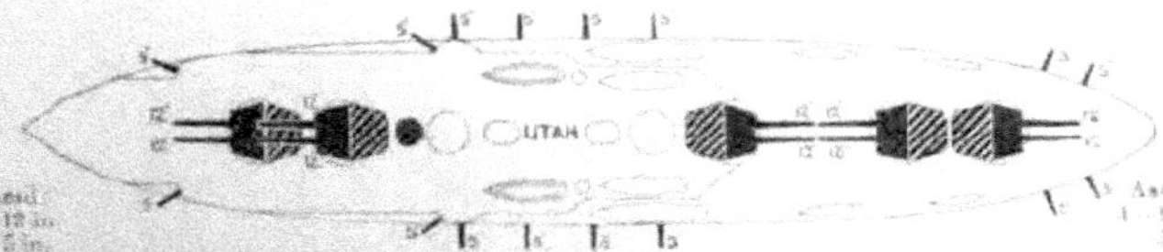

Ahead: 4—12 in. —5 in.

Astern: 4—12 in. 5 in.

Broadside: 10—12 in., —5 in., 1—21 in. torpedo tube.

Machinery: Parsons turbine. 4 screws. Boilers: 12 Babcock. Designed H.P. 28,000 = 20.75 kts. Coal: *normal* 1667 tons; *maximum* 2581 tons in *Utah*; 2560 tons in *Florida*. Oil, 400 tons. *Nominal* radius, 6720 miles at 10 kts., about 4600 at 19 kts. and 2000 at full speed.

Armour Notes.—Main belt 8½ feet wide; upper belt 8 feet wide. 2" splinter bulkheads between all 5 inch battery guns. 1" [illegible] Special sub-division and powerful pumping system against damage by mines or torpedoes.

Gunnery Notes.—Height of guns above water: In fore fore-turret, 33 feet; after fore-turret, 40 feet; amidship turret, 32 feet; in after pair of turrets, 25 feet. 16—5 inch officially listed, but by photographs only six mounted, 1920.

Engineering Notes.—Although turbines are adapted for superheated steam, no superheaters are installed. Have new type [illegible] Grate area, [illegible] sq. feet. Heating surface, 61234 sq. feet. Weight of Machinery: *Florida*, [illegible] tons; *Utah*, [illegible] tons. Electrical installation: 4 sets, each 300 k.w., 125 volts, 2400 amps., by General Electric Co. *Florida* on trials developed [illegible] shaft h.p. inclusive of main engines and all auxiliaries. In *Utah* [illegible] r.p.m. = [illegible] kts. [illegible]

Name	Builder	Machinery	Laid down	Completed	Trials: Full Power	Turbines	Boilers	Best recent speed
Florida Utah	New York Y. N.Y. Shipbuilding Co.	New York Yard N.Y. Shipbuilding Co.	Mar., '09 [illegible] Sep., '11 Mar., '09 [illegible] Aug., '11	40511 = 22.08 [illegible] = 21.04	Parsons Parsons	Babcock Babcock	

General Notes.—Authorised 1908, as No. 30 (*Florida*), 31 (*Utah*). Both fitted as flagships.

Continued Notes.

UTAH.

Aircraft View in earlier pages.

FLORIDA.

NEW U. S. BATTLE-CRUISERS. (Official Nos. 1—6.)

CONSTELLATION, CONSTITUTION, LEXINGTON, RANGER, SARATOGA, and UNITED STATES.

Normal displacement, 43,500 tons. *Full load,* tons. Complement, 1324.

Length (*w.l.*), 850 feet. Beam, (outside bulges, 105 ft. 5¼ in. / inside, 101 ft. 8¼ in.) (*Mean* draught, 31 feet / *Max.* " —) Length (*o.a.*), 874 feet.

Guns (**Dir. Con.**):
- 8—16 inch, 45 cal. Mk. I.
- 16—6 inch, 53 cal. Mk. XII.
- 4—3 inch (AA.) Mk. III.
- 4—6 pdr. (saluting)

Torpedo tubes (21 inch):
- 4 *submerged.*
- 4 *above water.*

Armour (*See Notes*):

(No details available.)

Anti-Torpedo protection: Bulges carried up to lowers.

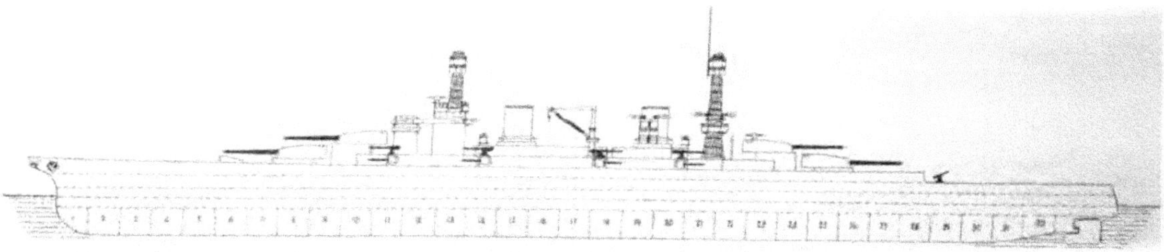

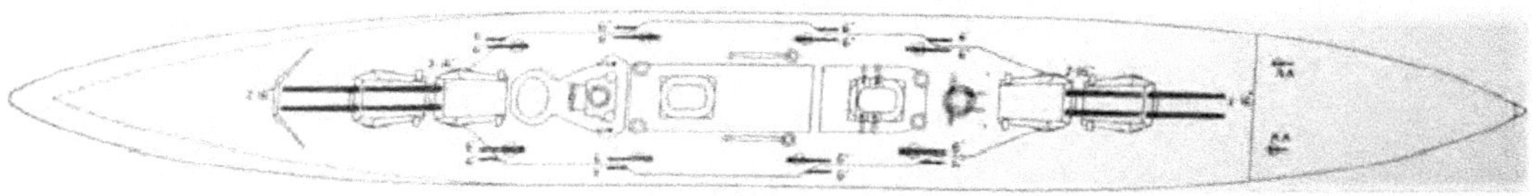

Ahead:
4—16 inch.
4—6 inch.

Broadside: 8—16 inch, 8—6 inch.

Astern:
4—16 inch.
4—6 inch.

Machinery: Turbines (see Table) and electric drive. Boilers: 16 (type unknown). Designed H.P. 180,000 = 33·25 kts. Fuel: oil only, *normal* tons, *maximum* tons.

Gunnery Notes.—Originally designed with 10—14 and 18—5 inch guns. Unofficial reports in 1920 said *Saratoga* and *U. States* would have 18 inch guns, but such statements have not been officially confirmed, and appear untrustworthy.

Engineering Notes.—Each ship to have 4 large steam turbines, each driving a 4000 KVA, 3-phase, 5000-volt generator at 1835 r.p.m. On each of the 4 propeller shafts, two 3-phase induction motors of 23,000 H.P., each having 22 poles and running at 331 r.p.m. These motors can also operate on 44 poles giving 6800 H.P. at 170 r.p.m., the condition for half-speed. Heating surface: 198,000 sq. ft. + 18,000 ft. superheating surface.

Armour Notes.—Armour contracted for by Bethlehem, Carnegie and Midvale Steel Companies.

Name.	*Builder.*	*Machinery.*	*Laid down.*	*Completed.*	*Trials.*	*Turbines.*	[illegible]
Constellation	Newport News	Newport News				W'stinghouse	
Constitution	Philadelphia, N.Yd					G.E.	
Lexington	Bethlehem (Quincy)	Beth'hem(Quincy)	1920			G.E.	
Ranger	Newport News	Newport News	1920			W'stinghouse	
Saratoga	New York S.B. Co.	N. York S.B. Co.				G.E.	
United States	Philadelphia, N.Yd		1920			G.E.	
			1920				

General Notes.—All to be fitted with Flag Officers' Quarters. Dates of completion uncertain; perhaps not until 1[illegible] Design largely revised 1919, to give extra protection to turrets, conning tower, magazines and communicati[illegible] at the expense of a small reduction in speed. Original design (which included 7 funnels) was given in 1919 "Fight[illegible] Ships," p. 195. Authorized under Three Year Programme thus: *Lexington* (1), *Constellation* (2), *Saratoga* [illegible] *Ranger* (4), 1916 Section; *Constitution* (5), 1917 Section; *United States* (6), 1918 Section.

VERMONT. 1919 *Photo,*

KANSAS (Sept., 1905), **VERMONT** (Aug., 1905), **MINNESOTA** (April, 1905), **NEW HAMPSHIRE** (—1906). *Normal* displacement, 16,000 tons. *Full load* displacement, 17,650 tons (*N. H.* 17,784 tons). Complement, 1206-9. Length (*waterline*), 450 feet. Beam, $76\frac{7}{8}$ feet. *Max.* draught, $26\frac{3}{4}$ feet. Length (*over all*), $456\frac{1}{4}$ feet. Guns (**Dir. Con.**): 4—12 inch, 45 cal., 8—8 inch, 45 cal., 12—3 inch, 13 pdr., 2—3 inch anti-aircraft and 4 or 2—6 pdr. saluting. Torpedo tubes (21 inch): 4 *submerged*. Armour (Krupp): 9" Belt (amidships), 4" Belt (ends), 3" Deck (slopes), 7" Lower deck side, 10" Barbettes, 12-8" Turrets to these, 7" Battery, 2" Casemates (for 14 pdrs.), $6\frac{1}{4}$"—6" Small turrets, 9" Conning tower, 5" Director station (near C.T.). Machinery: 2 sets vertical 4 cylinder triple expansion. 2 screws (outward turning). Boilers: 12 Babcock. Designed H.P. 16,500 = 18 kts. Coal: *normal* 900 tons; *maximum*, 2455 tons in *Kansas*, 2486 in *Vermont*, 2420 in *Minnesota*, 2582 in *N. Hampshire*. Built 1904-8.

Appearance similar to *Vermont* above.

LOUISIANA (Aug., 1904) & **CONNECTICUT** (Sept., 1904). As *Kansas*, &c., above, but differ as follows:—Displacement: *Full load*, 17,666 tons. Complement (as flagship), 1237. Armour (Krupp): 11"—9" Belt (amidships), 3" Deck (flat on belt), 12"—8" Turrets (N.S.), 10"—$7\frac{1}{2}$" Turret bases (N.C.), 7" Lower deck redoubt, 7" Battery, 2" Casemates (14 pdr.), 6" Secondary turrets (N.C.). Designed H.P. 20,525 = 18 kts. Coal: *maximum* 2446 tons, *Louisiana*; 2510 tons, *Connecticut*. Built 1903-6.

NEW JERSEY. (Main deck battery now removed). 1919 Photo, O. W. Wa

VIRGINIA (April, 1904), **NEW JERSEY** (Nov., 1904), **GEORGIA** (Oct., 1904), **NE** (Oct., 1904), **RHODE ISLAND** (May, 1904). *Normal* displacement, 14,948 tons. displacement, 16,094 tons. Complements, 1109-1130. Length (*waterline*), 435 feet. Be *Max.* draught, $25\frac{5}{8}$ feet. Length (*over all*), $441\frac{1}{4}$ feet. Guns (Model, '99): 4—12 inch, 8 inch, 45 cal., 8—3 inch, 13 pdr., 2—3 inch AA., and in *N.J.* 4—3 pdr., *Georgia Nebraska* 2—3 pdr., all saluting. Torpedo tubes (21 inch): 4 *submerged*. Armour H.N.): 11"—8" Belt (amidships), 4" Belt (ends), 3" Deck (flat on belt amidships), 10"—7 (H.N.), 12"—8" Turrets (H.N.), 6" Secondary turrets, 6" Lower deck side (H.N.), 6" Bat 9" Conning tower (K.N.C.). Machinery: 2 sets 4 cylinder vertical inverted triple ex screws. Boilers: 12 Babcock. Designed H.P. 25,463 = 19 kts. Coal: *normal* 900 tons 1961 to 2031 tons (average 1995). Built 1901-7.

MISSOURI. 1919 *Photo.*

OHIO (May, 1901), **MAINE** (July, 1901), **MISSOURI** (Dec., 1901). *Normal* displacement, 12,500 tons. *Full load* displacement, 13,500 tons. Complement, 1017—1053. Length (*waterline*), 388 feet. Beam, $72\frac{1}{4}$ feet. *Max.* draught, $25\frac{1}{2}$ feet. *Length over all*, $393\frac{3}{8}$ feet. Guns (M. '99): 4—12 in., 40 cal. (Mk. III, mod. 3), 8—6 inch, 50 cal., 2—3 inch AA., 4 or 2—6 or 3 pdr. Torpedo tubes (18 inch): 2 *submerged*. Armour (Krupp): $11''$—$7\frac{1}{2}''$ Belt (amidships), 4" Belt (bow), 10" Bulkhead (aft), $2\frac{3}{4}''$ Deck (on slopes), 4" Deck (aft), 12" Turrets (H.N.), 12"—8" Turret bases, 6" Lower deck side, 6" Battery, 6" Casemates (forward), 10" Conning tower. Machinery: 2 sets vertical inverted triple expansion 4 cylinder, except *Missouri*, which is 3 cylinder. 2 screws. Boilers: 12 Thornycroft, except *Maine*, 12 Babcock. Designed H.P. 16,000 = 18 kts. Coal: *normal* 1000 tons; *maximum* 2331 tons, *Ohio*; 1901 tons, *Maine*; 1933 tons, *Missouri*. Built 1900-1904. To be Training Ships for Engineering Personnel.

(*Illustration in preceding column.*)

ALABAMA (May, 1898), **ILLINOIS** (October, 1898), **WISCONSIN** (November, 189 displacement, 11,552 tons. Complement, 817. Guns: 4—13 inch, 35 cal. 8—6 inch, inch AA., 4 or 2—6 pdr. No torpedo tubes. Armour (Harvey-nickel): $16\frac{1}{2}$—$9\frac{1}{2}''$ Belt, 4" 4" Deck (aft., 12" Bulkheads, 14" Turrets, 15"—10" Turret bases, $5\frac{1}{2}''$ Upper belt and Casemates, 10" Conning tower. Machinery: 2 sets vertical triple expansion. 3 cylinders Boilers: 8 single-ended cylindrical, except *Illinois*, 8 Mosher. Designed H.P. *forced* 11,2 Coal: *normal* 850 tons; *maximum*, 1481 tons *Alabama*; 1522 tons *Illinois*; 1447 tons Built 1896-1901. To be Training Ships for Engineering Personnel.

1910 *P*

KENTUCKY (March, 1898). *Normal* displacement, 11,520 tons. Complement, 867. Guns inch, 35 cal., 4—8 inch, IV, 35 cal., 8—5 inch, 40 cal., 2—3 inch AA., and 2—6 pdr. tubes: none. Armour (Harvey-nickel): $16\frac{1}{2}''$—$9\frac{1}{2}''$ Belt, 4" Belt (bow), 5" Deck (aft), 2 (flat on belt) 4" aft, 10" Fore bulkhead, 12" After bulkhead, 17"-9" Turrets (for 13in.), 9" Turr 8 inch), $15\frac{1}{2}''$—$12\frac{1}{2}''$ Barbettes, $5\frac{1}{4}''$ Lower deck side, $5\frac{1}{4}''$ Battery, 7" & 6" Battery (bulkhea Conning tower. Machinery: 2 sets vertical triple expansion 3 cylinder. 2 screws. Boilers: 8 (fitted 1910-11). Designed H.P. 10,500 = 16 kts. Coal: *normal* 410 tons; *maximum* 1658 to

Undergoing alterations 1920.

KEARSARGE (March, 1898). Sister to *Kentucky* above. Converting 1920 for service as Crane at Navy Yards. Will retain 4—13 inch guns and turrets. 1 funnel. 4 1-2... removed. H.P. will be about 500...

Note.—As these ships are to be the subject of important experiments, they have been reinstated in this edition.

Photo added 1920.

(*To be Fleet Target Ship, controlled by W/T.*)

COAST BATTLESHIP No. 4 (ex-*Iowa*, 1896). 11,346 tons. Dimensions: 360 (*w.l.*) × 72¼ × 27¼ feet (*max.* draught). Guns: 4—12 inch, 8—8 inch. Armour: 14"—7" Belt, 17"—15" main turrets, 8"—7" secondary turrets, 15"—12½" main barbettes, 8"—6" secondary barbettes, 3"—2¾" decks. Designed H.P. 12,000 = 17 kts. Boilers: 3 D.E., 2 S.E.* Coal: 1682 tons.†

* To be adapted for burning oil } on conversion to Target Ship
† To have small quantity of oil }

C.B.S. No. 1. Photo added 1920.

COAST BATTLESHIP No. 1 (ex-*Indiana*, 1893). **COAST BATTLESHIP No. 2** (ex-*Massachusetts*, 1893). 10,288 tons. Dimensions: 348 (*w.l.*) × 69¼ × 26½ feet (*max.* draught). Guns: 4—13 inch, 8—8 inch, 4—3 inch. Armour: 18"—8½" belt, 15" (main), 6" (secondary) turrets, 17" (main), 8"—6" (secondary) barbettes, 3"—2¾" decks. Designed H.P. 9000 = 17 kts. Boilers: 8 Babcock. Coal: 1523-1526 tons.

Note.—Will probably be used as Target Ships for Aircraft, &c. *C.B.S. No.* 3 (ex-*Oregon*) presented to State of Oregon as a relic.

TONOPAH (190
CHEYENNE (19
OZARK (1901).
TALLAHASSEE

TONOPAH. 1919 *Photo, O. W. Waterman.*

3225 tons. Complements, 237. Dimensions: 252 × 50 × 12½ feet (*mean* draught). Guns: 2—1 cal.) 3—4 inch, 1—3 inch AA., 2—6 pdr., 2—1 pdr. Armour: Krupp, 11" belt a H.P. 2400 = 12 kts. Coal: *maximum, Ozark*, 301 tons, *Tallahassee*, 314 tons, *Tonopah*, *Cheyenne*, 132 tons coal + 60,816 gallons of oil fuel.

MONTEREY (1891). 4084 tons. Guns: 2—12 inch, 35 cal., in fore turret; 2—10 incl after turret; 1—3 inch AA., 6—6 pdr., 1—1 pdr. Armour (Harvey): 13"—5" belt Base of fore turret 13", of after turret 11½". Designed H.P. 5000 = 14 kts. Babcock bo 211 tons. Complement, 237.

MONADNOCK (1883). 3990 tons. Guns: 4—10 inch, 2—4 inch, 1—3 inch A/ 2—1 pdr. Armour: Iron belt 9", steel turrets 11½"—7½". H.P. (*on trials*) kts. Coal: 395 tons. Complement, 238.

SEATTLE (main deck battery now removed). 1919 Photo, O. W. Waterman

CHARLOTTE (ex-*North Carolina*, October, 1906), **MISSOALA** (ex-*Montana*, December, 1906), **SEATTLE** (ex-*Washington*, March, 1905). *Normal displacement*, 14,500 tons. *Full load*, 15,981 tons. *Seattle*, 15,712 tons. Complement, 890—899. Length (*waterline*), 502 feet. Beam, 72¾ feet. *Maximum* draught, 27 feet. Length (*over all*), 504¼ feet. Guns (M. '99): 4—10 inch, 40 cal., 4—6 inch, 50 cal., 10—3 inch, 14 pdr., 2—3 inch anti-aircraft, 4—6 or 3 pdr. (saluting). Torpedo tubes (21 inch): 4 *submerged*. Armour (Krupp, but part Midvale in *C.* and *M.*): 5"—3" Belt, 3" Deck (ends), 5" Lower deck side, 6" Lower deck bulkheads, 8"—4" Barbettes (N.C.), 9"—5" Turrets, 5" Battery, 2" Battery for 14 pdrs., 9" Conning tower (fore), 5" Tube (N.C.), 5" After C. T. (N.C.). Machinery: 2 sets 4 cylinder triple expansion. 2 screws, outward turning. Boilers: 16 Babcock. Designed H.P. 23,000 = 22 kts. Coal: *normal*, 900 tons; *maximum*, *Charlotte* and *Missoala*, 2164 tons; *Seattle*, 2062 tons. Built 1903-8. *Memphis* (ex-*Tennessee*) of this class wrecked 1916.

HUNTINGTON.* 1919 [illegible]

*Frederick and *Huron* believed to be the same, no cage mast close to fore funnel; no high ventilator between cage mast and first funnel; no ventilator just before mainmast; only two topmasts mainmast. Compare with *Pittsburgh* below.

PITTSBURGH (and *Pueblo*). 1919 Photo

HUNTINGTON (ex-*West Virginia*, April, 1903), **PUEBLO** (ex-*Colorado*, April, 1[illegible] **PITTSBURGH** (ex-*Pennsylvania*, Aug., 1903), **FREDERICK** (ex-*Maryland*, Sept., 1[illegible] and **HURON** (ex-*South Dakota*, July, 1904). Displacement, *normal*, 13,680 tons, *full load*, 15,[illegible] Complement, 640.* Length (*waterline*), 502 feet. Beam, 69½ feet. *Maximum* draught, 26 fee[t] *full load* displacement. Length (*over all*), 504 feet. Guns (model 1899): 4—8 inch, 45 cal., 4[illegible] inch, 50 cal., 2—3 inch anti-aircraft, 4—3 pdr. Torpedo tubes (18 in.): 2 *submerged*. Arm[our] (Krupp): 6" Belt, 3½" Belt (ends), 4" Deck (on slopes), 5" Upper belt, 4" Bulkheads, 6" Turrets (N.[C.]) 5" Battery, 6" Casemates (N.C.), 9" Conning tower (N.C.). Machinery: 2 sets 4 cylinder vert[ical] triple expansion. 2 screws. Boilers: 16 or 20 Babcock & Wilcox. Designed H.P. 23,000 = 22 [kts.] Coal: *normal*, 900 tons; *maximum*, *Huron*, 2233 tons; *Frederick* and *Huntington*, 2008 to[ns]; *Pittsburgh*, 1992 tons; and *Pueblo*, 1976 tons. Built 1901-8.

**Pittsburgh*, 926.

St. Louis. 1919 *Photo, O. W. Waterman.*

CHARLESTON (Jan., 1904), **ST. LOUIS** (May, 1905). Displacement, *normal*, 9700 tons, *full load*, 10,839 tons. Complement, 720. Length (*waterline*), 424 feet. Beam, 66 feet. *Maximum* draught, 24½ feet. Length *over all*, 426½ feet. Guns: 12—6 inch, 50 cal., 4—3 inch, 14 pdr., 2—3 inch AA., 4 (*Charleston*) 2 (*St. Louis*)—3 pdr. (saluting). Torpedo tubes: *nil*. Armour (Krupp): 4" Belt, 3" Deck, 4" Lower deck redoubt, 4" Battery, 5" Conning tower. Machinery: 2 sets vertical 4 cylinder triple expansion. 2 screws. Boilers: 16 Babcock. Designed H.P. 21,000 = 21·5 kts. Coal: *normal* 650 tons; *maximum* 1818 and 1793 tons. Begun 1902, completed 1905-6. *Milwaukee* of this class wrecked 1917.

(*Flagship, Destroyer Force, Pacific Fleet.*) 1920

BROOKLYN (1895). *Normal* displacement, 9215 tons; *full load*, 10,068 tons. Complem
Length (*waterline*), 400½ feet. Beam, 64¾ feet. *Maximum* draught, 26½ feet. Guns: 8
35 cal., 8—5 inch, 40 cal., 2—3 inch (anti-aircraft), 4—6 pdr. (saluting). Torpedo tul
Armour (Harvey): 3" Belt, 6" Deck (amidships), 3" Deck (ends), 8" Barbettes, 5½" Hoods to l
6" Hoists, 4" Shields (fixed), 5" guns, 7½" Conning tower. Machinery: 4 sets verti
expansion. 2 screws. Boilers: 7 cylindrical (5 double-ended and 2 single-ended). Desi
forced 18,000 = 21 kts. Coal: *normal* 900 tons; *maximum* 1449 tons. Laid down 1893. C
1896. Reconstructed 1909.

(*Flagship, Destroyer Force, Atlantic Fleet.*)

ROCHESTER (ex *Saratoga*, ex *New York*). (December, 1891.) (Reconstructed 1907-08
displacement, 8150 tons; *full load*, 8900 tons. Complement, 578. Length (*waterline*
Beam, 64⅚ feet. *Maximum* draught, 24½ feet. Guns: 4—8 inch, 45 cal., 8—5 in
2—3 inch AA. Torpedo tubes: *Nil*. Armour (Harvey nickel): 4" Belt, 6" Deck (
(Krupp) 6"—4" Barbettes, 6½" Turrets; 5" Hoists, 7" Conning tower. Machinery: 4
triple expansion. 2 screws. Boilers (1907): 12 Babcock. Designed H.P. *forced* 16,
Coal: *normal* 750 tons; *maximum* 1100 tons. Laid down at Philadelphia, Sept., 1890.
1893. Re-named *Saratoga*, 1911, and again re-named *Rochester*, 1917. Large refit i

NUMBER (4), NUMBER (5), NUMBER (6), NUMBER (7), NUMBER (8), NUMBER (9), NUMBER (10), NUMBER (11), NUMBER (12), and NUMBER (13).

Normal displacement, 7100 tons; *full load*, 7610 tons. Complement, 350.

Length (*waterline*), 550 feet. Beam, 55 feet. { *Mean* draught, 13' 6" / *Max.* " 14' 4" } Length over all, 555½ feet.

Guns: (**Dir. Con.**)
8—6 inch, 52 cal. Mk. XII.
2—3 inch anti-aircraft
2—3 pdr.
Torpedo tubes (21 inch):
4 in two twin deck mountings above water.

Armour:
(Not known.)

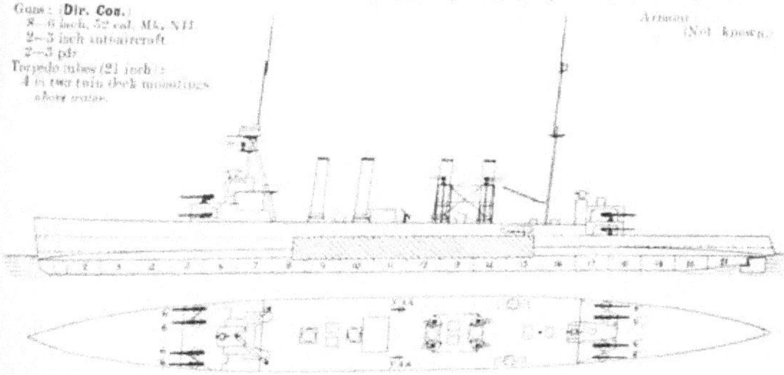

Ahead:
4—6 inch.

Broadside: 4—6 inch, 2—21 inch T.T.

Astern:
4—6 inch.

Machinery: Turbines (see Table for types) with reduction gears. Designed S.H.P. 90,000 = 35 kts. 4 screws. Boilers: see Table. Fuel: oil only; capacity unknown.

Note.—Navigation Direction Finder Tower on after bridge.

Name.	*Built and Engined by*	*Laid down*	*Completed*	*Trials*	*Turbines*	*Boilers*	*Heating Surface (sq. ft.)*
Number (4)	Todd Co., Tacoma	Dec. '18	Aug. '21 †		Westgh. Parsons	12 Yarrow	90276
Number (5)	" "	Dec. '18	Dec. '21 †		"	12 Yarrow	
Number (6)	" "	May '19	July, '22 †		"	12 Yarrow	
Number (7)	Union Ironworks	...			Curtis	12 Yarrow	
Number (8)	" "				"	12 Yarrow	
Number (9)	Wm. Cramp & Sons	Feb. '20	Jan. '21 †		Parsons	12 White-F	89840
Number (10)	" " "	Mar. '20	Sept. '21 †		"	12 White-F	
Number (11)	" " "	...	Nov. '21 †		"	12 White-F	
Number (12)	" " "	...	Dec. '21 †		"	12 White-F	
Number (13)	" " "	...	Feb. '22 †		"	12 White-F	

† Contract dates for completion.

General Notes.—Estimated cost (without guns) in 1916, five to six million dollars each; revised estimate 1919, 7½ million dollars each. Provided by the 1916 section of the three year programme.

Official Model (1).

Official Model (2).

1919 Photo.

OLYMPIA (November, 1892). *Normal* displacement, 5865 tons (*full load* 6588 tons). Complement, 433. Length (*waterline*), 340 feet. Beam, 53 feet. *Maximum* draught, 25 feet. Guns: 10—5 inch (51 cal.), 2—3 inch (AA.), 2—1 pdr. Armour: 4¾" deck on slopes, 2" flat.

Machinery: 2 sets vertical inverted triple expansion. 2 screws. Boilers: 4 double-ended and 2 single-ended. Designed H.P. *forced* 17,000 = 21·69 kts. Coal: *normal* 500 tons; *maximum* 1169 tons.

General Notes.—Laid down at San Francisco, June, 1891. Completed, 1895. Cost, *complete*, about £550,000. Flagship of the late Admiral Dewey, at battle of Manila. Reconstructed partially 1901-03, re-armed 1917-18.

1919 Photo, O. W. Waterman.

Columbia (see next column for details).

(*Flagship, Train, Atlantic Fleet.*)

Minneapolis. 19

(*Flagship, Train, Pacific Fleet.*)

COLUMBIA (July, 1892), & **MINNEAPOLIS** (August, 1893). Displacement, 7 Complement, 460 and 463. Length (*waterline*), 411¼ feet. Beam, 58¼ feet. *Maximum* 24½ feet. Guns: 3—6 inch, 45 cal. 4—4 inch, 40 cal., 2—3 inch AA., 2 (Co (*Minneapolis*)—3 pdr., (and 2—1 pdr. in *Minneapolis*).

Machinery: 3 sets vertical inverted triple expansion. 3 screws. Boilers: 8 doub single-ended cylindrical. Designed H.P. *forced* 18,500 = 22·80 kts. Coal: *normal* *maximum* 1561 tons in *Columbia*, 1433 tons in *Minneapolis*.

(*Station Ship, Honolulu.*)

CHICAGO (1885). Displacement, 4500 tons. Complement, 426. Armament: 5—5 2—3 inch AA., 2—3 pdr., 2—1 pdr., 2 M.G. Designed H.P. 9000 = 18 kts. Boilers + 4 S.E. Coal: 870 tons.

BIRMINGHAM. 1919 Photo.

(CHESTER CLASS—3 SHIPS.)

SALEM (July, 1907), **BIRMINGHAM** (May, 1907), **CHESTER** (June, 1907). *Normal* Displacement, 3750 tons; *full load*, 4687 tons. Complement, 367-378 (*B'ham* 377, flag 426). Length (*w.l.*), 420 feet. Beam, 47 feet. *Maximum* draught, 18¾ feet, *Mean* draught, 16¾ feet. Length (*over all*), 423 feet. Guns (**Dir. Con.**): 4—5 inch, 51 cal., 2—3 inch, 50 cal., 1—3 inch AA., 2—3 pdr. (saluting). Torpedo tubes (21 inch): 2 *above water*. Armour: 2" Nickel on *w.l.*, 2"—1" Steering gear protection, 2"—1" Protection to engines. No conning tower. Machinery: *Salem*, G. E. turbines with reduction gear (new 1916-17). *Chester*, Parsons' turbine. 4 screws. *Birmingham*, 2 sets 4 cylinder triple expansion. 2 screws. Boilers: various (*see Notes*). Designed H.P.: *Salem* (with new turbines) 20000. *Birmingham* and *Chester*, 16,000 = 24 kts. Coal: *normal* 475 tons; *maximum*, *Salem* and *Birmingham*, 1433 tons; *Chester*, 1408 tons.

Name.	*Builder.*	*Laid down.*	*Completed.*	*Refit.*	*Trials:— 4 hours full p.*	*Boilers.*	*Best recent speed.*
Salem (†)	Fore River Co.	Aug., '05	July, '08	1916-17	15,273 = 25·95*	12 Fore River	
Birmingham	Fore River Co.	Aug., '05	Apl., '08		15,476 = 24·32	12 Fore River	
Chester (†)	Bath Ironworks	Sept., '05	Apl., '08		25,400 = 26·52†	12 Normand	

* With original Curtis turbines. † Light on trials, 3673 tons.

General Notes:—Authorised 1904, re-armed 1917-18. Average cost, *exclusive of armament*, £325,000 each. These ships are good sea boats, but roll heavily. Serve as Flagships to Destroyer Forces, Atlantic and Pacific Fleet. Are very lightly built, and cannot maintain full power for any length of time.

CHATTANOOGA. (*Galveston*, no top on main mast). 1920 Photo, *Freund*.

(DENVER CLASS—6 SHIPS.)

CHATTANOOGA (March, 1903), **CLEVELAND** (September, 1901), **DENVER** (June **DES MOINES** (September, 1902), **GALVESTON** (July, 1903), & **TACOMA** (June, (Wood sheathed and coppered). *Normal* displacement, 3,200 tons. Complement, 302 Length (*waterline*), 292 feet. Beam, 44 feet. *Maximum* draught, 17 feet. Guns (model 189 5 inch, 50 cal., 1—3 inch AA.* 6†—6 pdr., §2 landing. No torpedo tubes. Armour (I nickel): 2½" Deck (on slopes), ½" (flat). Machinery: 2 sets 4 cylinder triple expansion. 2 Boilers: 6 Babcock & Wilcox. Designed H.P. 5,398 = 16·5 kts. Coal: *normal* 467 tons; *ma* 705 to 741 tons = 7,000 miles at 10 kts. *nominal*.

* Not in *Chattanooga*. † 4 in *Denver*. § Also 2—1 pdr. in *Cleveland*.

CHESTER. 1920 Photo.

(New Orleans Class—2 Ships.)

Albany.

Photo, O. W. Waterman.

NEW ORLEANS (Dec., 1896) & **ALBANY** (Jan., 1899). *Normal* displacement, 3430 tons (*full load*, 3954). Complement, 326-340. Length (*w.l.*), 326 feet. Beam, 43¾ feet. *Maximum* draught, 19 feet. Length *over all*, 354¼ feet. Guns (U.S. models): 8*—5 inch, 50 cal., 1—3 inch AA., 2—3 pdr. (+2 1 pdr. in *N.O.*). No torpedo tubes. Armour (Harvey-nickel): 3¼" Deck (slopes) and Cofferdam amidships, no gun shields, 1¼" Deck (flat). Machinery: 2 sets 3 cylinder triple expansion. 2 screws. Boilers: 4 double-ended. Designed H.P. *forced* 7500 = 20 kts. Coal: *normal* 450 tons; *maximum* 782 tons in *Albany*, 693 tons in *New Orleans*.

General Notes.—Laid down at Elswick (for Brazil) as the *Amazonas* (*New Orleans*) and *Abreu* (*Albany*). Purchased by U. S. (1898) just before war with Spain. ***Albany*** completed 1900, ***New Orleans*** 1898. Rearmed with U. S. guns, 1907.

* Believed reduced to 6—5 inch in 1920, first pair of guns in waist being removed and side plated up to improve weatherliness.

1919 Photo.

MARBLEHEAD (1892). 2072 tons. Complement, 252. Dimensions: 257 *w.l.* × 37 × *mean* draught. Guns: 8—4 inch, 1—3 inch AA. Designed H.P. 5400 = 17 kts. Mac sets inverted triple expansion. Boilers: 3 D.E. and 2 S.E. Coal: 340 tons.

Gunboats.

Building or Completing.

NUMBER (22). To be built by Charleston, N.Y., begun Dec., 1919. **ASHEVILLE** (Charl July, 1918). *Normal* displacement, 1575 tons. Complement, Length (*waterlin* Beam, 41¼ feet. *Mean* draught, 11⅓ feet. Guns: 3—4 inch, 50 cal., 2—3 pdr. Parsons turbine with reduction gear. 1 screw. Boilers: 3 Bureau (modified T Designed H.P. 800 = 12 kts. Fuel, tons.

1919 Photo.

SACRAMENTO (Feb., 1914). *Normal* displacement, 1425 tons. Complement, 163. l *line*), 210 feet. Beam, 50⅝ feet. *Mean* draught, 11¼ feet. Guns: 3—4 inch, 50 cal 2—1 pdr. Machinery: 1 set triple expansion. Boilers: 2 Babcock. H.P. (on trials kts. Coal: *maximum*, 428 tons. Built by Cramps. Completed, 1914.

Photo wanted.

SAMOA (ex German s.s. *Staatssekretaer Solf*) (1913, seized 1917). 550 tons. Complement, 30. Dimension: 131 (*waterline*) x 25½ x feet. Guns: 4—3 pdr., 2 M.G. Machinery: 1 set vert. compound. H.P., Speed and Fuel not known.

Photo wanted.

WILMETTE (ex s.s. *Eastland*, 1903, taken over 1918). 2,600 tons. Complement, 209. Length (*waterline*), 265 feet. Beam, 38½ feet. Draught, feet. Guns: 4—4 inch, 2—3 inch AA., 2—1 pdr. Machinery: 2 sets triple expansion. 2 screws. Boilers: 3 S.E. Designed I.H.P. 4000=16.5 kts. Fuel, tons.

Dubuque. 1920 Photo.

DUBUQUE (1904), **PADUCAH** (1904). *Normal* displacement, 1085 tons. Complement, 188 and 175. Length (*waterline*), 174 feet. Beam, 35 feet. *Mean* draught, 12¼ feet. Guns: *Dubuque*, 2—4 inch, 4—6 pdr., 2—1 pdr. *Paducah*, none. Machinery: 2 sets triple expansion. 2 screws. Boilers: 2 Babcock. Designed H.P. 1000=12.9 kts. Coal: *maximum* 252 tons.

1919 Photo, Favs.

WHEELING (1897). *Normal* displacement, 990 tons. Complement, 162. Length (*waterline*), 174 feet. Beam, 34 feet. *Mean* draught, 12 feet. Guns: 4—4 inch, 40 cal.; 2—3 pdr.; 2—1 pdr. Machinery: 2 sets vertical triple expansion. 2 screws. Boilers: 2 single-ended. Designed H.P. 1250=12 kts. Coal: *maximum*, 231 and 256 tons. *Nominal* radius: 4000 at 10 kts.

Mainmast of all removed. Photo, Abrahams & Sons.

ANNAPOLIS (1896). **VICKSBURG** (1896). *Normal* displacement, 1010 tons. Complement, 146—153. Length (*waterline*), 168 feet. Beam, 36 feet. *Mean* draught, 12 feet. Guns: *Annapolis*, none. *Vicksburg*, 6—4 inch, 2—1 pdr. Machinery: 1 set vertical triple expansion. 1 screw. Boilers: 2 Babcock. Designed H.P. 800=12 kts. Coal: 235 tons.

1920 Ph

HELENA (1896) & **WILMINGTON** (1895). Displacement tons. Complement, 193 to 207. Length (*waterline*), 250 Beam, 39¼ feet. *Mean* draught, 9 feet. Guns: 8—4 inch, 4 4—3 pdr. Machinery: 2 sets vertical triple expansion. 2 s Designed H.P. 1988=13 kts. Coal: *maximum*, 256 t *Wilmington*; 307 tons in *Helena*. *Nominal* radius: 2200 m 10 kts.

NASHVILLE (1895). 1371 tons. Guns: None. H.P. 2500= Coal: 381 tons. Complement, 175.

1919 *Photo.*

(*Used as Yacht by the Secretary of the Navy.*)

DOLPHIN (1884). 1486 tons. Complement, 161. Guns: 1—4 inch, 2—6 pdr. Speed, 15·5 kts. Coal: 225 tons.

No photo available.

NANTUCKET (ex *Rockport*, ex *Ranger*, 1876). 1261 tons. I.H.P. 500=10 kts. Coal, 182 tons.

Mainmast now removed.

(*S/M. Base, San Pedro, Cal.*)

ALERT (1875). 1100 tons. Complement, 149. Dimensions 177½ × 32 × 13 feet (*mean* draught). Guns: 4—6 pdr. H.P. 500=10 kts. 2 Babcock boilers. Coal: 202 tons.

Non-effective Gunboats used for Naval Reserve Duties.

(*Minnesota Naval Reserves.*)

***ESSEX** (1874). 1375 tons. Guns: 6—3 pdr. H.P. 800=10 kts. Coal: 155 tons.

(*Naval Reserves, 9th, 10th, 11th Districts.*)

***GOPHER** (1871). 840 tons. 5—3 pdr., 1—1 pdr. Speed, 9 kts. Coal: 80 tons.

(*Naval Reserves.*)

WOLVERINE (1842). 685 tons. H.P. 365=10·5 kts. Paddle wheel. Coal: 115 tons.

* *Rated as unserviceable for War Purposes.*

Small Gunboats for Service in the Philippines.

Monocacy. 1919 *Photo, Le*

MONOCACY (1914), **PALOS** (1914). 190 tons. C
Guns: 2—6 pdr., 6 machine. Machinery: 2 sets vert
Boilers: 2 Babcock. Designed H.P. 800=13¼ k
tons. Built at Mare Island Navy Yard and re-erect
Dock and Engineering Co.

No photo available.

CALLAO (1888). 243 tons. Guns: none. Spee
Coal: 33 tons.

259 + 60* = 319† U. S. DESTROYERS.—Tabulated Details.

*Building, completing, or under contract, Oct. 1st, 1920. †12 Boats not under contract or construction; excluded as they will probably be cancelled.

No.	Class.	First begun.	Last completed	Displacement.	Dimensions. Length (w.l.) × Beam × Mean draught	Designed H.P. = kts. speed.	Max. Fuel.	Guns. AA = Anti-Aircraft.	Torpedo tubes.	Complement. Peace/War
				tons.			tons.			
148	*Various Yards**	1918	1920*	1215	310 × 31 × 9¼	27,000t—25,000t = 35†	..	4—4 inch (50 cal.)§ 1—3 inch AA.	12—21 inch	122/139
3	*Bath I. W.**	1918	1920	1215						
5	*N. Y/S. B. Co.**	1918	1920	1215						
11	*Newport News**	1917	1919	1213						
10	*N. Y/S. B. Co.**	1917	1919	1211						
52	*Fore River** / *Union I.W.**	1917	1919	1191	310 × 31 × 9	27,000t—24,200t = 35†	..	4—4 inch (50 cal.) 2—3 inch AA.	12—21 inch‡	122/139
21	*Cramp**	1917	1919	1154						
17	*Bath I. W.** / *Navy Yards**	1917	1919	1154						
6	*Caldwell*	1916	1917	1125	310 × 30½ × 8	18,500t = 30 / 20,000t = 32	260 *oil* ?	5 or 4—4 in. (50 cal.) 1—3 inch AA.	12—21 inch	122/139
1	*Isabel*	—	1917	797	230 × 26¼ × 9¼	*trials)* 8400t = 28.8	..	4—3 inch 2 M.G. (AA.)	4—18 inch	115
6	*Allen*	1915	1917	1110	310 × 29⅝ × 9¼	17,000t = 29.5	290 *oil* ?	4—4 inch (50 cal.) 2—3 inch AA.	12—21 inch	103/136
5	*Conyngham*	1914	1916	1050	310 × 29⅝ × 9¾	17,000t = 29.5	305—310 *oil*	4—4 inch (50 cal.)	8—21 inch	106/132
6	*Cushing*	1913	1915	1050	300 × 30¼ × 9½	16,000tr = 29	307—311 *oil*	4—4 inch (50 cal.)	8—21 inch	106/133
8	*Aylwin*	1912	1915	1040					8—18 inch	
21	*Drayton*	1909	1912	742	289 × 26¼ × 8¼	12,000t = 29½	222—238 *oil*	Nil	6—18 inch‖	86/107

In above Table, displacements and dimensions are generally the average for each class. Dates refer to time first boat of class was begun and the last boat commissioned for service. Launching dates are given in description by classes on subsequent pages. In H.P. column, *t* = turbines, *tr* = turbines combined with reciprocating engines.

* Classified by Building Yards. †Many did not attain this speed on trials. §4—5 inch guns in the "5 N.Y./S.B. Co." boats. 2 of the "148 Various Yard" boats have 8—4 inch. ‡Possibly removed from 15 boats converted to Mine Layers and Aircraft Tenders. ‖Only 4 tubes in two boats.

148 boats, by various yards, (Flush Deckers).

BY PRIVATE CONTRACTORS.

10 *Bethlehem Fore River Plant, Quincy*: **Aulick, Bancroft, Belknap, Gillis, McCalla, McCook, Osmond Ingram, Rodgers, Turner, Welles.** Designed S.H.P. 27,000 = 35 kts. Curtis turbines and reduction gear. 2 screws. 4 Yarrow boilers—27,540 sq. ft. heating surface. Other details as "General Notes."

35 *Bethlehem Destroyer Plant (Squantum)*: **Bailey, Ballard, Billingsley, Breck, Case, Charles Ausburne, Converse, Dale, Delphy, Doyen, Edwards, Flusser, Greene, Henshaw, Isherwood, Lardner, Laub, McCawley, McDermut, McLanahan, Meade, Meyer, Moody, Morris, Osborne, Putnam, Reid, Sharkey, Shubrick, Sinclair, Swasey, Thornton, Tingey, Toucey, Worden.** Designed S.H.P. 27000 = 35 kts. Curtis turbines with reduction gear. 4 Yarrow boilers—27,540 sq. ft. heating surface. Other details as "General Notes."

39 *Bethlehem Union Plant (San Francisco)*: **Bruce, Chase, Coghlan, Farquhar, Farragut, Fuller, Hull, John Francis Burnes, Kennedy, Kidder, Lamson, La Vallette, Macdonough, Marcus, Mervine, Mullany, Nicholas, Paul Hamilton, Percival, Preston, Reno, Robert Smith, Selfridge, Shirk, Sloat, Somers, S. P. Lee, Stoddert, Sumner, Thompson, William Jones, Wood, Woodbury, Yarborough, Young, Zeilin,** *Nos.* [illegible]. Designed H.P. 27,000 = 35 kts. Machinery: G. E. Curtis turbines and reduction gear. 2 screws. 4 Yarrow boilers—27,540 sq. ft. heating surface. Other details as "General Notes."

25 *Cramp (Philadelphia)*: **Alden, Barker, Borie, Broome, Bulmer, Chandler, Edsall, Ford, Hovey*, John D. Edwards, Long*, MacLeish, McCormick, Parrott, Paul Jones, Peary, Pillsbury, Pope, Simpson, Smith Thompson, Southard, Stewart, Tracy, Truxton, Whipple.** Designed S.H.P. 27,000 = 35 kts. Parsons turbines with reduction gear. 2 screws. 4 White-Forster boilers—27,500 sq. ft. heating surface. Other details as "General Notes."

14 *Newport News*: **Abel P. Upshur, Branch, Clemson, Dahlgren, Dallas, George E. Badger, Goldsborough, Graham, Herndon, Hunt, Mason, Satterlee, Sommes, Welborn C. Wood.** (Contract for Nos. 200—205 has been cancelled). Designed S.H.P. 27,000 = 35 kts. Westinghouse turbines with reduction gear. 2 screws. 4 White-Forster boilers—27,500 sq. ft. heating surface. Other details as "General Notes."

15 *New York S.B. Co.*: **Barry, Bainbridge, Childs, Goff, Hopkins, Humphreys, James K. Paulding, King, Lawrence, McFarland, Overton, Reuben James, Sands, Sturtevant, Williamson.** Designed S.H.P. 27,000 = 35 kts. Machinery: Westinghouse turbines and reduction gear. 2 screws in all; also 4 White-Forster boilers—27,500 sq. ft. heating surface. Other details as "General Notes."

1 *Union I. W.*†: **Chauncey,** Designed S.H.P. 27,500 = 35 kts. Curtis (geared cruising) turbines. 2 screws. 4 Yarrow boilers—27,540 sq. ft. heating surface. Otherwise as "General Notes."

BY NAVY YARDS.

6 *Mare Island N. Yd.*: **Decatur, Litchfield, Perry, Trever, Wasmuth, Zane.** Designed S.H.P. 26,000 = 35 kts. 2 screws. 4 Normand boilers—27,000 sq. ft. heating surface. Otherwise as "General Notes."

3 *Norfolk N. Yd.*: **Hulbert, Noa, William B. Preston.** S.H.P., speed, turbines, boilers, as Mare Island, N. Yd., boats just above. Other details as "General Notes."

General Notes.—*Normal* displacement, 1215 tons (1308 *full load*). Dimensions: 310 (*w.l.*) × 30' 11½" × 9¼ feet mean (draught 10' 9¼" *full load*). Armament: 4—4 inch*, 50 cal., 1—3 inch, 23 cal. AA., 12—21 inch tubes in 4 triple deck mountings. (S.H.P., speed, machinery, boilers, as separately noted to each class). Oil fuel: — tons. Complement, 122 (peace), 130 (war).

Trials.—Very few boats have attained the designed speed of 35 kts. The average trial speed can be taken as about 34.5 kts.

Appearance Note.—All have stump mainmast, S.L. on tower between best skids and side under beam guns plated over. This note is only good until July, 1921, as all other Flush Deckers will eventually be altered to appearance of above boats.

**Hovey, Long*, have each 8—4 inch paired on twin mountings in same positions as other boats.

†Contract for *Chauncey* assigned to Union I. W., but while building the Yard became the Bethlehem Union Plant.

For illustrations v. opposite page.

Builder	Name	Begun.	Launch.	Comp.
Bethlehem (Quincy).	Aulick	3/12/18	11/4/19	26/7/19
	Bancroft	4/11/18	21/3/19	30/6/19
	Belknap	3/10/18	14/1/19	28/4/19
	Gillis	27/12/18	29/5/19	3/9/19
	McCalla	25/9/18	18/2/19	30/4/19
	McCook	10/3/18	31/3/19	19/5/19
	O. Ingram	15/10/18	28/2/19	27/6/19
	Rodgers	3/10/18	21/12/18	28/3/19
	Turner	21/12/18	17/5/19	24/3/19
	Welles	18/11/18	8/5/19	2/9/19
Bethlehem (Squantum).	Bailey	3/10/18	5/2/19	27/6/19
	Ballard	3/6/18	7/12/18	5/6/19
	Billingsley	28/4/18	10/12/19	1/2/20
	Breck	8/6/19	5/9/19	28/11/19
	Case	3/6/19	21/9/19	*29/11/19
	C. Ausburne	27/4/18	14/12/19	28/2/20
	Converse	18/3/18	28/11/19	27/4/20
	Dale	6/1/18	19/11/19	14/2/20
	Delphy	20/4/18	18/7/18	30/11/18
	Doyen	24/3/19	26/7/19	10/10/19
	Edwards	20/4/18	10/10/18	24/4/19
	Flusser	5/3/18	7/11/19	31/12/19
	Greene	3/6/18	2/11/18	9/5/19
	Henshaw	31/12/18	28/6/19	24/6/19
	Isherwood	24/5/19	10/9/19	26/11/18
	Lardner	10/6/19	29/9/19	10/12/18
	Laub	20/4/18	25/8/18	17/3/19
	McCawley	4/11/18	14/6/19	29/8/19
	McDermut	20/4/18	6/8/18	27/3/19
	McLanahan	20/4/18	22/9/18	5/4/19
	Meade	24/9/18	24/5/19	6/9/19
	Meyer	6/2/19	18/7/19	30/9/19
	Moody	10/12/18	28/6/19	25/9/19
	Morris	20/7/18	12/4/19	25/7/19
	Osborne	4/11/18	29/12/19	17/3/20
	Putnam	30/6/19	30/9/19	18/12/19
	Reid	18/3/18	15/10/19	6/11/19
	Sharkey	14/4/19	12/8/19	20/10/19
	Shubrick	3/6/18	31/12/19	3/7/19
	Sinclair	10/10/18	2/6/19	26/8/19
	Swasey	27/8/18	7/5/19	31/7/19
	Thornton	3/6/18	22/3/19	15/7/19
	Tingey	8/8/18	24/4/19	27/7/19
	Toucey	26/4/19	5/6/19	31/10/19
	Worden	30/6/19	24/10/19	21/12/19
Bethlehem, San Francisco.	Bruce	20/7/19	20/5/20	....
	Chase	5/5/19	2/9/19	....
	Coghlan	25/6/19	16/6/21	....
	Farquhar	13/8/18	18/1/19	....
	Farragut	4/7/18	21/11/18	4/6/20
	Fuller	4/7/18	5/12/18	28/2/20
	Hull	....	....	....
	J. F. Burnes	4/7/18	10/11/18	1/5/20
	Kennedy	25/9/18	15/2/19	....
	Kidder	5/3/19	10/7/19	....
	Lamson	13/8/19	....	....
	La Vallette	14/4/19	15/7/19	....
	MacDonough	24/5/20	....	....
	Marcus	20/5/19	22/8/19	....
	Mervine	28/4/19	11/8/19	....
	Mullany	3/8/19	....	....
	Nicholas	11/3/19	1/5/19	....
	P. Hamilton	25/9/18	21/2/19	....
	Percival	4/7/18	5/12/18	31/3/20
	Preston	19/7/18	....	....
	Reno	4/7/18	22/1/19	23/7/20
	R. Smith	13/5/19	19/9/19	....
	Selfridge	28/4/19	25/7/19	....
	Shirk	13/2/19	20/6/19	....
	Sloat	18/1/19	14/5/19	....
	Somers	4/7/18	28/12/18	21/6/20
	S. P. Lee	31/12/18	22/4/19	....
	Stoddert	4/7/18	8/1/19	30/6/20
	Thompson	14/8/18	15/1/19	....
	Wm. Jones	10/2/18	9/4/19	....

Builder	Name	Begun.	Launch.
Bethlehem, San Francisco, continued.	Wood	23/4/19	28/5/19
	Woodbury	3/10/18	6/2/19
	Yarborough	27/2/19	20/6/19
	Young	20/1/19	8/5/19
	Zeilin	20/1/19	28/5/19
	(322)	....	....
	Sumner	27/8/19	....
	(334)	....	....
	(335)	....	....
Cramp.	Alden	24/10/18	7/6/19
	Barker	30/4/19	11/9/19
	Borie	30/4/19	4/10/19
	Broome	8/10/18	14/5/19
	Bulmer	11/8/19	22/1/20
	Chandler	19/8/18	10/3/19
	Edsall	15/9/19	....
	Ford	11/11/19	....
	Hovey	7/9/18	26/4/19
	J. D. Edwards	21/5/19	18/10/19
	Long	23/10/18	26/4/19
	MacLeish	19/8/19	18/12/19
	McCormick	11/8/19	14/2/20
	Parrott	23/7/19	23/11/19
	Paul Jones	23/12/18	....
	Peary	9/9/19	6/4/20
	Pillsbury	23/10/19	....
	Pope	9/9/19	21/5/20
	Simpson	7/10/19	28/1/20
	S. Thompson	24/3/19	14/7/19
	Southard	18/8/18	31/3/19
	Stewart	9/9/19	4/1/20
	Tracy	3/4/19	12/8/19
	Truxtun	3/12/19	....
	Whipple	12/6/19	6/11/19
Newport News.	A. P. Upshur	20/8/18	14/2/20
	Clemson	11/5/18	5/9/18
	Dahlgren	8/6/18	20/11/18
	Dallas	25/11/18	31/5/19
	G. E. Badger	24/9/18	6/3/21
	Goldsboro'	8/6/18	20/11/18
	Graham	7/9/18	22/3/19
	Herndon	25/11/18	31/1/19
	Hunt	20/8/18	14/2/20
	Mason	10/7/18	8/3/19
	Satterlee	10/7/18	21/12/18
	Sommes	10/7/18	21/12/18
	W. C. Wood	24/9/18	6/3/20
N.Y. S.B. Co.	Barry	26/7/19	....
	Bainbridge	27/5/19	12/6/20
	Childs	19/3/19	— — 20
	Goff	16/6/19	2/6/20
	Hopkins	30/6/19	26/6/20
	Humphreys	31/8/18	28/7/19
	J. K. Paulding	31/7/18	20/4/20
	King	28/4/19	....
	Lawrence	14/8/18	....
	McFarland	31/7/18	30/3/20
	Overton	30/10/18	10/7/19
	R. James	2/1/19	4/10/19
	Sands	22/3/19	28/10/19
	Sturtevant	23/11/18	....
	Williamson	27/3/19	16/10/19
U.I.W.	Chauncey	17/6/18	29/9/18
Navy Yards.	Decatur	....	....
	Hulbert	18/11/18	28/6/19
	Litchfield	15/1/19	12/6/19
	Noa	18/11/18	28/8/19
	Perry	....	....
	Trever	12/8/19	....
	Wasmuth	12/8/19	9/9/19
	W. B. Preston	18/11/18	12/8/19
	Zane	15/1/19	

Illustrations for "148 boats by various Yards (Flush Deckers)" listed on opposite page.

ALDEN (and Cramp boats). *1920 Photo, Mr. Metcalf, Midshipman, R.N.*

CONVERSE (and Bethlehem Squantum boats). *1920 Photo.*

No photos available.

Generally as types illustrated elsewhere on this page, but boats launched from (about) middle of 1920 and onwards have guns arranged as *Preble, &c.*; also twin SL. like *Brooks, &c.*

BOATS BY NEWPORT NEWS, NEW YORK S.B. CO., UNION I.W., AND NAVY YARDS.

[illegible] (and Bethlehem Fore River boats). *1920 Photo.*

Note.—Bethlehem (San Francisco) boats same appearance as *Converse* and *Rodgers* above.

Notes for all "Flush-decker" Types.

Commencing with *Caldwell* (69), concluding with *Pruitt* (347), and after deducting 6 cancelled boats (200—205), in all there are 273 boats. Type is believed to have originated with the Bath Iron Works' design for *Manley*. Design is stand excepting that (a) the various Bethlehem Yards followed their own design, the other yards adopting the Bath Iron Works' plans, and (b) very minor variations were introduced to conform with constructional practice of each y Boats on this page and preceding two pages are distinguished at present by stump mainmasts and mounting of S. L. on tower between boat skids. However, the whole 273 "Flush Deck" T.B.D., whether in service, buildin completing, will be modified as *Pruitt* Class below, *i.e.*, (a) with stump mainmast, (b) S. L. on tower between boat skids, (c) after 4 inch gun raised on after deck-house and 3 inch AA. gun mounted on quarter deck. When t alterations are carried out, the whole series will be uniform in appearance, excepting perhaps such boats as may be modified for duty as Light Mine Layers, Aircraft Tenders, &c.

Twelve Boats, *Nos.* 348—359, not contracted for. Although not definitely cancelled, it is most unlikely they will be built now. They are therefore omitted from all and any totals.

3 Bath I.W. Boats (Flush Deckers).

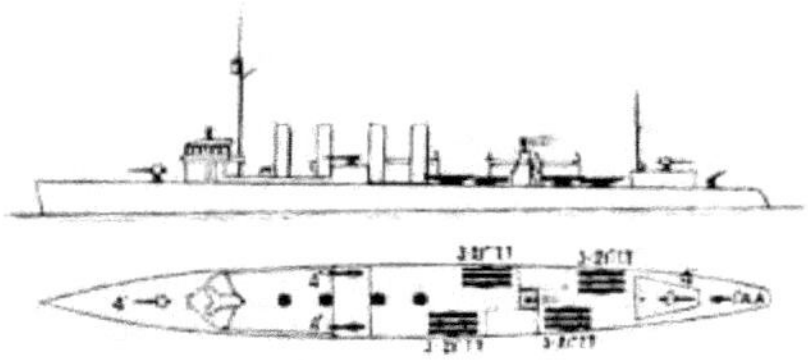

3 *Bath I.W.*: **Preble, Pruitt, Sicard.** Designed S.H.P. 26,000 = 35 kts. Parsons turbines and reduction gears. 2 screws. 4 Normand boilers—27,000 sq. ft. heating surface. Other details as "General Notes."

General Notes.—*Normal* displacement: 1215 tons (1308 *full load*). Dimensions: 310 (*w.l.*), 314 ft. 4½ in. (*o.a.*) × 30 ft. 11¼ in. × 9 ft. 4 in. mean draught, 9 ft. 0¾ in. *full load*. Armament: 4—4 inch, 50 cal., 1—3 inch, 23 cal. AA., 12—21 inch tubes in 4 triple deck mountings. Oil fuel: tons. Complement, 122. In these boats the after 4 inch gun is on deck-house and the AA. gun is on quarter deck—a modification which will be effected in all other "Flush Deck" T.B.D.

		Begun.	Launch.	Comp.
B.I.W.	*Preble*	12/4/19	8/3/20	19/3/20
	Pruitt	25/6/18	....	....
	Sicard	18/6/19	20/4/19	1/5/20

5 New York S.B. Co. Boats (Flush Deckers).

BROOKS. 1920 *Photo, by courtesy of Buil*

5 *New York S. B. Co.*: **Brooks, Fox, Gilmer, Hatfield, Kane.** Designed S.H.P. 27,000 = Westinghouse turbines with reduction gear. 4 White-Forster boilers—27,500 sq. ft. heating surface details as "General Notes."

General Notes.—Normal displacement, 1215 tons (1308 *full load*). Dimensions: 310 (*w.l.*) × 30' 11¼" × 9¼ f draught (9' 9¾" *full load*). Armament: 4—5 inch, 51 cal., 1—3 inch, 23 cal. AA., 12—21 inch tubes in 4 tr mountings. (S.H.P., speed, machinery and boilers as noted above.) Oil fuel: tons. Complement, same as the 148 boats previously described, but are more powerfully armed and have twin S. L. between boat skids.

		Begun.	Launch.	Comp.
N.Y.S.B.Co.	*Brooks*	11/6/18	24/4/19	18/6/20
	Fox	25/6/18	12/6/19	17/5/20
	Gilmer	25/6/18	24/5/19	30/4/20
	Hatfield	10/6/18	17/3/19	16/4/20
	Kane	3/7/18	12/8/19	11/6/20

11 Newport News boats. (Flush Deckers.)

LAMBERTON. 1919 *Photo.*

11 *Newport News*: **Abbott, Bagley, Breese, Gamble, Haraden, Hopewell, Lamberton, Montgomery, Radford, Ramsay, Thomas.**

General Notes.—Normal displacement 1213 tons (1306 *full load*). Dimensions: 310 (*w.l.*) × 30′ 11⅝″ × 9′ 3⅝″ mean draught (10′ 3¼″ *full load*). Armament: 4—4 inch, 50 cal., 2—3 inch, 23 cal. A.A. Designed S.H.P. 26,000 = 35 kts. 2 screws. Machinery: Curtis (geared cruising) turbines. Boilers: 4 Thornycroft—28,000 sq. ft. heating surface. Oil: — tons. Complement, 122 (peace), 139 (War).

	Begun.	Launch.	Comp.		Begun.	Launch.	Comp.
Abbot ..	5/4/18	4/7/18	18/7/19	*Lamberton* ..	1/10/17	30/3/18	22/8/18
Bagley ..	11/5/18	19/10/18	26/8/19	*Montgomery* ..	2/10/17	23/3/18	26/7/18
Breese ..	10/11/17	11/5/18	23/10/18	*Radford* ..	2/10/17	5/4/18	30/9/18
Gamble ..	12/11/17	11/5/18	29/11/18	*Ramsay* ..	21/12/17	8/6/18	14/2/19
Haraden ..	30/3/18	4/7/18	6/6/19	*Thomas* ..	23/3/18	4/7/18	25/4/19
Hopewell ..	19/1/18	8/6/18	21/3/19				

Trials.—None attained 35 kts. Majority made 33 to 34 kts. *Hopewell* only made 28.22 kts. on preliminary acceptance trials in shallow water. On later official trials made over 30 kts.

10 New York S. B. boats. (Flush Deckers.)

HERBERT. 1921 *Photo, by courtesy of Builders.*

10 *New York S. B. Co.*: **Babbitt, Badger, De Long, Dickerson, Herbert, Jacob Jones,* Leary, Schenck, Tattnall,* Twiggs.***

General Notes.—Displacements, 1211 tons normal (1304 or 1306 *full load*). Dimensions: 310 (*w.l.*) × 30′ 11⅝″ × 9¼ feet mean draught (9′ 9¼″ *full load*). Armament: 4—4 inch, 50 cal., 2—3 inch, 23 cal. A.A. Torpedo tubes: 12—21 inch, in 4 triple deck mountings. Designed S.H.P. 26,000 = 35 kts. 2 screws. Machinery: Parsons turbines with reduction gear. 4 Thornycroft boilers—27,648 sq. ft. heating surface. Oil only: — tons. Complement, 122 (peace), 139 (War).

	Begun.	Launch.	Comp.		Begun.	Launch.	Comp.
Babbitt ..	19/2/18	30/9/18	24/4/19	*Jacob Jones* ..	21/2/18	20/11/18	20/10/19
Badger ..	9/1/18	24/8/18	29/5/19	*Leary* ..	6/3/18	21/12/18	5/12/19
De Long ..	21/2/18	29/10/18	20/9/19	*Schenck* ..	26/3/18	23/4/19	30/10/19
Dickerson ..	25/5/18	12/3/19	3/9/19	*Tattnall* ..	1/12/17	5/9/18	26/6/19
Herbert ..	9/4/18	18/5/19	21/11/19	*Twiggs* ..	23/1/18	28/9/18	28/7/19

Trials.—Exceeded 35 kts. *Badger*, 35.35. *Dickerson*, 35.13. *Tattnall*, 35.88.

52 Fore River* & Union I. W. boats. (Flush Deckers.)

Luce (& Fore River boats). 1920 *Photo.*

Burns (& Union I. W. boats). 1920 *Pho*

26 *Fore River** : **Bell, Bush, Colhoun, Cowell, Crosby, Dyer, Foote, Gregory, Israel,† Kalk, Kimberly, Lansdale,† Little, Luce,† Maddox, Mahan,† Maury,† Meredith, Murray,† Palmer, Sigourney, Stevens, Stribling,† Stringham, Thatcher, Walker.**

Trials.—From 34.34 (*Israel*) to 35.77 (*Bush, Palmer*). Fifteen boats did not attain 35 kts.

26 *Union I. W.** : **Anthony,† Burns,† Champlin, Chew, Crane, Gridley, Harding, Hart,† Hazlewood, Hogan, Howard, Ingraham,† Ludlow,† Mackenzie,† McKean, McKee, Mugford,‡ O'Bannon, Renshaw, Ringgold, Rizal,† Robinson, Schley, Sproston†, Stansbury, Williams.**

Trials.—Best boats *Harding*, 35.02, *Schley*, 35.2. Others under 35 kts, down to *Hazlewood*, 33.56.

Displacement, 1191 tons *normal* (1284 tons *full load*). Dimensions: 310 (*w.l.*) × 30 ft. 11¼ in. × 8 ft. 1¾ in. (*mean*) draught (9¼ feet *full load*). Armament: 4—4 inch, 50 cal. and 2—3 inch, 23 cal. AA. Tubes: 12—21 inch in 4 triple deck mountings (*v. Footnotes* † ‡). Designed S.H.P. 27,000 = 35 kts. Machinery: *Gridley, Harding, McKee, McKean, Ringgold, Robinson*, G. E. turbines with reduction gear; all other boats, Curtis with reduction gear. 2 screws. 4 Yarrow boilers—27,540 sq. feet heating surface. Oil fuel: tons. Complement, 122 (peace), 139 (war). *Rizal* built at expense of Philippines Government, and manned by Filipino crew.

*While work was proceeding on these boats, both Yards were absorbed by the Bethlehem Shipbuilding Corporation.

†Are Light Mine Layers; some (or all) tubes removed.

‡Tender to Fleet Air Detachment. All tubes removed and replaced by 1,500 gallon petrol tanks.

FORE RIVER BOATS:—

	Begun.	Launch.	Comp.		Begun.	Launch.
Bell	15/11/17	20/4/18	1/8/18	*Luce*	9/2/18	29/6/18
Bush	4/7/18	27/10/18	19/2/19	*Maddox*	27/7/18	27/10/18
Colhoun	19/9/17	21/2/18	12/6/18	*Mahan*	4/5/18	4/8/18
Cowell	15/7/18	23/11/18	17/3/19	*Maury*	25/2/18	4/7/18
Crosby	23/6/18	28/9/18	24/1/19	*Meredith*	29/6/18	22/9/18
Dyer	28/9/17	13/4/18	2/7/18	*Murray*	22/12/17	3/6/18
Foote	7/8/18	14/12/18	21/3/19	*Palmer*	29/5/18	18/8/18
Gregory	25/8/17	25/1/18	31/5/18	*Sigourney*	25/7/18	16/12/17
Israel	26/1/18	22/6/18	20/8/18	*Stevens*	20/10/17	13/1/18
Kalk	19/8/18	21/12/18	29/3/19	*Stribling*	14/12/17	29/5/18
Kimberley	21/6/17	4/12/17	27/4/18	*Stringham*	19/10/17	30/3/18
Lansdale	20/4/18	21/7/18	26/10/18	*Thatcher*	8/6/18	31/8/18
Little	18/6/17	11/11/17	5/4/18	*Walker*	18/6/18	14/9/18

UNION I. W. BOATS:—

	Begun.	Launch.	Comp.		Begun.	Launch.
Anthony	18/4/18	10/8/18	19/6/19	*Mackenzie*	4/7/18	29/7/18
Burns	15/4/18	4/7/18	7/8/19	*McKean*	12/2/18	4/7/18
Champlin	31/10/17	7/4/18	11/11/18	*McKee*	29/10/17	3/3/18
Chew	2/1/18	26/5/18	12/12/18	*Mugford*	20/12/17	14/4/18
Crane	7/1/18	4/7/18	18/4/19	*O'Bannon*	12/11/18	28/2/19
Gridley	1/4/18	4/7/18	8/3/19	*Renshaw*	8/5/18	21/9/18
Harding	12/2/18	4/7/18	24/1/19	*Ringgold*	20/10/17	14/4/18
Hart	8/1/18	7/4/18	26/5/19	*Rizal*	28/6/18	21/9/18
Hazlewood	24/12/17	22/6/18	20/2/19	*Robinson*	31/10/17	28/3/18
Hogan	25/11/18	12/4/19	30/9/19	*Schley*	29/10/17	28/3/18
Howard	9/2/18	26/4/19	28/1/20	*Sproston*	20/4/18	10/8/18
Ingraham	12/1/18	7/7/18	15/5/19	*Stansbury*	9/12/18	16/5/19
Ludlow	7/1/18	9/6/18	23/12/18	*Williams*	25/3/18	4/7/18

(*Continued on next*

21 Cramp boats (Flush Deckers).

DENT. 1919 *Photo*.

1920 *Photo by courtesy of Builders*.

21 *Cramp*: **Barney, Bernadou, Biddle, Blakely, Breckinridge, Cole, Dent, Dorsey, Du Pont, Elliott, Ellis, Greer, J. Fred Talbott, Lea, Rathburne, Roper, Talbot, Tarbell, Upshur, Waters, Yarnall.**

Displacements: 1154 tons *normal* (1247 tons *full load*). Dimensions: 310 (*w.l.*) × 30′ 11¼″ × 9 feet mean draught. *Full load* draught, 9¼ feet. Armament: 4—4 inch, 50 cal., 2—3 inch, 23 cal. AA., 12—21 inch tubes in 4 triple deck mountings. Designed S.H.P. 26,000 = 35 kts. Machinery: Parsons turbines with reduction gear. Boilers: 4 White-Forster—27,500 sq. feet heating surface. Oil: 294* = 4960* miles at 14 kts.* Complement, 122 (peace), 139 (war). *Trials*: All made *about* 35.1 on *trials*, except *Upshur*, 34.61.

*Unofficial figures.

	Begun.	Launch.	Comp.
Barney	26/3/18	5/9/18	14/3/19
Bernadou ..	4/6/18	7/11/18	19/5/19
Biddle	22/4/18	3/10/18	22/4/19
Blakely	26/3/18	19/9/18	8/3/19
Breckinridge ..	11/3/18	17/8/18	27/2/19
Cole	25/6/18	11/1/19	19/6/19
Dent	30/8/17	23/3/18	9/9/18
Dorsey	18/9/17	9/4/18	15/9/18
Du Pont ..	2/5/18	22/10/18	30/4/19
Elliott	24/2/18	4/7/18	25/1/19
Ellis	8/7/18	30/11/18	7/6/19
Greer	24/2/18	1/8/18	31/12/18
J. F. Talbott ..	8/7/18	14/12/18	30/6/19
Lea	18/9/17	29/4/18	21/10/18
Rathburne ..	12/7/17	27/12/17	24/6/18
Roper	31/2/18	17/8/18	15/2/19
Talbot	12/7/17	20/2/18	20/7/18
Tarbell	31/12/17	28/5/18	27/11/18
Upshur	19/2/18	4/7/18	20/12/18
Waters	26/7/17	9/8/18	8/8/18
Yarnall	12/2/18	19/6/18	29/11/18

17 Bath I.W. & Navy Yard boats (Flush Deckers).

EVANS. 1919 Photo

4 *Bath I. W.*: **Aaron Ward, Buchanan, Crowninshield, Hale.** Designed S.H.P. 24,200 = 35 Otherwise as "General Notes" below.

4 *Bath I. W.*: **Evans, Philip, Wickes, Woolsey.** Designed S.H.P. 24,610 = 35 kts. *Trials*: *Philip* ... *Wickes* 35.34. *Woolsey* 24,575 S.H.P. = — kts. Otherwise as "General Notes" below.

6 *Mare Island N. Y.*: **Boggs, Claxton, Hamilton, Kennison, Kilty, Ward.** Designed S... 24,000 = 35 kts. Otherwise as "General Notes" below.

2 *Mare Island N. Y.*: **Fairfax, Taylor.** Designed S.H.P. 24,200 = 35 kts. Otherwise as "General N... below.

1 *Charleston N. Y.*: **Tillman.** As *Boggs, Claxton, &c.*, above.

General Notes.—Displacements: 1154 tons *normal* (1247 *full load*). Dimensions: 310 (*w.l.*) × 30′ 11¼″ × 9 feet ... draught (*full load* draught, 9¼ feet). Armament: 4—4 inch, 50 cal., 2—3 inch, 23 cal. AA., 12—21 tubes in 4 triple deck mountings. Designed S.H.P. and speed as separately noted. Machinery: Parsons tu... with reduction gear. 2 screws. 4 Normand boilers—27,000 sq. ft. heating surface. Oil fuel only: — Complement, 122 (peace), 139 (war).

	Begun.	Launch.	Comp.
A. Ward ..	1/8/18	10/4/19	21/4/19
Buchanan ..	29/6/18	2/1/19	20/1/19
Crowninshield ..	5/11/18	24/7/19	6/8/19
Hale	7/10/18	25/9/19	12/6/19
Evans	28/12/17	30/10/18	11/11/18
Philip	1/9/17	25/7/18	24/8/18
Wickes	26/6/17	25/6/18	31/7/18
Woolsey	1/11/17	17/9/18	30/9/18

	Begun.	Launch.	C...
Boggs	15/11/17	25/4/18	[illegible]
Claxton	25/4/18	15/1/19	
Hamilton ..	8/6/18	10/1/19	
Kennison ..	14/2/18	8/6/18	[illegible]
Kilty	15/12/17	25/4/18	[illegible]
*Ward**	15/5/18	1/6/18	[illegible]
Fairfax	10/7/17	15/12/17	[illegible]
Taylor	15/10/17	14/2/18	[illegible]
Tillman	29/7/18	7/7/19	[illegible]

*Note rate of building.

6 Caldwell class. (Flush Deckers).

STOCKTON (& CONNER). *Photo, Lieut. H. Reuterdahl, U.S.N.R.F.*

2 *Cramp*: **Conner, Stockton.** *Conner*, 1225 tons *full load*; *Stockton*, 1238 tons *full load*. Designed H.P. 20,000 = 30 kts. Parsons (geared cruising) turbines. 3 screws. Boilers: 4 White-Forster—21,620 sq. ft. heating surface. *Trials*: *Conner* 30·17, *Stockton* 30·12.

Note.—*Stockton* has 5—4 inch, 2—4 inch being paired on the fo'xle mounting. Otherwise as "General Notes" below.

Photo wanted.

Each has *four* funnels.

CRAVEN, CALDWELL, MANLEY.

1 *Norfolk N.Yd.*: **Craven.** 1187 tons *full load*. Designed H.P. 20,000 = kts. Parsons turbines, with reduction gear. 2 screws. Boilers: 4 Thornycroft—21,500 sq. ft. heating surface. *Trials*: 32·33.

1 *Mare Island N.Yd.*: **Caldwell.** As *Craven*, above, but engined with G. E. Curtis turbines and reduction gear. *Trials*: 31·7.

1 *Bath I. W.*: **Manley** (1917). As *Craven*, above, but boilers 4 Normand—21,500 sq. ft. heating surface. *Trials*: 32·23.

General Notes.—*Normal* displacement: 1125 tons, *full load* as separately noted. Dimensions: 310 (*w.l.*) × 30⅝ × 8 feet (mean draught). *Full load* draught: 9 feet 6½ inches. Armament: *Stockton* 5—4 inch (50 cal.), others 4—4 inch (50 cal.), (and in all) 2—3 inch anti-aircraft, 12—21 inch tubes in 4 triple deck mountings. 280 tons oil fuel. Complement 122 (139 War), *Stockton* 128 (134 War).

CALDWELL CLASS:—

Photo wanted.

Has *three* funnels.

GWIN.

1 *Seattle Con. & D. D. Co.*: **Gwin.** 1228 tons *full load*. Designed H.P. 18,750 = 30 kts. Parsons turb reduction gear. 2 screws. Boilers: 4 Yarrow—22,252 sq. ft. heating surface. Otherwise as "Genera in preceding column.

	Begun.	Launch.	Comp.		Begun.	Launch.
Conner	16/10/16	21/8/17	10/2/18	*Caldwell*	9/12/16	10/7/17
Stockton	13/10/16	17/7/17	26/11/17	*Gwin*	21/6/17	22/12/17
Craven	20/11/17	29/6/18	19/10/18	*Manley*	22/8/16	23/8/17

*The first boat completed of the long "Flush Decker" series.

1 Special Boat.

ISABEL. *Photo added.*

1 *Bath Iron Works*: **Isabel** (ex-yacht, launched, purchased and converted, 1917). 797 tons *normal* (930 load). Dimensions: 230 (*w.l.*), 245⅓ (*o.a.*) × 26⅔ × 9¼ feet mean draught. Complement, 115. Ar 4—3 inch (50 cal., S.A.), 2 machine AA. Tubes: 4—18 inch in two twin mountings. S.H.P. 4400 = *trials*). Parsons turbines and 2 Normand boilers. 2 screws. Oil fuel: tons.

(All on this page unofficially known as "Thousand Tonners.")

6 Allen class.

2 *Bath I.W.*: **Allen** (1916), **Davis** (1916). 1071 tons (1185 *full load*). Designed H.P. 17,500=30 kts. Parsons (geared cruising) turbines. Machinery weighs 350 tons. Boilers: 4 Normand—22,500 sq. ft. heating surface. Trials: *Allen*, 30·29; *Davis*, 30·36 kts.

2 *Fore River*: **Rowan** (1916), **Sampson** (1916). 1110 tons (1125 *full load*.) Designed H.P. 17,000=29·5 kts. Curtis (geared cruising) turbines. Machinery averages 385 tons. Boilers: 4 Yarrow—21,500 sq. ft. heating surface. Trials: *Rowan*, 29·57 kts.; *Sampson*, 29·52 kts.

1 *Mare Island Navy Yard*: **Shaw** (1916). Displacement, H.P. and speed as *Rowan*. Parsons (geared cruising) turbines. Boilers: 4 Thornycroft. Trials: 29·5 kts.

1 *Cramp*: **Wilkes** (1916). 1110 tons (1124 *full load*). Designed H.P. 17,000=29·5 kts. Parsons (geared cruising) turbines. Machinery weighs 361 tons. Boilers: 4 White-Forster. Trials: 29·58 kts.

General Notes.—Dimensions: 310 (*w.l.*) × 29½ × 9¼ to 9¾ feet *mean* draught. *Full load* draught of first two: 9 ft. 9½ in. Others, 10 ft. 8¼ in. Armament: 4—4 inch (50 cal.), 2—3 inch anti-aircraft, 12—21 inch deck tubes in 4 triple deck mountings. Designed H.P. 17,000=29·5 kts. Oil fuel: 290 tons (estimated). Complement: 122/136 war.

SAMPSON. *Davis* has after S.L. before mainmast on high lattice tower. 1919 *Photo, O. W. Waterman.*

5 Conyngham Class.

TUCKER (of Conyngham class). 1919 *Copyright Photo, O. W. Waterman.*

2 *Cramp*: **Conyngham** (1915), **Porter** (1915). 1090 tons (1205 *full load*.) Designed H.P. 18,000=29½ kts. Parsons (geared cruising) turbines. Machinery weighs 375 tons. Boilers: 2 White-Forster—24,000 sq. ft. heating surface. Oil: 308 tons. Trials: *Conyngham*, 29·63 kts.; *Porter*, 29·58 kts.

1 *New York S.B. Co.*: **Wainwright** (1915). 1050 tons (1265 *full load*). Designed H.P. 17,000=29½ kts. Parsons (geared cruising) turbines. Machinery averages 369 tons. Boilers: 4 Normand—21,500 sq. ft. heating surface. Trials: 29·67 kts. Oil: 308 tons. (*Jacob Jones* of this type lost during war.)

1 *Fore River*: **Tucker** (1915). Displacements as *Conyngham*. Designed H.P. 17,000=29½ kts. Curtis (geared cruising) turbines. Machinery weighs 369 tons. Boilers: 4 Yarrow—21,500 sq. ft. heating surface. Oil: 309 tons. Trials: 29·56 kts.

1 *Bath I.W.*: **Wadsworth** (1915). 1060 tons (1174 *full load*). Designed H.P. 17,000=29½ kts. Parsons turbines with reduction gear. Machinery weighs 323 tons. Boilers: 4 Normand—21,500 sq. ft. heating surface. Oil: 316 tons. Trials: 30·67 kts.

General Notes.—Dimensions: 310 (*w.l.*) × 29½ × 9¼ to 9¾ feet (*mean* draught). *Full load* draught 10 feet 1¼ inches to 10 feet 8¼ inches. Armament: 4—4 inch (50 cal.), 8—21 inch tubes in 4 twin-deck mountings. Designed H.P. 17,000=29·5 kts. Unofficial endurance figures: 5640 miles at 14 kts. Complement: 106, (132 war.)

6 Cushing Class.

NICHOLSON. 1919 *Copyright Photo, O. W. Waterman.*

1 *Fore River*: **Cushing** (1912). 1050 tons (1071 *full load*). Parsons (geared cruising) turbines. Machinery weighs 390 tons. Boilers: 4 Bureau Modified Yarrow—21,500 sq. feet heating surface. Trials: 29·18 kts.

1 *New York S.B. Co.*: **Ericsson*** (1914). 1090 tons (1211 *full load*). Parsons geared turbines with reciprocating machinery weighs 364 tons. Boilers: 4 Thornycroft—26,936 sq. ft. heating surface. Trials: 29·29 kts.

1 *Bath I.W.*: **McDougal** (1914). 1025 tons (1139 *full load*). Two sets Parsons turbines and two reciprocating. Machinery weighs 325 tons. Boilers: 4 Normand—21,500 sq. feet heating surface. Trials: 30·7 kts.

3 *Cramp*: **Nicholson** (1914), **O'Brien*** (1914), **Winslow** (1915). 1050 tons (1171 *full load*). 2 Cramp-Zoelly turbines with 2 reciprocating. Machinery weighs 351 tons. Boilers: 4 White-Forster—21,600 sq. feet heating surface. Trials: *Nicholson* 29·08 kts., *O'Brien* 29·17 kts., *Winslow* 29·05 kts.

General Notes.—Dimensions are about 300 (*w.l.*) × 30¼ × 9 feet 4 inches to 9 feet 9 inches (*mean*). *Full load* draught: 9 feet 8 inches to 10 feet 9 inches. Armament: 4—4 inch (50 cal.), 8—21 inch tubes in 4 twin-deck mountings. Designed H.P. 16,000=29 kts. Oil fuel: 305 to 311 tons. Complement: 106, (war 132.) Built to guaranteed radius at 15 kts.

* Have very low mainmasts.

Continued on next page.)

8 Aylwin Class.

(Unofficially known as "Thousand Tonners")

DUNCAN. 1919 *Copyright photo, O. W. Waterman.*

4 *Cramp*: **Aylwin, Balch, Benham, Parker** (1913). 1036 tons (1158 *full load*). Cramp-Zoelly turbines. Machinery: 347 tons. Boilers: 4 White-Forster—21,500 sq. feet heating surface. Oil fuel: *Aylwin*, 307 tons; *Balch*, 306 tons; *Benham*, 311 tons; *Parker*, 317 tons. Trials: *Aylwin* 29·6, *Balch* 29·62, *Benham* 29·50, *Parker* 29·58 kts.

2 *Bath I.W.*: **Cassin, Cummings** (1913). 1020 tons (1139 *full load*). Parsons turbines and reciprocating engines (on port shaft only). Machinery: 329 tons. Boilers: 4 Normand—21,500 sq. feet heating surface. Oil fuel: 312 tons. Trials: *Cassin* 30·14. *Cummings* 30·57 kts.

1 *New York S. B. Co.*: **Downes** (1913). 1072 tons (1190 *full load*). Turbines and reciprocating engines. Machinery: 386 tons. Boilers: 4 Thornycroft—26,456 sq. feet heating surface. Oil fuel: 308 tons. Trials: 29·07 kts.

1 *Fore River*: **Duncan** (1913). 1014 tons (1135 *full load*). Turbines and reciprocating engines. Weight: 348 tons. Boilers: 4 Yarrow—21,500 sq. feet heating surface. Oil fuel: 308 tons. Trials: 29·14 kts.

Note.—Dimensions *about* 300 (*w.l.*) × 30½ × 9¼ to 9¾ (*mean*). *Full load* draught 10¼ to 10¾ feet. *Designed* H.P.: 16,000 = 29 kts. Armament: 4—4 in. (50 cal.), 8—18 in. tubes in four twin-mountings. Complement 106, war 134. Built to specified radius at 15 kts. Reciprocating engines not used over 15 kts. Very economical steamers.

BENHAM (*Aylwin* class). 1920 Pho

Aircraft view showing disposition of guns and T.T. Note stump mainmast.

21 Drayton Class.

(Unofficially known as "Flivver Type.")

All 742 tons* (883 *full load*). Dimensions: 289 (*w.l.*) × 26½ × 8¼ feet (*mean*). Armament: *Nil* (1920). *6—tubes (in pairs). Designed H.P. 12,000 = 29½ kts. *Full load* draught, from 9¼ to 11 feet. Complement 90, w

**Mc Call, Sterett* have only 4—18 inch T.T. in pairs, 1920.

TRIPPE (all have *four* funnels). 1919

5 *Bath I.W.*: **Drayton** (1910), **Jenkins** (1912), **Jouett** (1912), **Trippe** (1910), **Paulding** (1910). Mach Parsons turbines. Weight of machinery: 263 tons in first three, 270 tons in *Trippe*, 269 tons in *Paulding*. B 4 Normand. Oil fuel: *Drayton*, 227 tons; *Jenkins*, 222 tons; *Jouett*, 225 tons; *Trippe*, 224 tons; *Pauldi* tons. Trials: *Drayton*, 30·83; *Jouett*, 32·27; *Jenkins*, 31·17; *Paulding*, 32·80; *Tripp*e, 30·89.

(*Continued on next pa*

U. S. DESTROYERS ("Flivver" Type)—*Continued.*

21 **Drayton class**—*Continued.*

(General details given on preceding page.)

No photo available.
All have *four* funnels like *Beale*

4 *New York S. B. Co.*: **Ammon** (1910), **Burrows** (1910), **Jarvis** (1912), **McCall** (1910). Parsons turbines* 3 screws. Machinery: 269 tons in *Jarvis*, 288 tons in others. Boilers: 4 Thornycroft. Oil fuel: **224** tons in *Burrows* and *McCall*, 223 in *Jarvis*, 227 in *Ammon*. Trials: *Ammen*, 30·48; *Burrows*, 30·41; *Jarvis*, 30·04; McCall, 30·66 kts.

*Geared cruising in *Jarvis* only.

BEALE (only). 1919 *Copyright photo, O. W. Waterman.*

No photos.
Have *three* funnels like *Roe* and *Perkins*.

MAYRANT, PATTERSON, WARRINGTON.

4 *Cramp*: **Beale** (1912), **Mayrant** (1910), **Patterson** (1911), **Warrington** (1910). *Mayrant* re-engined 1916-17 with Westinghouse turbines and mechanical reduction gear (284 tons). *Warrington* has Cramp-Zoelly turbines. 2 screws. Weight of machinery, 284 tons. Boilers: 4 White-Forster. *Beale* and *Patterson* have Parsons turbines. 3 screws. Machinery: 273 tons. Oil fuel: 227 tons *Patterson*, *Beale*: 236 tons in other two. Boilers: 4 White-Forster. *Trials*: *Beale*, 29.55; *Mayrant*, 30.22; *Patterson*, 29.69; *Warrington*, 30.12 kts.

No photo.

HENLEY only: *four* funnels, like *Beale*.

PERKINS (*Walke* and *Sterett* similar). 1920 *Photo.*

4 *Fore River*: **Henley** (1912), **Perkins** (1910), **Sterett** (1910), **Walke** (1911). *Henley* re-engined 1916-[illegible] with Westinghouse turbines and mechanical reduction gear (305 tons). Other three Curtis turbines only (290 tons) 2 screws. Boilers: 4 Yarrow. Oil fuel: *Walke*, 238 tons; others, 230 tons. Trials: *Henley*, 30·72; *Perkins* 29·78; *Sterett*, 30·37; *Walke*, 29·78 kts.

No photo.
Has *four* funnels like *Beale*.

FANNING only.

ROE (MONAGHAN and TERRY). 1919 *Photo.*

4 *Newport News*: **Fanning** (1912), **Monaghan** (1911), **Roe** (1909), **Terry** (1909). Parsons turbines. 3 screws. Machinery: 276 tons. Boilers: 4 Thornycroft. Oil fuel: 222-223 tons. *Trials*: *Fanning*, 29·99; *Monaghan*, 30.45; *Roe*, 29.60; *Terry*, 30.24 kts.

U. S. EAGLE BOATS & SUBMARINE CHASERS.

Fabricated Patrol Vessels.

EAGLE 17. — 1919 Photo by courtesy of the Ford Motor Co., Detroit.

54 "EAGLE BOATS":—

Eagle 1—15
Eagle 17—19
Eagle 23—24
Eagle 26—29
Eagle 31—60

Missing Numbers:—Transferred to Coast Guard: *16, 20, 21, 22, 30.* Capsized and lost, June, 1920: *25.*

All launched 1918.

Displacement, *normal* 500 tons, *full load* 615 tons.

Dimensions: Length, 200 feet (*p.p.* and *o.a.*). Beam, 25½ feet. *Mean* draught, 7¼ feet; *full load*, 8½ feet. (Tip of propeller blade passes 3 ft. 2½ ins. below draught line).

Armament: 2—4 inch (50 cal.), 2—3 inch AA., 2 M.G. (1 Y-gun in *Nos.* 4—7 only, for training purposes). Carry 12 depth charges.

Machinery: Poole geared turbine. 2 Bureau Express boilers. 1 screw. Designed H.P. 2500 = 18 kts. Fuel: 105 tons coal + 45 tons oil. Endurance: 3500 miles at 10 kts. Machinery weighs 93.7 tons. Heating surface, 3000 sq. feet.

Complement, 61.

Notes.—Assembled by Ford Eagle Boat Plant, River Rouge; machinery and fittings by Ford Plant, Highland Park, Detroit. Estimated cost, £275,000 per vessel, but this has been largely exceeded. Are the American equivalent of the British "P-boats." The first British P-boat was commissioned on January 1st, 1916; the first Eagle Boat on October 28th, 1918. A very large sum of money was spent on erecting the fabricating plant for the Eagle Boats. British shipbuilders, without any extraordinary preparations, additional plant or slips, produced the larger British P-boats in about the same time as the Ford Eagle Boats are built. Contract for *Nos.* 61—112 cancelled November, 1918.

Submarine Chasers ("SC." boats).

SC 200. — 1919.

115*—110 ft. boats:—Numbered between **SC. 23** and **SC. 444.**† Built 1917-19 by Navy Yards and private contractors.§

Designed displacement, 54 tons; *actual* displacement, owing to heavier guns and alterations, 77 tons *normal*, 85 tons *full load*. Wooden hulls. Length, 105 feet 110 feet (*o.a.*). Beam, 14 feet 8¾ ins. *Mean* hull draught, 5 feet 5¼ ins., *full load* aft 8½ ins. Machinery: three sets of 220 B.H.P. Standard petrol motors = 18 kts. speed. Owing to added weights, 16·85 kts. is reported to be actual full speed. 2400 petrol = 900 miles at 10 kts. Complement, 26. Have small radius W/T.

Armaments: Designed for 1—6 pdr. and 2 M.G., but these were only mounted in a few. Majority have 1—3 inch (23 cal.), and 2 Colt M.G. Some boats have an extra 6 pdr. depth charges.

Hydrophone Gear: Boats fitted out in this way have either (*a*) K-tube fish hydrophones of acoustic radius, or (*b*) SC and MB hydrophone tubes built into (and insulated from) 3 miles acoustic radius. All boats with hydrophones carry extra load of depth cha

Notes.—Quick rollers with period of about 5 secs.

*Latest official total available to February 1st, 1920; present total probably under 100, later Sales. 341 originally built (exclusive of 100 for French Navy). 11 lost, 12 tra to U.S. War Department, 22 transferred to U.S. Coast Guard, 8 sold to Italy, and 4 t 170 sold or placed on Sale List up to February 1st, 1920.

†Numbers are too irregular to be quoted in full; also, recent Sales have not been report is therefore deemed advisable only to give "limit numbers."

§For full List of Builders, Contracts Transferred to France, &c., v. "Fighting Ships," 1919 p. 225.

Official Summary to October 1st, 1920, of all Types.

(Exclusive of surrendered German boats.)

88 Completed, or nearly ready.

49 Building and completing.

7 Authorized, but not yet under contract or construction. (Are Fleet boats).

Notes.

The only official information published with regard to U.S. Submarines merely mentions contractors, place of construction and building dates.

Details in Table opposite and on subsequent pages, giving displacements, dimensions, H.P., speed, armament, endurance and complement, are entirely from unofficial sources, and must be regarded as only approximate.

No further Submarines are projected, other than those already authorized by Congress. But the following recommendations have been made, with regards to future types of Submarines :—

That three types of Submarines be built, as follows :—

A :—A large-radius, ocean-going type, with highest possible W/T. sending range ; to have heavy armament, moderate surface and submerged speeds, but considerable submerged radius. Surface displacement, *about* 2000 tons ; length, *about* 200 feet ; 2—5 inch (high velocity) and 1 AA. guns ; 6—21 inch torpedo tubes (4 bow, 2 stern) and 15—21 inch torpedoes to be carried.

B :—Large radius Mine-laying type on the lines of the German *U 117—126* Class (for details of this class, *v.* Japanese *O— 1* or French *René Audry*).

C :—A "General Purposes" type of *about* 800-1000 tons, 200-250 feet long.

No.	Type	Date	Displacement	H.P.	Max. speed	Fuel or Battery Endurance	Complement	T. Tubes	Max. draug[ht]
	Ocean-going :		tons		kts.				feet
9	*"Fleet 1—9"*	*Proj.*	1950-2025	...	22-24	...	...	...	...
3	*AA 2, AA 3 (H)* / *AA 1* (H)	'17-'20 / '15-'19	1106 / 1190	4000 / 1350	20 / 10-12	For 7000 miles.	...	4	15
10	*S 3—S 13 (BD)*	'17-'21	800-854 / 940-1090	1200	14-14 / 12-11	...	37	4	13-1
9	*S 2, 14—S 17 (L)* / *S 48—S 51 (L)*								
31	*S 1, S 18—S 47 (H)*								
	Coastal:								
7 / 20	*R 21—R 27* (L) / *R 1—R 20* (H)	'17-'20	500-520	...	...	...	...	...	...
6	*H 4—H 9 (H)*	1918	355 / 467	480 / 320	13 / 11	For 2800 miles at 11 kts. / For 30 miles at 5 kts.	25	4	13
6 / 10	*O 11—O 16* (L) / *O 1—O 10* (H)	'16-'18	485-520 / 570-630	1200 / 730	14 / 10-11	For 5200 miles at 11½ kts.	...	4	...
4 / 3	*N 4—N 7* (L) / *N 1—N 3* (H)	'15-'17	330-317 / 385-410	900 / 620	13 / 10-11	For 2500 miles at 11 kts. / 100-120 at 5 kts.	18	4	12
1	*M 1* (H)	'13-'16	490 / 680	1800	14 / 10-11	For 3000 miles at 14 kts. / For 140 miles.	...	4	14
4	*L 5—L 8* (L)	'13-'16	450-490 / 570-720	1300 / 750	14 / 10·5	For 4500 miles at 11 kts. / For 150 miles at 5 kts.	...	4	...
7	*L 1—L 4*, *L 9—L 11* (H)								
8	*K 1—K 8* (H)	'11-'14	380-390 / 520-530	1000 / 680	14 / 10·5	For 4500 miles at 10 kts. / For 120 miles at 5 kts.	18	4	12½
2	*H 2—H 3* (H)	'10-'13	350 / 470	950 / 620	14 / 10·5	For 2300 miles at 11 kts. / For 100 miles at 5 kts.	18	4	12
2	*F 2, F 3* (H)	'09-'12	330 / 400	480 / 350	12 / 10·5	For 2300 miles at 11 kts. / For 100 miles at 5 kts.	18	4	12
2	*E 1, E 2* (H)	'09-'12	287 / 340	500 / 250	12 / 11	For 2100 miles at 11 kts. / For 100 miles at 5 kts.	15	4	...
3	*D 1—D 3* (H)	'07-'10	228 / 340	600 / 330	12 / 9·5	For 1240 miles at 9-10 kts.	15	2	...

(BD) = "Bureau Design," *i.e.*, by Bureau of Construction, Navy Dept. (H) = Boats built by Electric Boat Co. ("Hollands" or to their designs. (L) = Boats built by Lake Co., or to their designs. (LI) = Plus-Laurenti boat built by Cramp

55* + 7 *projected* "Fleet Submarines."

*About 33 delivered or nearing completion; 21 building or completing.

Note.—Contracts with Electric Boat Co. for 9 Submarines (*Nos. 159—167*), and with Lake T. B. Co. for 4 Submarines (*Nos. 173—176*), cancelled December, 1918.

Projected.

9 *type unknown*: "*Fleet 1—Fleet 9*," authorized, but not yet contracted for or begun. Three will be built at Portsmouth Navy Yard (2 commenced 1920, 1 to be begun) and six by private firms. Tenders were invited from contractors for construction of six boats, each of 2025 tons surface displacement, 300 × 27½ × feet, 22 kts. surface speed. The lowest tender, submitted by Bethlehem S. B. Corporation, was, however, for a type of 1950 tons and 21 kts. Will be numbered in *AA* or *BB* series.

Note.—These boats may be similar to the recommended Type "A" outlined on preceding page.

AA 1 (after launch). *Photo, Kadel & Herbert.*

3 *Holland type*: **AA 1** (ex-*Schley*, Fore River Co., 1918), **AA 2, AA 3** (1919, by Fore River Co.). Dimensions: about 264 (*w.l.*), 270 (*o.a.*) × 23 × 14 feet. Machinery: 4 sets of 1000 H.P. Nlseco Diesel engines on *surface*, arranged in tandem pairs to 2 screws. Armament: 2—3 inch AA. guns and 4 tubes.

"Fleet Submarines"—*Continued.*

S 3. 1920 *I*

10 "*Bureau design*" boats: **S3, S4, S6—S13**; all by Portsmouth N. Y. *S 3—9* launched 1919-20, complete or completion. *S 10—13* begun; will be ready 1921. Dimensions: 231 × 21½ × 12½ feet. Some armed 1—5 inch gun. Tubes: 4 bow (12 torpedoes carried). *S 5* foundered 1920.

Building or Completing.

9 *Lake type*: **S2** (1919), **S14—S17** (1919-20), and *S 48* (1920), *S 49—S 51* (*building*). All by Lake T. B. Co. B to be similar to *S 1*, *S 18—S 47* given below. *S 2*, *S 14—17*, delivered or nearly finished, *S 48—51* completi building.

Photo wanted.

31 *Holland type*: **S 1** (1918), **S 18—47** (1919-21), viz., *S 1*, *S 18—S 29* by Fore River S. B. Co., *S 30*—S 41* by I. W., *S 42—S 47* (to be begun). Dimensions: about 230 × 21 × 12½ feet. Machinery: 2 sets of 600-7 Nlseco Diesel engines. Will have large radius of action on *surface*. Armament: 1—3 inch gun and 4 *S 1* delivered; *S 18—34* delivered or nearly ready; *S 35—41* building or completing; *S 42—47* building; building or completing.

**Note.*—*S 30* built at expense of Philippines Government and will be manned by a Filipino crew.

(*Continued on next*

81 Boats.

Note.—"*Number 108*" a small experimental boat, will not be built.

1920 *Photo.*

7 *Lake type*: **R 21—R 27** (1918-19), all built by Lake T. B. Co. No details known. Completed 1918—19.

R 5—R 10. 1920 *Photo.*

20 *Holland type*: **R 1—R 20** (1917-19), viz., *R 1—R 14* by Fore River S. B. Co., *R 15—R 20* by Union I. W., San Francisco. No details known. In above illustration R 5 and R 10 have 1—3 inch gun mounted; platform for guns provided in other boats. Completed 1918-19.

Photo wanted.

4 *Holland type*: **H 4—H 9** (1918), assembled by Bremerton N. Yd., Puget Sound. Dimensions: 150¼ × 15¾ × 12¼ feet. 4 bow 18 inch tubes. Nlseco Diesel engines. Ordered for Russian Government, but bought in knock-down condition by U.S. Government and assembled in 8½ months.

Photo wanted.

6 *Lake type*: **O 11—O 16** (1917-18), viz., *O 11—O 13* by Lake T. B. Co., *O 14—O 16* by California S. B. Co. Dimensions about 172 × 18 × — feet. 4 torpedo tubes. Machinery: 2 sets Sulzer Diesel engines. Radius of action O 1—O 10 below. Completed 1918. Believed to mount 1—3 inch A.A. gun.

1920 *Photo.*

10 *Holland type*: **O 1—O 10** (1917-18), viz., *O 1* by Portsmouth N. Y., *O 2* by Bremerton, Puget Sound, N. Y., *O 3—O 10* by Electric Boat Co. and Fore River Co. Dimensions: about 175 × 17 × feet. 4 torpedo tubes. Machinery: 2 sets of 4-cycle Nlseco Diesel engines. Electric batteries weigh 65 tons. Radius of action: 5000-6000 miles, at 11-12 kts. *on surface*. Completed 1918. Believed to mount 1—3 inch A.A. gun.

N 6. 1920 *Photo.*

4 *Lake type*: **N 4, N 5, N 6, N 7,** (launched 1916-17) by Lake T. B. Co. Dimensions: about 155 × 14½ × feet. torpedo tubes. Machinery: 2 sets Sulzer Diesel engines. Radius of action: about 2500 miles at cruising speed *on surface*; 100-120 miles at 5 kts. *submerged*. Completed 1918.

Continued on next page

Aircraft Carriers.

ne Layers **AROOSTOOK** and **SHAWMUT** (v. next column), may act, from time to time, as Tenders to Flying Boats, Aeroplanes, and as Kite Balloon Ships.

stroyer *Mugford*. Tender to Fleet Air Detachment; torpedo tubes replaced by tanks holding 1500 gallons petrol.

20 Report by Chief of Bureau of Construction & Repairs, says plans were prepared for conversion of Car Ferry *Flagler* to an Aircraft Carrier. This may be "Hull 180," referred to below.

Also v. Note under "Tenders to Torpedo Craft" for "7 Standard Destroyer and Aviation Tenders" reported building.

Building or Completing.

Hull delivered to Contractors for conversion, Oct. 1920.

WRIGHT (ex-Emergency Fleet Corporation Hull No. 180, "Type B"). 14,246 tons. Dimensions: 488 (*o.a.*) × × 29 feet. Designed speed: 15 kts.

Undergoing re-construction.

To be modified on lines of H.M.S. *Argus*, funnels and ventilators being led through side of hull. Flying Deck about 520 × 60 ft. 30 Battle, Bombing and Torpedo-dropping 'planes to be carried. Also machinery and repair shops for aircraft to be installed.

LANGLEY (ex-Fleet Fuel Ship *Jupiter*, 1912, converted 1920-21). 19,360 tons. Dimensions: 542 × 65 × 27 feet (*mean*). S.H.P. 7161. 15 kts. G. E. turbines with Melville-McAlpine electric drive.

Note.—All figured details as originally designed. Alterations caused by re-construction not yet known.

Mine Planters.

14 Destroyers to be converted into "Light Minelayers." Flush Deck T.B.D. *Anthony, Burns, Hart, Ingraham, Israel, Lansdale, Luce, Ludlow, Mahan, Maury, Murray, Rizal, Sproston, Stribling*, so converted.

Same as "Shawmut" below.

(Pacific Fleet.)

AROOSTOOK (Cramps, 1907. Ex-S.S. *Bunker Hill*, of Eastern Steamship Co., purchased 1917 and converted by Boston N.Yd. into Mine Planter). All details as *Shawmut*, below.

SHAWMUT (AROOSTOOK same). 1920 *Photo*.

(Atlantic Fleet.)

SHAWMUT (Cramps, 1907. Ex-S.S. *Massachusetts*, of Eastern Steamship Co., purchased 1917 and converted by Boston N.Yd. into Mine Planter). 3800 tons. Complement, 314. Dimensions: 395 (*o.a.*) × 52¼ × 16 feet (*mean*). Guns: 1—5 inch, 2—3 inch AA., 2 M.G. Reciprocating engines and 8 S. E. boilers. 2 screws. H.P. 7000 = 20 kts. Fuel: 400 tons coal + 160 tons oil.

Mine Planters—*continued*.

Photo, U. S.

(Atlantic Fleet.)

SAN FRANCISCO (1889). Displacement, 4083 tons. plement, 303 (319 as Flagship). Dimensions: 324½ × 49¼ × 18¾ (*mean*) feet. Armament: 5—5 inch, inch AA., 4—6 pdr., 2—1 pdr., 2 M.G. Carries 300 Ma mines. 4 searchlights. Designed H.P. 8500 = 19 8 Babcock boilers. Coal: 663 tons.

Photo, U. S

(Pacific Fleet.)

BALTIMORE (1888). Displacement, 4413 tons. Comple 332. Dimensions: 335 (*o.a.*) × 48⅗ × 19½ (*mean*) Guns: 4—5 inch, 51 cal., 2—3 inch AA., 2 M.G. Arm 4" deck. Designed H.P. 8500 = 20 kts. Boilers: 8 cock. Coal: maximum 1092 tons.

Fuel Ships.

(Oil Carriers.)

Photo wanted.

(Fleet assignments not known.)

ALAMEDA (1919), **RAMAPO** (1919), **SAPELO** (1920), **SEPULGA** (1920). 16,800 tons. Dimensions: 463¼ (*p.p.*) × 60 × 26¼ feet (*mean draught*). Quad. Exp. reciprocating engines. 3 S. E. boilers. I.H.P. 2900 = 10.5 kts. Guns: 2—5 inch, 51 cal. Are Shipping Board contracts taken over by Navy Department. (*Patoka* N.O.T.S.)

BRAZOS (with 2 T.B.D.). *1920 Photo.*

("Pecos" Atlantic Fleet, "Neches" & "Brazos" Pacific Fleet).

BRAZOS (1919), **NECHES** (1920), *PECOS*: all three by Boston N.Yd. 14,800 tons. Dimensions: 475½ (*o.a.*) × 56 × 26⅝ feet *mean*. Recipro. engines and 4 B. & W. boilers. Designed I.H.P. 5200 = 14 kts. Deadweight capacity 9600 tons. Own fuel 826 tons, drawn from cargo holds. Guns: 4—5 inch.

CUYAMA (v. next column). *1920 Photo.*

KANAWHA (& MAUMEE). *1919 Photo, O. W. Waterman.*

("Kanawha," "Cuyama" Pacific Fleet. "Maumee" Atlantic Fleet).

CUYAMA (1916) **MAUMEE** (1915), **KANAWHA** (1914). Oil Tankers. 14,500 tons. Dimensions: 475¼ × 56 × 26¾ (*mean*). *Cuyama* and *Kanawha* reciprocating engines, 5200 H.P. *Maumee* 2-cycle Diesel motors 5000 H.P. Speed: 14 kts. Fuel capacity: 7500 tons cargo oil, 1575 tons own bunker oil. Guns: *Cuyama* 4—5 inch; *Maumee* 4—4 inch; *Kanawha* 4—4 inch, 2 M.G.

Following 5 Ships are Colliers and Oil Carriers.

(Pacific Fleet).

JASON (1912), **ORION** (1912). 19,250 tons. Dimensions: 536 × 65 × 27⅔ (*mean*). Speed: 14 kts. Fuel capacity (deadweight to designed draft): 10,500 tons cargo fuel + 2000 tons own bunker fuel. *Maximum* cargo capacity (close stowage): 11,500 tons coal + 2575 tons oil. Guns: 4—4 inch.

Fuel Ships—*Continued.*

(Colliers and Oil Carriers).

PROTEUS. *Photo added 1*

(Atlantic Fleet).

NEREUS (1913), **PROTEUS** (1912). 19,080 tons. Dimen 522 × 62 × 27⅔ (*mean*). Speed: 15 kts. Fuel capacity (dead to designed draft): 10,500 tons cargo fuel + 2000 own bunke *Maximum* cargo capacity (close stowage): 11,800 tons coal tons oil or 10,100 tons coal + 3050 tons oil. Guns: 4—4 inc

(Pacific Fleet). *1919 Photo, O. W. Waterm*

NEPTUNE (1911). 19,480 tons. Dimensions: 542 × 65 × 27⅝ (*me* Speed: 13 kts. Fuel capacity (deadweight to designed dra 10,500 tons cargo fuel + 2000 tons own bunker fuel. *Maxi* cargo capacity (close stowage): 10,200 tons coal + 2925 tons o 11,700 tons coal + 1250 tons oil. Guns: 4—4 inch. Westinghouse geared turbines.

(Continued on next pa

Photo wanted.

3 *Holland* type: **N1–N3** (1916-17) viz., *N 1–N 3*, by Electric Boat Co., and Seattle Con. and D. D. Co. Dimensions: about 147 × 16 × 12 feet. 1—3 inch AA. gun. 4 torpedo tubes. Machinery: 2 sets 450 H.P. Nlseco Diesel engines. Radius of action: 2500 miles at cruising speed on *surface*, and 100-120 miles at 5 kts. *submerged.*

Photo, copyright by F. Muller, Junr.

1 *Holland* type: **M 1.** (Electric Boat Co., 1915.) Dimensions: 165 × 16 × 13¾ feet. Armament: 1—3 inch, 50 cal. gun (high angle anti-aircraft model) and 4 torpedo tubes. Machinery: 2 sets of 8 cylinder 600 H.P. Nlseco Diesel engines. Radius of action: 3000 miles at 14 kts. on *surface*. Has three periscopes.

Photo wanted.

4 *Lake* type: **L 5** (Lake T. B. Co., 1915), **L 6, L 7** (Craig S. B. Co., Long Beach, Cal., 1916), **L 8** (Portsmouth N. Y., 1917). Dimensions: about 165 × 14¾ × feet. 1—3 inch AA. gun. 4 torpedo tubes. Machinery: 2 sets of 600 B.H.P. Busch Sulzer Diesel engines. Radius of action: about 4500 miles on *surface* at cruising speed, 150 miles at 5 kts. *submerged.*

Note.—When in European waters, during the War, were marked as "AL" Class, to distinguish them from the British "L Class" Submarines.

L 3. 1916

7 *Holland* type: **L 1–4, L 9–11** (1914-15), viz., Electric Boat Co. Dimensions: about 169 × 17¼ × 3 inch AA. gun. 4 torpedo tubes. Machinery: 2 sets of 450 H.P. Nlseco Diesel engines in *L* — Radius of action: about 4500 miles on *surface* at cruising speed, and 150 miles at 5 kts. *submerged.* each.

Note.—*L 6* and *L 7* contracted for by Lake Co., but were built by the California S. B. Co., Long Beach, in European waters, during the War, were marked as "AL" boats, to distinguish them from the British Submarines.

K 5. 1916

8 *Holland* type: **K 1** (ex-*Haddock*), **K 2** (ex-*Cachalot*), **K 3** (ex-*Orca*), **K 4** (ex-*Walrus*), and **K 5–K 8.** (K 1, 2, 5, 6, by Fore River, K 3, 7, 8, by Union I. W., K 4 by Moran Co. Dimensions: about 153 × 1 4 bow torpedo tubes. 8 torpedoes carried. Machinery: 2 sets 6-cylinder 500 B.H.P. Nlseco Diesel en of action: 4500 miles at 10 kts. on *surface*. 120 miles at 5 kts. *submerged.*

(Continued on ne

1920 *Photo.*

H 1, H 2.

2 Holland type: **H 2** (ex-*Nautilus*), and **H 3** (ex-*Garfish*), (1913). H 2 by Union I. W., H 3 by Moran Co. Dimensions: about 150 × 16 × [illegible] feet. 4 torpedo tubes. Radii of action: about 2300 miles at 11 kts. on surface, [illegible] miles at 5 kts. submerged. H 1 wrecked, March, 1920.

No photo available.

2 Holland type: **F 2** (ex-*Barracuda*), and **F 3** (ex-*Pickerel*), (1911). F 3 by Union Ironworks, F 2 by Moran Co. Seattle. Dimensions: about 142 × 15 × [illegible] feet. 4 torpedo tubes. Radii 2300 at 11 kts. surface, 100 at 5 kts. submerged. Reported to have been re-engined during 1915-16 with Nlseco & Craig Diesel engines. F 1 (ex-*Skate*) foundered before the war, was salved and stricken from the Navy Register. F 4 (ex-*Carp*) lost during the war.

2 Holland type: **E 1** (ex-*Skipjack*), and **E 2** (ex-*Sturgeon*), (1911). Dimensions: 135 × 14 × [illegible] feet. Built by Fore River Co. 4 torpedo tubes. Radii 2100 at 11 kts. surface, 100 at 5 kts. submerged. E 2 experimentally equipped with Edison nickel batteries. Carry Mk. VII Mod. 1 and 18 inch torpedoes.

Photo, Boston News Co.

3 Holland type: **D 1** (ex-*Narwhal*), **D 2** (ex-*Grayling*), and **D 3** (ex-*Salmon*), (1909-10). Built by Fore River [illegible] Dimensions: 134½ × 14 × [illegible] feet. 4 torpedo tubes. Radius of action on surface, 1250 miles at cruising speed [illegible] 9-10 kts.

Note.

The Old Submarines A 1, A 3, A 5, A 7, B 2 and all "C" Class, placed on Sale List, 1919.

A 2, A 4, B 1, B 3, not yet officially notified as for sale, but will probably be disposed of at an early date.

G 2, G 3, G 4, unofficially reported as for sale. As these three boats were not very successful, it appears likely they will be disposed of. Accordingly they have been deleted.

Tenders to Torpedo Craft.

First class cruiser, **BROOKLYN,** is fitted for and serves as flagship to T.B.D. in full commission, Pacific Fleet. She is described on a previous page.

Third class cruiser, **SALEM,** serves as flagship to T.B.D. in reserve, Pacific Fleet. Is described on a previous page.

First class cruiser, **ROCHESTER,** is flagship to T.B.D. in full commission, Atlantic Fleet. Third class cruiser, **CHESTER,** is flagship to T.B.D. in Reserve, Atlantic. Both ships described on preceding pages.

Unofficially reported, 1920, that the Hog Island E.F.C. Yard is building hulls of 7 "Standard Destroyer and Aviation Tenders." When launched, to be fitted out at Philadelphia N.Yd.

For **BRIDGEPORT** v. "Repair Ships" p. 225.

Building

WHITNEY (Boston N.Yd., begun 1920), *DOBBIN* (Philadelphia N.Yd., begun 1919). 10,600 tons. Dimensions: 460 (*w.l.*), 483¾ (*o.a.*) × 61 × 21 feet (*mean* draught). Guns: 8—5 inch, 4—3 inch AA. Torpedo tubes, for testing purposes: 2—21 inch. Geared turbines. Boilers: 2 Bureau Modified Thornycroft. 1 screw. Designed S.H.P. 7000 = 16 kts. Coal: 1107 tons. Complement, 332. Equipped to serve as Depôt, Repair and Hospital Ship for 18 Destroyers. To have special anti-torpedo protection. Generally sister ships to *Holland*, Tender to Submarines.

1919 *Photo, C. E. Waterman.*

(*Pacific Fleet.*)

MELVILLE (1915). 7150 tons. Complement, 412. Dimensions: 400 × 54½ × 20 feet (*mean* draught). Guns: 2—5 inch (51 cal.), 1—3 inch AA., 2—3 pdr. Torpedo tubes: 1—18 inch. Machinery: Parsons turbines and Westinghouse reduction gear. 2 Babcock and Wilcox boilers. H.P. (*on trials*) 4006 = 15.09 kts. Fuel: 900 tons *oil*.

Tenders to Torpedo Craft—*Continued.*

(*Pacific Fleet Reserve Flotillas.*) 1920 *Photo.*

BLACK HAWK (Cramp, 1913, ex-Grace Steamship Co. S.S. *Santa Catalina*, taken over 1917). 13,500 tons. Dimensions: 404½ × 53½ × 28½ feet. Guns: 4—5 inch, 1—3 inch AA., 2 M.G. I.H.P. 3400 = 13 kts. Oil: 2108 tons. Complement, 471.

1919 *Photo.*

LEONIDAS (1898). 4023 tons. Dimensions: 274 × 39¼ × 17¾ feet. Complement, 88. Guns: 1—6 inch, 2—3 inch, 2—3 inch AA., 2 M.G. H.P. 1100 = 8½ kts. Coal: 205 tons.

(*Atlantic Fleet.*)

DIXIE (ex *El Sud*), (1893). 6114 tons. Complement, 446. Dimensions: 391 × 48¼ × 20 feet (*mean* draught). Armament: 4—3 inch, 1—3 inch AA., 2—6 pdr., 2—1 pdr., 2 M.G. H.P. 3800 = 14.5 kts. Coal: 1100 tons.

Tenders to Torpedo Craft—*Contin*

As "Dixie" in preceding column.

(*Pacific Fleet.*)

BUFFALO (*ex El Cid*), (1892). 6000 tons. Compl Dimensions: 391 × 48¼ × 19½ feet. Armament: 6—[illegible] pdr., 2—1 pdr., 2 M.G. H.P. 3600 = 14.5 kts. Coal:

PRAIRIE. 19[illegible]

(*Pacific Fleet.*)

PRAIRIE (*ex El Sol*), (1890). 6620 tons. Compl Dimensions: 404¾ × 48¼ × 20¾ feet (*mean* draught). 8—3 inch, 1—3 inch AA., 2—1 pdr., 2 M.G. H.P. 38[illegible] Coal: 1331 tons.

(*Atlantic Fleet.*)

PANTHER (1889). 3380 tons. Complement: 258. 304¾ × 40¾ × 15¾ feet (*mean* draught). Guns: 4—3 inc AA. H.P. 3200 = 13.5 kts. Coal, 691 tons.

Tenders to Submarines.

Building.

HOLLAND (Puget Sound N Yd. Begun 1929). 10,600 tons. Dimensions: 460 (w.l.), 483¾ (o.a.) × 61 (*extreme*) × 21 feet *mean draught*). Guns: 8—5 inch, 4—3 inch AA. Geared turbines. 1 screw. Boilers: 2 Bureau Modified Thornycroft. Designed S.H.P. 7000 = 16 kts. Oil: 1107 tons. Complement, 288. Generally sister ship to *Whitney* and *Dobbin*, Tenders to Torpedo Craft.

1912 *U. S. Navy Photo.*

BUSHNELL.

(*Atlantic Fleet.*)

BUSHNELL (1915). 1580 tons. Complement: 203. Dimensions: 350¼ × 45⅝ × 15 feet (*mean draught*). Guns: 4—5 inch (51 cal.), 2—3 pdr. (saluting). Machinery: Turbines with reducing gear. H.P. (*on trials*) 2617 = 14·15 kts. Fuel: 660 tons oil.

Tenders to Submarines—*Continued.*

Photo, Copyright, E. Muller, Jr.

(*Submarine Base, New London, Conn.*)

FULTON (ex-*Niagara*) (1914). 1408 tons. Complement: 144. Dimensions: 226¼ × 35 × 13 feet (*mean draught*). Guns: 2—3 inch (50 cal.), 1—1 pdr. auto anti-aircraft. Torpedo tubes: None officially listed, but from published plans and photos 4 tubes appear to be mounted on fox'sle in two twin deck-mountings. Machinery: 1100 B.H.P. 6-cyl. 2-cycle Nlseco Diesel engines. 1 screw. Designed speed: 12·25 kts.; 12·34 on trials. Fuel: 234 tons oil.

1920 *Photo.*

(*Pacific Fleet.*)

BEAVER (Newport News, 1910, purchased 1918). 5970 tons. Dimensions: 380 (o.a.) × 47 × 21 feet (*max. draught*). Guns: 4—5 inch, 1—3 inch AA., 2—1 pdr., 2 M.G. 1 screw. I.H.P. 4500 = 16·5 kts. Fuel: 2350 barrels oil fuel. Complement, 373.

Tenders to Submarines—*continued.*

1920 *Photo*

(*Atlantic Fleet.*)

CAMDEN (Flensburg, Germany, 1900, ex-German-Australian *Kiel*, seized 1917). 9000 tons (estimated). Dimensions: 4... (o.a.) × 48 × 24 feet. Guns: 4—4 inch, 2—1 pdr., 4 ... I.H.P. 2550 = 12 kts. Coal: 975 tons. Complement, 327.

1920 *Photo*

SAVANNAH (Flensburg, Germany, 1899, ex-Hamb... American S.S. *Saxonia*, seized 1917). 10,800 tons. Dim... sions: 414¼ × 46 × 26¼ feet. Guns: 4—5 inch, 2 M... I.H.P. 2600 = 10·5 kts. Coal: 743 tons + 531 ... additional stowage = 1274 tons. Complement, 295.

(*Asiatic Fleet.*)

RAINBOW (1890, purchased 1898). 4300 tons. Dimensi... 351½ × 41 × 17¼ feet. Guns: 4—5 inch, 2 M.G. I.H.P. 1800 = 12 kts. Coal: 1166 tons. Complement, 258.

Old Monitor *Monadnock* attached to Submarine Division Asiatic Fleet, and Gunboat *Alert* (Submarine Base, San Pe... Cal.), are not rated officially as Tenders to Submarines.

Fuel Ships—*Continued.*

All on this page Colliers —

(Pacific Fleet.)

VULCAN (1909). 11,250 tons. Dimensions: 403 × 53 × 24⅝ feet (*mean*). Speed: *about* 12 kts. Fuel capacity (dead-weight to designed draft): 7200 tons cargo fuel + 300 tons own bunker fuel. *Max.* capacity (close stowage): 7000 tons cargo fuel + 825 tons own bunker fuel. Guns: 4—4 inch.

Undergoing re-construction on Asiatic Station, 1920.

(Asiatic Fleet.)

***ABARENDA** (1892). 6680 tons. Dimensions: 325½ × 42 × 22⅝ feet (*Plimsoll mark*). Guns: 4—6 pdr. H.P. 1050 = 9 kts. *Max.* fuel capacity: 3500 tons cargo + 825 own bunkers.

Photo added 1920.

***AJAX.** 9250 tons *approx.* Dimensions: 387½ × 46½ × feet. H.P. 2000 = 10 kts. *Max.* fuel capacity: 4800 tons cargo + 500 own bunkers. Guns: 4—6 pdr.

*Bought during War with Spain.

Fuel Ships—*Continued.*

Photo added 1920.

***BRUTUS.** 6550 tons. Dimensions: 332½ × 41½ × 23 feet (*Plimsoll mark*). H.P. 1200 = 10 kts. *Max.* fuel capacity: 4850 tons cargo + 450 tons own bunkers. Guns: 4—6 pdr.

*Bought during War with Spain.

Also *Arethusa, Beaufort, Cæsar, Mars, Nero, Patoka, Robert L. Barnes, Sara Thompson,* all on Naval Overseas Transport Service and not assigned to Fleet Duties.

Ammunition Ships.

NITRO, PYRO, *completing.* *1920 Photo.*

NITRO, PYRO (both built by Puget Sound N.Yd.*). 10,600 tons. Dimensions: 460 × 61 × 21 feet. Guns: 4—5 inch, 2—3 inch AA. Designed S.H.P. 5300. Speed: 16 kts. Parsons (geared) turbines. 2 screws. Fitted with plant for powder testing and cooling, also large cold storage capacity for meat, in addition to ammunition-carrying spaces. Complement, 226.

*No launching date; built in dry dock and floated out. Completed 1920.

Supply Ships.

(Atlantic Fleet.)

BRIDGE (1916). 8500 tons. Complement, 92. 423 × 55½ × 20½ feet (*mean draught*). Gun 50 cal., 1—3 inch AA., 1 M.G. H.P. 4000 = ciprocating engines. 2 screws. Fuel: 1000 t

(Pacific Fleet.)

RAPPAHANNOCK (Vegesack, Germany, 1 German Lloyd S.S. *Pommern,* seized 1917) Complement, 119. Dimensions: 471½ × 5 Guns: 1—5 inch, 1—3 inch. I.H.P. 485 Coal: 3060 tons.

(Continued on

Supply Ships—*Continued.*

(Asiatic Fleet.) 1920 Photo, "—R."

OMPEY (Bought 1898). 3085 tons. Complement 71. Dimensions: $234 \times 33\frac{1}{4} \times 15\frac{5}{6}$ feet *(mean draught)*. Guns: 4—6 pdr. Speed: 10·5 kts. Coal: 205 tons.

(Refrigerator and Supply Ship, Pacific Fleet.)

CIER (1891). 8325 tons. Complement, 92. Dimensions: $355\frac{1}{3} \times 46 \times 29\frac{1}{3}$ feet *(mean draught)*. Guns: 1—5 inch, 2—3 inch, 1—3 inch AA. H.P. *(on trials)* 2127 = 12·3 kts. Coal: 39 tons

(Refrigerator and Supply Ship, Pacific Fleet.) ... Dimensions: 383×

Supply Ships—*Continued.*

1920 Photo.

(Supply Ship, Atlantic Fleet).

CULGOA (1889). 6000 tons. Complement, 92. Dimensions: $346\frac{1}{4} \times 43 \times 21\frac{3}{4}$ feet *(mean draught)*. Guns: 4—3 inch, 50 cal., 1—3 inch AA. H.P. 2350 = 13·25 kts. Coal: 980 tons.

Repair Ships.

Building.

MEDUSA (begun at Puget Sound N.Yd., Jan., 1920). 10,000 tons. Dimensions: 460 × 70 × 19 feet. Guns: 4—5 inch, 4—3 inch AA. Geared turbines. Designed S.H.P. 7000 = 16 kts.

1920 Photo, Seward, Weymouth.

*(U.S. Naval Forces, European Waters.)**

BRIDGEPORT (Vegesack, Germany, 1901, ex-North German Lloyd S.S. *Breslau*, seized 1917). 8600 tons. Dimensions: $429\frac{1}{2} \times 54\frac{1}{4} \times 28\frac{1}{4}$ feet. Guns: 8—5 inch, 1—3 inch AA., 2 M.G. I.H.P. 3600 = 12.5 kts. Coal: 1060 tons. Complement, 793.

*Will probably be assigned for duties as Tender to 2nd Destroyer Flotilla, Atlantic Fleet.

Repair Ships—*Continued.*

("Prometheus," Atlantic Fleet; "Vestal," Pacific Fleet.) 1919

PROMETHEUS (1908). **VESTAL** (1908). 12,585 tons. G
4—5 inch, 50 cal., 1—3 inch AA. H.P. 7500 = 16 kts.
Prometheus: 1331 tons; *Vestal*: 1422 tons. Babcock boil

No photo available.

(For Target Repairs: Pacific Fleet.)

NANSHAN. 5140 tons. Dimensions: $295\frac{3}{4} \times 39 \times 21\frac{1}{4}$ feet. I
1400 = $10\frac{1}{2}$ kts. *Max.* fuel capacity: 425 tons. Guns: 4—6

(For Target Repairs: Atlantic Fleet.) 1919 Photo.

LEBANON (1897). 3285 tons. Dimensions: 249 *(o.a.)* × $37\frac{1}{2}$ × 1
feet. Guns: 2—3 inch, AA. H.P. 1000 = 10 kts. Coal: 1
tons. Complement, 61.

(For Radio Repairs: Pacific Fleet.) Photo added 1920.

SATURN. 4840 tons. Dimensions: $297 \times 40\frac{1}{4} \times$ feet. H.P. 2500 = 11 kts. *Max.* fuel ...

Navy Transports.

Projected

HEYWOOD. (*Authorised: not yet begun*). 10,000 tons. Complement 258. Dimensions: 483¼ × 64 × 19 feet. Guns: 4—5 inch, 2—3 inch AA., 2—6 pdr., 2—1 pdr. Turbines and B. and W. boilers. Designed S.H.P. 5400 = 10 kts. Fuel, 1200 tons.

1920 Photo.

HENDERSON (1916). 10,000 tons. Complement: 133. Dimensions: 483¼ × 61 × 20 feet (*mean* draught). Guns: 8—5 inch, 50 cal., 2—3 inch, 2—1 pdr. H.P. 4000 = 14 kts. Reciprocating engines. 2 screws. Fuel: 1200 tons *oil*. Has Sperry gyro. stabilizers. To take 2000 men and 32 horses.

Photo, F. C. Bowen, Esq.

HANCOCK (1879). 8500 tons. Complement: 289. Dimensions: 465¼ × 45¼ × 21¼ feet. (mean draught). Guns: 4—3 inch, 2—3 inch AA., 2—1 pdr., 2 M.G. H.P. 3100 = 13 kts. Coal: 1425 tons.

Navy Transports—*Continued.*

1920 Photo, U.S.N.

(*Asiatic Fleet.*)

GENERAL ALAVA (bought 1898). 1115 tons. Complement, 97. Dimensions: 212½ (*p.p.*) × 29 × 11 feet. I.H.P. 770 = 10.5 kts. Coal: 246 tons.

Also about 20 or more other cargo ships employed on N.O.T.S.* to be disposed of as soon as circumstances permit.

*Naval Overseas Transport Service.

Hospital Ships.

RELIEF completing. 1920 Photo.

RELIEF (Dec., 1919, Philadelphia N.Y.). 9800 tons. Complement 272. Dimensions: 460 (*p.p.*) × 61 × 19½ feet. Designed H.P. 5250 = 16 kts. Parsons (geared) turbines, 2 screws. Babcock and Wilcox boilers. Oil: 2032 tons.

Hospital Ships—*Continued.*

MERCY.

(*Pacific Fleet.*)

COMFORT (ex-Ward Liner *Havana*, built by Cram... tons, and **MERCY** (ex-Ward Liner *Saratoga*, 1907). 10,112 tons. Complement 336. Dimensi... 24 to 26 feet. I.H.P. 8500 = 18 kts. Coal: ... Purchased 1918.

(*Atlantic Fleet.*)

SOLACE (1896). 5700 tons. Complement, 260. 377 × 44 × 22 feet (mean draught). H.P. ... Purchased during War with Spain.

Tugs and Mine-Sweepers.

1920 Photo.

("Bird" Class—54 Boats.)

LAPWING, OWL, ROBIN, SWALLOW, SANDERLING, AUK, CHEWINK, CORMORANT, GANNET (all built by Todd S. B. Co., Tebo Yacht Basin, Brooklyn, N. Y.).

TANAGER, CARDINAL, ORIOLE, CURLEW, GREBE, MALLARD, ORTOLAN, PEACOCK (all built by Staten Id. S. B. Co., N. Y.).

AVOCET, BOBOLINK, LARK, PIGEON, REDWING, RAVEN, SHRIKE (all built by Baltimore D. D. & S. B. Co.).

PELICAN, FALCON, OSPREY, SEAGULL, TERN (all built by Gas Engine & Power Co., Morris Heights, N. Y.)

TURKEY, WOODCOCK, QUAIL, PARTRIDGE (all built by Chester S. B. Co.)

SANDPIPER, WARBLER, VIREO, WILLET (all built by Philadelphia Navy Yard).

SWAN, WHIPPOORWILL, BITTERN (all built by Alabama S. B. & D. D. Co., Mobile).

WIDGEON, TEAL, BRANT (all built by Sun S. B. Co., Chester).

KINGFISHER, RAIL (both built by Puget Sound Navy Yard).

EIDER, THRUSH (both built by Pusey & Jones, Wilmington).

FINCH, HERON (both built by Standard S. B. Co., N. Y.).

FLAMINGO, PENGUIN (both built by New Jersey D. D. & T. Co., Elizabeth Port).

All built 1918-19. Displacement: 950 tons normal, 1009 *full load* at 10½ feet aft. Dimensions: 187¾ (o.a.) × 35½ × 9¾ feet (mean draught). Guns: 2—3 inch AA., 2 M.G. Machinery: 1 set triple expansion and 2 B. & W. boilers. Designed I.H.P. 1400 = 14 kts. Oil fuel only.

Note.—*Cardinal*, *Plover*, *Goshawk* and Nos. 55, 56, 57 stopped. These craft may also serve as Mine Layers, Tugs, and Salvage [illegible]

Fleet Tugs.

(*Atlantic Fleet.*)

ALLEGHENY (1917), **SAGAMORE** (1917). Ocean-going Tugs of 1000 tons. Dimensions: 149 × 30 × 14½ feet. Guns: 2—3 inch AA., 2 M.G. I.H.P. 1800 = 13 to 14 kts. Oil: 279 tons.

(*Pacific Fleet.*)

UNDAUNTED (1917). 450 tons. Guns: 1—3 pdr., 2 M.G. I.H.P. 1000 = 11.5 kts. Oil fuel only.

(*Atlantic Fleet.*)

CHEMUNG (ex-*Powhatan*, 1916), **WANDO** (1916). 575 tons. Dimensions: 122½ × 26½ × 11½ feet. 2—3 pdr., 1 M.G. H.P. 800 = 11 kts. Oil fuel only.

("Iroquois" Atlantic Fleet. "Mohave" Pacific Fleet.)

ARAPHO (1914), **MOHAVE** (1914). 575 tons. 2—3 pdr. H.P. 800 = 10 to 11 kts. Oil: 149 tons.

(*Atlantic Fleet.*)

LYKENS (1912). 625 tons gross. Dimensions: 157 × 29 × 15 feet. Guns: 1—3 inch, 2 M.G. H.P. 1000 = [illegible] kts.

ONTARIO. 1919 Photo.

(*Pacific Fleet.*)

SONOMA (1912) and **ONTARIO** (1912). 1120 tons. Dimensions: 175 (p.p.) × 34 × 12¾ feet (mean). Guns: 1—3 inch, 1—3 inch AA., 2 M.G. Speed: 13 kts. Coal: 440 tons (average). These two tugs are ocean-going and available as repair ships.

(*Atlantic Fleet.*)

PATAPSCO (1908), **PATUXENT** (1908). 755 tons. Dimensions: 148 × 29 × 12½ feet. Guns: 1—3 inch AA., 1 M.G. H.P. 1160 = 13 kts. Coal: 324 tons.

(*Pacific Fleet.*)

SEA ROVER (1908). [illegible] tons. Guns: 1—3 inch, 2 M.G. I.H.P. 700 = 10 kts. Fuel: 1200 barrels oil fuel.

(*Atlantic Fleet.*)

POTOMAC (1897). 785 tons. Guns: 2—3 pdr. H.P. 2000 = 16 kts. Coal: 200 tons.

(*Asiatic Fleet.*)

PISCATAQUA (1897). 854 tons. Guns: 2—3 pdr., 2 M.G. H.P. 2000 = 16 kts. Coal: 236 tons.

Fleet Tugs—*Continued.*

(*Asiatic Fleet.*)

WOMPATUCK (1896). 462 tons. Guns: 2—1 pdr., 1 M.G. H.P. 650 = 13 kts. Coal: 130 tons.

(*Atlantic Fleet.*)

UNCAS (bought 1898). 411 tons. Guns: 2—1 pdr., 1 M.G. I.H.P. 750 = 12 kts. Coal: 120 tons.

(*Pacific Fleet.*)

IROQUOIS (1892). 702 tons. Guns: 2—3 pdr. H.P. 1000 = 13 kts. Oil fuel.

(*Pacific Fleet.*)

CHALLENGE (1889, bought 1918). [illegible] tons. Guns: 2—3 pdr. Speed: 11 kts.

In addition, about 57 Tugs and Mine Sweepers assigned for Duty at Navy Yards, Naval Stations, Naval Districts, &c. (This total includes 19 Tugs of the 1000 ton "Allegheny Class" not assigned to Fleet Duties.)

Converted Yachts.

Note.—All purchased during War with Spain.

SYLPH (1898). 152 tons. Armament: Nil. H.P. 550 = 15 kts. Coal: 48 tons. Complement, 36.

(*Continued on next page.*)

Converted Yachts—*Continued.*

Photo, Lieut. F. S. Dowell, U.S.N.

(Presidential Yacht.)

MAYFLOWER (1896). 2690 tons. Armament: 2—3 inch, 1—3 inch AA., 4—6 pdr. H.P. 2400 = 14¼ kts. Coal: 537 tons. Complement, 185.

SCORPION (ex *Sovereign*, 1896). 775 tons. Armament: 4—6 pdr. H.P. 2800 = 17.8 kts. Coal: 136 tons. Complement, 91. Station Ship at Constantinople; interned during War. Reported relieved by *Galveston*, 1919.

VIXEN (ex-*Josephine*, 1896). 806 tons. Guns: 4—3 pdr., 2—1 pdr. H.P. 1250 = 16 kts. Coal: 190 tons. Complement, 105.

Converted Yachts—*Continued.*

YANKTON (ex-*Penelope*, 1893). 975 tons. Armament: 2—3 inch, 2—3 pdr., 1 M.G. H.P. 750 = 14 kts. Coal: 174 tons. Complement, 117. 1" deck amidships.

Photo: favour of Commissioner R. P. Forshew, N.Y.N.M.

GLOUCESTER (1891). 786 tons. Guns: 5—3 pdr., 2—1 pdr. H.P. 2000 = 17 kts. Coal: 120 tons. Boilers: Babcock.

About six or eight Yachts taken over during the War are still in service, but will be disposed of as soon as possible.

Surveying Service.

Photo added 1920.

HANNIBAL (1898). Ex Fuel Ship. 4000 tons. Dimensions: 274 × 39½ × 17¾ feet. Complement, 306. Coal: 491 tons.

For Duty with Naval Reser

Various ships—no exact list available, but are m very old vessels.

Rated as Unserviceable for War

Adams (1874).	*Hartford* (1858).
Boston (1883).	*Intrepid* (1904).
C.T.B. No. 6 (1897).	*Mohican* (1872).
Cumberland (1904).	*Old Constellation*
Essex (1876).	*Old Constitution*
Gopher (1871).	*Philadelphia* (18
Granite State (1818).	*Reina Mercédes*
	Southery (1889).
	Vesuvius (1887).

Ex-German Warships assigned t

Battleship (Dreadnought type):—*Ostfriesland*.

Light Cruiser:—*Frankfurt*.

Destroyers:—*S 132, G 102, V 43*.

Submarines:—*U 117, U 111, UB 148, UB 88*, boat.

All above will be broken up eventually, bu Experimental and Propaganda Duties. F ex-Austrian Battleships, v. page 381.

Notes.

[illegible] Revised, 1920, from materials furnished by courtesy of the Commandant, U.S. Coast Guard, Treasury Department, Washington, D.C. Photos also official unless otherwise acknowledged.

Re-classified 1920, as (a) Cruising Cutters, (b) Inshore Patrol Cutters, (c) Harbour Cutters, and (d) Harbour Launches.

Ships arranged alphabetically. Classes comprising several ships inserted under earliest name in alphabetical order (Ex Eagle Boats taken as "E" and Ex Submarine Chasers as "S").

Length is o.a. Beam is moulded. Draught is max. Tonnages are displacement.

1920 Stations in italics and 1920 Signal Letters in leaded type, usually after description of ships, e.g., *Acushnet* stationed at *Woods Hole, Mass.*, Signal Letters **GVHP**.

33 Cruising Cutters.

ACUSHNET (1908). Steel, 1 screw. 800 tons. Dimensions: 152 × 29 × 13¼ feet. Speed: 12 kts. Guns: 2—1 pdr. (*Woods Hole, Mass.* **GVHP.**)

ALGONQUIN (1898). Steel, 1 screw. 1119 tons. Dimensions: 205½ × 32 × 13½ feet. Speed: 16 kts. Guns: 4—6 pdr., 2 M.G. (*Astoria, Oreg.* **GVFL.**)

U. S. COAST GUARD. (Alphabetically arranged.)

Cruising Cutters—Continued

ANDROSCOGGIN (1908). Wood, 1 screw. 1605 tons. Dimensions: 210 × 35¼ × 17½ feet. Speed: 13 kts. Guns: 4—6 pdr., 2 M.G. Usually on Grand Banks for Fisheries Duty. (*Boston, Mass.* **GVHK.**)

1920 Photo.

APACHE (1891). Iron, 1 screw. 708 tons. Dimensions: 185½ × 29 × 9½ feet. Speed: 12 kts. Guns: 3—3 inch, 2 M.G. (*Baltimore, Md.* **GVBS.**)

BEAR (1874, transferred from Navy, 1885). Wood, 1 screw. 1700 tons. Dimensions: 198 × 29½ × 18¼ feet. Speed not known. Guns: 3—6 pdr. (*Seattle, Wash.* **GVDM.**)

Cruising Cutters—Continued

1920 Photo.

(Ex Eagle Boats.)

Name. (ex-Eagle No.)		Station.	Signal Letters
ROTHWELL	(21)	*San Diego, Cal.*	**GVJ[illegible]**
CARR	(30)	*Panama Canal Zone.*	**GVJ[illegible]**
EARP	(22)	*Honolulu, T.H.*	**GVJ**
McGOURTY	(16)	*New London, Conn.*	**GVJ[illegible]**
SCALLY	(20)	*San Juan, P.R.*	**GVJ**

All built 1919. Steel. 500 tons. Dimensions: 200¾ × 25½ × 11 feet. Guns: 2—4 inch, 1—3 inch AA., 2 M.G. Speed 18 kts. (geared turbine)

1920 Photo.

COMANCHE (ex-*Windom*, 1896). Iron, 2 screws. 589 tons. Dimensions: 170¾ × 27 × 8¾ feet. Guns: 2—6 pdr. Speed: 13 kts. (*Galveston, Tex.* **GVDT.**)

Best available scan

Cruising Cutters—*Continued.*

(For *Earp*, v. "Ex-Eagle Boats.")

GRESHAM (1897). Steel, 1 screw. 1090 tons. Dimensions: 205½ × 32 × 12½ feet. Speed: 14 kts. Guns: 4—6 pdr., 2 M.G. (*Boston, Mass.* **GVFD**).

Not completed

HAIDA, MODOC, MOJAVE, TAMPA (all projected or building). Steel. 1600 tons. Dimensions: 240 × 39 × 16½ feet. To have 5 guns each. Machinery: Turbo-electric. To be built by Union Con. Co., Oakland, Cal.

Photo, Second, Weymouth.

ITASCA (ex-*Bancroft*, 1893, taken over from Navy, 1906). Steel, 2 screws. 980 tons. Dimensions: 189½ × 32 × 13[illegible] feet. Speed: 14 kts. Guns: 2—3 inch, 4—6 pdr. (*New York, N.Y.* **GVHF**).

Cruising Cutters—*Continued.*

No photo.
Are Shallow-draught, Stern-Wheel Steamers.

Name.	*Station.*	*Signal Letters.*
KANKAKEE	*Evansville, Ind.*	....
YOCONA	*Vicksburg, Mass.*	....

Built 1919. Steel hull, wood deck-houses. 383 tons. Dimensions: 182 × 34 × 3½ feet. Guns: Unknown. Speed: 12 kts.

MANNING (1897). Composite, 1 screw. [illegible] tons. Dimensions: 205 × 32 × 13½ feet. Speed: 15 kts. Guns: 4—6 pdr., 2 M.G. (*Norfolk, Va.* **GVFK**).

(For *McGourty*, v. "Ex-Eagle Boats.")
(For *Modoc* and *Mojave*, v. *Haida*.)

MORRILL (1889). Iron, 1 screw. 420 tons. Dimensions: 145½ × 24 × 9½ feet. Speed: 13 kts. Guns: 2—6 pdr., 2 M.G. (*Detroit, Mich.* **GVDN**).

Cruising Cutters—*Continued.*

ONONDAGA (1898). Steel, 1 screw. 1192 tons. 205½ × 32 × 13½ feet. Speed: 14 kts. Gu[illegible]
(*South Baltimore, Md.* **GVFM**).

OSSIPEE (1915). Steel, 1 screw. 908 tons. 165¾ × 32 × 11¾ feet. Speed: 12 kts. Gu[illegible] 2—6 pdr., 2 M.G. (*Portland, Me.* **GVBW**).

PAMLICO (1907). Steel, twin screw. 451 tons. 158 × 30 × 5½ feet. Speed: 11 kts. Gu[illegible]
(*Newbern, N.C.* **GVHJ**).

(*Continued*

Best available scan

Cruising Cutters.—*Continued.*

No Photo.

PEQUOT (1911). Wood. 283 tons gross. Dimensions: 155 × 22 × 9½ feet. Guns: Unknown. Speed: Unknown. (*South Baltimore, Md.*, **GVCM**).

(For *Scally* v. "Ex-Eagle Boats.")

SEMINOLE (1900). Steel, 1 screw. [illegible] tons. Dimensions: 188 × 29½ × 12 feet. Speed: 14 kts. Guns: [illegible]—3 inch, 2 M.G. (*Wilmington, N.C.*, **GVFP**).

SENECA (1908). Steel, 1 screw. 1445 tons. Dimensions: 204 × 34 × 17½ feet. Speed: 13 kts. Guns: 4—6 pdr., 2 M.G. (*New York, N.Y.*, **GVHL**).

Not completed.

SHAWNEE (projected). Steel. 800 tons. Dimensions: 168 × 30 × 14 feet. No further details. To be built by Union Con. Co., Oakland, Cal.

1920 Photo.

SNOHOMISH (1908). Steel, 1 screw. 879 tons. Dimensions: 152 × 29 × 15½ feet. Speed: 12 kts. Guns: 2—6 pdr., 2 M.G. (*Port Angeles, Wash.*, **GVHN**).

1920 Photo.

TALLAPOOSA (1915). Steel, 1 screw. 912 tons. Dimensions: 165¾ × 32 × 11¾ feet. Speed: 12 kts. Oil fuel only. Guns: 2—3 inch, 2 M.G. Derelict Destroyer for Gulf of Mexico. (*Mobile, Ala.*, **GVDP**).

(For *Tampa*, v. *Haida*.)

Cruising Cutters—*Continued.*

TUSCARORA (1902). Steel, 1 screw. 739 tons. Speed: 14 kts. Dimensions: 178 × 30 × 11 feet. Guns: 2—6 pdr. (*Key West, Fla.*, **GVFS**).

1919 Photo, U.S. Navy Publicity Bureau.

UNALGA (1912). Steel, 1 screw. 1181 tons. Dimensions: 190 × 32½ × 14 feet. Speed: 13 kts. Guns: 3—6 pdr. (*San Francisco, Cal.*, **GVHS**).

YAMACRAW (1909). Steel, 1 screw. 1082 tons. Dimensions: 191⅔ × 32⅔ × 13 feet. Speed: 13 kts. Guns: [illegible] pdr. (*Savannah, Ga.*, **GVHR**).

27 Inshore Patrol Cutters.

Arcata (1903). Wood, 1 screw. 138 tons. Dimensions: 85 × 17½ × 10½ feet. Speed: 11 kts. (*Port Townsend, Wash.*, **GVHC.**)

No Photo.

Name.	*Station.*	*Signal Letters.*
Arrow*	*Key West, Fla*	**GVHQ.**
Kangaroo†	*Key West, Fla.*	**GVFT.**

Wood. 12 tons. Dimensions: 62½ × 11 × 2½-2¾ feet. *Arrow* 18 kts., *Kangaroo* 21 kts. (both twin-screw petrol motors).

*No date. †Built 1917

No Photo.

Messenger (1916). Wood. tons. Dimensions: 66 × 13 × 2¾ feet. Speed: 12 kts. (petrol). (*Key West, Fla.* Signal letters illegible in Official Tables)

Inshore Patrol Cutters—*Continued.*

As Illustration of SC boats, U.S. Navy pages.

(Ex-Submarine Chasers.)

Name (late *SC*).		*Station.*	*Signal Letters.*
Adams	(268)	*Boston, Mass.*	....
Boyce	(437)	*Chicago, Ill.*	**GVKF.**
Cook	(438)	*Sault Ste. Marie, Mich.*	**GVKC.**
Cygan	(335)	*Key West, Fla.*	**GVKR.**
Deering	(333)	....	....
Hahn	(415)	*New York N.Y.*	**GVKS.**
Hansen	(68)	*Key West, Fla.*	....
Johansson	(435)	*Philadelphia, Pa.*	**GVKB.**
Klingelhoefer	(433)	*Chicago, Ill.*	**GVKD.**
Knudsen	(431)	*Chicago, Ill.*	**GVJS.**
Larsen	(203)	*Miami, Fla.*	**GVKN.**
Melahatos	(197)	*S. Baltimore, Md.*	**GVKL.**
Newbury	(70)	*Key West, Fla.*	....
Ovesen	(199)	*S. Baltimore, Md.*	**GVKM.**
Quigley	(22)	*S. Baltimore, Md.*	**GVKH.**
Richards	(27)	*S. Baltimore, Md.*	**GVKJ.**
Smith	(155)	*Key West, Fla.*	**GVKQ.**
Stellenwerf	(417)	*New London, Conn.*	**GVJW.**
Talley	(334)	....	....
Taylor	(153)	*Key West, Fla.*	**GVKP.**
Tingard	(183)	*New London, Conn.*	**GVJT.**
Vaughan	(152)	*Key West, Fla.*	....

Built 1917-18, taken over 1919-20. Wood. 75 tons. Dimensions: 110 × 14¾ × 6 feet. Guns (if retained): Not known. Petrol engines (twin screw). Speed: 11 kts.

No Photo.

Swift (1917). Wood. tons. Dimensions: 66 × 13½ × 3½ feet. Speed: 16 kts. (twin screw, petrol). Guns: (*San Francisco, Cal.*, **GVHT.**)

26 Harbour Cutters.

No Photo.

ALERT (1907). Wood. 35 tons. Dimensions 5½ feet. Speed: 7 kts. (*Mobile, Ala.*, **GVF**

ARUNDEL (ex-*Manhattan*, 1873). Iron, 1 Dimensions: 102 × 20½ × 8½ feet. Spee *Baltimore, Md.*, **GVCD.**)

CALUMET (1894). Iron, 1 screw. 170 t 94½ × 20½ × 8½ feet. Guns: 2 M.G (*New London, Conn.*, **GVDR.**)

(*Conti*

No photo available.

Name.	*Station.*	*Signal Letters.*
CHATTAHOOCHE	*New York N.Y.*	....
CHAUTAUQUA	*New York N.Y.*	**GVJM.**
CHENANGO	*Philadelphia, Pa*	**GVJN.**
CHICOPEE	*Boston, Mass.*	**GVJL.**
CHILLICOTHE	*New York N.Y.*	....
CHINCOTEAGUE	*S. Baltimore, Md.*	**GVJQ.**
CHIPPEWA	*New York N.Y.*	**GVJK.**
CHOPTANK	*S. Baltimore, Md.*	**GVJP.**
CHOWAN	*Norfolk, Va.*	**GVJR.**
CHULAHOMA	*New York N.Y.*	....

All built 1910. Wood. 215 tons. Dimensions: 88 × 20 × 8¾ feet. Speed: 10 kts. Are Ex Navy Tugs taken over by Coastguard.

DAVEY (1908). Steel, 1 screw. 182 tons. Dimensions: 92½ × 19 × 10½ feet. Guns: 1—1 pdr. Speed: 10 kts. (*New Orleans, La.*, **GVHM.**)

Harbour Cutters—*continued.*

GOLDEN GATE (1896). Steel, 1 screw. 240 tons. Dimensions: 110 × 20½ × 9¾ feet. Speed: 12 kts. (*San Francisco, Cal.*, **GVFH.**)

No photo available.

GUARD (1914). Wood. 52 tons. Dimensions: 67 ft. 7 in. × 12½ × 6¼ feet. Speed: 9 kts. (*Friday Harb., Wash.*, ———)

GUTHRIE (1895). Iron, 1 screw. 149 tons. Dimensions: 88 × 17½ × 9 feet. Speed: 11 kts. (*Philadelphia, Pa.*, **GVBQ**)

Harbour Cutters—*continued.*

HUDSON (1893). Iron, 1 screw. 179 tons. Dim... × 20 × 9 feet. Speed: 12 kts. (*New York,* ...

MACKINAC (1903). Steel, 1 screw. 241 tons. Di... 110 × 20½ × 10½ feet. Guns: 1 M.G. Speed... (*Boston, Mass.* **GVHB.**)

1919 *Photo, "Syren & Ship...*

MANHATTAN (1918). Steel, 1 screw. Ice Breaker, ... Vessel, Tug and Fire Float. 400 tons. Dimensions ... × 24 × 11¾ feet. Speed: 12 kts. (*New York, N.Y.*...

Harbour Cutters—*continued*.

No photo available

PENROSE (1883). Wood. 41 tons. Dimensions: 67 × 12 × 5½ feet. Speed: 7 kts. (*Pensacola, Fla.* **GVDL**).

[illegible] Official Photo.

TIOGA (1916). Steel. 131 tons. Dimensions: 81½ × 20 × 8½ feet. Speed: 10 kts. (*Baltimore, Md.* [illegible]).

No photo available

TYBEE (1895). Steel. 43 tons. Dimensions: 63 × 11⅔ × 8 feet. Speed: 10 kts. (*Savannah, Ga.* **GVDW**).

Harbour Cutters—*continued*.

WINNISIMMET (1903). Steel, 1 screw. 182 tons. Dimensions: 96½ × 20½ × 9 feet. Speed: 12 kts. (*South Baltimore, Md.*, **GVFW**).

WISSAHICKON (1904). Steel, 1 screw. 194 tons. Dimensions: 96½ × 20½ × 9½ feet. Speed: 12 kts. (*New York N.Y.*, **GVHD**).

No photo available

YORK (1898). Wood. tons. Dimensions: 125 × 16½ × 7 feet. Speed: 14 kts. (*Norfolk, Va.*)

22 Harbour Launches.

Name.	*Signal Letters.*	*Built.*	*Ton.*
Advance	......	1917	..
Carolina	......	1906	..
Cossack	**GVFN**	1916	..
Dare	......	1913	..
Dart	......	1909	..
Dash	......	1912	32
Delmarva	......	1900	..
Guide	......	1907	41
Javelin	......	1907	.
Lookout	......	....	.
Patrol	......	1917	2
Pioneer	......	1915	3
Relief	......	1897	.
Scout	**GVFC**	1903	4
Search	......	1917	.
Sentinel	......	1918	
[illegible]	......	1917	
Valiant	......	1905	
Venture	......	1910	
Vidette	......	1913	
Vigilant	......	1910	
Voyager	......	1906	

*Tons *gross*.
†Displacement.
§Twin screw (others petrol motors).

Note.—All above wood.

1 Station Ship.

Colfax (1871). Iron. 486 tons. Dimensions: [illegible] feet. No engines.

IMPERIAL JAPANESE NAVY.

Revised, 1920, at the Navy Department, Tokyo, by courtesy of H. E. The Minister of Marine.

Errata.

On Map:—

For **KOREA** read **CHOSEN**.

For **YEZO** read **HOKKAIDO**.

For **MASANPHO** read **CHINKAI**.

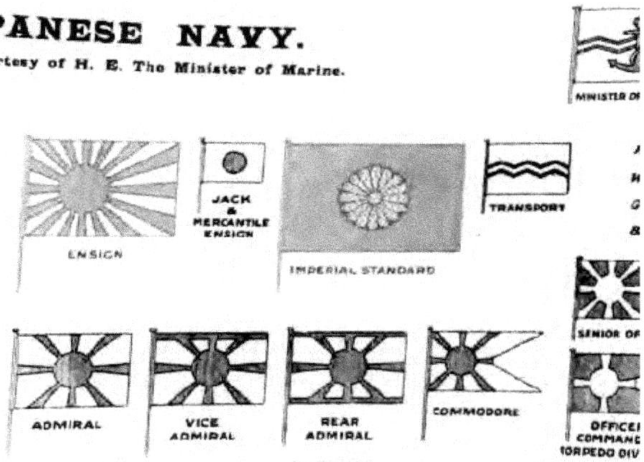

YOKOSUKA

to	miles.
Hakodate	450
Hong Kong	1580
Honolulu	3400
Kobe	300
Kure	400
Manila	1720
Nagasaki	700
Ryojun (*Port Arthur*)	1250
Saigon	2409
Shanghai	1050

NAGASAKI

to	miles.
Chemulpo	446
Foo-chow	650
Hong Kong	1067
Keilung (*Formosa*)	630
Kiao-chau	450
Kure	240
Kobe	340
Maidzuru	400
Manila	1300
Chinkai	170
Matsumai	750
Pescadores	820
Ryojun (*Port Arthur*)	585
Takeshiki	120
Saigon	1990
Shanghai	320
Vladivostock	620
Wei-hai-wei	480

TAKESHIKI

to	
Kiao-chau	
Maidzuru	
Chinkai	
Matsumai	
Nagasaki	
Ryojun (*Port Arthur*)	
Vladivostock	
Wei-hai-wei	

MAIDZURU

to	miles.
Matsumai	440
Nagasaki	400
Ominato	465
Takeshiki	340
Vladivostock	620

(RYOJUN) PORT ARTHUR

to	miles.
Hong Kong	1275
Kiao-chau	280
Nagasaki	585
Vladivostock	1170
Wei-hai-wei	100

MATSUMAI

to	m
Maidzuru	
Nagasaki	
Ryojun (*Port Arthur*)	1
Takeshiki	
Vladivostock	
Yokosuka	
Nearest Russian coast (150 English miles N.E. by E. of Vladivostock)	

JAPANESE DOCKYARDS.

Divided into 2000 yard squares. Uniform scale. Soundings in fathoms. Heights in feet.

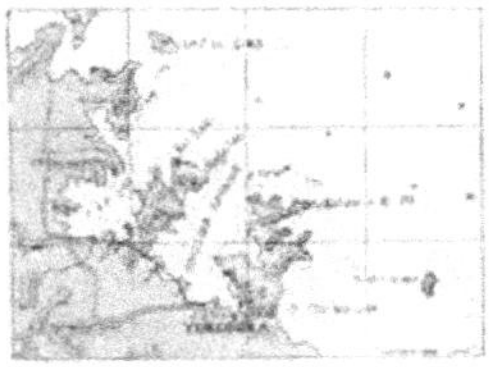

YOKOSUKA.

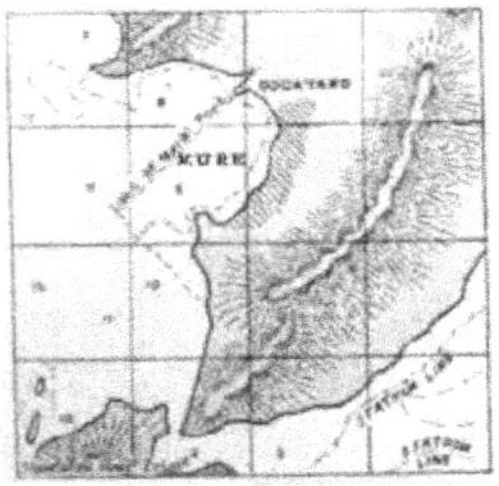

KURE.

MAIDZURU.

RIOJUN (PORT ARTHUR).

Imperial Dockyards. Principal Naval Harbours (*Ching-ju-fu*).

YOKOSUKA (in Sagami). Fleet Base, Dockyard, T.B.D., Submarine and Aircraft Stations. Fortified. Six slips, one for Battle-Cruisers, two for Dreadnoughts, three smaller. Docks No. 5, $747 \times 115 \times 41$ feet, able to take any ship, and No. 4, $538 \times 98\frac{1}{2} \times 32$. Also one dock (No. 2) $447 \times 94\frac{1}{2} \times 29$ feet; two others: No. 1, $298 \times 82 \times 21$ feet; No. 3, $265 \times 45 \times 18$ feet. Employees (in 1914) 8000. Naval Engineering Academy here. Naval Aviation Station at Oihama. Admiralty Chart No. 997.

KURE (in Aki). Fleet Base, Dockyard, T.B.D., and Aircraft Base. — slips; yard can build largest ships. Docks No. 1, $413 \times 58\frac{1}{2} \times 28$ feet; No. 2, $485\frac{1}{2} \times 81 \times 35\frac{1}{2}$ feet; No. 3, $666 \times 100 \times 34$ feet. Perhaps a fourth big dock. Ordnance and Armour Plate built here.* Cadets Academy at Etajima. Naval Air Station at Himonura. Admiralty Chart No. 3469.

SASEBO† (in Hizen). Fleet Base, Dockyard, T.B.D. and Aircraft Base. Slips for Light Cruisers, Torpedo Craft, Submarines, &c. There are four docks here: (1) $435 \times 94 \times 29$; (2) $475 \times 85 \times 32$; (3) $538 \times 93\frac{1}{2} \times 33$ feet; (4) $777 \times 111 \times 38$ feet. Perhaps a fifth big dock. There is also a floating dock for Torpedo Craft here (1500 tons). Admiralty Chart No. 359.

MAIDZURU (or MAIZURU) (in Tango). Admiral 2174. Destroyers built here. Light Cruiser sl 1919-20. One dock 450 feet long, one small for T. 1906), a third dock 540 feet long completed 1913 fourth dock.

RYOJUN (PORT ARTHUR). Dry docks: No. 1, 370 × No. 2, $500 \times 92\frac{1}{2} \times 32$ feet. Admiralty Chart No.

DAIREN (Dalny). Naval dock, $381 \times 43 \times 19\frac{3}{4}$ feet.

CHINKAI (Korea), previously known as Masampo.

TOKUYAMA BAY (150 miles N.E. of Nagasak Station and Fleet Anchorage being prepared, Sep

* Ordnance Factory somewhat damaged by explosion, June, 1920. † *See next page for Map.*

JAPANESE DOCKYARDS.

Divided into 2000 yard squares. Uniform scale. Soundings in fathoms. Heights in feet.

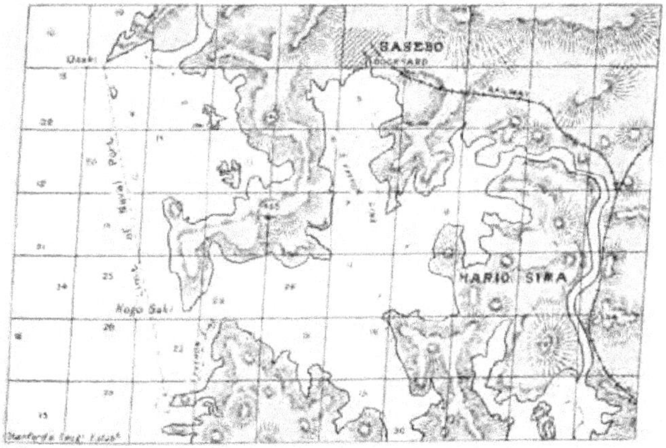

SASEBO.

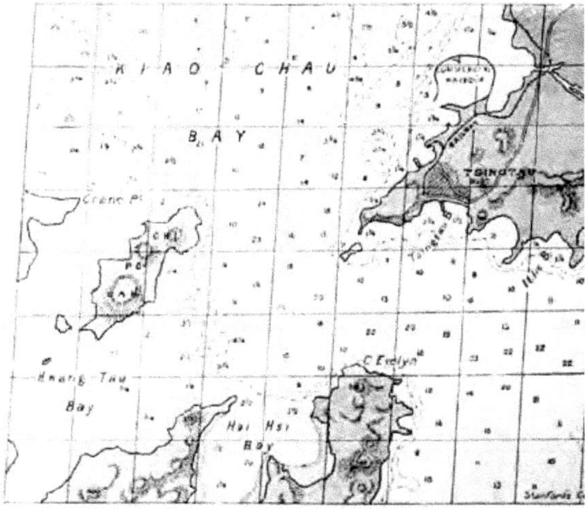

KIAO-CHAU and TSINGTAO (China). German naval base in the Far East, occupied by t Japanese Expeditionary Army, 1914. There used to be a floating dock here, 410 × 100 (12,000 tons), but it was probably destroyed by the Germans before surrender. Tsingta being used as a commercial harbour, but the naval base facilities, created during German oc are at the disposal of Japanese warships in any emergency.

VLADIVOSTOK (see Russian Navy section) also occupied temporarily by Japanese troops for tection of Japanese nationals and their property.

Lesser Naval Harbours (Yoko).

OMINATO. Torpedo base. | Repair Stations for T.B.D., &c. Small 1500 ton Floating Dock at each
RAKO (Pescadores). |

Mercantile Marine.

From "Bureau Veritas" 1919-20 figures. Ships under 100 tons *gross* excluded.

1642 steam and motor ships of 2,747,591 tons *gross*.
42 vessels (with auxiliary motors) of 11,256 tons *gross*.
1573 sailing vessels of 210,857 tons *gross*.

During 1919, Japanese Shipyards launched 133 vessels of 611,883 tons *gross*; some of these for foreign account.

Coinage.

Yen (100 sen) = 2s. 0½d. British, $0.50 U.S.A., *about*. (1914 rates.)

Principal Mercantile Ports.

Yokohama, Hakodate, Nagasaki, Moji, Kobe, Dairen (Dalny).

Overseas Possessions.

Formosa, Karafuto (Saghalien), Pescadores. Protectorate in Chosen (Korea). Kiaochau Peninsula and Tsingtao leased from China. Also Mandatory Power for administration of Marshall, Caroline and Marianne Islands.

Private Docks.

Details of Private Docks will be found on a later page (after the Silhouettes), with a full des of the Private Japanese Shipbuilding Yards.

Principal Naval Guns.

* = Q.F. with brass cartridge case.

Notation.	Calibre.	Length in calibres.	Model.	Weight of Gun.	Weight of A.P. shot.	Maximum Initial Velocity.	Maximum penetration firing A.P. capped at K.C. 5000 yards.	3000 yards.	Danger space against average ships at 10,000 yards.	5000 yards.	3000 yards.	Service rate of Fire. Rounds per minute.
	inches.			tons	lbs.	F.S.	inches.	inches.				
A^a	16	45	...	...	...		...	...	...	...	...	...
A*	14	45	V	82	1400		...	...	...	...	...	...
A^a	12·6	40	C	66	990	2306	12	15	...	...	...	0·2
A^a	12	45	(04)	58½	850	2800	16	20	185	500	780	...
A^a	12	40	E	49	850	2423	12	15½	120	450	720	1
A^a	10	45	E	32¼	500	2710	10½	13½	150	480	720	3
A^a	10	45	O	32	490	2500	9¾	12¼	130	450	700	...
B	9	35	O	19½	268	2400	5½	7¼	...		...	...
A*	8	45	O	15½	188	2800	7⅓	10½	105	430	625	1
A	8	45	E	17⅓	250	2740	7	10	110	425	600	1·2
B*	8	40	E	15¼	250	2580	5½	7¼	100	400	580	1·2
C	6	45	(04)	8¼	100	3000	4½	6½	75	250	475	...
C	6	50	V	8	100	3000	4½	6½	75	250	475	6
E*	6	40	E	6¼	100	2500	3	4½	65	210	425	7
F*	6	40	E	6	100	2220	2½	4	35	150	360	8
E*	5·5	...	...	...	...	...	...		...	...	...	...
F*	4·7	45	...	...		...	...	...	...	...		...
F*	4·7	40	E	2	45	2150	...	2½	...	...	...	8
F*	4·7	32	E	1¼	36	1938	...	...	...	...	...	8·6
F*	3		...		14	...	...	...	...	...	...	...
F*	3	40	...	2	12	2200		...	...	...	...	...

In the Model column C = Schneider-Canet; E, Elswick; O, Obukhoff (Russian); V, Vickers.

16 inch in *Kaga*, *Tosa*, *Nagato* and *Mutsu*.

14 inch, 45 cal. in *Ise* class (2), *Fuso* class (2), *Kongo* class (4).

12 inch, 50 cal. in *Settsu*.

12 inch, 45 cal. in *Settsu*, *Satsuma*, *Aki*, *Kurama*, *Ibuki*, *Ikoma*, *Katori*, *Kashima*, *Iwami*, *Mikasa*.

12 inch, 40 cal. in *Asahi*, *Shikishima*, *Fuji*.

10 inch, 45 cal. in *Satsuma*, *Aki*, *Kashima*, *Katori*, *Suwo*, *Okinoshima*, *Mishima*.

8 inch, 40-45 cal. in 1st Class Cruisers, *Kurama*, *Ibuki*.

6 inch, 50 cal. in *Fuso* class (2), *Kongo* class (4), *Settsu*.

6 inch, 40-45 cal. in *Aki* and older Battleships, 1st and 2nd Class Cruisers, &c.

5·5 inch in *Ise* class (2), *Tatsuta* class (2), *Kuma* class (5).

Note.

Kure Naval Arsenal has concluded experiments with anti-aircraft guns. Large numbers of these guns are now being manufactured for all classes of warships and defences of naval harbours.

Unofficially reported, 1920, that large sums have been spent on improved gun mountings and loading gear, fire controls and directors, &c. Large sums also said to have been laid out, 1917–20, on experiments in internal protection against torpedoes and mines. A very satisfactory system is said to have been arrived at.

(All details tabulated above are unofficial.)

Ordnance Factories.

Guns manufactured at Kure, N.Y. Also by Muroran Steel Works, Hokkaido, owned by Hokkaido Colliery & S.S. Co., Messrs. Vickers Ltd. and Sir W. G. Armstrong-Whitworth & Co., and licensed to build Armstrong and Vickers' types of guns. There are also Japanese "Yamanouchi" types of small guns.

Shells manufactured by Muroran Steel Works, Kobe Iron & Steel Works, Kawasaki Zosen Kaisha, Mitsubishi Zosen Kaisha, Nippon Heiki Kaisha, and other firms.

Armour Plates.

From Imperial Iron Works, Yawata, Kyushu and Kure D.Y.

Torpedoes.

Size	Name or Mark	Air Pressure lbs.	Charge lbs.	Max. Range at	kts.	Type o[f]
21"	...	...	...	...	...	...
18"	...	...	...	...	...	...
15"	...	...	...	...	...	...
14"	...	...	...	...	...	...

Torpedoes built by Mitsubishi Zosen Kaisha, Kobé Steel Works (output, 250 torpedo[es]) and by Navy Yards which also construct tubes. Deck T.T. in some of later T.B.[…] have folding lips.

Mines.—No details known.

Searchlights.—No details known.

JAPANESE NAVY.

Official Classification of Ships.

The official classification, instituted in 1912, is as follows:—

Battleships.	Battle-Cruisers.	1st class Cruisers (over 7000 tons).	2nd class Cruisers (under 7000 tons).	1st class Coast Defence Ships (over 7000 tons).	2nd class Coast Defence Ships (under 7000 tons).	1st Class Gunboats (over 800 tons).	2nd Gu (under
Kaga	*Atago*	Nisshin	*Yura*	Iwami	Okinoshima	Mogami	Saga
Tosa	*Takao*	Kasuga	*Isudzu*	Fuji	Mishima	Yodo	Uji
Mutsu	*Akagi*	Aso	*Nagara*	Suwo	Chiyoda	Chihaya	Fushi
Nagato	*Amagi*	Idzumo	*Natori*		Akitsushima		Toba
Hiuga	Kirishima	Iwate	*Oh-i*		Manshu		Sumid
Ise	Haruna	Yakumo	*Kitakami*		Matsue		
Yamashiro	Hi-yei	Adzuma	*Kiso*		Hashidate		
Fu-so	Kongo	Asama	Kuma		Karasaki		
Settsu	Kurama	Tokiwa	Tama		Komahashi		
Aki	Ibuki		Tatsuta		Wakamiya		
Satsuma	Ikoma		Tenriu		Musashi		
Kashima			Hirado		Yamato		
Katori			Chikuma				
Mikasa			Yahagi				
Asahi			Tone				
Shikishima			Niitaka				
Hizen			Tsushima				
			Tsugaru				
			Chitose				
			Suma				
			Akashi				

Destroyers:—1st class, over 1000 tons; 2nd class, 1000-600 tons; 3rd class, under 600 tons.
Torpedo boats:—1st class, over 120 tons; 2nd class, under 120 tons.

Minister of Marine: Admiral Baron Kato, G.C.M.G.
Naval Attaché (London): Captain S. Kobayashi, C.B.
Assistant Naval Attaché (London): Lieut.-Commr. K. Kurokawa.

Personnel: Active, in 1916, 63,225 all ranks. At end of war about 80,000. Now about 60,000—70,000. Yearly entries about 5,000—6,000 conscripts and 5,000—6,000 volunteers. Serve 4 years on active list and 7 years in reserve. Reported that 85% of complements are on board all ships and balance crews in immediate readiness at the depôts.
Reserves:—35,000 or more.

Personnel and Uniforms.

INSIGNIA OF RANK—EXECUTIVE OFFICERS—SLEEVES. (Changed to this: 1908).

Sho-i Ko-hos
Midshipma

Executive Branch:	Tai-sho.	Chu-sho.	Sho-sho.	Tai-sa.	Chu-sa.	Sho-sa.	Tai-i.	Chu-i.	Sho-i	Has a strip
Corresponding British or U.S.:	*Admiral*	*Vice-Ad.*	*Rear-Ad.*	*Captain.*	*Commander.*	*Lieut. Com.*	*Lieutenant.*	*Sub-Lieut.*	*Acting Sub-Lieut.*	half the wid of a Sho-i.

BAND between stripes	(BRANCHES, with but after Executive)
Violet	Kikwan (*Engineer*) (with executive rank and curl. Titles as above with the prefix Kikan).
Red	Gun-i (*Doctor*).
White	Shukei (*Paymaster*)
Brown	Zosen (*Constructor*).
"	Zoki (*Engineer-Constructor*). with curl, as Executive.
Purple-Brown	Zohei (*Gun Constructor*).
Blue	Suiro (*Hydrographer*).

CAP.

The cap is the same as the British (but without gold embroidery in the senior ranks).

CAP BADGE.

Small anchor, surrounded by cherry leaves.

The senior officer of any branch on board the ship always carries the affix "*cho.*" Thus: Ho-jitsu-cho (*Gunnery*), Sui-rai-cho (*Torpedo*), Ko-kai-cho (*Navigator*), Gun-i-cho (*Senior Doctor*), Shukei-cho (*Senior Paymaster*).

Undress is a military tunic (dark blue) with the sleeve insignia of rank in black braid only, with curl for executives and engineers. Other branches cannot be distinguished from each other in this tunic.

Scale : 1 inch = 160 feet.

JAPANESE WARSHIPS RECOGNITION SILHOUETTES. *(Re-arranged and extended 1920.)*

ONE FUNNEL.

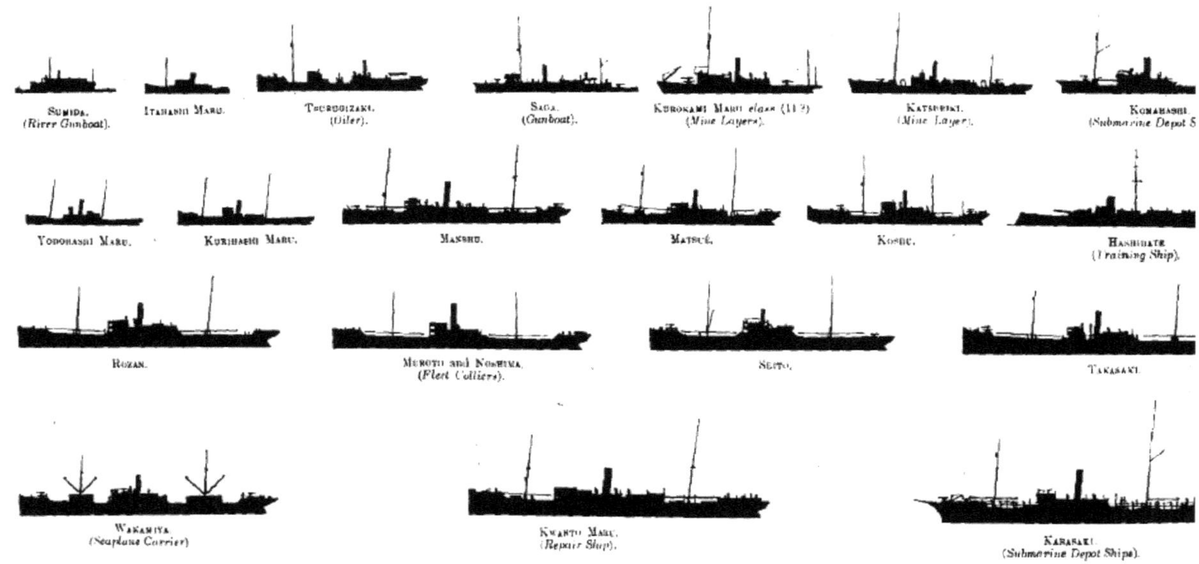

Note.

Appearance of following Special Service Ships ne
Information regarding their appearance would
preparation of their Recognition Silhouettes.

T.B.D. Depot Ships:—ISUSAN, KENKAI, REKISAN, SHIROGANE.

TWO FUNNELS.

FUSHIMI & TORA (*River Gunboats*).

YODO.

AKASHI.

SUMA.
(*Has mainmast now like "Akashi."*)

KASUGA *class* (2 ships).

FUSO *class* (2 ships).

SARUHASHI (MARU).

CHIHAYA.

ISE *class* (2 ships).

AKITSUSHIMA.

CHITOSE.

NAGATO *class* (2 ships).
(*Silhouette is only approximate, but from official drawing.*)

(Two Funnel types continued on next page

TWO FUNNELS—*continued.*

OKINOSHIMA *class* (2 ships).

FUJI.

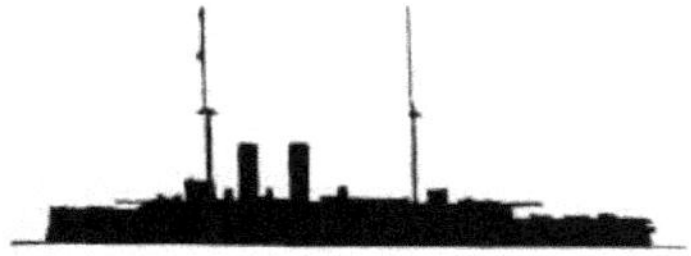

ASAHI.
(*Training Ship.*)

ASAMA *class* (2 ships).
("*Tokiwa*" *Training Ship.*)

MIKASA.

KATORI.

KASHIMA.

IWAMI.

SATSUMA.

IKOMA.

THREE FUNNELS.

MOGAMI.

TATSUTA *class* (2 ships).

TAMA *class* (5 ships).
(Distinguished by tripod foremast; two guns and two before mainmast.)

Three Funnel Types continued

THREE FUNNELS—*continued.*

NAGAURA MARU.
(*Submarine Depot Ship*).

NIITAKA *class* (2 ships).

TONE.

YAKUMO.
(*Training Ship*).

SUWO.
(*Training Ship*).

SHIKISHIMA.
(*Submarine School*).

IDZUMO *class* (2 ships).
"*Idzumo*" *attached to T.B.D. Flotillas.*

TSUGARU.
(*Mine Layer*).

HIZEN.

KURAMA
(*Ibuki* same without tripods).

SETTSU.

ADZUMA.
(*Training Ship*).

AKI.

KONGO *class* (4 ships)

Scale: 1 inch = 160 feet.

JAPANESE NAVY RECOGNITION SILHOUETTES.

FOUR FUNNELS.

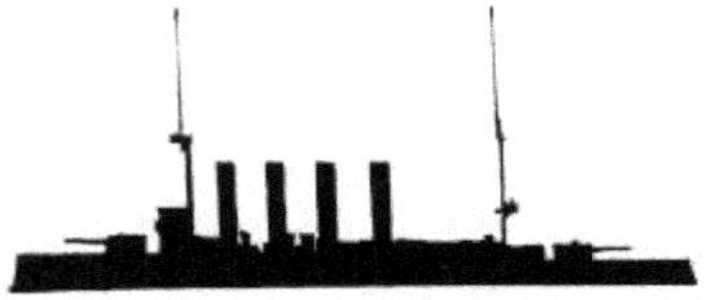

Aso.
(*Mine Layer*).

Hirado *class* (3 ships).

TORPEDO CRAFT. **Scale: 1 inch = 160 feet.**

Shirataka t.b.

Nos. 68–75 t.b.
(8 Schichau *type*)

(1)

(2)

Hayabusa *class* t.b. (11)

Kagero *class* t.b.d. (

Urakaze t.b.d.

Momo *class* t.b.d. (10).

Momi *class* t.b.d. (8).

Minekaze *class* t.b.d. (7).

Sakura *class* t.b.d. (2).
(May now have raised fore funnels.)

Kaba *class* t.b.d. (10).

Amatsukaze *class* t.b.d. (4).

Tanikaze *class* t.b.d. (2)

Akebono *class* t.b.d. (2).
Asashio *class* t.b.d. (2).

Later Arare *class* t.b.d. (33).
Earlier Arare *class* t.b.d. (2
(*Vary a little in details.*)

Yamakaze *class* t.b.d. (2).

JAPANESE SUBMARINES—RECOGNITION SILHOUETTES. *(Re-arranged 1920.)*

Scale: 1 inch = 80 feet. *Note.*—When running in surface trim, Nos. 6—13, 16, 17, do not show full length of hull, but position of stern in boats is indicated by small steel pennants as shown in drawing Removeable bridge screens are shown as shaded

Nos. 1—5. Nos. 6, 7. Nos. 8, 9. No. 13. No. 15.

Nos. 10—12. Nos. 16, 17.

All Boats given below are ex-German Types:-

O 7. O 6. O 4, O 5. O 2.

O 3. O 1.

PRINCIPAL PRIVATE ESTABLISHMENTS (including Private Docks).

Firms marked * are Warship Builders. Zosen Kaisha = Shipbuilding Co.

AIZAWA COMPANY, Osaka. Three slips.

ASANO ZOSEN KAISHA, Ushioda, Tsurumi. Large new Yard, covering 250 acres, completed 1918. Extensions over another 250 acres *projected*. Already has six concrete slips, 600 feet long. Four concrete slips of 800 feet length and two of 500 feet length *projected*. One 80-ton floating crane and many small travelling cranes for yard and fitting-out basin. Is to undertake warship construction.

*FUJINAGATA COMPANY, Osaka. Two small docks. Six slips. Builds destroyers.

HARADA DOCKYARD COMPANY, Osaka. Two small docks. Two slips.

HARIMA DOCK COMPANY, Aioi. Dock, 428 × 54 × 20 feet.

*ISHIKAWAJIMA COMPANY, Tokyo. Extended 1917. At Tokyo: graving dock, 270 × 41$\frac{3}{4}$ × 12 feet. Builds destroyers. Seven slips. Formerly had a Branch Yard at Uraga, but this has been sold to the Uraga Dock Co. There is a small Branch Yard now at Yawata (Kyushu).

*KAWASAKI DOCKYARD COMPANY, Hyogo Works, Kobé, and at Yoshiura, near Kure, and at Dairen (or Dalny), North China. Build, engine, and equip warships and mercantile vessels of all classes. At Kobé: Six building slips, *i.e.*, one of 720 feet, one of 550 feet, one of 530 feet, one of 475 feet, one of 420 feet, and one of 410 feet. One graving dock 407 × 64 × 24 feet. Also several patent slips. Floating cranes: one 200-ton, and several smaller. Yard covers 133 acres, and has a water-front of 1000 metres. Total number of persons employed 12,000. Licensees for the Curtis Marine Steam Turbine, Ansaldo-Laurenti types of submarines, and the Ansaldo-Fiat Diesel engines. Submarines built at Yoshiura D.Y. near Kure.

KUHARA ZOSEN KAISHA, Tokuyama, Shimonoseki. New large Yard, *completing*.

*MITSU BISHI ZOSEN KAISHA:—

(*a*) Kobé Yard.—Build, engine, and equip destroyers and mercantile vessels. Three building slips: two of 510 feet and another (length not known). Floating docks: one 12,000-ton, one 7000-ton and one 1600-ton. Large cranes: one 100-ton sheer-leg and one 40-ton floating crane. Yard covers 46 acres, with a water-front of 1700 metres. Number of persons employed 7500.

(*b*) Nagasaki Yard.—Warships and mercantile vessels of all types built here; also engines, boilers, equipment, &c. Seven building slips: one of 767 feet, one of 605 feet, one of 602 feet, one of 540 feet, one of 510 feet, one of 451 feet, and one of 418 feet. Three graving docks, *viz.*: No. 3, 728 × 96$\frac{1}{3}$ × 34$\frac{1}{2}$ feet (*Dreadnought*); No. 1 (Tategami), 523 × 89 × 26$\frac{1}{2}$ feet (*Dreadnought*); and No. 2 (Mukaijima), 371 × 66 × 23$\frac{1}{2}$ feet. Large cranes: one 150-ton hammer crane; one 60-ton; and one 34-ton floating cranes. Yard covers 127 acres, and has a water-front of 2750 metres. Number of persons employed 15,500. Licensees for the Parsons' Marine Steam Turbine.

(*c*) Hikoshima (Moji).—Docks here, (1) 360 × 60 × 25$\frac{3}{4}$ feet; (1) 150 × 74 × 25$\frac{3}{4}$ feet; (3) 1200 × 31 × 18$\frac{1}{2}$ feet.

MITSU BUSSAN KAISHA Yard at Hibiya. Five slips. Three graving docks for ships feet long.

NITTA COMPANY, Osaka. Five slips. Will undertake repair work, and is building a d that purpose. Shipbuilding Yard will be closed.

OKAMOTO Yard, Hyogo, Kobé Bay.

ONO ZOSEN KAISHA, Osaka. Four slips and three small dry docks.

*OSAKA IRONWORKS, Sakurajima, Osaka, Innoshima and Bingo. At Osaka: graving dock 12$\frac{1}{3}$ feet.

SUZUKI COMPANY. Harima Yard at Owe has five concrete ways (500 × 60 feet each), gı 440 × × feet. At the Toba Yard, two concrete ways (each 300 × 45 feet), and a sma × 42 × 13$\frac{1}{2}$ feet. Took over the Toba Shipbuilding Co. and Bingo Dockyard Co. d Licensees for Sulzer Diesel Engines.

UCHIDA COMPANY, Yokohama. Shipyard and engine works. Three slips. Dry dock pr

*URAGA DOCK COMPANY, Uraga. Builds light cruisers, destroyers, torpedo craft, anc vessels: also builds boilers and engines, &c. Five building slips, *viz.*: two of 500 feet 400 feet. Four dry docks: (1) 482$\frac{1}{4}$ × 70 × 35$\frac{1}{2}$ feet; (2) 448$\frac{1}{4}$ × 65$\frac{1}{4}$ × 25$\frac{1}{4}$ feet; (3) 469 × (4) 452 × (60-69) × 32 feet. Yard covers 37 acres, and has a water-front of 970 metres. T of persons employed 5000.

YOKOHAMA DOCK COMPANY, Yokohama. Up till recent times only undertook repa is now extended and equipped for shipbuilding. Graving docks: No. 1, 483$\frac{1}{2}$ × 93 No. 2, 375 × 60$\frac{3}{4}$ × 26$\frac{1}{4}$ feet; No. 3, 481 (*o.a.*) × 66$\frac{3}{4}$ × 21$\frac{1}{4}$ feet.

Other Private Docks are:—Hakodate, dry dock 480$\frac{1}{4}$ × 81$\frac{3}{4}$ × 30$\frac{1}{2}$ feet. At Nagasaki: Mat dry dock, 310 × 48 × 17 feet; and Messrs. Y. Iwasaki's patent slip (1200 tons). Temposan No. 3, 256$\frac{1}{2}$ × 44 × 13$\frac{3}{4}$ feet; No. 1, 245 × 38 × 12$\frac{1}{4}$ feet; also two old w In all there are twelve small docks at Osaka, and in the island of Innoshima, close t of which the largest is 425 × 59 × 27 feet.

Notes to Shipbuilding Establishments.

The Kawasaki and Mitsu Bishi Companies are, of course, the two great priv building establishments. Japanese shipbuilding is hampered to a large extent by steel supplies. The industry is experiencing a marked "post-war slump." At Ju shipyards were operating in the Osaka District, but only 12 are working now. Wag increased so much, the cheap labour, which gave Japanese shipbuilders advantage competitors, has largely disappeared. But, given adequate supplies of material a labour, the large Japanese companies guarantee that they could begin, launch and following warship types in the periods stated:—Largest type of Battleship or Batt 20 months; Light Cruiser, in 11 months; Large Destroyer, in 7 months (or less)

Nomenclature and "Kana" of Japanese Warships.

The system of nomenclature now being used is thus:—

Battleships and Battle Cruisers: Named after Provinces.
Light Cruisers: Named after Rivers.
Destroyers (First Class): Named after Winds.
Destroyers (Second Class): Named after Trees, Flowers and Fruits.

In the following Ship Pages, the "kana" (or Japanese characters) of the various Ship Names are given wherever possible. It must be pointed out that there are a fair number of variations in "kana," and there is no fixed rule to determine which variant is to be chosen for the name of a ship.

The names on sterns of Japanese Warships are, by Regulation, written in a script selected for each ship by the Technical Bureau. The exact "kana" for the latest Japanese Warships are not exactly known: and the Japanese characters in the following pages, for ships such as *Amagi, Kaya, Yura, Nagara, Kitakami, Tenryu* (or their sister ships), are not necessarily those which will eventually appear on the sterns of these ships.

The names, painted on the sides of Japanese Destroyers, are written, by Regulation, in a script known as "katakana," which differs entirely from the "kana" of the larger vessels, as given on subsequent pages.

Fleet Establishment.

In the middle of 1920, the constitution of the Imperial Navy was unofficially tabulated thus:—

	"First Period."	"Second Period."	"Third Period."
Battleships	6	5	4
Battle-Cruisers	4	3	—
Cruisers	7	4	14
Coast Defence Ships	1	—	5
Gunboats	—	16	2
Destroyers	45	35	—
Submarines	*	*	*
Totals	63	62	25

*No details published.

At the end of 1923 it is intended that the Battle Fleet should be composed of the following ships:—

8 DREADNOUGHTS: *Ise, Hiuga, Fuso, Yamashiro, Nagato, Mutsu, Kaga* and *Tosa.*

6 BATTLE-CRUISERS: *Haruna, Kirishima, Akagi, Amagi, Atago, Takao.* The last four have still to be built, and it is doubtful if they can be in service by the end of 1923.

Russian Warships.

Various Russian Warships (as detailed in Russian Navy Section) have been taken over by the Japanese at Vladivostok.

Also various Russian Amur River Gunboats of the **BURIAT** class appear to be in Japanese hands at ports in North Eastern Asia—probably Nicolaievsk and Khabarovsk. These Craft were disarmed by the Bolsheviks 1918-19; some have been partially dismantled; one or two have been damaged by shell and rifle fire.

Japanese Naval Programmes.

The aim of these has always been clearly defined, *viz.*, the establishment of a Fleet of one squadron of eight Battleships and one squadron of eight Battle-Cruisers, all sh of the "First Period" (or under eight years old). This scheme of naval expansion is ac known as the "8-8 Standard."

The three important Japanese Navy Laws are:—

A:—The "8—4 Unit Programme" of 1915, estimated to cost 354 million yer
B:—The "8—6 Unit Programme" of 1918-24, estimated to cost about 350 milli
C:—The "8—8 Unit Programme" of 1920-28, estimated cost uncertain.

Much time and patience has been expended in an attempt to elucidate the abo Programmes, in order to discover the identity of the ships begun under these three sche has been found impossible to get any clear understanding of the subject because (*a*) tl Programmes appear to overlap one another; and (*b*) published information describing t ship Construction, sanctioned by the three projects, is contradictory and unreliable.

On the next page will be found a Table giving a summary of Japanese Naval Cons for the past five or six years. This Table is entirely unofficial in origin, but endeavour h made to check the accuracy of the particulars given. Ships are apportioned to the years i credits were first asked for their commencement, and no attempt is made to classify then past Navy Laws.

The summary of the 1920-28 Programme is the most reliable that can be obtained at p This Programme was sanctioned by the Imperial Diet in July, 1920. The number of larg appears to be fixed, but variations may be made in the number of smaller ships, and parti in Submarines. In addition to amounts for Dockyard Extensions and subsidiary purp devotes 761, 112, 152 Yen to construction.

It must be pointed out that the 1920-28 Programme is entirely distinct from (and add to) any previous Programmes. The 1918-24 Programme has not yet run it's full course, for have still to be expended from 1920 onwards. The credits to be taken in successive ye construction *alone* are as follows:—

	Programmes previous to 1920–21. (Yen).	Programme approved in 1920–21. (Yen).
1920–21	105,434,509	89,391,224
1921–22	123,336,807	44,914,940
1922–23	121,160,589	79,765,070
1923–24	58,629,666	96,104,964
1924–25	———	109,811,444
1925–26	———	112,054,794
1926–27	———	113,692,857
1927–28	———	115,377,355

It is estimated that 3 Light Cruisers, about 14 Destroyers, and various Submarines and Auxiliaries remain to be allotted to 1920–21 under the programmes *previous* to that year. V so commenced will be entirely distinct from any ships begun under the 1920-21-22 Sectio the 1920-28 Programme.

It will be noticed from the Table that Japanese Construction is somewhat in arrears. example, it was intended that *Amagi* and *Akagi* should be commenced in 1919, but these (and other) ships are only authorised, and not yet commenced.

A key is appended at the foot of the Table. It only requires to be added that some Destro in the 1918 column were launched some months ago in a very forward state of construction, are assumed to be delivered now, although formal reports of their completion have not l received.

JAPANESE NAVAL PROGRAMMES, 1915—1920.

YEAR	1920-28 Programme.	1920	1919	1918	1917	1916	1915
BATTLESHIPS	4 Ships	(Nil)	(Nil)	*KAGA* (*B*) *TOSA* (*B*)	**MUTSU** (c)	**NAGATO** (c)	(Nil)
BATTLE-CRUISERS	4 Ships	ATAGO (a) TAKAO (a)	AMAGI (a) AKAGI (a)	(Nil)	(Nil)	(Nil)	(Nil)
LIGHT CRUISERS	12 Ships	YURA (a)	*ISUDZU* (*B*) NAGARA (a) NATORI (a)	**OH-I** (c) **KITAKAMI** (c) *KISO* (*B*)	**KUMA** (D) **TAMA** (D)	**TATSUTA** (D) **TENRIU** (D)	(Nil)
DESTROYERS (First Class)	37* boats of both classes. *Total open to future variation.	Namikaze (a) Nokaze (a) Numakaze (a)	Akikaze (a) Hokaze (a) *Shiokaze* (*B* ?) Tachikaze (a) Yukaze (a)	***Hakaze* (D)** *Nadakaze* (*B*) ***Okikaze* (D)** ***Shimakaze* (D)** ***Yakaze* (D)**	***Minekaze* (D)** ***Sawakaze* (D)**	***Kawakaze* (D)** ***Tanikaze* (D)**	***Amatsukaze*** ***Hamakaze* (*D*** ***Isokaze* (D)** ***Tokitsukaze***
DESTROYERS (Second Class)		Ashi (a) Hasu (a) Hishi (a) Sumire (a) Tade (a) Tsuta (a) Warabi (a) Yomogi (a)	*Aoi* (*B*) *Fuji* (*B*) *Hagi* (*B*) *Kiku* (*B*) *Susuki* (*B*)	***Momi* (D)** ***Kaya* (D)** ***Kaki* (D)** ***Kuri* (D)** ***Nashi* (D)** ***Nire* (D)** ***Take* (D)** ***Tsuga* (D)**	Special Programme ***Enoki* (D)** ***Maki* (D)** ***Keyaki* (D)** ***Kuwa* (D)** ***Nara* (D)** ***Tsubaki* (D)**	(Nil)	Special Progr ***Hinoki* (D)** ***Kashi* (D)** ***Momi* (D)** ***Yanagi* (D)**
SUBMARINES	70 or 75 boats ?	?	Nos. :— 44 (a) 48 (a) 45 (a) 49 (a) 46 (a) 50 (a) 47 (a)	Nos. : *27* (*B*) *33* (*B*) 39 (a) *28* (*B*) *34* (*B*) *40* (*B*) *29* (*B*) 35 (a) 41 (a) *30* (*B*) 36 (a) **42** (c ?) *31* (*B*) 37 (a) 43 (a) *32* (*B*) 38 (a)	Nos. :— **22** (D) **23** (D) *24* (*B*) **25** (c) **26** (D)	Nos. :— **19** (D) **20** (D) **21** (D)	Nos. **16** (D) **17** (D) **18** (D)
AUXILIARIES	18 Ships (4–5 Gunboats)	*IRO* (*B* ?) *SHIRIYA* (*B* ?) *TSURUMI* (*B* ?)	SATA (a) HOSHO (a)	**NINOSHIMA** (D) ERIMO (a) *NOTORO* (*B*) *SHIRETOKO* (*B*) **MUROTO** (D) **NOSHIMA** (D)	**YENOSHIMA** (D) (Special Programme)	**YENTO** (D) **KATASHIMA** (D) **KUROKAMI** (D) **SUNOSAKI** (D) **TSURUGIZAKI** (D)	**KUROSH** **TOSHIM** **KATSURI** **ASHIZAK** **KATOKU** (Shijiki—lo

Key.—(a) = Authorized, will be begun 1920-21. (*B*) = Building. (c) = Completing. (D) = Delivered and in service.

Future Construction.

Under the 1920-28 Navy Law, the construction of 4 Battleships and 4 Battle-Cruisers is projected.

ぎまあ *AMAGI*, ぎかあ *AKAGI*, *ATAGO & TAKAO*.

(all projected; not yet begun).

Displacement, *about* 43,500 tons. Complement, .

Length feet. Beam, feet. Draught, feet.

Guns: *not known*.

Armour: *not known*.

Note.

No official details communicated, and the few unofficial reports on these ships which have come to hand are not trustworthy. They say "*Akagi* and *Amagi* may have 12—16 inch guns, while *Atogo* and *Takao* may have 8—18 inch." Not certain that these four ships are of uniform design.

Machinery: screws. Boilers: Designed S.H.P. = kts. Fuel: *normal* tons, *maximum* tons.

Name	*Builder*	*Machinery.*	*Laid down*	*Completed*	*Trials.*	*Boilers*	*Best recent speed.*
Takao	Mitsubishi, Nagasaki	Mitsubishi	1920 or 1921				
Atago	Kawasaki, Kobe	Kawasaki					
Amagi	Yokosuka D.Y.						
Akagi	Kure D.Y.						

Notes.—*Amagi*, *Akagi*, authorised under 1919 Section of 1918-21 Navy Law; *Atago* and *Takao* under 1920 Section. The fourth ship was to have been named *Ashitaka*, but *Takao* was name adopted perhaps because *Ashitaka* might have been confused with Destroyers *Ashi* and *Take*.

(KAGA CLASS—2 SHIPS.)

がか *KAGA (building).* さと *TOSA (building).*

Normal displacement, 40,600 tons. Complement, .

Length, 700 feet. Beam, feet. Draught, feet.

Guns:
- 8 *or* 10—16 *inch* } Dir.
- 20—5.5 inch } Con.
- 4—12 pdr. AA.
- 4 M.G.
- 4 Landing

Torpedo tubes (*21* inch):
- 8, positions unknown

Armour:
- 12" Belt ......
- " Ends ......
- " Deck ......
- " Turrets ....
- " Gunhouses ..
- " Battery......
- " C.T. ....
- Special Anti-Torp

Note.

No plans available. All details given are entirely unofficial.

Machinery: Geared turbines. screws. Boilers: Kampon or Kansei. Designe unofficially reported as 60,000 = 23 kts. Fuel: *normal* tons; *maximum* tons

Name	*Builder*	*Machinery*	*Laid down*	*Completed*	*Trials*	*Boilers*
Kaga	Kawasaki, Kobe	Kawasaki	20th July, 1920			
Tosa	Mitsubishi, Nagasaki	Mitsubishi	16th Feb., 1920*			

General Notes.—Both ships authorised for commencement under 1918 Programme, but not begun until 1920 probably not be launched until 1922.

* Nominal date of commencement.

(NAGATO CLASS—2 SHIPS.)

とがな NAGATO (Nov. 9th, 1919). つむ MUTSU (May 31st, 1920).

Normal displacement, 33,800 tons. Complement, 1336.

Length, { 660' 7" *p.p.* / 700' 0" *o.a.* } Beam, 95 feet. Draught, { 30 feet *mean*. / ... feet *max*.

* Approximate.

Guns (Japanese):
- 8—16 inch, 45 cal. } **Dir. Con.** ?
- 20—5·5 inch, — cal. }
- 4—12 pdr. AA. ?
- 4 M.G.
- 4 Landing.

Torpedo tubes (21 inch):
- 8 (positions unknown).

Armour:
- " Belt ....
- " Ends ....
- " Deck.........
- " Turrets ....
- " Battery ....
- " Conning tow[er]

(Special ant[i-torpedo] protect[ion])

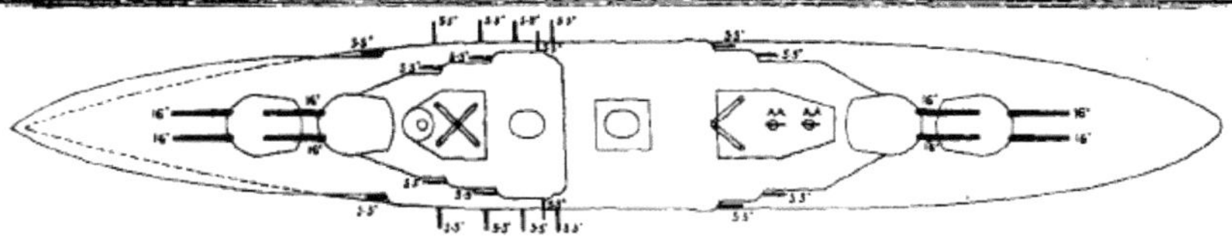

Ahead:
- 4—16 inch.
- (6?)—5·5 inch.

Astern:
- 4—16 inc[h].
- (4?)—5·5 inc[h].

Broadside: 8—16 inch, 10—5·5 inch.

Machinery: Geared turbines. Screws. Boilers: Kampon or Kansei. Designed H.P. unofficially reported as 46,000 = 23 or 23·5 kts. Fuel: *normal* tons, *maximum* tons.

Name	*Builder*	*Machinery*	*Laid down*	*Completed*	*Trials*	*Boilers*
Nagato	Kure D.Y.		28 Aug., 1917	Winter 1920	*	...
Mutsu	Yokosuka D.Y.		1 June, 1918	Summ'r 1921	*	

General Notes.—*Nagato* provided for by 1916 Programme, and laid down in dry dock at Kure D.Y. during the summer of 1917, floated out Nov. 9th, 1919. *Mutsu* provided for by 1917 Programme; estimated cost is about eight million pounds for *Mutsu*.

せい ISE (Nov., 1916). ひうが HIUGA (Jan., 1917).

Normal Displacement, 31,260 tons. Complement, 1,360. Length (*p.p.*), 640 feet.
Beam, 94 feet. *Max.* draught, 28½ feet. Length (*o.a.*), 683 feet.

Guns (Japanese **Dir. Con.?**):
12—14 inch, 45 cal.
20—5·5 inch, — cal.
4—12 pdr. (anti-aircraft).
Torpedo tubes (21 inch):
6 *submerged*.

Armour (Japanese):
12"—8" Belts
5"—3" Belt (Ends)
2—1½" Decks
12"—8" Turrets
6" Battery
12"—6" C.T.

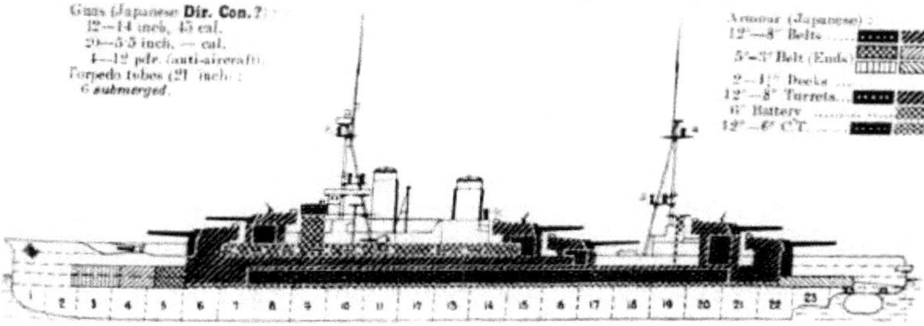

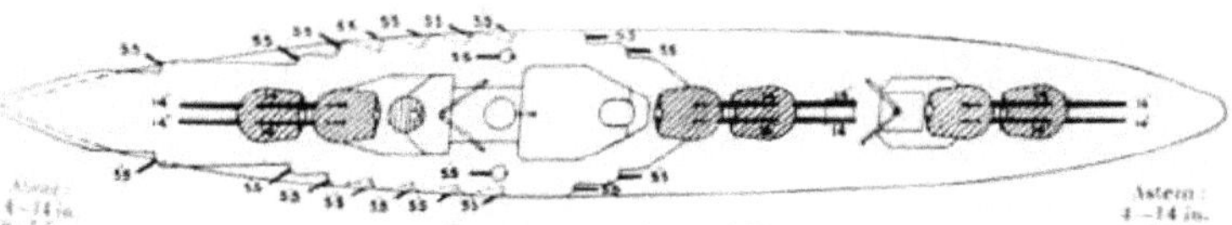

Ahead: 4—14 in. [illegible]—5·5 in.

Broadside: 12—14 in., 10—5·5 in.

Astern: 4—14 in. 6—5·5 in.

Machinery: Brown-Curtis turbines in *Ise*; Parsons turbines in *Hiuga*. 4 screws. Boilers: 24 Kansei. Designed H.P. 45,000 = 23 kts. Fuel: *normal* 1000 tons; *maximum* 4000 tons coal + 1000 tons oil.

Name	Built by	Machinery	Laid down	Com-pleted	Trials H.P. — kts.	Boilers	Turbines
Ise	Kawasaki Co.	Kawasaki Co.	May '15	Dec. '17	= 23	Kansei	Curtis
Hiuga	Mitsu Bishi Co.	Mitsu Bishi Co.	May '15	Mar. '18		Kansei	Parsons

Hiuga. 1918 *Photo by courtesy of the Navy D*[illegible]

Ise. 1918 *Photo by courtesy of the Navy Depa*[illegible]

Note.—In above illustrations of *Ise* and *Hiuga* running trials, chart-house and bridges are not [illegible] searchlights are not in position. The large tanks at base of after tripod mast, in view of *Ise*, are [illegible] shipped for trial purposes.

General Notes.—Built under the 1914 Naval Programme. These ships are an improved and slightly [illegible] type. It is unofficially reported that they are strongly protected against aerial attack by thick protective decks over vital parts of the hull. They may not have the 14—12 pdr. [illegible] attention is reported to have been paid to the rapid replenishment of fuel, stores, ammunition [illegible] internal protection against mine and torpedo explosions by minute sub-division, using [illegible] machinery magazine spaces, &c.

FU-SO (March, 1914). YAMASHIRO (Nov. 1915).

ふさふ FU-SO (March, 1914). ろしまや YAMASHIRO (Nov. 1915).

Normal Displacement, 30,600 tons. Complement, 1243 and 1272.

Length (*p.p.*) 630 feet. Beam, 94 feet. *Max. draught*, 28½ feet. Length *over all*, 673 feet.

Guns (Japanese **Dir. Con.?**):
- 12—14 inch, 45 cal.
- 16—6 inch, 50 cal.
- 4—12 pdr. (anti-aircraft).
- 4 machine.
- 4 landing.

Torpedo tubes (21 inch):
- 6 *submerged*

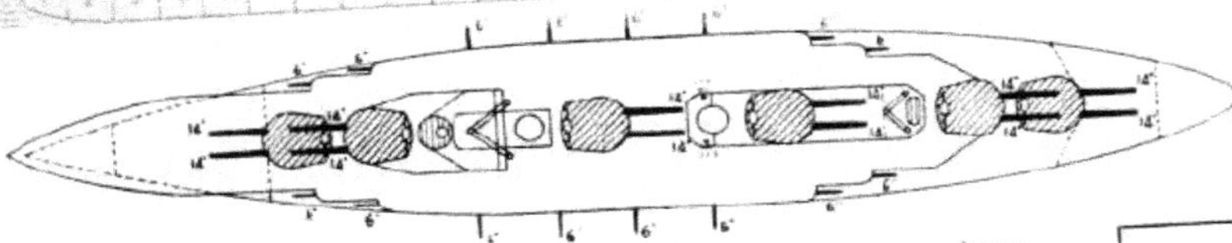

Broadside: 12—14 in., 8—6 in.

Ahead:
4—14 in.
4—6 in.

Astern:
4—14 in.
4—6 in.

Yamashiro. 1916 Photo by courtesy of the Navy.

Armour:
12"—4" Belt (amidships)
4"—4" Belt (bow)
4" Belt (stern)
8" Upper belt
6" Battery
12"—8" Barbettes
12"—8" Gunhouses
12" Fore C.T.
Bulkheads 12" to 4"
2" & 1½" Decks
Hoods { Gunhouses / C.T.

Gunnery Notes. (unofficial). R.F. in roof of each gun-house, over and Torpedo control tower (aft) and in fire-control top. In Yo... photo, note short calibre Q.F. fixed to chases of 14-inch guns ... calibre practice.

Torpedo Notes: 10 searchlights. Net defence right up to stern.

Machinery: Brown-Curtis turbines. 4 screws. **Boilers**: 24 Miyabara; *Yamashiro* may have Kansei. **Designed H.P.** 40,000=22·5 kts. in *Yamashiro*, 22 kts. in *Fuso*. Fuel: *normal*, 1000 tons; *maximum* 4000 tons coal + 1000 tons oil fuel.

Name.	*Builder.*	*Machinery*	*Laid down*	*Completed*	*Trials.* H.P. kts.	*Turbine*	*Boilers*	*Best recent speed*
Fu-So **Yamashiro**	Kure Yokosuka	Kawasaki Kawasaki	Mar.'12 Dec.'13	Nov.'15 Apl.'17	46,500 23	Curtis Curtis	Miyabara Miyabara	

General Notes.—*Fu-so* 1911 Naval Programme. *Yamashiro* 1913 Programme. *Fu-so* built in dry dock and floated out March, 1914.

Fu-So. 1916 Photo by courtesy of the Imper...

つつせ **SETTSU** (March, 1911).

Normal displacement, 21,420 tons. Complement, 1000.

Length (*p.p.*), 500 feet. Beam, 84½ feet. { *Maximum* draught, 28½ feet. / *Mean* „ 27½ „ } *Length over all*, 533 ft.

Guns (M. '04) (*See Notes*):
- 12—12 inch, 45 and 50 cal.
- 10—6 inch, 50 cal.
- 8—4.7 inch, 50 cal.
- 4—12 pdr. AA.
- 4 Machine.
- 4 Landing.

Dir. Con.

Torpedo tubes (18 inch):
- 4 *submerged* (broadside).
- 1 „ (stern).

(458 tons.)

Armour (Krupp):
- 12" Belt (amidships) ......
- 4" Belt (ends) (N.C.) ......
- 3" and 1½" Armour decks
- 7"-4" Lower deck side
- 11" Turrets ....
- 11" Turret bases ......
- 6" Battery ....
- 10" Conning tower........

(4960 tons.)

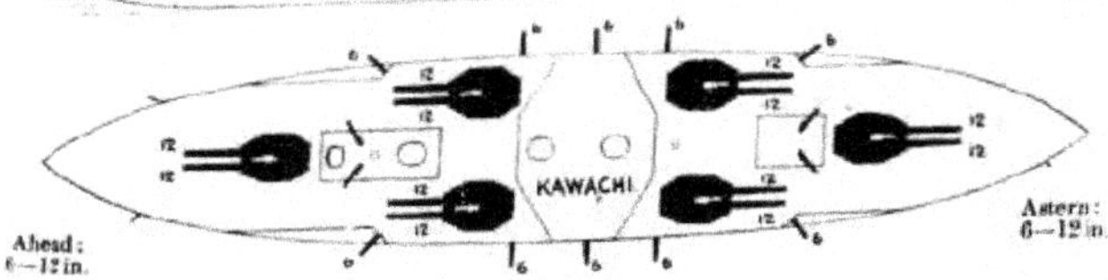

Ahead: 6—12 in.

Astern: 6—12 in.

Broadside: 8—12 in.

Machinery: Curtis turbine: 2 screws. Boilers: 16 Miyabara. Designed H.P. 25,000 = 20.5 kts. Coal: *normal*, 1100 tons; *maximum*, 2300 tons + 400 tons oil fuel.

Gunnery Notes.—The 12 inch guns in end barbettes are 50 calibres long; those in the four beam barbettes are only 45 calibres long. With the exception of 12 pdrs. (Japanese), all calibres are Elswick models.

Engineering Notes.—Weight of machinery, 2146 tons.

Name.	*Built at*	*Machinery*	*Laid down.*	*Completed.*	*H.P.*	*Trials. kts.*	*Turbines*	*Boilers.*	*Best recent speed.*
Settsu	Kure D.Y.	Kure D.Y.	Jan., '09	1912	15,000 =	18.2 =	Curtis	Miyabara	

Notes.—*Kawachi*, a sister ship to *Settsu*, destroyed by explosion in Tokuyama Bay, July, 1918.

SETTSU.

Photo by courtesy of the Navy Department, Toky

1911 JAPANESE BATTLE-CRUISERS. (Dreadnoughts Nos. 2, 3, 4, 5).

(Kongo Class—4 Ships.)

うかんこ **KONGO** (May, 1912). **HI-YEI** (Nov., 1912).

なるは **HARUNA** (Dec., 1913). ましりき **KIRISHIMA** (Dec., 1913)

Normal displacement, *Kongo, Hi-yei*, 27,500 tons; *Haruna, Kirishima*, 27,613 tons. Complement, 980.

Length (w.l.), 704 feet. { *Haruna, Kirishima*: Beam, 92½ feet. *Max.* draught, 27 feet. / *Kongo, Hi-yei*: " 92 " " " 27½ ".

Guns (*see Notes*):
8—14 inch, 45 cal. (A*)
16—6 inch, 50 cal.
8—12 pdr. (semi-automatic)
(Some anti-aircraft ?)
4 Machine.
4 landing.
Torpedo tubes (21 inch):
8 *submerged*.

Armour (Krupp):
Vertical.
8" Belt (amidsh
3" Belt (ends)
6" Upper belt
6" Battery ...
9", 6", 5" Bulk
8", 6" Bulkhea
10"— " Barbe
9"— " Gunho
[illegible]" C.T. base
10" C.T. (" b
" Fore comm
6" Torpedo c
" Tube (C.T.

Armour (H.T. ?):
Decks
" Forecastle
" Main......
" Middle ....
" Lower ....

Special Protection:
" Internal I

Ahead:
4—14 in.
2 to 4—6 in.

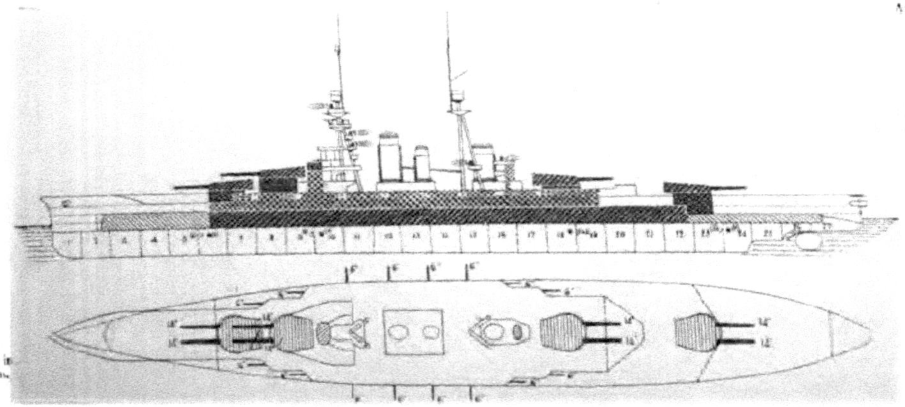

Astern:
4—14 in.
4—6 in.

Broadside: 8—14 in., 8—6 in., 4—21 in. T.T.

Machinery: Parsons 4-shaft (in *Haruna* only Curtis 4-shaft) turbines. Designed H.P. 64,000 = 27.5 kts. Boilers: See *Notes*. Fuel: *normal*, 1,200 tons; *maximum*, 4,000 tons + 1,000 tons oil.

Name	Builder	Machinery	Laid down	Completed	Trials	F.P.	Turbine	Boilers	Best recent speed
Kongo	Vickers	Vickers	Jan. '11	Aug. '13	4[illegible],000 = 25	78,000 = 27.5	Parsons	Yarrow	1916
Hi-yei	Yokosuka	Mitsu Bishi	Nov. '11	Mar. '14			Parsons	Kansei	
Haruna	Kawasaki	Kawasaki	Mar. '12	Mar. '15		82,000 = 27.7	Curtis	Yarrow	
Kirishima	Mitsu Bishi	Mitsu Bishi	Mar. '12	Apr. '15		80,000 = 27.5	Parsons	Yarrow	

General Notes.—*Kongo* 1910-11 Programme, others 1911-12 Programme.

Illustrations for "Kongo" class Battle Cruisers.

HIYEI. *Photo by courtesy of The Imperial Japanese Embassy*, 1917.

KONGO. This ship only has fore funnel set further forward than in other three ships.

KIRISHIMA, in action. (*Special Manœuvres*, 1919.) *From "The Navy," Tokyo*, 1910.

Note extra tops added to fore tripod mast.

These Notes are not from any official data.

Appearance Notes.—All four ships now have raised fore-funnels, as *Hiyei* opposite, but *Kongo* has her fore further forward and well away from the second funnel.

Gunnery Notes.—In *Kongo* guns are Vickers models; but in other three ships all calibres are of Japanese lacture. *Kongo* has combined Vickers (hydraulic) and Janney-Williams (electric) manœuvring systems barbettes; there is also a small auxiliary hydraulic installation, generally used for cleaning purposes, whi be used in emergency for working the 14-inch guns.

Armour Notes.—Main belt is 12' 5" deep, 8" thick, and extends between Barbettes Nos. 1 and 4.* Upper belt be Barbettes Nos. 1 and 3, 6" thick, and carried up to forecastle deck. **Bulkheads**: Main belt is closed by di bulkheads of 8"—6" aft and by a 6"—5" bulkhead forward. Upper belt closed by 6" bulkhead aft and bulkhead forward.

Anti-Torpedo Protection Notes.—Internal sub-division by longitudinal and cross bulkheads. Extra protection gi armour to all magazine spaces. Port and starboard engine rooms are divided by an unpierced longit bulkhead along keel-line. It is said contract for *Kongo* stipulated she should float with 50 feet of her side away, should not heel more than 11°, and *automatically* regain the vertical in a specified length of time, thou an increased draught.

Aircraft Notes.—Experiments with launching of seaplanes from *Kongo's* decks took place during September, Anti-aircraft guns mounted now in these ships.

Torpedo Notes.—Tubes are twin submerged type, at varying levels, some being only 6 feet below waterline. E in the case of the tubes in wake of No. 3 Barbette, starboard Tube is before Port Tube. *Kongo* has comb hydraulic and electrically-operated tubes. In *Haruna*, tubes are Elswick 21-inch side-loading, hydrauli operated.

Engineering Notes.—In *Kongo, Hiyei* and *Kirishima* Parsons turbines have H.P. rotors on outboard shafts and L.P inner shafts, with astern turbines aft and in same casing. *Kongo's* 36 Yarrow boilers are in five compartme Pressure: 205 lbs. per sq. inch.

*There is also a narrow 3-inch strip of armour, 2 feet 6 inch deep under whole length of main belt; this is not show plans opposite.

(See facing page for Plan and Details.)

AKI (April, 1907).

Displacement, 19,800 tons. Complement, 937.

Length (*o.a.*), 492 feet; (*p.p.*), 460 feet. Beam, 83¾ feet. *Maximum* draught, 28 feet.

Guns (Elswick M. '04):
- 4—12 inch, 45 cal. } Dir.
- 12—10 inch, 45 cal. } Con.?
- 8—6 inch, 45 cal.
- 8—12 pdr.
- 4 Machine
- 4 Landing

Torpedo tubes (18 inch):
- 4 submerged (broadside).
- 1 submerged (stern).

([illegible] tons)

Armour (Krupp):
- 9" Belt (amidships)
- 6" Belt (bow) (N.C.)
- 4" Belt aft
- 2" Deck
- 7" Upper belt
- 8" Turrets
- 9½" Turret bases
- 7" Bases to 10 in. turrets
- 7" Battery for Q.F. (N.C.)

([illegible] tons)

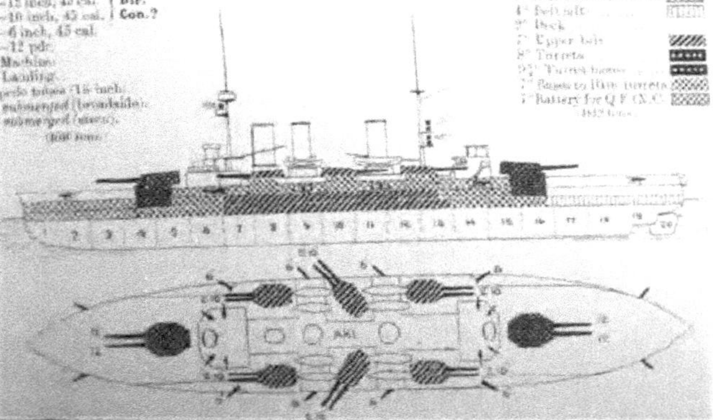

Machinery: Curtis turbine, by Fore River Co., U.S.A. 2 screws. Boilers: 15 Miyabara. Designed H.P. 24,000 = 20 kts. Coal: *normal*, 900 tons; *max.*, 2000 tons + 170 tons oil fuel. Begun Kure D.Y. March, 1905. Completed 1910. Trials: 28,740 S.H.P. = 20·25 kts. Refitted, 1915.

SATSUMA さつま (November, 1906).

Normal displacement, 19,350 tons. Complement, 937.

Length (*waterline*), 479 ft. Beam, 83¾ ft. *Maximum* draught, 28 ft. Length *over all*, 4

Guns (Elswick M. '04):
- 4—12 inch, 45 cal. } Dir.
- 12—10 inch, 45 cal. } Con.?
- 12—4·7 inch, 45 cal.
- 4 Machine
- 4 Landing

Torpedo tubes (18 inch):
- 4 submerged (broadside)
- 1 submerged (stern).

Armour (
- 9" Belt (
- 6" Belt (
- 4" Belt (
- 2" Deck
- 7" Uppe
- 7"—9"
- 9" Turr
- 5" Batte

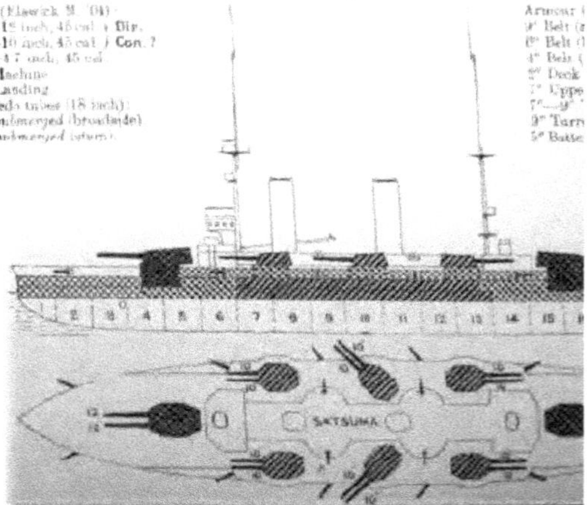

Machinery: 2 sets vertical triple expansion. 2 screws. Boilers: 20 Miyabara. Desig kts. Coal: *normal* 750 tons; *maximum* 2000 tons + 300 tons oil fuel. Built and engi Begun May, 1905. Completed April, 1910. Trials: I.H.P. 18,507 = 18·95 kts. Refitte

まらく **KURAMA** (Oct., 1907) & **IBUKI** (Nov., 1907). きふいっ

Normal displacement, 14,600 tons. Complement, 845.

Length (*p.p.*), 450 feet. Beam, 75½ feet. *Max.* draught, 26¼ feet. Length *over all*, 485 feet.

Guns (Japanese):
- 4—12 inch, 45 cal.
- 8—8 inch, 45 cal.
- 14—4·7 inch, 50 cal.
- 4 Machine
- 4 landing.

Torpedo tubes (18 inch):
- 3 *submerged*

(4158 tons)

Armour (Krupp):
- 7" Belt (amidships) ....
- 4" Belt (ends) (N.C.) ....
- 2" Armour deck ....
- 5" Lower deck redoubt ....
- 7" Turrets (N.C.) ....
- 7" Turret bases (N.C.) ....
- 5" Battery ....
- 6" Secondary turrets (N.C.)
- 8" Conning tower (fore) ....
- 6" " " (aft) ....

(2900 tons)

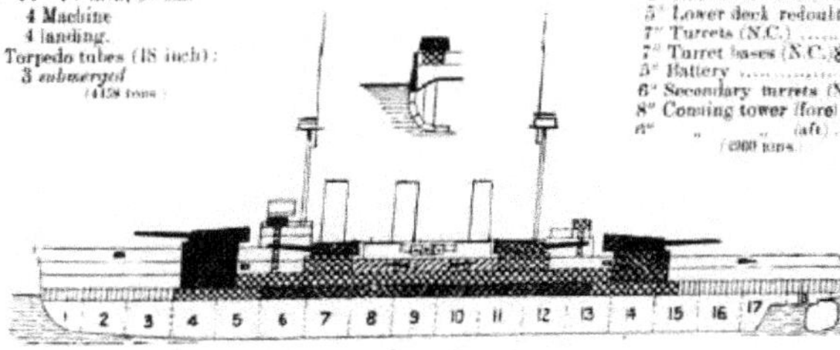

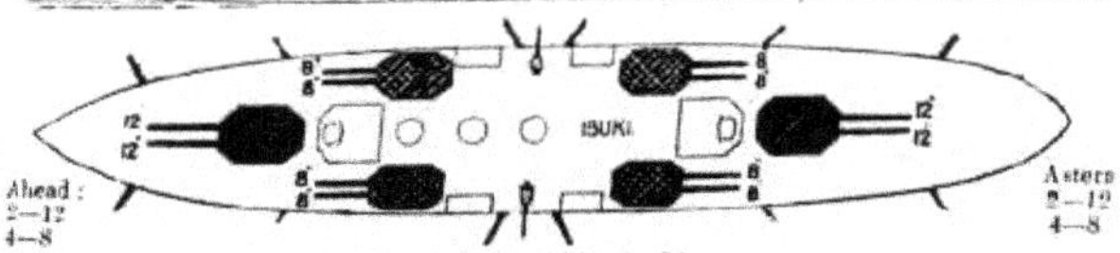

Ahead:
2—12
4—8

Astern:
2—12
4—8

Broadside: 4—12 in., 4—8 in.

Machinery: *Kurama*, 2 sets reciprocating. 2 screws. *Ibuki*, Curtis turbine. Boilers: Miyabara. Designed H.P., *Kurama*, 22,500 = 21·25 kts.; *Ibuki*, 27,000 = 22 kts. Coal: *normal*, 600 tons; *maximum*, 2000 tons + about 250 tons oil.

**Armour Notes*—Main belt is 7½ feet wide, thinning at ends.

**Gunnery Notes*—Loading positions, big guns: all round. Hoists: electric. Big guns: hydraulic and electric gears; secondary turrets, electric gear. Arcs of fire: 12 inch, 270°; secondary turrets, about 130°.

**Torpedo Notes*—Bullivant net defence to both ships.

Engineering Notes—Weight of machinery, 2078 tons.

Name	*Builder*	*Machinery*	*Laid down*	*Completed*	*Trials*		*Boilers*	*Best recent speed*
					at 4/5	*6 hrs f.p.*		
Kurama *Ibuki*	Yokosuka Kure	Yokosuka Fore River Co., U.S.A.	Aug.,'05 May,'06	Oct.,'09 1910	20,978 = 21·	27,353 = 20·87	Miyabara in both	

General Notes.—**Weights**: hull and fittings, 6105 tons, equipment 731 tons. *Ibuki* launched six months after laying down.

*These notes are unofficial.

KURAMA. *Photo, Symonds.*

IBUKI.

Differences.—*Kurama* has tripod masts, and slightly taller funnels. Yard on foremast is high up wi[th] searchlight top *under*. In *Ibuki*, yard is lower down, with searchlight top *above* yard.

Photo, Symonds

(New photo of this ship wanted.)

(Training Ship: Gunnery and Torpedoes.)

IKOMA (April, 1906). まこい

Displacement, normal 13750 tons; full load 15,150 tons. Complement, 817.

Length (p.p.), 440 feet; (o.a.), 475 feet. Beam, 75 feet. Mean draught, 26 feet.

Guns (see Notes):
4—12 inch, 45 cal.
10—6 inch, 45 cal.
8—4·7 inch.
6—12 pdr.
Torpedo tubes (18 inch):
2 submerged.
1 above water (stern).
(1719 tons.)

Armour (Krupp C. & N.C.):
7" Belt (amidships)
4" Belt (ends)
2" Armour deck
5" Lower deck redoubt
9-7" Turrets
7" Turret bases
5" Lower deck side
5" Casemates (N.C.)
8" Conning tower (fore)
6" „ „ (aft)
(3517 tons.)

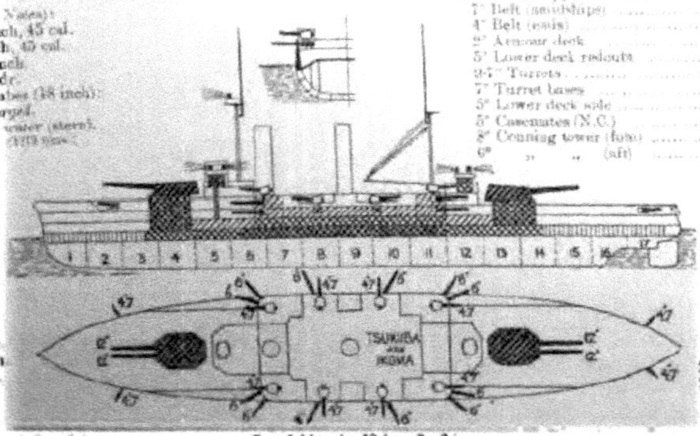

Ahead: 2—12 in. 2—6 in.

Astern: 2—12 in. 2—6 in.

Broadside: 4—12 in., 6—6 in.

Machinery: 2 sets reciprocating. 2 screws. Boilers: Miyabara. Designed H.P. 19,000 = 20·5 kts. Coal: normal, 600 tons; maximum, 2000 tons + 150 tons oil fuel. Weight of machinery, 2132 tons. Built and engined by Kure D.Y. 1905-7. On first trials made 21·9 kts. with 22,670 I.H.P. Refitted 1918-19. *Tsukuba*, of this class, destroyed by explosion, Jan., 1917.

1919 Corrections to Plans.—Main deck 6 inch battery removed and ports plated up. 4—4·7 inch over old two-tier sponsons removed. 6—6 inch now mounted on shelter deck, viz: 4 over old two tier sponsons (in old 4·7 inch positions) and 2

(Is second unit of the "Katori" Class.)

ましか **KASHIMA** (March, 1905). Displacement, 16,400 tons. Complement, 94[...] 470 feet. Beam, 78 ft. 1¼ in. *Max.* draught, 26 ft. 7½ in. Guns: 4—12 inch, 45 cal. ([...] 45 cal., 12—6 inch, 45 cal., 10—12 pdr., 3—3 pdr., 4 M.G., 2 field. Torpedo [...] 5 *submerged*. Armour (Krupp): 9" Belt (amidships), 6½" Belt (ends), 3" Deck on s[...] (K.C.), 9"—5" Turret bases, 8"—6" Secondary turrets (K.C.), 6" Lower deck (sid[...] Conning tower, 4" Upper deck battery. Machinery: 2 sets 4 cylinder vertical quac[...] 2 screws. Boilers: 20 Kansei. Designed H.P. 16,000 = 18·5 kts. Coal: n[...] *maximum*, 2,150 tons. Built at Elswick 1904-6. Refitted 1916.

りとか **KATORI** (July, 1905). Displacement, 15,950 tons. Comple[...] (o.a.), 455¾ feet. Beam, 78 feet. *Max.* draught, 27 feet. *Kashima* above, except *max.* coal, 1860 tons. Built by Vickers 1904-6. *distinguish*: Funnels closer together than *Kashima*; fore topmast is *before* m[...]

MIKASA さかみ (November, 1900). (Reconstructed 1907-08.) Displacement, 15,362 tons. Complement, 830. Length (*waterline*), 415 feet. Beam, 76¼ feet. *Maximum* draught, 26 feet. Length *over all*, 432 feet. Guns: 4—12 inch, (45) cal., 14—6 inch, 45 cal., 20—12 pdr., 2—3 pdr., 2—2½ pdr., 2 Maxims. Torpedo tubes (18 inch): 4 *submerged*. Armour (Krupp): 9" Belt (amidships), 4" Belt (ends), 12" Bulkheads, 3" Deck (slopes), 14"—10" Barbettes, 10"—8" Turrets to these, 6" Lower deck redoubt, 6" Battery (K.C.), 6" Casemates (N.C.). Machinery: 2 sets 3 cylinder vertical triple expansion. 2 screws. Boilers: 25 Belleville. Designed H.P. 15,000=18 kts. Coal: *normal* 700 tons; *maximum* 1500 tons. Built by Vickers, Barrow., 1899-1901. Sunk 1905. Salved and rebuilt 1906.

From "The Navy," Tokyo, 1918.

HIZEN (ex-Russian *Retvizan*) (October, 1900). (Reconstructed 1907-08.) Displacement, 12,700 tons. Complement, 791. Length (*waterline*), 376 feet. Beam 72½ feet. *Maximum* draught, 26 feet. Guns: 4—12 inch, 40 cal., (Obukhoff), 12—6 inch, 45 cal. (Japanese), 14—12 pdr., 4—2½ pdr., 2 M.G. Torpedo tubes (18 inch): 2 *submerged* (25° abaft). Armour (Krupp): 9"—6" Belt (amidships), 4" Belt (bow), 9" Bulkhead, 3" Deck (slopes), 6" Lower deck redoubt, 10" Turrets (N.C.), 9" Turret bases (N.C.), 5" Battery (N.C.), 10" Conning tower (N.C.) Machinery: 2 sets, 3 cylinder triple expansion. 2 screws. Boilers: 24 Niclausse. Designed H.P. 16,000=18 kts. Coal: *normal* 1016 tons; *maximum* 2000 tons. Built by Cramp, Philadelphia. Scuttled at Port Arthur, 1904, and captured on fall of fortress. Salved and repaired, 1905-6.

Asahi is something like Mikasa, but easily distinguished from her because her fore funnel is much neare than Mikasa's.

(Gunnery Training Ship.)

ひさあ **ASAHI** (March, 1899). *Normal* displacement, 14,765 tons. Complement, 836. *over all*, 425¼ feet. Beam, 75¼ feet. *Maximum* draught, 27¼ feet. Guns: 4—12 inch, 14—6 inch, 40 cal., 20—12 pdr., 2—3 pdr., 2—2½ pdr. Torpedo tubes (18 inch): 4 *su* Armour (Harvey Nickel): 9" Belt (amidships), 4" Belt (ends), 14" Bulkheads, 3" Deck (a slopes), 14"—10" Barbettes, 10"—8" Turrets to these, 6" Lower deck redoubt, 6" Casemat 14" Conning tower, 4" After tower. Machinery: 2 sets vertical triple expansion. 2 Boilers: 25 Belleville. Designed H.P. 16,000=18 kts. Coal: *normal* 700 tons; *maximum* 16 Built by Thomson, Clydebank, 1897-1900.

(Temporarily used as Submarine School.)

ましきし **SHIKISHIMA** (November, 1898). 14,580 tons. Complement, 836. Length (438 feet. Beam, 75½ feet. *Max.* draught, 27½ feet. Guns: As *Asahi* above. Torpedo tubes *Asahi* above, but has an additional 18 inch tube *above water*. Armour, machinery, screws boilers also as *Asahi*. Designed H.P. 14,500=18 kts. Coal (*max.*): 1500 tons. Built ... Ironworks, 1897-1900.

IWAMI みはい (July, 1902). (ex-Russian *Orel* reconstructed, 1907). *Normal* displacement, 13,516 tons. Complement, 806. Length (p.p.), 376½ feet. Beam, 76 feet. *Mean* draught, 26 feet. Length *over all*, 397 feet. Guns (Japanese): 4—12 inch, 45 cal., 6—8 inch, 45 cal., 16—12 pdr., 4—2½ pdr., 4 M.G. Torpedo tubes (15 inch): 2 *submerged* (broadside), 1 above water (*bow*). Armour (Krupp): 7½"–6" Belt (amidships), 4" Belt (ends), 3" Armour deck (slopes), 10" Turrets (N.C.), 10" Turret bases (N.C.), 6"—4" Upper belt, 10" Conning tower. Machinery: 2 sets vertical 4 cylinder triple expansion. 2 screws. Boilers: 20 Belleville. Designed H.P. 16000=18 kts. Coal: *normal* 750 tons; *maximum* 1520 tons. Laid down at Galernii Island, June, 1900, and completed for sea in September, 1904. Captured in Battle of Tsushima, 1905.

(*Cadets' Training Ship.*)

うはす **SUWO** (ex-Russian *Pobieda*. May, 1900). *Normal* displacement (after reconstruction), 13,500 tons. Complement, 787. Length (*waterline*), 424 feet. Beam, 71½ feet. *Maximum* draught, 25½ feet. Length *over all*, 435 feet. Guns (Obukhoff): 4—10 inch, 45 cal., 10—6 inch, 45 cal., 16—12 pdrs., 4 machine, 2 field. Torpedo tubes (15 inch): 2 *submerged* (at 20° abaft). Armour: 9"—6" Belt; 2¾" Deck (on slopes); 10" Turrets; 8" Turret bases; 5" Lower deck side; 5" Casemates (10); 10" Conning tower (K.N.C.) Machinery: 3 sets vertical 3 cylinder triple expansion. 3 screws. Boilers: 30 Bellevilles, placed fore and aft. Designed H.P. 14,500=18 kts. Coal: *normal* 1063 tons; *maximum* 2100 tons. Also 250 tons liquid fuel. Scuttled at Port Arthur; [illegible] on fall of fortress. Salved and refitted 1905-6.

(*Seamanship Training Ship.*)

じふ **FUJI** (March, 1896). 12,649 tons. Complement, 741. Length (*waterline*), 390 73 feet. *Maximum* draught, 27 feet. Guns: 4—12 inch, 40 cal., 10—6 inch, 40 cal., 16- pdr. 4—2½ pdr. Torpedo tubes (18 inch): 4 *submerged*, 1 *above water* (bow). Armour (Ha (amidships), 14" Bulkheads. 2½" Deck (flat on belt), 4" Lower deck redoubt. 6" Casemates. 14" Conning tower. Machinery: 2 sets 4 cylinder vertical triple expansi Boilers: 10 Miyabara. Designed H.P. *forced* 13,500=18 kts. Coal: *normal* 700 tons; tons.

Note.—There is a new Destroyer named *Fuji*, meaning "Wistaria." The above Battleship is named mountain Fujiyama.

***OKINOSHIMA** (ex-Russian *General Admiral Graf Apraxin*, 1896) 4126 tons, & ま [illegible] (ex-Russian *Admiral Seniavin*, 1894). 4160 tons. 3—10 inch, 45 cal. (*Mishim* with, in both 4—4·7 inch, 10—3 pdrs., 2—2½ pdr. Armour: 10"—8" belt amidsh Designed H.P. 5000=16 kts. Boilers: 8 Belleville. Coal: *maximum* 450 tons. Battle of Tsushima, 1905.

**Note.*—These ships serve as Barracks for Boys.

んしつに NISSHIN (Feb., 1903) & KASUGA (Oct., 1902). かすか

Displacement 7,750 tons. Complement 610 and 595.

Length (*waterline*), 357 feet. Beam, 61 feet 11 ins. *Maximum* draught, 25¼ feet.

Guns (Elswick):
4—8 inch, 45 cal.
14—6 inch, 45 cal.
10—12 pdr.
4—2½ pdrs.
2 Maxims.
2 Field guns.
Torpedo tubes (18 inch):
4 *above water* (in casemates).

Armour (Terni):
6" Belt (amidships)
4½" Belt (ends)
1½" Deck (on slopes)
5½" Turrets
5½" Turret bases (N.C.)
6" Lower deck side
4½" Lower deck bulkheads
6" Battery (N.C.)
4½" Battery (bulkheads)
4½" Conning tower

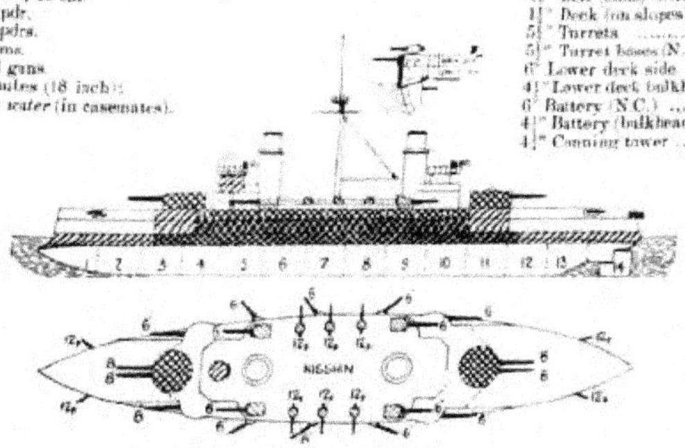

Ahead:
2—8 in.
4—6 in.

Astern:
2—8 in.
4—6 in.

Broadside: 4—8 in., 7—6 in.

Machinery: 2 sets 3 cylinder vertical triple expansion. 2 screws. Boilers: 12 Kansei. Designed H.P. 13,500 = 20 kts. Coal: *normal*, 650 tons; *maximum*, 1,200 tons.

General Notes.—Laid down for Argentina by Ansaldo, Genoa, in 1902, as the *Moreno* and *Rivadavia*. Purchased end of 1903 by Japan for £160,000, just before outbreak of Russo-Japanese War. Both had large refits 1911. *Kasuga* stranded at N.W. entrance to Banka Straits, Dutch East Indies, during January, 1918, and was not refloated till May. She has undergone extensive repairs in a Japanese dockyard.

(*Now fitted as Mine Layer*). *Photo, His Honour Mr. Justic*

そあ ASO (ex-Russian *Bayan*). (La Seyne, 1900).

Displacement, 7800 tons. Comple Length (*waterline*), 443 feet. Beam 57½ feet. *Max.* draught, 22 feet. Lengt 450 feet. Guns: 2—8 inch, 45 cal., 8—6 inch, 45 cal., 16—12 pdr., Torpedo tubes (at 20° abaft): 2 *submerged*. Tubes are 15 inch. Armour (Krupp): (amidships), 4" Belt (forward), 8" Bulkhead (aft), 2" Deck, 3¼"—2½" Upper belt, 7" Big gu Hoists, &c., 3¼" Battery redoubts, 6¼" Conning tower. (Total about 1500 tons. Machiner vertical triple expansion. 2 screws. Boilers: 24 Belleville. Designed H.P. 16,500 = 21 kt *normal* 550 tons; *maximum* 1100 tons. Scuttled at Port Arthur, 1905, and captured fortress. Re-floated and re-fitted in Japan, 1905–6.

Idzumo. 1919 *Photo, Seward*,

もつい IDZUMO (Elswick, 1899) & てはい IWATE (Elswick,

Displacement, 9750 tons. Complement, 658. Length (*p.p.*), 400 Beam, 68½ feet. *Maximum* draught, 24¼ feet. Length *over all*, 434 feet. Guns (Elswick): 4— 40 cal., 14—6 inch, 40 cal., 12—12 pdr., 4—2½ pdr., 2 M.G. Torpedo tubes: 4 *subm* Armour (Krupp): 7" Belt (amidship), 3½" Belt (ends), 2½" Deck (slopes), 5" Lower deck (red 6" Turrets and bases, 6" Casemates, 14" Conning tower. (Total 2100 tons.) Machine Humphry and Tennant: 2 sets 4 cylinder triple expansion. 2 screws. Boilers: Belleville H.P. 16,000 = 20·75 kts. Coal: *normal* 550 tons

(Cadets' Training Ship.) 1920 Photo, M. Bar, Toulon.

まつあ **ADZUMA** (St. Nazaire, 1899). Displacement, 9126 tons. Complement, 644. Length (*waterline*), 430 feet. Beam, $59\frac{1}{2}$ feet. *Maximum* draught, 25 feet. Length *over all*, $452\frac{1}{2}$ feet. Guns (Elswick): 4—8 inch, 40 cal., 12—6 inch, 40 cal., 16—12 pdr., 4—$2\frac{1}{2}$ pdr. Torpedo tubes: 4 *submerged*. Armour (Krupp mostly): 7" Belt (amidships), $3\frac{1}{2}$" Belt (ends), $2\frac{1}{2}$" Deck (on slopes), 6" Turrets and bases (H.S.), 6" Casemates (H.S.), 5" Side above belt. (Total weight, 2000 tons). Machinery: 2 sets vertical triple expansion. 2 screws. Boilers: 24 Belleville. Designed H.P. 17,000 = 21 kts. Coal: *normal* 600 tons; *maximum* 1200 tons.

(Cadets' Training Ship.)

もくや **YAKUMO** (Vulcan Co., 1899). Displacement, 9,735 tons. Complement, 698. Length (p.p.), 390 feet. Beam, $64\frac{1}{4}$ feet. *Mean* draught, $23\frac{3}{4}$ feet. Length *over all*, 434 feet. Guns: 4—8 inch, 40 cal., 12—6 inch, 40 cal., 16—12 pdr., 4—$2\frac{1}{2}$ pdr. Torpedo tubes (18 inch): 4 *submerged*. Armour (Krupp): 7" Belt (amidships), $3\frac{1}{2}$" Belt (ends), $2\frac{1}{2}$" Deck (slopes), 6" Turrets (N.C.), 6" Turret bases (N.C.), 5" Lower deck side, 6" Casemates (8), 10" Conning tower. (Total weight, 2040 tons) Machinery: 2 sets vertical triple expansion. 2 screws. Boilers: 24 Belleville. Designed H.P. 15,500 = $20\frac{1}{2}$ kts. Coal: *normal* 550 tons; *maximum* 1200 tons. *Armour Notes:* Belt is 7 feet wide.

Tokiwa (Coast Training Ship). 1920 Photo, M. B

まさあ **ASAMA** (March, 1898). はきと **TOKIWA** (July, 1898). Displacemen[illegible] Complement 648. Length (o.a.), 442 feet. Beam, $67\frac{1}{4}$ feet. *Maximum* draught, $24\frac{1}{4}$ [illegible] (Elswick): 4—8 inch, 40 cal., 14—6 inch, 40 cal., 12—12 pdr., 4—$2\frac{1}{2}$ pdr. Torpedo [illegible] inch): 4 *submerged*. Armour (Harvey-nickel): 7" Belt (amidships), $3\frac{1}{2}$" Belt (ends), 2" De[illegible] 5" Upper belt (amidships), $3\frac{1}{2}$" Bulkheads to it, 6" Turrets and bases, 6" Casemates (10), 1[illegible] tower. (Total, 2100 tons.) Machinery: 2 sets 4 cylinder triple expansion. 2 screws. Miyabara. Designed H.P. *forced* 18,000 = $21\frac{1}{2}$ kts. Coal: *normal* 550 tons; *maximum* [illegible] Both built at Elswick. Very 'handy' ships. *Asama* was badly damaged by groundin[illegible] Coast of Central America, in December of 1914. Was salved and has been repaired and [illegible]

1889 SECOND CLASS COAST DEFENCE SHI[illegible]

たよち **CHIYODA** (June, 1890). Displacement 2439 tons. Complement 324. L[illegible] Beam, $42\frac{1}{2}$ feet. *Maximum* draught, $15\frac{1}{2}$ feet. Guns (Elswick): 10—4[illegible] 15—3 pdr. Torpedo tubes: 2 *above water*. Armour (chrome steel): $4\frac{1}{2}$" Belt, 1" [Deck flat on belt.] Machinery: 2 sets triple expansion. 2 screws. Boilers (1898) without economisers. Designed H.P. 7000 = 19 kts. Coal: *normal* 330 tons; *maxim*[illegible]

Future Construction (*Unofficial Notes*).

Under the 1920-28 Navy Law, it is proposed that 12 Light Cruisers should be built.

Possible that three further Light Cruisers may be authorized under 1921-22 Sections of the 1918-24 Navy Law.

Projected.

らゆ *YURA.* Builder unknown. Unofficially reported to be "*about* 7,000 tons displacement." Authorized under the 1920 Section of the 1918-24 Navy Law.

Projected or Building.

づすい *ISUDZU* — To be begun 1920-21 by Sasebo D.Y., Mitsubishi Co., and Uraga Dock Co.

らがな *NAGARA*

りとな *NATORI* — Laid down 10th August, 1920.

No details known. Authorized under 1919 Section of the 1918-24 Navy Law.

1918-19 "Middle Light Armoured Cruisers."

TAMA. *1920 Photo.*

まく KUMA (July, 1919). **また TAMA** (10th Feb., 1920). **ゐほお O**...

みかたき KITAKAMI (July, 1920). **そき** *KISO* (...

Normal displacement, 5500 tons. Complement, 439.

Length (*w.l.*), 500 feet (*o.a.*), 535 feet. Beam, 46¾ feet. *Mean* draught, 15¾ f...

Guns (*unofficial*):
7—5.5 inch, — cal.
3—12 pdr. A.A.
— M.G.
Torpedo tubes (21 inch):
4 *above water*
2 *submerged*

Armour (*u*...
2" Belt ...
" Deck ...
" C.T. ...

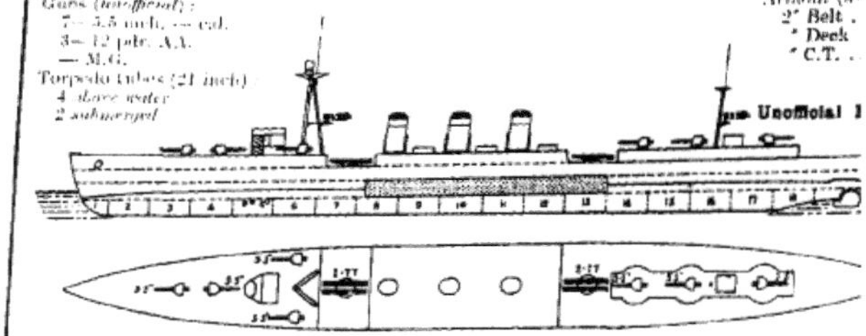

Ahead: 3—5.5 inch. Broadside: 6—5·5 inch. Astern: 1—5.5 ...

Machinery: Turbines. Boilers: screws. Designed H.P. 33 kts. Fuel: *normal*, tons; *maximum*, tons.

Name.	*Builder*	*Machinery*	*Laid down.*	*Completed.*	*Refit.*	*First Trials.*	*Boilers*
Oh-i	Kawasaki	Kawasaki	24/11/19				
Kitakami	Sasebo D.Y.		Sep., '19				
Kiso	Mitsubishi Nagasaki	Mitsubishi	10/6/19				
Kuma	Sasebo D.Y.		Aug. '18	Aug. '20			
Tama	Mitsubishi Nagasaki	Mitsubishi	Aug. '18	1920			

General Notes.—*Kuma* and *Tama* begun under the 1917 Naval Programme; *Oh-i, Kitakami, Kiso* under 1... Programme.

JAPANESE "SMALL LIGHT ARMOURED" CRUISERS.

(Tenryu Class—2 Ships.)

たつた TATSUTA (29th May, 1918) & うりんて TENRYU (11th March, 1918).

Normal displacement, 3500 tons. Complement, 332

Length (*o.a.*): 440 feet. Beam: 40¾ feet. *Mean* draught: 13 feet.

Guns:
4—5·5 inch, — cal.
1—12 pdr. AA.
2 M.G.
Torpedo tubes:
6 *above water* in two triple U.D. mountings.

Armour (unofficial):
1½" Belt .....
—" Deck .....
—" C.T. .....

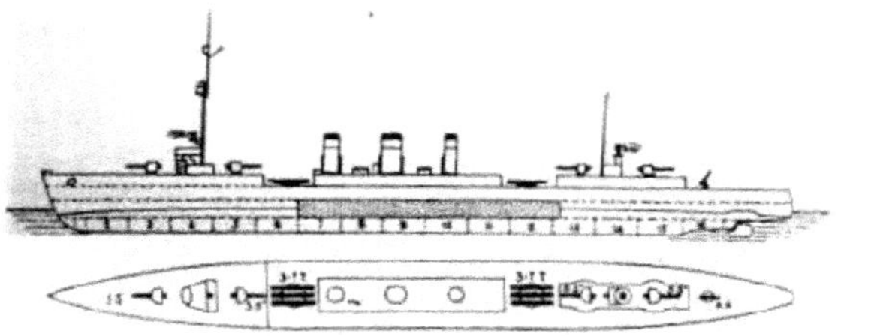

Ahead:
1—5·5 in.

Broadside: 4—5·5 in.

Astern:
1—5·5 in.

Machinery: Parsons or Curtis turbines. 3 screws. Boilers: 10 Kansei. Designed H.P. (unknown) =31 kts. Fuel: *normal*, tons; *maximum*, tons.

Name	*Builder*	*Machinery*	*Laid down*	*Completed*	*Trials*	*Boilers*	*Best recent speed.*
Tatsuta	Sasebo D.Y.		July, '17	31 Mar. '19	[illegible] kts. (unofficial)	Kansei	
Tenryu	Yokosuka D.Y.		May, '17	20 Nov. '18		Kansei	

General Notes.—Begun under the 1916 Naval Programme.

Tatsuta. 1920 *Photo, Navy D*

Tenryu (on trials; guns not mounted). From "The Navy

CHIKUMA. 1916 Photo by courtesy of the Imperial Japanese Embassy.

とらひ HIRADO (June, 1911). きはや YAHAGI (Oct., 1911). まくち CHIKUMA (April, 1911).

Normal displacement 4950 tons. Complement 403.

Length (*p.p.*), 440 feet; *o.a.*, 475 feet. Beam, 46½ feet. *Max.* draught, 16½ feet.

Guns:
8—6 inch, 50 cal.
4—12 pdr.
2 M.G.
Torpedo tubes (18 inch):
3 *above water*.
(361 tons.)

Armour:
3" Deck (amidships)
2" Deck (ends)
4" Conning tower
(423 tons.)

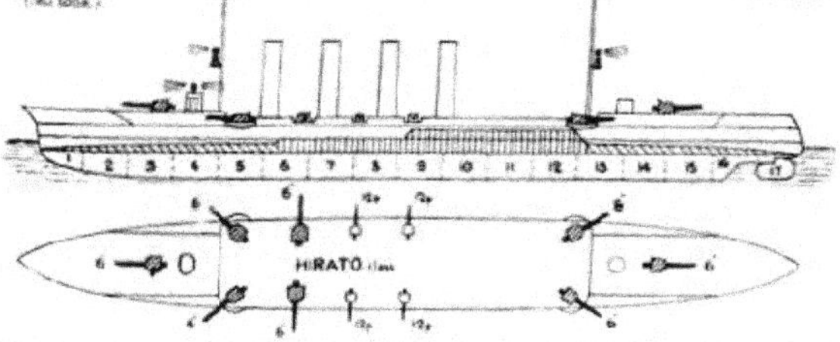

Machinery: Curtis turbines; (2 screws), except Yahagi, Parsons turbines (4 screws). Boilers: 16 Kansei. Designed H.P. 22,500 = 26 kts. Coal: *normal* 500 tons; *maximum* 900 tons + 300 oil.

Name	*Builder*	*Machinery*	*Laid down*	*Completed*	*Trials H.P. = kts.*	*Boilers*	*Best recent speed*
Hirado	Kawasaki Co.	Kawasaki Co.	Aug. '10	Oct. '12		Kansei in all.	
Yahagi	Mitsu Bishi	Mitsu Bishi	June '10	July '12	27,408 = 26.8		
Chikuma	Sasebo Yard	Kawasaki Co.	May '10	May '12	24,974 = 26.8		

Weight of hull, etc., 2278 tons. *Hirado* returned to Japan early in 1918 after having steamed 26,000 miles in nine months on active service.

ねと TONE (Oct., 1907).

Normal displacement 4105 tons. Complement 401.

Length (*p.p.*), 360 feet; *o.a.*, 406 feet. Beam, 47½ feet. *Mean* draught, 16¾ feet.

Guns:
2—6 inch, 50 cal.
10—4·7 inch, 50 cal.
2—12 pdr.
2 M.G.
Torpedo tubes (18 in.):
3 *above water*.
(261 tons.)

Armour:
3" Deck (amidships)
2" Deck (ends)
4" Conning tower
(42 tons.)

[illegible]

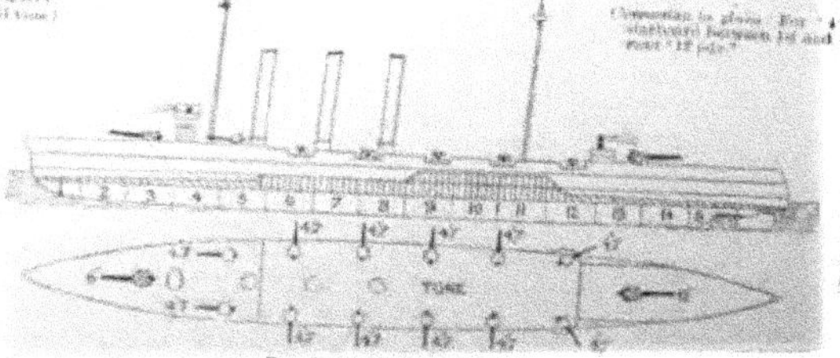

Ahead:
1—6 in.
2—4·7 in.

Broadside: 2—6 in., 6—4·7 in.

Machinery: 2 sets triple expansion. 2 screws. Boilers: 16 Miyabara. Designed H.P. 15,000 = 23 kts. Coal: *normal* 300 tons; *maximum* 850 tons + 125 tons oil.

Name.	*Builder*	*Machinery*	*Laid down*	*Completed*	*Refit*	*Trials*	*Boilers*
Tone	Sasebo	[illegible]					

NIITAKA has foremast aft of fore bridges; TSUSHIMA has foremast stepped through charthouse.

かたいに NIITAKA. ましつ TSUSHIMA.

NIITAKA (1902) & **TSUSHIMA** (Dec., 1902).

Displacement, 3420 tons. Complement, 307.

Length, 334½ feet. Beam, 44 feet. *Maximum* draught, 16 feet.

Guns (Elswick):
6—6 inch, 40 cal.
10—12 pdr.
4—2½ pdr.
Torpedo tubes:
None.

Armour (steel):
2½" Deck
(Cellulose belt.)
4" Conning tower
(K.N.C.)

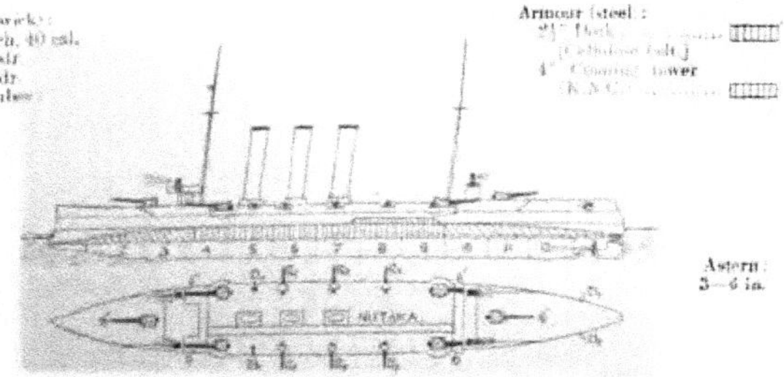

Ahead:
3—6 in.

Astern:
3—6 in.

Broadside: 4—6 in.

Machinery: 2 sets triple expansion. 2 screws. Boilers: 16 Niclausse. Designed H.P. 9400 = 20 kts. Coal: *normal* tons; *maximum* 600 tons. Laid down 1900 at Yokosuka and Kure.

A similar ship, *Otowa*, wrecked 1917. *Niitaka*, on patrol service during war, has steamed 50,000 miles in two years.

(*Fitted as Mine Layer.*)

るがつ **TSUGARU** (ex Russian *Pallada*). (Aug., 1899.) Displacement 6630 ton ment 532. Length (*waterline*), 410 feet. Beam, 55 feet. *Maximum* feet. Length (*over all*), 415½ feet. Guns (Russian): 10—6 inch, 45 cal., 12—12 p Torpedo tubes (18 inch): 2 *above water*. Armour: 2½" Deck, 4" Hoists, 6" C. T. Engine hatches. Machinery: 3 sets horizontal 3 cylinder triple expansion. 3 screws Belleville. Designed H.P. 11,600 = 20 kts. Coal: *normal* 900 tons; *maximum* 1430 at Port Arthur, October, 1904. Salved August, 1905. Used as a Training Ship.

せとち **CHITOSE** (January, 1898). Displacement 4992 tons. Complement 43 (*waterline*), 396 feet. Beam, 49½ feet. *Maximum* draught, 18 feet. inch, 40 cal. (*B*)., 10—4·7 inch, 12—12 pdr., 2—2½ pdr. Torpedo tubes (18 *above water*. Armour: 4½" Deck (amidships). Machinery: 2 sets vertical triple 2 screws. Boilers: 12 Miyabara. Designed H.P.: *forced* 15,500 = 22·5 kts. Co 350 tons; *maximum capacity* 1000 tons. Built and engined by Union Ironworks, San Laid down 1896, completed 1899. Made 22·87 kts. on first trials with 12,500 H.P. A *Kasagi*, wrecked in Tsugaru Straits, 1916.

AKASHI. 1913 Photo.

しかあ **AKASHI** (1897). Displacement 2800 tons. Complement, 310. Length, $295\frac{3}{4}$ feet. Beam, $41\frac{3}{4}$ feet. *Maximum* draught, $16\frac{1}{2}$ feet. Guns (Elswick): 2—6 inch, 40 cal., 6—4·7 inch, 40 cal., 12—3 pdr. Torpedo tubes (14 inch): 2 *above water*. Armour: 2" Deck, $4\frac{1}{2}$" Shields to 6 inch guns. Machinery: 2 sets vertical triple expansion. 2 screws. Boilers: 8 cylindrical (probably replaced now by Kansei or Miyabara). Designed H.P. 7600 = 19·5 kts. Coal: *normal* 200 tons; *maximum* 600 tons.

Note.—Was intended to be a sister ship to *Suma* (given in next column), but as *Suma* proved a poor sea boat, *Akashi* was built up amidships.

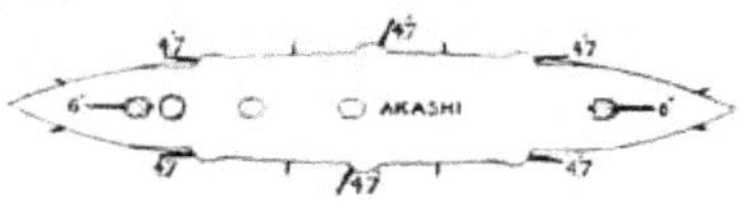

Now has fire-control top on foremast and a very small mainmast.

ます **SUMA** (1895). 2700 tons. Comp. 310. Dimensions: $305\frac{1}{4} \times 39\frac{1}{4} \times$ (*max.*) $16\frac{1}{2}$ feet. Armament (Elswick): 2—6 inch, 40 cal., 6—4·7 inch, 40 cal., 12—3 pdr. 2 tubes. 2" Deck. $4\frac{1}{2}$" Shields to 6 inch guns. Machinery: 2 sets vertical triple expansion. 2 screws. Boilers: 4 Miyabara. Designed H.P. *forced* draught 8500 = 20 kts. Coal: *normal* 200 tons; *maximum* 600 tons. (See note to *Akashi* above).

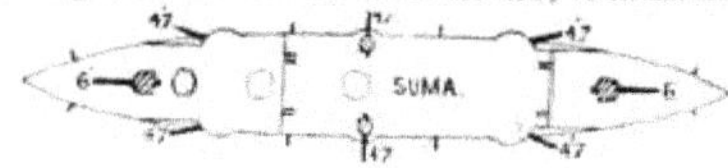

まきつきあ **AKITSUSHIMA** (1892). 3172 tons. Comp. 314. Dimensions:—302 × 43 × (*Ma*
$18\frac{1}{4}$ feet. Armament: 4—6 inch, 6—4·7, 10—3 pdr. 2—14 inch tubes. 3" Deck. Machiner
2 sets vertical triple expansion. 2 screws. Boilers: 8 Miyabara. Designed H.P. *natu*
8500 = 19 kts. Coal: 550 tons.

(*Continued on next pag*

(*Conscripts' Training Ship.*)

てたしは **HASHIDATE** (1891). 4277 tons. Comp. 360. Dimensions: 295 × 51 × (*max.*) 21¼ feet. Armament: 1—12·6 inch Canet, 11—4·7 inch, 6—12 pdr., 2—3 pdr., 2 M., 2—14 in. tubes. Armour (steel): 12" Barbette with 4" shield, 1½" deck. Machinery: 2 sets triple expansion. 2 screws. Boilers: 8 Miyabara (1903). Designed H.P. 5400 = 16 kts. Coal: 680 tons.

1st Class Gunboats (*Itto Hokan*).

(Over 800 tons).

みかも **MOGAMI** (Nagasaki, 1907). 1350 tons. Comp. 178. 316 × 31½ × 11¼ feet. Guns: 2—4·7 inch (50 cal.), 4—12 pdr. Armour: 2¼" deck. Torpedo tubes: 2—18 inch. Machinery: Parsons turbine. 3 screws. No cruising turbines. Designed H.P. 8000 = 23 kts. Boilers: 6 Miyabara. Coal: normal, tons; *maximum*, 350 tons + 70 tons oil. First turbine-engined ship of Japanese Navy.

と よ **YODO** (Kobe, 1907). 1250 tons. Comp. 193. 305½ × 32 × 11 feet. Guns: 2—4·7 inch, 50 cal., 4—12 pdr. Armour: 2½" deck. Torpedo tubes: 2—18 inch. Machinery: 2 sets 4 cyl. triple expansion. Designed H.P. 6500 = 22 kts. Boilers: 4 Miyabara. Coal: *normal*, tons; *maximum* 340 tons + 80 tons oil.

1st Class Gunboats—*continued.*

やはち **CHIHAYA** (1901). 1263 tons. Comp. 16
2—4·7 inch, 4—12 pdr. 2—18 inch tubes. Designed H
21 kts. Boilers: Thornycroft. Coal: *normal* 200 tons;
450 tons.

Note.—Five Gunboats reported to be included in 1920-28
Three further boats may be sanctioned under 1919 or 19
of the 1918-24 Navy Law. No details being available, t
cannot be placed either to 1st, 2nd or River Classes.

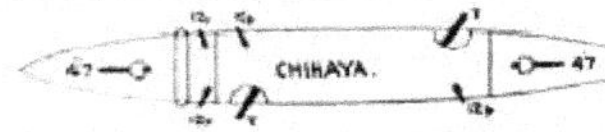

2nd Class Gunboats (*Nito Hokan*). (Under 80

Photo by courtesy of the Navy Depa

SAGA (Sasebo, 1912). 785 tons. Complement 98. 210
feet. Guns: 1—4·7 inch, 2—12 pdr. 1 machine. 3
Triple expansion. 3 screws. Boilers: 2 Kansei. Des
1600 = 15 kts. Coal: *maximum* 400 tons.

(*Continue*

2nd Class Gunboats—*Continued.*

UJI (1903). 620 tons. Complement 86. 180½ × 27½ × 7 feet. Armament: 4—12 pdr., 3 M.G. H.P. 1000 = 13 kts. Belleville boilers. Coal: 150 tons.

River Gunboats.

FUSHIMI (1906). 180 tons. Complement 45. 160 × 24½ × 2¼ feet. Guns: 2—6 pdr., 4 M. 800 H.P. = 14 kts. Yarrow boilers. Coal: 25 tons.

Also **TOBA** (1911). 250 tons. Complement 59. 180 × 27½ × 2½ feet. Guns: 2—12 pdr., 6 M. 1400 H.P. = 15 kts. Coal: 80 tons.

SUMIDA (1903). 126 tons. Complement 40. 145 × 24 × 2 feet. Guns: 2—6 pdr., 4 M. 550 H.P. = 13 kts. Thornycroft boilers. Coal: 40 tons.

89* + *11*† + 11§ Destroyers. (*Kuchikukan*).

* In service. † Building or completing. § Authorised, not begun.

Boats over 1000 tons are officially rated first class; 1000—600, second class; and under 600 tons, third class.

Totals.	Class.	Date.	Displacement.	H.P.	Max. Speed.	Fuel. Coal Oil.	Complement.	Tubes.	Max. Draught.
	First Class:—		tons.		kts.	tons.			feet.
3	*Namikaze*	*Proj.*	...	...	...	...	...	...	...
5	*Akikaze*	'20—?	...	...	...	...	...	...	...
4	*Okikaze*	'18—'21	1345	28,500*t*	31	...	148	6	10¼
2	*Minekaze*	'18—'20				...	145		
2	*Tanikaze*	'16—'19	1300	28,000*t*	34	...	128	6	10½
4	*Amatsukaze*	'15—'17	1227	27,000*t*	34	145 + *195*	145	6	10½
2	*Umikaze*	'08—'11	1150	10,500*t*	31·5	250 + *180*	139	3—1	9½
	Second Class:—								
8	*Ashi*	*Proj.*	...	...	...	...	...	...	...
5	*Ae-i*	'19—?	...	...	...	...	...	...	...
8	*Momi*	'18—'20	850	21,500*t*	31·5	...	110	4	9¾
6	*Tsubaki*	'17—'18	835	16,000*t*	31·5	90 + *210*	109	6	7¾
4	*Momo*	'15—'16							
10	*Kaba*	'14—'15	665	9,500	30	99 + *125*	92	4	8¼
1	*Urakaze*	'13—'15	955	22,000*t*	28	*248*	117	4	10½
2	*Sakura*	'11—'13	605	9,500*t*	30	123 + *50*	92	4	7¼
	Third Class:—								
33	*Arare*	'05—'08	375–381	6,000	29	{100 or 95 + *15*}	61	2	8¼—9
2	*Asagiri*	'02—'04							
2	*Asashiho* (T)	'01—'04	333	7,000	31	96	60	2	8½
2	*Akebono* (Y)	'97—'00	306	6,000	31	90-95	55	2	8
3	*Kagero* (T)	'97—'00	326	5,400	30	81	61	2	7¼

(T) = Thornycroft. (Y) = Yarrow type. *t* = Turbine.

Future Construction.—(*Unofficial Notes*).

Reported that 37 New Destroyers of 1st and 2nd classes are projected under the 1920-28 Navy Law. There may be 14 further boats to be put in hand under the final years of the 1918-24 Navy Law.

(*Continued on next page.*)

4* + 6† + 3§ 1st Class Destroyers (over 1000 tons).

* In service. † Building or Completing. § Projected.

3 Namikaze class.

Authorized.

Name.	*Builder.*	*Begun.*	*Name.*	*Builder.*	*Begun.*
Namikaze	Mitsubishi Co.	....	*Nokaze*	Maidzuru D.Y.	....
Numakaze	Nagasaki or Kobé	....			

No details known. Authorized under 1920 Section of 1918-24 Navy Law.

5 Akikaze class.

Projected or Begun.

Name.	*Builder.*	*Begun.*	*Name.*	*Builder.*	*Begun.*
Akikaze	Mitsubishi Co.	1920 or	*Hokaze*		
Yukaze	Nagasaki or Kobé	1921	*Shiokaze*	Maidzuru D.Y.	1920 or 1921
			Tachikaze		

No details known. Believed authorized under the 1919 section of the 1918-24 Navy Law.

7 Minekaze class. (2 Minekaze + 5 Okikaze.)

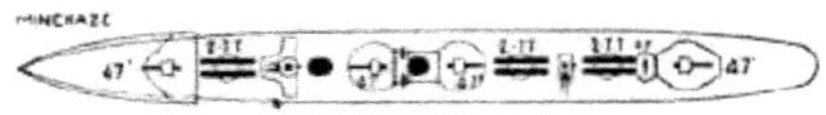

Sawakaze (on trials; no guns or T.T.). 1920 *Photo, Navy Dept., T*

7 *Minekaze* class: **Minekaze, Okikaze, Nadakaze, Shimakaze,** (4) by Maidzuru D.Y.: **Saw** **Hakaze** and **Yakaze** all built by Mitsubishi Co., at Nagasaki or Kobé. *Normal* displacement 1 Dimensions: 320 (*p.p.*), 336½ (*o.a.*) × 29½ × 9½ feet (*mean* draught). Armament: 4—4.7 inch, 45 cal and 6—21 inch torpedo tubes. 2—30 inch searchlights. Designed S.H.P. 28,500 = 34 kts. Machinery: 2 shaft turbines and 8 Kampon or Kansei boilers. Fuel: tons. Complement: *Minekaze, Sawa* others 148. *Minekaze* and *Sawakaze* belong to the 1917 Programme; *Okikaze, Nadakaze, Shimakaze, H* *Yakaze* to the 1918 Programme.

Name.	*Begun.*	*Launch.*	*Comp.*	*Name.*	*Begun.*	*Launch.*
Minekaze	20/4/18	8/2/19	Yes.	*Sawakaze*	7/1/18	7/1/19
Okikaze	22/2/19	3/10/19	Yes.	*Hakaze*	11/11/18	21/6/20
Nadakaze	9/1/20	....	....	*Yakaze*	15/8/18	10/4/20
Shimakaze	5/9/19	31/3/20	Yes.			

2 Tanikaze class.

KAWAKAZE. 1920 *Photo, Navy Dept., T*

2 *Tanikaze* class: **Tanikaze** (Maidzuru D.Y., July, 1918) and **Kawakaze** (Yokosuka D.Y., 1917). Dimensions: 320 × 29½ × 9¼ feet. Armament: 3—4.7 inch, 45 cal., 6 tubes in three twin deck m 2—30 inch searchlights. Designed H.P. 28,000 = 34 kts. Machinery: Parsons turbines. Boilers Complement, 128. Begun under 1916 Programme.

(*Continued on n*

1st class Destroyers—*continued.*

4 Amatsukaze class.

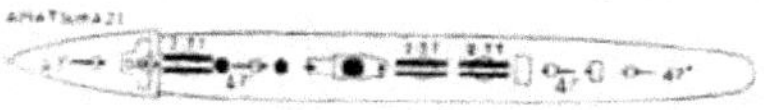

AMATSUKAZE. 1917 *Photo by courtesy of the Navy Department, Tokyo.*

4 *Amatsukaze* class: ***Amatsukaze*** and ***Isokaze*** (both launched at Kure, Oct., 1916), ***Hamakaze*** (Nagasaki, Oct., 1916), and ***Tokitsukaze*** (Kobé, Dec., 1916).* 1227 tons. Dimensions: 310 (*p.p.*), 326½ (*o.a.*) × 28 × 9¼ feet. Armament: 4—4.7 inch (40 cal.), 2 M.G., and 6—18 inch tubes in 3 twin deck mountings. 1 searchlight. Machinery: 3 sets Parsons turbines and Kanpon boilers. Designed H.P. 27,000 = 34 kts. Fuel: 245 tons coal + 195 tons oil. Complement, 145. Built under the 1915 Naval Programme.

**Note.*—*Tokitsukaze* wrecked: salved in three sections, March, 1918, and practically rebuilt at Maidzuru D.Y.

---

2 Umikaze Class.

Fore-funnel now raised.

2 *Umikaze* class: ***Umikaze*** (Maidzuru, 1910) and ***Yamakaze*** (Nagasaki, 1911). 1150 tons. Dimensions: 310 (*p.p.*) 323½ (*o.a.*) × 27½ × 9 feet (*mean draught*). Armament: 2—4.7 inch (40 cal.), 5—12 pdr. Torpedo tubes: in *Umikaze*, 4—18 inch in two twin deck mountings; in *Yamakaze*, 3—18 inch in single deck mountings. One searchlight. Machinery: 3 sets Parsons turbine and Kanpon boilers. Designed H.P. 20,500 = 33 kts. Fuel: 250 tons coal + 180 tons oil = 2700 miles at 15 kts. Complement, 120. Trials: *Umikaze* 33.46 kts.

33* + 5† + 8‡ 2nd class Destroyers. (900 to 665 tons).

*In service. †Building or completing. ‡Projected.

8 Ashi Class.

Projected.

Name.	Builder.	Begun.	Name.	Builder.	Begun.
Ashi	..	1920 or 1921	*Tade*	..	1920 or 1921
Hasu	..		*Tsuta*	..	
Hishi	..		*Warabi*	..	
Sumire	..		*Yomogi*	..	

No details known. Begun under 1920 section of 1918-24 Navy Law.

---

5 Ao-i Class.

Building.

Name.	Builder.	Begun.	Name.	Builder.	Begun.
Aoi	Kawasaki Co., Kobé	1.4.20	*Hagi*	Uraga Co., Uraga	29.2.20
Kiku		20.1.20	*Susuki*	Ishikawajima Co., Uraga	8.3.20
*Fuji**	Fujinagata Co., Osaka	9.12.19			

No details available. Begun under the 1919-20 section of the 1918-24 Navy Law.

*Means "Wistaria"; not to be confused with Battleship *Fuji*, named after mountain.

(*Continued on next page.*)

8 Momi class.

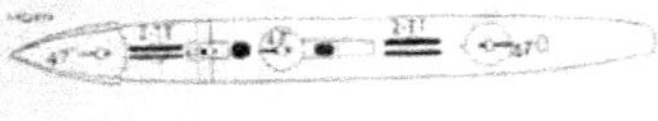

MOMI. (Tank for trials shipped before mainmast). 1920 *Photo, Navy Dept., Tokyo.*

8 *Momi* class: ***Momi*** and ***Kaya*** (Yokosuka D.Y.), ***Kuri*** and ***Nire*** (Kure D.Y.), ***Nashi*** and ***Take*** (Kawasaki Co., Kobe), ***Kaki*** (Uraga Co.), and ***Tsuga*** (Ishikawajima Co.). 850 tons. Dimensions: 275.5 (*p.p.*) × 27.4 × 7¾ feet. Armament: 3—4.7 inch, 45 cal., 2 M.G., 4—21 inch torpedo tubes in two twin deck mountings. 1—35 inch searchlight. Designed S.H.P. 21,500 = 31.5 kts. Machinery: turbines, + Kansei boilers. Fuel: tons. Complement, 110. Built under 1918-19 Programme.

Name.	*Begun.*	*Launch.*	*Comp.*	*Name.*	*Begun.*	*Launch.*	*Comp.*
Momi	23/12/18	10/6/19	27/12/19	*Nashi*	2/12/18	26/8/19	10/12/19
Kaya	13/12/18	10/6/19	28/3/20	*Take*	2/12/18	26/8/19	25/12/19
Kuri	5/12/19	19/3/20	30/4/20	*Kaki*	27/2/19	20/10/19	Yes.
Nire	5/9/19	22/12/19	31/3/20	*Tsuga*	5/3/19	17/4/20	Yes.

10 Momo class (4 Momo + 6 Tsubaki).

1920 *Photo, Seward, Weymouth.*

10 *Momo* class: ***Maki*** and ***Keyaki*** (Sasebo D.Y., 1918), ***Kuwa*** and ***Tsubaki*** (Kure D.Y., 1918), ***Enoki*** (Maidzuru D.Y., 1918), and ***Nara*** (Yokosuka D.Y., 1918). ***Yanagi*** (1917) and ***Momo*** (1916), both built at Sasebo; ***Kashi*** (1916) and ***Hinoki*** (1916), both built at Maidzuru. 835 tons. Dimensions: 275 (*o.a.*) × 25 × 7¾ feet. Armament: 3—4.7 inch, 2 M.G., and 6—18 inch tubes in two twin and two single deck mountings. One searchlight. Machinery: 3 sets Curtis turbines and 4 Kansei boilers. Designed H.P. 16,000 = 31.5 kts. Fuel: 90 tons coal + 218 tons oil. Complement, 109. The first six provided for by Special 1917 Programme. The last four built under the 1915 Naval Programme.

10 Kaba class.

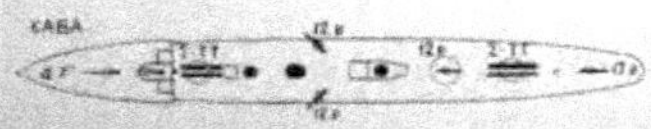

UME (foreground) KUSUNOKI (behind). 1919 *Photo, Comm. Hollerton, R.N.*

Note.—Some boats of *Momo* and *Kaba* classes had no forward tubes during War, but they are probably re-mounted now.

10 *Kaba* class:—***Kaba*** (Yokosuka), ***Kaede*** (Maidzuru), ***Kashiwa*** and ***Matsu*** (Nagasaki), ***Katsura*** (Kure), ***Kiri*** (Uraga), ***Kusunoki*** and ***Ume*** (Kobe), ***Sakaki*** (Sasebo), and ***Sugi*** (Osaka). All launched February–March, 1915. 665 tons. Dimensions: 260 (*p.p.*) 274 (*o.a.*) × 24 × 7.9 feet. Armament: 1—4.7 inch, 4—12 pdr. (2 anti-aircraft model) and 4—18 inch tubes. Machinery: 3 sets, 4-cylinder triple expansion and 4 Kansei boilers. Designed H.P. 9,500 = 30 kts. Fuel: 90 tons coal + 136 tons oil. Complement, 92. These boats are said to have been built in seven months. Built under 1914 Naval Programme, and served in Mediterranean during war. 12 *Kaba*s built in Japanese Yards for French Navy—see French *Tribal* class, *Algérien, Arabe, &c.*

(*Continued on next page.*)

2nd Class Destroyers—*Continued.*

1 Special Boat.

1917 *Photo, O. Parker, M.B.*

1 Yarrow type: ***Urakaze*** (Yarrow, Scotstoun, 1915). 955 tons. Dimensions: 275 (*p.p.*) 288 (*o.a.*) × 27·5 × 9·5 feet. Armament: 1—4·7 in., 4—12 pdr., 4—21 inch tubes in two twin deck mountings. Machinery: 22,000 H.P. turbines = 28 kts. Yarrow large tube boilers. Fuel: 248 tons oil only. Endurance about 1800 miles at 15 kts. Complement, 117. Trials: 30·26 kts. at 1082 tons load displacement.

Note.—It was intended that this boat should have a Diesel engine for cruising speed, combined by Föttinger hydraulic transmission to the turbines. Owing to the War, the Föttinger gear and Diesel engine were never delivered. They were replaced by a big oil tank. *Kawakaze* (sister-boat) ceded to Italy, and now Italian *Audace*.

2 Sakura Class.

Fore funnel may now be raised.

2 *Sakura* class:—***Sakura*** (1911), ***Tachibana*** (1912). Both built at Maidzuru. 605 tons. Fuel: 125 tons coal + 30 tons oil: otherwise as 10 *Kaba* class, on preceding page.

42 3rd Class Destroyers (under 600 tons).

33 Later *Arare* type, viz.:—5 built at Maidzuru: ***Ayanami*** (1909), ***Isonami*** (1908), ***Uranami*** (1907), ***Oit*** ***Yunagi*** (1906); 2 built at Uraga: ***Kikudzuki*** (1907), and ***Nagatsuki*** (1906); 4 built at Naga ***Matsukaze, Minadzuki, Shirayuki*** and ***Shiratsuya*** (all 1906); 1 built at Osaka: ***Ha*** (1906); 5 built at Kobe: ***Hatsuharu, Shigure, Udzuki*** (all 1906), ***Asakaze*** and ***Harukaze*** 1905; 4 built at Sasebo: ***Mikadzuki, Nowaki, Yudachi*** (all 1906), and ***Yugure*** (1905); 12 at Yokosuka: ***Hatsuyuki, Hibiki*** (both 1906), ***Arare, Fubuki, Hatsushimo, Kamik*** ***Kisaragi, Nenohi, Ushiho, Wakaba, Yayoi*** (all 1905), and ***Ariyake*** (1904). 375—381 Dimensions: 227 to 234½ (o.a.) × 21½ × 6 feet. Armament: 6—12 pdr., and 2—18 inch tubes. 1 searcl Designed H.P. 6,000 = 29 kts. Coal: 100 tons, but *Ayanami, Isonami*, and *Uranami* have 95 tons coal + 1 Endurance about 1,200 miles at 15 kts. Complement 61.

Also

2 Earlier *Arare* type: ***Asagiri*** (1903), ***Murasame*** (both built at Yokosuka, 1902). Generally the same as above 33 boats, but have 2—12 pdr., and 4—6 pdr. guns.

Note to "Arare" type Plans.—2—12 pdr., not shown, are mounted abeam of 2nd and 3rd funnels.

(6 pdr. on forecastle now replaced by a 12 pdr.)

2 *Thornycroft* (1902):—***Asashiho*** (1902), ***Shirakumo*** (1901), 333 tons. Guns: 2—12 pdr., 4—6 pdr. Torpedo t 2—18 inch. H.P. 7000 = 30 to 31 kts. Boilers: Thornycroft. Coal: 96 tons. Complement, 60.

42 3rd Class Destroyers—*Continued.*

Note.

All T.B.D. listed below now used only as Tenders to Torpedo Schools. Will probably be removed from the Navy in the near future.

Note to Plans.—The 6 pdr. on forebridge has been replaced by a 12 pdr.

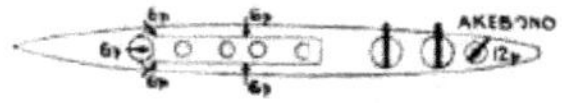

2 Yarrow type:—**Akebono, Oboro** (both 1899). 306 tons. Dimensions: 220 × 20½ × 5 feet. Guns: 2—12 pdr., 4—6 pdr. Torpedo tubes: 2—18 inch. H.P. 6000=31 kts. Coal: 90—95 tons. Complement, 51.

Note to Plans.—The 6 pdr. on forebridge has been replaced by a 12 pdr.

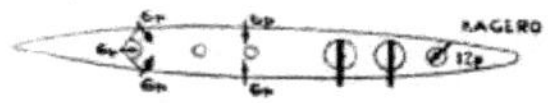

3 *Thornycroft* type:—**Kagero** (1899) **Shiranui** (1899) **Usugumo** (1900). 326 tons. Dimensions: 210 × 19¼ × 6 feet (mean draught). Guns: 2—12 pdr., 4—6 pdr. Torpedo tubes: 2—18 inch. H.P. 5400=30 kts. Coal: 81 tons. Complement, 61. *Murakumo* and *Yugiri* of this class also exist, but are only used for Subsidiary Duties.

20 Torpedo Boats.

No.	Type.	Date.	Displacement.	H.P.	Maximum Speed.	Coal.	Complement.	Tubes.
			Tons.		kts.	tons.		
	First class:—							
11	*Hayabusa* (N)	'00-'04	150-152	4200	28½-29	24-26	30	3
1	*Shirataka* (S)	1900	115	2600	25	24	30	3
	Second class:—							
8	{67—68} {70—75} (S)	1903	87-89	1200	23½-24	26·5	23	3

(S)=Schichau type. (N)=Normand type.

1st Class Torpedo Boats

(*No Shirataka*)

(A) *Hayabusa* class (*Normand* type). Note position of mast and bridge, and compare with (B).

(B) *Hayabusa* class: Compare with (A).

11 *Hayabusa* class (*Normand* type), viz.:—9 *Kure D.Y.*: **Aotaka, Hato, Hibari, Kari, Kiji, Sagi, T.** (all 1903), **Kamome** and **Udzura** (1904); 2 *Kawasaki Co., Kobe*: **Hashitaka** (1903), and **Ot** 150 to 152 tons. Dimensions: 147 × 16 × 4½ feet. Armament: 1—4 pdr., 2—3 pdr., 3—14 in Machinery: 2 sets triple expansion. Designed H.P. 4200=28·5 to 29 kts. Coal: 26 tons. *Hayabusa, Manadzuru* also exist, but only used for Subsidiary Purposes.

(*Continued on next*

1st Class Torpedo Boats

Continued.

1920 *Photo, Navy Dept., Tokyo.*

1 Sohichou type: **Shirataka** (1898). 127 tons. Dimensions: 151 × 16·8 × 4·5 feet. Armament: 2—3 pdr., 3—14 inch tubes. 1 Searchlight: Designed H.P. 2600=28 kts. originally, 25 kts. now. Coal: 24 tons.

2nd Class Torpedo Boats.

(2d Suiraitei.)

Are only used now for Harbour Duty.

1920 *Photo, Navy Dept., Tokyo.*

8 Sohichou type: **Nos. 70–75, 67–68** (built at Yokosuka, Sasebo and Kure, 1902-3). 81 to 89 tons. Dimensions: 126 × 16½ × 3¼ feet. Armament: 2—6 pdr., (Yamanouchi) and 3—14 inch tubes. Designed H.P. 1200=23½ to 24 kts. Coal: 26½ tons.

A*—23† + *12*§ + 15 *(or more?)*‡ Submarines. *(Sensuikan).*

* 7 surrendered German Submarines described on later pages and separately totalled and tabulated in Table B.
† Completed. § Building and completing? ‡ Authorized, but not begun.

No.	Type	Date	Displacement	H.P.	Max. speed	Fuel.	Complement	T. Tubes	Max. draught
	Ocean-going:—		tons		kts.	tons.			
?	*Nos. 51—?*	?	...	...	...	...	...	...	...
7	*Nos. 44—50*	'20—?	...	...	...	...	...	...	...
17	*Nos. 27—43*	'18—?	...	...	...				
8	*Nos. 14, 19 & 20* / *Nos. 22—26* (ND)*	'17—'21	750-800 / 1000-1200		...		...	...	...
9	*Nos. 18 & 21* (AL?)*	'15—'20	700 / 1070	...	...	...	...	...	...
	Coastal:—†								
1	*No. 15* (SL)	'12—'19	450 / 675	2000 / 800	17 / 10	...	...	6	10
2	*Nos. 16, 17* (VJ)	'16—'19	290 / 320	600 / 350	14 / 8	10	26	4	11·5
1	*No. 13* (VJ)	'10—'12	300 / 330	1000 / 300	14 / 8	About 15			
3	*Nos. 10—12* (VJ)	'10—'11	290 / 320	600 / 300	13 / 8	About 15	26	2	11·5
2	*Nos. 8, 9* (V)	'07—'08	285 / 315	600 / 180	12 / 8	About 15			
2	*Nos. 6, 7*	'05—'06	73 / 86	300 / 30	8·5 / 5	1½	16	1	7·5
5	*Nos. 1—5* (H)	'05—'06	105 / 125	160 / 70	9 / 7	2	17	1	9

(ND)—Navy Department design. (AL)—Ansaldo-Laurenti design. (SL)—Schneider-Laubeuf design. (V)—Vickers type. (VJ)—Vickers type, modified by Japanese Navy Department. (H)—Holland type.
* Classification } Revised by Navy Dept., Tokyo, 1920.
† Details, all types }

New Construction. *(Details unofficial).*

Unofficially reported that 70 or 75 Submarines are projected under the 1920-28 Navy Law. In addition, further [illegible] may be authorized under the 1918-24 Navy Law, but not yet commenced.

U.S. and German Naval Journals say Japan contemplates building one or two Submarine Cruisers, each to be [illegible] with 3 sets of 3000-4000 B.H.P. Sulzer-Diesel engines. These boats are to be larger than the German *U 142* class of 2158/2785 tons. Implicit reliance need not be placed in these reports, but the surrendered German [illegible] *O 1—O 7* (described later) will probably have a marked effect on future Japanese Submarine designs.

Projected.

Total unknown: Nos. 51—? Authorized for commencement under 1920 Programme.

7 boats (*types unknown*): *Nos. 44—50.* Believed authorized under 1919 Programme and to be begun 1920-21.

Notes.—Nothing known as regards design of above boats. May be divided into Ocean-going, Coastal and Misc[illegible] types.

(Continued on next pa[illegible]

8 + (*12 building and completing?+7 projected?*) **Ocean-going boats.**

Building, completing, and to be begun?

17 boats; *types unknown*: ***Nos. 27—33, 34—37, 38—41, 42*** and ***43.*** Building or completing by Navy Yards and private shipyards, under 1918-19 Naval Programme. No details known.

**Note.*—*Nos.* [illegible] and *No.* [illegible] to be launched 1920-21, to be completed 1920-21 (or later). *No.* [illegible] Progress with *Nos.* [illegible] not known; will perhaps be begun 1920-21.

Photo wanted.

8 *Navy* type: ***Nos. 14, 19, 20*** and ***Nos. 22—26.*** Built by Navy Yards and Private Yards, about 1917-21. Displace 700-900 tons on *surface*, 1000-1200 tons *submerged*. No other details available.

**Notes.*—*Nos. 14, 19, 20, 22, 23, 26* believed to be in service or nearly finished. *No. 25* launched at Kobé October 18th, 1919 (now completing ?). *No. 24* to be launched 1920-21.

Photo wanted

2 (type uncertain): ***No. 18*** and ***No. 21,*** built about 1915-20 by Kawasaki Co., Kobé. Believed to be Ansaldo-Laurenti boats, displacing 700 tons on *surface* and 1070 tons *submerged*. *No. 18* belongs to the 1915-16 Programme; *No. 21* to the 1916-17 Programme. Completed 1919-20.

*Are unofficial Notes.

1 **Schneider-Laubeuf boat.**—(*Coastal*).

1920 *Photo*, *Navy Dep*

1 *Schneider-Laubeuf* type: ***No. 15*** (Chalon-sur-Saône, France, 1914). Displaces 450 tons on *surface*, 670 t[illegible] Dimensions: *about* 182 × 19 × 10 feet. Machinery: 2 sets of 1000 B.H.P. Schneider-Carels Diese[illegible] kts. on *surface*. *Submerged* speed: *about* 9½-10 kts. Armament: 1 small Q.F. gun, 2 bow torpe[illegible] tubes in w.t.c. under superstructure.

Note.—A similar boat, *No.* [illegible], ceded to French Navy in 1914, and now French *Armide*.

2 **Modified Vicker-Japanese type**—(*Coastal*).

No 17. 1920 *Photo*, *Navy*

2 *Modified Japanese-Vickers* type: ***Nos. 16*** and ***17,*** built at Kure D.Y., 1916-19, under 1915-16 Pro[illegible] 290 tons on *surface*, 320 tons *submerged*. *About* 11¼ feet draught. In these two boats two bow t[illegible] when in surface-running trim. There are also two other tubes (position uncertain, but probabl[illegible] 4 tubes in all. Other details as Table A.

(*Conti*[illegible]

A—JAPANESE SUBMARINES—*Continued.*

1 Modified Vickers-Japanese Boat (*Coastal.*)

1920 *Photo, Navy Dept., Tokyo.*

1 *Modified Japanese-Vickers type*: **No. 13** (Kawasaki Co., Kobe, 1911). Displace 300 tons *on surface*, 330 tons *submerged*. Otherwise much the same as *Nos.* 8, 9, below.

2 Vickers Type—(*Coastal.*)

No. 8.

1920 *Photo, Navy Dept.*

2 *Vickers type*: **Nos. 9** and **8** (Vickers, Barrow, 1908). Dimensions: *about* 135 × 13½ × 11½ feet. 2 Other details as Table.

3 Vickers-Japanese Type—(*Coastal.*)

No. 12.   No. 11.

From "The Navy" (Tokyo), 1920.

3 *Vickers-Japanese type*: **No. 12—10** (Kure D.Y., 1910-11). Dimensions: *about* 135 × 12½ × 11½ feet. 2 bow tubes. Radius of action: *about* 1000 miles *on surface* at 10 kts. Other details as Table A.

2 ——— *Boats—(*Coastal*).

*Type unknown; probably experimental improvements on Old *Holland* boats.

O 7.

1920 *Photo, Navy Dept.*

(For description, see next page.)

2 —— Boats. *Coastal—continued.*

O 6. (No. 7 illustrated on preceding page.) 1920 *Photo, Navy Dept., Tokyo.*

2 type unknown: **No. 7** (Kawasaki Co., 1906-7). **No. 6** (Kawasaki Co., Kobé, 1906-7). Displacement: 79 tons on surface, 96 tons submerged. Dimensions: 45 × 8 × 7½ feet. 1 bow torpedo tube. 1 set petrol motors.

5 Old Holland Boats.

No. 2. 1920 *Photo, Navy Dept., Tokyo.*

5 early Holland type: **Nos. 5—1** (1905-6, imported from U.S.A. about time of Russo-Japanese War). Dimensions: [illegible] × 11½ × [illegible] feet. 1 bow torpedo tube. Petrol engine of 160 H.P. for surface running. 1 screw. Probably of little effective value now.

B.—7 ex-German Submarines.

Assigned to Japan for experimental and propaganda duties. To be broken up eventuall[illegible]

Type	No.	Class	Completed	Displacement	B.H.P. / E.H.P.	Speed	Endurance miles (at kts.)	Complement
Ocean	1	O 2 [illegible]	1916	715 [illegible]	2400 / 1180	16 / 9.5	9500 (8)	35
	1	O 2 [illegible]	1915	7[illegible]	2000 / 1000	15.5 / 9.5	11250 (8)	36
Small cruiser	2	O 4 / O 5	1918 / 1918	515 / 630 / 516 / 620	1300 / 500	11 / 8	9000 (4) / 55 (4)	34
Mine-laying	1	O 3	1918	1163 / 1501	2800 / 1200	15 / 7.5	11500 (8)	40
	1	O 7, O 3 [illegible]	1918	491 / 746	600 / 620	11.5 / 6.6	9850 (7) / 56 (4), 35 (5)	32

Notes.

For dimensions, armament, &c., see descriptions under photographs. All are double-hull boats. G[illegible] Krupp 30 or 45 cal. "Ubts. u. Tpbts." marks. Weight of shells: 5.9 inch, 101 lbs.; 4.1 inch, [illegible] 22 lbs. Torpedoes: All 19.7 inch G.VII*** steam-heater, 430 lbs. charge, 11,706 yds. at 28 k[illegible] 35 kts.

2 Ocean-going Boats.

1920 *Photo, Na[illegible]*

1 "Mittel U" type: **O 2** (ex *U* [illegible]), Krupp, Germania Yard, Kiel. Dimensions: about 213¾ × 20¾ × [illegible] —4.1 inch, 1—3.4 inch. Torpedo tubes: 4—19.7 inch (2 bow, 2 stern). 8 torpedoes carried. German *U* [illegible]—[illegible] class.

(Continue[illegible]

JAPANESE (ex-German) SUBMARINES.

(All 1920 Photographs, by courtesy of the Navy Dept., Tokyo).

1 "Mittel-U" type: **O 2** (ex U 46), by Danzig D.Y. Dimensions: about 213¼ × 21¼ × 11¾ feet. Gun: 1—4.1 inch Krupp. Torpedo tubes: 6—19.7 inch (4 bow, 2 stern). 6 torpedoes carried. Periscope depth: 45 feet. (U 43—50 class of German Navy).

2 Small Ocean-going Boats.

O 7.

O 6.

2 "UB III" type: **O 6** (ex UB 125), **O 7** (ex UB 143), both by Weser Co. Dimensions: 181¾ × 19¼ × 12 feet. Guns: In O 6, 1—3.4 inch "Flak" (AA.); in O 7, 1—4.1 inch Krupp. Torpedo tubes: 5—19.7 inch (4 bow, 1 stern). 10 torpedoes carried. Oil: 35/71 tons + 8 tons lubricating. Max. dive: 245 feet. Crash dive: 29.5 feet in 40 secs. Periscope depth: 42 feet. (UB 48—249 classes, German Navy).

3 Mine-laying Boats.

O 1

1 "UE" type: **O 1** (ex U 125), by Blohm & Voss, Hamburg. Dimensions: 267 ft. 5 in. × 24 ft. Gun: 1—5.9 inch (Krupp). Torpedo tubes: 4—19.7 inch (all bow). Mines: 42, laid through 2 port... Reported that, owing to defective stability only about 32 mines can be carried. Designed to carry (12 outside hull in pressure-tight cylinders, and 12 within), but doubtful if whole 24 were ever carried... UE class, German Navy).

O 5

O 4.

2 "UC III" type: **O 4** (ex UC 90), **O 5** (ex UC 99), by Blohm & Voss, Hamburg. Dimensions: 184 × 18... feet. Gun: 1—4.1 inch (Krupp), 130 rounds carried. 3—19.7 tubes, 2 amidships (above water in surfa... 1 stern. 7 torpedoes carried. 14 mines carried in 6 laying chutes forward. Periscope depth: 42 feet... dive: 246 feet. Can dive to 29½ feet in 45 secs. (UC 90—105 type, German Navy).

Note.

Under the 1920-28 Navy Law, it is proposed that 15 Fleet Auxiliaries should be built. Also some further vessels may be authorized under the 1918-24 Navy Law. Information is not available for the classification of these ships.

Aircraft Carrier.

Projected.

HOSHO. Authorized under the 1919 Naval Programme. No details available.

(Also v. third column for *Wakamiya*)

Destroyer Depot Ships.

Photo wanted.

ISUSAN (Maru). (——). 217 tons. Dimensions: 131 × 21 × 6.2 feet. Speed: 8 kts. No other details known.

Photo wanted.

KENKAI (Maru). (Kure D.Y., 1902). 315 tons. Dimensions: 128 × 22 × 8½ feet. Speed: 12.7 kts. No other details known.

Destroyer Depot Ships—*continued.*

Photo wanted.

REKISAN (Maru). (——). 450 tons. 129½ × 22¼ × 8½ feet. Guns: Not known. I.H.P. 202 = 8.7 kts. No other details known.

Photo wanted.

SHIROGANE. (——). 413 tons. Dimensions: 128 × 23 × 8.3 feet. Speed: 10.5 kts. No other details known.

Submarine Depot Ships.

Battleship **SHIKISHIMA** is being used as Submarine School, pending completion of buildings ashore.

1917 *Photo by courtesy of the Navy Department, Tokyo.*

KOMAHASHI. (Sasebo, 1913). 1230 tons. 227 × 35 × 17½ feet. Guns: 3—12 pdr. H.P. 1824 = 13.9 kts. Was originally built as a Naval Transport.

Submarine Depot Ships—*contin*

KARASAKI (ex *Ekaterinoslav*, 1896, captured 1904 Dimensions: 440 × 49½ × 15¾ feet (max. draw ment: 1—12 pdr. H.P. 3200 = 13 kts.

NAGAURA (Maru) (ex-Gunboat *Tatsuta*, Elsw tons. 240 × 27½ × 9¼ feet. Guns: Not k 5000 = 21 kts. Coal: 155 tons.

Seaplane Depot Ship.

1917 *Photo by courtesy of the*

WAKAMIYA (1901). 7600 tons. Dimensions: 19 feet. Armament: 2—12 pdr. H.P. 1 Carries 4 seaplanes. Was originally built as a

Note.—Now has hangars at foot of each mast, as shown by S

Mine Layers.

Note.—Possible that under the 1918-21 and 1920-28 Navy Laws further Mine Layers may be authorised, building or completing. No details are available of such vessels at present. 1st Class Cruiser **ASO** and 2nd Class Cruiser **TSUGARU** are fitted as Mine Layers.

1918 Photo, Navy Dept., Tokyo.

KUROKAMI.

(KUROKAMI CLASS—11 SHIPS.)

NINOSHIMA (Maru). Building or completing at Kure D.Y. 1918-19 Programme.

YENOSHIMA (Maru).* Built under 1917 Special Programme, at Maidzuru D.Y.

YENTO* (Maru,) Mar., 1917; **KATASHIMA (Maru)** Feb., 1917), both built at Maidzuru D.Y., and **KUROKAMI (Maru)** Kure D.Y., Feb., 1917. Built under 1916-17 Programme.

ASHIZAKI (Maru) and **KATOKU (Maru).** Both launched at Maidzuru D.Y., Oct., 1915. Built under 1915-16 Programme.

KUROSHIMA (Maru) (Oct., 1914) and **TOSHIMA (Maru)** (Oct., 1914). Both built at Sasebo D.Y.

SOKUTEN (Maru). March, 1913, built by Maidzuru D.Y.

NATSUSHIMA (Maru). Yokosuka D.Y., March, 1911.

Details of all above ships: 450 tons. Dimensions: 150 × 25 × 7.5 feet. Guns: Not known. H.P. 600 = 12 kts.

*May also be referred to as *Enoshima Maru* and *Eno Maru*.

Mine Layers—*continued.*

1920 Photo, Navy Dept., Tokyo.

KATSURIKI. (Kure D.Y., Oct., 1916). 2000 tons. Dimensions: 290 × 39 × 14 feet. Guns: 2—12 pdr. H.P. = 13 kts. Built under 1915-16 Programme. Fitted with 4 galleys for sweeps.

No Photographs or further details available of the following nine Old Mining Vessels.

Name (Built)	*Tons.*	*Speed. (kts.)*
DAI SHICHI* SHINTEN (1901) ..	320	11
DAI NI† OMINATO (Maru) (1900) ..	392	12
DAI NI† KWAITEN (Maru) (1900) ..	284	12.8
TAKESHIKI (Maru) (1899) ..	360	13
DAI ICHI‡ TENKYO (1899) ..	304	12.6
NASAMI (Maru) (1898) ..	287	12.4
DAI GO§ SHINTEN (1898) ..	315	11
DAI SAN‖ SHINTEN (1888)	295	10.4
KEIKWAN (Maru) (——) ..	135	10

In English, these first two words mean:—* No. 7. † No. 2. ‡ No. 1. § No. 5. ‖ No. 3.

Mine-Sweepers.

Some of the Mine Layers, detailed above, probably equipped also as Sweepers. No exact details known.

Oilers & Tank Vessels.

Building?

IRO, SHIRIYA, TSURUMI.* Authorised under 1920 Programme. Possibly building by Asano Co's. Yard at Tsurumi.

*Name also given as *Irosaki*.

Building?

SATA. Authorised under 1919 Naval Programme. No [illegible] available.

Building.

ERIMO (authorised), *NOTORO* (begun November 20) and *SHIRETOKO* (in building). All by Kawasaki, Kobé. 8,000 tons. Belong to 1918-19 Programme. [illegible] other details known.

[Continued on next

Oilers & Tank Vessels—*Continued.*

1920 Photo, Navy Dept., Tokyo.

SUNOSAKI (Maru) (Yokosuka D.Y., June, 1918). 5800 tons. Dimensions: 400 × 50 × 25.2 feet. Guns: 2—12 pdr. H.P. 6000 = 14 kts. Carries 5000 tons oil. Built under 1916-17 Programme.

1920 Photo, Navy Dept., Tokyo.

TSURUGIZAKI (Kure D.Y., June, 1917). 1970 tons. 220½ × 31 × 14 feet. H.P. 900 = 9 kts. Cargo: 1,100 tons oil. Built under 1916-17 Programme.

Also 10 (or more) other small Tank Vessels; names and details not known.

Fleet Colliers.

Muroto. *1920 Photo, Navy Dept., Tokyo.*

NOSHIMA (Feb., 1919), **MUROTO** (Oct., 1918). Both built by Mitsubishi Co., Kobé. 8750 tons. Dimensions: 345 × 50 × 23.2 feet. Guns: Unknown. I.H.P. 2640 = 12.5 kts. Coal: 877 tons. Built under Special 1918-19 Programme.

Naval Auxiliaries and Transports.

KOMAHASHI (Submarine Depot Ship) and **WAKAMIYA** (Seaplane Carrier), both listed on preceding page. Were both originally built as Naval Transports, and can be so used if necessary.

1920 Photo, Navy Dept., Tokyo.

ROZAN (ex German S.S. *Ellen Rickmers*, built at Geestemünde, 1906, and captured 1914). 4117 tons *gross*. Dimensions: 367.7 × 47.7 × 27.4 feet (*depth of hold*). H.P. 1600 = 11 kts.

Naval Auxiliaries and Transports.—C

1920 Photo, Navy

SEITO (ex-German S.S. *Durendart*, built at Flensbu captured 1914). 3844 tons *gross*. Dimension 49.2 × feet. Guns: 2—12 pdr. H.P. 152

1920 Photo, Navy I

KOSHU (ex-German S.S. *Michael Jebsen*, built Kiel, 1904, and captured 1914). 1521 tons g sions: 251.1 × 36.1 × 18.9 feet (*depth of hol* —12 pdr. H.P. 800 = 9.5 kts. Ice-breaking

1920 Photo, Nav

TAKASAKI (1902). 4746 tons. Dimensions: × 15¾ feet. Guns: 2 small Q.F. Speed: other details known.

(Continued o

Naval Auxiliaries & Transports—*continued.*

Now has fore topmast.

MANSHU (ex-*Manchuria*, 1901). 3916 tons. Complement, 193. Guns: 2—12 pdr., 2—3 pdr. I.H.P. 5000 = 17.6 kts. Coal: 400 tons.

1920 Photo, Navy Dept., Tokyo.

MATSUE (ex-*Sungari*, 1898). 2550 tons. Complement, 126. Guns: 2—12 pdr., 2—3 pdr. I.H.P. 1500 = 13 kts. Coal: 139 tons and some oil fuel.

Repair Ship.

1920 Photo, Navy Dept., Tokyo.

KWANTO (Maru) (1898). 10,000 tons. 410 × 49½ × feet.

Salvage Ships.

Note.—Silhouettes given of all four ships.

Photo wanted.

KURIHASHI (Maru) (——). 1040 tons. 182 × 80¼ × 11⅞ feet. I.H.P. 1200 = 12·5 kts.

Photo wanted.

YODOHASHI (Maru) (——). 870 tons. 159 × 27½ × 12¼ feet. I.H.P. 718 = 11·8 kts.

Photo wanted.

SARUHASHI (Maru) (——). 600 tons. 140 × 26¼ × 10½ feet. I.H.P. 885 = 13·1 kts.

Photo wanted.

ITAHASHI (Maru) (——). 333 tons. Speed 12 kts. No other details known.

Photo by courtesy of the Minist

MUSASHI (Yokosuka, 1886). **YAMATO** (Kobé, 1 posite built. 1502 tons. Guns: 4—12 pdr., H.P. 1622. Original speed was 13½ kts. Coal:

Training Ships.

(All described on preceding pages):—

For Gunnery:
ASAHI. Battleship.

For Gunnery and Torpedoes:
IKOMA. 1st Class Cruiser.

For Seamanship:
FUJI. Battleship.

For Engineer Branch:
SUWO. Coast Defence Battleship.

For Cadets:
ADZUMA., **TOKIWA.**, **YAKUMO.** 1st Class Cruisers.

For Conscripts:
HASHIDATE. 2nd Class Coast Defence Ship.

For Boys:
Old 2nd Class Coast Defence Ships, **OKINOSHI** **MISHIMA**, are used for this service as Barrac

Ex-German Warships Assigned to Jap

For Experimental and Propaganda Duties only. No incorporated in the Imperial Navy:—

Dreadnoughts:—**OLDENBURG,* NASSAU.***
Light Cruisers:—**AUGSBURG.**
Destroyers:—**V 127*, V 80,* S 60*, T 181.**
Submarines:—7 boats (c. Submarine Paquel.

FRENCH FLEET.

(Officially Revised, 1920, by order of H.E. The Minister of Marine.)

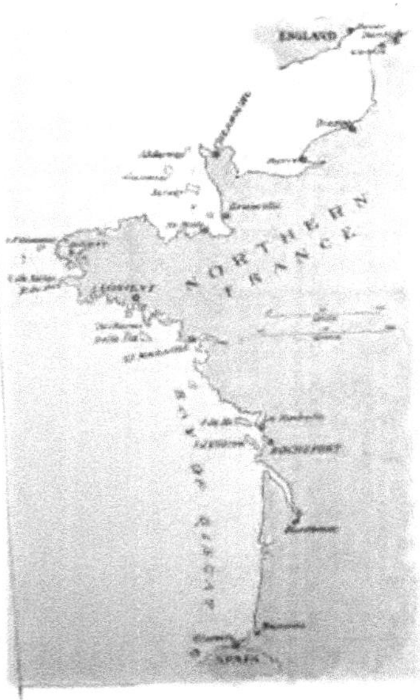

CHERBOURG

to	knots.
Bermuda	2933
Brest	210
Copenhagen	834
Devonport	111
Dunkerk	186
Lorient	300
Kronstadt	1500
Portsmouth	77
Rochefort	410
Stockholm	1258
Toulon	1817
Wilhelmshaven	476

BREST

to	knots.
Algiers	1369
Bermuda	2811
Cherbourg	210
Devonport	130
Gibraltar	955
Liverpool	407
Malta	1954
Kronstadt	1672
New York	2954
Oran	1676
Portsmouth	226
Rio de Janeiro	4837
Rochefort	300
Toulon	1659
Wilhelmshaven	785

TOULON

to	knots.
Aden	2882
Ajaccio	160
Algiers	405
Bizerta	450
Brest	1659
Constantinople	1336
Genoa	164
Gibraltar	713
Kronstadt	3280
Lorient	1656
Port Mahon	297
Malta	612
Naples	407
Oran	542
Port Said	1485
Portsmouth	1804
Saigon	7150
Sevastopol	1654
Spezia	195
Tangiers	738
Taranto	706

ORAN

to	knots.
Algiers	400
Bizerta	650
Gibraltar	225
Toulon	542

BIZERTA

to	knots.
Algiers	400
Gibraltar	875
Malta	240
Oran	650
Spezia	430
Toulon	450

SAIGON

to	knots.
Hong Kong	930
Manila	875
Port Arthur	2065
Singapore	640

MERCANTILE MARINE:

From "Bureau Veritas" 1919-20 figures. Ships under 100 tons gross exclude[d]
1259 steam and motor ships of 1,919,588 tons gross; also 131 ex-enemy steam and mot[or]
178,909 tons gross under French management.
101 vessels (with auxiliary motors) of 116,898 tons gross.
748 sailing vessels of 306,201 tons gross.
During 1919, French Shipyards launched 34 vessels of 32,663 tons gross; some of these
account.

COINAGE: 1 franc (100 centimes) = 5d. British, — cents American (rate varies).

TRADE PORTS (in order of importance): Marseilles, Havre, Dunkirk, Bordeaux, Calais, Dieppe, Rouen.

OVERSEA POSSESSIONS: Corsica, Algeria, Tunis, Senegal, etc. Madagasc[ar], Martinique, S. Pierre, New Caledonia, Pondicherry, Annam, Cochin China, Tonquin

FRENCH FLEET.

ADDENDA AND DELETIONS.

Sloops.

ARRAS. 1920 Photo, H. Ferand, Brest.

MONTMIRAIL (*of "Arras Type"*) launched September, 1920. The launching dates for *Reims, Remiremont, Vimy*, are those communicated by the Ministry of Marine, but more correct dates appear to be **REIMS** (13th March, 1920), **REMIREMONT** (31st July, 1920), **VIMY** (December, 1919).

1920 Photo.

SCARPE, SUIPPE. Seen from quarter, both ships have same appearance as above photo.

Destroyers and Torpedo Boats.

Delete following boats, all condemned and to be sold :—

2 *Hussard* type : ***Cavalier, Tirailleur.***

2 *Claymore* type : ***Massue, Pierrier.***

2 *Mousquet* type : ***Belier, Epieu,***

1 *Normand* type : ***Borée.***

3 *Normand* type : ***Nos. 288, 281, 271.***

Submarines.

1920 Phot

AMARANTE class :—1 Boat (name unknown) with large and high bridge screen.

Miscellaneous Ships.

Oil Tanker **DURANCE** launched September, 1920.

Minister of Marine—Mons. Landry.

Naval Attaché (London)—Captain Blanpré. *Assistant Naval Attaché*—Le Comte de la Rochefoucauld.

Personnel (1920-21).

31,500 officers and men for ships in commission and reserve; foreign service, O.D. (offensive-defensive) coastal flotillas; submarines and aviation + 53,500 officers and men for Ministry of Marine, prefectures and training stations, and in barracks = 85,000 total.

(Flags 1919).

Drawings prepared from information furnished by courtesy of the Ministry of Marine, Paris.

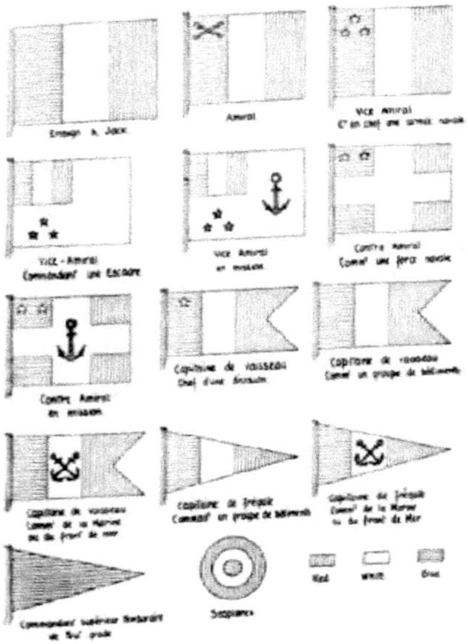

(Charts (a) uniform scale. Soundings in fathoms. Heights in feet.)

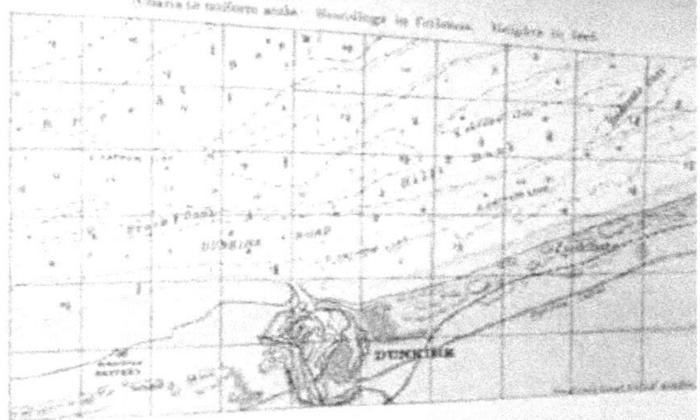

Dunkirk. For details see next page.

Calais. For details

FRENCH NAVAL BASES.

(To Uniform Scale: Soundings in fathoms. Heights in feet.)

CHANNEL.

For Charts, **see preceding page.**

DUNKIRK. Fine harbour. Breakwater. Torpedo, Submarine and Aircraft base. There are four docks here—one 622 × 69 × 26¼, two 357½ × 46 × 26 or 21 feet, and one smaller.

CALAIS. Torpedo and submarine base. Dock, 426¼ × 69 × 28½.

Private Docks.

BOULOGNE. Floating Dock, 236¼ × 62¼ feet (1000 tons).
DIEPPE. Dry Dock, 361 × 59 × 27 feet.
HAVRE. (No. 4). 674¼ × 98¾ × 28½ feet; also 2 large and 3 small docks.
GRANVILLE. Dry Dock, 225 × 46½ × 23 feet.
PAIMPOL. Dry Dock, 229½ × 36 × feet.
Private Oil Tanks at Dunkirk, Calais, Rouen, Havre, Cherbourg.

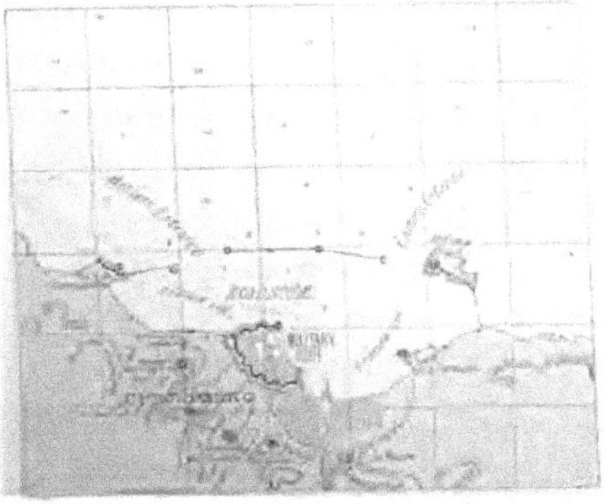

CHERBOURG. Admiralty Chart No. 2602. Rise of Spring Tide, 17¾ feet.

CHERBOURG. Base for Submarines and Torpedo Craft. Protected by a breakwater about two miles long. Heavily fortified.

Dry Docks.

No. 1	360 × 62¼ × 32¼ feet	No. 5	656 × 93½ × 39¼ feet.
„ 2	393 × 62¼ × 32¼ „	„ 6	606 × 69½ × 29¼ „
„ 3	390½ × 62¼ × 32¼ „	„ 7	265 × 59 × 20¼ „
„ 4	360 × 62¼ × 32¼ „	„ 8	257 × 59½ × 20¼ „

Ex-German Floating Dock towed here 1920, about 459¼ × 98·4 × ——feet. Floating Submarine Salvage Dock, about 200 × 28 feet. Also ex-German Pressure-testing Dock for submarines. 5 Naval Oil Tanks, 11,000 tons capacity total.

T.B.D., T.B., and S/M. built here.

ATLANTIC COAST.

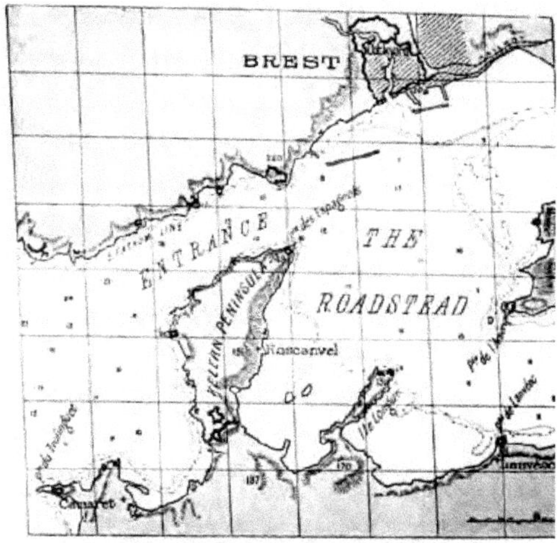

BREST. Admiralty Chart No. 3427. Rise of Spring Tide, 19½ feet.

BREST. Forts, very numerous, but only armed with old guns of 9·4 inch calibre and sm Oil Tanks 27,552 tons capacity total. Very big dock in town, 737 × 114 × 34. Partic docks here:

No. 1	380½ × 70¼ × 35½ feet.	No. 6	226¼ × 46 × 16 fe
„ 2	584 × 72¾ × 35½ feet.	„ 7	187 × 53¾ × 16 fe
„ 3	584 × 93¾ × 35½ feet.	„ 8	820 × 121½ × 35½ fe
„ 4	672½ × 129¼ × 35½ feet.	„ 9	820 × 121½ × 35½ fe

LORIENT. Dockyard and Base closed 1920. Still H.Q. for Coastal Defence. Prov Naval Ordnance at Gâvres. Docks here are:—No. 1, 380¼ × 56½ × 20½ feet; No. 2, (feet; No. 3, 646 × 118 × 29 feet *building*. 2 Oil Tanks, 1070 tons capacity total.
ROCHEFORT. Is situated 25 Km. up river, and is only accessible to Torpedo Craft. repairs to small craft. The three docks here cannot accommodate modern big ships. 489 × 67½ × 25½. Fuel Station for Flotillas and Dirigibles. 12 small Oil Tanks, 200 ton There is a Government dock at LA PALLICE (ROCHELLE), 553 × 72 × 30½ feet, and another sm

Big Private Docks.

BREST. (Town) 737 × 114 × 34 feet. | BORDEAUX. 538 × 109½ × 33 feet;
ST. NAZAIRE. (C.G.T.) 334½ × 98½ × 29 feet (enlarged). | LA ROCHELLE. 590 × 77 × 30 feet
Private Oil Tanks at La Pallice, Bordeaux, St. Loubès, Blaye and Cette.

FRENCH NAVAL BASES—MEDITERRANEAN.

(To uniform scale. Divided into 2000 yard squares. Soundings in fathoms. Heights in feet.) **Dockyards named in heavy type.**

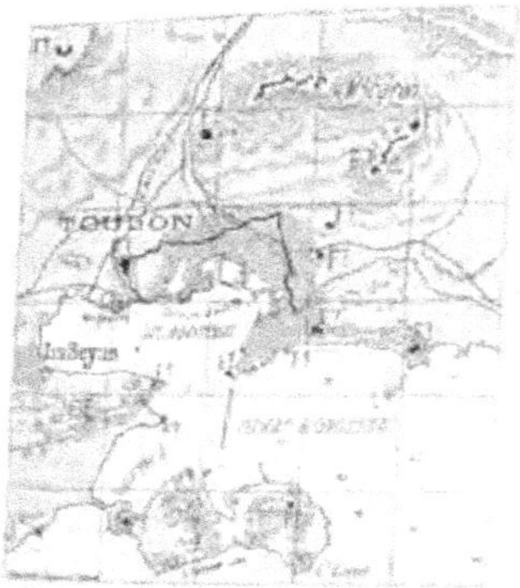

Toulon. Admiralty Chart No. 151-2608.

TOULON. Fortified. Fleet Aircraft and Submarine base. Destroyers, T.B. and Submarines built here in Dockyard. 4 Naval Oil Tanks of 9350 tons capacity (total). Also large private storage at Marseilles and small tanks at La Seyne.

Dry Docks.

MISSIESSY.	No. 1	427 × 93¼ × 30¼	CASTIGNEAU. No. 1	325 × 70½ × 25¼
"	" 2	427 × 93¼ × 30¼	" " 2	385½ × 70¼ × 27¼
"	" 3	585 × 93¼ × 30¼	" " 3	535 × 75½ × 27¼

Floating dock for torpedo craft. Submarine Floating Salvage Dock, "Atlas" 300 × 42¾ feet, 1000 tons lift. Also 3 small docks in the Arsenal Principal.

Naval Aviation Centres.

Very large establishment at Fréjus, St. Raphael, near Toulon. Also Training School at Berre and Istres, near Marseilles. Stations at Aubagne and Cuers for Naval Airships.

Naval Bases.—(*Continued*).

AJACCIO. | BONIFACIO. | Fortified harbours used by the French fleet. No docks. At Ajaccio, 4 Oil Tanks of 3750 tons capacity (total).

ALGIERS. Coaling Station. Two dry docks (No. 1) 455 × 74 × 28 feet; (No. 2) 268 × 54 × 19 feet. Average depth of harbour, 8 fathoms.

ORAN. Torpedo base. Excellent harbour, with an average depth of five fathoms. Two breakwaters. Fortified. 2 Oil Tanks of 1340 tons capacity (total). Floating Submarine Salvage Dock 210 × 64 feet.

BIZERTA. Destroyer, Submarine and Aircraft base. Lies inside a narrow channel. To be fortified and defended by AA. guns. Dry Docks: (No. 1) 656 × 90½ × 33½; (No. 3) 295 × 46¼ × 17¼; (No. 4) 656¾ × 90½ × 33¼. New docks 985 feet long and slipways for small craft projected. Floating dock for torpedo craft. Submarine Salvage Ship *Vulcain* stationed here. 6 Oil Tanks 6300 tons capacity (total). Subsidiary Stations at CAP BON, BONE and PHILLIPE-VILLE.

BEYROUT. Base for French Syrian Squadron.

Oran. Admiralty Chart No. 812.

Bizerta. Admiralty Ch

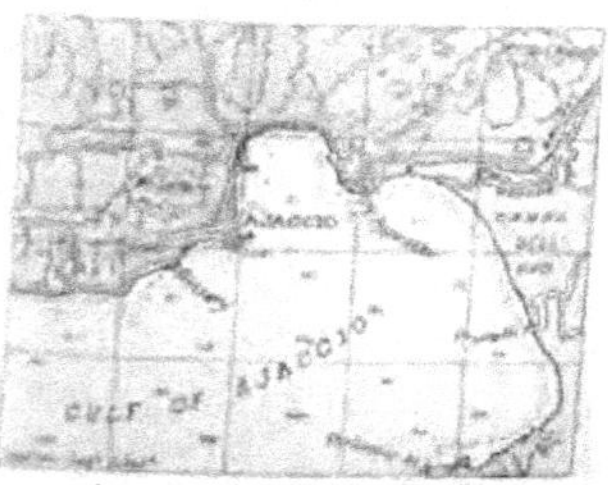

Ajaccio. Admiralty Chart No. 429.

Large Private Docks (*Mediterranean*).

MARSEILLES—(No. 1) 595½ × 61 × 22¾ feet, and No. 7, 669¼ × 82 × 29¼ feet.

Dakar (Senegal). Admiralty Chart No. 1001.

FRENCH COLONIAL NAVAL BASES.

African W. Coast.

DAKAR (Senegal). Naval dry dock here, 629¼ × 92¾ × 30¼ feet. Aircra Subsidiary Stations at RABAT and CASABLANCA.

African E. Coast.

DIEGO-SUAREZ, TAMATAVE. Naval Stations, Madagascar. Practic facilities at latter harbour.

In the Far East.

SAIGON. Situated 42 miles up the Donnai River. Anchorage average, 9 f One large dry dock 518 × 72 × 30 feet, and two small ones. Floating D × 66 feet, 10,000 tons lift. Slip for building ships up to 300 feet long; Also small Stations at ALONG (Tonkin), CAP S. JACQUES (Cochin-Ch

In the West Indies.

FORT DE FRANCE (Martinique).

Naval Ordnance (converted from official 1919 Table furnished by courtesy of the Ministry of Marine).

Calibre (ins.)	13.4		12					9.4			7.6			6.5				5.5			
Do. (c/m)	34		30					24			19			16				14			
Mark	1912		1906-10	1906	'93-96 M	1893-96	1893	1902-6	1893-6	1893	1902	1893-6	1893	'93-'96M	1893-6	1893	'91 (11)	1918	1910	1893	1916
Length (calibres)	45		45	45	40	40	40	49.5	40	40	50	40	40	45	45	45	45	—	55	45	..
Weight of Gun (English tons)	63.33		54.7	54.7	47.3	47.3	43.4	29.1	23.6	20.4	14.9	12.5	10.2	7.96	7.96	6.52	6.94	..	5.17	4.01	..
Weight of Shell (capped) in lbs.	A.P. 1100.5	Ogival 1256.6	922.3	900.4	770.3	770.3	770.3	487.2	374.8	374.8	199	199	199	121	121	121	121	..	80.5	84	..
Weight of Charge (lbs.)	330.7	337.3	267.4	282.2	284.4	244.7	217.2	148.1	147.4	127.9	84	74.5	49.2	45.6	43.7	28.9	28.9	..	22.7	16	..
Max. Pressure (lbs.)	59,525	57,320	57,320	57,320	61,729	63,934	51,808	57,320	63,934	52,910	59,083	61,739	47,400	47,400	61,739	44,092	44,092	..	55,155	53,706	..
Muzzle-Velocity (ft.-secs.)	2624.7	2605	2559	2559	2839	2674	2559	2625	2723	2625	3117	2756	2526	2953	2838	2527	2527	..	2723	2395	..
Remaining Velocity at 21,872 yds. / 20,000 m.	..	1300	..	..	..	..	..	..	..	..	..	..	..	..	..	..	..	..	..	..	..
Remaining Velocity at 16,404 yds. / 15,000 m.	1178	1462	..	..	..	..	..	1102	..	..	..	..	..	..	..	..	..	..	..	..	..
Remaining Velocity at 10,936 yds. / 10,000 m.	1473	1762	1529	1529	1411	1325	1260	1348	1135	1112	1106	1010	992	928	918	922	922	..	860	905	..
Remaining Velocity at 5468 yds. / 5000 m.	1640	2162	1992	1992	2028	1893	1801	1896	1755	1654	1860	1571	1447	1417	1351	1257	1257	..	1230	1108	..
Angle of descent at 20,000 m.	..	25° 10'	..	..	..	..	..	..	..	..	..	..	..	..	..	..	..	..	..	..	..
Angle of descent at 15,000 m.	21° 10'	14° 50'	..	..	..	..	..	25° 50'	..	..	..	..	..	..	..	..	..	..	..	..	..
Angle of descent at 10,000 m.	9° 30'	7° 35'	9° 25'	9° 25'	9° 50'	11° 20'	12° 40'	11° 10'	14° 40'	16° 10'	14° 10'	18° 10'	20° 50'	23° 20'	24° 30'	26° 30'	26° 30'	..	27° 30'	28° 40'	..
Angle of descent at 5000 m.	3° 10'	2° 53'	3° 14'	3° 14'	3° 0'	3° 20'	3° 40'	3° 20'	3° 40'	4° 0'	3° 10'	4° 10'	5°	4° 20'	5° 10'	6° 20'	6° 20'	..	6° 10'	7° 20'	..

Torpedoes. *(Details unofficial.)*

Whitehead pattern, made at Toulon Torpedo Factory; also Schneider pattern at Schneider & Cie's works.

French Designation.	Diameter.	Length.	Charge.	Pressure in chamber.	*Maximum range.*
	inches	feet	lbs	lbs per sq. in.	
1919 55 c/m	21·7	...	...	...	
1904 model	18	$19\frac{1}{4}$	198	2150	6000
45 c/m *long*	18	$16\frac{1}{4}$	198	1270	2000
45 c/m *short*	18	$13\frac{3}{4}$	185	1000	1500
38 c/m *long*	15	$18\frac{3}{4}$	100		1000
38 c/m *short*	15	$16\frac{1}{4}$	85		1000

Notes.—All fitted with gyros. Mines: 9 marks in use up to 1914. Include 360-kilo Sautter-Harlé spherical and 700-kilo Breguet spherical types. Paravanes British marks.

General Notes.

Colour of Ships.

Big ships.—Grey all over, but some have grey-green or greenish-brown turrets, or black barbette shields.

Destroyers, etc.—Light grey; have identity letters, taken from their names, painted on bows—*e.g.*, CL = *Claymore*, FF = *Fanfare*, EH = *Enseigne Henry*.

Submarines.—Sea green or light grey; have identity letters painted on bows or "air rudders" on the same system as destroyers.

Uniforms.

	Vice-Amiral.	Contre-Amiral.	Capitaine de vaisseau.	Capitaine de fregate.	Lieutenant de vaisseau.	Enseigne.	Enseigne 2e classe and Aspirant.
Corresponding British or U.S. Navies.	"	*Rear-Ad.*	*Captain.*	*Commander.*	*Lieut.*	*Lieut.* (junior.)	*Sub-Lieut.* and *Midshipman.* (U.S. Ensign.)

Note.—New rank of Capitane de corvette, introduced June, 1917, with four gold stripes on sleeve, and intermediate between ranks of Capitane de fregate and Lieutenant de vaisseau. Equivalent to British or U.S.A. rank of Lieut.-Commander.

Only three of the Capitaine de fregate's upper stripes are gold: the other two are white.

Epaulettes with parade uniform are of the usual sort, *except* that—
A Vice-Admiral's epaulettes have the usual anchor and 3 stars.
A Rear-Admiral's " " " " 2 "

Caps are the usual shape, but have gold bands round. The cocked hat carries the tri-colour.

(Warship Builders only, arranged alphabetically.)

CHANTIERS DE BRETAGNE (formerly Etab. F. De la Brosse & Fouché), NANTES. D t. b., etc. Slips: Two, 328 feet long; two, 260 feet long; one, 230 feet long; one long. Build turbines, recipro. and Diesel engines, boilers, etc.

CHANTIERS DE LA GIRONDE, BORDEAUX. All kinds of warships. Water front: 6 Area of works: about 48 acres. Four slips: (1) 597 feet, (2) 515 feet, (3) 436 feet, feet. Titan cranes to first two. One dock, which can be used for building: 594 × 32¾ feet, with 140 ton electric crane. Special submarine workshop: 141 × 49 fee allied to Schneider et Cie.

CHANTIERS DE LA LOIRE. Works at ST. NAZAIRE (all kinds of warships) and NAN ST. NAZAIRE, 2 slips, 575 feet long: 1 slip, 555 feet long; 1 slip, 505 feet long; an 460 feet long. Machinery works at SAINT DENIS.

CHANTIERS DE PROVENCE. Yard at Port de Bouc.

CHANTIERS DE ST. NAZAIRE (Penhoet). Principal yard ST. NAZAIRE, with slips, and 590 feet long. Works also at ROUEN. All kinds of warships.

CHANTIERS NAVALS FRANÇAIS, CAEN. 10 slips, 400-500 feet; 4 slips, 300 feet. 1250 acres.

DYLE & BACALAN, BORDEAUX. Destroyers, t.b., etc. Five slips, longest of which is 3 Served with two 40 ton cranes. Machinery: 7000 h.p.

FORGES ET CHANTIERS DE LA MEDITERRANÉE. Principal yard LA SEYNE, near T Other yards at HAVRE, GRANVILLE and MARSEILLES. 3 slips, (*a*) 492 feet, (*b*) 590 feet, feet. All kinds of warships built at La Seyne, smaller craft at other yards. Total em circa 6500.

AUGUSTIN NORMAND & Cie. Destroyer, torpedo boat and submarine works at I Build ships up to 4000 tons. Yard covers 4 hectares. Large water front. Empl 1500. Build submarine mine-layers, Diesel engines, Normand and Normand-Sigaudy l etc.

SCHNEIDER & Cie. (formerly Schneider-Canet). LE CREUSOT WORKS: Ordnance, Mun Armour Plate. HAVRE, HONFLEUR AND HOC WORKS: Aviation and Diesel engines, p range. CHALON-SUR-SAÔNE YARD: Destroyers, torpedo boats and submarines. CHAMP SUR-SEINE WORKS: Electric equipment for submarines. LA LONDES-LES-MAURES V (WAR): Torpedoes. STATION DE CREUX ST. GEORGES (near Toulon): For testing subma

SOCIETÉ PROVENCALE DE CONSTRUCTIONS NAVALES (La Ciotat) MARSEILLES.*

SOCIETÉ NORMANDE DE CONSTRUCTIONS NAVALES, HAVRE and CHERBOURG.*

* Firms allied to Schneider & Cie.

Besides firms named above, machinery and boilers built by Etab. Indret (Nantes), Dela Belleville & Cie., and other firms.

Diesel Engine Builders:—	*Type of Engine:—*
F. et C. de la Med.	Sulzer types.
Ch. de la Loire	Sulzer types.
Comp. de Con. Mecanique Procedes Sulzer	Sulzer types.
Delaunay-Belleville	Sulzer. Burmeister & Wain.
Etab. de Material Naval & Industriel	(Own design.)
Thompson-Houston Co.	Tosi.
Dujerdin & Cie.	Werkspoor.
Brissoneau & Lotz	Werkspoor.
Weyner & Richemond	Werkspoor.
Ch. de St. Nazaire	Werkspoor.
Soc. de Moteurs Chaleassene	Sabathé.
Augustin Normand & Cie.	Normand.
Sautter-Harlé & Cie.	M.A.N.
Lefiaive & Cie.	M.A.N.

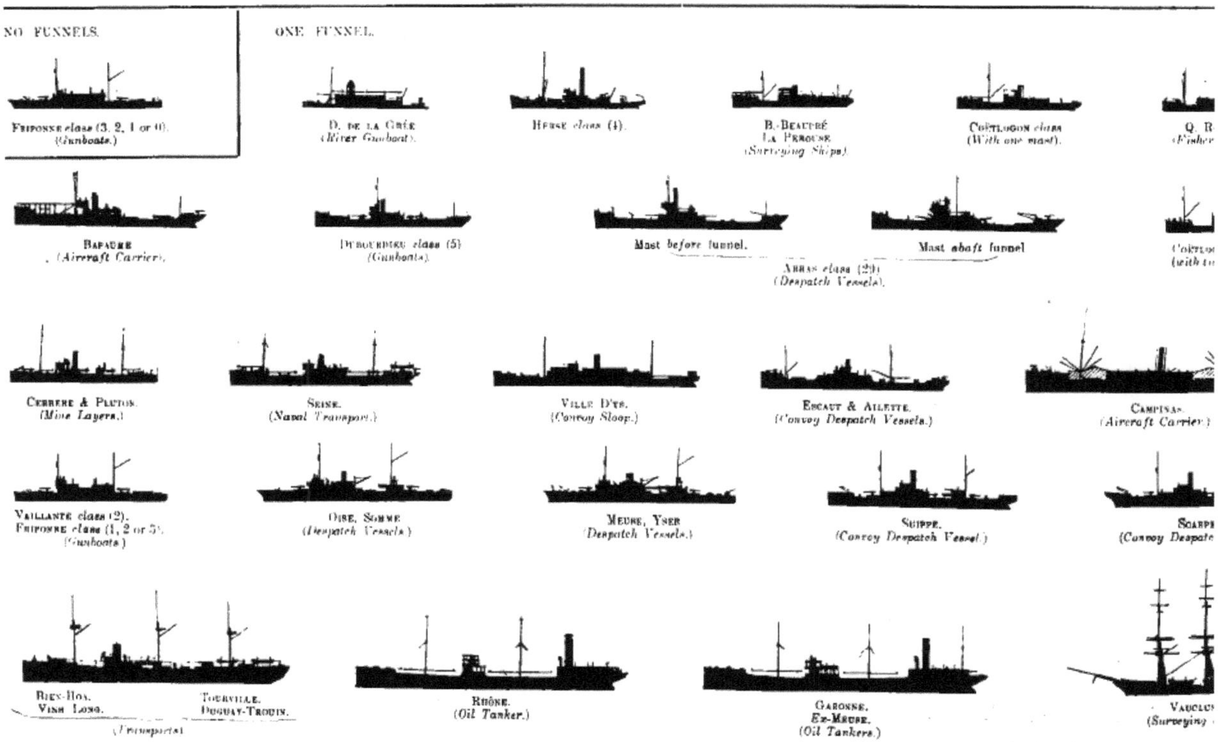
NO FUNNELS.
FRIPONNE class (3, 2, 1 or 0).
(Gunboats.)
ONE FUNNEL.
D. DE LA GRÈE
(River Gunboat).
HERSE class (4).
B.-BEAUPRÉ
LA PEROUSE
(Surveying Ships).
COËTLOGON class
(With one mast).
Q. R
(Fisher
BAPAUME
(Aircraft Carrier).
DUBOURDIEU class (5)
(Gunboats).
Mast before funnel.
Mast abaft funnel
ARRAS class (29)
(Despatch Vessels).
COËTLO
(with tw
CERBERE & PLUTON.
(Mine Layers.)
SEINE.
(Naval Transport.)
VILLE D'YS.
(Convoy Sloop.)
ESCAUT & AILETTE.
(Convoy Despatch Vessels.)
CAMPINAS.
(Aircraft Carrier.)
VAILLANTE class (2).
FRIPONNE class (1, 2 or 3).
(Gunboats.)
OISE, SOMME
(Despatch Vessels.)
MEUSE, YSER
(Despatch Vessels.)
SUIPPE.
(Convoy Despatch Vessel.)
SCARPE
(Convoy Despatc
BIEN-HOA.
VINH LONG.
TOURVILLE.
DUGUAY-TROUIN.
(Transports)
RHÔNE.
(Oil Tanker.)
GARONNE.
EX-MEUSE.
(Oil Tankers.)
VAUCLU
(Surveying

TWO FUNNELS.

ARDENT *type* (20).
(Gunboats.—Vary a little in details.)

AISNE, MARNE
(Despatch Vessels.)

ALTAIR *type* (5).

CASSIOPÉE, REGUL

"Star Class" (7).
(Sloops.)

LA HIRE *(Training Ship.)*

D'ESTRÉES.

BRETAGNE.

LORRAINE } *May be rebuilt 1920-21*
PROVENCE } *as "Bretagne."*

THREE FUNNELS.

POTHUAU
(Gunnery Training Ship.)

COURBET.

JEAN BART *(Has cranes as above Silhouette.)*
FRANCE } *(Have derricks as "Courbet.")*
PARIS }

(Three Funnel Types continued on next page

T

THREE FUNNELS.

Republique*
Patrie. } (*Training Ships.*)

* *Has no T.T. at stern. Also has Y-derricks instead of cranes abeam of 2nd funnel.*

D'Entrecasteaux.
(*Training Ship.*)

Du Chayla.
Cassard.

Colmar.

Strasbourg.

Metz.

Foudre.
(*Aircraft Carrier*

FOUR FUNNELS

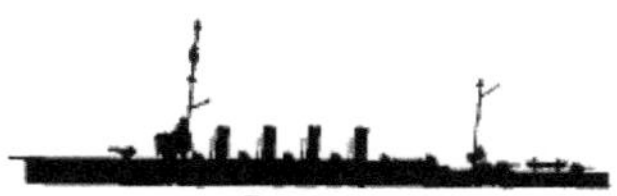

Thionville.

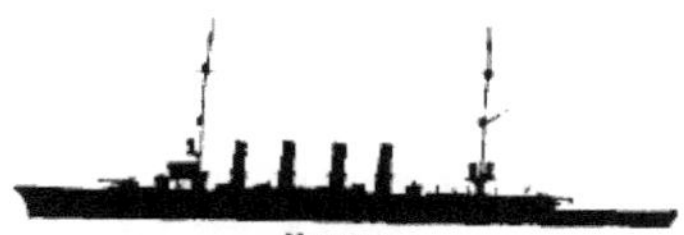

Mulhouse.

Four Funnel Types continued on next pag

FRENCH WARSHIPS (four to six funnels), RECOGNITION SILHOUETTES.

FOUR FUNNELS—(*continued.*)

GUICHEN

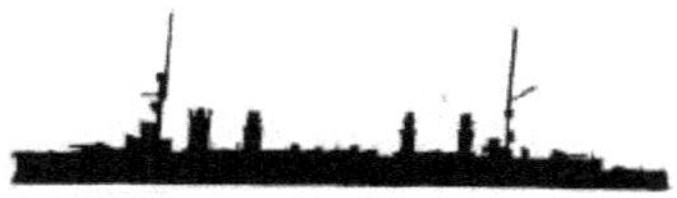

JURIEN DE LA GRAVIÈRE

DESAIX.

JULES MICHELET.

VICTOR HUGO.
JULES FERRY.*

* *Has cranes between 1st and 2nd funnels instead of derricks.*

AUBE AND MONTCALM *classes* (6 ships).
(Impossible to differentiate these classes except quite clo

FIVE FUNNELS.

CONDORCET.

VERGNIAUD.
VOLTAIRE.

DIDEROT.

SIX FUNNELS.

ERNEST RENAN.

EDGAR QUINET

JEANNE D'ARC (*Training Ship.*)

Scale: 1 inch = 160 feet.

FRENCH TORPEDO CRAFT—RECOGNITION SILHOUETTES.

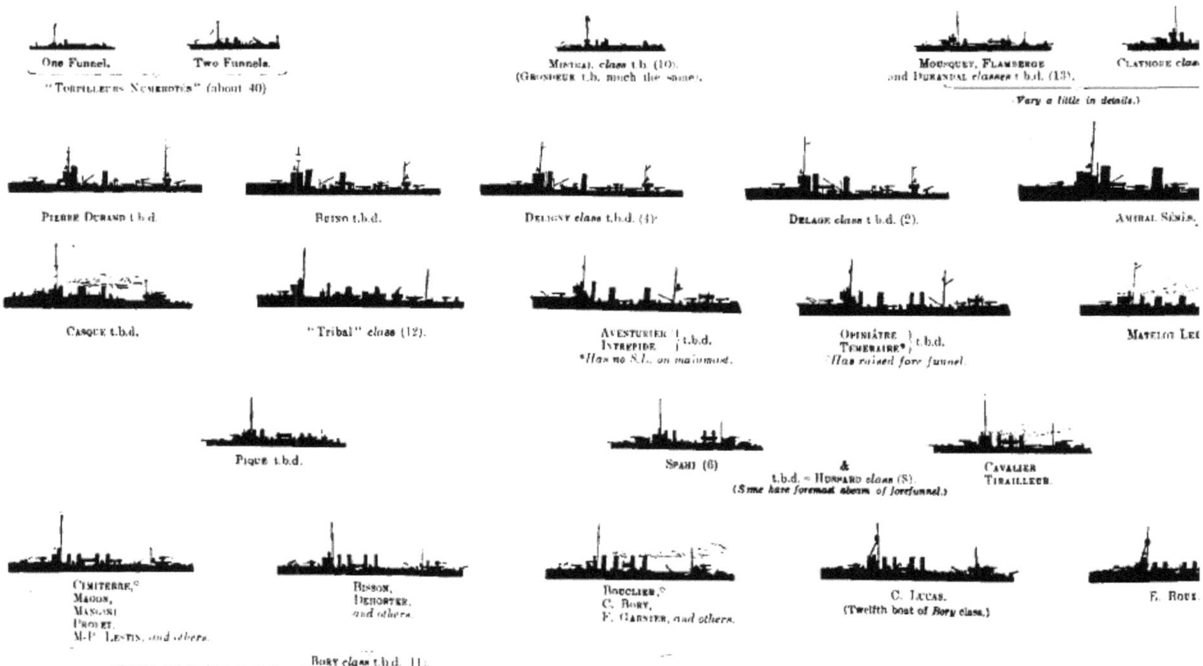

*Bouclier and Cimiterre nominally units of "Casque type," and not included in total for "Bory class."

SUBMARINE CHASERS.—**Scale: 1 inch = 80 feet.**

V1—V73 *classes.*

C 1—C100 *class.*

C101—C116 *class.*

SUBMARINES.—**Scale: 1 inch = 80 feet.**

Note.—**Except "R. Morillot" (Coastal Boat), all boats given below are "Sea-going" Types. For "Ocean-going" Types (and ex-German "UB" boats), v. next page.**

ROLAND MORILLOT.

COULOMB.
(Modified Pluviôse class.)

AMAZONE.
ANTIGONE.
ARMIDE.

LIEUT. O'BYRNE.
DUPETIT-THOUARS.
HENRI FOURNIER.

CURIE.*
(Modified Pluviôse class.)

*Rebuilt by Austrians during the War. Accuracy of this Silhouette very doubtful now.

Modified Pluviôse class:—
ARAGO. BRUMAIRE. EULER. FARADAY. FRANKLIN. FRIMAIRE. MONTGOLFIER. NEWTON. NIVÔSE.

VOLTA.

LE VERRIER.

(Modified Pluviôse class.)

CLORINDE.
CORNELIE.

AMARANTE.° AMPHITRITE. ANDROMAQUE. ARÉTHUSE. ARTÉMIS. ASTRÉE.* ATALANTE.
(°There are Mine Laying boats of this class and have no guns.)

BELLONE.
GORGONE.
HERMIONE.

DAPHNÉ.

(Continued on next

NOTE.—All types on this page are Ocean-going. The "Coriolan Class" boats are rated as Ocean-going by convention only. During their career as German "UB 88 Class" bo they were more correctly rated as a small Ocean-going type.

Fulton.
Joessel.

Dupuy de Lôme.
Sané.

Gustave Zédé.
Néréide.*
(*Has thin telescopic funnel.)

Laplace. Regnault.
Lagrange. Romazotti.

Coriolan.
Jean Corre.
Trinité Schillemans.

Victor Reveillé.
(Mine Layer.)

Jean Roulier. Jean Autric.*
Marrast. Leon Mignot.*
* *Periscope housing does not rise above bridge screen in these two boats.*

René Audry.
(Mine Layer.)

Halbronn.

(NORMANDIE CLASS—4 SHIPS.)

NORMANDIE (Oct., 1914), **GASCOGNE** (Sept., 1914), **LANGUEDOC** (May, 1915), **FLANDRE** (Oct., 1914).

Displacement, 25,230 tons. Complement,

Length (*waterline*), 576 feet. Beam, feet. *Maximum* draught, feet

Guns (Mod. '11):
12—13·4 inch, 45 cal (A[7]).
18—5·5 inch, 55 cal. (M. '10).
(Some anti-aircraft.)
4—3 pdr.
Torpedo tubes: 6.

Armour:
" Belt (amidships) ......
" Belt (bow) ....
" Belt (aft) ....
" Deck ....
" Side above belt }
" Big gun turrets ......
" Barbettes ....
" Battery......................
" Conning tower ......
(Special anti-torpedo protection.)

No Plans available.

Machinery: Turbines. screws. Boilers: various types. Designed H.P. =25 kts. Fuel tons *normal*, tons *full load*. Endurance: miles at kts.

Gunnery Notes.—13·4 inch guns originally built were used by French Army during the war, and some were re-bored to 15·75 inches. Some of the 13·4's intended for *Gascogne* are said to have been captured by the Germans at Lille.

Name	*Builder*	*Machinery*	*Laid down*	*Completed*	*Trials*	*Turbines*	*Boilers*	*Best max. speed*
Normandie	S. Nazaire (Lre)	S. N. (Loire)	July, '13	?				
Gascogne	Lorient	A. & C. de Bretagne	Oct., '13					
Languedoc	Gironde	Gironde	July, '13					
Flandre	Brest	...	Oct., '13					
Bearn	La Seyne	La Seyne	Jan., '14					

General Notes.—Designed by Chief Constructor Doyère. Fitted very wide bilge keels. Freeboard forward, 23 feet; aft, 16½ feet. Belong to 1913 programme.

Note.

All work on these ships was stopped soon after the outbreak of war. In 1918- design was considerably altered. Full details of the changes effected are not yet a but the Ministry of Marine communicated these points in 1919:—

(*a*) Main battery of 12—13·4 inch guns retained;
(*b*) Number of 5·5 inch guns reduced to 18 from 24;
(*c*) Six torpedo tubes to be retained;
(*d*) Speed raised from 21·5 to 25 knots;
(*e*) Special anti-torpedo protection (*soufflage contre-torpilles*) given to hull.

During 1919-20 various unofficial projects were discussed by French publicists f verting these ships to Battle Cruisers, Liners, Cargo Vessels, Target Ships, &c. The M of Marine informs "Fighting Ships" that up to July 8th, 1920, no decision had been at as regards the future of these ships. It has been subsequently decided that *Bearn*, class, shall be finished as an Aircraft Carrier.

(Bretagne Class—3 Ships.)

BRETAGNE (April, 1913), **LORRAINE** (September, 1913), **PROVENCE** (April, 1913).

Normal displacement, 23,230 tons. Complement, 1109.

Length (*waterline*), 541⅓ feet. Beam, 88⅓ feet. *Maximum* draught, 29 feet. Length *over all*, 541⅓ feet.

Guns (M. '12. **Dir. Con.** *in Bretagne*):
- 10—13·4 inch, 45 cal.
- *22—5·5 inch, 55 cal. (M. '10).
- 4—3 pdr. (may be AA.)†
- 2—1 pdr.
- 2 landing.

Torpedo tubes (18 inch) M. '12:
- 4 *submerged*.

* 18 in *Bretagne*

† = 2—14 pdr. AA. in *Bretagne*?

Armour (chromo-nickel):
- 10⅝" Belt (amidships) ....
- 7" Belt (ends) ....
- Deck (*see Notes*) ....
- 10"—6" Lower deck side
- 17"—10" Turrets..........
- 11" Turret bases (N.C.)
- 7" Secondary battery ....
- 12½" Conning tower ......

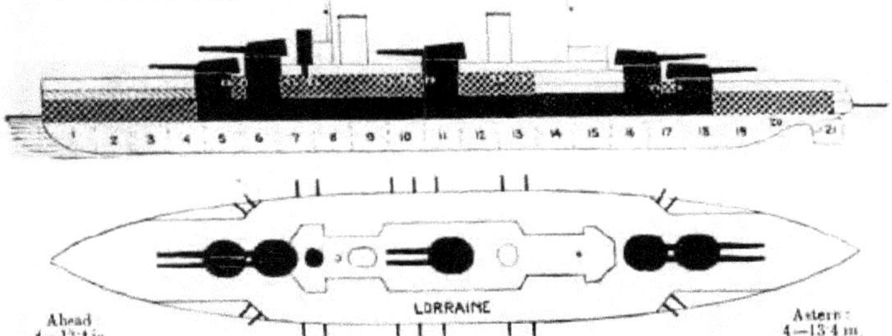

Ahead: 4—13·4 in. 3 or 4—5·5 in.

Astern: 4—13·4 in. 4—5·5 in.

Broadside: 10—13·4 in., 9 or 11—5·5 in.

Machinery: Parsons turbine. 4 screws. Boilers: see *Notes*. Designed H.P. 29,000 = 20 kts. Coal: *normal* 900 tons; *maximum* 2680 tons. Also 300 tons oil.

Gunnery Notes.—Carry 100 rounds per gun for 13·4 inch and 275 per gun for the 5·5 inch. Janney electro-hydraulic mountings to big guns. Special cooling for magazines—temperature 77° Fahr. Magazines can be completely flooded inside ten minutes. Height of guns above l.w.l.: 1st turret, 30½ feet; 2nd, 3[illegible] feet; 3rd, 33½ feet; 4th, 28½ feet; 5th, 21½ feet. Arcs of training: Nos. 1 and 5, 270°; Nos. 2 & 4, 280°; No. 3, 120° either beam. Arcs of secondary guns: 120°.

Torpedo Notes.—Bullivant net defence. 8—36" and 2—30" searchlights. 4—200 k.w. dynamos. Torpedoes: 1912 Schneider, 6000 yards range. 24 carried. Also 30 blockade mines.

Armour Notes.—Turrets of maximum thickness at ports, instead of uniform thickness as in *J. Bart*. (See Addenda Notes.) The double bottom is carried to the under side of protective deck. Main belt is 13¼ feet wide, 5 7/12 feet below and 7½ above l.w.l. Battery 197 feet long with 7 in. bulkheads. Protective decks: lower 2" slopes, 1½" flat. Upper, 1¼" flat on top of belt.

Engineering Notes.—Boilers: *Bretagne*, 24 Niclausse; *Lorraine*, 24 Belleville; *Provence*, 18 Guyot du Temple. Grate area *Bretagne*, 2,000 sq. feet; *Lorraine*, 2,050 sq. feet; *Provence*, 1,492 sq. feet. Heating surface: *Bretagne*, 64,600 sq. feet; *Lorraine*, 63,700 sq. feet; *Provence*, 42,[illegible] sq. feet. Pressure, 256 lbs. per sq. in. 300 turbine r.p.m. 2,700 tons coal in 13 bunkers, 250 tons oil in 4 tanks. Endurance: 4,700 miles at 10 kts., 2800 at 18½ kts.

Name	*Builder*	*Machinery*	*Laid down*	*Completed*	*Trials*	*Turbines*	*Boilers*	*Best recent speed*
Bretagne	Brest	La Seyne	July '12	Sept '15		Parsons	Niclausse	
Lorraine	S. Nazaire (Pen.)	S. Nazaire (Pen.)	Nov. '12	July '16		Parsons	Belleville	
Provence	Lorient	La Seyne	June '12	June '15		Parsons	D.T. Guyot	

General Notes.—B., L. and P. belong to the 1912 programme, one of them being a replace ship for the *Liberté*, blown up September, 1911. Estimated cost, £2,900,000 per ship = £126 per ton. Bilge keels: 245 × 2½ feet. Bulkheads are longitudinal and transverse; latter solid amidships.

*Unofficial.

Bretagne (as rebuilt 1919-20). 1920 *Copyright Photo, M.*

Notes.

During 1919-20 refit, following alterations were effected in *Bretagne*: Director Fire Control installed; elevation of main guns largely increased; forward pair of 5·5 inch guns removed and ports plated up; nets removed; tripod foremast added; mainmast cut down to stump; fore-funnel raised. Flying-off platforms are to be added to the superfiring turrets.

When trials of *Bretagne* are finished, similar alterations will probably be carried out in *Provence* and *Lorraine*.

Provence (Lorraine similar).

Addenda Notes.—According to the 1917 Edition of "Flottes de Combat," the barbette shields are [illegible]. Those for the end barbettes are 13·4" thick, for the superfiring barbettes 9½" thick and for the centre [illegible]

1910 FRENCH DREADNOUGHTS. Nos. 4—1 (*Cuirassés de 1er rang*) (20 knots).

COURBET (Sept., 1911), **JEAN BART** (Sept., 1911), **PARIS** (Sept., 1912) & **FRANCE** (Nov., 1912).

Normal displacement, 23,120 tons. *Full load*, 25,850 tons. Complement, 1108.

Length (*p.p.*), 541⅓ feet. Beam, 88½ feet. *Maximum* draught, 29 feet. Length *over all*, 544½ feet.

Guns (M. '06—10):
- 12—12 inch, 45 cal.
- 22—5·5 inch, 55 cal. (M. '10).
- 4—3 pdr. (may be A.A.)
- 2—1 pdr.
- 2—landing.

Torpedo tubes (18 inch):
- 4 *submerged*.

Armour: (Chrono-nickel):
- 11¾" Belt (amidships)......
- 7" Belt (bow) (N.C.) ...
- 7" Belt (aft) (N.C.) ......
- 2¾" Deck ....
- 10"—6" Lower deck side
- 12½" Turrets (N.C.) ......
- 11" Turret bases (N.C.) ..
- 7" Secondary battery ...
- 11¾" Conning tower (N.C.)

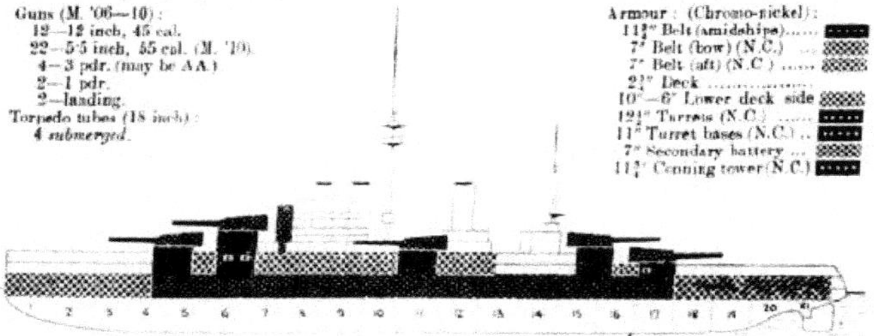

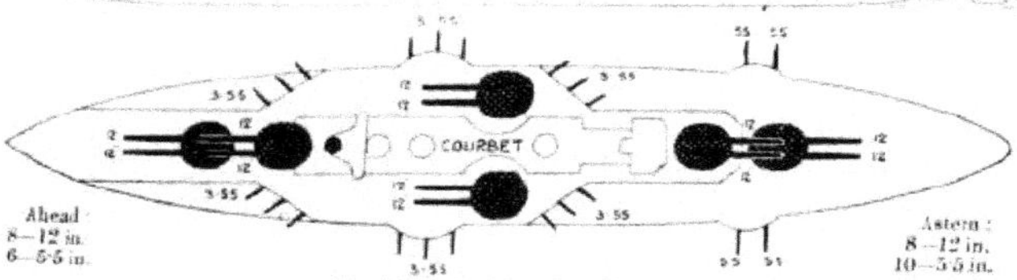

Ahead: 8—12 in. 6—5·5 in.

Astern: 8—12 in. 10—5·5 in.

Broadside: 10—12 in., 11—5·5 in.

Machinery: Parsons turbine. 4 screws. 24 Boilers: Belleville or Niclausse (large tube). Designed H.P. 28,000 = 20 kts. Coal: *normal* 906 tons; *maximum* 2452 tons; also 250 tons oil (may be increased to 450 tons). Endurance —— at 10 kts., 2700 at 18¾ kts.

Armour Notes.—Belt, 13¾ feet wide; 7¼ feet of it above water, 5½ feet of it below. For about 325 feet it is 10¾". Upper belt, 7" thick forms redoubt for secondary battery. The 4—5·5 inch aft are in casemates. Prot. decks: 2¾" curved from lower edge of belt to l.w.l. level. Above that a flat 1¾" deck from end to end on top of main belt. Above again is a 1¾" splinter deck against aerial attack. Conning tower is in three stories. 10¾" communication tube. Two armoured fire control stations on top.

Gunnery Notes.—Amidship 12 inch 180°, the centre line turrets 270° each. The 5·5 inch 120°.

Height of guns above water: Turrets (1) 30½ feet, (2) 37¾ (amidships) 25 feet, (5) 28½ feet, (6) 21¼ feet; upper deck battery, 5·5 inch, 21½ feet; main deck, 5·5 inch aft, 11¼ feet. Ammunition carried: 100 for each 12 in., 275 for each 5·5 inch, 300 for each 3 pdr. Westinghouse refrigerators in magazines. 77°F.

Torpedo Notes.—8 searchlights (36 inch), also 2—30 inch. Carries 12 torpedoes and 30 blockade mines.

Engineering Notes.—Pressure: 256 lbs. [illegible] shafts: H.P. turbines; inner L.P. [illegible] Condensers (2): 32,676 sq. feet cooling surface. [illegible] 300 r.p.m. Belleville oil fuel system.

Name	Builder	Machinery	Laid down	Completed	Trials Full power	Boilers	Best recent speed
J. Bart	Brest		Nov., '10	June '13	= 22·04	Belleville	22·6
Courbet	Lorient		Sept., '10	Sept. '13	= 20·14	Niclausse	
Paris	La Seyne		Nov., '11	Aug. '14	32,010 = 21·8	Belleville	
France	St. Nazaire		Nov., '11	June '14		Belleville	

General Notes.—Designed by M. Lyasse. [illegible] Average [illegible] [illegible] height of [illegible] feet [illegible] in a head sea. On the other hand, J. Bart [illegible]

COURBET. [illegible] Photo, M. Bar.

Note.—Courbet only has stump mainmast as above view. Other three ships rigged as J. Bart.

JEAN BART. This ship only has "[illegible]" [illegible] in place of derricks. Photo, [illegible]

(DANTON CLASS—4 SHIPS.)

DIDEROT (April, 1909), **CONDORCET** (April, 1909), **VERGNIAUD** (April, 1910), and **VOLTAIRE** (Jan., 1909).

Normal displacement, 18,890 tons. Complement, [illegible].

Length (*waterline*), 475¾ feet. Beam, 84⅔ feet. { *Mean* draught, 27 feet. / *Max.* " 28½ " } Length (*over all*), 481 feet.

Guns:
4—12 inch, 45 cal. (M '06)
12—9·4 inch, 49·5 cal. (M '02-'06)
16 or 14—3 in. (14 pdr.), 65 cal.
(2—3 pdr., S.A.)
4—3 pdr. (AA.)
2—3 pdr. (landing).
2—1 pdr.
Torpedo tubes (M '04, 18 inch):
2 *submerged* (broadside).

Armour:
10¼" Belt (amidships)
6" Belt (ends)
2" Deck (flat on upper belt)
2" Deck (below belt)
9"-6" Lower deck side
2¼" Upper belt (bow)
12¼" Turrets (big guns) (N.C.)
11" Turret bases (N.C.)
8¾" Secondary turrets (6) (N.C.)
12" Conning tower (N.C.)
Total weight about 4600 tons

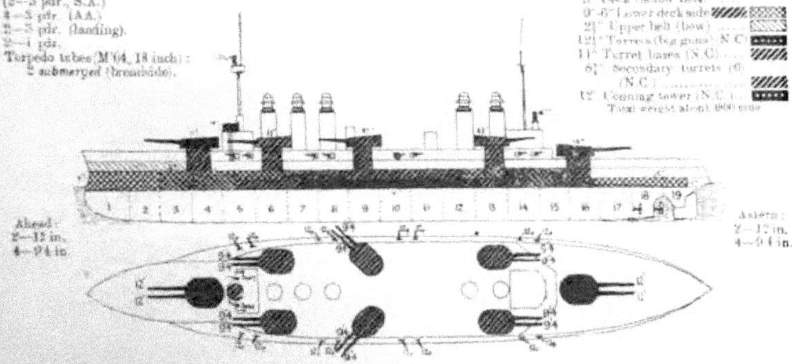

Ahead: 2—12 in. 4—9·4 in.

Astern: 2—12 in. 4—9·4 in.

Broadside: 4—12 in., 6—9·4 in.

Machinery: Parsons turbine. 4 screws. Boilers: 26 Belleville or Niclausse (large tubes). Designed H.P. 22,500 = 19·25 kts. Coal: *normal* 965 tons; *maximum* 2100 tons. Endurance: 3750 miles at 10 kts., 2300 at 18¾ kts.

Armour Notes.—Main belt is about 6½ feet wide, about 2½ feet of it below waterline. Lower edge is 3" thick. There is a deck on top of the main belt as well as below it. Fore bulkhead 7", aft ditto, 8"–7".

Gunnery Notes (unofficial).—Arcs of fire: Big guns, 260°. Secondary guns about 140° or more. Rounds carried: 12 in., 74 per gun; 9·4 in., 100 per gun; 12 pdrs., 300 per gun. The 14 pdr. guns are a Schneider-Canet model. Some 14 pdr. and 3 pdr. removed during the War; may not be remounted. 14 pdr. positions on plans not correct.

Engineering Notes.—Bellevilles of these ships have 332 elements and 3852 tubes. Niclausse 339 elements and 3814 tubes. Both types equipped for burning oil fuel.

Name	Builder	Machinery	Laid down	Completed	Trials: 24 hrs.	Trials: 10 hrs.	Trials: 3 hrs.	Boilers	Best recent speed
Diderot	St. Nazaire	St. Nazaire	[illegible]	Sept. '11	[illegible]	[illegible]	[illegible]	Niclausse	
Condorcet	St. Nazaire	St. Nazaire	Aug. '06	June '11	[illegible]	= 19·31	[illegible]	Niclausse	
Vergniaud	Bordeaux	F. & Ch. Medit.	Nov. '06	Jan. '11	[illegible]	[illegible]	[illegible]	Niclausse	
Voltaire	La Seyne	F. & Ch. Medit.	July '06	Aug. '11	[illegible]	[illegible]	[illegible]	Belleville	

[illegible]

CONDORCET. 1919 *Photo*, Comm. B[illegible]

Notes.

Vergniaud and *Voltaire* have had mainmasts cut down to height of fun[illegible] photo available of these ships.

Mirabeau of this class stranded on Crimean Coast, Feb., 1919. Salved Ap[illegible] after removal of forward 12 inch barbette and side armour. Wi[illegible] as a Target Ship.

DIDEROT.

PATRIE. 1920 *Photo*

Training Ships.—"Republique" for Gunnery; "Patrie" for Torpedoes.

REPUBLIQUE (Sept., 1902) & **PATRIE** (Dec., 1903). Displacement: 14,865 tons. Complement: 789. Length (*waterline*), 439 feet. Beam, 79½ feet. *Maximum* draught, 27½ feet. Length (*over all*), 443¾ feet. Guns (M '93-'96): 12—6·5 inch, 45 cal., 4 or 6—14 pdr. in *Republique*, 25—3 pdr. (4 may be AA.), 2—1 pdr. Torpedo tubes (18 in.): 2 *submerged*, and in *Patrie* only, 1 *above water* at stern and on quarter-deck for long range firing. Armour: 11¾" Belt (amidships), 7" Belt (ends), 3" Deck (flat below belt), 10"—5" Upper belt, 3" Upper Belt (forward), 11" Turret (bases), 5¼" Casemates, 5½" Small turrets, 4" Bases to them, 13" Conning tower. **Machinery**: 3 sets vertical inverted triple expansion. 3 screws. Boilers: 24 Niclausse. Designed H.P. 18,000 = 18 kts. Coal: *normal* 905 tons; *maximum* 1825 tons. *Nominal* radius 6,000 at 10 kts.

Notes.—*Republique* built at Brest, 1901-6, machinery by Indret; *Patrie* built at La Seyne, 1902-6, machinery by F. & C. de la Med. Both made 19 kts. on trials. In 1917-18, 6—6·5 inch were removed for defence of Salonica and never re-mounted. 4—12 inch removed 1920, on conversion to training ships.

Notes.

Following Old Battleships were removed from the Active List between Jan. and June, 1[illegible]

St. Louis. Kept for Subsidiary Duties (Accommodation Ship, &c.).

Charlemagne. Sold; being broken up.

Jaureguiberry. Accommodation Ship.

Henri IV. W.T. School. Reported to have no guns.

Amiral Trehouart.

Bouvines. Sold; being broken up.

The following three Battleships are to be removed from the Active List before the end of [illegible] and are therefore deleted.

Democratie.

Justice.

Verité.

Note: Mainmast of E. Quinet cut down; rig now as E. Renan opposite.

Waldeck-Rousseau. *Photo, Geiser, Alger.*

(Waldeck Rousseau Class—2 Ships).

EDGAR QUINET (Sept., 1907), WALDECK ROUSSEAU (March, 1908).

Normal displacement, 14,100 tons. Complement, 892.

Length (*waterline*), 515 feet. Beam, 70¾ feet. *Max.* draught, 27½ feet. Length *over all*, 521½ feet.

Guns—(M. '02):
- 14—7·6 inch, 50 cal.
- (20)—9 pdr.*
- (2)—9 pdr.*
- 4—3 pdr. AA.
- 2—1 pdr. or M.G.

Torpedo tubes (18 inch):
- 2 *submerged*.

* *See Notes.*

Armour:
- 6¾" Belt (amidships)
- 3" Belt (ends)
- 6" Bulkhead (aft)
- 2½" Deck (slopes)
- 6" Turrets (N.C.)
- 5" Turret bases
- 6" Casemates (4)
- 10" Conning tower

Total weight 2400 tons

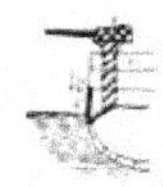

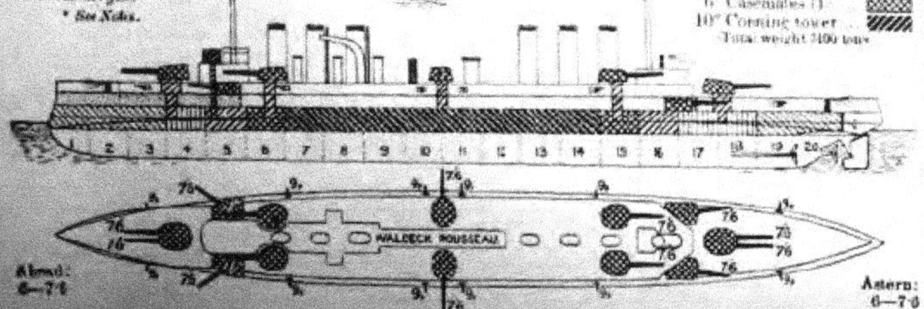

Ahead: 6—7·6

Astern: 6—7·6

Broadside: 9—7·6 in.

Machinery: 3 sets vertical triple expansion. 3 screws. Boilers (largetube): *See Notes.* Designed H.P. 36,000 = 23 kts. Coal: *normal* 1,242 tons; *maximum* 1,200 tons. *Nominal endurance:* 6560 miles at 10 kts.

Gunnery Notes.—9 pdr. positions on plans not correct. Reported in 1919 that W. Rousseau had only 12—9 pdr., 2—9 pdr. AA.

[illegible]	[illegible]	[illegible]	Laid down	Completed	Trials: 30 boilers 10 hours	Trials: Full power 3 hours	Boilers	Best recent speed
[illegible]	[illegible]		Nov., 96 [illegible]	1911 [illegible]	35,[illegible] = 23·[illegible]	40,[illegible] = 23·[illegible] 36,[illegible] = 23·1	40 Belleville 42 Niclausse	23

E. Renan. 1920 Photo, M. Ba[illegible]

ERNEST RENAN (March, 1906).

Normal displacement, 13,640 tons. Complement, 824.

Length (*waterline*), 515 feet (o.a. 521½ feet). Beam, 70 feet. *Max.* draught, 27¼

Guns:
- 4—7·6 inch, 50 cal. (M '02)
- *12—6·5 inch, 45 cal. ('93-'96 M
- 16—9 pdr.
- 2—9 pdr.
- 8—3 pdr. (4 are AA.)
- 2—1 pdr.

Torpedo tubes (18 inch):
- 2 *submerged*.
- (1904 Model torpedoes)

Armour
- 6¾" Bel
- 3¼" Bel
- 6" Bulk
- 2¼" Dec
- 6" Main
- 4"–5" T
- 4" Case
- 4" Secon
- 5" Base
- 8" Conn
- (Total w

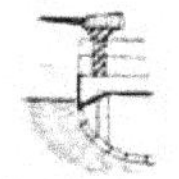

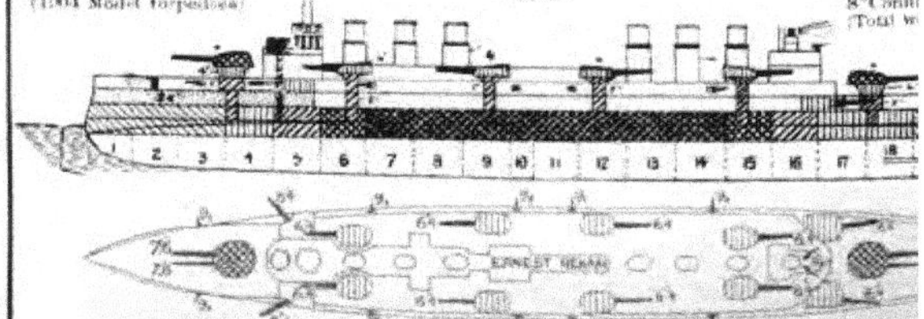

Machinery: 3 sets vertical triple expansion. 3 screws. Boilers: 42 Niclausse (large tube 36,000 = 23 kts. Coal: *normal* 1354 tons; *maximum* 2300 tons. *Nominal endurance:* 6700

Armour Notes.—Belt is 12½ feet wide, 4¾ feet of it below waterline. Upper edge 5" thick; b

**Gunnery Notes.*—Loading positions, big guns: all round. Hoists: electric to all guns. A electrically and hand. Arcs of fire: Big turrets, 240°; secondary turrets, 150°; casemate guns above water: Bow and secondary turrets, 29½ feet; after turret, 22 feet. F 20 feet: after casemates, 12 feet.

General Notes.—Laid down at St. Nazaire, Aug., 1904; completed, 1909. Made 24·24 with 37,685 H.P. Best speed now 21 kts. Mainmast removed 1918.

* Gunnery Notes unofficial

Photo, Geiser.

JULES MICHELET (Aug., 1905). Displacement 13,370 tons. Complement 770. Length (*waterline*), 489 feet. Beam, $70\frac{1}{4}$ feet. *Max.* draught, $27\frac{3}{4}$ feet. Guns: 4—7·6 inch, 50 cal., 12—6·5 inch, 45 cal., 2—9 pdr., 24—3 pdr. (4—3 pdr. are AA.), 2 M.G. Torpedo tubes (18 inch): 2 *submerged*. Armour: $6\frac{3}{4}$" Belt (amidship), 3" Belt (ends), 6" Bulkhead (aft), $2\frac{1}{2}$" Deck (slopes), 6" Turrets, 5" Turret bases, 4" Casemates (4), 4" Small turrets (8), 5" Bases to these, 8" Conning tower. Machinery: 3 sets vertical triple expansion. 3 screws. Boilers: 28 Du Temple-Guyot (small tube). Designed H.P. 29,000 = 22 kts. (Trials, 30,438 = 22·9.) Coal: *normal* 1330 tons; *maximum* 2100 tons. Begun at Lorient D.Y. 1904, completed 1908.

JULES FERRY. *Photo, Geiser, Algiers.*

JULES FERRY* (August, 1903), & **VICTOR HUGO** (March, 1904). *Average* displacement 12,518 tons. Complements, 737—731. Length (*o.a.*), *J.F.* $486\frac{1}{2}$ feet, *V.H.* 489 feet. Beam, $70\frac{1}{4}$ feet. *Mean* draught, $26\frac{1}{2}$ feet. Guns (M. '93): 4—7·6 inch, 45 cal., 16—6·5 inch, 45 cal., 24—3 pdr. (4—3 pdr. are AA.), 2—9 pdr., 2 M.G. Torpedo tubes (18 in.): 2 *submerged*. Armour: $6\frac{3}{4}$" Belt (amidships), 3" Belt (ends), 5" Bulkhead (aft), $2\frac{1}{4}$" Deck (slopes), 8" Turrets, 8" Turret bases, $5\frac{1}{4}$" Casemates (4), $5\frac{1}{4}$" Small turrets, 8" Conning tower. Machinery: 3 sets vertical 4 cylinder triple expansion. 3 screws. Boilers: 28 Guyot in *J. Ferry*, 28 Belleville in *V. Hugo*. Designed H.P. 27,500 = 22 kts. Coal: *normal* 1330 tons; *maximum* 1900 tons. Endurance at 10 kts., 6630; at 20 kts., 2600 miles.

GLOIRE. (All rated in "2e Catégorie," 1920.) *Photo, Geiser, A*

GLOIRE (June, 1900), **MARSEILLAISE** (July, 1900), **CONDE** (March, 1902), and **AM** **AUBE** (May, 1902). *Average displacement* 10,376 tons. Complement, 630. Length (*wate* $452\frac{1}{4}$ feet. Beam, $66\frac{1}{2}$ feet. *Maximum* draught, $25\frac{1}{4}$ feet. Length *over all*, $458\frac{3}{4}$ feet. Guns (M 2—7·6 inch, 45 cal., 8—6·5 inch, 45 cal., 6—3·9 inch, 2—9 pdr., 18—3 pdr. Torpedo tub in.): 2 *submerged*. Armour: $6\frac{3}{4}$"—4" Belt, $2\frac{1}{2}$" Deck (slopes), 8" Turrets (N.C.), 4" Hoists, 5" deck side, $2\frac{1}{2}$" Bow, $4\frac{1}{4}$" Casemates (4), $4\frac{1}{4}$" Small turrets (4), $3\frac{1}{4}$" Bases to these, 8" Conning t Machinery: 3 sets vertical triple expansion. 3 screws. Boilers: 28 Belleville or Nicl Designed H.P. 20,500 = 21 knots. Coal: *normal* 970 tons; *maximum* 1700 tons.

Notes.—*Gloire* and *Condé* built at Lorient, 1899-1904. *Marseillaise* at Brest, 1900-03; *A. Aube* at St. Nazaire 1900-01. 21·5, 21·3, 21·6, and 22·8 kts. on trials respectively.

DESAIX (March, 1901). *Displacement* 7,700 tons. Complement 520. Length (*waterline*), feet. Beam, $58\frac{1}{2}$ feet. *Mean* draught, $24\frac{1}{2}$ feet. Guns—(M. '93): 8—6·4 inch, 45 cal., 4—3·9 in 2—9 pdr., 10—3 pdr. (4—3 pdr. are AA.), 4 M.G. Torpedo tubes (18 in.): 2 *submerg* Armour: 4" Belt (amidships), $3\frac{1}{4}$" Belt (bow), $2\frac{1}{4}$" Armour deck, 4" Turrets, 4" Turret bases, Conning tower. Machinery: 3 sets 3 cylinder vertical triple expansion. 3 screws. Boilers: Belleville. Designed H.P. 17,100 = 21 kts. Coal: [illegible]
at St. Nazaire [illegible]

MONTCALM. (Mainmast is in one piece.)

(Rated 1920 in "2e Categorie")

QUEYDON (September, 1899), **MONTCALM** (March, 1900). Displacement, 9102 tons. Complement, 600. Length (*waterline*), 452¼ feet. Beam, 63¾ feet. *Max.* draught, 25 feet. Length (*over all*), 459 feet. Guns (M. '93 & '96): 2—7·6 inch, M. 93, 8—6·5 inch, M. '93-'96, 2—9 pdr., 20—3 pdr. Torpedo tubes (18 inch): *2 submerged*. Armour (Harvey Nickel): 6¼" Belt (amidships), 4" Belt (ends), 3" Forecastle, 2" Deck (on slopes), 3"-2¼" Lower deck side, 8" Turrets, 5" Turret bases, 4" Casemates (8), 6" Conning tower. Machinery: 3 sets vertical triple expansion. 3 screws. Boilers: *Gueydon*, 20 Niclausse; *Montcalm*, 20 Normand-Sigaudy (small tube). Designed H.P. 19,600=21 kts. Coal: *normal* 1030 tons; *maximum* 1650 tons. Begun at Lorient and La Seyne, 1898: completed, 1902. *Montcalm* refitted 1929.

(Cadets' Training Ship.)

JEANNE D'ARC (June, 1899). Displacement 11,200 tons. Complement 657. Length (*waterline*), 475¼ feet. Beam, 63½ feet. *Maximum* draught, 26½ feet. Guns—(M. '93): 2—7·6 inch, 40 cal., 16—3 pdr. (4—3 pdr are AA.), 8—1 pdr. Torpedo tubes (18 inch): 2 *submerged*. Armour (Harvey): 6" Belt (amidships), 4" Belt (bow), 3" Belt (aft), 2¼" Deck (on slopes), 7¼" Turrets, 4" Turret bases, 3" Lower deck, 5" Casemates (8), 6" Conning tower. Machinery: 3 sets vertical inverted triple expansion. 3 screws. Boilers: 48 Du Temple (small tube). Designed H.P. 33,000 =23 kts. (Best actual: 29,690 I.H.P.=21·7. In 1920, best speed 19 kts. *or less*). Coal: *normal* 1400 tons; *maximum* 1900 tons. Laid down at Toulon, October, 1896. Completed 1902.

(Sea-going Gunnery School Ship.)

POTHUAU (1895). Displacement 5460 tons. Complement 424. Length (*wateri* Beam, 50⅓ feet. *Maximum* draught, 22¼ feet. Guns—(M '93): 10—5·5 inch, 45 ca AA., 12—3 pdr., 8 M.G. Torpedo tubes: 4 *above water*. Armour: 3½" Belt (amids (ends), 3¾" Deck (*reinforcing belt*), 7" Turrets, 3" Hoists to these, 2" Casemates, 2" F 9½" Conning tower. Machinery: 2 sets horizontal triple expansion. 2 screws. Belleville. H.P. 10,400=19 kts. Coal: *normal* 538 tons; *maximum* 790 tons.

Projected.

(——— CLASS.—6 SHIPS.)

——— (1). ——— (4).
——— (2). ——— (5).
——— (3). ——— (6).

Displacement, tons. Complement, .

Length (*w.l.*), feet. Beam, feet. Draught, feet.

Guns:
(*Not known*).
Torpedo tubes:
(*Not known*).

Armour:
(*Not known*).

Machinery: Turbines, probably "all-geared" type. Boilers: Designed S.H.P. = kts. Fuel: tons *normal*, tons *full load* = miles at kts.

Notes.—Provided by Law of 30th March, 1912, and intended that they should be begun 1917-19, but none so laid down. Design not decided on, but plans have been considered of the following type:—About 5270 tons. Guns: 8—5·5 inch, 55 cal. (M. 1918), mounted in 4 turrets on centre line; 4—14 pdr. AA., 3 M.G., 12—21.7 inch tubes in 4 triple deck mountings. Armour: 2" Belt for 4 of length. 1" Deck, also some form of anti-torpedo protection. S.H.P. 50,000 = 30 kts. To be fitted for carrying aircraft.

1914 (ex-German) **Light Cruiser.**

METZ. 1920 Photo, Freund, Bre

METZ (ex-German *Königsberg*, December, 1915).
Displacement, *about* 4,200 tons. Complement, —.
Length, 450 feet. Beam, 43½ feet. Draught, 16 feet.

Guns (*see Notes*):
8—5·9 inch, 50 cal. (semi-auto.)
2 or 3—22 pdr. anti-aircraft.
Torpedo tubes (19·7 inch):
2 *above water*.
2 *submerged*.
200 mines can be carried.

Armour:
2" Belt (amidships
" Belt (ends) ...
" Deck (amidships
1½-¾" Gun Shields
4" Conning Tower
" Conning Tower

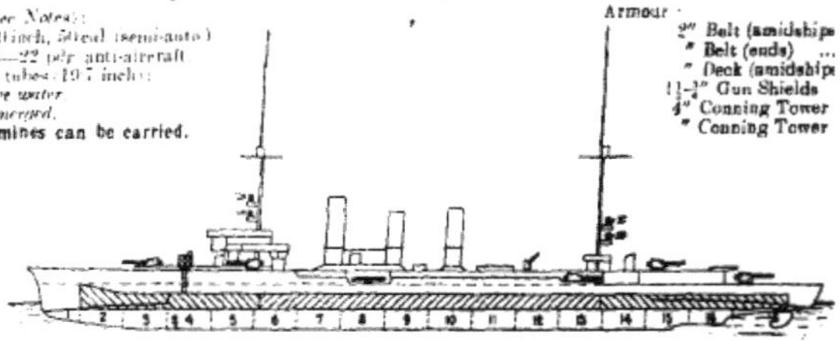

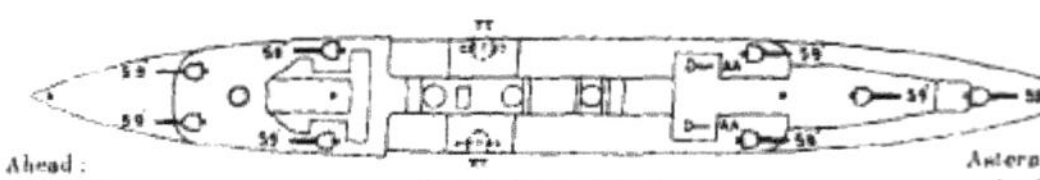

Ahead:
4—5·9 in.

Broadside: 5—5·9 in.

Astern:
4—5·9 in.

Machinery: German "Marine type." screws. Boilers: Schulz-Thornycroft. Designed H. — = 28·5 kts. Coal: *normal*, tons; *maximum*, 1240 tons (+500 tons oil fuel).

Gunnery Notes.—Will probably be re-armed with French 5·5 inch guns.

Name	*Builder*	*Machinery*	*Laid down*	*Completed*	*To be re-commissioned*
Metz	Weser Co., Bremen	Weser Co.	1914	Aug '16	October—December, 1921

General Notes.—Was begun as *Ersatz Gazelle* of German 1914-15 Pro... added to French N...

1920 Photo, H. Freund, Brest.

STRASBOURG (ex-German *Regensburg*, April, 1914).

Normal displacement, 4900 tons. Complement, 364 (as German Ship).

Length (*w.l.*), 456 feet. Beam, 45 feet. Draught, 16 feet *mean*, 17 feet *max*.

Guns (German):
- 7—5.9 inch
- 2—22 pdr. AA.
- 2 machine

Torpedo tubes (19.7 inch German):
- 2 *submerged*.

Mines:
- Fitted to carry 120.

Armour (Krupp):
- 4" Belt (amidships) ....
- 2½" Belt (ends) ....
- 2" Deck (amidships) ....
- 4" C.T. ....

Note.—Is being refitted at Brest D.Y. for service in French Navy. Will probably be re-armed with French Marks of 5.5 inch (14 cm.) guns. To be re-commissioned April—June, 1921.

Machinery: German "Marine type" turbines. Boilers: 14 Schulz-Thornycroft. Designed S.H.P. 26,000 = 27.25 kts. Coal: *normal*, 470 tons; *maximum*, 1300 tons. Oil: 220 tons. Endurance: 5000 to 6000 miles at 15-17 kts.

Notes.—Begun under German 1912 Programme as *Ersatz* ———, by Weser Co., Bremen. Completed, 1914. Surrendered at Cherbourg, June, 1920. Armaments mounted in this ship, when in German Navy, were (a) 12—4.1 inch, 1914-15, (b) 2—5.9 inch, 8—4.1 inch, about 1915-17, (c) 7—5.9 inch, at end of war. *Ancona* (ex-German *Graudenz*) of Italian Navy, is practically a sister ship.

Now has small crow's nest high on foremast, and short mainmast.

1914 *Official* Aus[...]

(IMPROVED SPAUN CLASS).

THIONVILLE (ex-Austrian *Novara*, Feb., 1913).

Normal displacement, 3500 metric tons. Complement, 318.

Length (*w.l.*), 410½ feet. Beam, 42 feet. *Mean* draught, 15 feet. Length *over all*, 4[...]

Guns: (Austrian).
- 9—4·1 inch (50 cal. semi-auto).
- 1—11 pdr. AA.
- 1 machine.
- 1 landing.

Tubes: (Austrian)
- 4—20·7 inch, *U.D.*, twin mounts p. and s.
- 1—*U.D.* over stern for D.C. or Torpedo-Mines.

Armour (steel[...]
- 2½" Belt (a[...]
- 2" Bulkhead[...]
- ¾" Deck.....

Note.—Re-fitting at Bizerta, 1920, for service in French Navy. Will probably be re-armed with F[...]

To be re-commissioned July—September, 1921.

Machinery: Turbine (*see Notes*). 4 screws. Boilers: 16 Yarrow. Designed H.P. [...] Coal: *normal*, 450 tons; *maximum*, 750 tons. Radius of action: 860 miles at 27 kts., [...] 24 kts.

Name	*Builder*	*Machinery*	*Laid down*	*Completed*	*Trials*	*Turbines*
Thionville	Fiume	Ganz-Danubius	Feb.'12	Oct.'14		A.E.G.

Notes.—Built for Austro-Hungarian Navy. Interned at Cattaro, Nov., 1918. Towed to Bizerta, 1920, T[...] near Brindisi while on passage, but she was refloated. *Venezia* (ex-*Saida*) and *Brindisi* (ex-*Helg*[...]) added to Italian Navy.

MULHOUSE. [illegible] Photo, [illegible]

MULHOUSE (ex-German *Stralsund*) (Nov., 1911).

Normal displacement, 4550 tons; 5100 tons, *full load*. Complement, 370.

Length (*w.l.*), 446¼ feet. Beam, 43·6 feet. *Mean* draught, 16¾ feet.

Guns (German):
- 7—5·9 inch, 50 cal.
- 2—22 pdr. anti-aircraft
- 2 machine

Torpedo tubes (19·7 in. German):
- 2 *submerged.*
- 2 *above water.*

Carries 120 mines.

Armour:
- 3¼" Belt (amidships)
- 2¼" Belt (ends) .....
- 2" Deck (amidships)
- 4" Conning tower...

Note.

Will probably be re-armed with French 5·5 inch guns.

To be re-commissioned January—March, 1922.

Machinery: Turbine. 4 screws. Boilers: 16 Schulz-Thornycroft. Designed H.P.: 24,200 = 26·75 kts. Coal: *normal* 750 tons; *maximum* 1200 tons. Oil: 130 tons.

Armour Notes.—Belt is very narrow, and at full load practically submerged.

Engineering Notes.—4 boilers are oil burning.

Name	*Builder*	*Machinery*	*Laid down*	*Completed*	*Trials (mean): 6 hours*	*Trials (mean): Best hour*	*Turbines*	*Best speed*
Mulhouse	Weser, Bremen	Weser	Ap. '10	June '12	27,062 = 26·9	35,515 = 28·27	Bergmann	...

Begun under German 1910 Programme. *Magdeburg* of this class sunk 1914 in the Baltic. **Breslau** mined and sunk off Imbros, 1918. Sister ship *Taranto* added to Italian Navy, 1920. They were very fast ships, but vibrated greatly over 22 kts. Proved bad sea boats on first trials, and large bilge-keels were fitted to lessen rolling. All were heavily forced on trials. Very lightly built, extensive use being made of steel castings and aluminium. Engine-rooms very cramped. Double-bottom shallow, and only carried up amidships to under side of protective deck. Above protective deck are coal-bunkers, about 12 feet wide. *Mulhouse* surrendered at Cherbourg and added to French Navy, Sept., 1920.

COLMAR. 1920 Photo, *Freund.*

COLMAR (ex-German *Kolberg*, Nov. 1908).

Normal displacement, 4350 tons. Complement, 379.

Length (*waterline*), 426½ feet. Beam 46 feet. *Maximum* draught, 17¾ feet.

Guns (German):
- 6—5·9 inch.
- 2—22 pdr. anti-aircraft
- 2 Machine.

Torpedo tubes (17·7 in., German):
- 2 *submerged.*
- 2 *above water.*

Can carry 120 Mines.

Armour (Krupp):
- 2" Deck (amidships)
- ¾" Deck (ends) ...
- 4" Conning tower

Note.

Will probably be re-armed with French 5·5 inch guns.

To be re-commissioned April—June, 1922.

Machinery: Turbines: Melms & Pfenninger. 4 screws. Boilers: 15 Schulz-Thornycroft. Designed H.P. 19,600 = 25 kts. Coal: *normal* 400 tons; *maximum* 900 tons, and some oil fuel.

Name	*Builder*	*Machinery*	*Laid down*	*Completed*	*Trials*	*Turbines*
Kolberg	Schichau	Schichau	Mar., '07	Dec. '09	30,400 = 26·3	Melms & P.

Note.—Surrendered at Cherbourg and added to French Navy, Sept., 1920.

JURIEN DE LA GRAVIÈRE (July, 1899). Displacement, 5690 tons (sheathed Complement 469. Length (*waterline*), 449¼ feet. Beam, 49¼ feet. *Mean* d Guns: 8—6·5 inch, 2—9 pdr., 45 cal., 10—3 pdr. (4—3 pdr. are AA.), 6 M.G. Torped 2 *submerged*. Armour (Harvey Nickel): 3" Deck (cellulose belt), 6½" Conning t shields. 1¾" Casemates. Machinery: 2 sets vertical triple expansion. 3 screws. Bo H.P. *full power*, 18,400 = 22·9 kts. Present speed is poor. Coal: *normal* 560 1000 tons. Laid down at Lorient, Nov. 1897. Completed, 1903. Best recent spee

D'ESTRÉES (Rochefort, Oct., 1899). Displacement 2460 tons (sheathed and coppe 250. Length (*waterline*), 311¾ feet. Beam, 39⅝ feet. *Mean* draught, 17½ feet. Gu 2—5·5 inch, 45 cal., 5—3·9 inch, 45 cal., 1—9 pdr., 8—3 pdr. (6—3 pdr. are AA (*removed*). Armour (hard steel): 1½" Deck (cellulose belt). Machinery: 2 sets 2 screws. Boilers: 8 Normand. Designed H.P. 8500 = 20 to 21 kts. Coal: *maximum* 550 tons. Begun, 1897; completed, 1899.

GUICHEN. 1920 *Photo, H. Freund, Brest.*

GUICHEN (May, 1897). Displacement, 8,280 tons (sheathed). Complement, 560. Length (*waterline*), 436¼ feet. Beam, 54¾ feet. *Maximum* draught, 24½ feet. Guns—(M. '91): 2—6·5 inch, 45 cal., 12—3 pdr.,* 5—1 pdr.* Torpedo tubes: (*removed*). Armour (Harveyed): 2¼" Deck (cellulose belt), 1½" Casemates, 2" Gun shields, 6¼" Conning tower (fore). Machinery: 3 sets 4-cylinder triple expansion. 3 screws. Boilers: 28 Lagrafel d'Allest. Designed H.P. 24,000 (*f.d.*) = 23 kts. Coal: *normal* 1,460 tons; *maximum* 2,030 tons. Laid down at St. Nazaire, 1895; completed, 1899. Placed in Special Reserve, Brest, for refit, 1919-20.

* Possibly all removed.

(*Illustration at head of next column.*)

D'ENTRECASTEAUX (June, 1896). Displacement, 8,123 tons (sheathed and coppered). Complement 521. Length (*waterline*), 393¾ feet. Beam, 58½ feet. *Maximum* draught, 26 feet. Guns—(M. '93): 2—9·4 inch, 40 cal., 2—3 pdr., 4 Maxims. Torpedo tubes (18 inch): 2 *submerged*. Armour (Harvey): 3¼" Armour deck, 9" Turrets, 4" Turret bases, 2¾" Casemates (8, all chrome steel), 10" Conning tower. Machinery: 2 sets triple expansion. 2 screws. Boilers: Cylindrical (5 double-ended). Designed H.P. 14,500 = 19·5 kts. (*never made*); present speed 18 kts. or *less*. Coal: *normal*, 650 tons; *maximum*, 950 tons. Laid down at La Seyne, 1894, completed, 1899. Now serves as Flagship, Atlantic Training Ships, and as Training Ship for W.O., C.P.O., &c. Is probably only used for Harbour Duties now, and is no longer a Sea-going Vessel.

D'ENTRECASTEAUX. 1920 *Photo, H. Freund, B*[illegible]

(*Described at foot of preceding column.*)

CASSARD (1896), **DU CHAYLA** (1895). Old Cruisers of about 3,950 tons. Complement, [illegible] Length (*waterline*), 324½ feet. Beam, 45 feet. *Maximum* draught 22 feet. Guns (M. '87): [illegible] 6·5 inch, 45 cal., 4—3·9 inch, 8—3 pdr., 2—14 inch tubes above water. *Armour*: 2" Deck [illegible] Sponsons to the four main deck 6·4 inch guns. H.P. 10,000 = 19·8 kts. *Du Chayla* could [illegible] make 13—14 kts. 1920. Coal: *maximum*, 624 tons. Boilers: Lagrafel d'Allest. *Cassard* [illegible] Rochefort D.Y. 1920. Both begun at Cherbourg D.Y. 1893.

CALAIS (mast *before* funnel).

Appearance Note.—Mast may be either before or abaft funnel (stepped to starboard of centre line), as above views of *Calais* and *Dunkerque*. All ships by one yard have same appearance, *e.g.*, all La Seyne boats as *Calais*; all Brest D.Y. boats as *Dunkerque*.

COUCY.

DUNKERQUE (mast *abaft* funnel).

("ARRAS" CLASS—29* SHIPS.)

*In July, 1920, 14 in service, 12 *completing*, 3 *building*.

7 built at Navy Yards:—

ARRAS (July, 1918), **DUNKERQUE** (July, 1918), **REIMS** (November, 1918), and (November, 1918), all four built by Brest D.Y. **NANCY** (Cherbourg D.Y., M **BAR-LE-DUC** (November, 1918) and **BELFORT** (March, 1919), both by L'

22 built by Private Yards:—

LASSIGNY (July, 1919) and **LES EPARGES** (September, 1919), both by Ch. Nantes. **REMIREMONT** (July, 1919) and **REVIGNY** (September, 1919), bo la Gironde, Bordeaux. **TAHURE** (——, 1919) and **TOUL** (April, 1919), bot la Loire, St. Nazaire. *BACCARAT* (*building*) and *BETHUNE* (*building*), bot Provence, Port du Bouc. **COUCY** (June, 1919), **EPINAL** (August, 1919), V (August, 1919), **VIMY** (December, 1919) and **VITRY-LE-FRANCOIS** (Mar five by Ch. de St. Nazaire, Penhoet. **AMIENS** (May, 1919), **CALAIS** (Nove **CRAONNE** (January, 1920), and **LIEVIN** (March, 1920), all four by F. et C La Seyne. **EPERNAY** (September, 1919), **LUNEVILLE** (Jan, 1920), **MON** (June, 1920), *MONTMIRAIL* (*building*) and **PERONNE** (March, 1920), all C. de la Med., Graville.

Despatch and Anti-Submarine Vessels of 700 tons. Dimensions: 246 ft. 0 in. × 2 9 ft. 2 in. Guns: 2—5.5 inch, 1—14 pdr. AA., 4 M.G. Carry 20 depth cha with four searchlights; also have hydrophones. Designed S.H.P. 5000 = 20. burning boats; = about 18-19 kts. in coal-burning boats. Machinery: 2 (geared) turbines. 2 screws. Boilers: 2 Normand or Guyot du-Temple oil-b tube). Fuel: 200 tons oil = Endurance: 1000 miles at 17 kts. *Craonne*, *Li* others for Foreign Stations) will burn coal (about 175 tons carried). Comple

Notes.—5.5 inch guns reported to be M.1910 (and perhaps M.1916), but some said armed with old 138 m/m. guns re-bored to 140 m/m. 5.5 inch have large degree *max.* range 17,000 metres. They are very roomy and comfortable ships for their high bows make them very dry in head seas. But with sea on the beam, to roll "like old boots" (*comme veritable sabots*), owing to heavy topweight o structure, etc. Oil-burning boats can do 21-22 kts. on light draught in good in a seaway they drop down to 13 kts. Built under Programme VI and VII following thirteen were cancelled 1918-19:—*Betheny*, *Compiegne*, *Chalons*, *Ch* *Douaumont*, *Gerbevillers*, *La-Fère Champenoise*, *Noyon*, *Roye*, *St. Dir*, *Senlis*, *Soi* *Bapaume* (30th ship of class), converted to Aircraft Carrier, and listed on a la

SCARPE (with Yacht bow). 1920 Photo, M. Bar, Toulon.

ESCAUT (AILETTE same). The 3·9 inch guns are concealed within lidded ports at angle superstructure.

ANCRE. Appearance uncertain.

SUIPPE (straight stem). 1920 Photo, M. Bar, Toulon.

AILETTE (Brest D.Y., 1918). **ANCRE** (Lorient D.Y., 1918). **ESCAUT** (Brest D.Y., 1918). **SCARPE** (Lorient D.Y., 1918). **SUIPPE** (Lorient D.Y., 1918).

Despatch Vessels of 650 tons. Dimensions: 256 feet 2 inches × 26 feet 5 inches × 8 feet 10 inches. Guns: 4—3·9 inch, 2—9 pdr. A.A., one or two M.G. Designed H.P. 5000 = 21 kts. Machinery: reciprocating. Boilers: two (type unknown). 2 screws. Oil fuel: 135 tons = 4000 miles at 10 kts. Complement 65.

Note.—Modifications of *Aisne*, *Marne*, *Meuse*, &c., described on a later page. Externally they are disguised by varying appearances as mercantile vessels, like convoy sloop *Ville D' Ys* below. Built under 1917 War Programme.

Photo wanted.

(See Silhouette for appearance.)

VILLE D' YS (ex-*Andromède*), Swan Hunter, Wigham Richardson & Co., Wallsend-on-Tyne, 1917. Convoy sloop of 1490 tons. Dimensions: 255 feet 3 inches (*p.p.*) 276 feet (*o.a.*) × 3[illegible] × 12 feet 3 inches. Guns: 2—5·5 inch, 3—14 or 9 pdr. Designed I.H.P. 2500 = 17 [illegible] Machinery: 1 set 4-cyl. triple-exp. Boilers: 2-cyl. 1 Screw. Coal: 270 tons = 2400 mil[illegible] 12 kts. Complement, 168 (war, probably inclusive of "panic" parties), about 90 (peace).

Note.—Generally same build as British convoy sloops. Built under 1916 War Programme. Repor[illegible] have been begun for British Navy as *Andromeda* and turned over to French Navy.

(*Continued on next p*[illegible]

ALGOL.

ANTARÈS.

("ETOILE" or "STAR" Class—7 Ships.)

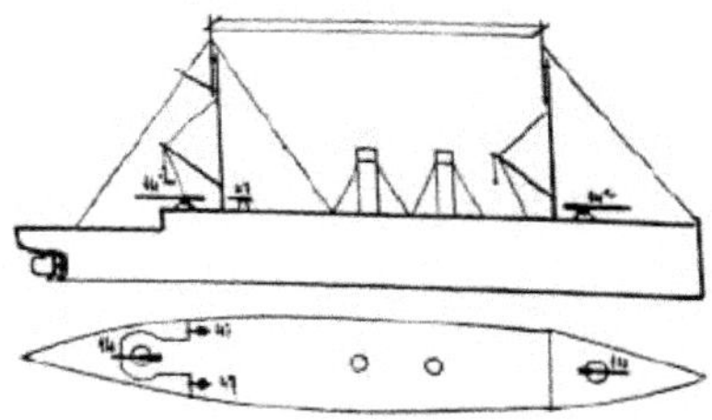

(Plans* by courtesy of Ministry of Marine.)

* On these 14 cm. = 7·5 inch, 47 mm. = 6 pdr.

ALDEBARAN, **ALGOL** (Barclay, Curle, Whiteinch, N.B., 1916).
ALTAIR, **ANTARÈS** (Wm. Hamilton, Port Glasgow, N.B., 1916).
BELLATRIX (Henderson, Glasg
CASSIOPÉE, **REGULUS** (Barclay, Curle,

Sloops, displacing 1250, 1420 and 1470 tons. Dimensions: 225¼ × 33½ × 11- 2—5.5 inch, 2—6 pdr. A.A. Are fitted as Fleet Sweeping Vessels. Designe 17 kts. (All made 17.2-17.6 kts. on trials.) Machinery: 1 set, 4 cyl. Boilers: Not known. 1 screw. Coal: 260 tons = 2400 miles at 12 kts. or 80.

Notes.—These sloops are practically replicas of the British "Flower Class" Sl in pairs by British shipyards. *Rigel*, the companion boat to *Bellatrix* by M lost during the War. All built under 1916 War Programme.

(Contin

OISE (& SOMME). 1919 *Photo.*

YSER (& MEUSE). 1920 *Photo. M. Mar.*

Note.

Though the design of these Despatch Vessels is uniform, the latitude allowed to the bui has resulted in differing appearances. Vessels built by same yard have the same a

AISNE (Lorient D.Y., 1916).
MARNE (Lorient D.Y., 1916).
MEUSE (Rochefort D.Y. 1918).,
SOMME (Brest D.Y., 1917).
OISE (Brest D.Y., 1917).
YSER (Rochefort D.Y., 1917).

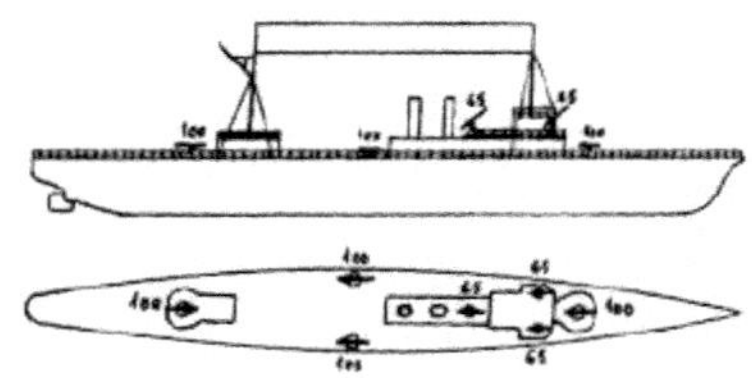

(Plans by courtesy of the Ministry of Mari

MARNE (& AISNE). 1919 *Photo.*

Despatch Vessels, displacing 650-680 tons. Dimensions: 256¾ × 26$\frac{1}{5}$ × 8¾ feet. Ar 4—3·9 inch (M. 1916, semi-auto), 3—9 pdr. AA., 2—6 pdr. Designed I.H.P. 4000 kts. Machinery: Reciprocating. Boilers: 2 (types unknown). 2 screws. Oil fuel tons = 4000 miles at 10 kts. Complement, 97. All built under 1916 War Program

*On Plans, 100 mm. = 3·9 inch; 65 mm. = 9 pdr.

Continued on next

"Avisos"—Despatch Vessels or Sloops (*Continued*).

No Photo available.
Clipper bow, 1 funnel, 2 pole masts.

CHAMOIS (Port de Bouc, 1905). Despatch Vessel of 431 tons. Dimensions: 168¾ × 25¼ × 9¼ feet. Guns (*not known*). H.P. 600 = 12 kts. Boilers: du-T. Guyot. Coal: 70 tons. Training Ship for Navigation.

Despatch Vessel *Quentin Roosevelt*, built 1915-18, has been assigned to Fisheries Protection Duty, and is listed with Miscellaneous French Ships on a later page.

The Ministry of Marine says the old Sloop *Décidée* still exists, but "*Fighting Ships*" has been informed from other sources that she was sunk during the War and present *Décidée* is only a Trawler.

Gunboats (*Cannonières*).

Note.—Gunboats *Coquette, Guerrière, Heroine, Joyeuse, Mutine*, were cancelled 1918-19.

1919 *Photo.*

DUMONT D'URVILLE. "DUBOURDIEU" TYPE.—5 steam-engined boats.

DUBOURDIEU (April, 1918). **DUCHAFFAULT** (September, 1918). **DUPERRÉ** (December, 1918). **DUMONT-D'URVILLE** (November, 1918). **DU COUËDIC** (April, 1919). *Normal* displacement, 500 tons. Dimensions: 210.8 (*o.a.*) × 26.2 × 8.75 feet (*max.* draught). Guns: 1—5.5 inch, 1—3.9 inch. S.H.P. 2000 = 17.4 kts. Machinery: 2 sets of Breguet de Woech turbines. 2 screws. Boilers: . Oil: 143 tons = 1985 miles at full speed. Complement, 72.

Notes.—Built under 1917-18 War Programme. Completed June, 1919—May, 1920. Sixth boat of this class, *Dècres*, cancelled. Are also rated as "Anti-Submarine Gunboats" (*Cannonières Contre Sous-marins*). In build they are very like *Arras* class Sloops.

Gunboats (*Continued*).

CONQUERANTE.

VAILLANTE type:—2 Diesel-engined boats.

CONQUERANTE (1917), **VAILLANTE** (1917). Both by Brest D.Y. 457 tons. *about* 211 × 22$\frac{7}{12}$ × 8¼ feet. Guns: 2—3.9 inch. Carry D.C. B.H.P. 18[illegible] Machinery: 2 sets of 900 B.H.P. Sulzer-Diesel engines. Oil fuel: 30 tons. Co[illegible]

Notes.—Are a modified *Friponne* design, for which see below. Both built and [illegible] Programme. Fitted as Mine Sweepers.

ENGAGEANTE. 1921 *Photo.*

FRIPONNE type:—3 Diesel-engined boats.

DILIGENTE (Brest D.Y., 1916), **ENGAGEANTE** (Brest D.Y., 1917), **SU**[illegible] (Brest D.Y., 1916). Displacement: 355 tons. Dimensions: 199⅞ × 22$\frac{7}{12}$ × [illegible] 2—3.9 inch. Carry D.C. and may also be equipped as Mine Sweepers. B.H.[illegible] Machinery: 2 sets 450 B.H.P. Sulzer-Diesel engines. Oil: 30 tons = 3000 [illegible] 1600 miles at 15 kts. Complement, 50.

Notes.—Generally the same design as the *Ardent* type (steam-driven) Gunboat[illegible] but above three Gunboats have Diesel engines. All built under 1916 W[illegible] *Chiffonne, Friponne, Impatiente* and *Mignonne* sold to Rumania, January.

(*Continue*

(ARDENT type—20 Steam-engined boats.)

ARDENT.

1919 *Photo.*

DEDAIGNEUSE.

1919 *Copyright Photo, M. Ba*

AGILE (Brest D.Y., 1916).
ALERTE (Rochefort D.Y., 1916).
ARDENT (Brest D.Y., 1916).
AUDACIEUSE (Port de Bouc, 1917).
BATAILLEUSE (Port de Bouc, 1917).
BELLIQUEUSE (Bordeaux, 1917).
CAPRICIEUSE (Nantes, 1916).
CURIEUSE (Lorient D.Y., 1916).
DEDAIGNEUSE (Bordeaux, 1916).
EMPORTÉ (Ch. de la Loire, St. Nazaire, 1916).
ETOURDI (Lorient D.Y., 1916).
EVEILLE (La Seyne, 1917).
GRACIEUSE (Lorient D.Y., 191
IMPETUEUSE (Ch. de la Giron deaux, 1917).
INCONSTANT (Brest D.Y., 191
LURONNE (Brest D.Y., 1917).
MALICIEUSE (Chantiers de I Port de Bouc, 1916).
MOQUEUSE (Lorient D.Y., 191
SANS SOUCI (Lorient D.Y., 19
TAPAGEUSE (Port de Bouc, 191

Displacement: 350 tons. Dimensions: 197¼ × 22$\frac{1}{?}$ × 8½ feet. Guns: 2—3.9 inch, D.C. Designed I.H.P. 1800-1200 = 17-15 kts. Machinery: Reciprocating. Various water-tube types. Fuel: 85 tons coal = 3000 miles at 10 kts., 1600 miles speed. Complement, 55.

Notes.—Built as Anti-Submarine Gunboats or "Cannonières Contre-Sousmarins." At t of the War, many of these boats were converted for Mine Sweeping Service. Som been engined with machinery stripped from old Coastal Torpedo Boats ("Torpille Défense Mobile"). They vary a little in build, but the illustrations given are belie be typical of these boats. Note that masts are stepped well to starboard of the cent *Audacieuse, Batailleuse, Impetueuse* and *Luronne*, 1917 War Programme; all others War Programme. As they were only built for the duration of the war, they are not ex to last long. Four were scrapped in 1920, after a life of only two or three years.

BOUDEUSE.

1919 *Copyright Photo, M. Bar, Toulon.*

River Gunboats (*Canonnières Fluviales*).

Note.—All boats in this column built for service abroad.

Projected.

TWO NEW RIVER GUNBOATS (not yet named). To be built at Toulon D.Y. About 190 tons. To replace *Argus* and *Vigilante*, condemned, 1920.

Completing.

BALNY (1914). Built by Chantiers de Bretagne, Nantes. 244 tons. Dimensions: 167¼ × 22 × 3¼ feet. H.P., 800 = 14 kts. Guns: 1—12 pdr., 2—1 pdr., 4 M.G. Fouché w.t. boilers. Coal: 27 tons.

Photo, Ch. de Bretagne.

DOUDART DE LA GRÉE (1909). 214 tons. 167¼ × 22 × 3¼ feet. Guns: 6—1 pdr. H.P. 920 = 14 kts. Coal: 27¼ tons. Complement, 57. Built by Chantiers de Bretagne, Nantes.

River Gunboats (*Continued*).

For service abroad.

No photo available.

PEI-HO (1901). 123 tons. Speed, 10 kts. Guns: 3—1 pdr. May be condemned soon, if not already scrapped.

Rhine and Danube Flotillas.

On Danube two Submarine Chasers and three Vedette Boats.

4 small "Chaloupes à moteur" *P I—P IV* were completed 1920, for service on the Rhine. Each armed with 2 M.G. Have petrol motors. No other details known.

Old Torpedo Gunboat (*Contre-Torpilleur*).

Copyright photo, G.S.G.

LA HIRE (1898). Displacement, 900 tons. Complement, 148. Length, 264½ feet. Beam, 27⅞ feet. *Maximum* draught, 12⅞ feet. Guns: 2—3·9 inch, 4—9 pdr., 2—3 pdr. No tubes. Designed H.P. 6400 = 23 kts. (never made), now 19-20 kts. Coal: 160 tons. Boilers: Normand.

Note.—Serves as Gunnery Training Ship. Armament given that mounted during 1920, but she has no definite armament, as guns may be changed from time to time for instructional purposes.

1 + 6 projected **Destroyers.***

	Class.	First Begun.	Last Completed	Normal Displacement	I.H.P.‡	Max. Speed.‡	Max. Coal or oil ‡	Complement.	Max. Draught.	Tubes
				tons.		kts.	tons.		feet.	
6	*(New)*	*Pro.*		...	...	...	...	...	...	...
1	*Amiral Sénès*†	1916	1918	2458	55000 *t*	36·9	720 *oil*	...	$11\frac{3}{4}$	4
2	*Delage class*†	1917	1919	990	23800	33·3 *t*	332 *oil*	...	$9\frac{3}{4}$	6
4	*Deligny class*†	1916	1918	910	23800	33·74 *t*	305 *oil*	...	$8\frac{3}{4}$	6
1	*Bainot*†	1916	1918	907	25150	36·6 *t*	330 *oil*		$9\frac{1}{2}$	6
1	*Pierre Durand*†	1915	1917?	950	22600	34·4 *t*	317 *oil*	...	$9\frac{1}{4}$	6
1	*Matelot Leblanc*†	1915?	1917?	836	17000	32·5 *t*	95 *coal and oil*	102	9	4
12	*Tribal class*	1917	1917	690	10000	29	100+120	87	$9\frac{7}{8}$	4
4	*Temeraire class*	1910	1914	950	18000	32 *t*	240+92	102	$10\frac{1}{2}$	4
1	*E. Gabolde*	1913	1921§	905	20000	31 *t*	200	...	10	6
12	*Bory class*	1910	1915	880-780	17000-13500	31-30 *t*	200-140	83	10	4
3	*Casque class*	1908	1912	820-790	14400-13000	31 *t*	140	83	10	4
8	*Hazzard class*	1906	1912	514-407	8000-7200	28 *t & tr*	100-110	77-69	10	3
10	*Claymore class*	1903	1911	415-340	6400	$30\frac{1}{2}$-28	80	71	$11\frac{1}{2}$	2
8	*Mousquet class*	1900	1904	370-310	7300-6000	$30\frac{3}{4}$-28	75	71	$10\frac{3}{4}$	2
3	*Flamberge class*	1900	1902	310	5600-4800	27-26	70	71	$10\frac{3}{4}$	2
2	*Durandal class*	1899	1901	305					$8\frac{1}{2}$	2
1	*Pique class*	1900	1902	310	5700	26	70	71	$8\frac{1}{2}$	2

t = turbines. *tr* = turbines and reciprocating.

* Owing to prospective condemnation of boats in last six classes, total of completed boats is only approximate.

† Ex-Enemy boats taken over 1920.

§ *E. Gabolde* not yet finished.

‡ Details for ex-German boats in these columns are *trial* figures.

Note to ex-Enemy Destroyers.

(v. next two pages.)

On completion of refits, these boats will be commissioned for service in the French Navy on the following dates:—

Amiral Sénès	April—June, 1921
Delage / *Rogout de la Touche* }	April—June, 1921
Chastang	October—December, 1921
Deligny	July—September, 1921
Mazaré	October—December, 1921
Vesco	July—September, 1921
Bainot	January—March, 1922
Pierre Durand	January—March, 1922
Matelot Leblanc	July—September, 1921

6 Boats: Type unknown.

(May be contracted for as *M* [illegible]—*M* [illegible].)

These boats are projected, but not yet begun. No decision appears to have been arrived at, as regards their design. Alternative plans have been prepared of three types displacing 1800, 2000 and 2300 tons. Maximum speed: 36 kts. Armament: 4—3.9 inch and 9—21.6 inch tubes (in three triples), all guns and tubes being on the centre line. Design is said to have been adopted by the Conseil Supérieur, September, 1919, but may be changed or abandoned before the boats are begun.

Plans have also been considered for a type of "Contre-Destroyer" of 2500 tons armed with 5—5.5 inch, 4—3 [illegible] pdr. AA. and 4—21.6 inch tubes. Speed: 30 or 36 kts.

Possible that both types may be built, the "Contre-Destroyers" for Ocean Duties and the smaller boats for Fleet and Coast Service.

1 ex-German Boat.

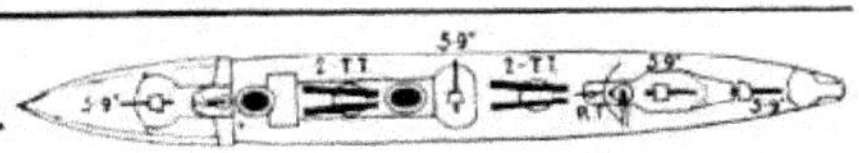

1 *Schichau, Danzig*: **Amiral Sénès** (ex German *S 113*, completed May, 1918). 2485 tons. Dimensions: 360 [illegible] × 14[illegible] feet draught aft. Guns: 4—5.9 inch (45 cal. "Flak"). Torpedo tubes: 4—21.6 inch, in two twin deck mountings. Machinery: Turbine. Boilers: 4 "Marinetyp" (Schulz-Thornycroft). Designed S.H.P. 55,000 = 36.9 kts. (trials, 34 (at sea)). 2 screws. Oil fuel: 210 tons *normal*, 720 tons *max.* [illegible] miles at [illegible] kts.

Gunnery Notes.—5.9 inch have large degree of elevation and can be used as A.A. guns. May be replaced by French 5.5 inch models. German directors, fire controls, R.F. and electric transmitters, etc., stripped before surrender.

Torpedo Notes.—Tubes 32 feet long, electrically operated. Axes diverge 15°.

General Notes.—Dimensions given above are open to question. French reports give 334.6 × 34.1 × 12.8 feet mean draught. Originally a unit of the German "1916 Design" Class (*S 113*—*B 123*). A sister boat (*V 116*) now Italian *Premuda*. The class represents the largest and most powerful Destroyer type extant, but their sea performance is said to be very bad indeed. Owing to the tremendous topweight of guns and tubes, they are said to roll almost to a capsizing angle. The Germans are said to have lengthened these boats by about 10 feet at bows in an attempt to improve stability. French reports also say they never got beyond 30 kts. on trials.

2 Delage Class (ex-German "Mobilisation" type).

R. DE LA TOUCHE. From "Die Technik im [illegible]"

2 *Howaldt* boats: **Delage** (ex-German *H 147*) and **Rageot de la Touche** (ex-German *H 146*). 279.8 × 27.4 × 9.3 feet. Guns (German)*: 3—4.1 inch, 45 cal. "Flak" (AA. mountings). Torpedo tubes*: 6—19.7 inch, in two single tubes to port and starboard, under bridges, and two twin mountings a[illegible] [illegible] fuel: 166 tons *normal*, 320 tons *max.* = 1840 miles at 20 kts. Surrendered at Cherbourg and ad[illegible] Navy, September, 1920. (Other details as Table.) Were units of German *H* [illegible] class.

* Note to *Delage* Class (and the *Deligny*, *Bisson*, *Pierre Durand* Classes hereafter):—All these bo[illegible] surrendered without torpedo tubes and with some (or none) of the guns listed above.

To distinguish.—Boats carried "flying" derrick post close to *after* funnel with boom slung forw[illegible] platform before amidships bandstand.

4 Deligny Class (ex-German "Mobilisation" type).

1920 Photo.

4 *Schichau* (Elbing) boats: **Chastang** (ex-German *S 156*), **Deligny** (ex-German *S 1[illegible]*), **Mazé** [illegible] *S 1[illegible]*) and **Vesco** (ex-German *S 1[illegible]*). Dimensions: 272.3 × 27.3 × 8.4 feet. Guns and to[illegible] *Delage* class above. Oil fuel: 162 tons *normal*, 305 tons *max.* = 1530 miles at 20 kts. Surrende[illegible] and added to French Navy, September, 1920. Other details as Table. Were units of German [illegible]

To distinguish.—Derrick post close to *fore* funnel with boom slung aft. No compass platform.

(Continued

1920 *Photo, G. Dott, Esq.*

1 Vulcan (Stettin) boat: **Bulno** (ex-German *V 130*). Dimensions: 269 × 27·2 × 9·2 feet. Guns and torpedo tubes: As *Delage* class on preceding page. Oil: 167¼ tons *normal*, 330¼ tons *max.* = 1695 miles at 20 kts. Other details as Table. Surrendered at Cherbourg and added to French Navy, September, 1920. Was unit of German *V 125*—*130* class.

To distinguish.—Derrick-post *midway* between funnels, with boom sling aft. Ventilator and compass platform just aft of second funnel, *plus* long unbroken forecastle.

1 Vulkan Boat (ex-German "Mobilisation" type.)

1920 *Photo, G. Dott, Esq.*

1 Vulkan boat: **Pierre Durand** (ex-German *V 79*). Dimensions: As *Bulno*, above. Guns and torpedo tubes: as *Delage* class, on preceding page.* Oil: 160¼ tons *normal*, 317¼ tons *max.* = 1810 miles at 20 kts. Other details as Table. Surrendered at Cherbourg and added to French Navy, September, 1920. Was unit of German *V 67*—*84* class.

*Was also fitted to carry 24 mines when first completed.

To distinguish.—Single forward tubes are mounted in well before bridges, and, between these tubes, arched ventilating trunk. Ventilator and compass platform just aft of second funnel, *plus* short forecastle. Derrick-post *midway* between funnels, with boom sling aft.

1 Ganz-Danubius Boat.

1 *Ganz-Danubius* (Fiume) boat: **Matelot Leblanc** (ex-Austrian *Dukla*). Dimensions: 275 × 25·7 ... tons (Austrian): 2—3·9 inch, 4—11 pdr., 2—11 pdr. AA. Torpedo tubes (Austrian): 4—20·8 in ... details as Table. Towed from Cattaro to Bizerta and added to French Navy, September, 1920. W... Austrian *Tatra* class. Seven sister boats incorporated in Italian Navy.

12 Tribal Boats.

(1917 War Programme)

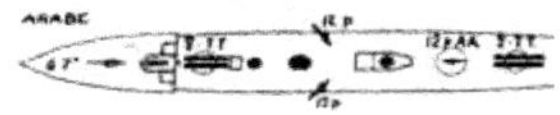

TONKINOIS 1920 *Copyright Photo, M. Bar, Toulo...*

12 built by Japanese Yards: ***Algerien, Annamite, Arabe, Bambara, Hova, Kabyle, Maro... Sakalave, Senegalais, Somali, Tonkinois, Touareg.*** Built by Kawasaki Co., Mitsu Bishi Co... other Japanese Yards. All begun March, 1917, launched May and July, 1917, and completed July and Septe... 1917. Average time of construction 5 months. 690 tons. Dimensions: 260 (*p.p.*), 271¼ (*o.a.*) × 27¼ × 7... *Normal* draught, 9⅞ feet; *load* draught, Armament: 1—4·7 inch, 4—12 pdr. (one 12 pdr. is AA.... 4—18 inch torpedo tubes in two twin deck mountings. Designed H.P. 10,000 = 29 kts; made about 30 on... Machinery: 3 sets 4-cyl. triple expansion. 4 Kansei boilers. Fuel: 100 tons coal + 120 tons oil = 3000 mi... 15 kts. and 950 miles at full speed. Complement, 87.

Note.—These boats are practically replicas of the ...

4 Temeraire type:—Purchased Boats.

INTRÉPIDE (*Aventurier* also has 3 funnels, but no S.L. on mainmast). 1920 *Photo, Freund, Brest.*

4 *Chantiers de Bretagne* (*Nantes*): **Aventurier** (ex-*Salta*), **Intrépide** (ex-*Rioja*), **Opiniâtre** (ex-*San Juan*), and **Temeraire** (ex-*Mendoza*) (all 1911). *Normal* displacement: 950 tons; *full load*, 1180 tons. Dimensions: 281½ × 28½ × 8¾ feet *normal* draught. *Full load* draught, 10¼ feet. Armament (Schneider-Canet): 4—3·9 inch, 2—3 pdr. AA., 4—18 inch tubes. Designed H.P. 18,000 = 32 kts. Machinery: Rateau-Bretagne turbines + 5 White-Forster boilers; 4 burn coal or oil and 1 burns oil only. Fuel capacity: 232 tons coal + 82 oil. Radius of action: 3,000 miles at 15 kts.

Note.—Built for Argentina and purchased by France immediately before outbreak of war.

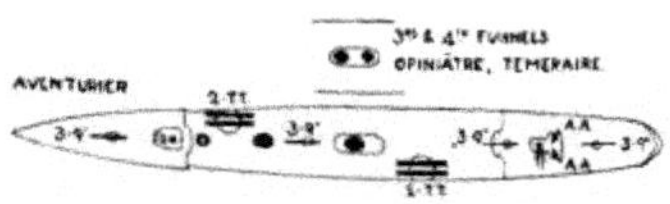

OPINIÂTRE. 1920 *Photo, Ba*

Note.—TEMERAIRE 4 funnels, as *Opiniâtre*, but also has raised fore funnel like *Aventurier* and *Intré*

OPINIÂTRE

1 Special Boat (1913 Programme)

(Building or Completing).

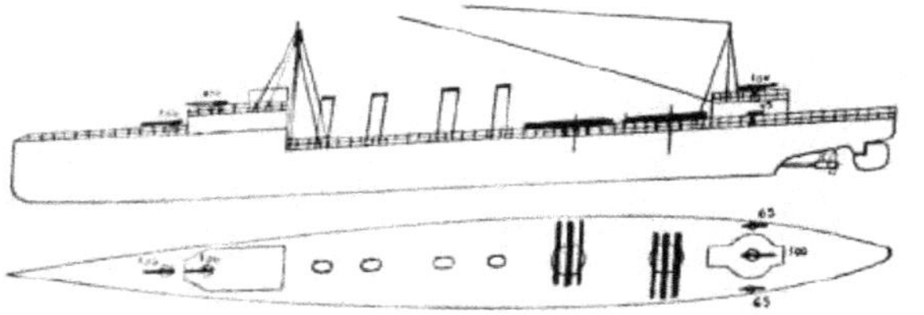

Plans by courtesy of French Ministry of Marine.

1 *Normand, Havre:* **Enseigne Gabolde** (1921). 900 tons. Dimensions: 269 × 26·9 × 10 feet. Armament: 3—3·9 inch, 2—9 pdr. (or 1—14 pdr. AA.), 6—18 inch torpedo tubes in two triple deck mountings. Designed H.P. 20,000 = 21 knots. Parsons geared turbines. Normand boilers. Fuel: 200 tons.

Note.—On plans 100 = 3·9 inch and 65 = 9 pdr. guns, but 9 pdr. guns may be replaced by 1—14 pdr. AA. gun mounted between funnels. Work on this T.B.D. was suspended during the War. Begun in December, 1913; launch not yet reported. Will run trials about April, 1921.

12 Bory Type.

General Note.—All these boats are armed with 2—3·9 inch, 4—9 pdr., 1—2 pdr. AA., 4—18 inch tubes in pairs. All adapted for mine sweeping and carry D.C. Endurance: about 1,500 miles at 15 kts. and 400–500 at full speed. Complement, 85. They differ according to builders in appearance, dimensions, H.P., turbines, &c.

War Losses: *Boutefeu, Faulx, Fourche, Renaudin.*

12 Bory Class (*continued.*)

(1913 Programme).

Mec. Lestin. (Note pennant ? and small double foremast.) 1920 Photo, *M. Bar, T.*

2 *Rochefort Dockyard:* **Enseigne Roux** (1915), **Mecanicien-Principal Lestin** (1915). 880 tons. Dimensions: 271 × 29·9 × 10 feet. Designed H.P. 17,000 = 30 kts. Fuel: 200 tons oil. Endurance for these two boats:—about ... miles at 15 kts. and 550 miles at full speed. Begun as M86 and M87.

**Lestin* has normal foremast.

(1911 Programme).

Note strutted foremast. 1919 Photo, *Commr. R. D. Holberton, R.*

1 *Toulon Dockyard:* **Commandant Lucas** (1914). 800 tons. Dimensions: 272·4 × 26 × 10 feet. Laval Turbines. Du Temple-Guyot boilers. Designed S.H.P. 15,000 = 31 kts. Fuel: 140 tons oil. Practically sister-boat to *Protet* and *Bisson* on next page. Begun as M78.

(*Continued on next.*)

12 Bory Type.—*Continued.*

All armed with 2—3·9 inch, 4—9 pdr., 1—2 pdr. (A.A.), and 4—18 inch tubes in pairs. Endurance: about 1500 miles at 15 kts., and 400-500 miles at full speed. All equipped for Mine Sweeping, and carry D.C.
(With *C. Lucas* on preceding page, three boats below built under 1911 Programme.)

1919 *Photo, Comm. B. D. Holbertson, R.N.*

1 *Rochefort Dockyard*: **Protet** (1913). 800 tons. Dimensions: 272.4 × 26 × 10 feet. Laval turbines. 2 du Temple Guyot boilers. Designed H.P. 15,000 = 31 kts. Fuel: 140 tons oil. Practically sister-boat to *C. Lucas* and *Bisson*. Begun as M87.

1918 *Photo.*

1 *Ch. de Bretagne (Nantes)*: **Magon** (1913). 851 tons. Dimensions: 272.4 × 25.6 × 10 feet.* Rateau-Bretagne turbines. Du Temple Guyot boilers. Designed S.H.P. 15,000 = 30 kts. (32.2 on trials). Fuel: 150 tons oil. Begun as M 79.

* As given by Ministry of Marine, 1919. Builders say dimensions are 243 · 27 · 8 ft.

1919 *Photo, Comm. B. D. Holbertson, R.N.*

1 *Schneider & Cie (Creusot)*: **Mangini** (1913). 800 tons. Dimensions: 253½ × 25.7 × 10 feet. Designed H.P. 15,000 = 30 kts. Zoelly turbines. Du Temple-Guyot boilers. Fuel: 160 tons oil. Begun as M 80.

12 Bory Type.—*Continued.*
(1910 Programme Boats).

No Photo available.

1 *Normand (Havre)*: **Francis Garnier** (1912). 780 tons. Dimensions: 244½ × 25 × 10½ feet. 13,500 = 31 kts. Turbines: Parsons. Boilers: Normand. Fuel: 148 tons oil.

1919 *Photo, Comm. B. D. Ho*

1 *Toulon D. Y.*: **Bisson** (1912). 800 tons. Dimensions: 272.4 × 26 × 10 feet. Laval turbines Guyot boilers. Designed S.H.P. 15,000 = 31 kts. Fuel: 140 tons oil. Practically sister-boat *Protet.*

DEHORTER.

1918 *Copyright Photo*

1 *Ch. de Normandie (Rouen)*: **Dehorter** (1912). 870 tons. Dimensions: 248½ × 25½ × 10 fee 14,000 = 31 kts. Turbines: Parsons. Boilers: Normand. Fuel: 140 tons oil.

(Contin

12 Bory Type—*Continued.*

(1910 Programme Boats.)

All boats given on this page are armed with 2—3·9 inch, 4—9 pdr., 1—2 pdr. AA. guns, 4—18 inch torpedo tubes, in pairs. All adapted for mine-sweeping and carry D.C. For other details see general notes on preceding page.

No photo available.

1 *Soc. de la Loire (St. Nazaire)*: **Capitane Mehl** (1912). 800 tons. Dimensions: 256½ × 25¼ × 10 *feet.* Turbines: Parsons. Boilers: Normand. Designed S.H.P. 15,000 = 31 kts. Fuel: **140** tons oil.

No photo available.

1 *Soc. de la Gironde (Bordeaux)*: **Commandant Riviere** (1912). 800 tons. Dimensions: 256½ × 25¼ × 10 feet. Breguet turbines, Du Temple-Guyot boilers. Designed S.H.P. 14,500 = 31 kts. Trials: 15,760 = 32·5 kts. Fuel: 140 tons oil.

1920 Photo, M. Bar, Toulon.

1 *Dyle & Bacalan (Bordeaux).* **Commandant Bory** (1912). 750 tons. Dimensions: 253¼ × 25¼ × 10 feet. Schneider-Zoelly turbines, Du Temple-Guyot boilers. Designed S.H.P. 14,1[illegible] = 31 kts. Fuel: 140 tons oil.

3 Casque Type (1908 Programme).

Generally identical with preceding *Commandant Bory* type. All armed with 2—3·9 inch, 4—9 pdr., 1 [illegible] guns, with 4—18 inch torpedo tubes in pairs. All three boats equipped for mine sweeping. Complement, [illegible] fuel 140 tons oil only. Endurance: about 1500 miles at 15 kts. 400 miles at full speed.

BOUCLIER.

1918 Copyright Photo, M. Bar, T[illegible]

1 *Normand (Havre).* **Bouclier** (1911). 780 tons. Dimensions: 237 × 24·9 × 9·4 feet. Designed H.P. 14,000 = 31 [illegible] on trials). Turbines: Parsons. Boilers: Normand. Fuel: 140 tons oil.

1917 Photo, Commr. B. D. Holbertson, R[illegible]

1 *Ch. de la Gironde (Bordeaux).* **Cimeterre** (1911). 894 tons. Dimensions: 213·3 × 26 × 10 feet. Designed H.P. 1[illegible] kts. Breguet turbines. Du Temple Guyot boilers. Trials: 31·2.

(Continued on next page.)

3 Casque type (1908 Programme)—*Continued.*

Armed with 2—3·9 inch, 4—9 pdr., and 4—18 inch tubes in pairs; also adapted for mine-sweeping and carry D.C.
For other details see Note on page preceding this.

Copyright Photo, M. Bar, Toulon.

1 *Forges et Ch. de la Méditerranée (Havre)*: **Casque** (1910). 830 tons. Dimensions: 246·5 × 21·8 × 10 feet. Designed H.P. 14,400 = 31 kts. (34·9 on trials). Parsons turbines. Normand boilers.

8 Hussard type.

(Can also be divided as 2 *Cavalier* type on this page + 6 *Hussard* type on next page.)

Differ in design according to builders, but all armed with 6—9 pdr. guns and 3—18 in. tubes (2 deck and 1 bow). Designed H.P. ranges from 8000 S.H.P. down to 7200 I.H.P. for speed of 28 kts., but *Cavalier* and *Mameluk* made over 30 kts. on trials. Radius of action, about 1000 miles at 15 kts., but *Cavalier* (oil fuel) about 1300 miles at 15 kts. Complements, 77 to 79. Maximum fuel in all, about 100 tons. Plans opposite for two boats on this page, remaining six on next page as plans for *Hussard*.

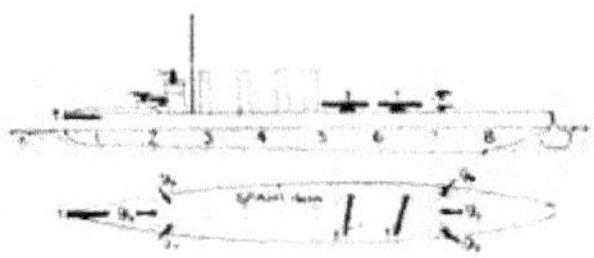

For description by classes, see next column on this page and first column on next page.

8 Hussard type (*Continued*).

CAVALIER.

1918 Copyright photo, M.

1 *Normand (Havre)*: **Cavalier** (1910). 527 tons. Dimensions: 222 × 21 × 10 feet. Turbines: Parsons. Bo oil-burning. Fuel: 130 tons oil only. Trials: *Cavalier*, 31·1.

1918 Copyright photo,

1 *Ch. de la Gironde (Bordeaux)*: **Tirailleur** (1908). 470 tons. Dimensions: 207 × 22·8 × 2½ feet. Re reciprocating engines. Du Temple-Guyot boilers. Coal: 100 tons. Trials: 28·82.

Continue

8 Hussard type—(*continued*).

All armed with 4—9 pdr. and 3—18 inch tubes. For other details, see Note on page preceding this.

1917 *Photo.*

2 *Rochefort* [illegible]: **Enseigne Henry** and **Aspirant Herbert** (1911). 450 tons. Dimensions: 2?1 × 21 × 10½ feet. Designed H.P. 7500 = 28.5 kts. Coal: 100 tons. Trials: *E. Henry* 29.5 kts., *A. Herbert* 29.? kts.

Note.—These boats have *four* funnels in pairs, the foremost funnel being close up against the chart house and almost abeam of funnel.

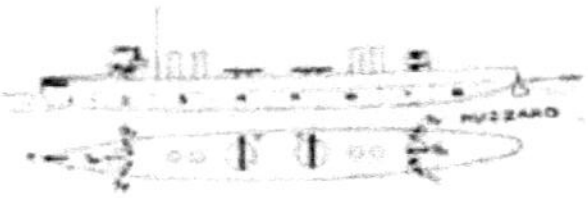

HUSSARD (and other boats).

Copyright photo, M. Bar, Toulon.

2 *Ch. de la Loire* (*Nantes*): **Hussard** (1908), 4?0 tons, and **Mameluk** (1909), 4?? tons. Dimensions: 2?6½ × 2?½ × 9½ feet.
1 *Dyle & Bacalan* (*Bordeaux*): **Lansquenet** (1909), ??? tons. Dimensions: 221 × 2?½ × 10 feet.
1 *F. & C. de la Med.* (*Havre*): **Spahi** (1908), 45? tons. Dimensions: 221 × 21½ × 9½ feet.
Machinery of all, reciprocating. Boilers: *Spahi*, Normand; others, Du Temple-Guyot. Coal: 100 tons in all. Trials: *Hussard*, 2?1; *Mameluk*, 2?6; *Spahi*, 2?7. [illegible] during the War.

33 Boats of Older Types (built 1896-1911).

Commonly known as "Hurricane Deck Type" (*Type Pont à Caillebotis*).

All are armed with 1—9 pdr., 6—3 pdr. and 2 tubes, and complement, 71. Boats on this page: 18 [illegible] tubes; on next page, 15 inch. Plans for all types as shown on this page. [illegible] of all boats on this page also [illegible] at 11 kts.

FANFARE. Some boats have mast very close to or almost abeam of fore funnel, as above. *Photo,* [illegible]

HACHE. Some boats have mast before funnel, as above.

1920 *Photo, M. Bar, Toulon.*

1? (Claymore class): **Carquois, Coutelas, Glaive, Mortier, Obusier, Pierrier, Poignard, St[illegible] Trident** and **Tromblon** (Rochefort D.Y., 1905-8); **Cognee, Hache** and **Massue** (Toulon D.Y., 19[illegible]); **Claymore** and **Fanfare**† (Normand, Le Havre, 1903-06); **Oriflamme** (Ch. de la Bretagne, Nantes, [illegible]); **Fanion** (Dyle and Bacalan, Bordeaux, 1908); **Gabion** and **Sape** (Soc. de St. Nazaire, Penhoët, 1907). [illegible] 3?5 tons. Dimensions: 19?8 × 21·? × 1?·5 feet. Designed H.P. 6400; all made 28 to 30½ kts. on trials. Boilers: Du Temple-Guyot or Normand. Coal: 90 tons.

* May also be divided as 14 Claymore type and 5 Fanfare type. † Has [illegible]

Older Types—(*Continued*): Armament: 1—9 pdr., 6—3 pdr. Boats on this page only, 2—15 inch tubes. Radius of action, 950 to 920 miles at 15 kts.

MOUSQUET class 1919 *Copyright Photo, M. Bar, Toulon.*

(13 boats, viz., 8 *Mousquet* class + 3 *Flamberge* class + 2 *Fauconneau* class.)

8 *Mousquet* class:—**Épieu, Harpon,** and **Sagaie** (Normand, Le Havre, 1902-3). **Francisque** and **Sabre** (Rochefort D.Y., 1902-3), **Bélier** (Ch. de la Bretagne, Nantes, 1902-3), **Bombe** (F. & C. de la Med., Havre, 1903), **Baliste** (Soc. de St. Nazaire, Rouen, 1903). 310-370 tons. Dimensions: 18[illegible] × 20[illegible] × 10[illegible] feet. H.P. 6000-7300 = 29 to 30·8 kts. on trials. Coal: 75 tons.

Also

3 **Escopette, Pertuisane,** and **Rapière,** (Rochefort D.Y. 1900-01). H.P. 4800-5200 = 26-27 kts. on trials, otherwise as *Mousquet* class above.

Also

2 **Fauconneau** and **Hallebarde** (Normand, Le Havre, 1899-1900). 310 tons. Dimensions: 180·4 × 20·6 × 7·8 feet. H.P. 5600 = 27 kts. Coal: 67 tons.

1 boat: **Pique,** (F. & C. de la Mediterranée, Granville, 1900). 310 tons. Dimensions: 180½ × 20·6 × 9⅚ feet. H.P. 5800 = 26 kts. Normand boilers. Coal: 67 tons. Has *four* funnels in pairs, the foremost two being higher than the aft pair.

1917 *Photo.*

11 "SEA-GOING TORPEDO BOATS."

Torpilleurs de haute mer.

2 screws. 15 inch torpedoes. Boilers: Normand, Thornycroft, and Du Temple G[illegible]

Totals.	Type.	Date.	Displacement.	I.H.P.	Max. Speed.	Coal.	Complement.	Tubes
10	*Mistral*	1901	150—186	4018	kts. 28·2	tons 25	32	3
1	*Grondeur*	1892	130	1750	22·8	10	28	2

All are Normand type. The *Forban* and *Mistral* class considerably exceeded 30 knots [illegible]

1917 *Copyright* [illegible]

10 Normand type: **Bourrasque, Mistral,* Rafale, Siroco*** (Normand, Havre); **Simo**[illegible] (F. & C. de la Med., Granville); **Audacieux,* Trombe*** (F. & C. de la Bretagne, N[illegible]) **Tramontane** (Dyle & Bacalan, Bordeaux). All launched 1900-01. 2—3 pdr. guns and 3 t[illegible] action: 2500 miles at 10 kts. * These boats have 1" armour amidships.

1 F. & C. de la Mediterranée, Granville: **Grondeur** (1892). 2—1 pdr., 2 torpedo tubes.

(Also old T.B. *Chevalier* for Auxiliary Duties.)

About 40 "Torpilleurs Numerotés."

Totals	Numbers	First Begun	Last Completed	Displacement	I.H.P.	*Max.* Speed	Coal	Complement	Tubes	*Max.* Draught
				tons		kts.	tons			
40	369—295	'03	'07	96-101	2000	26	14	23	3—1	8¼
	292—243	1899	1904	86-94	1500-2000	25	11	22	2	9

Notes.

About 40 of these boats are now in service; exact numbers uncertain, but numbered between [illegible] and [illegible].
Following boats have double numbers of number and name:—[illegible] (*[illegible]S*), [illegible] (*[illegible]S*), [illegible] (*[illegible]S*), [illegible] (*[illegible]S*), [illegible] (*[illegible]S*), [illegible] (*[illegible]S*), [illegible] (*[illegible]S*).
All boats are *Normand types*. Several exceeded 27 kts. on original trials, but few can exceed 25 kts. now.
All originally armed with 2—1 pdr. guns and 2 or 3—15 inch tubes, but these have been removed from some boats—see *Notes* below.

No. 339. 1920 *Photo, Bar, Toulon.*

Total number unknown: Are *Normand* type boats numbered between **369** and **295**, of 96 to 101 tons. *Nominal* full speed radius 200 miles. At 10 kts., 1800 miles.

Notes.—A large number of these Torpedo Boats were modified during the War for service as Coastal Patrol Boats, Guardships over Net Barrages, Boom Defences, &c., and for short-distance convoy work. The two amidships 15 inch torpedo tubes were removed and replaced by either a 14 pdr. gun (75 mm. field gun on naval mounting) or by a 9 pdr. gun or 2—3 pdr. guns, as above illustration. The stern 15 inch tube was retained, although it was of little use at full speed or in any seaway.

"Torpilleurs Numerotés"—(*continued.*)

Old *Normand* type boats numbered between **292** and **243**, of 86-94 and 80 tons. From *No.* 202 and [illegible] funnels—appearance more or less as [illegible] in preceding column.

Notes.—Uncertain how many of the above boats exist now. A few (including 243) appear to be in service [illegible] T.B. numbered between 35 and 275, condemned 1920.

Submarine Chasers (*Chasseurs de Sousmarins*)

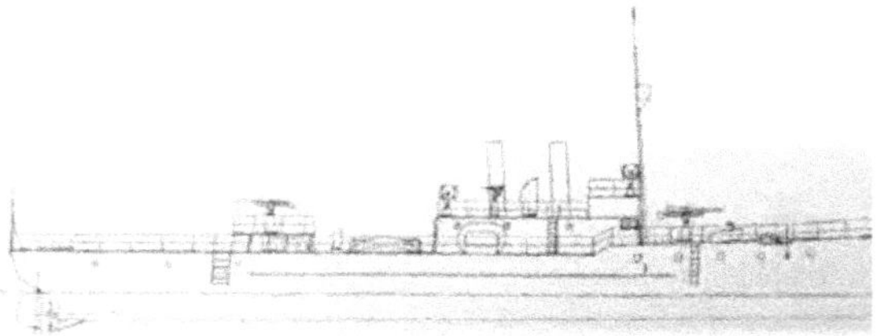

16 *Normand type* boats: ***C101—C106*** (Normand, Le Havre, 1918-20), ***C107—C116*** (builder unknown [illegible] tons. Dimensions: 135¾ (*p.p.*), 142 (*o.a.*) × 17¼ × 4¼ feet (draught amidships), 7¼ feet (draught [illegible] support). Guns: 2—14 pdr., 2 M.G. Designed I.H.P. 1300 = 16.5 kts. Triple expansion engines. 2 boilers. Coal: 32 tons. Complement, . Contract for 2 other boats with Normand cancelled 1918.

Note.—10 Niclausse type Chasers, *Fanfaron*, *Fantasque*, *Furieuse*, etc., have been converted into small Nav[illegible] their guns have been removed.

*Perhaps built by Chantiers de Bretagne, Nantes.

(*Continued on ne[illegible]*

Submarine Chasers (*continued.*)

C50. — *1920 Photo, H. Freund, Brest.*

92 (or less)*—*110 foot* type: Numbered between **C 1,** and **C 100.** Built by U.S. Navy Yards and smaller shipbuilding firms, 1917–18, for U.S. Navy as *SC5—SC404*, but contracts transferred to French Government. *Designed* displacement, 51 tons; *actual* displacement, 77 tons. Dimensions: 105 ft. (*p.p.*), 110 ft. (*o.a.*) × 14 ft. 8¾ in. × 5 ft. 5½ in. (*mean* hull draught). Machinery: 3 sets of 220 B.H.P. Standard petrol motors, totalling 660 B.H.P. = 17 kts. Petrol: 2400 gallons = 900 miles at 10 kts. **Armament:** *C 1—C 50* were sent across the Atlantic with 2—6 pdrs. only and may still retain these guns; *C 51—C 100* have 1—3 inch (23 cal.) gun. M.G. may have been added by French Navy. Carry depth charges. Complement, 26.

*Total number now in service uncertain. 4 condemned 1920, C 8 sunk by *Rivoa*, April, 1920.—Boats sold out unknown.

M. L. (*Vedettes*).

V82. — *1920 Copyright Photo, M. Bar, Toulon.*

51 (or less)*—*80 foot* type: Numbered between **V 1** and **61** (built in England 1917) and **V 62—73** (built in Canada 1917–18). About 35–38 tons. Complement, 8–10. Length, 80–85 feet. Beam, [illegible] feet. Draught, [illegible] feet. Armament: 1—75 mm. (14 pdr.) field gun converted to naval mounting, [illegible]. Machinery: 2 sets petrol motors. 20–22 kts. nominal, 19 kts. above best maximum at sea for any length of time. Radius of action: about 500 miles at full speed and 1000 miles at 12–15 kts. or less. Fuel: about 2000 galls. petrol. *V 21*, *V 25* and *V 27* lost during the War.

*Total number now in service uncertain; 30 were condemned between January 1st, 1920 and July 1st, 1920.

50* + *5* (+ *12 proj.*) = 67 **Submarines** (*Sousmarins*).

No. in Class	Class, Type (Design) †	Date.	Displacement. *Surface* / *Submerged* (tons)	H.P. ‡ *Surface* / *Submerged*	Speed ‡ *Surface* / *Submerged* (kts.)	Endurance. *Surface* / *Submerged*
	(Ocean-going)					
6	(*Q115—120*)	'21—?	1100 / —	...	...	...
2	*Marcast* class§	'17—'18	821 / —	2400 / 1100	16·2 / 8·5	8500 miles at 8 kts. / 65 miles at 5 kts.
1	*Halbronn*§	'16—'17	1250 / —	5500 / 1780	17·75 / 8	18000 miles at 8 kts.
2	*Leon Mignot* class§	'16—'17	798 / —	2300 / 1100	16·5 / 9·1	8500 miles at 8 kts. / 100 miles at 5 kts.
2	*Jossel* class (S)	'11—'20	900 / 1250	2700 / 1650	16·5 / 11	1200 miles at 10 kts. / 115 miles at 5 kts.
4	*Lagrange* class (H)	'14—'20	830 / 1130	2600 / 1650	16·5 / 11	1200 miles at 10 kts. / 115 miles at 5 kts.
2	*Dupuy de Lôme* class (H)	'15—'16	833 / 1000	4000 / 1700	18 / 11—12	1250 miles at 14 kts. / 130 miles at 5 kts.
2	*Néréide* (S)	'13—'16	787 / 1000	2400 / 1510	16·8 / 10·5	2500 miles at 12 kts. / 110 miles at 5 kts.
	Gustav Zédé (S)	'12—'14		2800 / 1540	15·8 / 10·5	2300 miles at 10 kts. / 110 miles at 5 kts.
	(Sea-going)					
6	(*Q121—136*)	'21—?	...	...	...	...
3	*Lieut. O'Byrne* class (SL)	'17—'20	350 / 500	1000 / 460	14 / 8·5	...
3	*Carrican* class§	'17—'18	514—521 / 730	1100 / 760	14 / 8	6000 miles at 5 kts. / 55 miles at 4 kts.
3	*Armide* class (L)	'14—'16	460 / 665	2200 / 850	17·5 / 11	2600 miles at 11 kts. / 150 miles at 5 kts.
1	*Daphné* (H)	'14—'16	640 / 945	1600 / 1400	15 / 10	2500 miles at 14 kts. / 100 miles at 5 kts.
3	*Bellone* class (H)	'14—'18	520 / 780	2100 / 1400	17 / 11	2300 miles at 10 kts. / 115 miles at 5 kts.
5	*Amarante* class (H)	'13—'18	410 / 550	1200 / 700	14 / 8	1300 miles at 10 kts. / 80 miles at 2 kts.
2	*Clorinde* class (R)	'13—'17	410 / 570	1300 / 800		
13*	Modified *Pluviôse* class (L)	'07—'14	400 / 565 ¢	840 / 600 ¢	13 / 9 ¢	1150 miles at 10 kts. / 80 miles at 5 kts.

Table continued on next page.

*6 "Modified *Pluviôse*" class will be condemned soon, reducing (*a*) total of completed boats to 44; (*c*) Class total to 7.

† (H) = Hutter. (L) = Laubeuf. (R) = Radiguet. (S) = Simonot. (SL) = Schneider-Laubeuf.

‡ As designed; not attained in some cases.

§ Ex-German boats, taken over 1920. The *Carrican* class boats can also be classified as a "small [illegible]".

¢ *Curie* of this class re-engined by Austrians. Present displacement, H.P. and speeds unknown.

Summary Table of Submarine Types—*continued.*

No. in Class	Class, Type (Design)*	Date	Displacement *Surface* / *Submerged*	H.P.† *Surface* / *Submerged*	Speed † *Surface* / *Submerged*	Endurance. *Surface* / *Submerged*	Tubes or Gear *Mines*	Complement
	Mine Layers.		tons					
1	*M. Callot* (SL)	'17—'21	930 / 1300	...	...	...	...	...
1	*P. Chailley* (NF)	'17—'21	830 / 1130	...	...	...	...	...
1	*René Audry*‡	'17—'18	1163 / 1501	2400 / 1200	13 / 7·5	11,500 miles at 8 kts. / ...	4/42§	40
1	*Victor Réveille*‡	'15—'16	532 / 833	330 / 300	10·5 / 8	5000 miles at 7 kts. / 75 miles at 5 kts.	2/38§	32
2	*Amarante* class (H)	'13—'15	410 / 550	1200 / 200	14 / 8	1300 miles at 10 kts. / 50 miles at 5 kts.	...	...
	Coastal.							
1	*Roland Morillot*¶	'16—'17	263 / 324	264 / 280	8½ / 6	3000 miles at 5 kts. / 70 miles at 3 kts.	2	21

* (SL) = Schneider-Laubeuf. (NF) = Normand-Fenaux. (H) = Hutter.

† H.P. and speeds *as designed*; not attained in some cases.

‡ German boats taken over 1920.

§ Mines laid by 2 mine-discharging tubes.

¶ German boat captured 1916.

6 Boats (Class un-named—Ocean-going).

To be begun 1921.

6 boats, type unknown: *Q 115*—*Q 118* to be begun at Cherbourg D.Y.; *Q 119*, *Q 120* to be begun at Toulon D.Y. Reported to be uniform design, 1100 tons *on surface*. No other details known.

Note.—Replace 6 boats, ordered 1914 and cancelled February, 1915.

2 ex-German Boats (Ocean-going—*Diesel*).

Photo, New Illustrated Co.

2 ex-German "Mittel U" type: **Marrast** (ex-German *U 165*), **Jean Roulier** (ex-German *U 166*), Bremer-Vulkan, Vegesack, and completed 1918. Displacements: 521 tons *surface*; [illegible] Dimensions: 235 × 21 ft. 1 in. × 12 ft. 9 in. Machinery: 2 sets 6-cylinder 2 cycle M.A.N. Dies[illegible] = 14·2 kts. *on surface*. *Submerged speed*: 8·5 kts. Oil fuel: 47/100 tons. Endurance: 8500 mil[illegible] *surface*; about 90-100 miles at 3 kts. *submerged*. Guns: 2—4·1 inch (Krupp). Torpedo tubes: [illegible] bow, 2 stern). Stowage for 12 torpedoes. Complement, 36.

Notes.—Units of German *U 160*—*172* class and among last submarines completed for the German Navy [illegible] design.

1 ex-German Boat (Ocean-going—*Diesel*).

Photo wanted.

(Photographie demandée).

1 ex-German "UK-Hilder" type: **Halbronn** (ex-German *U 139* (*Schwieger*), built by Krupp-Germania, Kiel 1917). Displacements: 1930 tons *on surface*, [illegible] tons *submerged*. Dimensions: 302¼ × 29¾ × 15 feet. 2 sets Krupp or M.A.N. 6 cylinder 4 cycle Diesel engines. B.H.P. 3500 = 17·75 kts.* *on surface*. Also [illegible] auxiliary engine. Speed *submerged*: 8 kts.* Oil fuel: 104/295 tons. Endurance: 18,000 miles [illegible] *surface*. Guns: 2—5·9 inch (Krupp). Torpedo tubes: 6—19·7 inch (4 bow, 2 stern). Stowage for [illegible] Complement, 83. [illegible]

Notes.—Unit of German *U 139*—*141* class. Double-hull type, fitted with anti-rolling tanks. Found deficient by Germans; *U 141* of the class had wood filling along upper deck to increase stability. This "Cru[illegible] was a very qualified success.

*One boat of this class on trial for German Navy could only make 15·8 kts. *on surface* and 7·4 kts. *submerge*[illegible]

(Continued on [illegible]

FRENCH SUBMARINES—OCEAN-GOING BOATS.

2 ex-German Boats (Ocean-going—*Diesel*.)

Photo, News Illustrations Co., 1920.

2 *ex-German* "Mittel-U" type: ***Jean Autric*** (ex-German *U 105*), ***Leon Mignot*** (ex-German *U 108*). Both built by Krupp-Germania, Kiel, and completed 1917. Displacements: 798 tons *on surface*,— tons *submerged*. Dimensions: *about* 235 × 21 × 12½ feet. Periscope depth: 45 feet. Machinery: 2 sets 6 cylinder 4 cycle M.A.N. or Krupp Diesel engines. B.H.P. 2300 = 16.5 kts. *on surface*.* Speed *submerged*: 9.4 kts.* Endurance: 8500 miles at 8 kts. *on surface*; *about* 100 miles at 3 kts. *submerged*. Guns: 1—4.1 inch and 1—3.4 inch (Krupp "Ubb and Tybts" models). Tubes: 6—19.7 inch (4 bow, 2 stern). Stowage for 12 torpedoes. Complement 36.

Notes.—Units of the German *U 105—114* class. Are double-hull type.

*Under trial by U.S. Navy, *U 111* (sister boat to above) could not exceed 12.8 kts. *on surface* and 7.8 kts. *submerged*.

2 Joessel Class (Ocean-going—*Diesel*.)

JOESSEL. *1920 Photo by courtesy of the Ministry of Marine.*

2 *Simonot type*: ***Fulton*** (1919) and ***Joessel*** (1917). Both built at Cherbourg D.Y. Displacements: 900 tons *on surface*, 1250 tons *submerged*. Dimensions: × × feet. Machinery: *on surface*, 2 sets Schneider-Carels Diesel engines, 2700 B.H.P. = 16½ kts. *Submerged* speed: 11 kts. Endurance: *on surface* and *submerged* about the same as *Lagrange* class in next column. Armament: 1 or 2—14 pdr. guns and 8 or 10 tubes, and/or torpedoes carried in external cradle gears. Stowage for 10 torpedoes. Complement, 40.

Notes.—Begun as *Q 109—110* of 1914 Programme. Undergoing extended trials and experiments, 1920, but can be regarded as completed. *Joessel* reported to have made over 18 kts. *on trials*.

4 Lagrange Class (Ocean-going—*Diesel*.)

LAGRANGE. *1920 Copyright Photo, M. Bar*

4 *Hutter type*: ***Lagrange*** (1917), ***Regnault*** (1919), and ***Romazotti*** (1917). All built at ***Laplace*** (Rochefort D.Y., 1919). Displacements: 840 tons *on surface*, 1130 tons *submerged*. 246¾ × 21 × 13 feet. Machinery: *on surface*, 2 sets 1300 B.H.P. Sulzer-Diesel engines, totallin for 16½ kts. *Submerged* speed: 11 kts. Endurance: *on surface*, 2500 miles at 14 kts., 4200 miles miles at 8 kts. When *submerged*, 115 miles at 5 kts. Armament: either (a) 1—3.9 inch and 1— (b) 2—14 pdr. guns. 8 or 10—18 inch tubes or torpedoes carried in external cradle gears. Stowage f and 440 rounds of ammunition for guns. Complement, 46.

Notes.—Begun as *Q 111—114* of 1914 Programme. Have been running extended trials 1919-20, attaini Can be regarded as completed.

2 Dupuy de Lome Class (Ocean-going—*Steam*.)

SANE.* *1919 Copyright photo,*

2 *Hutter type*: ***Dupuy de Lome*** (1915) and ***Sane*** (1916). Both built at Toulon D.Y. 19 tons *on surface*, 1100 tons *submerged*. Dimensions: 246¾ × 21 × 13 feet. Machinery: *on surfac* and oil-fired boilers: I.H.P. 4000 = 18 kts. *Submerged* speed: 11 to 12 kts. Endurance: 19 *on surface*, 150 miles at 4 kts. *submerged*. Armament: 2—14 pdr. guns and 8—18 inch torpedo tu cradle gears. Complement, 40.

Notes.—Begun as *Q 105* and *Q 106* of 1912 Programme. Both completed 1916. Are enlargements successful design for the submarine *Archimède* (built 1907-11 condemned 1920).

*Now has bridge screen and guns, as Silhouette.

(*Ocean-going Types conti*

2 Gustave Zédé Class.

(Ocean-going, one boat *Diesel*, one boat *steam*.)

General Notes.—These two Submarines are usually rated together as a class. They have the same dimensions, displacement, and armament but for comparative purposes, they were built with different systems of surface propulsion. It was originally intended that each boat should have 2 sets of 2000 B.H.P. Diesel engines totalling 4000 B.H.P. for a surface speed of 20 kts. After two years' experimental work, the Diesel engines of *G. Zédé* were given up and steam engines of about 50% less power adopted. Normands, after several years' research, produced reliable Diesel engines for *Néréide*, but of 40% less power than originally contemplated. These boats are expensive examples of the risks incurred, in trying to force Diesel engines to powers beyond the limits of contemporary knowledge.

Néréide (without guns). 1920 *Photo by courtesy of the Ministry of Marine.*

1 Simonot type: **Néréide** (Cherbourg D.Y., 1914). Displacements, dimensions, armament and complement, as *Gustave Zédé* below. Machinery: 2 sets 2 cycle 8 cyl. 1200 B.H.P. Schneider-Carels Diesel engines (built by Normand, le Havre), 2400 H.P. = 16.8 kts. *on surface*. Electric motors + batteries 1540 H.P. = 11.5 kts. *submerged*. Radius of action: *about* 3500 miles at 12 kts. *on surface*.

Notes.—Ordered January, 1911, as Q [illegible] of 1911 Programme. Completed, end of 1916.

1917 *Photo, Oscar Parker, M.B.*

Has been somewhat altered since this photo was taken; [illegible] Silhouette for present build.

1 Simonot type: **Gustave Zédé** (Cherbourg D.Y., 1913). Displacements: 787 tons *on surface*, 1000 tons *submerged*. Dimensions: 243 × 19½ × 12¼ feet. Machinery: 2 sets reciprocating, triple expansion + 4 Normand-Du Temple oil-burning boilers, 2200 H.P. = 15.8 kts. *on surface*. Electric motors + batteries, 1540 H.P. = 10.5 kts. *submerged*. Radii of action: 2500 miles at 10 kts. and 1800 miles at 14 kts. *on surface*. Armament: 1—14 pdr. and 1—3 pdr. Q.F. guns. Torpedo tubes: 2—18 inch [illegible] + 6—18 inch torpedoes carried in external dropping gears. Complement, 40.

Notes.—Begun as Q [illegible] of 1911 Programme. Completed October, 1914. Sunk 1916, after accumulator explosion, but was salved and repaired.

6 Boats. (Class unknown. Sea-going—*Diesel*.)

6 boats, *type unknown*; projected. Unofficially reported that these will displace *about* 500 to 600 tons

Notes.—Begun under 1920 Programme. Until named, they may be known as Q [illegible].

3 Lieut. O'Byrne Class (*Sea-going—Diesel*).

Lieut. O'Byrne.

3 Laubeuf type: **Lieutenant O'Byrne** (June, 1920), **Dupetit-Thouars** (——) and **Henri Fourn**[illegible] all by Schneider, Chalon-sur-Saône. Displacements: 350 tons *on surface*, 500 tons *submerged*. Dimensions: 173·6 × 15·4 × feet. Machinery: 2 sets Schneider-Carels Diesel engines, 1000 B.H.P. = 14 kts. *Submerged speed*: 8½ kts. Armament: 1 small Q.F. gun and 4 torpedo tubes. Complement, 24.

Notes.—Originally ordered for Rumanian Navy but taken over under French 1917 Programme. Are doubl

3 ex-German Boats. (Sea-going*—*Diesel*.)

Photo, U.N. [illegible]

3 ex-German "*UB III*" type: **Jean Corre** (ex-German *UB 155*), **Carissan** (ex-German *UB 99*) and **Schlllemans** (ex-German *UB 94*), all completed 1918. Builders uncertain, but probably Vulkan, H Weser, Bremen. Displacements: *on surface* (*a*) 523 tons for *ex-UB155*; (*b*) 516 tons for other two; *submerged* 733 tons. Dimensions: 181½ × 19½ × 12 feet. Periscope depth: 42 feet. Machinery: 2 sets 6 c engines, B.H.P. 1100 = 14 kts. *on surface*. *Submerged speed*: 8 kts. Oil fuel: 35/71 tons. Lubricat tons. Endurance: 9000 miles at 6 kts. *on surface*; 55 miles at 4 kts. *submerged*. Gun: 1—4·1 inc "Ubts. u. Tbts" model. Torpedo tubes: 5—19·7 inch (4 bow, 1 stern). Stowage for 10 torpedoes. hull type. *Max. dive*: 245 feet. "Crash-dive" to 29.5 feet in 40 secs. Complement, 34.

Notes.—Were units of German *UB* [illegible] and *UB* [illegible] classes.

*The rating of these boats as "Sea-going" is purely conventional. During the War, the Admiralty rated th as a "Small Ocean-going" type, which is a more correct description of their capabilities.

3 Armide Class (Sea-going—*Diesel*)

ARMIDE

1920 *Photo, by courtesy of the Ministry of Marine.*

3 *Laubeuf* type "De": **Amazone,** (1916), **Antigone** (1916), and **Armide** (ex-Japanese *No.* 14, 1915). Built by Schneider et Cie, Chalon-sur-Saône. Displacements: 460 tons on *surface*, 665 tons *submerged*. Dimensions: 184¼ × 17 × 10½ feet. Machinery: 2 sets, Schneider-Carels Diesel engines, 2200 H.P. = 17½ kts. *on surface*. *Submerged speed*: 11 kts. Endurance: 960 miles at 13 kts., 2600 miles at 11 kts. on *surface*. Armament: 1—1 pdr. Q.F. gun. Torpedo tubes: 2 internal + 4 torpedoes carried in external dropping gears. Double-hull type.

Note.—*Amazone* and *Antigone* begun for Greek Navy; *Armide* for Japanese Navy. All three taken over for French Navy during War.

1 Daphné Class (Sea-going—*Diesel*)

DAPHNÉ*

1920 *Photo by courtesy of the Ministry of Marine.*

1 *Hutter* type: **Daphné** (Toulon D.Y., 1915). Displacements: 640 tons *on surface*, 945 tons *submerged*. Dimensions: 223 × 18¼ × 12½ feet. Machinery: 2 sets 800 B.H.P. Sulzer Diesel engines, 1600 H.P. = 15 kts. *on surface*. *Submerged speed*: 11 kts. Endurance: *on surface* (*a*) 2300 miles at 14 kts., (*b*) 4900 miles at 11-12 kts., 100 miles at 5 kts. when *submerged*. Armament: 2—14 pdr. gun. 10—18 inch torpedo tubes or torpedoes carried in external dropping gears. Complement, 55. Is an enlarged *Archimède* design.

Note.—Begun as *Q* [illegible] of 1913 Programme. *Diane* of this type lost during the War.

*Guns (*v.* Silhouette) not shown in this photograph. May have steel bridge screen now.

3 Bellone class (Sea-going—*Diesel*).

GORGONE*

1920 *Photo by courtesy of the Ministry*

3 *Hutter* type: **Bellone** (Rochefort D.Y., 1914) and **Hermione** (Toulon D.Y., 1913), **Gorgone** (
1915). Displacements, 520 tons *on surface*; 780 tons *submerged*. Dimensions: 196·8 × 17·7 × 12 feet.
2 sets Sabathé-Diesel engines, 2100 H.P. = 17 kts. *on surface*. *Submerged* speed, 9 kts.
on surface, 2300 miles at 10 kts., 115 miles at 5 kts. *submerged*. Armament: 1—3 pdr. Q.F. gun. 8—1
tubes or torpedoes carried in external dropping gears. Complement, 29.

Notes.—*Bellone* begun as *Q* [illegible] of 1912 Programme and completed 1917. *Hermione* (completed 1918
(completed 1916), begun as *Q* 102 and *Q* 104 of 1913 Programme, as a "Modified *Archimède*" desi

*Steel bridge screen and gun abaft C.T. (*v.* Silhouettes) not shown in this view.

5 Amarante class (Sea-going—*Diesel*)

ATALANTE

1919 *Copyright Photo, M.*

5 *Hutter* type: **Arethuse, Artemis** and **Atalante** (Toulon D.Y., 1913), **Andromaque** (
1915), **Amphitrite** (Rochefort D.Y., 1914). Displacements: 410 tons *on surface*, 52
Dimensions: 174 × 16¼ × 10-11 feet. Machinery: 2 sets 6 cyl., 2 cycle Schneider-Carels or Sabat
1300 B.H.P.† = 14 kts. *on surface*.† *Submerged* speed: 8 kts. Radii of action: 1950 miles at 1
80 miles at 5 kts. *submerged*. Armament: 1—3 pdr. Q.F. gun, 8—18 inch torpedo tubes or to
external dropping gears. Complement, 30. Are an improved and faster type derived from La
Pluviose" boats on next page.

Notes.—Begun under 1912 Programme as *Q* [illegible], *Q* [illegible]. *Amphitrite* completed 1918, the oth
of this class lost during the War.

**Amarante* and *Astrée* of this class adapted for mine-laying; described on a later page.

†Only *Arethuse*, *Artemis*, *Atalante* developed this B.H.P. on *trials*; other boats could not exceed 8
13.3 kts. speed.

[*Continu*

2 **Clorinde class** (*Sea-going Diesel*)

CLORINDE. 1920 *Photo by courtesy of the Ministère of Marine.*

2 *Bourdelles* type: ***Clorinde*** (Rochefort D.Y., 1913), ***Cornelie*** (Rochefort, 1913). Displacements: 410 tons on surface; 570 tons submerged. Diesel engines are 2 sets, 6 cyl. 2 cycle M.A.N. type. Electric motors + batteries of 410 H.P. All other details as the *Amarante* class on preceding page.

Notes.—Begun as Q 89 and Q 91 under 1909 Programme. *Clorinde* completed 1917 and *Cornelie* 1916.

13 **"Modified Pluviôse" class** (*Sea-going—Diesel*)

(Also known as "Brumaire" class).

La Vendée. 1920 *Copyright photo, M. Bar, Toulon.*

13 *Laubeuf* type viz:—

5 *Cherbourg Dockyard*: ***Brumaire**** and ***Frimaire*** (1911), ***Euler*** and ***Nivôse*** (1912), ***Franklin*** (1913).
4 *Rochefort Dockyard*: ***Faraday**** and ***Volta*** (1911), ***Montgolfier*** and ***Newton*** (1912).
4 *Toulon Dockyard*: ***Arago,* Coulomb,* Curie†*** and ***Leverrier*** (1912).

Displacements: About 400 tons *on surface*, 505 tons *submerged*. Dimensions: Length 166½ to 172 feet. Beam: 16½ to 17¼ feet. Draught: 10½ feet. Armament: 1—3 pdr. gun, and 7—18 torpedoes carried in external cradle gears. Machinery of these boats: 2 sets Diesel engines, 6 cyl. 4 cycle M.A.N. type by Cie. de la Loire; Sautter-Harlé; Normand & Cie.; Lefalve & Cie. B. H.P. 840 = 13 kts. *on surface*. Electric motors and batteries, 600 H.P. = 9 kts. *submerged*. Endurance: 1450 miles at 9-10 kts. *on surface*. 80 miles *submerged* at 4½ kts. Complement, 26.

Notes.—Begun as Q 60—Q 88. *Brumaire, Frimaire, Nivôse*, 1905 Programme; others 1906 Programme. Double-hull design.

*Six of these will be condemned very soon, leaving 7 boats as total of class.

†*Curie* captured by Austro-Hungarian Navy, Dec. 1914, and during 1915-18 was the Austro-Hungarian Submarine *XII*. Recovered by French Navy in 1918. She was re-engined and re-built by the Austrians, these alterations having a marked improvement on her design. The above figures for displacements, H.P. and speeds are probably no longer accurate for *Curie*.

1 **Special Boat** (*Mine-laying—Diesel.*)

Building or Completing.

(Begun 1917: to be completed 1921).

1 *Schneider-Laubeuf* type: ***Maurice Callot*** (Ch. de la Gironde, Bordeaux: launched 1919 ?). 920 tons *on surface*; 1200 tons *submerged*. No further details known. Mines reported to be carried abaft C.T. Begun under 1917 Programme. Double-hull design.

1 **Special Boat** (*Mine-laying—Diesel.*)

Building or Completing.

(Begun 1917: to be completed 1921).

1 *Normand-Fenaux* type: ***Paul Chailley*** (A. Normand & Cie., Le Havre, launched 1919 ?). Displ. 850 tons *on surface*, 1150 tons *submerged*. No other details available. Begun under 1917 Programme.

(*Mine-laying Types continued on next page.*)

2 ex-German Boats (Mine-laying—*Diesel*)

1920 *Copyright photo, M. Bar, Toulon.*

1 *ex-German* "UE II" type: ***René Audry*** (ex German *U 119*, built by Vulkan, Hamburg, completed 1918). Displacements: 1163 tons *on surface*, 1501 tons *submerged*. Dimensions: 267 ft. 5 in. × 24 × 13 ft. 10 in. Periscope depth: *about* 45 feet. Machinery: 2 sets of 6 cyl. 4 cycle M.A.N. Diesel engines *on surface*. B.H.P. 2400 = 1[illegible] kts. *Submerged speed*: 7.5 kts. Endurance: 11,500 miles at 8 kts. Oil capacity: 93.5/194 tons. *on surface* Guns: 1—5.9 inch (Krupp). Torpedo tubes: 4—19.7 inch (bow). 24 torpedoes carried: 12 in pressure-tight cylinders alongside upper deck, rest within hull.* 2 horizontal mine-laying chutes and ports under stern. Stowage for 42 mines, but owing to deficient stability, it is reported that only 32 mines can be carried. Originally unit of German *U 117—126* class, designed for mine-laying off U.S. coast. Were double-hull boats. Complement, 40.

*Doubtful if whole 24 were ever carried.

Photo wanted.

1 *ex-German* "UE I" type: ***Victor Réveille*** (ex-German *U 79*, built by Vulkan, Hamburg, completed 1916). Displacements: 755 tons *on surface*, 832 tons *submerged*. Dimensions: 186 ft. 4 in. × 19 ft. 16 in. × 15 ft. 3 in. Machinery: 2 sets of Diesel engines. B.H.P. 900 = 10.8 kts. *on surface*. *Submerged speed*: 8 kts. Endurance: 8000 miles at 7 kts. *on surface*, 75 miles at 3 kts. *submerged*. Oil capacity: 77/87 tons. Guns: 1—4.1 inch (Krupp). Torpedo tubes: 2—19.7 inch, bow tube to port, stern tube to starboard, and both *above water* when in surface trim. 2 horizontal mine-laying tubes under stern. Stowage for 36 or 38 mines and 2 torpedoes. Originally unit of German *U 71—80* class. Saddle-tank type of hull. Complement, 32.

2 of Amarante class (Mine-laying—*Diesel*)

Astrée.

1920 *Photo by courtesy of the Ministry o[f]*

2 *Hutter* type: ***Astrée*** (Rochefort D.Y., 1918), ***Amarante*** (Toulon D.Y., 1915). Displacements: [illegible] *surface*, 530 tons *submerged*. Dimensions: 171 × 16½ × 10–11 feet. Machinery: 2 sets of 6 cyl. 2 [illegible] or Carels Diesel engines, 1200 B.H.P.* = 14* kts. *on surface*. *Submerged speed*: 8 kts. Endurance [illegible] at 10 kts. (*on surface*), 80 miles at 5 kts (*submerged*). Mines carried: Not known.

Notes.—Built as Q 95 and Q 96 (?) under 1912 Programme. Completed 1918. *Amarante* reported to [illegible] type laying gear, mines being stowed on each beam.

**Max.* on trials, 900 000 B.H.P. = 13.3 kts.

1 Captured German boat (Coastal—*Diesel*)

1918 *Photo by courtesy of Senor Pascual [illegible]*

Above view of German *UB 23* is generally correct for appearance of ROLAND MORILLOT (ex-*UB 26*).

1 *ex-German* "*UB II*" boat: ***Roland Morillot*** (ex-German *UB 26*, launched by Weser Co., Breme[n] during 1916 and repaired at Cherbourg D.Y.). Displacements: 263 tons *on surface*, 321 tons *su[bmerged]* [Dimen]sions: 118½ × 14½ × 12 feet. Machinery: 2 sets, 142 B.H.P. 6 cylinder oil motors = 8½ kts. *on su[rface]* [Submerged] speed: 5½ to 6 kts. Endurance: 9000 miles at 5 kts. *on surface*, 70 miles at 3 kts. *submerged*. [Guns:] 3 pdr., 1 M.G., 2 (bow) 18 inch torpedo tubes. Carries 4 torpedoes. Oil: 22/28 tons. Lubric[ating] [illegible] Single-hull type. *Max.* dive: 165 feet. Crash dive to 20 feet in 40 secs. Complement, 21.

... so far as can be ascertained, there are no Sea-going Depot Ships for Destroyers, Submarines or Submarine Chasers in the French Navy.

Aircraft Carriers.

(Portes-Avions.)

Launched and to be converted.

BEARN (late Dreadnought of *Normandie* class, launched at La Seyne, April, 1920). Length (*w.l.*) 576 feet. Alterations to displacement, armament, armour, H.P., speed, etc., effected on alteration are not known yet. Reported that she will be modified on lines of British *Eagle*, *Hermes*, etc.

1920 *Photo, M. Bar, Toulon.*

BAPAUME (L'Orient D.Y., August, 1918). Converted "Aviso" of *Arras* class. 700 tons. Dimensions: 246 (*o.a.*) × 28.5 × 9.2 feet. Guns: 1—14 pdr. AA. aft. S.H.P. 5000 = 20.5 kts. Geared turbines. Oil: 200 tons = 1400 miles at 17.5 kts. Is attached to St. Raphael Naval Aviation Station.

1919 *Photo, M. Bar, Toulon.*

CAMPINAS (ex-liner of Chargeurs Reunis, built ——, and taken over during war). 3319 tons. Dimensions: 357·6 × 42·7 × 23·2 feet. Guns: not known. I.H.P. 1460 = 11·8 kts. on trials; present best speed, 10 kts. 1 set triple-exp. engines. 2 Boilers. Coal: 988 tons. Two hangars for 6-10 machines and K.B. gear.

Aircraft Carriers *(continued)*.

(Also Aviation School Ship.)

FOUDRE (October, 1895). Displacement, 6100 tons. Complement, 328. Length, 380½ feet. Beam, 51¼ feet. *Maximum* draught, 23¼ feet. Guns: 8—3·9 in., 4—9 pdr., 2—3 pdr. Armour: Deck 3¼", 4" casemates. Designed H.P. 11,500 = 19 kts. *Trials* (1897): I.H.P. 11,930 = 19·57 kts.

Mine Layers *(Mouilleurs de Mines.)*

PLUTON. *Photo by courtesy of Messrs. A. Normand.*

PLUTON (1913), 660 tons, & **CERBERE** (1912). Both built by Messrs. Augustin Normand, Havre. 566 tons. Dimensions: 193½ × 27½ × 10¾ feet. Guns: 2—3·9 inch and 1—3 pdr. AA. Designed H.P. 5000 = 20 kts. Coal: 150 tons. Carry 120 mines.

Destroyers *(used for Mine-Laying)*.

(Not known.)

Submarine Mine Layers.

Paul Chailly and *Maurice Callot*, —— *Victor Réveille* carries 36 mines. *René Audry* carries 42 mines. All listed on earlier pages. Two submarines, *Astrée* and *Amarante* (also previously described) have been adapted for mine laying; number of mines carried not known.

Mine Sweepers *(Drageurs de Mines)*.

Note.—All destroyers of the *Bouclier* class, and nearly all gunboats of the *Diligente*, *Friponne* and *Vaillante* types are fitted as mine sweepers, also various Torpedo Boats and Submarine Chasers.

Mine Sweepers *(continued)*

PIOCHE (towing K.B).

CHARRUE (1914), **HERSE** (1914), **PIOCHE** (191 (1914). 255 tons. Guns: 2—3 pdr. Designed H. Coal: 35 tons.

(No photo available.)

ORIENT (1908), 380 tons. **IROISE** (1907), 240 tons (1897), 155 tons. **ALCYON I** (——), 300 t 8-10 kts. Above ships are Trawlers purchase 1910-12. Further Mine-Sweeping Trawlers, wh be permanently retained for naval service, are l page but one.

Transports *(Transports)*, Supply Ships *(Soutie* and Refrigerator Ships *(Navires Frigori*

J. COEUR. 1920 *Photo,*

CHAMPLAIN (April, 1919), **JACQUES COE** 1919), **PRIMAGUET** (April, 1919), all three by
COETLOGON (June, 1919), **FORFAIT** (— **HAMELIN** (——), **LAMOTTE-PICQUET** (- Ch. de Bretagne, Nantes. 700 tons. Dimensio (*p.p.*), 167.2 (*o.a.*) × 25.9 × 11 feet (*mean* draugh (*max.* draught). Guns: 2—3.9 inch, 1 M.G. 1. = 12.5 kts. (light) and 1100 = 8 kts. (load draught expansion engines. 1 screw. 2 cylindrical boile 80 tons.

Notes.—Champlain, J. Coeur and *Primaguet* fitted as Re Ships. Completed from February, 1920 onwards. as a Class of 13 Patrol Vessels. Above 7 con converted to Surveying Ships *B-Beaupré* and *La* other 4 stopped.

Supply Ships, &c.—*continued.*

1920 Photo, French Govt.

SEINE (1913). 3160 tons. Designed H.P. 1950 = 13 kts. Guns: 2—3·9 inch, 1—9 pdr. Coal: 295 tons.

No Photo available.

LOIRET (ex S.S. *Paris*, purchased 1909). 2200 tons. Guns: 1—9 pdr. H.P. 1060 = 11 kts.

VINH-LONG (1881), **BIEN HOA** (1880). 5500 tons. Speed: nominal 13 kts. Draught: about 22 feet.

DUGUAY-TROUIN (ex-*Tonquin*, 1878), **TOURVILLE** (ex-*Gironde*, 1878). 5115 tons. Designed H.P. 2700 = 14 kts.

Note.—Mostly served as Hospital Ships during the War; being overhauled for further service, 1920. Usually described as "Indo-China Transports" or "Annamite Class."

Supply Ships, &c.—*continued.*

Also *Arcturus, Evangéline, Gâvres* and *Marguerite VI.*

Following were Supply Ships to Syrian Squadron, 1920, but are probably only Trawlers:—

Apache, *Jeanne Genévière*, *Meg*, *Mohican*, *René*, *Sidi-ferruch*, *Truite*

Oil Tankers (*Pétroliers.*)

Building and Completing.

AUBE (July, 1920), *DURANCE* (on stocks), both by L'Orient D.Y. 2500 tons. Dimensions: 242.8 (o.a.) × 38 × 15.7 feet. S.H.P. 1000 = 10 kts. Breguet turbines. Carry 1500 tons oil. Endurance: 1580 miles at 10 kts.

Same appearance as "Rhône" below.

———— (ex-*Meuse*, 1915, present name not known). 7500 tons. Speed, 13 kts. Carries 6400 tons of oil as cargo.

1920 Photo, B. Frères.

GARONNE (1911). Oil Tanker. 10,800 tons. H.P. 2900 = 13 kts. Carries 7900 tons oil as cargo.

Oil Tankers—*continued.*

1920 Photo, F

RHÔNE (1913). Oil Tanker. 7830 tons. H.P. 2[illegible] Carries 4500 tons oil as cargo.

No Photo available.

CZAR NICOLAI II (1895, captured 1914). 5300 tons. 18.7 feet. Guns: 1—14 pdr., 2—3 pdr. Speed: [illegible] Carries 2500 tons oil.

Salvage Ship.

(For appearance, see Silhouettes.)

VULCAIN (1908). 229 tons. Speed: 12 kts.

Tugs. (*Remorqueurs.*)

HIPPOPOTAME, MAMMOUTH, MASTO[illegible] RHINOCEROS. Ocean-going Tugs of 900 tons.

GLADIATEUR, ATHLETE. No details.

CENTAUR (1915). **TRAVAILLEUR, INDEFATIG[illegible]** others.

Following Tugs of 360 tons built during the War:—

Clameur, Fracas, Lutteur, Tapage, Tintamare, Tumulte, Crabe, Calmar, Homard, Tourteau, Michel-et-Renée.

Also many other smaller tugs and tanks.

Yachts.

Photo wanted.

ALBATROS (*ex-Bernida*, ——). 567 tons. Guns: 3—14 pdr. Speed: 8.72 kts. (*on trials*, January, 1920). Serves as Yacht for the Governor of Syria.

Photo wanted.

DIANA (——). 1200 tons. Draught (aft): 15.8 feet. Guns: 2—14 pdr. Speed: 12 kts. Coal: 200 tons. Serves as Yacht for the Governor of Morocco.

Note.—A Naval Yacht Reserve is to be registered and enrolled August, 1919-1920.

Navy Trawlers (*Patrouilleurs*).

A large number of Trawlers were built during the War to the order of the Ministry of Marine, for service as Patrol Boats, Mine Sweepers, &c. So many were sold out to private owners, 1919-20, it is almost impossible to say how many are still in the French Navy. The following are believed to be still in service, 1920:—

"*Bird*" *Class*:—

Canard, Colombe, Coq, Faisan, Goëlette, Paon, Perdreau, Pigeon, Pingouin, Pintade, Ramier, Tourterelle, Vanneau, all 677 tons.

Bergeronnette, Engoulevent, Fauvette, Loriot, Martinet, Merle, Passereau, Perruche, Pie, Pierrot, Roitelet, all of 460 tons.

Navy Trawlers—(*continued*).

"*Fish*" *Class*:—

Ablette, Anguille, Brochet, Brème, Brochet, Carpe, all 460 tons; *Perche, Tanche, Truite*, 410 tons.

"*Insect*" *Class*:—

Cigale, Coccinelle, Libellule, all 450 tons.

"*Flower*" *Class*:—

Clematite, Coquelicot, Perce-Neige, Tulipe, Violette. Have Diesel engines and no funnels.

"*Tree*" *Class*:—

Cèdre, Chêne, Erable, Frêne, Hêtre, Orme, Platane, Peuplier, all 360 tons.

"*Rock*" *Class*:—

Granit, Grès, Marbre, Mica, Porphyre, Pyrite, Silex, Quartz, all 390 tons.

Miscellaneous:—

Autruche, Cerbère, Eole, Orage, all 471 tons; *Mistral* and *Trompette*, with Sulzer Diesel engines, *Amiral, Elan, Zebu*, &c., &c.

Surveying Ships. (*Navires Hydrographiques*).

B.-Beaupré. 1920 Photo, H. Pernot, Brest.

BEAUTEMPS-BEAUPRÉ (*ex-D'Estaing*) and **LA PÉROUSE** both launched 21st November, 1919. 1200 tons. Dimensions: 196.8 (*o.a.*) × 28.9 × 18 feet. Guns: 1—3 pdr. AA. H.P. 1100 = 10.67 (*on trials*). 2 W.T. boilers. Coal: 80 tons. Complement, . Are Navy Transports, completed as Surveying Ships, 1920.

Surveying Ships—(*continued*).

Photo, N. D.

VAUCLUSE (1880, refitted 1920). About 1600 tons. Armament: 2—3 pdr. Speed: about 10-11 kts.

UTILE (1894). 400 tons. Speed: 13 kts. Guns: 1—3 pdr.

Also:—

Alidade (*ex-Martin-Pêcheur*), *Astrolabe* (*ex-Mauviette*), *Boussole* (*ex-Pinson*), *Gustave Zédé* (*ex-Ortolan*), *Octant* (*ex-Pivert*). Are "Bird Class" Trawlers built during the War and adapted for Surveying Service, 1920.

And *Sonde* (1911) of 51 tons. 10 kts. speed.

Fishery Protection. (*Garde-pêche*).

Note.—Convoy Sloop *Ville d'Ys* (previously described) is on the Newfoundland Banks.

Q. Roosevelt. 1920 Photo, by courtesy of "Le Yacht."

QUENTIN ROOSEVELT (*ex-Flamant*, Rochefort D.Y., December, 1916). Despatch Vessel of 610 tons. Dimensions: 154 × 28 × 13 feet. Guns: 1—14 pdr., 1—1 pdr. I.H.P. 1100 = 13 kts. Coal: 100 tons = 1500 miles at 10 kts. Was begun 1913, stopped 1914-17, completed April, 1918.

Also:—

Trawlers *Sentinelle, Sajou, Pervenche*, and *Escafaut*, tender to *Q. Roosevelt*.

Training Ships and Establishments.

Flagships, R.A. :—**PATRIE** (Mediterranean) **D'ENTRECASTEAUX** (Atlantic).

FOR GUNNERY.

Battleship	*République*
Cruiser	*Pothuau**
Gunboat	*La Hire*
Submarine Chaser	*C 79*

*Sea-going Gunnery Ship.

FOR TORPEDOES.

Battleship	*Patrie*
T.B.D.	*Hoche*
T.B.	*No. 369*
Submarine Chaser	*C 86*

FOR MINES AND SWEEPING.

(*Nil.*)

FOR SUBMARINE EXPERIMENTS.

Yacht	*Henriette**
Drifters	*Vigoureux** *Primivere†*
P.W. Steamer (old)	*Guêpe†*
Hulk	*Pescurt†*

* Probably for Hydrophones. † For Depth Charge Experiments.

FOR AVIATION.

Aircraft Carrier	*Bapaume**
Submarine Chasers	*C 81** *C 25**
Drifter	*Humard†*

*Attached to St. Raphael Naval Aviation Station. † For Kite Balloons.

FOR NAVIGATION

Sloop	*Chamois*
T.B.	*Audacieux*
Sailing Vessels	*Railleur* *Mutin*

FOR W/T. W/TELEPHONE EXPERIMENTS.

Hulk	*Henri IV*

FOR ENGINEERS.

T.B.D.	*Cavalier*
T.B.	*Nos. 360, 332, 328, 310.*
Trawlers or Patrol Boats	*Aster* *Perce-Neige*
Hulks	*St. Louis* *Jauréguiberry*

FOR CADETS.

Cruiser	*Jeanne d'Arc*
Sloops	*Arras* *Meuse*
T.B.D.	*Aventurier*
T.B.	*Typhon*
Gunboat	*Sans Souci*

FOR W.O. AND P.O.

Cruiser	*D'Entrecasteaux*

FOR BOYS.

.. ..	*Armorique*
.. ..	*Magellan*

MOROCCO.

(French Protectorate.)

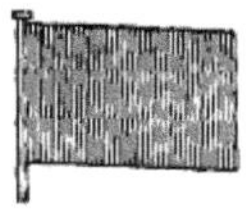

SIDI-EL-TURKI.

SIDI-EL-TURKI (1898). 450 tons. 154 × 2
Guns: 2—12 pdr. *Maximum speed*: 12.5 kt

1 funnel, 3 masts.

EL HASSANEH (1882). 1164 tons. 240¾ ×
Guns: 1 old 6 inch, 4 machine. *Nominal s*
Coal: 100 tons.

NOOR-EL-BAHR (1898) and two Coastguard S
Germany about 1906, also reported to exist.

SURRENDERED EX-ENEMY WARSHIPS ASSIGNED TO FRANCE.

Note.—All are Ex-Enemy Ships which have been definitely assigned to France for "Propaganda purposes," *i.e.*, for public exhibition, experiments, etc. They cannot be added to the French Navy, and must be broken up or destroyed by a specified date. *Ships whose names are in light type are at French Ports, but have not been assigned to any country.*

Nationality.	Battleships.	Light Cruisers.	Torpedo Craft (c).	Submarines (d).	Remarks.
GERMAN	**THURINGEN** (a)	**EMDEN** (b)	***V 126*** ***V 100*** ***V 46***	***U 166*** ***U 57*** ***UB 6*** ***U 160*** ***U 35*** ***UB 4*** ***U 157*** ***UB 142*** ***UC 104*** ***U 151*** ***UB 126*** ***UC 103*** ***U 113*** ***UB 84*** ***UC 100*** ***U 94*** ***UB 73*** ***UC 74*** ***U 91*** ***UB 24*** ***UC 58*** ***U 61*** ***UB 14*** ***UC 27*** ***U 79*** ***UB 8*** ***UC 23*** (Another boat)	(a) For gunnery experiments and investigation of propagation of fires. (b) Depth Charge Target. (c) All useful equipment to be stripped for re-fitting Ex-Enemy Destroyers added to Navy. (d) Several of these foundered on passage from England to France.
AUSTRIAN ..	**PRINZ EUGEN** (a) **ERZHERZOG KARL** **ERZH. FRIEDRICH**		***Satellit*** ***Roka*** ***Pandur*** And four Torpedo Boats :— ***Nos. 13, 15, 16, 17***	? (b)	(a) Target Ship for Aircraft Bombs and sub-surface attack. (b) Numbers cannot be discovered; 3 Austrian boats reported towed to Bizerta. Add : Minelayers *Basilisk*, *Dromedar*.
TURKISH ..	?	?	?	?	No information available.
BULGARIAN ..	——	——	——	*XVIII*	Taken over by France, 1919. Present disposal unknown.

(ROYAL) ITALIAN NAVY.

Revised 1919, from Official List, "Regia Marina Italiana, Maggio 1919." Corrected by courtesy of the Ministry of Marine, 1920.

Correction to above Map: *delete* AUSTRIA. For Triest read Trieste.

Principal Mercantile Ports.

Genoa (*Genova*), Naples (*Napoli*), Leghorn (*Livorno*), Palermo, Venice (*Venezia*), Catania (Sicily).

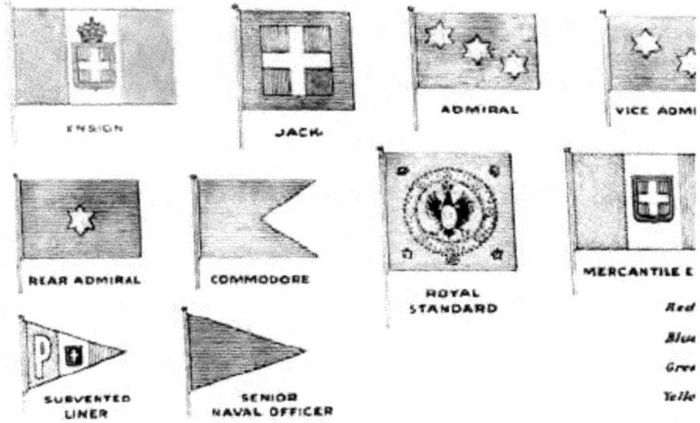

Private Docks.

GENOA: Three graving docks, Bacino di Darsena, 267¼ × 78¾ × 22 feet; No. 1, 571 × 81¼ ×
 No. 2, 702 × 60½ × 28 feet. Floating dock, 257 × 46 × 17 feet, 2900 tons lift.
LEGHORN: Orlando Bros., 451 × 71¾ × 23 feet. 2 patent slips for 1500 tons.
NAPLES: Municipal No. 1, 659¼ × 95 × 33½; No. 2, 348¾ × 60 × 23 feet.
PALERMO (Sicily): Dock, 565 × 85¾ × 27¼ feet. Patent slip, 1500 tons.
TRIESTE: Two Private docks here, 456 × 73 × 19 feet and 360 × 54 × 18¾ feet; there is a do
 Rocco, across the bay, 414 × 66 × 21 feet. Two floating of 2,000 and 1,500 tons.
Also various Austrian Docks taken over. No details available.

Colonies: Eritrea, Italian Somaliland, Tripoli and Cyrenaica (Libia Italiana).
Minister of Marine: Contr' Ammiraglio Sechi (June, 1919). *Naval Attaché, London*: C
 Bianchi.
Personnel: About 40,000 *active* personnel *before war*.

Mercantile Marine (*1916*).

From "Bureau Veritas" 1919-20 figures. Ships under 100 tons *gross* excluded.

711 steam and motor ships of 1,861,247 tons *gross*.
22 vessels (with auxiliary motors) of 14,307 tons *gross*.
694 sailing vessels of 160,759 tons *gross*.

During 1919, Italian Shipyards launched 32 vessels of 82,713 tons gross; some of these account.

Coinage.

Lira (100 centesimi) = 4½*d.* British, about ... cents U.S.A. Value varies.

ITALIAN DOCKYARDS, WEST & SOUTH COAST.

In 2000 yard squares. Uniform scale. Soundings in fathoms. Heights in feet.

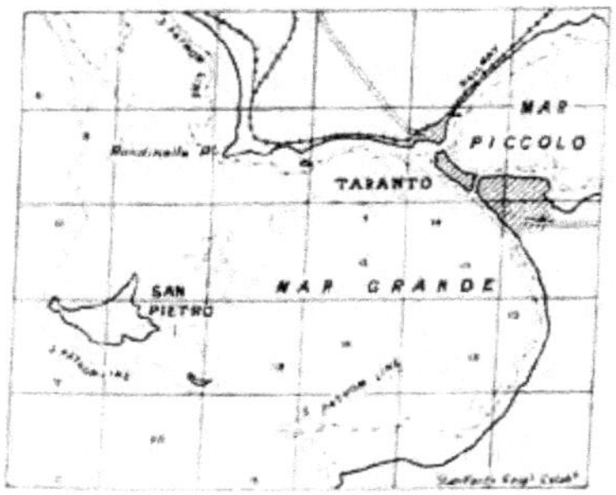

TARANTO.

SPEZIA.

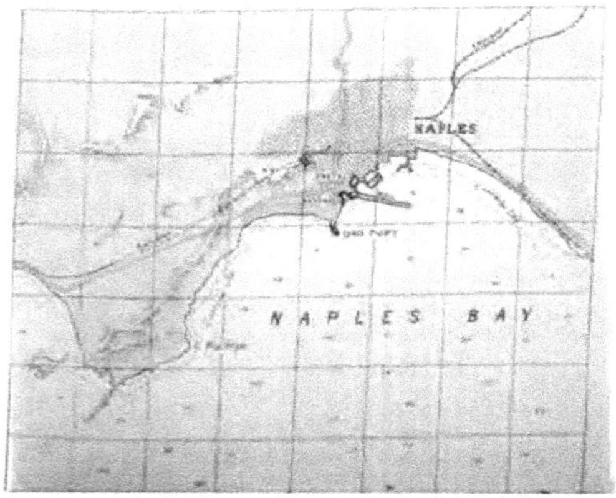

TABLE OF DISTANCES.

MADDALENA

to	knots.
Ajaccio	60
Bonifacio	20
Naples	210

SPEZIA

to	knots.
Ajaccio	160
Bizerta	430
Genoa	50
Leghorn	40
Maddalena	175
Malta	580
Naples	290
Taranto	620
Toulon	195
Venice	1000

NAPLES

to	knots.
Ajaccio	265
Bizerta	315
Bonifacio	225
Maddalena	210

TARANTO

to	knots.
Bizerta	550
Malta	310
Messina	220
Naples	380
Venice	575

VENICE

to	knots.
Pola	75
Taranto	575
Toulon	1197

TRIESTE to

	MILES.		MILI
Brindisi	372	**Pola**	1
Cattaro	337	Smyrna*	9
Constantinople*	1050	„ †	10
„ †	1152	Salonika*	9
Corfu	480	„ †	10
Fiume	110	Toulon	11
Gibraltar	1648	Wilhelmshaven§	32
Malta	730	„ ‡	374
Piraeus*	708	Zeebrugge§	305
„ †	839	„ ‡	375

* By Canal of Corinth
† By Cape Matapan.
§ By English Channel
‡ By North of Scotland and West of Ireland.

Royal Dockyards.

SPEZIA. Docks: No. 5, 702 × 105½ × 33 feet; No. 6, 508 × 90 × 33 feet; No. 4, 354¾ × 71½ × 29 feet; Nos. 3 and 2, 430 × 77 × 30 No. 1, 358 × 71 × 30 feet. Two large slips. (At Muggiano, there is a special testing dock for submarines belongi Fiat-San-Giorgio Co.) Total employees about 5300.
Admiralty Chart No. 155.

NAPLES (NAPOLI). One small Government dock, 247 × 62 × 19½ feet. New large dock, *projected*, 1918. One small slip. T employees 3029. Good anchorage. Commercial harbour and various new harbour works *projected*. There are two mon docks (*v.* preceding page). Admiralty Chart No. 1728.

CASTELLAMARE. No docks. Building yard. Two large slips, two small. Total employees 1920.

TARANTO. One dock, "Principe di Napoli," 709 × 108 × 32½ feet (2 sections). Floating dock, 365 × 61 × 25 feet (4800 t and another smaller (2900 tons). One large slip. Total employees 1561. Admiralty Chart No. 1643. Two 120 ton Krupps are mounted in a turret on San Pietro.

MADDALENA (SARDINIA). No docks. Total employees 113. **MESSINA** (SICILY): Government Dock, 243 × 71½ × 26 feet.

Torpedo Craft, Submarine and Naval Aircraft Stations: Spezia, Naples, Taranto. Also minor N Stations at Genoa, Maddalena, Gaeta, Messina.

ITALIAN DOCKYARDS.—ADRIATIC.

(To Uniform Scale. Divided into 2000-yard squares. Soundings in fathoms. Heights in feet).

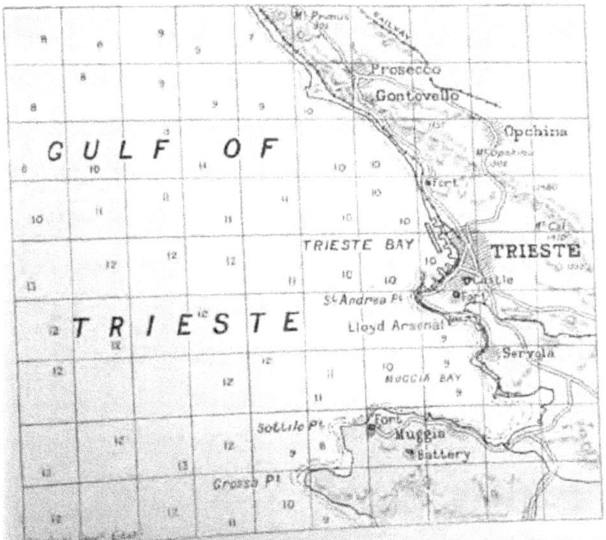

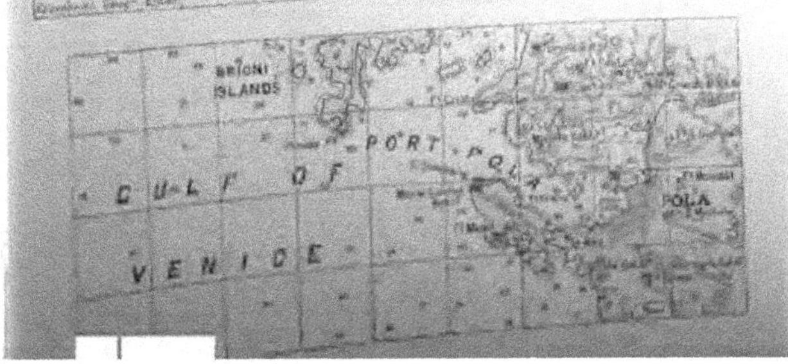

Arsenals.

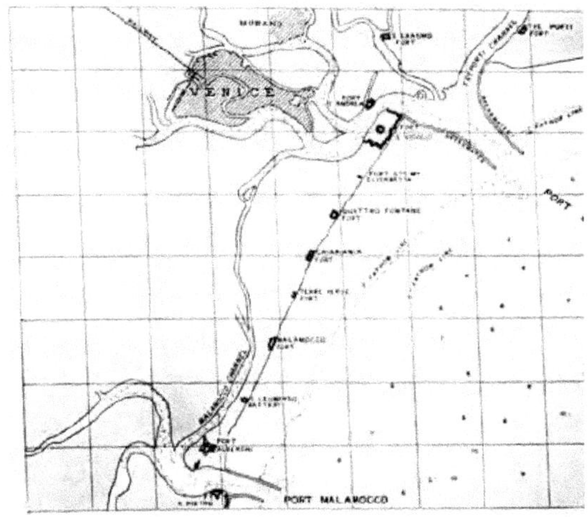

Royal Dockyards.

VENICE (Venezia). One dock, No. 4, 656 × 106 × 38½ feet; one dock for long c 525 × 78 × 25½ feet; one small dock No. 2, for third-class cruisers, 295 × 59 × 1 large slips, one small slip. Submarines built in this yard. Employees 27 Port Lido, 29 feet. Malamocco, 28 feet at low water. Commercial harl Admiralty Chart No. 1483.

TRIESTE. No details of Dockyard known. In D.Y. Basin there used to be at Ber Dock, 225 × 60 feet to lift 3750 tons. One building slip here. Admiralty C Rise of Spring Tide, 2 feet.

POLA (Late Austro-Hungarian Naval Base). Heavily fortified. Slips: One 690 fe smaller. Two dry docks: (a) 467 × 85 × 27½ feet and (b) 397 × 86 × 31½ feet. (c) 585 × 111 × 37 feet (22,500 tons); (d) 459½ × 90 × 32¾ feet (15,000 tons); (e feet; (f) small, only used for torpedo craft, 1,000 tons lift, completed June, 1500, 3 slips. Naval hospital. Admiralty Chart No. 202. Rise of Spring Tid

Torpedo Craft, Submarine & Naval Aircraft Stations: V Also minor Naval Station at Ancona.

(For BRINDISI, see next page).

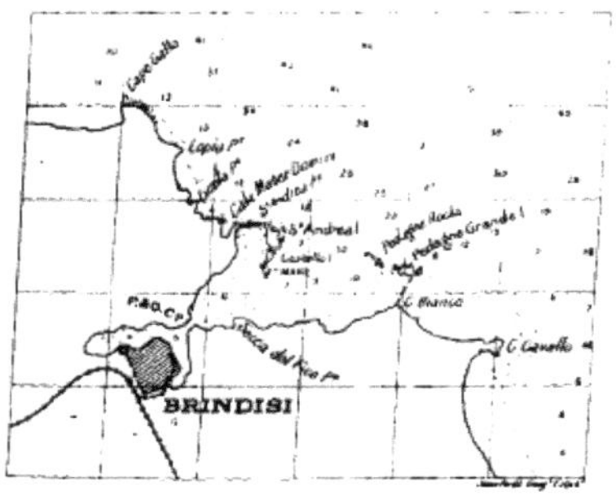

BRINDISI.

BRINDISI. No docks known to exist. Admiralty Chart No. 1492.

Uniforms.

(*Note.*—A five pointed silver star is worn on lapel of coat.)

INSIGNIA OF RANK ON SLEEVES. *NOTE.*—STAFF OFFICERS HAVE NO CURL.

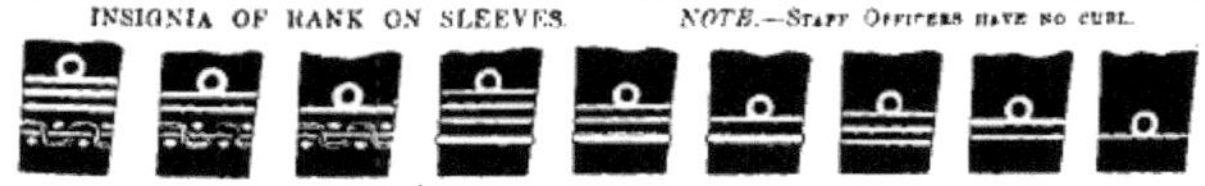

Ammiraglio	Vice-Ammiraglio.	Contr' Ammiraglio.	Capitano di Vascello.*	Capitano di Fregata.	Capitano di Corvetta.	Tenente di Vascello.	Sottotenente di Vascello.	Guardia-marina.
Admiral.	*V.-Admiral.*	*R.-Admiral.*	*Captain (senior).*	*Captain (junior).*	*Commander.*	*Lieut.*	*Sub-Lieut.*	*Midshipman.*

SOTTO-AMMIRAGLIO

(*) Sotto-Ammiraglio: New rank introduced, equivalent to British Commodore, and intermediate between Contr' Ammiraglio and Capitano di Vascello.

Lesser rank Allievo di marina (*Naval Cadet*).

Note.—Primo tenente has a gold band on shoulder strap for 12 years seniority. All lieutenants have stars on their shoulder straps corresponding to their stripes.

Other branches: Ingegnere (*Constructor*), Macchinista (*Engineer*), Medico (*Doctor*), Commissario (*Paymaster*)—with relative rank with the executives, without curl.

Colours: *Constructors*, dark purple; *Engineers*, black; *Doctors*, blue; *Paymasters*, red.

Note.—All officers under arms on duty wear a blue sash over right shoulder, ending in a blue knot at left hip; worn with belt. Officers on staff duty wear it on opposite shoulder, and without belt. Undress tunic has insignia of rank on shoulder straps, with stars as stripes. Senior lieutenants wear a piece of gold under the stars of shoulder strap.

Charts to Uniform Scale, divided into 2,000-yard squares. Soundings in fathoms. Heigh

Private Yards.

(*All undertake Naval Construction.*)

GIO. ANSALDO & C. (GENOA, COTRONA & SESTRI PONENTE, also at CORNIGLIANO & SA &c., consists of 28 establishments in various parts of Italy.) Build, engine, and com warships and mercantile vessels of all types. Slips: 3 large and 8 smaller at Sestri P three slips at Genoa Harbour, also other slips at Cantiere Savoia, Cornigliano a Area of works: hectares. Number of Employees, 60,000. (Also see ANSALDO-SAN GIORG

FRATELLI ORLANDO & C. (LEGHORN). Build and engine warships and mercantile ve types. Are affiliated to Acciaierie di Terni (Armour Plate and Shell Foundry) and Vi (Ordnance Works). No. of slips: Docks, one, 451 feet (*o.a.*) × 71¾ feet × 23 fee works: hectares. Number of employees: .

ODERO & C. (SESTRI PONENTE, near Genoa). Build and engine warships of all classes. N . Area of works: . Number of employees: . (Build battleships at Foce, n

(Above are the three greatest Italian warship-building firms).

ANSALDO-SAN GIORGIO, LTD. (MUGGIANO, SPEZIA). Build warships and mercantile vessels. in construction of Laurenti type submersibles and Diesel engines.

C. & T. T. PATTISON (NAPLES). Build destroyers, torpedo boats, mine sweepers, &c., al turbines, &c.

CANTIERE NAVALE FRANCO TOSI (LEGHORN & TARANTO). Build destroyers, torpedo b marines and mine layers or sweepers. Also boilers, turbines ("Tosi" type), and Diesel en

CANTIERE NAVALE RIUNITI (SPEZIA & PALERMO). Build torpedo craft and submarines.

OFFICINE MECCANICHE (MILAN & NAPLES). Build Belluzo turbines and other machinery.

OFFICINE INSUBRI (MILAN). Construct large Diesel engines for naval purposes.

STABILIMENTO TECNICO TRIESTINO (TRIESTE). Frontage, 2,100 feet. Area 123,000 2 Dreadnoughts' slips. 1 Patent slip, 450 tons capacity. 1 Floating Dock, 183 × 68 feet. Yard at Monfalcone for light cruisers, T.B.D., submarines, destroyed during war. Also San Rocco.

SOCIETA ANONIMA CANTIERI NAVALI DI QUARNARO (late Ganz-Danubius Co.) 5 Slips (2 up to 600 feet). Branch Yard at Port Rè for torpedo craft and submarines.

SOCIETA ANONIMA WHITEHEAD (FIUME). Slips for building torpedo craft and sub Large torpedo factory.

SOCIETA ESERCIZO BACINI (RIVA TRIGOSA).

S. A. V. I. N. E. M. (Societa Anonima Veneziana Industrie Navale e Meccaniche), GIUDECA, Build boilers, machinery, &c.

Fleet Establishment.

By Royal Decree the following vessels will be sold or disposed of before 31st Dec., 1920:—

(1) Second and third battleships laid down in 1903 or before.
(2) Light cruisers or scouts laid down in 1903 or before.
(3) Destroyers laid down in 1903 or before.
(4) Auxiliaries of all classes laid down in 1895 or before.
(5) Small craft for special services or local use no longer required.

OFFICIAL TABLE OF NAVAL ORDNANCE.

Official Designation :— Calibre mm/length cal.	381/40	305/46	305/40	254/45	254/45	254/40	203/45	190/45	152/50	152/45	152/40	120/50	120/45	102/45	102/35	76/50
Mark A = Armstrong, V = Vickers S =	A. V. e S.	A. e V.	A.	A.	V.	A.	A.	A. e V.	A.	S.	A.	A. e V.	A.	S.-A.	S.	A. e V.
Date of introduction.	1914	1909	1900-04	1907	1906	1899-03	1897	1908-1906	1918-1913	1911	1899-1916	1909	1913-18	1917	1914-15	1909
Designation by Calibre. c/m.	38.1	30.479	30.479	25.4	25.4	25.4	20.32	19.05	15.24	15.24	15.24	12	12	10.2	10.2	7.62
Calibre, in inches	15	12	12	10	10	10	8	7.5	6	6	6	4.75	4.75	4	4	3
Lengths: Total, in feet	51.87	47.77	41.707	39.07	38.715	34.609	31.126	29.22	25.94	23.42	20.77	20.38	18.38	15.715	12.247	13.271
Lengths: Rifled Bore, in inches	511.7	477.9	381.42	358.4	370.5	341.1	308.9	281.7	256.8	21.92	199.8	204.64	174.64	150.74	114.29	126
Lengths: Powder Chamber, in inches	84.5	97.7	92.12	74.94	74.91	55.10	47.4	51.65	44.6	44.6	37.7	28.64	35.03	27.16	23.50	22
Lengths: Bore, in calibres	34.11	37.3	31.95	35.84	37.05	34.11	38.41	37.5	42.77	36.54	33.31	43.31	36.98	37.53	28.46	42
No. of Grooves	120	72	48	60	70	42	32	44	36	58	28	36	36	40	32	28
Twist of Rifling, in calibres	44.927	30	40–600 600–30	30	60–30	30	60–20	90–30	33	36	30	30	30	—	—	30
Total Weight, in tons	83.56	62.99	51.77	34.40	35.339	30.511	19.586	14.478	8.100	7.025	6.503	3.662	4—035	2.327	1.900	1.122
Firing Charge: Armour-piercing projectile, lb	—	346	194	185	185	90.38	51.807	70.987	—	—	18.077	—	—	—	—	—
Firing Charge: Common Shell H.E., lb.	428.27	270.9	194	185	185	90.38	51.807	70.987	32.79	30.64	18.077	11.66	9.589	9.470	6.50	3.02
Weights: Armour-piercing projectile, lb.	—	907.2	942.7	404	494	494	269	201.39	—	—	107.07	—	—	—	—	—
Weights: Shell H.E., lb.	1925	884.4	881.74	469.8	489.8	189.8	256.23	498.5	110.22	103.61	102.36	48.74	48.74	30.31	30.31	14.05
Weights: Shrapnel, lb.	—	—	—	—	—	—	—	—	—	—	—	55.33	55.33	33.28	33.28	15.08
Bursting Charge: Armour-piercing projectile, lb.	—	16.63	9.79	4.37	4.37	4.37	2.003	2.332	—	—	1.014	—	—	—	—	—
Bursting Charge: Shell H.E., lb.	84.85	53.13	53.13	29.86	29.86	29.86	14.704	11.708	5.906	7.528	6.90	2.711	2.711	2.866	2.866	1.102
Bursting Charge: Shrapnel, lbs.	—	—	—	—	—	—	—	—	—	—	—	0.65	0.65	0.474	0.474	0.236
Muzzle Velocity, in ft. secs.	2206.6	2755.9	2346.8	2788.77	2788.77	2411.46	2.559	2788.77	2854	2723	2280	2788	2460	2788	2460	2460
Muzzle Energy —Total tons per sq. inch	20	18.63	15.75	17.71	17.71	15.09	16	17.98	18.37	16.86	13.73	18.37	15.75	18.37	18.37	18.37

15" 40 cal. in *Faa' di Bruno.*
12" 46 .. *Duilio* class (2 ships), *G. Cesare* class (2 ships), and *D. Alighieri.*
12" 40 .. *V. Emanuele* class (3 ships).
10" 45 .. *S. Marco* class (2 ships) and *Pisa.*
8" 45 .. *V. Emanuele* class (3 ships).
7.5" 45 .. *San Marco* class (2), and *Pisa.*
6" 50 .. *Duilio* class (2 ships).
4.7" 50 .. *Dante Alighieri, Marsala* class (2), *Quarto,* and *Libia.*

Note.—Above list of guns mounted in various classes is unofficial. 12 inch and 7.5 inch guns "Mobile Batteries" are dismountable.

Torpedoes, Mines, D.C., Air Bombs, &c.

No reliable details available. Torpedoes manufactured by State Arsenals and Whitehead & C., Societa Italiana, Naples.

Colour of Ships.

Battleships, Cruisers, Light Cruisers, Destroyers, are all very light grey. Destroyers have identity letters, taken from their names, painted on bows, *e.g.*, AC = *G. Acerbi*, AU = *Audace*, BR = *P. Bronzetti*, IV = *Impavido*, PL = *R. Pilo*, OR = *V. Orsini*, MT = *A. Mosto*, etc.

Official Classification of Ships. (C)—for Colonial Service.

"Navi di Battaglia."

1st Class: A. Doria, C. Duilio, C. di Cavour, G. Cesare, Dante Alighieri

2nd Class: V. Emanuele, R. Elena, Roma, San Giorgio, San Marco, Pisa

3rd Class: Varese, F. Ferruccio

Esploratori :—
Bari
Ancona
Taranto
Brindisi
Venezia
N. Bixio
Marsala
Quarto
Premuda
C. Rossarol
Leone
Tigre
Pantera
Aquila
Falco
C. Mirabello
A. Riboty
A. Poerio
G. Pepe
Agordat

1st Class: Sicilia, Bronte, Trinacria, Eritrea, Europa

S. Caboto (C)
E. Carlotto (C)
Minerva
Archimede

"Navi Sussidiarie."

2nd Class: Campania (C), Libia, Etna, Vulcano, Bengasi, A. Vespucci

3rd Class.

4th Class.

Guiliana
Cunfida
Misurata
Tobruk
C. Verri

Note.—Destroyers, "Sea-Going Torpedo Boats," Coastal T.B. and Submarines are as listed on subsequent Ship P[...] Above Table merely shows how Official Classification departs from order of Classification adopted for "Fig[...] Ships." On Ship Pages, tonnages and fuel capacities are stated in metric tons.

NO FUNNELS. **Scale : 1 inch = 80 feet.**

APE. VESPE. *(Lagoon G. B.)*

CARSO.

MONTE SANTO *class* (2). GRAPPA *class* (4)

(Mobile Batteries).

ONE FUNNEL. **Scale : 1 inch = 160 feet.**

BRONDOLO. MARGHERA. *(Lagoon G. B.)*

FAA' DI BRUNO. *(Monitor.)*

ANTEO. *(Salvage Ship.)*

ACHERONTE *class* (5).

TEVERE.

ATT

CAMPANIA.
Basilicata salved but non-effective until re-fitted.

S. CABOTO. *(Colonial G.B.)*

ERITREA. *(Transport).*

BRONTE. *(Fuel Ship).*

C. VERRI.

ARCHIMEDE.

GIULIANA.

TOBRUK.

A. MAGNAGHI. *(Surveying Ship).*

VULCANO. *(Repair Ship)*

CALABRIA.

GARIGLIANO.

A. VESPUCCI. *(Training Ship).*

TWO FUNNELS.

VARESE class (2 ships).
(Training Ship).

SICILIA (*Repair Ship*).
Note.—Two funnels forward closely paired. (Funnels half cased.)

AGORDAT.

MINERVA.

Torpedo Gunboats, used as Mine Layers.

ETNA.

PUGLIA.
Mine Layer.

TRINACRIA (*Royal Yacht*).

BENGASI (*Transport*).

A. DORIA.
C. DUILIO.

C. DI CAVOUR.
G. CESARE.
(No longer have searchlights before and abaft funnels.)

Scale: 1 inch = 160 feet.

ITALIAN WARSHIPS—RECOGNITION SILHOUETTES.

(Re-arr

THREE FUNNELS.

Quarto.

Bari.

Ancona.

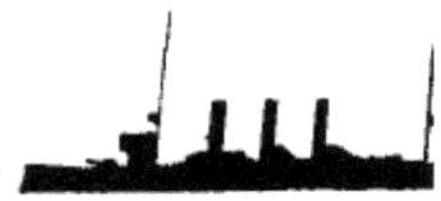

Libia.

Pisa.

Roma.

V. Emanuele. *
R. Elena.

* Differs slightly, c. Photo on later page.

FOUR FUNNELS.

Marsala.
N. Bixio.

Brindisi.
Venezia.

Taranto.

San Giorgio.
San Marco.

D. Alighieri.

TORPEDO CRAFT. **Scale: 1 inch = 160 feet.**

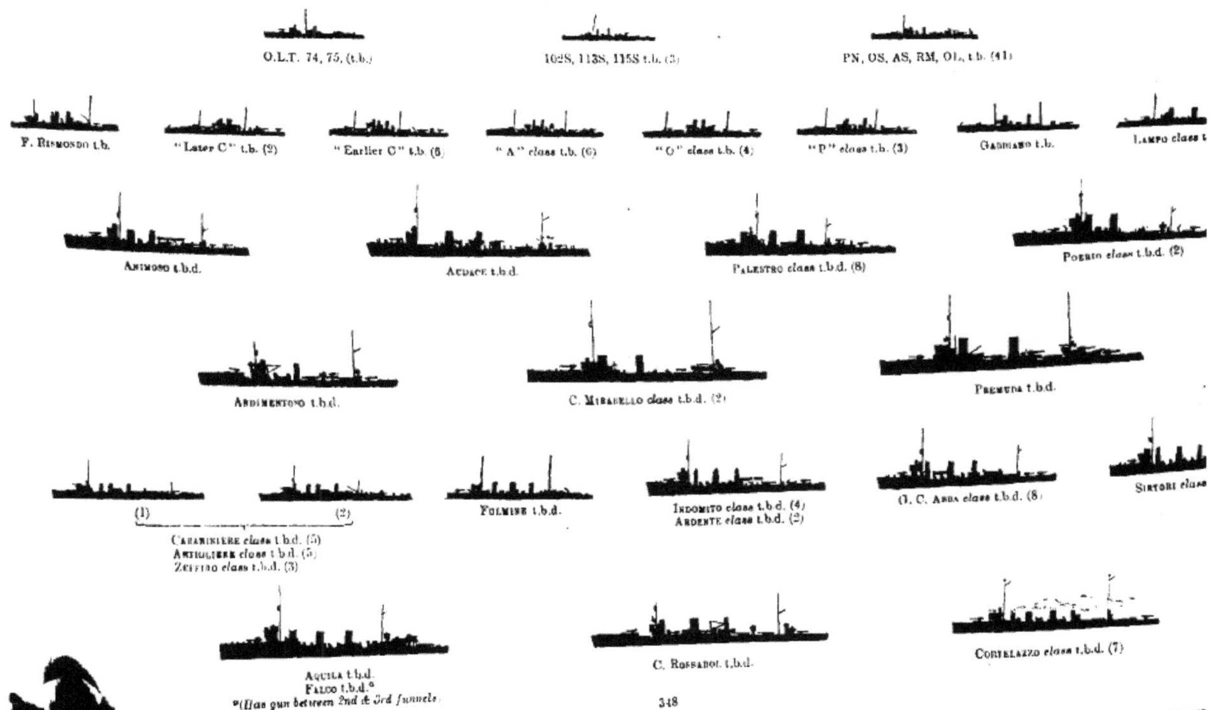

ANTI-SUBMARINE TYPES. **Scale: 1 inch = 80 feet.**

SUBMARINES. **Scale: 1 inch = 80 feet.**

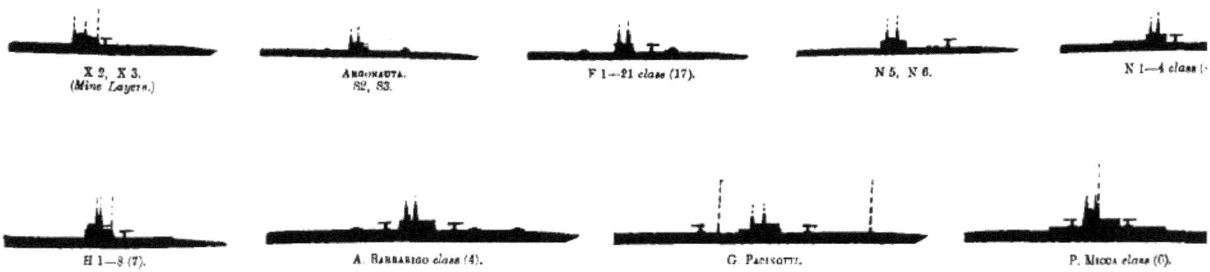

(Duilio Class—2 Ships).

CAIO DUILIO (April, 1913) and ANDREA DORIA (March, 1913)

Normal displacement, 24,500 tons. Complement, 1074.

Length (*p.p.*) 551·3 feet. Beam, 91·8 feet. *Mean* draught, 30·7 feet. Length *over all*, 575 feet, 9ins.

Guns (Armstrong or Vickers):
- 13—12 in., 46 cal. (A[5])
- 16—6 inch (45 cal.)
- 13—14 pdr.*
- 6—14 pdr. (A.A.)
- 2 M.G.
- 4 landing.

Torpedo tubes (18 inch):
- 2 broadside (*submerged*)
- 1 stern (*submerged*)

*3—14 pdr. temporarily removed during war

Armour (Terni):
- 9¾"–8" Belt (amidships)
- 5" Belt (ends)
- 1¾" and 1 ⅝" Decks
- 6¾" Lower deck side
- 9½" Barbettes
- 11 Turrets to these
- 6¾" Battery
- 11" Conning tower (fore)
- 6¾" Conning tower (aft)

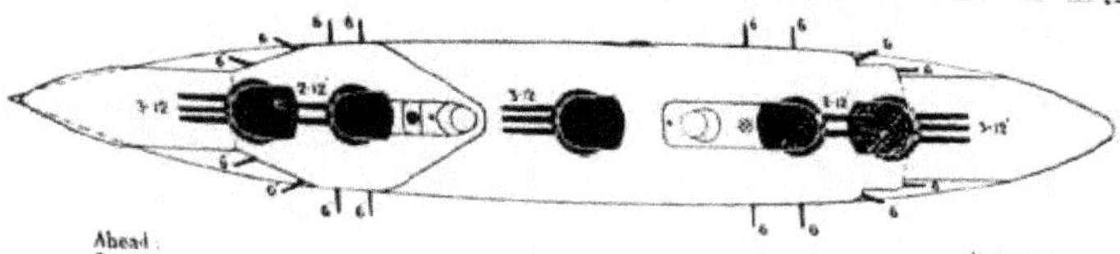

Ahead:
5—12 in.
4—6 in.

Astern:
5—12 in.
4—6 in.

Broadside: 13—12 in., 8—6 in.

Machinery: Parsons turbine. 4 screws. Boilers: 20 Yarrow. Designed S. H.P. 24,000 = 22 kts., but actually need 33,000–35,000 S.H.P. for this speed. Coal: *normal* 1000 tons; *maximum* 1200 tons and 800 tons oil fuel = about 4000 miles at 10 kts. and 2500 miles at 19 kts.

Gunnery Notes: / *Torpedo Notes:* As *C. di Cavour* class on next page. Range-finders mounted to the rear of roof of each barbette shield in armoured hoods and over each C.T. 3 inch (14 pdrs.) are 50 cal. Armstrong or 60 cal. Vickers. 8 searchlights. †All details unofficial.

Name	*Builder*	*Machinery*	*Laid down*	*Completed*	*Trials H.P.=kts.*	*Boilers*	*Best recent speed*
Caio Duilio	Castellamare D.Y.	Ansaldo	Apr.'12	1915	33,000 = 22	Yarrow	
Andrea Doria	Spezia	Ansaldo	Mar '12	May '15	32,000 = 21.5.	Yarrow	

General Notes.—Are improved *C. di Cavour* design, 16—6 inch being substituted for 18—4.7 inch. Otherwise—and excepting altered appearance—there is very little difference between the classes. Machinery, main guns, hull form and internal sub-division, exactly as *Cavours*, and also with same form of triple bottom. Designed by Engineer-Gen. Masdea. These ships exceed their designed displacement of 22,500 tons in the same manner as *Cavour* and *Cesare*, and require to develop 10,000 S.H.P. in excess of designed S.H.P., to reach their designed speed.

ANDREA DORIA. 1917 *Photo, O.*

Note.

To distinguish from *C. di Cavour* class, note that *both* tripod are before funnels. Funnels are also *equal* in height. Amidships barbette is on same level as *quarter-deck* triple—compare *Cavour* where it is on level with forecastle triple. General appearance lighter than *Cavours*.

1917 P

CAIO DUILIO. Has had a circular F.C. or R.F. top added to fore tripod mast, fore topmast.

[illegible]E DI CAVOUR (G. Cesare Class—2 Ships).

CONTE DI CAVOUR (Aug., 1911), **GIULIO CESARE** (Oct., 1911).

Normal displacement, 24,500 tons. Complement, 1074.

Length (*p.p.*) 554·3 feet. Beam, 91·8 feet. *Mean* draught, 30·7 feet. Length *over all*, 557 feet 9 inches.

Guns (Elswick):
13—12 in., 46 cal.
18—4·7 inch, 50 cal.
12—14 pdr., 50 cal.*
6—14 pdr. AA.
2 machine.
4 landing.

Torpedo tubes (18 inch):
2 broadside (*submerged*).
1 stern (*submerged*).

* 3—14 pdr. temporarily removed during war.

Armour (see *Notes*):
$9\frac{3}{4}$"–8" Belt (amidships)
5" Belt (ends)
$1\frac{1}{2}$" & $1\frac{1}{4}$" Decks
$6\frac{1}{4}$" Lower deck redoubt
$9\frac{1}{2}$" Barbettes
11" Turrets to these
5" Battery
11" Conning tower (fore)
$6\frac{1}{4}$" Conning tower (aft)

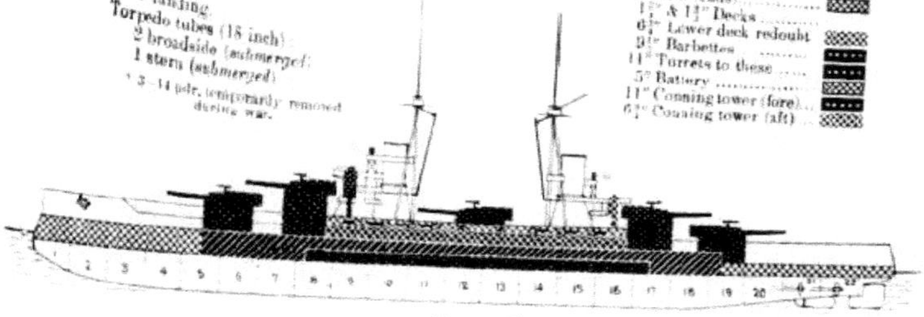

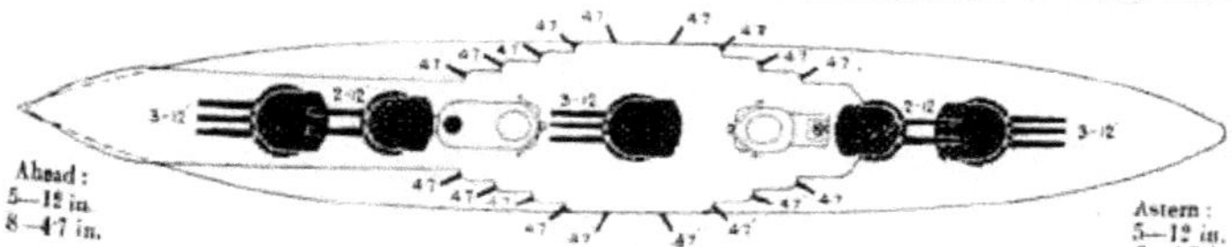

Ahead:
5—12 in.
8—4·7 in.

Astern:
5—12 in.
6—4·7 in.

Broadside: 13—12 in., 9—4·7 in.

Machinery: Parsons turbine. 4 screws. **Boilers:** various, see *Notes*. Designed S.H.P. 24,000 = 22 knots, but actually require 32,000 to 32,500 S.H.P. for this speed. Coal: *normal* 1000 tons; *maximum*, in *C. di Cavour* 1300 tons coal + 930 tons *oil*; in *G. Cesare* 1450 tons coal + 850 tons oil. Endurance: about 1540 miles at 10 kts. and 2900 miles at 19 kts.

Gunnery Notes.—Arcs of fire: End triples 300°, super-firing twin barbettes 310°, amidships triple 130° each beam. 14 pdr. can be transferred to different mountings in various positions in these ships. These notes are unofficial.

Armour Notes. *C. di C.*, Terni; *G. C.*, Bethlehem. Hull has triple bottom around keel, but not near or above turn of bilges.

Name	*Builder*	*Machinery*	*Laid down*	*Completed*	*Trials Full Power.*	*Boilers*	*Best recent speed*
					H.P. / *kts.*		
C. di Cavour	Spezia D.Y.	Orlando	Aug. '10	Jan. '15	32,300 = 22	20 Yarrow	
G. Cesare	Ansaldo	Ansaldo	June '10	Mar. '14	30,740 = 21·7	24 Babcock	

General Notes.—Designed by Engineer Gen. Masdea. Were designed to displace 22,500 tons at 27½ feet mean draught. Alterations during construction added 2000 tons to displacement and 3 feet to mean draught. They can only reach their designed speed with difficulty. Third ship of class, *Leonardo da Vinci*, blew up at Taranto, August, 1916; salved and docked upside down, 1920. Hull to be used as Oil Pontoon.

Conte di Cavour. 1910 *Photo*, [illegible]

[illegible]

Giulio Cesare. 1910 *Photo*, [illegible]

[illegible]

DANTE ALIGHIERI (August, 1910).*

Normal displacement, 20,500 tons. Complement, 991.

Length (p.p.), 519·7 feet. Beam, 87·4 feet. *Mean* draught, 30·1 feet. Length *over all*, 549·5 feet.

* Official Italian List gives 20 Oct., 1910, as launching date.

Guns (Elswick):
- 12—12 in., 46 cal. (A*)
- 20—4·7 inch, 50 cal.
- 12—14 pdr., 50 cal.*
- 6—14 pdr., AA.
- 2 M.G.
- 4 landing.

Torpedo tubes (18 inch.):
- 2 broadside (*submerged*).
- 1 stern (*submerged*).

* 3—14 pdr. temporarily removed during war.

Armour (Terni):
- 9¾" Belt (amidships)
- 6" Belt (forward) ......
- 4¾" Belt (aft) ......
- 1¼" & 1½" Decks ......
- 8¾" Lowerdeck redoubts
- 9" Barbettes ....
- 11" Turrets to these
- 4" Secondary Turrets (4)
- 4" Batteries ....
- 11" Conning tower......
- 8" C.T. (aft) ....

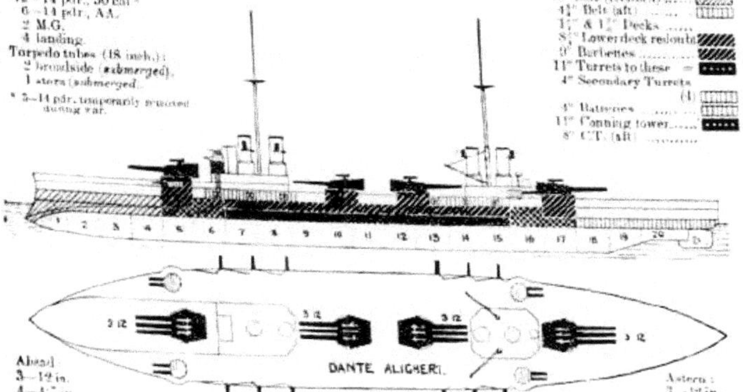

Ahead:
3—12 in.
4—4·7 in.

Astern:
3—12 in
4—4·7 in

Broadside: 12—12 in.
10—4·7 in.

Machinery: Parsons turbine. 4 screws. Boilers: 23 Blechynden. Designed S.H.P. 24,000=23 kts. but actually requires about 31,000 H.P. for this speed. Coal: *normal* 1000 tons; *maximum* 2200 tons. + 300 tons oil fuel=4520 miles at 10 kts. and about 5000 miles at 19 kts.

*Gunnery Notes.—Triple turrets are to Armstrong design, hydraulic power with special Ansaldo electric gear as reserve. 4·7 inch reported to have been fitted with improved breech mechanism in 1913-14 to give faster rate of fire.

* Not from any official source.

Name	Builder	Machinery	Laid down	Completed	Trials H.P. = kts.	Boilers	Best recent speed
D. Alighieri	Castellamare	Ansaldo	June '09	Sep. '12	31,290 = 23·8	Blechynden	

General Notes.—Designed by Engineer-Gen. Masdea. Was originally designed for 19,300 tons displacement, with a draught of 28½ feet mean, but alterations made during construction resulted in an addition of 1000 tons to displacement and about 2½ feet to draught. Consequently, she only attains her designed speed with difficulty. Triple steering gear, steam, special Ansaldo electric gear and manual power

1920 *Photo. O.*

At Fiume, with D'Annunzio's Forces.

Photo

Detail view of amidships triple barbette.

(V. Emanuele Class—3 Ships.)

EMANUELE (Oct., 1904), REGINA ELENA (June, 1904), ROMA (April, 1907).

Length (*p.p.*), 435 feet. Beam, 73½ feet. { *Maximum* draught, 27½ feet. / *Mean* „ 25¾ „ } Length *over all*, 475 feet 5 in.

Displacement, 12600 tons. Complement, 743.

Guns:
2—12 inch, 40 cal.
12—8 inch, 45 cal.
16—14 pdr., 40 cal.
4—14 pdr., A.A.
2—3 pdr.
2 Maxims.
2 landing.

Torpedo tubes (18 in.):
2 *submerged* (broadside).

Armour (Terni):
9·8" Belt
4" Belt (ends)
2·4" Deck (reinforcing)
·8" Bulkheads
8" Barbettes (N.C.)
8" Turrets
8" Lower deck (side)
2" Lower deck (bow)
6" Secondary turrets (6)
7½" Tertiary battery
10" Conning tower (N.C.)

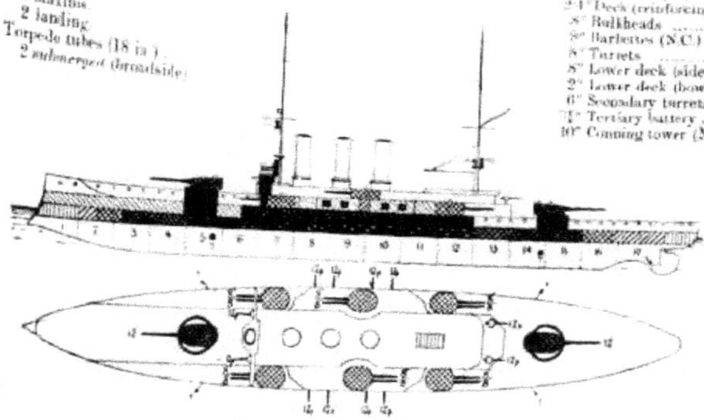

Ahead:
1—12 in.
5—8 in.

Astern:
1—12 in.
8—8 in.

Broadside: 2—12 in., 6—8 in.

Machinery: 2 sets 4 cylinder vertical inverted triple expansion. Designed H.P. 19,000 = 21·5 kts. 2 screws. **Boilers**: *V. Emanuele* and *R. Elena*, 28 Belleville; *Roma*, 18 Babcock & Wilcox. **Coal**: *normal* 1000 tons; *maximum* 2000 tons.

Aircraft and Gunnery Notes.—14 pdr. AA. are mounted singly on crowns of 12-inch turrets and central 8-inch turrets, but *Roma* has no AA. guns or crown of fore 12-inch turret. Nearly all R.F. positions over fore C.T. and behind each central 8-inch turret, 2—14 pdr. on each 8 in. turret, the 14 pdr. battery ports being plated up now. 1 large SL. in 2 tops on each mast and small SL. at after end of boat deck.

Name	*Builder*	*Machinery*	*Laid down*	*Completed*	*Last Refit*	*First Trials:* 24 hours	6 hours	*Boilers*	*Best recent speed*
V. Emanuele	Castellamare	Orlando	Sept.'01	Sept.'07		16,000 = 21	19,424 = 21·5	Belleville	
R. Elena	Spezia	Odero	Sept.'01	May '07		15,673 = 20·33	19,298 = 22·7	Belleville	21·2
Roma	Spezia	Ansaldo	Sept.'03	Dec.'08		16,000 = 20·5	21,000 = 22	Babcock	22

General Notes.—Designed by the late General Cuniberti. Average cost per ship, £1,000,000. *Roma* has less superstructure and is somewhat lighter, but has the same horse-power (not fitted as flagship).

V. Emanuele. *Note crows' nests high on each mast.* 1919 *Photo, Comdr. Hollert*

Note.

R. Elena as V. Emanuele, but differs thus: (*a*) No crows' nests high up on masts (*b* SL. platforms rather higher up foremast. (*c*) Flying bridge over C.T. as shown by Sil *V. Emanuele* and *R. Elena*. *Napoli* sold 1920. These three ships will probably be within the next twelve months.

Roma. 1919 *Photo, Comdr. Hollerton.*

Note different shape of funnels. Foremast is against fore funnel, and has F.C. top. "Starfish" low on mai

(San Giorgio Class—2 Ships.)

SAN GIORGIO (July, 1908), SAN MARCO (December, 1908).

Displacement { *San Giorgio*, 10,200 tons / *San Marco*, 11,000 tons } Complement { *S.G.* 689 / *S.M.* 691 }

Length (*p.p.*) 429 feet 11 in. Beam, 69 feet. { *S.G.* *mean* draught 24 feet, *max.* 26¼ feet. / *S.M.* *mean* „ 25¼ feet, *max.* 27 feet. }

Length *over all*, 462¼ feet.

Guns:
- 4—10 inch, 45 cal.
- 8—7·5 inch, 45 cal.
- 10—14 pdr.
- 6—14 pdr. AA.
- 2—3 pdr.
- 2 machine.
- 2 landing.

Torpedo tubes (18 inch):
- 2 *submerged* (broadside).
- 1 *submerged* (stern).

Note to Plans.
Now rigged with *two* masts.

Armour (*S.M.*, Midvale; *S.G.*, Terni):
- 8" Belt (amidships) ...
- 2¾" Belt (ends) ...
- 1½" & 1½" Decks...
- 7" Lower deck redoubt
- 8" Main barbettes ...
- 5½" Turrets to these
- 7" Citadel
- 6¼" Secondary turrets
- 9¾" Conning towers (f)

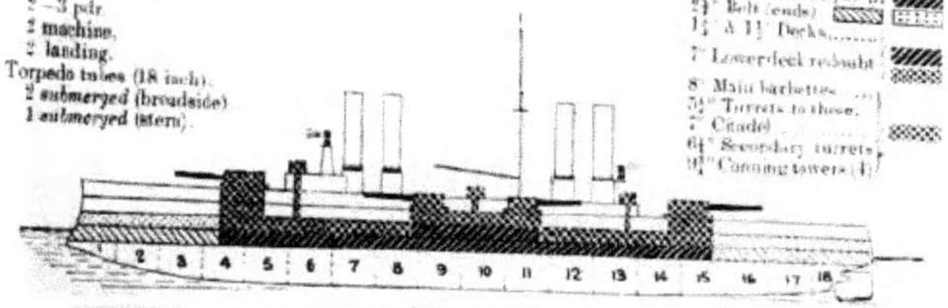

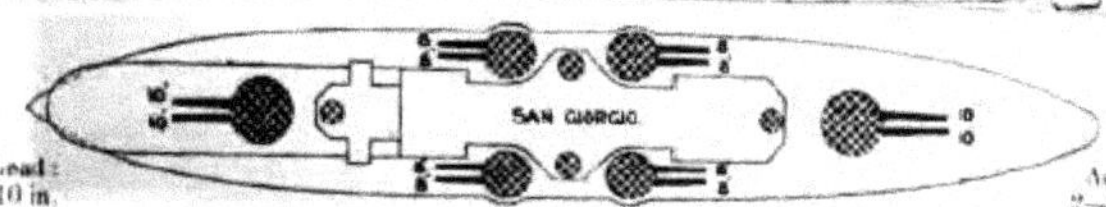

Ahead:
3—10 in.
4—7·5 in.

Astern:
2—10 in.
4—7·5 in.

Broadside: 4—10 in., 4—7·5 in.

Machinery: *S. Marco*, Parsons turbine; 4 screws. *S. Giorgio*, 2 sets 4 cylinder inverted triple expansion; 2 screws. Boilers: various. See *Notes*. Designed H.P., *S. Giorgio*, 18,000 I.H.P. = 22·5 kts.; *S. Marco*, 20,000 S.H.P. = 23 kts. Coal: *normal* 700 tons; *maximum* 1570 tons. Endurance: at 10 kts., 6270 miles for *San Giorgio*, 4800 miles for *San Marco*; 2640 miles at 20 kts. for *San Giorgio*, about 2500 miles at 21¼ kts. for *San Marco*.

Gunnery Notes.—All guns electrically controlled. Central pivot mountings. Fore 10 inch, 31 feet above waterline; after 10 inch, 22 feet; 7·5 inch guns 22 feet above water. Armoured fire-control towers between 7·5 inch barbettes. (These details are not official).

Name	*Builder*	*Machinery*	*Laid down*	*Completed*	*Trials: full power*	*Boilers*	*Best recent speed*
					H.P. — *Kts.*		
S. Giorgio	Castellamare D.Y.	Ansaldo	Mar.,'05	June '10	19,595 = 23	14 Blechynden	
S. Marco	Castellamare D.Y.	Ansaldo	Jan.,'07	July '10	23,000 = 23·75	14 Babcock	

General Notes.—Designed by Engineer Lieut-Gen. Masdea. In 1911, *S. Giorgio* grounded badly off C. Posillipo (Bay of Naples) but was salved and repaired. She was again badly damaged by stranding in Straits of Messina early in 1914.

San Marco. 1919 *Photo, Comm. Hollerton* [illegible]

San Giorgio. 1919 *Photo, Comm. Rolle* [illegible]

To distinguish between these two ships note the following differences:

SAN GIORGIO.

Side built up under fore bridge into superstructure.

FOREMAST: has small top with wide yard under same. Topmast is *abaft*. Both yards wide and far apart. This mast is rather shorter than main mast.

Searchlights on bridge between foremast and 1st funnel.

1ST FUNNEL: has no searchlight platform.

2ND FUNNEL: searchlights *low* on after side.

MAINMAST: has box top.

3RD FUNNEL: bridge across fore side.

4TH FUNNEL: searchlights *low* (on after side).

ALL FUNNELS: *half-cased* for about quarter of height.

SAN MARCO.

Side not built up under fore bridge.

FOREMAST: has no top. Two narrow [illegible] together and high up on topmast [illegible] *before* mast. Mast is equal in height to [illegible]

1ST FUNNEL: Searchlights built off fore sid[e] [illegible]

2ND FUNNEL: searchlights *high* on *fore* side [illegible]

MAINMAST: has no top.

3RD FUNNEL: no bridge.

4TH FUNNEL: searchlights *high* (on after si[de]) [illegible]

ALL FUNNELS: *plain* with low bonbinrs

1905. ITALIAN ARMOURED CRUISER.

PISA (Sept., 1907)

Displacement, 10,600 tons. Complement, 685.

Length (*p.p.*), 426½ feet. Beam, 68 feet 11 ins. *Mean* draught 24½ feet. *Maximum* draught, 25½ feet.
Length *over all*, 460 feet 11 ins.

Corrections to Plans.

Now has foremast with R.F. top and searchlight bridge across fore side of first funnel.

Guns (Vickers):
- 4—10 inch, 45 cal.
- 8—7·5 inch, 45 cal.
- 14—14 pdr., 50 cal.
- 6—14 pdr. (AA)
- 2—3 pdr.
- 2 M.G.
- 2 landing.

Torpedo tubes (18 inch):
- 2 *submerged* (broadside)
- 1 *submerged* (stern)

Armour (Vickers):
- 8" Belt (amidships)
- 3¼" Belt (ends)
- 1½" & 1¾" Decks
- 7" Lower deck redoubt
- 6¾" Main barbettes
- 8" Turrets to these
- 7" Citadel
- 7" Secondary turrets
- 7" Conning tower

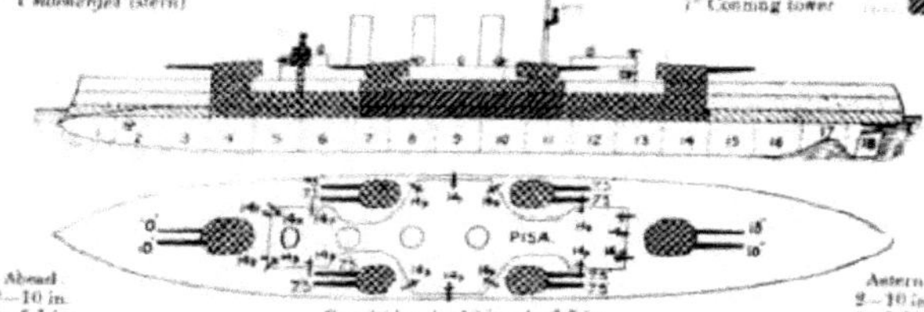

Ahead: 2—10 in. 4—7·5 in.

Broadside: 4—10 in., 4—7·5 in.

Astern: 2—10 in. 4—7·5 in.

Machinery: 2 sets 4 cylinder inverted triple expansion. 2 screws. Boilers: 22 Belleville. Designed H.P. 20,000=23 kts. Coal: *normal* 680 tons; *maximum* 1570 tons. Endurance: 6,270 miles at 10 kts., 2,300 miles at 21¼ kts.

Gunnery Notes.—All guns electrically controlled. Central pivot mountings. Guns 22 feet above water.

Name	*Builder*	*Machinery*	*Laid down*	*Completed*	*Trials*	*Boilers*	*Best recent speed*
Pisa	Orlando, Leghorn	Orlando	[illegible] '06	Dec., '07	24 hrs. = 21·4 kts. 4 hrs. 20,[illegible] = 23·[illegible] "	Belleville	

General Notes.—Designed by Engineer-General Orlando. The *San Giorgio* class (*see* preceding page) is a slightly enlarged *Pisa*. **A sistership, *Amalfi*, lost in the war.**

1919 Photo, Commr. R. D. Hollerton.

Note.

The addition of a foremast to this Armoured Cruiser, during the War, has considerably changed her appearance. It is possible for her to be mistaken now for one of the two Battleships, *Victor Emanuele* and *Regina Elena*. *Pisa* can easily be picked out by these features: (*a*) flush-decked hull from end to end; (*b*) hull not recessed for end-on fire of beam turrets; (*c*) main barbettes close to central superstructure; (*d*) only two small turrets slightly sponsoned, on each beam; (*e*) foremast well away from fore funnel; (*f*) searchlight bridge across fore funnel; (*g*) funnels thicker and not so high as those in *V. Emanuele* class; (*h*) small space between mainmast and third funnel; (*i*) superstructure abaft mainmast very long; (*j*) sloping stern with sternwalk.

2 Old Armoured Cruisers.

F. Ferruccio.

1929 Photo, Mr. H. C. Locker, Cadet, R.N.

(Both now serve as Cadets' Training Ships.)

VARESE (Aug., 1899), **FRANCESCO FERRUCCIO** (April, 1902). Displacement, 7350 tons. Complement 532. Length (p.p.), 344 feet. Beam, 59¾ feet. Maximum draught, about 24 feet. Guns: 1—10 inch, 45 cal., 2—8 inch, 45 cal., 2 or 4—5 inch, 40 cal., 6—14 pdr., 5—14 pdr. AA., 1 machine, 2—3 pdr., 1 landing. Torpedo tubes (18 inch): 4 *above water*, in 6 inch casemates. Armour (Terni): 6" Belt (amidships), 3" Belt ends, 1½" Deck, 6" Lower deck (redoubt), 6" Barbettes (N.C.), 6" Hoods to these (N.C.), 6" Battery (N.C.), 5" Battery bulkheads, 6" Conning tower (N.C.). Machinery: 2 sets triple expansion. 2 screws. Boilers (24): *Varese*, 24 Belleville; *F. Ferruccio*, 24 Niclausse. Designed H.P. 13,500 = 20 kts. Coal: *normal* 650 tons; *maximum* 1215 tons.

General Notes.—*Varese* built by Orlando, 1898—1901. *F. Ferruccio* at Venice, 1899—1904. Made about 20 kts. on trials, but present speed about 17 kts. *G. Garibaldi* of this class lost during the war.

1 ex-German Light Cruiser.

From a drawin[g]

BARI ex-German *Pillau*, ex-Russian *Muravieff Amurski*, April, 1914.

Displacement, 4350 tons. Complement, 372.

Length (p.p.), 403 feet. Beam, 46 feet. { *Mean* draught, 16 feet. *Max.* " 19 feet. } Length (o.a.

Guns (German):
8—5·9 inch, 50 cal., S.A.
2—22 pdr. anti-aircraft
Torpedo tubes (19·7 inch, German): 2 *above water*.

Armour:
1½" De
¾" Dec
3" Dec
3" Con

Note.

Will probably be re-armed with Italian guns and torpedo tubes. Plans accordingly deferred for the present.

Machinery: Schichau (Melms & Pfenninger) turbines. 3 screws. Designed H.P. Boilers: Yarrow (may be altered to Schulz Thornycroft). Coal: *normal* 500 tons; *max* and 250 tons oil fuel.

Name	*Builder*	*Machinery*	*Laid down*	*Completed*	*Trials: F. P.*
Bari	Schichau, Danzig	Schichau	Apl., '13	Dec. '15	

General Notes.—A sister ship, *Elbing*, sunk in the Battle of Jutland. Both vessels were seized by G[ermany on outbreak] of war with Russia, for whom they were originally laid down. *Bari* taken over by Italy, 192[0].

1912 Ex-German Light Cruiser.

GRAUDENZ. 1919 *Illustration.*

ANCONA (Ex-German *Graudenz*), October, 1913.

Normal displacement 4900 tons. Complement 364.

Length (*waterline*) 456 feet. Beam 45 feet. *Mean* draught 16 feet. (*Max* 17 feet).

Guns (German):
- 7—5.9 inch, 50 cal.
- 2—22 pdr. anti-aircraft
- 2 machine

Torpedo tubes (19.7 inch)
German: 2 *submerged* and 2 *above water.*

Can carry 120 mines each

Armour:
- 4" Belt (amidships)
- 2¼" Belt (ends)
- 2" Deck (amidships)
- 4" Conning tower

Note.

Will probably be rearmed with Italian guns and T.T.

Machinery: "Marine Type" Turbine. 4 screws. Boilers: 14 Schulz-Thornycroft. Designed H.P. 26,000 = 27.25 kts. Coal: *normal* 470 tons; *maximum* 1300 tons. Oil: 220 tons.

Armour Notes.—Belt is rather deeper than in *Taranto.*

General Notes.—Belongs to German 1912 Programme. Built by Kiel D.Y. Laid down 1912 and completed 1914. Radius of action: *about* 5,500 to 6,000 miles at cruising speed. Taken over by Italy, 1920. *Strasbourg* (ex-*Regensburg*), of French Navy, more or less a sister ship to above, *q. v.* for notes on various armaments mounted 1914-18.

1911 Ex-Austrian Light Cruisers.

Present appearance as silhouette.

MODIFIED "ADMIRAL SPAUN" TYPE—2 SHIPS.

BRINDISI (ex-Austrian *Helgoland*, Nov., 1912). **VENEZIA** (ex-Austrian *Saida*, Oct., 19

Normal displacement, 3500 metric tons. Complement, 318.

Length (*w.l.*), 410¼ feet. Beam, 42 feet. *Mean* draught, 15 feet. Length *over all*, 428¼ fe

Guns (Austrian):
Brindisi
- 9—4.1 inch.
- 1—3 inch (anti-aircraft).
- 1 machine.
- 1 landing.

Torpedo tubes (in all) 6—18 inch:
3 pairs *above water.*

Guns (Austrian):
Venezia
- 8—3.9 inch (50 cal. semi-auto).
- 1 machine.
- 1 landing.

Armour (steel):
- 2½" Belt (amidships
- 2" Bulkheads ......
- ¾" Deck ....

Note.

Will probably be re-armed with Italian guns and T.T.

Machinery: Turbine (*See Notes*). 4 screws. Boilers: 16 Yarrow. Designed H.P. 25,000 =
Coal: *normal*, 450 tons; *maximum*, 750 tons. Radius of action as *Ad. Spaun.*

Name	*Builder*	*Machinery*	*Laid down*	*Completed*	*Trials*	*Turbines*	*Boilers*
Saida	Montfalcone	Prager Masch	Sep.'11	May,'14		Pfennigér	Yarrow
Helgoland	Fiume	Ganz-Danubius	Oct.'11	Oct.,'14		A.E.G.	Yarrow

General Notes.—Ex-Austrian ships, assigned to Italy, 1920. Sister ship, *Thionville* (ex-*Novara*), now in French

1910 Ex-German Light Cruiser.

TARANTO (ex-German *Strassburg*, Aug., 1911).

Normal displacement, 4550 tons; 5100 tons *full load*. Complement, 370-373.

Length (*w.l.*), 446¼ feet. Beam, 43·6 feet. *Mean* draught, 16¾ feet.

Guns (German):
- 7—5·9 inch, 50 cal.
- 2—22 pdr. anti-aircraft } Had only 1
- 2 machine } gun, 1919.

Torpedo tubes (19·7 inch, German):
- 2 *submerged*.
- 2 *above water*.

Can carry 120 mines.

Armour:
- 3½" Belt (amidships)......
- 2¼" Belt (ends) ....
- 2" Deck (amidships)......
- 4" Conning tower ....

Note.

Will probably be re-armed with Italian guns and T.T.

Machinery: Turbine. 4 screws. Boilers: 16 Schulz-Thornycroft. Designed H.P. 26,000 = 27 kts. Coal: *normal*, 750 tons; *maximum*, 1200 tons. Oil: 130 tons.

Armour Notes.—Belt is very narrow, and at full load practically submerged.

Engineering Notes.—4 boilers are oil burning.

Name	*Builder*	*Machinery*	*Laid down*	*Completed*	*Trials (mean): 6 hours*	*Trials (mean): Best hour*	*Turbines*	*Present best speed*
Taranto	Wilhelmshaven Yard		Ap. '10	Dec '12	25,647 = 26·9	33,712 = 28·28	Parsons	

***Notes.*—Begun for German Navy under the 1910 Programme. Taken over by Italy, 1920. Sister ship, *Mulhouse* (ex-*Stralsund*), now in French Navy, to which refer for any further notes.**

1913 Italian (Colonial) Cruiser.

1919 *Photo by courtesy of the Minis*

CAMPANIA (July, 1914).

Displacement, 2480 tons (sheathed and coppered). Complement, 240.

Length (*p.p.*) 249·4 feet. Beam, 41¾ feet. *Max.* draught, 16¾ feet. Length *over all*

Guns:
- 6—6 inch, 45 cal.*
- 2—14 pdr.*
- 3—14 pdr. AA.

Torpedo tubes:
- 2 *above water*.

* *These guns temporarily removed during war. May not yet be remounted.*

Ahead:
3—6 in.

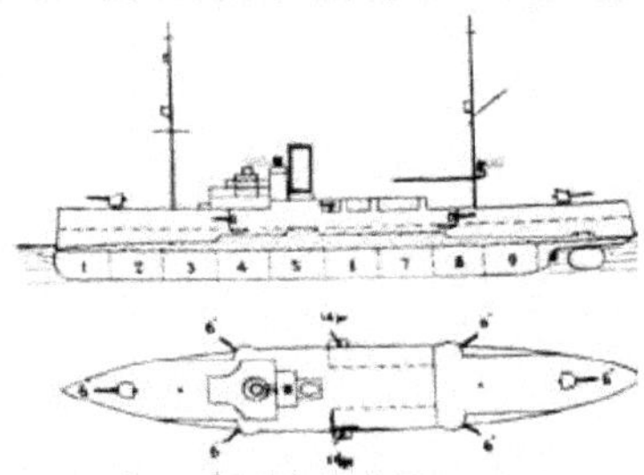

Broadside: 4—6 in.

Machinery: 2 sets vertical triple expansion. 2 screws. 4 cylindrical boilers. De = 16·5 kts. (*not attained on trials*). Coal: *normal* 200 tons, *full load* 500 tons = 4300 m about 2000 miles at 15 kts.

Name	*Builder*	*Machinery*	*Laid down*	*Completed*	*Trials:*
Campania	Castellamare D.Y.	Off. Mec., Naples	Sept. '11	Aug. '16	About 500 H.P. = 15·7 kts.

General Notes.—Built for Colonial Service. An improved *Calabria* type. *Basilicata* sunk by boiler ex Suez Canal, 1919. Re-floated 1920, but non-effective until repaired.

Note.—**These two Ships and "Quarto" (v. next page) are officially rated as "Esploratori" (Scouts) with the Destroyers (or Flotilla Leaders) of over 1000 tons.**

(Nino Bixio Class—2 Ships).

MARSALA (March, 1912), **NINO BIXIO** (Dec., 1911).

Displacement, 3,500 tons. Complement, 298.

Length (*p.p.*), 429·8 feet. Beam, 42·8 feet. *Mean* draught, 13·4 feet. Length *over all*, 460 feet.

Guns:
6—4·7 inch (50 cal.)
6—3 inch (14 pdr.)
2—2 pdr. AA.
Torpedo tubes (18 inch):
2 *above water*.
(Also mine-laying gear.)

Armour:
1½" Deck (amidships)
¾" Deck (ends) ......

Machinery: Curtis turbine. 3 screws. Boilers: 14 Blechyndes (12 oil-burning, 2 coal-burning). Designed S.H.P. 22,500 = 28 kts. Fuel capacity: 130 tons coal + 520 tons oil = 3000 miles at 10 kts.

Name	*Builder*	*Machinery*	*Laid down*	*Completed*	*Trials kts.*	*Boilers*	*Best recent speed*
Marsala	Castellamare D.Y.	Off. Mecc. Naples	Feb. '11	Sept. '14	= 27·?	Blechynden	
Nino Bixio	Castellamare D.Y.		Feb. '11	May '14	=	Blechynden	

General Notes.—Carry 200 blockade mines. Designed by Engineer-General Rota.

1919 *Photo, Comsnr. H. D. Holbarton.*

Marsala. To distinguish, note (a) only one top on foremast; (b) only one yard, high up on fore top ...

Nino Bixio. To distinguish note the following ...

1919 *Photo, Comsnr. H. D. Holbarton.*

1909 ITALIAN SCOUT CRUISER (Mine Layer).

1919 Photo, Commr. Holberton, R.N.

QUARTO (Aug. 1911).

Displacement, 3,280 tons. Complement, 253.

Length (p.p.), 413·4 feet. Beam, 42·1 feet. *Mean* draught, 13 feet. Length *over all*, 431¼ feet.

Guns:
6—4·7 inch (50 cal.)
6—3 inch (14 pdr.)
2—2 pdr. AA.
Torpedo tubes (18 inch):
2 *above water*.
(Also mine-laying gear.)

Armour:
1½" Deck (amidships)
¾" Deck (ends)

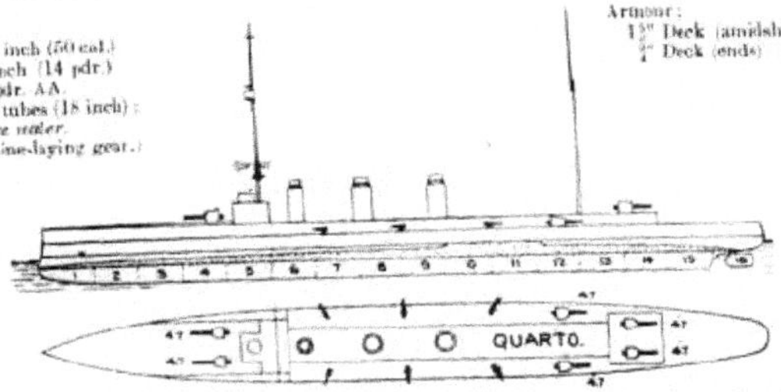

Machinery: Parsons turbine. 4 screws. Boilers: 10 Blechynden (8 oil-burning, 2 coal-burning). Designed S.H.P. 25,000 = 28 kts. Fuel capacity: *maximum*, 50 tons coal + 490 tons oil = 3270 miles at 10 kts., or 1220 miles at 25·7 kts.

Name	*Builder*	*Machinery*	*Laid down*	*Completed*	*Trials H.P.*		*Trials kts.*	*Boilers*	*Best recent speed*
Quarto	Vsdee Y.	Odero	Oct.,'09	Sep.,'12	28,400	=	28·6	Blechynden	29·5

General Notes.—Carries 200 blockade mines. Designed by Engineer-Col. Truccone.

1907 ITALIAN CRUISER.

1920 Photo, Ministe

LIBIA (ex Turkish *Drama*, Nov., 1912). 4500 tons. Complement, 317. D (*o.a.*) × 47½ × 16 feet. Armament: 2—6 inch, 45 cal., 8—4·7 inch, 50 cal., 2—14 pdr 2—18 inch tubes (*above water*). Deck, 4". Designed H.P. 12,500 = 22 kts. 16 N Coal: *normal* 275 tons, *maximum* 640 tons. Endurance: 4500 miles at 10 kts. B by Ansaldo. Was begun by Ansaldo for Turkish Navy in 1911 and appropriat outbreak of Turko-Italian War. Not completed until 1913.

OLD ITALIAN COLONIAL CRUISER.

CALABRIA (1894). 2492 tons. Sheathed and zinced. Complement, 208. Dimensi 41¾ × 17¼ feet (*max.*) Armament: 4—6 inch, 6—6 pdr., 2—1 pdr., inch tubes (*above water*). 2" Deck. Designed H.P. 4000 = 16 kts. Boilers: 4 *maximum* 500 tons. Built at Spezia.

PUGLIA. Launched 1898. Served as Mine layer during war, carrying 60 mines.

Displ. (tons)	3100	Guns—		Designed I.H.P.	7500
Length (ft. *o.a.*)	289¾	7—14 pdr. AA.		Speed (knots)	19.5
Beam (ft.)	40¼	Torpedo Tubes:		Coal (tons)	480
*Max.*draught (ft.)	19¼	2—18 inch (*above water*)		Complement	164

Photo, Abrahams, Devonport.

ETNA (1885). 3740 tons. Complement, 239. Dimensions: 283½ × 43¼ × 20 feet (*maximum*). Armament: 2—6 inch,* 1—14 pdr.,* 2—6 pdr.* Deck, 2". Designed H.P. 6400 = 17 kts. 4 cyl. boilers. Coal: 630 tons.

* Disarmed during later stages of the War and served as a Harbour Flagship. Guns given as officially listed by Ministry of Marine.

Plans of Flotilla Leaders.

(See following pages).

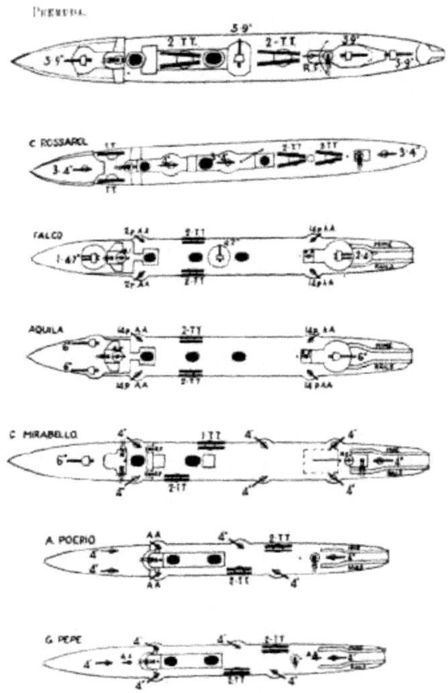

Summary Table.

8 + *3* **Flotilla Leaders** } **67** *completed.* }
= 70.
59 Destroyers } *3 building.* }

Total	Type.	Date	Displacement metric	S.H.P.*	Max. speed*	Fuel†	Complement (men)	Tubes Mines	Max. Draught
	Flotilla Leaders:—		tons		kts.	tons.			feet
1	*Premuda*	'16-'18	§2485	55,000 *t*	35·3	*210/720*	..	4	14¼
1	*C. Rossarol*	'14-'16	§1354	40,217 *t*	37·5	*150/327*	..	4/24	..
3	*Leone* (AS)	*Bldg.*	2200	42,000 *t*	*34*	*400*	..	6/*100*	..
2	*Aquila* (PN)	'13-'19	1556	44,000 *t*	*34*	*260*	139	4/50	..
2	*Mirabello* (AS)	'14-'16	1521	35,000 *t*	35	*350*	149	4/100	..
2	*Poerio* (AS)	'13-'15	1030	20,000 *t*	32	*270*	99	4	..
	Destroyers:—								
1	*Ardimentoso*	'14-'15	‖ 905	25,000 *t*	35	*162/305*	..	4/24	..
7	*Curtatone*	'12-'17	‖ 850	17,000 ?	32·5	95 (Coal & Oil)	102	4	..
8	*Palestro* (OL)	'17-'19	880-904	24,000 *t*	32	*170*	..	4	..
11	*Sirtori* (OS)	'16-'18	847	18,000 *t*	33	*170*	100	4/*10*	8¾
1	*Audace* (Y)	'13-'16	955	21,500 *t*	30	*245*	118	4	9¼
8	*Abba* {(PN) (OS)}	'13-'16	770	18,000 *t*	30	*150*	71	4	..
4	*Indomito* (PN)	'12-'14	690	15,000 *t*	30	*130*	71	4—2	..
1	*Animoso* (OS)	'11-'13	700	14,000 *t*	30	*130*	75	4	..
2	*Ardente* (OL)	'11-'12	686	15,000 *t*	30	*130*	71	2	..
10	*Ascaro* (AS)	'12-'13	398	6000	28.5	50+*35*	53	3	..
	Carabiniere (AS)	'09-'10	420			90	50	3	..
	Artigliere (AS)	'06-'07	390			80			
3	*Zeffiro* (PN)	'02-'04	380	5200	30	80 / *65*	58	4	7¼
2	*Lampo* (Sc)	'99-'01	360	5200	30	80 (coal)	58	2	8¾
1	*Fulmine* (AS)	'97-'00	340	4800	26	60 (coal)	46	2	8¾

(AS)=Ansaldo, Sestri-Ponente. (OL)=Orlando, Leghorn. (OS)=Odero, Sestri-Ponente. (PN)=Pattison, Naples. (Sc)=Schichau. (t)=Turbines.

*Details of ex-German boats in these columns are trial figures.
†In Fuel column, figures in italics denote oil fuel. *Ascaro* has 50 tons coal + 35 tons oil. *Zeffiro* class have 80 tons coal or 65 tons oil.
§Ex-German } Boats taken over 1920.
‖Ex-Austrian }

Flotilla Leaders.*

*Note.—There is no official Italian rating corresponding to this. All the large T.B.D. on this page, *A*... *Poerio* class in next page, are officially rated as *Esploratori* (Scouts) with Light Cruisers of *Bixio* and [illegible]

1 Ex-German Boat ("1916 Design")

1920 *Photo, Kruger,* [illegible]

1 Vulcan boat: **Premuda** (ex-German *V 116*, built 1916-18). Displacement, 2485 tons. Dimensions [illegible] 301½ × 13 feet. Guns and tubes (German)*: 4—5.9 inch (45 cal. "Flak"), 4—23.6 inch, in 2 twin [illegible] Machinery: A. E. G.-Vulkan turbines. Boilers: 4 "Marinetyp" (Schulz-Thornycroft). Designed S.[illegible] 15 kts. Oil fuel: 210 tons *normal*, 720 tons *max.*

Note.—German "1916 Design" boat, taken over by Italy, 1920. For other remarks, refer to *Amiral* [illegible] boat), in French Navy Section.

*Will probably be re-armed with Italian Guns and Tubes.

Special Note to "C. Rossarol" (*v.* next page).

The Germans say the contract for this boat (and *B 98*) was signed on August 5th, 1914, the [illegible] being ready in 12 months. They also say the sister boats *B 109—112* were built in five or six months [illegible] similar boats, *V 99*, *100*, in 1½ months. These German building periods, for large Destroyers, in [illegible] stages of the late war, place too great a tax on our credulity. It is believed all these big Destroyers were begun about Sept., 1913—March, 1914, as contracts for the Chinese Navy. Taken over by Germany, engined with machinery intended for the Russian "Improved *Novik*" types of t.b.d. Whether the "contract" was genuine or spurious is an open question. The Russian contract was probably [illegible] *C. Rossarol* appears to be a German modification of the Russian *Garibaldi* type.

(Continued

1 Ex-German boat.

Photo added 1919.

1 *Böhm & Voss boat*: **Cesare Rossarol** (ex-German *B 97*, 1914). 1254 tons. Dimensions: 321.5 × 30¾ × 9.4 feet. Guns and tubes (German) * 4—4.1 inch "Flak" (? on H.A. mountings), 4—19.7 inch tubes, in 2 twin and 2 single mounts. Turbines and Schulz-Thornycroft boilers. Made 37¼-38 kts. on *trials*. Oil: 150 tons *normal*, 526¼ tons *full load* = 2620 miles at 20 kts. Taken over by Italy, 1920.

*Will probably be re-armed with Italian guns and tubes. When first completed could carry 24 mines.

All units of "Improved Aquila" class, given in 1919 "Fighting Ships," have been cancelled.

---

3 Leone Class.

All work suspended.
(Projected date of completion, 1922.)

3* *Ansaldo type*: *Leone*,† *Pantera*,† *Tigre*.† *Building* or completing. 2200 tons. Dimensions: 359.6 (*p.p.*) × 34 × 11.5 feet (*mean*). Guns: 8—4.7 inch (45 cal.) 2—14 pdr. (40 cal.) AA., 2—2 pdr. AA. Tubes: 6 in two triple deck mountings. Designed S.H.P. 42,000 = 34 kts. 2 Parsons turbines. 4 Yarrow oil-burning boilers. 2 screws.

**Leopardo* and *Lince* cancelled.

†All will probably carry 100 mines.

2 Aquila class.

Photos wanted.

FALCO.

AQUILA. *Photo, Commr. E. D. Holberton, R.N., add.*

4 *Pattison type*: **Falco*** (1919), **Aquila*** (1916). 1550 tons *normal*, 1723 tons *deep load*. Dimensions: [illegible] × 31 × 10.7 feet (*mean*). Guns: (*a*) *Falco*, 5—4.7 inch, 40 cal.; 4—14 pdr., 40 cal.; 2—2 pdr. AA. [illegible] 3—6 inch, 40 cal.; 4—14 pdr., 40 cal. AA.; 2—2 pdr. AA. Tubes: 4—18 inch, in two twin deck [illegible] Designed S.H.P. 44,000 = 34 kts. 2 Tosi turbines. 4 Thornycroft oil-burning boilers. 2 screws. 1700 miles (15 kts.), 380 miles (full speed).

Note.—These two destroyers were to have been built for the Rumanian Navy. Appropriated for Italian N[illegible] break of war with Austria-Hungary. Two boats, *Nibbio*, *Sparviero*, sold back to Rumania, 1920.

*Are Mine Layers, carrying 50 mines each.

Flotilla Leaders (*continued*).

2 Mirabello class.

C. MIRABELLO. 1919 *Photo, Commr. Holbertan, R.N.*

2 *Ansaldo* type: ***Carlo Mirabello**** (1914), ***Augusto Riboty**** (1915). 1521 tons *normal*, 1785 tons *deep load*. Dimensions: 311.4 (*p.p.*) × 32 × 10.6 feet (*mean*). Guns: 1—6 inch (45 or 40 cal.), 7—4 inch, 2—2 pdr. AA. and 4 tubes in two twin deck mountings. Made 33.75 kts. on *trials*. 4 Yarrow oil-burning boilers. 2 Parsons geared turbines. 2 screws. Endurance: 2840 miles (15 kts.), 500 miles (full speed). *C. A. Racchia* mined off Odessa, 1920.

*These two boats are Mine Layers, each carrying 100 mines.

2 Poerio class.

G. PEPE. 1919 *Photo, Commr. Hallerton, R.N.*

2 *Ansaldo* type: ***Alessandro Poerio*** (1914), ***Guglielmo Pepe*** (1914). 1030 tons. Dimensions: 272.6 (*p.p.*) × 26.3 × 9.3 feet. Armament: 4 inch, 5 in *Poerio*, 6 in *Pepe*, 2—2 pdr. AA. 4—18 inch torpedo tubes in two twin-deck mountings. H.P. 20,000 = 30 kts. 3 Yarrow oil-burning boilers. 2-shaft Parsons turbines. Endurance: 2000 miles (15 kts.), 745 miles (full speed). *Trials*: 33.4 kts. Fitted as Minelayers.

Destroyers.

1 ex-German boat.

1920 *Photo, G. Doll, Esq.*

1 *Schichau* boat: ***Ardimentoso*** (ex-German S63, 1915). 965 tons. Dimensions: 271.4 × 25¼ × [illegible] and tubes (German)*: 3—4.1 inch ("Flak" = on high-angle mounts), 6—19.7 inch tubes, in two single mountings. Turbines and Schulz-Thornycroft boilers. German S 63—66 class made about [illegible] *trials*. Oil: 162 tons *normal*, 305 tons *max.* = 1960 miles at 20 kts. Taken over by Italy, 1920

*Will probably be re-armed with Italian guns and T.T. Was fitted to carry 24 mines when first compl[eted]

7 Cortelazzo class (ex-Austrian).

7 *Ganz-Danubius* type: ***Cortelazzo*** (ex-*Lika*), ***Grado*** (ex-*Triglav*), ***Monfalcone*** (ex-[illegible]) 1914-17; ***Pola*** (ex-*Orjen*, 1913), ***Fasana*** (ex-*Tatra*), ***Muggia*** (ex-*Csepel*), ***Zenson*** [illegible] launched 1912. 850 tons. Dimensions: 274 × 25¼ × 8¼ feet. Guns and torpedo tu[bes]: 2—3.9 inch, 4—11 pdr., 2—11 pdr. AA., 1 M.G., and 4—20.8 inch tubes. A. E. G. turbines and [illegible] On *trials*, made 32.5 to 33 kts. Other details as Table. Originally Austrian "Tatra Class." Ta[ken over] 1920. Sister boat, *Mateśit Lrblanc* (ex-*Dukla*), now in French Navy.

Note.—Another boat of this type, the unfinished Austrian *Huszar*, lies at Porto Rè, near Fiume, in area administered by D'Annunzio's Government.

*Will probably be re-armed with Italian guns and T.T.

(*Continued*

Plans of Destroyer Types.

11 Sirtori class.

V. ORSINI. *1919 Photo, Commr. Holberton*

11 *Odero type*: **Agostino Bertani,* Giacomo Medici,* Giuseppe la Farina,*** and **Nico** (all 1917): **Angelo Bassini,* Francesco Stocco, Giacinto Carini,* Giovan** **Giuseppe la Masa,* Giuseppe Sirtori,** and **Vicenzo Orsini** (all 1916). 847 tons. Dim (*p.p.*) × 24 × 9.2 feet (*mean*). Guns: (*a*) *A. Bertani, G. Medici, G. la Farina*, 6—4 inch, 2—14 pdr. *A. Bassini, G. Carini, G. la Masa, G. Sirtori*, 4—4 inch, 2—12 pdr. (*c*) *F. Stocco, G. Acerbi, V. Orsi* 2—2 pdr. AA. Tubes (in all): 4—18 inch, in two twin-deck mountings. Made about 34 kts. on t turbines. 4 Thornycroft oil-burning boilers. 2 screws. Endurance: 1700 miles (15 kts.), 470 mile *Benedetto Cairoli*, of this class, lost during War.

* These 7 boats are Mine Layers, each carrying 10 mines.

8 Palestro class.

PALESTRO. *Photo by courtesy of Flli. Orlando.*

8 *Orlando type*: **Calatafimi, Castelfidardo, Confienza, Curtatone, Monzambano, Palestro, San Martino, Solferino.** Launched 1918-20. 880 tons.* Dimensions: 256.6 (*p.p.*)* × 24.9 × 8.7 feet (*mean*). Guns: 4—4 inch, 45 cal., 2—14 pdr., 30 cal. AA., 2—2 pdr. AA. Tubes: 4 in two twin-deck mountings. Designed S.H.P. 24,000 = 32 kts. 2 Zoelly turbines. 4 Thornycroft oil-burning boilers. 2 screws.

* In *Calatafimi, Castelfidardo, Curtatone* and *Monzambano*, length *p.p.* is about 262.5 feet, and displacement is 904 tons.

1 Special Boat.

1919 Photo, Commr. Holberton

1 *Yarrow type*: **Audace** (ex-Japanese *Kawakaze*, Scotstoun, N.B., 1915). 955 tons. Dimensions: × 27¼ × 9¼ feet (*max.*). Guns: 7—4 inch, 2—2 pdr. AA. Tubes: 4—18 inch, in two twin mounti 34.5 kts. on trials. 2-shaft Parsons turbines and 3 Yarrow large-tube, oil-burning boilers. Endura miles (15 kts.), 560 miles (full speed). Trials, 35·2 (1682 tons load displacement).

Note.—This boat must not be confused with *Audace* built in 1912-13 by Orlando, as a sister boat to *Ardimo* page). The Orlando *Audace* was sunk during the War, and the Japanese *Kawakaze* (closely resemblin *Audace*) was purchased by Italy and re-named *Audace*.

8 Abba class.

G. ABBA. 1919 *Photo, Commr. Helberton, R.N.*

2 *Pattison* boats: ***Francesco Nullo,**** ***Antonio Mosto,*** (both 1914).
6 *Odero* boats: ***Rosolino Pilo, Giuseppe Abba,* Ippolito Nievo, Simone Schiaffino, Pilade Bronzetti*** (all 1914), ***Giuseppe Missori*** (1915). 770 tons. Dimensions: 236.2 (*p.p.*) × 24 × 8.8 feet (*mean*). Guns: 5—4 inch, 2—2 pdr. A.A. Tubes: 4 single 18 inch. Made about 31.8 kts. on trials. 2 Tosi turbines and 4 Thornycroft oil-burning boilers. 2 screws. Endurance: 1700 miles (15 kts.), 440 miles (full speed).

*At Fiume, attached to D'Annunzio's Forces.

1 Special boat.

1919 *Photo, Commr. Helbe*

1 *Orlando* type: ***Animoso*** (1913). 700 tons. Dimensions: 245.4 (*p.p.*) × 24.6 × 8.4 feet (*mean*). inch, 4—14 pdr., 2—2 pdr. AA. Tubes: 4 single 18 inch. On *trials*, made 36.1 kts. with (*about*) 2-shaft Zoelly turbines and 4 White-Forster oil-burning boilers. 2 screws. Endurance: 1650 m 430 miles (full speed). Her sister boat, *Audace*, was lost during the War.

4 Indomito class.

INDOMITO. 1919 *Photo, Commr. Holberton, R.N.*

4 *Pattison* type (Thornycroft design): ***Indomito*** (1912), ***Impavido, Insidioso, Irrequieto*** (1913). 690 tons, Dimensions: 238 (p.p.) × 24 × 8.4 feet. Guns: 1—4.7 inch, 4—14 pdr., 2—2 pdr. AA. Tubes: 2—18 inch, but *Indomito* only has 4 single 18 inch. On *trials*, 18,200 S.H.P. = 35.8 kts. 2 Tosi turbines and 4 oil-burning Thornycroft boilers. 2 screws. Endurance: 1520 miles (15 kts.), 364 miles (full speed). *Impetuoso* and *Intrepido* of this class lost in the War.

2 Ardente class.

ARDITO. 1919 *Photo, Commr. Holbe*

2 *Orlando* type: ***Ardito*** (1912), ***Ardente*** (1912). 686 tons. Dimensions: 238 (*p.p.*) × 24 × 8.4 feet (1—4.7 inch, 4—14 pdr., 2—2 pdr. AA. Tubes: 2 single 18 inch. On *trials*, 15,150 S.H.P. = 33 Parsons turbines and 4 oil-burning Thornycroft boilers. 2 screws. Endurance: 1520 miles (15 (full speed).

10 Carabiniere/Artigliere class.

ASCARO. Some boats rigged like this. 1919 *Photo.*

1917 *Photo, O. Parkes, Esq., M.B.*

GARIBALDINO. Has been lost, but some of the boats detailed below are rigged like this, with very short pole mast right at stern.

10 *Ansaldo* (Improved *Nembo*) type:—

5 *Carabiniere* class: **Ascaro** (1912), **Pontiere** (1910, reconstructed 1913), **Alpino, Carabiniere, Fuciliere** (1909). *Ascaro* 398 tons, others 420 tons. Dimensions, H.P., speed as *Artigliere* class below. Guns as *Artigliere* class below, except *Ascaro*, with 2—14 pdr. and 4—6 pdr. Carry 99 tons oil fuel, except *Ascaro*, with 50 tons coal + 33 tons oil. Endurance: 915 miles for *Ascaro*, 1250 miles for others, all at 15 kts. *Garibaldino* lost during War.

5 *Artigliere* class: **Corazziere** (1909), **Lanciere, Artigliere** (1906), **Bersagliere, Granatiere** (1906), 390 tons. Dimensions: $211\frac{1}{4}$ (*p.p.*) × 20 × $6\frac{3}{4}$ to 7 feet (*mean*). Guns: 4—14 pdr. Tubes: 3 single 18 inch. On trials, 6500 I.H.P. = 29·1 kts. 3 Thornycroft boilers and 2 screws. 80 tons coal. Endurance: 1000 miles at 15 kts.

2 Lampo class.

2 *Schichau* type: **Euro, Strale** (1899-1901). 320 tons. Dimensions: $196\frac{3}{4}$ × $21\frac{1}{2}$ × $8\frac{1}{2}$ feet. Armament: 4—14 pdr., 2—18 inch tubes. All have Schulz boilers. 31·2 kts. on trials. Endurance: 1100 miles at 15 kts.

3 Zeffiro class.

ZEFFIRO. 1919 *Illustrati*

Espero, Zeffiro 3 funnels as above.

3 *Nembo* class (Thornycroft pattern): **Aquilone** (1902), **Espero, Zeffiro** (both 1904). Built by P. Naples. 380 tons. Dimensions: 208 (*p.p.*) × $19\frac{1}{2}$ × $7\frac{1}{2}$ feet. Armament: 4—14 pdr., 4—18 inch tubes. I 3 Thornycroft, oil-burning in *Aquilone* and *Espero* (80 tons coal), coal-burning in *Zeffiro* (60 tons oil). Or 5350 I.H.P. = 30·3 kts. *Turbine* and *Nembo* and *Borea* of this class have been lost during the Endurance: 1000 miles at 15 kts.

1 Old Boat.

1 *Odero* type: **Fulmine** (1898). 310 tons. Dimensions: 200 × 21 × $8\frac{1}{2}$ feet. Armament: 4—6 pdr., 2— tubes. 4 Blechynden boilers. Designed for 28 kts, but never exceeded 26 kts. Endurance: 820 miles at

(Also see next page for other T.B.)

22 Sea-going Boats *(Torpediniere di alto mare).*

No.	Class.	Date.	Displacement.	H.P.	Max. Speed.	Fuel.	Complement.	Tubes.	Max Draught
			tons.		kts.	tons.			feet.
2	*"Later C"* (PN)	1908–9	216	3000	25	30 tons oil	33	3	$7\frac{1}{4}$
6	*"Earlier C"* (PN)*	1906–7	216	3000	25	40 tons coal	35	3	$7\frac{1}{2}$
6	*"A"* (OS)*	1906–7	216	3000	25	{ 40 tons coal / 30 tons oil	35 } / 33 }	3	$7\frac{1}{2}$
1	*Gabbiano* (SY)	1907	162	2000	25	18 tons oil	27	2 or 3	$7\frac{1}{4}$
4	*"O"* (OS)*	1905–7	210	3000	25	40 tons coal	35	3	$7\frac{1}{4}$
3	*"P"* (PN)*	1905–6	210	3000	25	30 tons coal	32	3	$7\frac{1}{4}$

(PN) = Pattison, Naples. (OS) = Odero, Sestri-Ponente. (S) = Schichau. (SY) = Spezia D.Y.

* *Note.*—All belong to a common design and are armed with 2—14 pdr., 3—18 inch tubes, arranged as *Pegaso* plan. The 3—3 pdr. guns, shown on plans, were replaced during war by 2—14 pdr. guns.

The *Andrea Bafile* type Convoy Gunboats, described on a later page, are of 23 kts. speed, carry 2—18 inch tubes, and can be taken as a Torpedo Boat type.

CALIPSO. 1929 *Photo, Ministry of Marine.*

2 *"Later C"* class: **Calipso, Climene** (Pattison, Naples, 1908). 216 tons. Dimensions: 164'3 × 17'4 × 5'7 feet (mean). Guns: 1—14 pdr., 1—14 pdr. A.A. Tubes: 3 single 18 inch. On trials, 3153 I.H.P. = 26 kts. 2 oil-burning Thornycroft boilers and 2 screws.

CASSIOPEA. 1920 *Photo, Ministry o*

AIRONE. (Mast stepped close to bridge.) 1920 *Photo, Ministry o*

Mainmast removed from some boats.

6 *"A"* class: **Airone, Ardea, Albatros, Astore, Arpia** (all 1907), **Alcione** (1906). B Sestri-Ponente. 216 tons. Dimensions: 164'3 × 17'4 × 5'7 feet (mean). Guns: 2—14 pdr. A.A. 3 single 18 inch. On trials, 3260 I.H.P. = 25'8 kts. 2 Thornycroft coal-burning boilers, except Thornycroft oil-burning boilers and 30 tons oil instead of 40 tons coal. 2 screws in all.

(Continued on

Sea-going Torpedo Boats (*continued.*)

Now rigged with 2 pole masts.

1 *Navy type*: ***Gabbiano*** (Spezia D.Y., 1907). 162 tons. Dimensions: 150 × 17·5 × 5·5 feet. Guns: 2—3 pdr. Tubes: 2 or 3 single. On trials 2200 I.H.P. = 26 kts. 2 coal-burning Normand boilers and 2 screws.

4 "O" class: ***Orfeo*** (1907), ***Olimpia, Orsa*** (both 1906), ***Orione*** (1905). Built by Odero, Sestri-Ponente. 216 tons. Dimensions: 167·3 × 19·7 × 5 feet (*mean*). Guns: 1—14 pdr., 1—14 pdr. AA. Tubes: 3 single 18 inch. On trials, 3300 I.H.P. = 25¼ kts. 2 coal-burning Blechynden boilers and 2 screws.

3 "P" class: ***Pegaso, Procione, Pallade*** (all 1905). Built by Pattison, Naples. 210 tons. Dimensions: 164.3 × 17.4 × 5.7 feet. Guns: 2—14 pdr. AA. Tubes: 3 single 18 inch. On trials, 3160 I.H.P. = 26.6 kts. 2 Thornycroft coal-burning boilers and 2 screws.

75 Coastal Torpedo Boats. (*Torpediniere Costiere.*)

No.	Class.	Date.	Displacement tons.	H.P.	Max. Speed. kts.	Fuel. tons.	Complement.	Tubes.
2	74 O.L.T., 75 O.L.T.	1917-'18	200	4500 *t*	29	20 oil		
4	70 C.P.—73 C.P.	1917-'18	169	3500	28—31	.. oil		
6	58 O.L.—63 O.L.	1915-'16	160	3700	28.1	.. oil	..	2
24	52 A.S.—57 A.S., 46 O.S.—51 S.O., 40 P.N.—69 P.N	1915-'18	160—165	3500	28—29	12 oil		
35	1 P.N.—38 P.N., 13 O.S.—24 O.S., 25 A.S.—32 A.S., 39 R.M.	1910-'14	125	2700—3200*	27.5—31.5	15 tons oil	23	2
1	*P. Rismondo*†	1910	110	2400	26	28 tons oil	20	2
3	115S., 113S., 102S.	'87-'93	79	700—1000	19	19 tons oil *or* 17 tons coal	17	2

t=turbines. * 33 boats recipro. engines, 2 turbines, 1 turbine and recipro. combined. † Captured Austrian

Note:—

The letters, following numbers, indicate builder and location of Yard in the following manner:—

A.S.=Gio. Ansaldo & C., Sestri-Ponente.
C.P=Cantiere Navale Riuniti, Palermo.
O.L.=Fratelli Orlando & Co., Leghorn.
O.L.T. „ „ „ , (T. added to turbine-driven boats.)
O.S.=Odero & C., Sestri-Ponen
P.N.=Pattison & C., Naples.
R.M.=Regia Marina (built in N
S.=Schichau (Germany).

(2 funnels.) 1919 *Photo by courtesy of Flli. Orlando.*

2 *Orlando* boats: ***74 O.L.T., 75 O.L.T.,*** Built at Leghorn. Begun 1917. Launched Oct., 1917–Jan., 1918., June-Sept., 1918. 200 tons. Guns: 1 or 2—14 pdr. AA. Torpedo tubes: 2—18 inch. S.H.P. 4500 = 2-shaft turbines and 2 water-tube boilers.

(*Continued on ne*

Photo wanted.

4 *Cantiere Navale Riuniti* boats: **70 C.P.-73 C.P.** Built at Palermo. Begun 1917, launched 1918, finished 1919. Details as 12 Pattison boats in next column, except that these boats are (*a*) 169 tons displacement, and have (*b*) *two* 14 pdr. AA. and (*c*) 2—18 inch torpedo tubes in *two single* deck mountings. (*d*) Speed: 28 to 31 kts.

Photo by courtesy of Fili. Orlando, 1919.

6 *Orlando* boats: **58 O.L.-63 O.L.** Built at Leghorn. Begun 1915-16, launched April-Sept., 1916. Completed May-Oct., 1916. Details generally as 12 Pattison boats in next column. But I.H.P. 3700 = 28.1 kts.

52 A.S. 1920 *Photo, Gio. Ansaldo & C.*

6 *Ansaldo* boats: **52 A.S.-57 A.S.** Built 1915-16 at Sestri-Ponente. Details as 12 Pattison boats below, but 2—14 pdr. guns and 1 M.G.

Note to Illustration.—14 pdrs. not combined " Y " type. Guns are in echelon and happened to coincide at the moment the photo was taken.

48 O.S. (2 funnels). 1920 *Photo, Ministry of*

6 *Odero* boats: **46 O.S.-51 O.S.** Built 1915-16 at Sestri-Ponente. Details as 12 Pattison boats belo
pdr. guns.

64—68 P.N. (2 funnels). 1920 *Illust*

40—45 P.N. (2 funnels). 1920 *Photo, Ministry*

12 *Pattison* boats: **40 P.N.-45 P.N.** (begun 1915, launched and completed 1916). **64 P.N.-69 P.N** and completed 1917-18). All built at Naples. About 160-165 tons. Dimensions: 139½ × 15 × 1—14 pdr. AA. Tubes: 2—18 inch in one twin mounting. I.H.P. 3500 = 28 to 29 kts. Mac reciprо. triple expansion. Boilers: Thornycroft.

(Continued

Coastal Torpedo Boats (*Continued*).

Appearance of all boats as above. *Official photo.*

1 *Navy Yard boat*: **39 R.M.** (Spezia Navy Yard)
15 *Pattison boats*: **33 P.N.—38 P.N.**, **6 P.N.—12 P.N.**, **1 P.N.—4 P.N.**
11 *Odero boats*: **18 O.S.—24 O.S.**, **13 O.S.—16 O.S.**
8 *Ansaldo boats*: **25 A.S.—32 A.S.**

36 boats in all. Launched 1911-13. Completed 1911-14. Average 125 tons. Dimensions: 139½ × 15 × 5 feet. Guns: 1—9 pdr. Tubes: 2—18 inch. I.H.P. 2700-3200 = 27.5-31.5 kts. Oil: 15 tons.

Boilers: 2 Thornycroft oil-burning, except 39 *R.M.* with special type. Machinery: 2 sets triple expansion, except (*a*) 39 *R.M.* with 3 sets triple expansion and 1 B.P. turbine to 3 screws; (*b*) 31 *A.S.* with 3-shaft Parsons turbines; (*c*) 32 *A.S.* with 2-shaft Bergmann turbines. 5 *P.N.*, 17 *O.S.*, 37 *P.N.*, lost in War, 38 *P.N.*, sold.

Now has pole mast just abaft fore bridge.

1 *ex-Austrian boat*: **Francesco Rismondo** (ex-Austrian *No. 11*). Armed with 2—3 pdr. and 2—17.7 inch tubes. Oil fuel only. White-Forster boilers. Deserted to Italian Navy, October, 1917.

(*All relegated to Harbour Duties.*)

3 *Schichau type*: **115 S, 113 S, 102 S.** 80-85 tons. Speed: 17-19 kts. Armament: 2—1 pdr., 2—14 inch tubes. (*Several have been re-boilered.*)

(M. A. S. Boats with torpedo tubes described on a later page.)

46* Submarines (*Sommergibili*).

* 11 Ocean-going, 33 Coastal, 2 Mine-laying boats; 2 Ocean-going boats reported not ready yet.

No.	Type.	First begun.	Completed.	Displacement	B.H.P. Kw.	Speed kts.	Endurance.	Tubes.	Complement.
*	*Ocean-going*:			*	*	*		*	
6	*P. Micca* (C)	1914	'17-'19	810 / 1211	2600 / 900	17 / 10	...	6	32
4	*A. Barbarigo* (FL)	1913	'17-'18	747 / 925	2600 / 900	17.5 / 10	2000 at 12 kts.	6	32
1	*G. Pacinotti* (FL)	1913	1916	700 / 870	2600 / 900	18 / 10	5000 at 12 kts.	5	32
	Coastal:—								
2	*N5 N6* (B)	1917	'18-'19	276 / 352	650 / 350	13.5 / 8	...	2	21
4	*N1—N4* (B)	1915	'17-'18						
7	*H1—H8* (H)	1916	1917	361 / 441	480 / 500	12.5 / 11	2000 at 13 kts. / 30 at 8 kts.	4	22
17	*F1—F21* (FL)	1915	'16-'17	260 / 318	700 / 250	13 / 8	1600 at 8.5 kts. / 90 at 4 kts.	2	21
2	*S2, S3* (SG)	1913	1915	255 / 390	600 / 400	13 / 8.5	1600 at 8.5 kts. / 12 at 5 kts.	2	21
1	*Argonauta*	1914	1915	255 / 300	600 / 400	15 / 8.5	950 at 12 kts. / 100 submerged	2	21
	Mine Laying								
2	*X2, X3* (B)	1916	1917	403 / 468	660 / 325	9.2 / 6.3	1200 at 12 kts.	(†) 6 or 9	..

(C)—Cavallini design; (FL)—Fiat-Laurenti design; (B)—Bernardis design; (H)—"Holland" (Electric B[oat]) design.

*Details in these columns are all official.

† Total of mine-discharge chutes; figures are not official.

(*Description by classes on following pages.*)

Note.—Ex-Austrian Submarine *W* is at Fiume, where there are said to be also several Italian Submarines under [the flag] of D'Annunzio's Government. Also at Fiume, 5 unfinished boats (2 almost ready), originally begun for the [Austrian] Navy.

ITALIAN SUBMARINES.

11 Ocean-going Boats *(Sommergibili d'alto mare).*

5 "P. Micca" class.

A. Emo. *1920 Photo by courtesy of the Ministry of Marine.*

6 *Carollini* type: ***Angelo Emo,* Lorenzo Marcello,* Lazzaro Mocenigo,* Galvani, Pietro Micca, Torricelli*** (Spezia Navy Yard). Dimensions: 206¾ × 19¾ × 14 feet. Guns: 2—14 pdr. (40 cal.) AA. 6—17·7 inch torpedo tubes. Machinery: In first three, 2 Tosi Diesel motors on surface and 2 Savigliano electric motors *when submerged*; last three have 2 Ansaldo Diesel and 2 Ansaldo electric motors. Carry 60 tons oil fuel. Other details as Table. Completed 1917-1919. Two boats unofficially reported as still on stocks 1920.

*These 3 boats begun at Venice D.Y., Aug., 1914. They were dismantled and taken to Spezia in the summer of 1915 and laid down for a second time.

4 "A. Barbarigo" class.

A. Barbarigo. *1920 Photo, Ansaldo San*

4 *Fiat-Laurenti* type: ***Agostino Barbarigo,*** (1914), ***Andrea Provana*** (—), ***Sebastiano*** [illegible] ***Giacomo Nani*** (Fiat-San Giorgio Co., Spezia). Dimensions: 213¼ × 19¾ × 13¼ feet. Guns: 2 [illegible] 6—17·7 inch torpedo tubes. Machinery as *G. Pacinotti* below. Carry 60 tons oil. Completed 1917 and Dec., 1918. Other details as Table.

1 "G. Pacinotti" class.

Photo, Ansaldo-San [illegible]

(Expected for Sale or Disposal, 1920.)

1 *Fiat-Laurenti* type: ***G. Pacinotti*** (Spezia, 1915 or 1916). Dimensions: × × feet. [illegible] pdr. (40 cal.) AA., 5—17·7 inch torpedo tubes (3 bow, 2 stern). Machinery: 2 Fiat Diesel engin[illegible] Savigliano electric motors *when submerged*. Other details as Table. A sister boat, *Alberto Guglie*[illegible] taken for a German U-boat by a British Destroyer and sunk.

33 Coastal Boats—(*Sommergibili Costieri*).

Note.—The Ansaldo San Giorgio Co. laid down two Coastal Boats in 1917 as a speculation. Were *about* 335/660 tons, 13/9 kts. speed, and had 4 bow tubes. Subsequent disposal of these boats unknown. They may have been abandoned in 1918-19.

N 5. *Photo by courtesy of the Ministry of Marine.*

2 *Bernardis* type: **N 5** (1917), **N 6** (1918). Both built by Tosi, Taranto. Dimensions: 152[illegible] × 13[illegible] × 10 feet. Machinery: 2 Tosi Diesel engines *on surface*. Guns, tubes and electric motors as *N 1—4* below. Other details as Table.

N 1—4. *Photo, Ansaldo-San Giorgio Co.*

4 *Bernardis* type: **N1—4** (Ansaldo-San Giorgio Co., Spezia, 1916-17). Dimensions: 150[illegible] × 14[illegible] × 10 feet. Guns: 1—14 pdr. (30 cal.) AA. 2 bow 18 inch tubes. Machinery: 2 Sulzer-Ansaldo Diesel engines *on surface*; 2 Tecnomasio Italiano electric motors *when submerged*. Oil: 9-10 tons. Are an "Improved *Nautilus*" class. Other details as Table.

Coastal Boats (*Continued*).

H 2. 1919 *Photo, Commr. Holbert*

7 *Holland* type: **H1, H2** (1916) **H3—H8** (1917). All built by Canadian Vickers Co., Montreal. Dime… 15[illegible] × 12[illegible] feet. Guns: 1—3 pdr. (30 cal.) AA. 4—18 inch bow tubes. Machinery: 2 Nlseco… *on surface*; 2 Dynamic Electric Co. motors *when submerged*. Oil: 14 tons. Other details as Table.

F 7. 1919 *Photo, Commr. Hollerto…*

17 *Fiat Laurenti* type: **F1, F2, F5, F7, F9, F10, F12, F14, F16, F18—21,** built by Fiat-San… Spezia and Odero, Sestri Ponente; **F6, F13, F15, F17,** built by Orlando, Leghorn. Launched an… 1916-17. Dimensions: 149[illegible] × 13½ × 10½ feet. Guns: 1—14 pdr. (30 cal.) AA. 2—17.7 inch… Machinery: 2 Fiat Diesel *on surface*; 2 Savigliano electric motors *when submerged*. Oil: 12 tons. A… proved *Medusa*" type. Other details as Table.

(Continued on n…

Coastal Boats—*Continued*

1919 *Photo, Commr. Holbertos, R.N.*

2 *Scott-Laurenti type*: **S2, S3** (1915), built by Scotts S. B. & E. Co., Greenock, N.B., for British Navy as S. 2, 3, ceded to Italy about 1916. Dimensions: 148 × 13½ × 9½ feet. 1 Q.F. gun. 2—18 inch bow tubes. Fiat type Diesel engines. Endurance: 1600 miles at 8½ kts. on surface; 75 miles at 5 kts. *submerged*.

1 *Fiat-Laurenti type*: **Argonauta** (ex-Russian ——*; Spezia, 1914). Dimensions: 147½ × 13¾ × 8½ feet. 1 Q.F. gun. 2—18 inch torpedo tubes. Fiat-Diesel engines. Radius of action: 950 miles at 12 kts. on *surface*; 57 miles *submerged*.

2 Mine Laying Boats (*Sommergibili—Affondamine*).

Photo wanted.

2 *Bernardis type*: **X2, X3** (Ansaldo, Spezia, 1916). Dimensions about 140 × 18 × 13 feet. Guns: 1—14 pdr. (30 cal.) A.A. Carry 18 mines in 6 or 9 discharge chutes. Machinery: 2 Ansaldo Diesel on surface; 2 Ansaldo electric *submerged*. Carry 11 tons of fuel. X 2, completed June, '17, and X 3, May, '18. (Other details as Table.)

"Mobile Batteries." (*Batterie Semoventi.*)

GRAPPA (on Gunnery Trials). *Photo by courtesy of the Ministry of Mar*

(GRAPPA CLASS—4 SHIPS.)

Name.	*Begun.*	*Launch.*	*Ready*
GRAPPA*	May 12th, 1918	Sept. 21st, 1918	Dec
MONTELLO	May 10th, 1918	Dec. 31st, 1918	July
MONTE NOVEGNO ..	May 30th, 1918	May 31st, 1919	Con
MONTE CENGIO ..	July 31st, 1918	Dec. 10th, 1918	Con

* May also be referred to as *Monte Grappa*.
† At Naples.

All begun and launched at Castellammare D.Y. 575 tons. Dimensions: 131¼ (*p.p.*) feet. Guns: 1—15 inch, 40 cal. (dismountable), 1 Colt M.G. Armour: " Belt. 2 sets Fiat San Giorgio Diesel, type "2 C. 216 F.M." B.H.P. 700 (designed), 6 = 7 kts. Oil fuel: 40 tons. Complement, . Designed by Major-General G.

(*Continued on next*

SABOTINO (*Monte Santo same*). *1920 Photo by courtesy of the Ministry of Marine.*

(MONTE SANTO CLASS—2 SHIPS.)

Name.	*Begun.*	*Launch.*	*Ready for Trials.*
MONTE SANTO ..	Sept. 10th, 1917	March 24th, 1918	July 23rd, 1918
SABOTINO	Sept. 25th, 1917	Feb. 20th, 1918	Sept. 2nd, 1918

Both by Venice D.Y. 570 tons. Dimensions: 122 (*p.p. Monte Santo*), 123 (*p.p. Sabotino*) × 28.4 × 6.2 feet. Guns: 1—15 inch, 40 cal. (dismountable). Armour: 3.9" Belt. Machinery: 1 Tosi Diesel. B.H.P. 350 = kts. Oil fuel: 5 tons. Complement, 45. Designed by Engineer Primeri.

Notes.

Pasubio, Vodice, given in 1919 "Fighting Ships," are not mentioned at all in the Official Tables of "Batteria Semoventi" furnished by the Italian Ministry of Marine. They were laid up in 1920 and will probably be disposed of. *Pasubio* was a unit of *Monte Santo* class. *Monfalcone* (already disposed of) was similar to *Monte Santo*, but smaller, and mounted a 12 inch gun. *Vodice* was on lines of *Carso*. *Cucco* (already disposed of) was similar to *Carso* but mounted a 12 inch gun. All these craft were built for defence of Venice. Another Mobile Battery of special design, *Alfredo Capellini*, capsized and sank off Ancona, Nov., 1918.

1920 Photo by courtesy of the Ministry

CARSO (Venice D.Y. Begun March 25th, 1917; launched June 9th, 1917; rea August 8th, 1917). 360 tons. Dimensions: 118.1 (*p.p.*) × 30.1 × 4.2 feet. G inch, 45 cal. Armour: 3.9" Belt. Machinery: 2 "P.4.1" Polar Diesel. B.I kts. Oil fuel: 5 tons. Complement, 30. Designed by Engineer Primeri

1919 Photo by courtesy of the Ministry of

FAA' DI BRUNO (Venice D.Y., 1916-17). 2,500 tons. Dimensions: 220 (*o.a.*) 6 feet. Complement, —. Guns: 2—15 inch, 40 cal., 2—14 pdr. AA., 2—1½ Armour: —" Barbette, —" Deck. H.P. 100 = 4 kts. Has only a small engine at to raise enough speed to move ship. Hull is box-shaped, without bow or stern.

Lagoon Gunboats (*Cannoniere lagunari*).

APE. 1920 *Official Photo.*

APE, VESPE. Built at Venice D.Y. 1918. About 50 tons. Dimensions: 60 × 10 × 2½ feet. Guns: 1—14 pdr. (field gun) and 10 machine. H.P. and speed unknown. Motor engines.

BRONDOLO. 1920 *Official Photo.*

(For Venetian Lagoons.—Have served as Mine Layers.)

BRONDOLO (1909), **MARGHERA** (1909). 120 tons. Dimensions: 125 × 11 × 2½ feet. Guns: 1—14 pdr. During War, each fitted to carry and lay 60 mines. Designed H.P., 450 = 13 kts. Coal: 12 tons. Complement, 17.

There are also about 62 small ships under 200 tons *gross* for service in Venetian Lagoons.

Colonial Gunboats.

1920 *Photo, A. Bo, Chefoo.*

SEBASTIANO CABOTO (Palermo, 1913). Colonial gunboat (*Cannoniera per Servizio Coloniale*). 1050 tons. Complement, 111. Dimensions: 196.8 × 31.8 × 9.2 feet. Guns: 6—14 pdr., 4 machine. Designed H.P., 1200 = 13 kts. 2 screws. 2 cyl. (direct flame) boilers. Coal: 100 tons. Was built for S. American Station.

River Gunboat.

Completing.
About 70% advanced.

(For service on Chinese Rivers.)

ERMANNO CARLOTTO (Shanghai, 1914). Shallow-draught river gunboat (*Cannoniera fluviale per la Cina*). 220 tons. Dimensions: 149 × 24½ × 3 feet. Guns: 2—14 pdr., 4 machine. Designed H.P., 1100 = 14 kts. 2 Yarrow boilers. Complement, 73.

Convoy Gunboats (*Cannoniere di s*

Not yet ready.

ALESSANDRO VITTURI	*ERNESTO GIO*
ANDREA BAFILE	*EMANUELE*
CARLO DEL GRECO	*TOLOSETTO F*

Building or completing by Pattison, Naples. 250 ... sions: 170.6 × 19 × 5.6 feet. Guns: 2—... Torpedo tubes: 2—18 inch in one twin mou... 2800 = 23 kts.

Submarine Chasers.

As Illustration for "SC boats," U.S. Navy S

SC 78	**SC 128**	**S**
SC 82	**SC 179**	**S**
SC 94	**SC 215**	

Above are numbers of U.S. series. New Italian ... unknown. Boats may be re-numbered in the ... series. About 77 tons. Dimensions: 105 (... 5½ feet. Guns: 1—3 inch (23 cal.), 2 Colt ... 660 = 16¾ to 18 kts. Fuel: 2400 gallons per ... 3 sets Standard motors. Wooden hulls.

(Continued

M. A. S. 1—434 (*Motobarche Armate*).

Note.—Correspond to British C.M.B. and M.L. types. Many sold out, scrapped, etc., 1919-20. Present total unknown, but far less than 434 boats.

Photo, Messrs. Ansaldo & Co.

"TIPO VELOCISSIME." 16 BUILT (1917-19). 18 tons. Length: 50 feet. Motors (Isotta Fraschini): 1000 B.H.P. = 35 kts. Guns: 3 M.G. Torpedo tubes: 2.

"TIPO VELOCI." 15 BUILT (1917-19). 30 tons. Length: 72.2 feet. Motors (Isotta Fraschini): 1000 B.H.P. = 30 to 33 kts. Guns: 1—14 pdr., 40 cal., 1 M.G. Torpedo tubes: 2—18 inch in some.

Same appearance as British M.L.

TYPE C: 102 BUILT (1916-19). 40 tons. Length: 80 feet. Motors (Standard): 450 B.H.P. = 19 kts. Guns: 1—14 pdr., 40 cal. (A.A.) and 1 M.G. Are much the same as British *M.L. 1—151*.

1919 Photo.

TYPE B: 54 BUILT (1916-19). 12 tons. Length: 59 feet. Motors (Isotta Fraschini): 500 B.H.P. = 18 kts. Guns: 1—14 pdr. 30 cal. and 1 or 2 M.G.

M. A. S. (*Continued*).

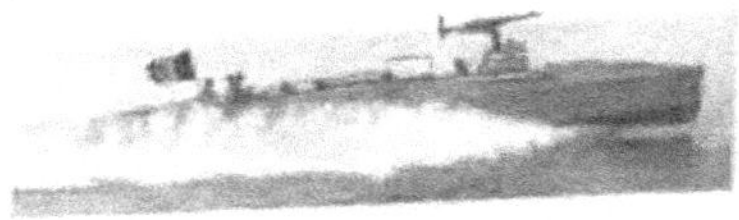

1919 Photo.

M.A.S. 91—102, 218—232 (built by Orlando).

1919 Photo.

With two torpedoes.

TYPE A: 183 BUILT (1916-19). 12 tons. Length: 53½ feet. Motors (Isotta Fraschini or Sterling or F.I.A.T.): 400 to 500 B.H.P. = 24 to 26 kts. Guns: (a) 1—3 pdr., 40 cal. and 1 M.G., or (b) 1—6 pdr., 43 cal. and 1 M.G. Several have 2 torpedo tubes or dropping gears.

"Grillo" Boats.

No photo available.

TOTAL NUMBER NOT KNOWN. About 20 tons. Dimensions: 39 × 6 × — feet. 1—15 H.P. electric motor and tunnel screw = 5 kts. 1—20 H.P. electric motor driving endless climbing chains for surmounting booms. 2—18 inch torpedoes carried in dropping gears. Complement, 3. Designed by Major General Pruneri.

"Subsidiary Ship" (*Nave Sussidiaria*).

Not completed by June, 1919.

NAME AND TYPE UNKNOWN. Is building at Riva T... 10,300 tons. Dimensions: 328.1 × — × — feet. Est. speed: 16 kts. Has special underwater protection.

Depot Ship (Red Sea).

ALULA (Motorship). No details or photograph of this ...

Submarine Salvage Ship.

(*Pontone per ricupero Sommergibili.*)

ANTEO (Smulders, Schiedam, 1914). Special Salvage S... Submarines. 2100 tons. Dimensions: 164 × 73¾ × ... I.H.P., 750 = 8 kts. 2 screws. 2 cyl. boilers. 2 ... each lifting 200 tons.

Salvage Ship.

Photo wanted.

TESEO (ex-Austrian *Herkules*, 1910; taken over 1920) ... tons. Dimensions: 210 × 331 × 13 feet. ...

Mine Layers (*Navi Affondamine*).

Note.—All vessels in section A—F fitted and employed as Minelayers during the war are fully described and illustrated on preceding pages.

A :—Light Cruisers.

NINO BIXIO, MARSALA QUARTO. } Carry 200 mines each.

B :—Old Cruisers.

PUGLIA. Carries 60 mines.

C :—Flotilla Leaders.

2 Aquila class: ***Aquila, Falco.*** Carry 50 mines.

2 Mirabello class: ***Augusto Riboty, Carlo Mirabello.*** Carry 100 mines each.

D :—Destroyers.

7 boats of G. Sirtori class, viz.: ***Angelo Bassini, Agostino Bertani, Giacinto Carini, Giacomo Medici, Giuseppe la Farina, Giuseppe la Masa, Nicola Fabrizi.*** Each carries 10 mines.

E :—Submarines.

X2, X3, described in Submarine Section. 18 mines each.

F :—Lagoon Gunboats.

BRONDOLO, MARGHERA. 60 mines each.

G :—Old Torpedo Gunboats.

(May no longer be Mine Layer.)

AGORDAT (1899). 1313 tons. Complement, 173. Dimensions: 287½ × 30½ × 12 feet. Armament: 2—4.7 inch, 4—14 pdr. AA., 2—14 inch tubes (above water). Deck, ¾". Designed H.P. 8000 = 23 kts. 8 Blechynden boilers. Coal: *normal* 190 tons; *maximum* 300 tons.

No photo available.

MINERVA (1892). 940 tons. Complement, 83. Length: 246 feet. Beam: 27 feet. *Max.* draught: 12½ feet. Guns: 3—14 pdr., 1—14 pdr. AA., 2—1 pdr. AA. Carries 60 mines. Armour: 1½" deck. Designed H.P. 4000 = 19 kts. 4 Blechynden boilers. Oil fuel: 80 tons.

Fuel Ships (*Navi rifornimento carbone e nafteline*).

Oil Tankers (*Cisterne per nafta*).

Not completed, May, 1919.

ARSA (*Vesuvo D.Y.*) and *FIVE OTHER OIL TANKERS building, completing or in service, of "Improved Acheronte" class.* No details available.

ACHERONTE, COCITO, FLEGETONTE, LETE, STIGE (all 1914-15). Oil tankers. 1284 tons. Armament: 1 H.P., 400. Fiat Diesel engines = kts. Cargo: 760 tons oil fuel.

No photo available.

GIOVE* (——). **NETTUNO*** (——). Oil tankers. 10,310 tons. Dimensions: 416¾ × 56¾ × 24 feet. Armament: 6—14 pdr. H.P. and speed not known. Cargo: 8000 tons oil fuel.

*These two Ships are not mentioned in Official List furnished by the Ministry of Marine, Rome. They may have been (*a*) sunk during the War, or (*b*) transferred to Mercantile Marine.

Fuel Ships (*Continued*).

BRONTE (1904). Fleet Collier, Oil Tanker a[...] 9460 tons. Complement, 79. Guns: 4—6 p[...] cargo, 3700 I.H.P. = 11·5 kts.; with reduc[...] I.H.P. = 14·5 kts. 2 screws. 4 cyl. boiler. [...] tons coal or 4000 tons oil + 2000 tons coa[...] 550 tons. *Sterope* (sister ship) lost during Wa[...]

Note.—*Garigliano* (see third column) can also carr[...] of oil fuel.

Repair Ships (*Navi Officina*).

Photo, Mr. Stewart M[...]

SICILIA (1891). 13,298 tons. Old Battleship. 4 old 13.5 inch and 4—14 pdr. AA. guns. 4¼"—2" Deck, 14¼" Barbettes. Speed: 20 kt[...] probably 12 kts. best speed now. Coal: 140[...]

VULCANO (ex *Savona*, 1885). 3260 tons. Co[...] Guns: 4—6 pdr. I.H.P., 2310 = 15·2 [...] 8 loco. boilers. Coal: 600 tons.

Transports (*Trasporti*).

Note.—*Brondo* (see Fuel Ships) is rated as a Transport.

1920 Official Photo.

ERITREA (ex *Kaiseri*, captured from Turkey, 1911). 1350 tons. Guns: 6—6 pdr. I.H.P., 1000 = kts. 1 screw. 2 cyl. boilers. Can carry 1700 tons cargo.

BENGASI (ex-*Derna*, captured from Turkey, 1911). Built 1890. 3560 tons. Complement, 64. Guns: 2—6 pdr., 4—3 pdr. I.H.P., 2000 = 13.6 kts. 1 screw. 4 cyl. boilers. Can carry 1050 tons as cargo.

(*Also Oil Tanker.*)

GARIGLIANO (1887). 1300 tons. Complement, 42. Armament: 2—12 pdr. H.P., 537 = 11 kts. 1 screw. 1 cyl. boiler. 120 tons coal. Can carry 260 tons cargo and 85 tons oil (as cargo).

Small Mine-Sweeping Tugs.

(*Rimorchiatori-Dragamine.*)

No Photo available.

Fourth and Fifth Series: **RD 18—RD 29** (builders unknown, 1916-18), **RD 30—RD 44** (builders unknown, 1918-19), **RD 45—49** (Pattison, Naples, 1918-19). No details known of these vessels. Generally as *RD* 1—18 below, but a little larger, so as to use engines stripped from other ships.

Appearance much as "Tevere" in next column.

Third Series: **RD 13, RD 15—RD 18** (Castellamare D.Y., 1916-17). As *RD* 1—18 below.

Second Series: **RD 7—RD 9, RD 11, RD 12** (Castellamare D.Y., 1916-17). As boats below.

First Series: **RD 1, RD 3—RD 6** (Castellamare D.Y., 1915-16). 200 tons. Dimensions: 115 × 19 × 6½ feet. I.H.P. 750 = 12 kts.

Tugs (*Rimorchiatori*), Tanks (*Cisterne*), Distilling Ships (*Navi Distillatrice*).

Salvage Ship *Teseo* (previously listed) is also a powerful Tug.

No photo available.

TREMITI. No details of this ship received.

Armed Tugs, Tanks, &c.—*Contin*

No photo available.

TITANO (1913). High Sea Tug. 970 tons. Gun H.P., 1970 = 15 kts. 2 screws. 2 Genna Coal: 100 tons.

No Photo available.

ERIDANO (1911). Tank. 1260 tons. Guns: 2— signed H.P., 1200 = 12.7 kts. 2 screws. 2 Ya Coal: 120 tons. Carries 400 tons water.

TEVERE (1897). Tank. 960 tons. Guns: 2—1 p 620 = 11 kts. 1 screw. 2 Thornycroft boilers. tons. Carries 320 tons water.

ATLANTE (1892), 776 tons, and **ERCOLE** (1892). Tugs. 853 tons. Armament: 2—6 pdr. Design 1830 and 1700 = 13 kts. 2 screws. 4 cyl. boiler 150 tons.

Also many other Tugs, 83 Harbour Tanks and 12 Lagoon

Royal Yacht.

Photo, Sword, [illegible]

TRINACRIA (ex-*[illegible]*, [illegible]). [illegible] tons. Complement, [illegible]. Dimensions: [illegible] × [illegible] × [illegible] feet. Armament: [illegible] 2—3 pdr. Designed H.P., [illegible] = [illegible] kts. [illegible] screw. [illegible] boilers. Coal: [illegible] tons.

Note.—Loaned 1920, for Service as a "Floating Fair Ship" for Trade Exhibitions and Propaganda purposes.

Armed Yachts (*[illegible]*).

Official Photo, 1920.

TOBRUK ([illegible]). [illegible] tons. Complement, 28. Dimensions: [illegible] × 25 × [illegible] feet (mean). Guns: 2—11 pdr. (A.A.). I.H.P., [illegible] = [illegible] kts. 1 screw. 1 cyl. boiler. Coal: 170 tons.

No photo available.

MISURATA ([illegible]). [illegible] tons. Complement, 33. Dimensions: 161[illegible] × 25[illegible] × 14 feet (mean). Guns: 2—14 pdr. (A.A.). I.H.P. 760 = 12.5 kts. 1 screw. 1 cyl. boiler. Coal: 120 tons.

No photo available.

CUNFIDA (ex-*[illegible]*, 189[illegible], captured during Turko-Italian War). [illegible] tons. Complement, 26. Dimensions: [illegible] × [illegible] × — feet. Guns: 1—14 pdr., 1—14 pdr. (A.A.). I.H.P. [illegible] = [illegible] kts. 1 screw. 2 cyl. boilers.

Armed Yachts (*continued*).

Official Photo, 1920.

GIULIANA (1893). 1100 tons. Complement, 46. Dimensions: 222[illegible] × 28 × 15 feet (mean). Guns: 4—6 pdr. I.H.P. 700 = 11.9 kts. 1 screw. 1 cyl. boiler. Coal: 180 tons.

Official Photo, 1920.

ARCHIMEDE (——). 815 tons. Guns: 4—6 pdr. I.H.P. —— = 13 kts. 1 screw. 1 cyl. boiler.

Note.—Captured by Germans at Odessa, June, 1918. Re-captured December, 1918.

Official Photo, 1920.

CAPITANO VERRI (ex-*Thetis*, captured during Turko-Italian War). 640 tons. Guns: 2—6 pdr. I.H.P., 355 = 10.5 kts. 1 screw. 1 cyl. boiler.

Surveying Ship (*Servizio Idrografi*

Official

AMMIRAGLIO MAGNAGHI (Ansaldo, —(8/14)) Complement, 137. Dimensions: 229[illegible] × 37[illegible] Guns: 4—14 pdr. I.H.P., 2000 = 14 kts. 2 sets vertical triple expansion. Boilers: 4 c[illegible]

Training Ships (*Navi Scuola*,

For Cadets:

Cruisers **F. FERRUCCIO** and **VARESE** described [illegible] page.

Also

AMERIGO VESPUCCI (1882). 2660 tons. [illegible] 279. Armament: 4—4·7 inch.* Armour: [illegible] ships. H.P., 3340 = 14 kts. 1 screw. 8 locom[illegible] Coal: 500 tons.

*Removed during War.

EX-ENEMY WARSHIPS ASSIGNED TO ITALY.

All to be broken up or otherwise disposed of.

NATIONALITY.	BATTLESHIPS.	CRUISERS.	TORPEDO CRAFT.	SUBMARINES.	REMARKS.
GERMAN	(Nil)	(Nil)	(Nil)	U 163 UB 102 UC 98 U 120 UB 95 UC 94 U 114 UB 80 UC 93 U 54 (All *B*)	All to be broken up under these conditions :— *Ships marked A* :—To be put out of commission within 18 months and to be destroyed or broken up within 5 years.
AUSTRIAN	TEGETTHOF (*B*) RADETZKY (*A*)* ZRINYI (*A*)* E. FRANZ FERDINAND (*A*)	(Nil)	T.B.D. :— Varasdinier (*A*), Csizkos (*B*), Dinara (*A*), Huszar (*A*), Scharfschutze (*B*), Turul (*A*), Velebit (*A*), Uskoke (*B*) T.G.B. :— Magnet (*A*), Trabant (*A*), Planet (*A*), Blitz (*A*), Komet (*A*), Meteor (*A*), Sebenico (*A*), Spalato (*A*), Zara (*A*) T.B. :— 52 T (A), No. 1 (*A*), No. 2 (*A*), No. 4 (*A*), No. 5 (*A*), No. 6 (*A*), No. 12 (*A*)	9 boats—numbers not reported (*A*)	*Ships marked B* :—May be kept in full commission for 1 year; but must be paid off at the end of 30 months and destroyed or demolished within 5 years. All periods reckoned from date of assignment to Italy. *List communicated to "Fighting Ships" by request of H.M. Under-Secretary for Foreign Affairs, November, 1920, gives these two ships as assigned to Italy. List furnished by the Ministry of Marine, Rome, August, 1920, also gives them as assigned to Italy but with the reservation: "The two ships of the *Radetzky* Class at Spalato (*Zrinyi*, *Radetzky*) will be transferred to United States on ratification of Peace Treaty with Austria." Possibly the U.S. Government has foregone claim on these two, and they are now Italian under Regulation (*A*).
TURKISH	—	—	—	—	No details yet available.

ESTHONIAN FLEET.

Compiled, 1919, from details supplied by courtesy of the Naval Staff, Reval ; Corrected 1920 by favour of Messrs. J. Pitka & Sons. Reval.

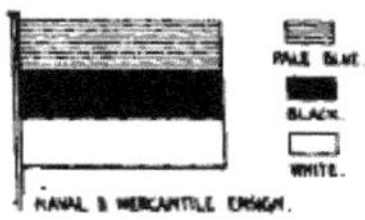

NAVAL & MERCANTILE ENSIGN.

Note.—The top bar of Ensign is pale blue.

Commander-in-chief of the Navy and Naval Defences : Rear-Admiral Johann Pitka, K.C.M.G.

Mercantile Marine consists of (*a*) 80 steamers, total gross registered tonnage 30,000 tons ; and (*b*) 177 Sailing ships, total gross registered tonnage 20,000 tons.

Coinage : Marks and pennies (Finnish system). Present rate of exchange, 45 marks = £1. (Normal rate of exchange, 25 marks = £1.)

Naval Officers and Men : 2,100.

RECOGNITION SILHOUETTES.

Scale : 1 inch = 160 feet

AHTI.

TAARA.

WANEMUINE.

LEMBIT (*Gunboat*).

Lake Peipus Gunboats.

TORPEDO CRAFT.

LENNUK (*T.B.D.*)

WAMBOLA (*T.B.D.*)

Naval Base, Ports, Harbours, etc.

ROHUKULA (*Naval Base*). (About 3 miles to the S.W. of Hapsal.) Extended
1914-16. Entrance to harbour, 18-20 feet. Depth in harbour, up to about 18 f
running on quays. Capable of accommodating at present 5 steamers, when c
repaired, 15 to 20 steamers. Good anchorage in the roadstead. Warehouses, S
Shops. Was partly destroyed by fire and bombardment when the Russian
but now being rebuilt and repaired by Esthonians.

TALLINN (Reval). (A late Naval Base of the Russian Navy, dredged to take l

A : HARBOUR. Depth at the entrance of the port, up to 35 feet. Depth in the
There is room to berth in the harbour, under present conditions, 30 large st
smaller steamers, besides sailing coasters. Good anchorage in the roads, no ti
modern mole in the place of the old wooden jetty, is in the course of complet
to the disturbances which Esthonia has experienced, this work has been tempo
The Esthonian Government proposes to finish this work on the first possib
and contemplates making the Port of Reval a modern free port, with all
appliances, to accommodate the enormous amount of shipping which will ha
Reval in the future, when Reval (being the nearest point to the Russian
and Siberia) will be used as a transit port for Russian trade. Large cold st
are in the course of construction. Port of Reval has 8 large grain elevator
for 750,000 poods (about 12,000 tons) and one small elevator. Also three
to lift from 40 to 60 tons, and one wooden crane. Besides this, 4 floating dry
of lifting 2,000, 1,250 and 3,000 tons are available. The cranes and floati
to the Esthonian Government, while the elevator, to the town. No
Floating Ships' Repairing Workshops. Also Salvage Steamers belonging to
Baltic Salvage Association. Three extensive Shipbuilding Yards (belongin
Shipbuilding Co., Ltd., Bekker & Co., Ltd., and the Peter Wharf (late Nob
Telliskoppel). The Telliskoppel Bay being an excellent site for a harbour
Government has decided to build an extensive harbour there.

B : NEW HARBOUR or MINE HARBOUR (MIINI Sadam). This harbour
a short distance from the Harbour of Reval (approximately one mile), is also
tion, and can already accommodate several small coasting vessels. There i

PERNAU. The normal depth on the bar and thence up to 2 miles above floati
to not being dredged during the War, is about 15 feet. This depth can ea
by dredging up to 20 feet, that is, level with the roadstead. Accommodatio
in the harbour. Three berths, with electric cranes, at the Waldhof Works
left side of the river. There is also a private quay on the right side of the

BALTIC PORT. Depth at entrance to port, 16 to 20 feet. Depth in port, 16 t
is room for two steamers alongside the quay, but as many as four have be
at one time. No tides. Difference in depth caused by the winds. The
is narrow and difficult. The best anchorage is under the Island of R
School.

NARWA. Average depth on the bar : in spring, 12 feet ; in summer, 18-21
entering harbour, 18-21 feet ; for coming to the town, 9-10 feet. Length :
2,850 feet ; of public quays, 1,600 feet ; of quays at the mouth of river, 1,10
in the roads outside the bar. Four Mooring Buoys. Esthonian Govern
improve the harbour arrangements in Narwa, especially for the timber t
possible. *Navigation School.*

(*Contin*

HAPSAL. Depth at entrance to port (average), 10 feet. No docks. One wooden jetty, about 250 feet long, with the same depth alongside.

PORT LOKSA. Open coast port, protected by breakwater. Slipway capable of lifting 6-700 tons, also at the same time two or three 400 ton vessels. Depth at entrance, 20 feet. Inside breakwater, 17-18 feet. Alongside quay, 14 feet. Length of quayside, about 500 feet. Capable of accommodating 3 coasting steamers and at the same time small sailing craft. Good anchorage in the roads.

HEINAST (Hainasch). Port protected by breakwater. Draft, 14 feet. Railway. Open port all year round practically. Proposed to be extended to cope with future traffic of Esthonia. *Navigation School.*

KURESAAR (Arensburg). On Island Esel (Oesel). Draft, 12 feet. Quay about 1 mile long. Good anchorage on roads. Protected by Isle Abroka. About 1½ miles from town is a small harbour for small coasting vessels. This will be rebuilt and extended to accommodate larger ships. *Navigation School.*

KERDLA (Kertel). Draft, 12-14 feet. Situated on the Isle of Dago in Bay. Goo[d] harbour. Capable of accommodating alongside quay 2 or 3 coasting steamers.

PORT KUNDA. Natural harbour in Bay. Good anchorage. Depth, 10-30 feet. [A]long quay, up to 19 feet.

WIRTS (Werder). On the N.W. Coast of Isle Wirts (Schildau). Quay for coasting [...] steamers, also for small sailing craft.

KIHELKOND. Very good natural site for a Naval Base and Wireless Station. Open p[ort] all year round. There is a small harbour for local vessels, also the harbour especi[ally built] by the Russians to accommodate torpedo boats.

SPITHAM. A harbour of refuge under construction.

KAESMU (Kaspewik). Good anchorage. *Navigation School.*

ASSERIN. Quay for small coasting steamers and sailing vessels.

ESTHONIAN DESTROYERS. (*Miini Ristlejad.*)

1919 *Photo.*

1 *Gavriil* type: **Wambola** (ex-Bolshevist *Spartak*, ex-Kerensky *Mikula Maklai*, ex-Czarist *Kapitan [Ko]r Kingsbergen*). Launched by Putilov Works, Petrograd, 1915 and completed 1918. 1585 tons. Dimensions: 314⅔ × 30½ × 9¾ feet. Guns: 4—4 inch, 1—2 pdr. anti-aircraft, 2 M.G. Torpedo tubes: 9—18 inch, in 3 triple-deck mountings. Also designed to carry and lay 80 mines. Designed S.H.P. 30,000 = 29 kts. A.E.G. Curtis turbines. Oil fuel: about 400 tons. Captured by British Light Cruisers and Destroyers in the Baltic, 1919 (with *Lennuk*) and transferred to Esthonian Navy.

1919 *Pho[to].*

1 *Izyaslav* type: **Lennuk** (ex-Russian *Avtroil*, launched by Reval S.B. Co., 1915, completed (1917). 18[..] Dimensions: 344½ × 31¼ × 9¾ feet. Guns: 5—4 inch, 1—9 pdr. anti-aircraft, 2 machine. Torpedo t[ubes: 9] —18 inch, in 3 triple-deck mountings. Also designed to carry and lay 80 mines. Designed S.H.P. 32[,000 =] 32 kts. A.E.G.-Curtis turbines. Oil fuel: 400 tons = 2400 miles at 15 kts. Complement, about [...] Captured (with *Wambola*) from Bolshevists by British Light Cruisers and Destroyers in the Baltic, 19[19, and] transferred to Esthonian Navy.

Gunboats.

Note.—Mainmast has been removed.

LEMBIT (ex German *B*[...], ex Russian *Bobr*, 1907). [...] tons. Dimensions: [...] (*o.a.*) × 36 × 8 feet [...] draught). Guns: 2[...] inch, 1—11 pdr. Us[ed to] carry about 50-60 [mines]. I.H.P. 800 = 12 kts. (In 1029 = 11·5 kts.) [...] 130 tons.

Note.—Above Gunboat [was] captured by Germans at [...] in 1918, and served [a] short time in German N[avy].

(Gunboats continued on next page.)

Gunboats—*(continued).*

No photo available.

LAINE (——). About 400 tons. Guns: 2—3 pdr. and some machine. Speed: 12 kts.

No photo available.

MEEME (ex German ———, captured by Esthonian warships from German "Landwehr" troops, at mouth of Duna, July, 1919; since refitted by Esthonian Government.) 77·3 tons. 90¼ × 21 × 7 feet. H.P. 225 = 13 kts. Guns: 1—3·4 inch, 1—4 pdr., 2—1 pdr., 6 M.G. (2 M.G. are in steel-plated turret). 1 torpedo tube being fitted.

Mine Layers.

No photo available.

KALEW, OLEW (ex-Russian Motorships *No. 7* and *No. 8*). 50 tons *gross*. Guns: 1—3 pdr. Speed: 9 kts. Have motor engines.

Mine Sweepers.

No photo available.

ALICE. **EDITH.** **KAETHE.** All three are Converted Tugs.

Also 10 small Mine Sweepers with motor engines.

Ice Breaker.

No photo available.

TARUJA (ex Russian *Herkules*). 900 tons. Guns: 1—5·1 inch, 1—11 pdr. Speed: 12 kts.

Peipus Lake* Flotilla.

*Lake Peipus is referred to in Bolshevist Naval Staff reports as Lake Chudskoe, Tchudakoe, &c. Photos below of *Taarta* and *Athi* taken during naval action with Bolshevist vessels, October 12th, 1919.

Gunboats.

[illegible] Photo received 1920.

WANEMUINE. UKU. 1920 Photo.

TAARA. **TAARTU.** **UKU.** **WANEMUINE.**

All above ships converted from Passenger Steamers during the War by the Russian Navy. Displacement: 100-200 tons. Speeds: 12·9 kts. Guns: *Taara*, 3—11 pdr.; *Tartu*, 2—1 pdr. machine; *Uku*, 1—11 pdr.; *Wanemuine*, 2—11 pdr.

Peipus Lake Gunboats—

ATHI. No details known. Armed with 2 Q.F

Transport (on Lake Peipus

No photo available.

KAJAK (ex-*Severnaja Komuna*, ex-*Gogare*, c Bolshevists). 5,100 tons *d.w.c.* Speed:

Small Craft (on Lake Peipus

Lotti. Armed Motor Patrol Boat.

Flirt. Communication vessel.

Also various tugs, small launches and motor exemplified by above illustrations; some on L

POLISH FLEET.

Personnel, in 1919, reported to be 1500 ex-officers and bluejackets from the late Austrian and Russian Navies.

Note.—During 1919, a British Naval Mission, under Commander Eric L. Wharton, R.N., visited Warsaw, to advise the Polish Government on the organisation of a small Preventive and Police Force.

Scale: 1 inch = 160 feet. RECOGNITION SILHOUETTES.

(*Not yet selected.*)

"A" *type* t.b. (5).

V 105-108 *type* t.b. (1).

6 Torpedo Boats.

(All for Police and C. G. Duties. To mount only one small gun.)

Photographs reserved for 1921 Edition.

5 *ex-German* "A" Class: To be assigned to Poland from boats surrendered by Germany. Identity of these boats not yet known, but the following are *assumed* details of the German T.B. numbered between *A 70* and *A 80*, from which selection will probably be made: Built 1916-18. 235 tons. H.P. 4700 = 25.5 kts. Oil: 71 tons = 836 miles at 20 kts. Guns: 2—3.4 inch, 30 cal. Krupp, "Tbts" model. Torpedo tubes: 1—17.7 inch. Can carry 10 mines.

6 Torpedo Boats—*Continued.*

V 106. 1920 *Photo, G. Dal*

1 *ex-German* boat, to be selected from "V 105—108" type. Built by Vulkan, 1913-15. 340 tons. Dimensi . . . 20.4 × 9.4 feet. H.P. 6000 = 30 kts. Fuel: 50.7 tons coal + 16.2 tons oil = 640 miles at 20 kts. . . . 3.4 inch, 45 cal. Krupp "Tpbts" model. Tubes: 3—17.7 inch, in one triple mounting in V 105; in t . . . mountings in V 106, 108. Can carry 12 mines. Original four boats of this Class begun for Netherlan . . . *Z 1—Z 4*, taken over by Germany, August, 1914, renumbered *V 105—108*. V 107 War Loss. Other . . . of this Class to be assigned to Brazil.

River Flotilla.

Not completed yet.

FOUR SMALL GUNBOATS, building at Danzig, 1920. No details known.

On the Pripet, Vistula and other rivers, there appear to be nondescript flotillas of Armed Steamers, Launches, &c. Some of these vessels have taken part in Polish Army Ope . . . against the Soviet Armies. During April, 1920, the Poles claimed to have captured Bolshevist River Monitors and various Armed Steamers near Czernobyl." During May the Bolshevists reported they had blown up "an Enemy Armoured Steamer" near the . . . of the Privzhna River. A photograph has been seen of a "Bolsherist Armoured Mo . . . captured on the Pripet. She is an antiquated Paddle-wheel River Steamer, the "arm . . . consisting of a "stockade" of wooden logs and iron rails, built up round the paddle . . . Within this "stockade," four or six machine guns were mounted, while two field gun . . . lashed to the forecastle and quarter-deck. "Warships" of this description can or . . . compared with the fighting craft extemporised during the U.S. Civil War.

2a

FINNISH FLEET

Revised 1920.

Scale: 1 inch = 160 feet. RECOGNITION SILHOUETTES.

Information not available for the preparation of Silhouettes.

Personnel.—Not known. Fleet is administered by a Captain-Commander and his Staff.

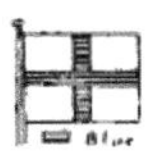

Mercantile Marine.

From "Bureau Veritas" 1919-20 figures.

Ships under 100 tons *gross* excluded.

111 Steam and Motor Ships of 75,963 tons *gross*.
17 Vessels (with Auxiliary Motors) of 3027 tons *gross*.
1959 Sailing Vessels of 279,601 tons *gross*.

Ports.—Abo, Hango, Helsingfors. Sveaborg is fortified.
Lapvik was an anchorage used by the Russian (Imperial) Navy during the War, and may possess defences.

Shipyards.

MASKIN OCH BROBYGGNADS A.B. (Helsingfors). 3 slips to build ships up to 350 feet long, and a fourth smaller slip. Floating dock to lift 4000 tons. 60 ton floating crane over fitting-out basin. Also build engines.

OSAKKEYHITO LEHTONIEMI AKT. WARKHUS (Warkhus). Small yard.

SANDVIKENSDOCKA OCH MECKANISKA VERKSTED (Helsingfors).

New Yard projected at Bornholmen, near Abo. At Abo, the Crichton Yard used to build Torpedo Craft for the (Imperial) Russian Navy.

At Helsingfors, "Oskar Eklund" Dry Dock 330 × 50 × 18½ feet.

At Nystad, hydraulic slipway to haul 1200 tons.

Minister of War: Col. Yolander (March, 1920).

2 Destroyers.

Photo wanted.

2 type unknown: ***Matti Kurki*** and ————. Complement, 52.

Note.—Above boats cannot be identified with any certainty; it can only be conjectured that they are t[illegible] Russian T.B.D. captured during 1918, of which the following are details:—

5 Yarrow type: ***Podvishni, Poslushni, Prosorlivi*** (Izhora Works, Petrograd, 1899-1900). ***R*** (Nevski Works, Petrograd, 1901). 220 tons. 190 × 18½ × 7½ feet. Guns: 2—11 pdr., 2 machine. T[illegible] Carry 14 mines. I.H.P. 3800 = 27 kts. Boilers: 4 Yarrow. Coal: 60 tons. Complement, 52.

Three Finnish "Torpedo Boats" reported to have been lost in ice in Gulf of Finland, Dec., 1919, but n[illegible] have not been reported. Possibly they were three of the above five ex-Russian boats; the se[illegible] forming the Finnish *Matti Kurki* class.

10 Torpedo Boats.

Photo wanted.

6 "*Type G*": Names or numbers not known. Complement, 51.

(*Continue*

Torpedo Boats—*continued*

Photo wanted.

4 "*Type C*": Names or numbers unknown. Complement, 35.

Note.—No information available for identification of above 10 T.B. Previous information was that the following 5 ex-Russian boats formed part of Finnish Fleet:—

5 *Normand* type: ***Nos. 212, 215, 216, 218, 222*** (1902-4). 150 tons. 147½ × 16 × 4½ feet. Guns: 2—3 pdr., 1 M.G. Tubes: 2—15 inch in twin mounting. I.H.P. 2200 = 29 kts. Coal: 30 tons. Complement, 28. Captured from Russia, 1918. Fitted as Minesweepers.

The total of 5 boats neither accords with "Type G" nor "Type C." Also the Normand boats cannot be divided into two classes. A warning must be added that the numbers and total of these 5 ex-Russian boats are now suspected as being untrustworthy.

2 Submarines.

Photo wanted.

2 *type unknown*: ——— and ———, under repair, 1920.

Note.—It can only be suggested that there are two of the Russian boats *AG 11, 12, 13, 15*, scuttled at Han salved. For details of Russian "*AG*" boats, *v. AG 22, 23*, in Russian Black Sea Section.

2 Gunboats.

No photo available.

KARJALA and ——— (ordered by Russian Government, 1916, completed at Crichton's Yard, Abo, 1918, and taken over by Finnish Government). 350 tons. Speed, 14 kts. Complement, 48.

1 Mine Layer.

Photo wanted.

——— (bought from Italy, 1920). Complement, 51. No other details known.

2 Motor Launches.

No photo available.

——— and ———, bought from Italy, 1920.

Also a number of small Patrol Boats.

Lake Ladoga Flotilla.

Various armed vessels are in service on Lake Ladoga, of wh details are available.

(ROYAL) SWEDISH NAVY.

Revised 1920 by courtesy of the Ministry of Sea Defence. Also corrected from the 1920 Edition of the Official Handbook "Svenska Marinens Rulla."

Illustrated from photos taken or collected by S. Anderson, Esq.

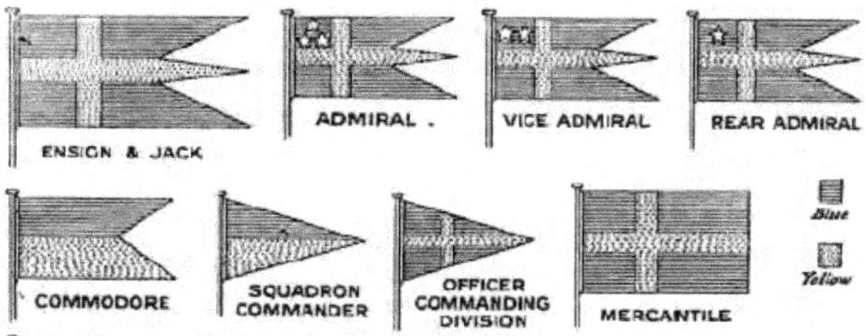

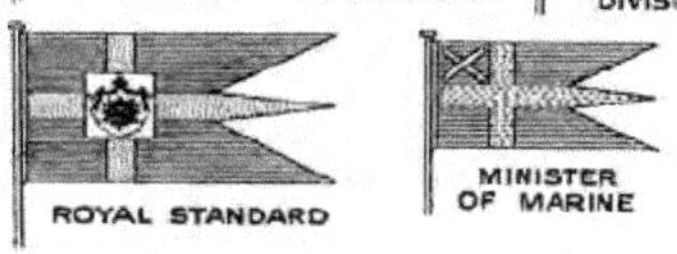

Minister of Sea Defence: General Hammerskjöld (Nov., 1920).

Naval Attaché (London): Captain D. Tiselius

Modern Swedish Guns (Bofors).

(All details are unofficial).

Notation	Designation		Length in calibres	Model	Weight of Gun	Weight of A.P. shot	Initial velocity	Max. penetration firing A.P. capped at K.C. 8000 yds.	5000 yds.	3000 yds.	Danger Space against average warship, at 10,000 yards	5000 yards	3000 yards	Nom. per minute
	c/m	inches			tons	lbs.	ft. secs.	in.	in.	in.				
A ⁵	28	11	45	'11	41	760½	2576†	10½	15	18	..	..	..	1·2
AA	25.4	10	42	'94	29	450	2362	..	8	11	125	340	672	1
A	21	8.3	44	'98	17	275½	2460†	..	7	9½	100	425	600	2
C	15.2	6	50	'03	7½	100	2789†	..	..	7	75	230	460	6
E	15.2	6	45	'98	6	100	2460	..	..	6	60	200	410	6
	12	4.7	50	'11	3½	46½	2822	..	..	..	..	..	..	8
I	12	4.7	45	'94	2	46½	2430	..	..	3½	..	..	..	8

Guns marked † in the velocity column fire Bofors special nitro-compound.

Torpedo.

Whiteheads. *Submerged* tubes (Elswick pattern), except in *Oden & Svea* class. Torpedoes, 18 in., 15 in. and 14 in. Since 1911 torpedoes have all been manufactured at Karlskrona.

Wireless. Marconi.

INSIGNIA OF RANK ON SLEEVES.

Vice-Am.	Konter-Am.	Kommendör.	Kommendörkapten 1 graden.	2 graden.	Kapten.	Löjtnant
British or U.S.A. ,,	*Rear-Ad.*	*Commodore.*	*Captain.*		*Commander & Lieut.-Com.*	*Lieutenan*

Note.—Full Admirals have 3 stars (as for Vice and Rear combined).

Reserve officers have *blue* above the stripes. All civilian officers have stripes without colour between stripes as follows:—Constructors and Engineers, *violet blue*; Doctors, *red*; *white*.

Swedish Arsenals.

KARLSKRONA. One t.b. slip. Dry docks: (Oscar II.), 410 × 68 × 24½; (No. 1), 20 (No. 2), 265 × 51¾ × 19½; (No. 3), 209 × 46½ × 20; (No. 4), 226 × 46¾ × 20; (No. 5), 2 (Old Dock), 233 × 46 × 17½. Submarine Station. Is fortified.

STOCKHOLM. One t.b. slip. One dry dock, 410 × 58½ × 22¾ feet. One dry dock la 79 × 29¾ feet. There are also two private docks at the Finnboda Yard (Beckholmen) below. (No. 1), 328 × 55 × 18 feet; (No. 2), 321½ × 35½ × 11½ feet. Mining and Station. Is fortified.

Swedish Private Yards

(that build warships).

A. B. LINDHOLMEN-MOTALA (Lindholmens Mekaniska Verkstad och Varf) Gothenbur 220 and 187 feet long. One dry dock, 402 × 58½ × 20 feet. One 60-ton crane. pneumatic plant. Total employees 900.

GÖTAVERKEN. (Gothenburg). One or two slips (up to 8000 tons), and two slips (2500 a No docks. Employees 700. 12,000-ton Floating Dock at this port, 159¾ × 82 × 2 by Götaverken Co.

BERGSUND (Bergsunds Mekaniska Verkstad och Varf) Stockholm. One slip, 151 feet lon One 40-ton crane. Employees 600.

FINNBODA. (Stockholm). One slip, 226 feet long. 60-ton crane. Employees 700. (325 × 52 × 17 feet; (2) East Dock, 321½ × 35 × 13 feet. One large floating dock, on

KOCKUM (Kockums Mekaniska Verkstads Aktiebolag) Malmo. One slip, 400 feet another, 275 feet long. Two dry docks (Harbour Commissioners) 502 × 66 × 23¾ feet a 12½ feet. 45-ton crane. Electric, pneumatic, and hydraulic power throughout yard. En

ÖRESUND SVARVET (Landskrona). One dry dock, 621 × 81 × 25 feet.

Mercantile Marine.

"Bureau Veritas" figures 1919-20. Ships under 100 tons gross excluded.

1010 steam and motor ships of 920,817 tons *gross*.
158 vessels with auxiliary motors, of 17,881 tons *gross*.
917 sailing vessels of 138,219 tons *gross*.

During 1919, 55 ships of 50,971 tons *gross* were launched by Swedish Shipyards. Som foreign owners.

Principal Trade Ports.

Gothenburg, Stockholm, Malmö, Norrköping, Helsingborg, Gefle and Sundsvall.

Coinage. (1914 rate of exchange).

1 krona = about 1s. 1½d. British, 25 cents U.S.A. (18 kronor = £1).

General Notes.—Personnel: About 4,000. *Colour of Ships*: Grey.

Scale: 1 inch = 160 feet.

SWEDISH RECOGNITION SILHOUETTES.

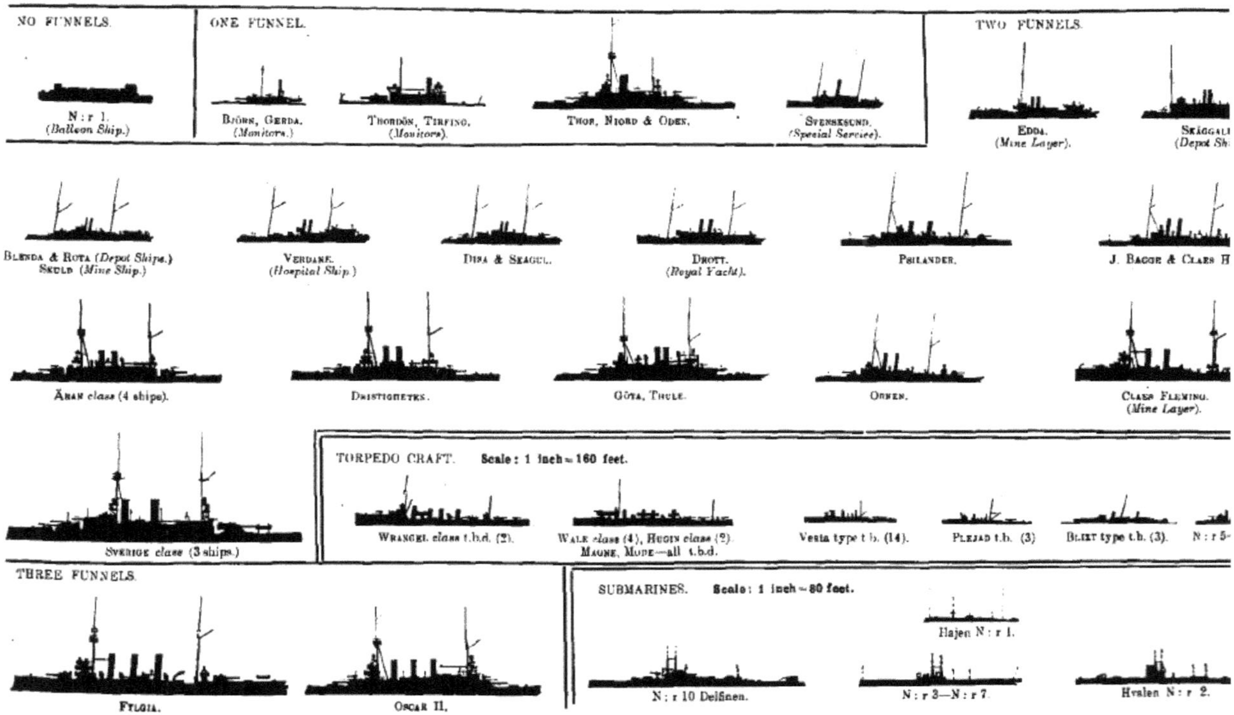

SWEDISH COAST DEFENCE BATTLESHIPS. (*Kustpansarfartyg*).

From *Sverige* class, down to and including *Svea* class, officially rated as *I kl. Pansarbåtar.*

(Sverige Class—3 Ships.)

SVERIGE (May, 1915), **DROTTNING VICTORIA** (Sept., 1917) and **GUSTAV DEN FEMTA** (Feb., 1918).

Displacement: *Sverige*, 7695 tons; *D. Victoria* and *Gustav V* about 7,300 tons. Complement, 408.

Length (*w.l.*) { *Sverige*, 392·8 feet; *D.V.* & *G.V.* 396·6 feet } Beam, 61 feet. *Max.* draught, 22 feet.

Guns (Bofors):
- 4—11 inch 45 cal.
- 8—6 inch, 50 cal.
- 4—14 pdr.
- 2—14 pdr. A.A.
- 2—6 pdr.
- 2 machine.

Torpedo tubes (18 inch):
- 2 *submerged* (broadside).

Armour: (Carnegie).
- 8" Belt (amidships) .....
- 6" Belt (ends) ....
- 1¾" Deck (slopes) ......
- 4" Redoubt ....
- 8" Big gun turrets .....
- 5"—3" Small turrets
- 8" Conning tower .....

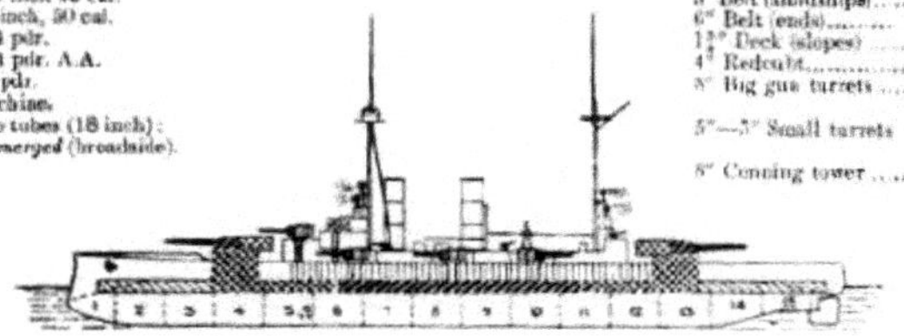

Ahead: 2—11 in. 4 to 6—6 in.

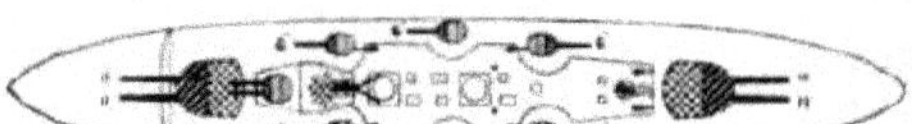

Astern: 2—11 in. 2 to 4—6 in.

Broadside: 4—11 inch, 5—6 inch.

Machinery: In *Sverige*, Curtis turbines by Bergsund Co., 4 screws; in *Gustav V* and *Drottning Victoria* Westinghouse geared turbines by the Motala Company. 2 screws. Boilers: 12 Yarrow. Designed S.H.P. 20,000 in *Sverige*, 22,000 in *D. Victoria* and *Gustav V* = 22·5 kts. Coal: *Normal* 350 tons; *maximum* 700 tons + 100 tons oil.

Engineering Notes.—In *D. Victoria* and *Gustav V*, Westinghouse turbines consist of 2 sets on each shaft, each containing two turbines of the "divided-flow" type, driving propellers through double-pinion reduction gears (total weight of turbines and reduction gears about 120 tons):

Shaft H.P.	22,000	15,000	3,150
R.P.M. (turbines)	3,600	3,186	2,070
R.P.M. (propellers)	200	177	115
Speed (knots)	22·5	20	12·5

General Notes.—*Sverige* was sanctioned 1911; but upon a change of Government was cancelled by the Liberals. The nation then voluntarily subscribed a sum which at the end of April, 1912, amounted to nearly £670,000 for the building of this ship, which was estimated to cost £670,000; the balance was devoted towards commencement of *D. Victoria* and *Gustav V* under the 1915-19 Naval Programme. Work on *Gustav V*. was stopped during the war, owing to non-delivery of armour plates from the U.S. Building was resumed in 1919.

191

1918 Photo by courtesy of S. Ans

Name.	Builder.	Machinery.	Laid down.	Completed.	Trials.	Boilers.
Sverige	Götaverken, Gothenburg	Bergsund Co.	Feb. '12	May, '17		Yarrow in all
D. Victoria	Götaverken, Gothenburg	Motala Co.	July '15	1920		
Gustav V	Kokum Co., Malmö	Motala Co.	Dec. '14	1921		

OSCAR II: Present appearance. *1913 Photo.*

OSCAR DEN ANDRA (1906).

Displacement, 4080 tons. Complement, 329.
Length (*w.l.*) 316½ feet. Beam, 50½ feet. *Maximum* draught, 18 feet.

Guns:
- 2—8·3 inch, 45 cal.
- 8—6 inch, 50 cal.
- 10—6 pdr.
- 1—1 pdr.

Torpedo tubes (18 inch):
- 2 *submerged* (Elswick).

Armour (Krupp):
- 6"—4" Belt (amidships)
- 2" Deck (slopes) ....
- 6" Bulkheads ....
- 4" Lower deck redoubt ......
- 7½"—5" Big gun turrets (N.C.) ......
- 7" Hoists to these..............
- 5"—3" Small turrets (N.C.)...
- 7" Conning tower..............

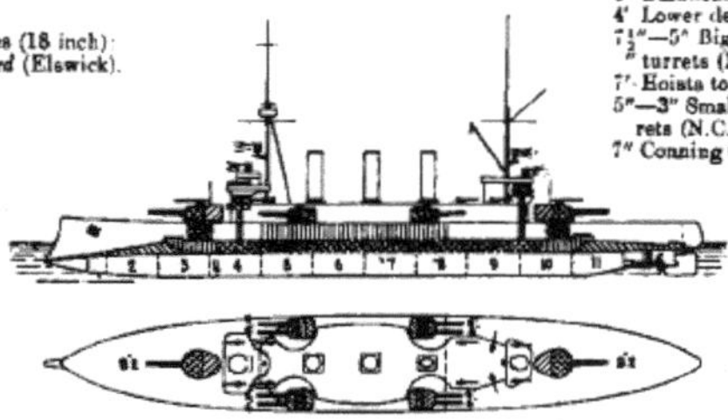

Ahead:
1—8·3 in.
4—6 in.

Astern:
1—8·3 in.
4—6 in.

Broadside: 2—8·3 in., 4—6 in.

Machinery: 2 sets triple expansion. 2 screws. Boilers: 12 Yarrow. Designed H.P. 9000=18·3 kts. Coal: *normal* 350 tons; *maximum* 500 tons=2950 miles at 10 kts.

Notes.—Four searchlights carried—one on each bridge, and one on each mast. Built by Lindholmen Co. On *trial*: 9400=18·96 kts.

TAPPERHETEN. *1918 Photo, Karlsson, Kar*

(ÄRAN CLASS—4 SHIPS).

Note.—Appearance now as plans below and silhouettes. Searchlights now on masts.

ÄRAN (1902), VASA (Dec., 1901), TAPPERHETEN (1902) & MANLIGHETEN (Dec

Displacements: *Aran*, 3650 tons; *Vasa*, 3745 tons; *Tapperheten*, 3990 tons; *Manligheten*, 384
Length, (*w.l.*) 287 feet. Beam, 49½ feet. *Maximum* draught, 16¾ to 17¼ feet.
Complement, 301. (*Manligheten* 285).

Guns:
- 2—8·3 inch, 45 cal.
- 6—6 inch, 45 cal
- 10—6 pdr.*
- 1—1 pdr.

Torpedo tubes (18 inch):
- 2 *submerged* (Elswick).

**Tapperheten* and *Manligheten* have only 8—6 pdr.

Armour (Krupp):
- 7" Belt (amidships) ....
- 2" Deck (flat on belt) ....
- 7" Bulkheads ....
- 7½"—5" Turrets ......
- 8" Supports ....
- 5½"—2½" Small turrets (N
- 4" Hoists, etc. ....
- 8" Conning tower ....

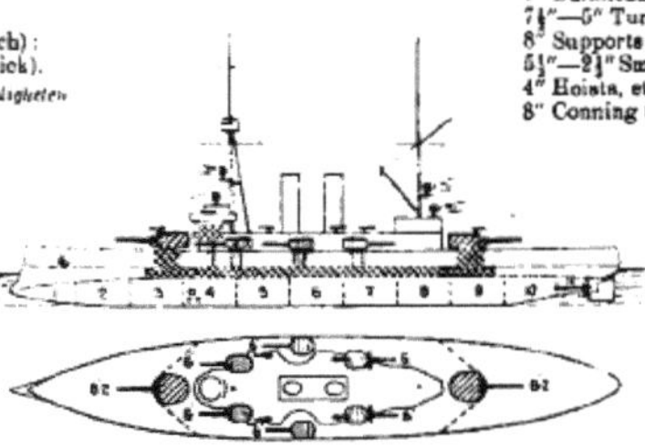

Ahead:
1—8·3 in.
4—6 in.

Astern
1—8·
4—6

Broadside: 1—8·3 in., 3—6 in.

Machinery: 2 sets triple expansion. 2 screws. Boilers: 8 Yarrow. Designed H.P. *Manligheten*, 6500 for others=17 kts. Coal: *maximum*, 300 tons. Endurance: 2000 at 10 kt
Notes.—Where built; *Aran*, Lindholmen; *Vasa*, Bergsund; *Tapperheten* and *Manligheten*.

Searchlights between funnels now in tops on each mast. 1914 *Photo, Karlsson, Karlskrona.*

DRISTIGHETEN (1900).

Displacement, 3620 tons. Complement, 262.

Length (*w.l.*), 285 feet. Beam, 48½ feet. *Maximum* draught, 17 feet.

Guns:
- 2—8·3 inch, 45 cal.
- 6—6 inch, 45 cal.
- 10—6 pdr.
- 1—1 pdr.

Torpedo tubes (18 inch):
- 2 submerged (Elswick).

Armour (Krupp):
- 8" Belt (amidships) ......
- 2" Deck (flat on belt) ......
- 8" Turrets ......
- 8" Turret supports ......
- 3½" Battery ......
- 3½" Hoists ......
- 8" Cotming tower ......

Ahead:
1—8·3 in.
2—6 in.

Astern:
1—8·3 in.
2—6 in.

Broadside: 2—8·3 in., 3—6 in.

Machinery: 2 sets triple expansion. Designed H.P. 5000 = 16 kts. natural draught. 2 screws.

Boilers: 8 Yarrow. Coal: normal 280 tons; *maximum* 310 tons = 2040 miles at 10 kts.

First trials: I.H.P. 5557 = 16·8 kts.

Built by Lindholmen Co. 4 searchlights. Fire control and tripod fore-mast, fitted 1911-12.

Thor, as rebuilt, 1916. 1916 *Photo, Karlsson*

THOR (1898), & NIORD (1898).

Displacement: *Thor*, 3650 tons; *Niord*, 3700 tons. Complement, 254.
Length, (*w.l.*) 278¼ feet. Beam, 48½ feet. *Maximum* draught, 18 to 18½ feet.

Guns (see Notes):
- 2—10 inch, 42 cal.
- 6—4·7 inch, 45 cal.
- 8—6 pdr.
- 1—1 pdr.

Armour (Harvey-ni
- 9½" Belt ......
- 2" Deck (flat on
- 8" Turrets ......
- 8" Supports and
- 3½" Battery......

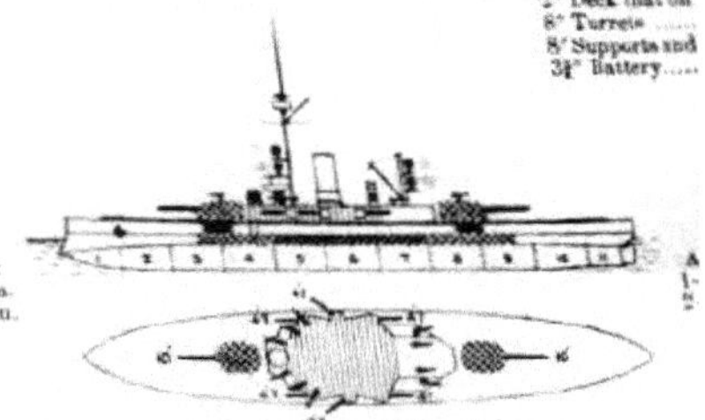

Ahead:
1—10 in.
2—4·7 in.

Broadside: 2—10 in., 3—4·7 in.

Machinery: 2 sets, vertical triple expansion, 2 screws. Boilers: 6 Cylindric
5000 = 16 kts. Coal: normal, 280 tons; *maximum*, 300 tons = 2530 miles at 10 k

Name	Builder	Machinery	Laid down	Completed	Last refit	First trials (f.d.)
Thor Niord	Bergsund Lindholmen	By builders	1896 1896	1899 1900	1916 1917	= 16·7 = 16·5

Notes.—The big guns of *Thor* are Canet, those of *Niord* are Bofors. 4 searchligh
in 1916 and *Niord* in 1917, their appearance being greatly altered. As orig
two masts and two funnels, like *Svea* class, on next page. Now they are al
re-built *Oden*, also given on next page.

1918 Photo, Karlsson, Karlskrona.

ODEN as re-built 1915. Appearance is now almost the same as that of *Thor* and *Niord* on preceding page, but note that there is no ventilator abeam of the conning tower, chart house is slightly deeper fore and aft, and small topmast to stump of old mainmast is much higher. As will be seen by comparing photos and plans the 4·7 inch guns are differently arranged. Has plain top on tripod mast. *Thor* and *Niord* have a ringed top.

ODEN (March, 1896).

Normal displacement, 3715 tons. Complement, 243.

Length (*o.a.*), 278¾ feet. Beam, 18½ feet. *Maximum* draught, 18½ feet.

Guns (Canet and Bofors):
- 2—10 inch, 42 cal. Canet.
- 6—4·7 inch, 45 cal. Bofors.
- 8—3 pdr.
- 1—1 pdr.

Armour (Harvey-nickel):
- 10" Belt (amidships)......
- 2" Deck (flat on belt)
- 10" Turrets ....
- 10" Supports and hoists
- 3½" Battery.................
- 9½" Conning tower ......

Ahead:
1—10 in.
2—4·7 in.

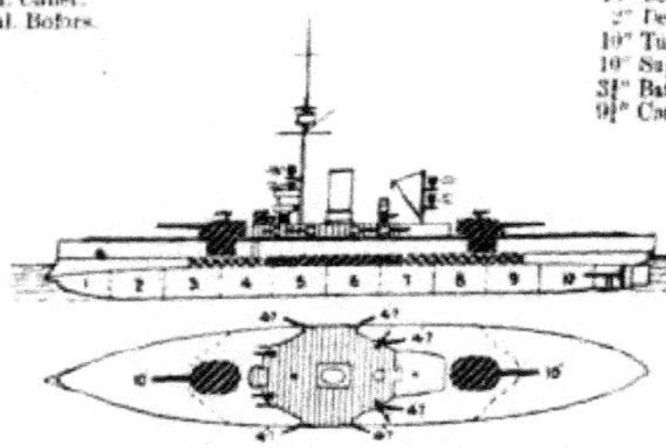

Astern:
1—10 in.
4—4·7 in.

Machinery: 2 screws. Boilers: 6 cylindrical. Designed H.P. 5000 = 16 kts. Coal: *normal* 275 tons; *maximum*, 300 tons. Endurance: 2530 miles at 10 kts.

Laid down by Bergsund Co., 1894. Completed 1898. 4 searchlights. Re-constructed 1915, when appearance was greatly changed. Appearance was originally like *Svea* class in next column.

GÖTA. THULE. *1914 Photo by courtesy of S. An* (SVEA).*

Forebridges run to tripod legs; no funnel caps. (*Old Gunboat with raking masts and funnels.*) Still has derricks just before fo

SVEA* CLASS—2 SHIPS.

*Svea of this class transferred to Depot Ships.

GÖTA (1889), & THULE (1893).

	Length (*w.l.*) (feet)		Beam (feet)		Draught (feet)		Displacement.
Göta......	258·5	×	47·9	×	17	=	3393 tons.
Thule ...	260·8	×	47·9	×	18·1	=	3305 tons.

Complement 265 and 269.

Guns:
- 1—8·3 inch, 45 cal.
- 7—6 inch, 45 cal.
- 9—6 pdr.
- 1—1 pdr.

Torpedo tubes: (15 inch):
1 *submerged* (bow), but *Thule* has none.

Armour (Creusot):
- 11½"—8" Belt.........
- 2" Deck (flat on bel
- 2" Deck (at ends) ..
- 10½" Conning tower

Armour (Krupp):
- 7½" fore turret (N.C
- 5" Hoist to this (N
- 5" Secondary turret
- 4" Hoists to these ..

Ahead:
1—8·3 in.
4—6 in.

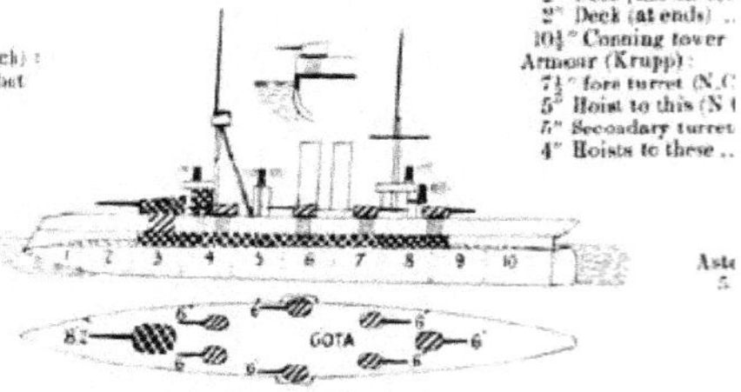

Aste
5

Broadside: 1—8·3 in., 4—6 in.

Machinery: 2 sets horizontal triple expansion. 2 screws. Boilers: 6 cylindrical. Designe 4650 = 15 kts. Coal: *normal* 200 tons; *maximum* 288 to 300 tons.

Notes.—Completely reconstructed, 1901–04. *Göta* built by Lindholmen Co., *Thule* by Bergsund C has square funnels, *Thule* round ones. 4 searchlights. Tripods fitted 1911-12.

THORDON AND TIRFING. *1911 Photo.*

THORDON, 1505 tons, & **TIRFING,** 1524 tons (launched 1866—67, and reconstructed 1903—1905). Complement, 91.

Guns: *Thordön,* 2—4·7 inch, 8—3 pdr. Armour (iron), $4\frac{1}{2}$" belt, 18" turret
Tirfing, 2—4·7 inch, 8—6 pdr. „ $4\frac{1}{2}$" „ $10\frac{3}{4}$" „

Both have a $9\frac{1}{2}$" conning tower. 1 screw. H.P. 380 = 6—7 kts. Coal: 115—120 tons.

1911 Official Illustration

BJORN & GERDA (built 1873—74, and reconstructed 1909—1910). 459 tons. Complement, 48. Guns: 1—4·7 inch, 5—6 pdr., Armour (iron), $2\frac{1}{2}$" belt, $16\frac{1}{2}$" casemate. H.P., 155 = 8 kts., but *Gerda* only, 133 H.P. for 7 kts. 2 screws. Coal: *maximum* 25 tons.

FYLGIA (Dec., 1905).

Displacement, 4980 tons. Complement, 362.

Length (*w.l.*), $377\frac{1}{4}$ feet. Beam, $48\frac{1}{2}$ feet. *Max.* draught, $20\frac{5}{8}$ feet.

Guns (M. '00):
8—6 inch, 45 cal.
14—6 pdr.
2—1 pdr.
Torpedo tubes (18 inch):
2 *submerged.*

Armour (Krupp)
4" Belt ....
2" Deck (slope
5"—2" Turret
4" Hoists and s
4" Conning to

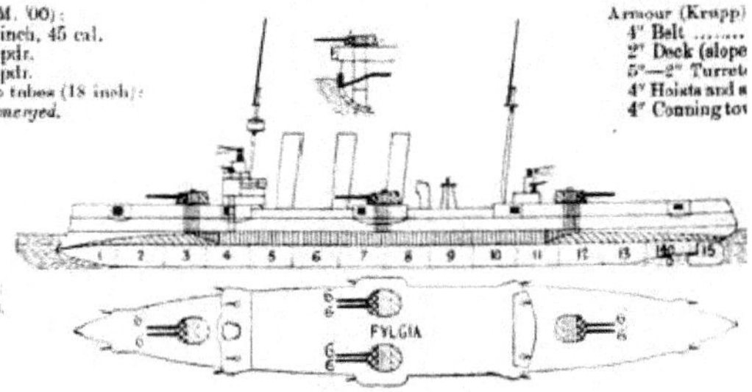

Ahead:
6—6 in.

Broadside: 6—6 in.

Machinery: 2 sets triple expansion. 2 Screws. Boilers: 12 Yarrow. Designed H.P. 12,0
Coal: *normal* 350 tons; *maximum* 900 tons = 5770 miles at 10 kts.

General Notes: Laid down by Bergsund Co., at Finnboda in 1903 and completed 19
12,440 H.P. = 22·7 kts. Machinery by builders. (There are no sponsons under beam 6-in
incorrectly shown on elevation, plan and section above.)

10 Destroyers.

No.	Type	Date	Displacement	H.P.	Max. speed	Fuel	Complement	T. tubes	Max. draug't feet.
			tons		Kts.	tons			
2	*Wrangel*	'15—'18	550	13000*t*	34			6	9¼
2	*Hugin*	'10—'12	460	8000*t*	30	80 + 3 oil	69	2	8¾
4	*Wale*	'06—'10	460	8000	30	80	65	2	8½
1	*Magne* (T)	1905	460	7700	30·7	80	63	2	9½
1	*Mode* (Y)	1902	453	6500	31	86	63	2	9

T = Thornycroft. Y = Yarrow. *t* = turbines.

WRANGEL AND WACHTMEISTER completing. *1918 Photo, S. Anderson, Esq.*

2 *Wrangel* class: **Wrangel** and **Wachtmeister** (Lindholmen Co., Gothenburg, both launched 1917.) Dimensions: 236¼ × 21½ × 9¼ feet. Armament: 4—14 pdr., 2 machine guns, 6—18 inch torpedo tubes. Designed S.H.P., 11,000 (*n.d.*), 13,000 (*f.d.*) = 34 kts. Reported to have attained 34.8 kts. on trials. Have De Laval geared turbines. Completed, early 1918.

No photo available.

Appearance same as *Wale* class in next column.

2 *Hugin* class: **Hugin** (1911), **Munin** (1912), 460 tons. Dimensions: 216 × 21 × 8 feet. Guns: 4—14 pdr., 2 machine. Torpedo tubes: 2—18 inch. 4 Yarrow boilers. Curtis turbines. S.H.P., 8,000 = 30 kts. Fuel: 80 tons coal + 3 tons oil. Endurance: 800 miles at 15 kts. *Hugin* built by Goteborg Co., and *Munin* by Kockum Co., Malmö

191

4 *Wale* class: **Wale** ('06), **Ragnar** ('07) **Sigurd** ('08), **Vidar** ('09). 460 to 462 tons. Dimensions: 214 × …
Armament: *Wale*, 2—14 pdr., 4—6 pdr., 2 machine; other boats, 4—14 pdr., 2 machine; 2—18 inch t…
4 Yarrow boilers. Hull on Thornycroft lines. All built in Sweden. Coal: 80 tons. Radius of action … at 15 kts.

Has mainmast now as *Wale* class above.

1 *Thornycroft*: **Magne** (1905). Dimensions: 215¼ × 20¼ × 9 feet. Armament: 6—6 pdr., 2 machine, 2 sin… tubes aft. 4 Thornycroft boilers. Endurance: 930 miles at 15 kts.

Has mainmast now as *Wale* class above.

1 *Yarrow*: **Mode** (1902). Dimensions: 220 × 28¼ × 8¼ feet. 4 Yarrow boilers. Armament: 6—6 pdr., Tubes: 2 single, aft. (18 inch). Endurance: 980 miles at 15 kts.

38 Torpedo Boats.

No.	Type	Date	Displacement	H.P.	Max. speed	Fuel	Complement	T. tubes	Max. draug't
	(First class)		tons		kts	tons			feet
17	*Plejad* (N)	'05–'10	106-122	2000	25-26	18	18	2	8½
11	*Blixt* (S)	'96–'04	94-117	1300	23	20 to 13	18-15	2	7½
	(Second class)								
10	*N:r* 5–15	'04–'06	58-61	800	21	22	14	2	6½

(N)=Normand type. (S)=Schichau type.

FIRST CLASS:—

Photo wanted.

(For appearance, see Silhouettes).

CASTOR, PLEJAD, POLLUX.

Other 14 boats (fore funnel lengthened now).

17 *Plejad* class: ***Astrea*** ('07), ***Castor*** ('07), ***Iris, Plejad*** ('05), ***Pollux*** ('07), ***Thetis*** ('07), ***Spica*** ('07), ***Vega*** ('07) ***Vesta, Argo, Antares, Arcturus, Altair, Polaris, Perseus, Regulus, Rigel*** ('09-'10). Armament: (except for *Plejad*, *Castor* and *Pollux*, which have 2–3 pdr.) 2–6 pdr., 2–18 inch tubes. Except *Plejad* (by Normand, Le Havre), all built in Sweden.

FIRST CLASS (*continued*):—

BLIXT *Photo, S. And*

11 *Blixt* class: ***Blixt*** ('98), ***Bris*** ('00), ***Kapella*** ('04), ***Meteor**** ('99), ***Mira*** ('02), ***Orion*** ('03), ***O Stjerna*** ('99), ***Sirius*** ('03), ***Virgo*** ('02), ***Vind*** ('00). Guns: 2–1 pdr. Torpedo tubes: 2–15 in in Sweden.

* In 1920 attached to 6th Mining Division.

SECOND CLASS:—

No. 11.

10 second class boats: ***Nr 5–15*** (No. 13 vacant). Armament: 1–1 pdr., 2–18 inch tubes.

(Vedette Boats of 23-83 tons are listed on a later page, with Swedish Miscellaneous Ships.)

10 (*+6 ? building*) **Submarines.**

Note. Much secrecy is observed with regard to Swedish Submarines, of which no particulars are published by the Ministry of Sea Defence. The following table and description by classes are not from any official data and must be accepted with caution.

No.	Type	Date	Displacement	H.P.	Max. speed	Fuel or Battery Endurance	Complement	T. tubes	Max draught
			tons		kts.				feet
6	First Class:— *Nr 11—Nr 18* (L)	15—?	...	...	...		...	...	...
1	*Nr 10* (N)	1915	230 / 320	800	15 / 9		17	2	
2	*Nr 9, Nr 8* (L)	1914							
6	Second Class:— *Nr 3—Nr 7* (N)	'09—'12	180 / 230	1000 / 200	15 / 8	1800 miles at 10 kts. / 40 miles at 5 kts.	17	2	...
1	*Nr 2* (L)	1909	180 / 230	800 / 200	15 / 8	1800 miles at 10 kts. / 40 miles at 5 kts.	17	2	7
1	*Nr 1* (H)	1904	107 / 127	200	8.5 / 7	...	8	1	9½

(L)=Laurenti type. (N)=Navy type. (H)=Holland type.

1st Class Submarines.

Uncertain how many of these boats have been completed.

6 *Laurenti* type: ***Nr 11 Aborren*** (1917), ***Nr 12 Gaddan*** (1918), ***Nr 13 Laxen*** (1918), ***Nr 17 Otter*** (——), ***Nr 18 Salen*** (——), all built by Kockum Co., Malmö; ***Nr 14 Bavfer*** (——), ***Nr 15 Hvalrossen*** (——), ***Nr 16 Iller*** (——), all built by Karlskrona D.Y. No details yet known. The conjunction of numbers and names given above is more or less conjectural and open to future correction.

Photo, Karlsson, Karlskrona.

1 *Navy* type: ***Nr 10 Delfinen*** (Bergsund Co., Stockholm, 1915). Armament: 1 machine gun, and 2—18 inch bow tubes.

Photo wanted.

2 *Laurenti* type: ***Nr 9 Svärdfisken, Nr 8 Tumlaren*** (Kockum Co., Malmö, 1914). 2—18 inch bow tubes.

2nd Class Submarines.

5 *Navy* type: ***Nr 3, Nr 4,*** both built by Stockholm D.Y., ***Nr 5, Nr 6,*** Kockum Co., ***Nr 7,*** Co., all built 1910-11. Dimensions: 139¾ × 11¾ × 9¼ feet. 2—18 inch tubes. These are improved *Hajens*. action at full speed (15 kts.), 475 miles.

1 *Laurenti-Fiat* type: ***Hvalen Nr 2*** (Spezia, 1909). Dimensions: 139 × 14 × 9¼ feet. 2—18 inch bow tubes. Sexl Fiat petrol engines. 3 screws.

When running on surface may have two masts raised, one right in bow and other right at stern.

1 *Holland* type: ***Hajen Nr 1*** (Stockholm D.Y., 1904). Dimensions: 65 × 12 × 11 feet. 1—18 inch bow tube. 3 carried. Petrol motor. 1 screw.

SWEDISH MISCELLANEOUS.

Torpedo Gunboats (*Torpedkryssare*).

PSILANDER.

JAKOB BAGGE and CLAS HORN.

ORNEN.

PSILANDER (1900), 814 tons, **JAKOB BAGGE,** 835 tons, **CLAS HORN** (1898), 846 tons, **ORNEN** (1896), 841 tons. Designed H.P. 4000 = 20 kts., *natural draught*. Coal, 100 tons. Armament: 2—4·7 inch, 4—6 pdr., 1—15 inch *submerged* bow tube.

General Notes: To distinguish between these ships compare relative positions of foremast, bridges, and fore funnel. In 1917, some (or all) of these four ships had extremely high top-gallant masts for W/T., and appeared to have a single stiffening strut to foremast, giving the appearance of a foretripod mast when seen full broadside.

Gunboat (*Kanonbat*).

1918 Photo, S. Andersson, Esq.

SKAGUL (1878). 589 tons. Guns: 2—4·7 inch, 4—6 pdr. Speed: 13 kts. Coal: 94 tons.

1918 Photo, S. Anderson, Esq.

DISA (1877, rebuilt 1908). 512 tons. Guns: 1—5·9 inch, 1—4·7 inch, 2—6 pdr., 2 machine. Speed: 11 kts. Coal: 98 tons.

Depot Ships (*Depåfartyg*).

Note.—Gunboat *Edda*, given on next page as a "Mine Ship," is officially rated as a "Depot Ship." Other old Vessels used as Parent Ships are *Stockholm* (for T.B. at Karlskrona), *Saga* (for Submarine at Karlskrona) and *Freja* attached to Flying Station.

Depot Ships—*continued*.

Undergoing re-construction 1920.

SVEA. Old Coast Defence Battleship, built by Lindhol[m] rebuilt 1904. —— tons. Dimensions: 248·4 × 4[...] Guns: 4—4·7 inch, 2—6 pdr. Armour: *11½"—8" [...] 10½" C.T. 2 screws. 6 cyl. boilers.† I.H.P. 46[...] Coal: 290 tons.

* May be removed. † May be reduced.

As rebuilt. (*Submarine Depot Ship.*) *1918 Photo, Karlskro[na]*

SKAGGALD (1879, rebuilt 1913). 670 tons. Guns[...] Speed: 13 kts. Coal: 94 tons.

1918 Photo, S. An[derson]

BLENDA (1874, re-built 1908). 464 tons. Guns: 2—[...] 11 kts. Coal: *about* 95 tons.

Balloon Ship *(Ballongfartyget).*

No Photo available.
No funnels; no masts.
See Silhouettes.

N : r 1 (1902). 264 tons. Assigned to Coastal Defence.

Mine Ships *(Minfartyg).*

1917 Official photo.

***CLAES FLEMING** (1914). 1,748 tons. Dimensions: 262 × 32·8 × 13·7 feet. Guns: 4—4·7 inch (50 cal.), 6 machine. Speed: 20 kts. Parsons turbines.

EDDA (1885) (1907). 662 tons. Guns: 4—6 pdr. Speed: 13·5 kts. Coal: 80 tons.

*These are the only two Swedish warships officially rated as "Mine Ships." *Skuld* and *Rota* are not so rated officially.

Mine Ships—*Continued.*

1918 Photo, Karlsson, Karlskrona.

SKULD (1879). Old Gunboat, 604 tons. Guns: 4—6 pdr. Speed: 13½ kts. Coal: 94 tons.

Note.—In 1919, the 2nd Mining Division consisted of *Skuld* (above), *Svensksund* (v. next page), and Vedette Boats *N : r 23, 24, 25* (see *N : r 20—25* on next page).

1918 Photo, S. Anderson, Esq.

(Is also rated as a "Depot Ship.")

ROTA (1878). 540 tons. Guns: 2—6 pdr. Speed: 13 kts. Coal: *about* 95 tons.

Note.—In 1920, the 3rd Mining Division consisted of *Rota* (above) and Vedette Boats *Sökaren, Sceparen, Sprengaren* (all listed on next page). The 6th Mining Division consisted of T.B. *Orken* and *Meteor*.

Hospital Ship *(Lasarettsfartyg).*

VERDANDE (1879). 514 tons. Speed: 13¼ kts. Coal:

Royal Yacht.

(Is officially rated as "Chefsfartyg" or "Flagship

DROTT (1877). 633 tons. Guns: 4—3 pdr. Desig 260 = 15 kts. Coal: 100 tons.

Surveying Ships *(Sjökarteverkets fartyg)*

SVENSKUND (see next page) is occasionally used as a Ship.

Five small vessels are employed on this service, viz.:—*Egl* 93 tons; *Tärnan* (1889, rebuilt 1909), 54 tons; *Falken* (tons; *Svalan* (1881), 127 tons, and *Ran* (1857) of 199 to

Pilot Vessels *(Lotsverketsfartyg).*

There are ten of these vessels, the largest and most modern *Vega* (1901), of 285 tons; *Malmö* (1897), of 138 tons; *Göteb* of 132 tons; and *Gäfle* (1892), of 110 tons. The rema built 1885—1886, are under 100 tons.

(Continued on n

Patrol or Coastguard Ships *(and Mine Sweepers).*

Photo wasted.

SÖKAREN, SVEPAREN, SPRÄNGAREN (1918, Motala Co.). 227 tons. Dimensions: 85·3 × 23 × 9·8 feet. Speed: 11 kts. Guns: 1—6 pdr.

Note.—These ships are fitted out as Mine Sweepers and attached to 3rd Mining Division. *Sprängaren* is officially rated as a "Vedette Boat."

Vedette Boats *(Vedettbatar).*

No photo available.

N : r 19 (1918). 67 tons. No guns. Speed: 10 kts. (oil motor). Officially rated as "Lysmaskinbat." ("Searchlight Boat.")

Photo, S. Anderson, Esq.

N : r 26—23 (ex-T.B. *79, 81, 83, 85*, 1903). 62—55 tons. Speed, 20 kts. Guns: 1—1 pdr. May still have 2—15 inch tubes.

No photo available.

N : r 20 (ex-T.B. *Komet*, 1896). 104 tons. Speed, 23 kts. Guns: 1—1 pdr. May still have 2—15 inch tubes.

Vedette Boats—*Continued.*

No photo available.

N : r 22 (ex-T.B. *Gudur*, 1894). 73 tons. **N : r 21** (ex-T.B. *Gondul*, 1894). 83 tons. Speed, 19½ kts. Guns: 2—1 pdr. May still each have 2—15 inch torpedo tubes.

1915 Photo, Karlsson, Karlskrona.

N : r 18—N : r 15 (1887-4), 86—72 tons; **N : r 14—N : r 9** (1891-1886), 55—51 tons; **N : r 8—N : r 5** (1885-1880), 40—43 tons; **N : r 5—N : r 1** (1880-1879), 23 tons. Speeds, 19 to 17 kts. All armed with 1—1 pdr. gun. *N : r 5—N : r 1* for Coastal Defence. *N : r 14—9* and *N : r 8—5* are ex-Torpedo Boats *77—67* (alternate numbers); *N : r 18—15* are ex-T.B. *4, 3, 2* and *1*.

For Various Special Services.

SVENSKSUND (1891). 394 tons. Guns: 2—6 pdr. Speed: 12½ kts. Coal: 50 tons.

Note.—This ship is employed as (*a*) a Mine Layer, (*b*) a Surveying Ship, (*c*) a Fisheries Protection Vessel, (*d*) an Ice Breaker, (*e*) a Repair Ship, and (*f*) a Training Ship, according to season of the year and when necessity arises for her use in any of these duties.

Training Ships *(Övningsfarty*

Cruiser **FYLGIA** (previously described) and gunboat
(*v.* preceding column) employed as Training Ship

JARRAMAS. 1917 Ol

JARRAMAS (1900). 337 tons. Speed, 11 kts.
NAJADEN (1897). 355 tons. Speed, 11 kts.

GLADAN. 1917 Photo,

GLADAN (1857). 327 tons. Speed, 10 kts.
FALKEN (1877). 138 tons. Speed, 8½ kts.

The old Corvette *Saga* (1878), of 1,777 tons, is also a Training Ship, and is attached to the Karlskrona S

Lastly, there are five very old "Accommodation a (*Logements- och andra fartyg*), all probably fighting purposes.

(ROYAL) NORWEGIAN NAVY.

(Revised, 1920, from information supplied by courtesy of H. E. the Minister of Defence.)

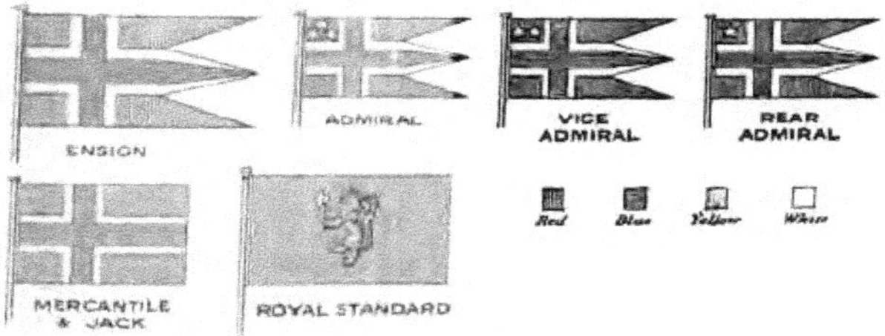

Arsenals.

HORTEN. One dry dock, 346 × 60 × 20¼ feet.
BERGEN (Marineholmen) Small dockyard. No building slips. No naval docks, but various private docks in the vicinity (for which see below).

Principal Private Yards.

FREDRIKSSTAD: Fredrikstad Mek. Verksted. Docks: 1 small floating.
KRISTIANIA: (a) Nylands Verksted. Docks: 1 floating (320 × 61 × 19 feet), 2 smaller floating.
(b) Akers Mek. Verksted. Docks: 1 dry, 1 floating (both small) and 1 dry, 586 × 82 × 23 feet (to be completed 1920).
SANDEFJORD: Framnes Mek. Verksted. Docks. 1 floating (360 × 65 × 21 feet), 1 smaller floating.
KRISTIANSAND: Kristiansands Mek. Verksted. Docks: 1 dry, 330¼ × $\frac{56}{32\frac{1}{2}}$ × 18 feet. 3 small slips.
BERGEN: (a) Laxevaags Dokkompagni. Docks: 1 dry, 365 × 55 × 18 feet, 1 slip.
(b) Bergens Mek. Verksted. Docks: 1 small dry.
(c) Bergens Dokkompagni. Docks: 1 small dry.
(d) Laxevaags Maskin-og Jernskibsbyggeri. Docks: none.
(e) Mjellem & Karlsen Mek. og Patentslipper. Two slips. Docks: none.
TRONDHJEM: Trondhjems Mek. Verksted. Docks: 2 small dry.
STAVANGER: Stavanger Stöberi og Dok. Docks: 2 small dry.
PORSGRUND: Porsgrund Mek. Verksted. One small floating dock. One slip.
MOSS: Moss Varft A/S. Docks, 3 small floating (can be used together for ships up to 300 ft.)

Mercantile Marine.

Bureau Veritas 1919-20 figures. Ships under 100 tons *gross* excluded.
1,380 Steam and Motor Ships of 1,044,260 tons *gross*.
162 Vessels (with Auxiliary Motors) of 25,738 tons *gross*.
581 Sailing Vessels of 270,657 tons *gross*.
During 1919 Norwegian Shipyards launched 82 ships of 52,578 tons *gross*, some of these for foreign account.

Principal Trade Ports.

Kristiania, Bergen, Trondhjem.

Coinage.

Same as Swedish.

Modern Guns (Elswick and Bofors).
(Details officially revised 1920.)

Notation.	Designation		Length in calibres.	Model.*	Weight of Gun.	Weight of A.P. shot.	Initial velocity.	Max. penetration firing A.P. capped at K.C. 5000 yards.	3000 yards.	Danger Space against average warship at 10,000 yards.	5000 yards.	3000 yards.
	c.m.	inches.			tons.	lbs.	ft. secs.					
MEDIUM.	*20.9*	8.2	44	...	18½	308	2300	6¼	9¼	100	405	594
	14.91	5.9	46	...	7	100	2625	3	4½	67	215	440
	12	4.7	44	...	2¼	44	2570	...	...	...	...	...
LIGHT.	*10.16*	4	40	B	2	32	2545	...	...	...	...	...
	7.6	3	50	B	1	12	2820	...	...	...	...	...
	7.6	3	40	E	¾	12	2210	...	...	...	...	...
	7.6	3†	25	B	¼	12	1630	...	...	...	...	...

* B = Bofors. E = Elswick. † Anti Aircraft. There is also a 3 pdr. AA. gun.

Minister of Defence.—Major-General Aratemark. *Naval Attaché, London.*—Captain J. Scott

Colour of Ships: Coast Defence Ships, Destroyers, Torpedo Boats, Gunboats, &c., all light gr

INSIGNIA OF RANK ON SLEEVES

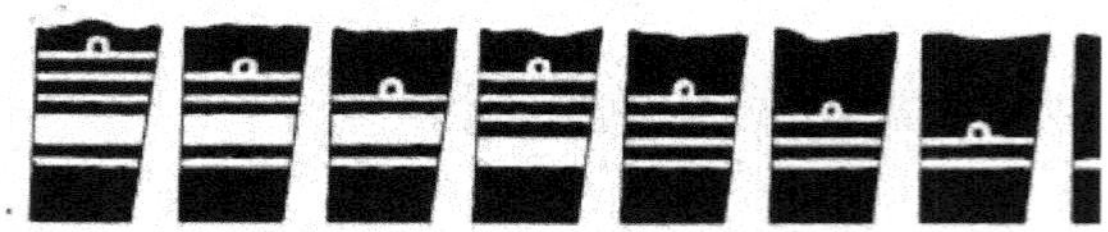

	Admiral.	Vice-Am.	Kontre-Am.	Kommandör.	Kommandör-Kaptein.	Kaptein.	Premier-Löitnant.	L
British or U.S.A.	"	*Vice-Ad.*	*Rear-Ad.*	*Commodore.*	*Captain.*	*Commander.*	*Lieutenant.*	*Su*

In relative rank Engineers have the same *without curl*.
" " Doctors " " " and a *red* passepoil above upper stripe.
" " Paymasters " " " " *blue* " " "

General Notes.—Personnel: About 1400 permanent, 1000 yearly conscripts, all seafaring men in

NORWEGIAN RECOGNITION SILHOUETTES.

Scale: 1 inch = 160 feet.

ONE FUNNEL.

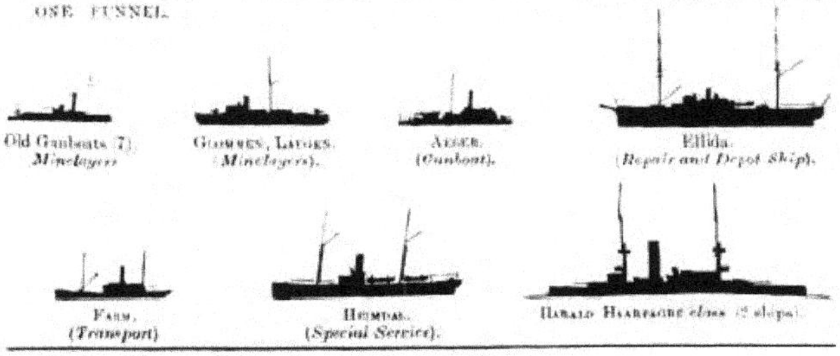

TWO AND THREE FUNNELS.

TORPEDO CRAFT. **Scale: 1 inch = 160 feet.**

SUBMARINES. **Scale 1 inch = 80 feet.**

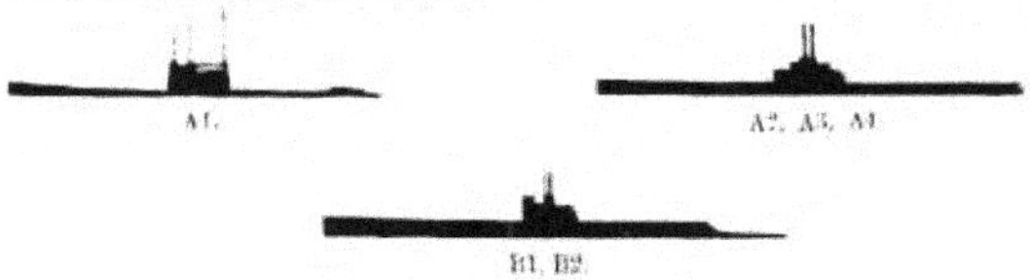

NORWEGIAN COAST DEFENCE BATTLESHIPS. *(Panzers*

Photo, Wils

NORGE (March, 1900), EIDSVOLD (June, 1900).

Displacement, 4166 tons. Complement, 270.

Length (*p.p.*), 290 feet. Length (*over all*), 301¼ feet. Beam, 50½ feet. *Max.* draught,

Guns:
2—8·2 inch, 44 cal.
6—5·9 inch, 46 cal.
8—12 pdr.
2—3 pdr. AA.
Torpedo tubes (18 inch):
2 *submerged*.

Armour (Krup
6" Belt......
2" Deck slo
8" Turrets
6" Bases
5" Casemat
6" Conning

Ahead:
1—8·2 in.
2—5·9 in.

Machinery: 2 screws. Boilers: 6 Yarrow. Designed H.P. 4500 = 16·5 kts. Coal: *maximum* 550 tons.

Notes.—Built at Elswick. Completed 1900-1901. Machinery by Hawthorn, Leslie & seaboats. On plans C = 8·2 inch guns; D = 5·9 inch; F = 12 pdr. guns.

Coast Service Battleships (*Panserskibe*)

TORDENSKJOLD. (Both exactly alike.)

HARALD HAARFAGRE (Jan. 1897), TORDENSKJOLD (March, 1897).

Displacement, 3858 tons. Complement, 249.
Length (*p.p.*), 279 feet. Length (*over all*), 291 feet. Beam, 48½ feet. *Maximum* draught, 17¾ feet. *Mean* draught, 16½ feet.

Guns (Elswick):
- 2—8·2 inch, 44 cal.
- 6—4·7 inch, 44 cal.
- 6—12 pdr.
- 2—3 pdr. AA.
- 2—1 pdr. (Hotchkiss)

Torpedo tubes*:
- *2 submerged*

Armour (Harvey):
- 7" Belt (amidships)
- 4" Belt ends
- 2" Deck (flat on belt)
- 8" Bulkheads
- 8"—5" Turrets
- 6" Bases of turrets
- 6" Conning tower

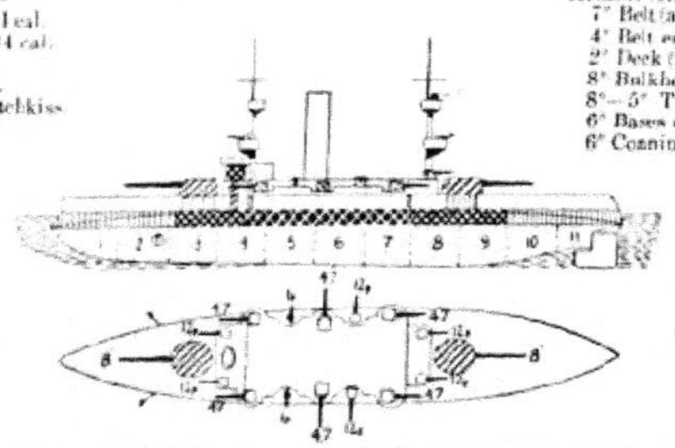

Ahead:
1—8 in.
2—4·7 in.

Astern:
1—8 in.
2—4·7 in.

Machinery: Boilers: 3 cylindrical. 2 screws. Designed H.P. 4500 = 16·9 kts. (made 17·2 on trial, 1897-8.) Coal: *normal* 400 tons; *maximum* 540 tons.

Notes.—Built at Elswick, and completed 1897–8. Engines by Hawthorn, Leslie & Co. Belt is 174 feet long by 6½ feet deep. Excellent seaboats.

* Tubes are 18 inch in *H. Haarfagre* and 17·7 inch in *Tordenskjold*.

4 Destroyers (*Torpedobaatsjagare*).

No.	Type	Date	Displacement (tons)	H.P.	Max. speed (kts.)	Coal or Oil (tons)	Complement	T. tubes
3	*Draug*	'08-'14	540	7500	27	95 coal	71	3
1	*Valkyrjen** (S)	1896	410	3300	23	95 coal	59	2

(S) = Schichau.

* *Valkyrjen* is not officially listed with Destroyers, but separately as a "Torpedodivisionsbaat" or Torpedo Di

Fore funnel reported lengthened in one or more boats.

Garm (1913), **Draug** (1908), and **Troll** (1910). All built at Horten. 540 tons. Dimensions: 2[illegible] × 8½ feet. Armament: 6—12 pdr., 3—18 in. tubes. H.P. 7500 (8000 in *Garm*) = 27 kts. Recip. engines in last two, turbines in *Garm*. Coal: 95 tons. Complement 71.

1918 Photo, Wilse, Chr

1 Schichau type: **Valkyrjen** (1896). 410 tons. Dimensions: 190 × 21 × 8 feet. Complement 59. 4—12 pdr., 2 torpedo tubes (18 in.). 2 screws. Boilers: Thornycroft. Designed H.P. 3300 = 23 kts. Coal:

30 Torpedo Boats.

(Details in table are by averages.)

No.	Type	Date	Dis-placement	H.P.	Max Speed	Coal	Com-plement	T. Tubes	Max. draught
3	Sea-going— *Snogg*	'16–'17	tons 220	3300	kts. 25	tons 30 coal+oil	32	4	feet 5½
9	1st class— *Delfin*	'96–'02	90–79	1000	20	17	19	2	7
3	2nd class— *Skarv*	'07–'12	92	1700	25	16	18	2	7½
5	*Grib*	1906–?	70	1100	22½	15	15	2	6½
	Ravn								
2	*Hauk*								
2	*Hvas*	'95–'03	65	500	19½	12–14	14	2	6
3	*Djerv*								
3	*Lyn*	'96–'02	40	650	18½	12–14	14–15	2–1	5½

"Large Torpedo Boats" (*Större Torpedobaater*).

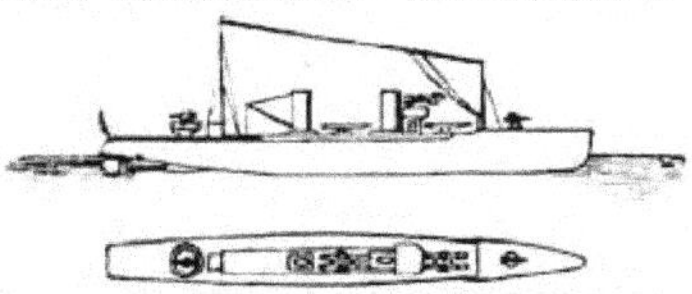

Plans by courtesy of Ministry of Defence, 1919.

3 new boats: ***Snogg, Stegg, Trygg*** (1916-17). 220 tons. Dimensions: 173.9 × 18 × 5.2 feet. Guns: 2—12 pdr. Tubes: four (size not known). Designed H.P. 3300 = 25 kts. Fuel capacity: 30 tons coal and oil. Complement, 32.

1st Class Torpedo Boats.

9 boats: ***Laks, Sild, Sael, Skrei,*** (1900). 90 tons and 21 kts. — *1918 Photo, Wilse, Christiania.*
Brand, Storm, Trods (1899). 20 kts. } 79 tons.
Delfin, Hval (1896). 19 kts. }
All armed with 2—1 pdr. guns and 2 tubes on deck. Complement, 19. Last two built by Schichau, Elbing.

2nd Class Torpedo Boats.

Grib, Jo, Lom, Orn, Ravn, as above photo. — *1918 Photo, Wi*
Appearance of Kjell, Skarv, Teist, not known.

8 { ***Kjell*** (1912), 94 tons; ***Skarv, Teist*** (1907-08), 98 tons. *Skarv, Teist* 2—3 pdr. guns. *Kjell* 1— of all 3 boats, 25 kts. Complements, 18.
Grib, Jo, Lom (1906) and ***Orn, Ravn*** (1903). All 70 tons, and about 22—23 kts. Guns: 2—1 has 2—3 pdr. Complements, 15.

Note.—All above 8 boats have 2 tubes, viz., 1 bow and 1 deck.

1918 Photo, Wilse

4 boats: ***Hauk, Falk*** (1903), 63 tons; ***Hvas, Kjæk*** (1900), 64 tons. All have 2—1 pdr. and 2 tu deck). First two 20 kts., others 19 kts. Complements, 14.

(Same appearance as above but shorter funnels.)

6 boats: ***Djerv, Dristig, Kvik*** (1897-98), 67 tons. ***Blink, Glimt, Lyn*** (1896), 45 tons. Al speed, guns: 2—1 pdr. and 2 tubes (1 bow 1 deck). Complements: 14 for all.

4 + ? **Submarines** *(Undervandsbaater).*

Note.—The only information officially published concerning Norwegian Submarines is: the ***A*** class, built between **1909** and **1914**, and the ***B*** and ***C*** classes, not yet completed. Excepting details of *B* class given below, the figures given in following table and description of classes are ***not*** from any official data.

No.	Type	Date	Displacements	H.P.	Max. speed	Fuel or Battery Endurance.	Complement	T. tubes	Max. draught
			tons		kts.	miles			feet
2?	*C2? C1?*	*Bldg.*	420 / 515	...	14 / 9	...	...	4	10
2?	*B2? B1?*	*Bldg.*	412 / 516	650	14	...	—	...	—
2	*A4–A2* (K)	1913	250 / [illegible]	900 / 630	14 / 9		15	3	9½
1	*A1* (K)	1909	205 / 257	400 / 250	11½ / 9	1450 at 9 kts. / 45 at 6.5 kts.	11	3	9½

(K) = Krupp-Germania type.

Building?

2 *Holland* type: *C2, C1,* ordered from Electric Boat Co., U.S.A., about 1915. Dimensions: about 180 × 17 × 10 feet. Armament: 1—12 pdr., 4—18 inch tubes.

Note.—Doubtful if either of the above boats has ever been commenced. The contract may have been cancelled.

Building?

For appearance, see silhouette.

2 *Type unknown*: *B2, B1,* assigned to Karljohans Yard, Tønsberg, for construction. For details, as communicated by the Ministry of Defence, see Table above.

Note.—Four further boats of this class, *B6, B5, B4, B3,* reported to be projected. It is doubtful if any of these six boats have yet been begun.

Submarines—*Continued.*

Photo Wanted.

(For appearance, see Silhouettes.)

3 Krupp-Germania type: ***A4–A2*** (all 1913). Dimensions: 152½ × 16½ × 9½ feet. Surface engines: [illegible]
3—18 inch tubes. 4 torpedoes carried.

Note.—*A4* completing at Kiel, was appropriated by Germany on outbreak of war and became the Germa[illegible]
surrendered at Harwich with other German Submarines.

1 Krupp-Germania type: ***A1*** (ex-*Kobben*) (1909). $\frac{205 \text{ tons.}}{255 \text{ tons.}}$ Speed: $\frac{11.7 \text{ kts.}}{9 \text{ kts.}}$ 3—18 inch tubes. 4 torpedoes carri[illegible]
paraffin engines. Dimensions: 129 × 12 × 9½ feet.

Gunboats.

(Class not known.)

Photo wanted.

ORKLA (1917), **RAUMA** (1917). 270 tons. Complement —. Dimensions: 124·7 × 23 × 5·8 feet. Guns: 3—4 inch. Designed H.P. 360 = 10 kts. Oil: tons.

1st Class Gunboats (Kanonbaater av I. Kl.).

1918 Photo, Wilse, Christiania.

FRITHJOF (1895). 1382 tons. Complement 166. Dimensions: 223¾ × 32¾ × 13¼ feet. Guns: 2—4·7 inch, 4—12 pdr., 4 M.G. Tube (18 inch), 1 sub. (bow). Designed H.P. 2800 = 15 kts. Coal: 160 tons.

2nd Class Gunboat (Kanonbaat av 2 Kl.).

Photo, Wilse, Christiania.

AEGER (1893). 383 tons. Complement 45. Guns: 1—8·2 inch, 1—10 ... 4 pdr. H.P. 370 = 9·8 kts. Coal: 25 tons. Max. draught: ...

Mine Layers (Minelæggere).

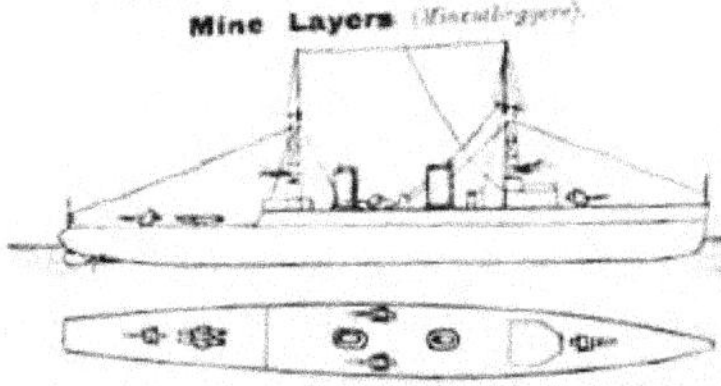

Plans by courtesy of the Ministry of Defence.

FRÖYA (1916). 735 tons. Complement, 79. Dimensions: 248·7 × 27 × 8·2 feet. Guns: 4—4 inch q.f. Designed H.P. 7000 = 22 kts. Coal: 95 tons + 60 tons oil. Carries about 200 mines.

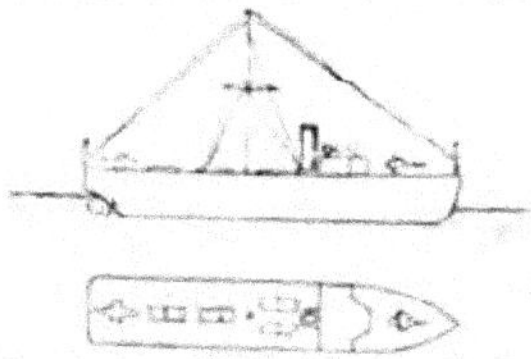

Plans by courtesy of the Ministry of Defence.

GLOMMEN (1917), **LAUGEN** (1917). Both built at Kaldnaes Yard, Tonsberg. 335 tons. Complement 35. Dimensions: 137·8 × 27·9 × 6·2 feet. Guns: 2—12 pdr. Designed H.P. 170 = 9·5 kts. Coal: 21 tons. Each carries about 50 mines.

VALE (as Minelayer) on right. 1918 Photo, S. Andersen, Esq.

Name.	Date.	Tons.	Crew.	Guns.	Speed kts.
Tyr	1888	281	44	1—10·2 in., 1—6, 2—3 pdr.	10·5
Gor	1885	276	44	1—10·2 in., 3—1 pdr.	10·5
Vidar / Brage	1881 / 1878	254	41	1—4·7 in., 3—1 pdr.	7·5 / 8·5
Nor	1878	254	39	1—4 in., 3—1 pdr.	9·5
Vale / Uller	1878 / 1874	255	41	1—4·7 in., 3—1 pdr.	8

(All fitted as Minelayers.)

For Special Service.

Photo, ...

HEIMDAL (1892). 640 tons. Complement 62. G... 2—3 pdr. H.P. 625 = 12 kts. Coal: 92 tons.

Transport (Transportskib.)

1918 Photo, ...

FARM (1900). 300 tons. Complement 32. Gu... 2—1 pdr. Speed: 10 kts.

Workshop and Mother Ship.

(Verkstedsog Moderskib.)

1918 Photo, S. A...

ELLIDA (1880). 1,045 tons. Guns: 4—9 pdr., 2—... 10·5 kts.

(ROYAL) DANISH NAVY.

Officially revised, 1920, by courtesy of H.E. the Minister of Marine.

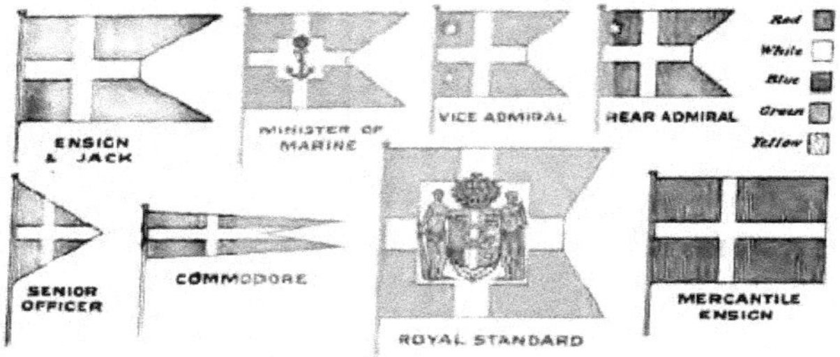

Note.—Division flag pennant, otherwise as commodore.

INSIGNIA OF RANK ON SLEEVES.

	1	2	3	4	5	6	7	8	9
	Admiral.	Vice-Admiral.	Kontre-Admiral.	Kommandør.	Kaptajn.	Premierløjtnant.	Løjtnant.	Sekond-løjtnant.	Kadet.
British or U.S.A.	*Admiral.*	*Vice-Admiral.*	*Rear-Admiral.*	*Captain.*	*Commander.*	*Lieut-Commander.*	*Lieutenant.*	*Sub-Lieut.*	*Midshipman.*

Band between stripes:— *Branches (curls, but after Executive).*

Crimson	Ingeniør (*Engineer*). Officers of the Technical Department have small V-shaped pieces of black braid sewn horizontally along gold stripes.	Without curl, but with Corps emblem over top stripe.
Blue	Intendant (*Paymaster*).	
Peachy Red	Læge (*Surgeon*).	

The broad stripes in ranks 1—3 are 2·7 cm. wide.
The narrow „ „ 7 & 9 are 0·7 „
The other stripes are 1·1 cm. wide. Distance between stripes is 0·5 cm.

Personnel: About 4000, all ranks.

Oversea Possessions: Iceland, Greenland, Faroes Islands.

Coinage: Krone (100 ore) = about 1s. 1½d. British; 30 cents U.S.A.

Minister of Marine: M. Klaus Berntsen.

Mercantile Marine.

From "Bureau Veritas" 1919-20 figures. Ships under 100 tons gross excluded.

447 Steam and Motor Ships of 641,567 tons gross.
99 Vessels (with Auxiliary Motors) of 13,958 tons *gross*.
584 Sailing Vessels of 102,376 tons *gross*.
During 1919, Danish shipyards launched 46 vessels of 37,766 tons *gross*; some of these for foreign account.

Naval Guns. (All details unofficial.)

Bore		Length in Calibres	Make and Date of Model		Weight of Gun	Weight of Projectile	Muzzle Velocity	Mo[unted in]
in.	cm.				tons	lbs.	ft. secs.	
9·4	24	43	Bofors	M. '06	21·5	352	2800	*P. Skram*
9·4	24	43	Bofors	M. '05	21·5	353	2477	*O. Fischer*
9·4	24	40	Canet	M. '96	22·0	353	2362	*H. Trolle*
9·4	24	40	Krupp	M. '93	23·4	353	2362	*Skjold*
5·9	15	50	Bofors	——	7·5	112	2690	*P. Skram*
5·9	15	43	Bofors	M. '96-'04	5·5	112	2296	*O. Fischer,*
4·7	12	40	Krupp	——	2·3	44	2362	*Skjold, He*

Arsenal.

COPENHAGEN. One slip. One dock (lengthened, 1912), 308¼ × 59 × 19¼ feet. Floa[ting docks:] No. 1, 115½ × 29 × 12¼ feet. No. 2, 113 × 39¼ × 12½ feet. Coast Defence Battleships, Cruise[rs, Torpedo] Boats, and Submarines built here.

BRAMSNÆSVIG. Mine and Torpedo Station; Torpedo Proving Range.

Private Establishments. (Revised 1920.)

COPENHAGEN. BURMEISTER & WAIN. Area covered by Shipyard: Length of water frontage: 4500 feet = about ⅚ of a mile. Shipbuilding Berths: Si[x] up to 600 feet long. Graving Dock: 475 × 64¼ × 23½ feet at entrance. Float[ing Dock:] 600 × 77 × 27¼ feet, to lift 11,500 tons. Patent Slip with maximum hauling capacity of [...] dw. Large fitting-out quay on water frontage with stationary cranes lifting 10, 15 an[d ...] respectively. Total number of persons employed: 5000. *Remarks.*—Patente[es of] Burmeister & Wain types of Diesel engines. Specialise in the construction of lar[ge ...] engined passenger and cargo vessels. Build hulls, machinery, boilers, turbines, re[...] engines and equipment for large mercantile ships, ferry boats, etc. All foregoing det[ails] furnished by Messrs. Burmeister & Wain, 1920.

COPENHAGEN FLOATING DOCK & SHIPYARD CO. 4 dry docks, viz.:—No. 1, [...] × 9 feet; No. 2, 445 × 116 × 9 feet; No. 3, 480 × 65 × 12 feet; No. 4, 510 × 68 [...] 3 floating docks, viz.:—No. 1, 170 × 45 × 12 feet (2700 tons). No. 2, 315 × 56 [...] (600 tons); No. 3, 350 × 55 × 15 feet (3900 tons). Also smaller building and patent [...]

HELSINGOR SHIPBUILDING & ENGINEERING CO., of HELSINGOR (ELSINOR[E) ...] docks, viz.:—No. 1, 341 × 46 × 14 feet; No. 2, 381 × 54 × 18 feet. One patent [...] tons capacity, and some building slips.

Other private docks in Denmark are located thus:—
AALBORG. Dry dock No. 1, 373 × 62¼ × 18 feet; Dry dock No. 2, 250 × 32¼ × [...]
FREDERIKSHAVN. Floating dock for vessels 200 feet long, drawing up to 16 fe[et ...] concrete floating dock, completed 1920.
NEXO (BORNHOLM). Dry dock, 170 × 30 × 12¼ feet.
SVENDBORG. Floating dock, 129 × 42 × 11½ feet.

Note to pages following this.

All displacements given in metric tons. Where measurement of length is not stat[ed, it is to] be taken as between perpendiculars (*p.p.*). All draughts *max. load aft*, unless otherwise [...]

DANISH WARSHIPS: RECOGNITION SILHOUETTES.

* These Silhouettes are from drawings furnished by courtesy of the Ministry of Defence.

Scale: 1 inch = 160 feet.

NO FUNNELS.

MINEKRAN V. MINEKRAN VI. (*Mining Vessels*).

GRØNSUND. (*Repair Ship*).

ONE FUNNEL.

MINEKRAN IV. (*Mining Vessel*)

FENRIS. (*Fisheries and C.G.*)

LOSSEN. (*Mine-Layer.*)

SKJ

KRIGER class (3). (*Surveying Ships.*)

DIANA. (*Fisheries and C.G.*).

HJÆLPEREN. (*Mine Layer.*)

GULDBORGSUND. (*Gunboat.*)

BESKYTTEREN.

ABSALON.

ISLANDS FAL

(*Fisheries and Coast Guard Vessels.*)

HEJMDAL. (*Training Ship*).

VALKYRIEN.

O. FISCHER & H. TROLLE.

P. SKRAM.

INGOLF (*Training Ship*).

TWO FUNNELS.

DANNEBROG (*Royal Yacht*).

FYLLA.

GEJSER.

TORPEDO CRAFT. **Scale: 1 inch = 160 feet.**
*Now have mainmasts.

P8, P9 (*Patrol Boats*).

Tumleren *class* t.b.* (3).

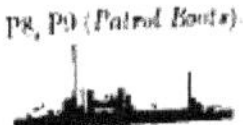
Hvalrossen *class* t.b. (3).

Havörnen *class* t.b. (3)
P2, P3 (*Patrol Boats*).

Ormen t.b.

Improved Ormen *class* t.b. (10).

Springeren *class* t.b. (3)

SUBMARINES. **Scale: 1 inch = 80 feet.**

Aegir *class* (5).

Havfruen *class* (6).
(One folding mast before C.T.

(HERLUF TROLLE CLASS—3 SHIPS.)

The three differ in dimensions and armament as follows:—

	PEDER SKRAM.	**OLFERT FISCHER.**	
Launched	May, 1908	1903	
Displacement (tons)...	3800	3700	
Complement	257	255	
Length (ft.)	275½ (*p.p.*)	271½ (*p.p.*)	
Beam (ft.)	51½	50½	
Max. draught (ft.) ...	16½	16¾	
Armament	2—9·4 inch, 43 cal. 4—5·9 inch, 50 cal. 8—14 pdr. 2—14 pdr. AA. 2—1 pdr. 2 Machine	2—9·4 inch, 43 cal. 4—5·9 inch, 43 cal. 6—14 pdr. 2—6 pdr. AA. 2—1 pdr. 2 machine	2— 4— 6— 2— 2 r
All 18 inch tubes.	4 tubes (*submerged*)	3 tubes (*submerged*)	3 t
H.P. } Trials	5400	4800	
Speed }	15·9	15·8	

(For plans and illustr

P. SKRAM, O. FISCHER, H. TROLLE.

For Tabulated details see preceding page.

Armour (*see Notes*):
7" Belt (amidships)
4" Belt (aft)
3" Deck (forward)
(Deck flat on belt.)
7" Turrets
6" Casemates (KNC)
7" Bulkhead
7" Conning tower

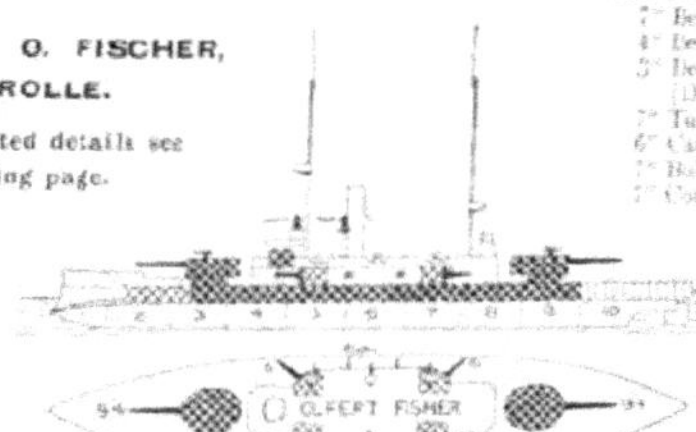

Ahead:
1—9·4 in.
2—5·9 in.

Astern:
1—9·4 in.
2—5·9 in.

Machinery: 2 sets triple expansion, 2 screws. All are fitted with 6 Thornycroft boilers. Coal (*maximum*): 250 tons in *O. Fischer* and *P. Skram*, 245 tons in *H. Trolle*.

Gunnery Notes.—Bofors 1906 models in *P. Skram*. Bofors 1903 models in *O. Fischer*. In *H. Trolle* 9·4 inch are Canet 1896 model, and 5·9 inch are Bofors 1896 model.

Armour Notes.—Krupp armour in *P. Skram*, Creusot in other ships. Belt is 7 feet deep, 3½ feet above, and 3½ feet below waterline.

Torpedo Notes.—All tubes 18 inch and submerged: 1 bow, 1 stern, 2 broadside in *P. Skram*, 1 bow and 2 broadside in *O. Fischer* and *H. Trolle*.

Engineering Notes.—Trial I.H.P. and speeds given in Table on next page as *O. Fischer* and *H. Trolle* did not reach designed speed of 16 kts. with 4200 I.H.P. on trials. Endurance is about 2000 miles at 10 kts. and 1050 miles at 14½ kts.

General Notes.—All built at Copenhagen Dockyard. [illegible] 2—14 or 6 pdr. anti-aircraft guns mounted on crowns of 9·4 inch barbettes.

1920 Photo, Lieut. Stern

OLFERT FISCHER. Fore chart house aft of bridges, also chart house at foot of mainmast.

1920 Photo, Ministry of Defence.

PEDER SKRAM. (Cleared for action: topmasts struck). No ventilators. Funnel rather shorter and thicker than in other 2 ships. Three searchlights round foot of foremast, and only one (on tower) abaft mainmast. Also note 14 pdr. guns on roof of 9·4 inch barbettes (are anti-aircraft type).

1917 Photo, Sérensen.

HERLUF TROLLE. Two searchlights at feet of masts, now as *O. Fischer* above. Ventilators funnel low.

DANISH COAST SERVICE SHIP. (*Kystforsvarsskib.*)

SKJOLD. *(In Reserve 1920.)* 1920 Photo, Lieut. ...

SKJOLD (1896). Laid down at Copenhagen 1893, and completed 1899. Displacement, 2200 tons. Complement, 137. Dimensions: 227¼ × 38 × 13½ feet. **Guns** (Krupp): 1—9·4 in., 40 cal., 3—4·7 in., 40 cal., 4—6 pdr., 1 Machine. **Armour (Harvey):** 10" Belt, 3" Belt (aft), 2" Deck (bow) (Deck flat on belt amidships), 8" Big turret, 5" Small turrets, 7" Bulkhead, 8" Conning tower. Machinery: I.H.P. 2400 = 13·4 kts. Boilers: 4 Thornycroft. Coal capacity: 110 tons (*maximum*).

DANISH CRUISER.

Officially rated as "Coast Defence Ship" (*Kystforsvarsskib* as *Skjold* above).

Note.

This ship was laid down in February, 1914, and was only launched in 1918. Work on her was stopped in February, 1920, but has been resumed. To be completed 1922.

NIELS JUEL (July, 1918.)

Displacement, 4,100 tons. Complement, —.

Length (*o.a.*), 296·1 feet. Beam, 53·5 feet. Draught (full load, aft), 15·8 feet. Length (*p.p.*), 285·1 feet.

Guns (Krupp):
- 10—5·9 inch, 45 cal.
- 3—6 pdr. (anti-aircraft).

Torpedo tubes:
- 2 submerged (broadside).

Armour (Krupp):
- 7" Belt (amidships) ...
- 4" Belt (ends)..........
- 7" Bulkhead ....
- 7" Conning Tower......

Machinery: 2 screws. Boilers: Thornycroft. Designed H.P. 5500 = 17 kts. Oil: tons.

Notes.—Laid down at Copenhagen Royal Yard in Feb., 1914. Originally designed as a Coast Defence Battleship, with 2—12 inch and 8 or 10—4·7 inch guns. Design altered 1920 as above official details: *viz.*, displacement raised by 200 tons; armament altered: 2 submerged tubes, turret armour and casemates dropped; speed raised by one knot; fuel altered to oil only.

DANISH CRUISERS (*Krydsere*).

VALKYRIEN, as re-constructed. *1920 Photo, Lieut. R. Steen Ste...*

VALKYRIEN (1888, re-built 1913 & 1920). **3020** tons. Complement 282. Dimensions 43½ × 18 feet (maximum draught). Guns: 2—5·9 inch (Bofors, 50 cal.), 6—12 pdr. (... pdr. AA., 2 machine, 3—15 inch above water tubes (2 bow, 1 stern). 2½ inch deck. ... 5200 = 17·4 kts. Coal: 496 tons. Sometimes used as a Training Ship in place of *Heim...*

1920 Photo, Lieut. K. Steen St...

HEJMDAL (1894, rebuilt 1909). 1313 tons. Complement, 156. Dimensions: 231¼ (*p.p.*) ... feet. Guns (Krupp): 2—4·7 inch, 4—20 pdr., 2—6 pdr., 2—6 pdr. AA., 2 machi... tubes. (1—18 inch, 1—15 inch). 1½" deck amidships. 6" engine hatches. H.P. ... Thornycroft boilers. Coal: 150 tons. Generally employed as Training Ship for Mid...

(*Continued* ...

DANISH CRUISERS—(continued).

GEJSER. 1920 Photo, Ministry of Defence.

GEJSER (1892, rebuilt 1907). 1300 tons. Complement, 156. Dimensions: 231¾ (p.p.) × 34 × 11¼ feet. Guns (Krupp): 2—4·7 inch, 4—20 pdr., 2—6 pdr., 2—6 pdr. AA., 2 machine. 2 torpedo tubes (1—18 inch, 1—15 inch) *above water*. 1¾" deck amidships. 6" engine hatches. H.P. 3100 = 17·1 kts. Coal: 150 tons. Boilers, 8 Thornycroft (*first ship to be so fitted*). Sometimes used as Training Ship for Boys.

Sloop.

Photo wanted.

Appearance as British "Flower Class" Sloops with two masts.

(Used as Fisheries Protection Vessel, Iceland).

FYLLA* (ex-British Fleet Sweeping Vessel *Asphodel*, launched December, 1915, by Messrs. D. W. Henderson, Glasgow. Sold to Danish Navy, June, 1920). 1250 tons. Dimensions: 255¼ (*p.p.*), 267¾ (*o.a.*) × 33½ × 11 (*mean*); 11¾ feet (*max. draught*). Guns: Sold without armament.* Designed I.H.P. 1400 = 17 kts., but actually requires about 2400 H.P. for this speed. Machinery: 1 set triple expansion, 1 screw. Boilers: 2 cylindrical. Coal: 130 tons *normal*; 260 tons *max.* (about 2000 miles at 15 kts. Complement (as British Ship), 79.† Built for British Navy under Emergency War Programme as a unit of the *Arabis* group of "Flower Class" Sloops.

* May be rearmed with 2—3·4 inch (Krupp), 2—3 or 1 pdr., 2 M.G. † Over 100 (Danish Navy).

23 Torpedo Boats. (*Torpedobaade*).

("Improved Ormen" class, single screw; next three classes twin screw; last two classes

No.	Class	Date.	Displacement	H.P.	Max. speed	Fuel	Complement
			tons		kts.	tons	
10	*Improved Ormen* (Ny)	15—19	108	2000	24·3	15 tons coal	22
3	*Hvalrossen* (Ny)	1912	187	2500	26	29 tons coal	30
3	*Søridderen* (Y)	1911	231	3000	27·5	80 tons coal	33
3	*Tumleren* (S)	1911	235	3000	27·5	" "	33
1	*Ormen* (No)	1907	105	2100	26·2	17 tons coal	21
3	*Havørnen* (Ny)	'98—'97	138	2100	23·5	16 tons coal	24

(Ny) = Navy type. (S) = Schichau type. (Y) = Yarrow type. (No) = Normand ty
* Also have two depth charge tubes.

Note.—In 1920 all Danish Torpedo Boats were numbered as follows:—

No.	Name	No.	Name	No.	Name
1	*Sælen*	9	*Støren*	17	*Spækhuggeren*
2	*Havkatten*	10	*Springeren*	18	*Vindhunden*
3	*Nordkaperen*	11	*Søridderen*	19	*Tumleren*
4	*Makrelen*	12	*Delfinen*	20	*Ormen*
5	*Narhvalen*	13	*Hvalrossen*	21	*Søbjørnen*
6	*Havhesten*	14	*Søulven*	22	*Havørnen*
7	*Søhunden*	15	*Flyvefisken*	23	*Hajen*
8	*Søløven*	16	*Søridderen*		

STØREN. 1920 Photo, Lieut. R. S. Sørensen, I

10 *Improved Ormen* class: **Havkatten No. 2, Saelen No. 1** (both launched 1912), **Havheste** **Makrelen No. 4, Narhvalen No. 5, Nordkaperen No. 3, Søhunden No. 7** (all lau **Søløven No. 8, Springeren No. 10, Støren No. 9** (all launched 1916). All by Copen Dimensions: 126·3 (*o.a.*) × 13·9 × 8 feet *normal* draught (9 feet *full load* draught). Armament: 2—4 pdr., [illegible] aircraft), 2—18 inch tubes (one bow, one deck, both above water), and two depth charge tubes.

Note.—For these boats, the Normand "Ormen" design was adopted, but various modifications were introdu Ministry of Defence. They have not proved very good sea boats.

(Continued on ne

DELFINEN. (*Hvalrossen* has raised fore funnel.) 1917 *Photo, Ebesen, Aarhus.*

3 *Hvalrossen* type: **Delfinen, No. 12, Hvalrossen, No. 13, Svaerdfisken No. 11,** (all 1913), built at Copenhagen Dockyard. Dimensions: 184¼ × 17 × 7 feet. Armament: 1—11 pdr. (aft), 4—18 inch tubes, viz.: one bow, one twin, and one single deck mounting, all being *above water*.

Note.—*Svaerdfisken* was badly damaged by a mine in Dec., 1918. Repairs were not begun till 1920.

FLYVEFISKEN. 1917 *Photo, Ebesen, Aarhus.*

3 *Yarrow* type: **Søridderen, No. 16, Soulven, No. 14, Flyvefisken, No. 15,** (all 1911). First built by Yarrow, others by Burmeister & Wain, Copenhagen. Dimensions: 181½ × 18 × 6¼ feet. Armament: 2—11 pdr., 5—18 inch tubes. There are 10 w.t. compartments. On trial did 5500 I.H.P. = 28.2 max. Curtis turbines in *Søridderen*. Yarrow boilers in all.

(Plan as Schichau class on previous page.)

VINDHUNDEN. 1917 *Photo, Ebbe*

3 *Schichau* type: **Tumleren, No. 19, Vindhunden, No. 18, Spækhuggeren, No. 17** (a by Schichau, others by Copenhagen Dockyard. Dimensions: 184¾ (o.a.) × 19¼ × 7.3 feet. Armame 5—18 inch tubes, 1 in bows *above water*, other 4 on deck. Turbine engines and Normand boilers i of action: 1400 miles at 14 kts.

Note.—The "Later A" types of Torpedo Boats, built by the Schichau and Vulkan Yards for the Germa 1916-18 were similar to the above Danish boats.

1917 *Photo, E*

1 *Normand* type: **Ormen, No. 20** (Copenhagen D.Y. 1907). 127¾ (o.a.) × 13.9 × 8.5 feet. Arm 1—18 inch tube (bow, *above water*), 2—18 inch amidships.

(*For old T.B., s*

Old Torpedo Boats.

HAJEN. All now have mainmasts. *1917 Photo, Ebbesen, Aarhus.*

3 *Havornen* class named **Havørnen, No. 22** (1897, rebuilt 1902), **Hajen, No. 23** (1896, rebuilt 1908), **Søbjørnen No. 21** (1899, rebuilt 1908). All built at Copenhagen D.Y. Armament: 1—3 pdr., 1 M.G., 4 tubes (all 15 inch in *Havørnen* and *Søbjørnen*; 2—18 inch + 2—15 inch in *Hajen*).

(Patrol Boats listed on next page.)

11 + 1 Submarines (*Undervandsbaade*).

No.	Type	Date	Displacement	H.P.	Max. speed	Fuel or Battery Endurance.	Tubes	Complement	Max. draug't
			tons						feet
1	(*New*) (HW)	*Bldg*	...	...	...	...	...	...	...
3	*Bellona* (N)	'15-?	301 / 369	1000 / 630	15 / 10½	...	4	...	8¼
5	*Aegir* (HW)	'12-'15	185 / 235	450 / 340	13.5 / 8	1400 miles at 10 kts.	3	11	8
6	*Havmanden* (HW)	'11-'14	164 / 201	450 / 275	13½ / 8	For 220 at 12 kts. ...	2	11	8

(N)=Navy type. (HW)=Hay-Whitehead type, designed by Marley F. Hay and Messrs. Whitehead, Fiume, Austria.

Note.—Danish Submarines numbered thus in 1920. *Havfruen* 2 —, *Havmanden* 3 —, *Thetis* 4 —, *2 den April* 5 —, *Najaden* 6 —, *Nymfen* 7 —, *Aegir* 8 —, *Ran* 9 —, *Triton* 10 —, *Neptune* 11 —, *Galathea* 12.

Begun May, 1920.

1 New boat, *not named*, building at Copenhagen D.Y. Dimensions: 134½ × 12¼ × 7⅞ feet. Will be of much the same design as *Aegir* class and will have Burmeister & Wain Diesel motors.

Completing.

Rota to run trials during Summer, 1920.

3 Navy type: **Bellona** (1919), **Flora** (1920), and **Rota** (1918). All built at Copenhagen D.Y. Burmeister & Wain 6-cylinder, 4-cycle Diesel engines. Dimensions: 155.7 × 14.4 × 8.8 feet. Armament: 1—6 pdr. anti-aircraft gun; 4—18 inch torpedo tubes.

GALATHEA. *1920 Photo, Lieut. R. Steen S.*

Hay-Whitehead type: **Aegir** (1914), **Galathea** (1914), **Neptun** (1915), **Ran** (1915), **Triton** ... built at Copenhagen D.Y. ... Dimensions: ... × 15½ × 12 ... feet. Armament: 1—6 pdr. anti-air ... inch torpedo tubes. These boats are an enlarged and slightly improved *Havmanden* design.

HAVFRUEN. *1917 Photo, Ebbesen*

6 Hay-Whitehead type:—**Havmanden** (1912) and **Havfruen** (1912), **Thetis** (1912), built by White... Fiume, and **Najaden** (1912), **Nymfen** (1914), **2 den April*** (1913), built at Copenh... 1 set 6-cyl. Diesel engines. Dimensions: 118⅛ × 11.9 × 8 feet. 2—18 inch torpedo tubes. Radius on *surface* (additional to that given in table) 100 miles at 10 kts; *submerged* 2 hrs. 10 mins. continuous 8 kts.

* In Danish, pronounced *Anden April*.

Patrol Boats. [illegible]

[illegible] Photo, [illegible]

P1. [illegible] British M.L. [illegible] built 1915, [illegible] Danish Navy 1920. [illegible] tons. Dimensions: [illegible] Guns: 1—3 pdr. [illegible] Fuel: 2000 gallons petrol. Complement, [illegible]

Note: British M.L. [illegible] and [illegible], also purchased 1920, but [illegible] in Danish Navy.

1920 Photo, Lieut. [illegible]

P8, P9. Built 1894-95. 47 tons. Guns: 1—3 pdr., 1 M.G. Tubes: 2—14 inch. I.H.P. 350=14 kts. Now serve as Tenders to Submarines and Coastal Defence Vessels.

P[illegible] Built [illegible] Photo, [illegible]

P2, (ex-Torpedo Boat [illegible] ex-[illegible]). **P3,** (ex-Torpedo Boat 77, ex-[illegible]). Built at Copenhagen D.Y. 1905. 125 tons. I.H.P. 1460=20 kts. (first year). Guns: 2 machine, 4 tubes ([illegible] and 2—15 inch). Coal: 12 tons. Complement, 21.

Patrol Boats—continued.

No photo available.

P4, P5, P6, P7. Built 1889-90. 22 tons. I.H.P. 180=12 kts. Each armed with 2 M.G.

Depot Ships.

Old Cruiser *Hekla* serves as a Receiving Ship for crews of Submarines and Seaplanes. Has neither boilers nor guns now and is therefore non-effective.

For Gunboat **GRONSUND,** see next page.

Mine Ships (*Minelægger* or *Minekrydser*).

1920 Photo, Lieut. Stensen, R.D.N.

MINEKRAN V* (1917), **MINEKRAN VI†** (1913), both built at Copenhagen D.Y. 200 metric tons. Dimensions: 88.5 (*p.p.*) × 20.7 × 6.6 feet (mean draught). Guns: 2—1 pdr. Machinery: in each, 2 sets of Bergsund surface-ignition, heavy-oil motors and electric drive. B.H.P. 290=10 kts. Complement, 26. Carry 50—60 mines each. In Danish * *V = Fem*, † *VI = Seks*.

1920 Photo, Lieut. Stensen, R.D.N.

LOSSEN (1910). 630 tons. Complement, 53. 149.3 (*p.p.*) × 28 × 9.5 feet. Guns: 2—6 pdrs. I.H.P. 900=15 kts. Coal: 29 tons. Carries about 150—180 mines.

No photo available.
Appearance similar to "Minekran IV" (next column).

MINEKRAN I* (1896). 110 tons. Dimensions: 72.2 × 16.4 × 6.6 feet. Guns: 2 machine. Speed: 8 kts. Complement, 18. In Danish, * *I = Et*.

Mine Ships—Continued.

1920 Photo, Lieut. [illegible]

HJÆLPEREN (1890). 300 tons. Complement, 10. 22.2 × 7.5 feet. 3 M.G. I.H.P. 330 = 11 kts.

1920 Photo, Lieut. [illegible]

MINEKRAN IV* (1884). 166 tons. Dimensions: [illegible] 5.9 feet. Guns: 2 M.G. I.H.P. 160=8 kts. Re[illegible] Complement, 29.

In Danish * *IV = Fire*.

No photo available.
Appearance similar to "Minekran IV" abov[e]

MINEKRAN II* (1878), **MINEKRAN III†** (18[illegible]) Guns: 2 M.G. Speed: 7 kts. Complement, 16.

In Danish * *II = To*; † *III = Tre*.

Also 2 Tugs (built 1878) used for Mining Duties, a[nd] Barges.

Torpedo Transport *(Torpedo Transportbaad)*

1920 photo, *Lieut. Steensen, R.D.N.*

SLEIPNER (1882) 80 tons. 73 × 14·8 × 5·9 feet (*mean draught*). Guns: none. I.H.P. 110 = 9 kts. Sometimes used as Tender to Submarines.

Repair Ship *(Værkstedsskib)*

1920 *Photo, Lieut. Steensen, R.D.N.*

GRONSUND (1883, re-built 1905-6 and 1917-18). Old gunboat of 260 tons. Complement, 20. 119 (*p.p.*), 125·3 (*o.a.*) × 20 × 8·1 feet. Guns: 2—6 pdr. A.A. B.H.P. 250 = 11 kts. Has internal combustion engines with "electric drive." Serves as Depot, Training and Repair Ship to Submarines, particularly those operating in the Great Belt.

Surveying Ships *(Opmaalingsskibe)*.

Krieger. 1920 *Photo, Lieut. Steensen, R.D.N.*

KRIEGER, MARSTRAND, 170 tons, **WILLEMOËS,** 160 tons. (All built 1861, re-built 1896). 114 × 17·5 × 6·4 feet (*mean draught*). All launched 1861. Complements, 28 to 32. I.H.P. 250 = 11 kts. Guns: 2 machine.

Fishery Cruisers *(Fiskeriinspektionsskibe)* and **Coastguard Vessels** *(Inspektionsskibe)*.

For Sloop *Fylla* used as Fisheries Protection Vessel, Iceland (see an earlier page).

1920 *Photo, Lieut. Steensen, R.D.N.*

(Fisheries Duties, Faroes Islands).

DIANA (Holland, 1916). 260 tons. 116·5 (*p.p.*), 124·8 (*o.a.*) × 21·5 × 8·8 feet. Guns: 2—3 pdr. I.H.P. 400 = 9·5 kts. Complement, 27.

Fishery Cruisers.—*Continued*

1920 *Photo, Lieut. Steens*

FENRIS (Holland, 1915). 190 tons. 86·4 (*p.p.*) × 2 Guns: 2—3 pdr. I.H.P. 420 = 12 kts. Complement

1920 *Photo, Ministry of* I

ISLANDS FALK (Elsinore 1906). 755 tons. Compl 170·6 (*p.p.*), 183·6 (*o.a.*) × 29·3 × 14·7 feet. Guns: 2—3 pdr. I.H.P. 1100 = 11·7 kts. (*f.d.*)

(*Continued on*

Fishery Cruisers (*Continued*).

1920 Photo, *J. Dannsø*.

(Now painted light grey).

BESKYTTEREN (1900). 447 tons. Comp. 56. 134 (*p.p.*), 142·5 (*o.a.*) × 24·8 × 11·5 feet. Guns: 1—6 pdr., 2—3 pdr. I.H.P. 620 = 12·5 kts. (*f.d.*). Reboilered 1920.

Gunboat. (*Kanonbaad*.)

1920 Photo, Lieut. Dannsø, R.D.N.

GULDBORGSUND (1884, re-built 1908). Old Gunboat of 270 tons. Complement, 40. 110 × 26 × 8·1 feet. Guns: 4—6 pdr., 2 M.G. H.P. 400 = 11·0 kts.

Note.—The S.S. *Thor* belonging to the Government of Iceland is also employed on Fisheries Protection Duties.

Lighthouse Tenders (*Fyrtransportskibe*) and Beacon Inspection Ships (*Vagerinspektionsskibe*).

Photo, S. Anderson, Esq.

ABSALON (1877). 292 tons. Complement, 26. 125 (*p.p.*), 134½ (*o.a.*) × 29·1 × 8·8 feet. Guns: 2—3 pdr. H.P. 390 = 11·6 kts. (*f.d.*).

Also—

		Tons.		I.H.P.		Kts.
C. F. Grove	...	465	...	450	...	9
Kattegat	...	263	...	200	...	9
Lövenörn	...	397	...	500	...	—
Nordsøen	...	268	...	260	...	—

Royal Yacht (*Dampskib til Hs. Majestæt Kongens Brug*).

1919 Photo, *Ebbesen, Aarhus*.

DANNEBROG (1879). 1190 tons. 235 (*p.p.*) × 26·3 × 10·4 feet. Guns: 2 M.G. I.H.P. 940 = 13 kts.

Training Ships (*Skoleskibe*).

Cruisers **VALKYRIEN** and/or **HEJMDAL** (d… previous page) used sometimes as Training S… shipmen, and **GEJSER** for Boys. Gunboat **G…** (… preceding page) sometimes used as a Trai… Submarine Service.

(*Boys' Training Ship*)

INGOLF (1876). 1000 tons. Complement, 125. 1… (*o.a.*) × 27·9 × 13·6 feet. Guns: 4—6 pdr., 1 … 900 = 11 kts. (*f.d.*).

Also four Cutters, *Bacchus*, *Seanes*, *Saga* and *Th…*

Non-Effective.

Fyen (Receiving Ship).

Esbern Snare (Store Ship).

Fremad I., *Fremad II.* } Tugs for Dockyard Duties.

Lille Belt, *Store Belt*, *Öresund* } Old Gunboats used as Store Hulks fo…

GERMAN FLEET.

		miles.
WILHELMSHAVEN to	Cherbourg	510
	Dover	330
	Gibraltar	1565
	Havre	450
	Heligoland	48
	Hull	330
	Kiao-chau	11100
	Malta	2555
	New York	3570
	Port Said	3530
	Portsmouth	440
	Rio de Janeiro	5470
	Rosyth	450
	Sheerness	280

		miles.
KIEL to	Copenhagen	160
	Danzig	280
	Libau	360
	Memel	330
	Kronstadt	670
	Skaw	240
DANZIG to	Libau	120
	Memel	113
	Kiel	280

Wilhelmshaven to *Kiel*, 80 miles (by canal much of the way).
" " 530 " (via Great Belt).

Flags.

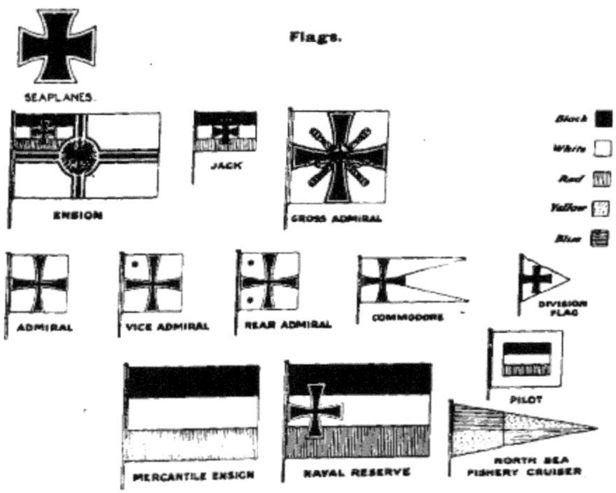

Mercantile Marine.

From "Bureau Veritas" 1919-20 figures.
Ships under 100 tons *gross* excluded

1021 Steam and Motor Ships of 712,432 tons *gross*.

116 Vessels (with Auxiliary Motors) of 17,491 tons *gross*.

1261 Sailing Vessels of 366,495 tons *gross*.

Coinage.

Mark (100 pfennige) *nominal* value = 11¾*d*. British, 24 cents U.S.A., but rate of ex depreciated to *about* ¾*d*.

Colour of Ships.

Big ships: Light grey all over.

Torpedo craft: Varies from black to dull brown.

Active Naval *Personnel*.

By Treaty of Versailles, limited to following *maximum* figures, inclusive of Officers ashore:—1500 Officers and Warrant Officers, 13,500 men. Total, 15,000 *Active* I

Administration, 1920.

Fleet to be controlled by the Minister of Defence, who is to be advised by the Ch "Admiralstab." To be three Administrative Departments:—(*a*) Naval Staff; (*b* Fleet Service (Naval Construction, Ordnance, Hydrography, &c.); (*c*) Victuallin Personnel, &c. Three Inspecting Departments:—(*a*) For Training Stations and (*b*) for Torpedoes and Mines; (*c*) Naval and Coastal Ordnance. Coastal Defences a

Minister of Defence: Herr Gustav Noske.

(Divided into 2000 yard squares. Uniform scale. Soundings in fathoms. Heights in feet.)

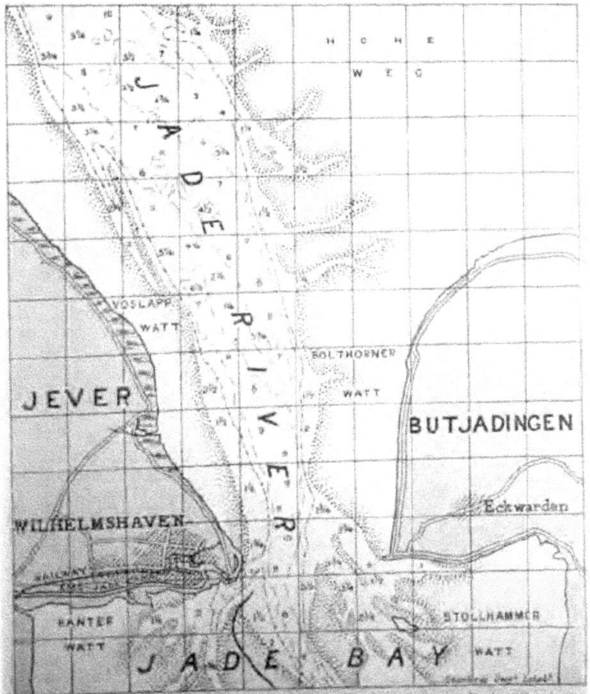

WILHELMSHAVEN. **North Sea Naval Station.** Two large basins. Docks (Graving): Nos. (1), (2) and (3), 625 × 99½ × 36 feet; No. (4), 584 × 101 × 35 feet; No. (5), 584 × 101 × 35 feet; No. (6), 584 × 94 × 35 feet; No. (7), 822 × 131 × 34 feet; No. (8.) 822 × 114¾ × 34 feet. Floating docks uncertain, owing to surrenders made at British and French ports, 1920. Four building slips, (1) 600 feet, (2) 345 feet, (3) 300 feet, (4) *about* 600 feet. Men employed, *about* 11,500. Machinery: 38 of 9300 H.P. Admiralty Chart No. 3346. Rise of Spring Tide, 11¼ feet. Harbour has to be dredged continually on account of the sand which silts up very badly.

NORTH SEA (*Nordsee*).

Principal Private Yards:

BLOHM & VOSS (HAMBURG). Ten *Slips* (four 600 feet long, five each about 500 fe[et], one slip 328 feet). Five *Docks* (floating). (1) 320 × 52 × 18. (2) 355 × 52 × [...] × 88 × 25. (4) 595 × 88 × 25. (5) 728 × 123 × 36 feet (46,000 tons). Pos[sibly] (40,000 tons) intended originally for Pola D.Y. (Austria-Hungary). Probably [sub]marines and destroyers now. *Employees*, 10,000. *Machinery*, 10,900 H.P. Bou[ght] & Schmilinsky Yard for extensions, 1917, with dock 400 × 50 × 16 feet.

A. G. WESER (BREMEN). Five *Slips* (one of 650 feet, one of 490 feet, two of 3[...] small). Three *Docks* (floating) of small size. Build submarines. *Empl[oyees]* [...] *Machinery*, 4220 H.P.

BREMER VULKAN (VEGESACK). Six *Slips* (one of 650 feet, two about 500 feet, th[...]. *Docks*, non floating. Probably build submarines and destroyers now. *Empl[oyees]*, 3100. *Machinery*, 2800 H.P.

VULCAN (HAMBURG). Two *Slips*, 650 feet or so. Four floating docks: No. 4, 6[...] 28 feet (17,500 tons); No. 3, 610 × 106 × 28 feet (27,000 tons). Another d[ock] 108¼ × 32¾ feet (32,000 tons) may exist. Two smaller floating docks, 433 × 7[...] (5000 tons) and 510 × 82 × [...] feet (11,000 tons). *Employees*, about 9000. I[...]

JOHANN C. TECKLENBORG (GEESTEMÜNDE). Six *Slips* (two of 600 feet, tw[o ...] feet, two smaller). 2600 men. H.P. 2600. Two dry docks.

REIHERSTIEG SCHIFFSWERFTE (HAMBURG). One *Slip* about 550 feet, one [...] five small. *Docks*, one (floating) 511½ × 97 × 26 feet (20,000 tons), t[...] *Employees*, about 2000. H.P. about 4000. Bought J. & N. Wichorst Yar[d for] extensions.

Largest Private Docks.

BREMERHAVEN. Port for Bremen, which is 50 miles inland of river mouth, and [...] from Bremerhaven. At Bremen are the headquarters of the N. Deutscher Ll[oyd and] big shipbuilding firm, Weser Act Gesellschaft. Kaiser Dry Dock No. 1, 741 × [...] can take ships up to 30,000 tons. Another (N.D.L. floating) 450 × 58 × [...] Bremerhaven, Seebeck Co. has one dock (No. 1) 550 × 63 × 21 feet, and three [...] Geestemünde, the same firm has two docks, each 547 × 80 × 18 feet. Also at [...] two smaller dry docks; and at Hammelwarden, one small dry dock. At Bremen [...] dock, 490 × 90 × 23 feet (12,000 tons), and another smaller (3000 tons).

EMDEN. New and large W/T. Station. One floating dock (Inner Harbour) 284½ [...] feet (3500 tons). 1 smaller (2500 tons). Lock between Inner Harbour an[d ...] 853 × 131 × 43 feet.

Note.

Torpedo Craft and Submarine Stations at Wilhelmshaven assigned to Ministry of commercial use.

(Divided into 2000 yard squares. Uniform Scale. Soundings in fathoms.)

KIEL CANAL.

Length: 61¼ miles. Width at surface: 394 feet and 144 feet at bottom. New double locks: 1082 × 147 × 36 feet. Old locks (at Holtenau and Brunsbüttel): 482 × 82 × 29½ feet. Sidings: 4 of 3600 × 540 feet, 11 other bays of 984 feet for turning. Speed of ships: 10 kts. for large, 15 kts. for small. Time of passage: 7 to 10 hours, according to size of ships; but 16 battleships can, in emergency, pass in 6 hours.

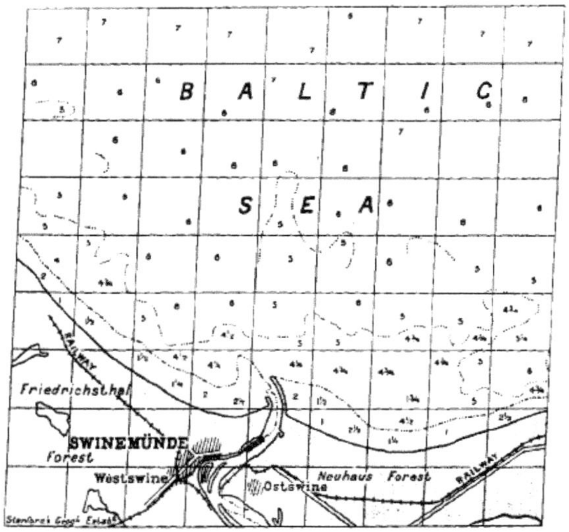

SWINEMÜNDE. To be Baltic Naval Station.

BALTIC (*Ostsee*).

Principal Private Yards.

HOWALDT (KIEL). Nine *Slips* (two over 700 feet, one about 500 feet, six smalle one floating (11,000 tons). *Employees*, about 3500. *Machinery*,

KRUPP'S GERMANIA (KIEL). Nine *Slips* (five of 700 to 520 feet, the rest sma marines built here for which 10-12 slips were available in 1914. *Employees*, 6400. 4000 H.P. *Docks*, None.

SCHICHAU (ELBING). Twenty-seven *Slips* (six over 650 feet long, the others all sn (six ?) *Docks* (floating) for torpedo craft. *Employees*, about 10,000. *Machinery*,

VULCAN (BREDOW-STETTIN). Seven *Slips* (two over 650 feet, two about 500 feet, t 350 feet or so). Two *Docks* (floating). (1) 416 × 58 × 17 feet. (2) 300 × 37 *Employees*, 7000. New Shipbuilding and Repair Yard projected at Königsberg.

State Yard.

KIEL. Late Naval Dockyard to be converted to Mercantile Construction and Repairs of slips unknown. Docks: (1) 423 × 71 × 28 feet. (2) 382 × 70 × 25½ feet. 64½ × 22½ feet. (4) 344 × 67 × 16 feet. (5) 570 × 94 × 30 feet. (6) 570 × 94 Floating docks: Number uncertain, owing to surrenders made 1920.

Fleet Establishment, 1920.*

(The following are the Vessels selected by the German Government, in accordance w Clauses of Treaty of Versailles.)

6 Battleships.

Not to be replaced until 20 years old, and any new ships built in replacement are not 10,000 metric tons displacement. Dates when ships arrive at age limit (reckoned of laying down) given in brackets after each name:—

Braunschweig (1921). *Elsass* (1921). *Hessen* (1922). *Hannover* (1924). *Schleswig-Holste Schlesien* (1925).

6 Light Cruisers.

Age limit, 20 years. New ships not to exceed 6,000 metric tons displacement. Dates in on same system as for Battleships:—

Amazone (1919). *Thetis* (1919). *Medusa* (1920). *Arkona* (1920). *Berlin* (1922). *Hamburg* (

12 Destroyers.

Age limit, 15 years. New boats not to exceed 800 metric tons displacement. Can be at dates given in brackets:—

T 185 (1924). *T 190, 196* (1925). *V 2, 3, 5, G 7, 8, 10, 11* (1926). *S 18, 23* (1927).

12 Torpedo Boats.

Age limit, 15 years. New boats not to exceed 200 metric tons displacement. Dates as for Destroyers above:—

T 90, 101 (1918). *T 102—7* (1920). *T 108—110, T 113* (1921).

General Note.—Any accidental losses may be replaced, irrespective of age limits, but must not exceed displacements indicated.

* "For reasons of economy and shortage of personnel, the German Naval Authorities have decided to comm the following ships in the near future:—Battleship *Hannover*, Light Cruisers *Hamburg*, *Medusa*, *Arkona*. vessels in commission at present are those of Mine Sweeping Flotillas—about 30 vessels in all." ("Schiffb 1920.)

GERMAN UNIFORMS, GUNS AND TORPEDOES.

INSIGNIA OF RANK ON SLEEVES FOR EXECUTIVES.

Gross-Admiral (= Admiral of the Fleet) as Admiral but 4 upper stripes.

Admiral. Vize-Admiral. (Vice-Ad.) Kontre-Admiral. (Rear-Ad.) Kommodore (Commodore). Kapitän zu See. (Captain) Fregatten-Kapitän & Korvetten-Kapitän (Commander) Kapitän-Leutnant. (Lieut.-Comm.) Oberleut z. See. (Lieut.) Leutenant z. See. (Sub-Lieut.)

Flaggoffiziere. Stabsoffiziere. Subalternoffiziere.

Note.—Torpedo officers are without the crown on sleeve, as also are all civilian branches. Engineer officers given same official status as Executive, July, 1908.

Torpedo officers have between the gold stripes, *brown*; Engineers, *black*; Doctors, *blue*; Paymasters, *light blue*; Constructors, *black*.

(The colour of the branch is also worn on the epaulettes. full dress, and worked into the shoulder straps.)

Paymasters, constructors and legal officers have silver buttons, badges, &c., as well as epaulette fringes, instead of gold, and cloth instead of velvet between the stripes. They are *officials* and not officers.

Torpedoes (1919).

Note: The following details are understood to be based on very sound information.

Size.	Name or Mark.	Air Pressure.	Charge. (lbs.)	Max. Range. (yds.) at kts.	Type of Heater.
19·7	G VII.*** (23 feet)	2275	430	11700 at 28 5500 at 35	Steam
19·7	G VI. AV **	2275	441	5500 at 27 2200 at 40	Dry
17·7	G/125	2133	308	6500 at 24 2200 at 36	Dry

Mines (Notes unofficial).

Date	Pre-War.	War.	—	"UC"† War.	"UE"† War.	A/S‡ War.
No. of Types	3		—	—	—	—
Weight of Charge (lbs.)	115—330	—	210*	265—440	Specially large	44
Laying Limit (ft.) ...	295—337	1130	650*	1200		11

* Torpedo-Mine. † For laying by S/M. ‡ Possibly Depth Charge, improved to 110 lb. type at end of War.

Spherical horned type said to have charge of 220-240 lbs. T.N.T. Pear-shaped "Carbonit" mines said to have charge of 186 lbs. wet guncotton. Marks I and II, soluble safety plugs. Later marks have safety pin attached by rod to sinker. Horns said to be glass tubes (containing battery acid) and covered with lead or soft metal. Bending of horn breaks glass tube and acid runs to battery, which at once becomes active and fires detonator. Also a horned A.E.G. type of which no details are known.

Naval Guns (Details unofficial).

Notation.	Calibre.	Usual Naval Designation.	Length in cals.	Date of Model.	Weight of Gun.	Weight of A.P. shot.	Initial Velocity.	Maximum penetration, direct impact against K.C. at 9000 yds.	6000 yds.	3000 yds.	Danger against ave[rage] 10,000	50[00]
	inch.	c/m.			tons.	lbs.	foot-secs.	in.	in.	in.	yards	yar[ds]
A**	11	28	40	'01	32·2	661·4	2756	6	10	14	150	45[0]
A**	9.4	24	40	'99	20·3	418·9	2756	4	8	12	130	40
A*	9·4	24	40	'98	20·3		2625	3½	7	10	110	35
B*	6·7	17	40	'01	7·5	154·5	2756	...	3	5	80	24
C*	5·9	15	50	...	5·5	101·4	3084	...	...	...	...	
C*	5·9	15	45	'09	5		2920	...	...	5	...	2[illegible]
E*	5·9	15	40	'01	4·9		2756	...	..	3½	..	1[illegible]
E*	5·9	15	35	...	4·4		2231	...	...	...	...	
F*	4·1	10·5	40	...	1·7	35·2	2756	...	...	...	...	
F*	4·1	10·5	35	...	1·2	38·2	2000	...	...	...	...	

* Brass cartridge cases to all guns.

Lesser guns: 3.4 inch (88 m/m) firing 22 lb. projectiles in modern and 15 lb. in ol 2 inch (4 pdr.) of 55 and shorter calibres.

A.A. guns: 5·9 inch, 4·1 inch, 3·4 inch War Marks on H.A. mounts ("Flak").

Projectiles: Guns of 11 inch and over fire A.P., Ersatz A.P., H.E., and comm

The 1899 and later models have the recoil utilized to return the gun to firing po over 5.9 inch. In 5.9 in. springs are employed. German guns have a lower than normally obtains, but mountings are arranged to give large degree of

11 inch, 40 cal., M. '01 in *Deutschland* and *Braunschweig* classes (removed from so

Scale: 1 inch = 160 feet.

GERMAN **SPECIAL SERVICE SHIPS**—RECOGNITION SILHOUETTES.

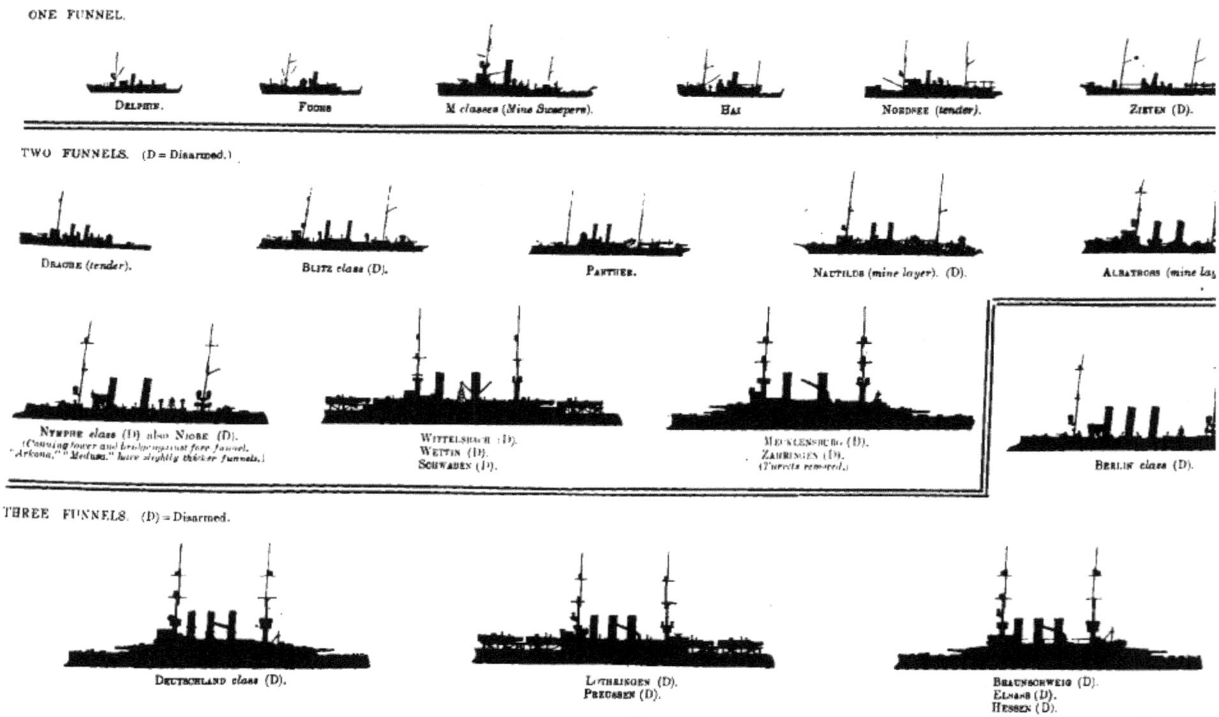

GERMAN DESTROYERS & TORPEDO BOATS—RECOGNITION SILHOUETTES.

Notes.—Majority disarmed. *Denotes types selected for Post-War Fleet.

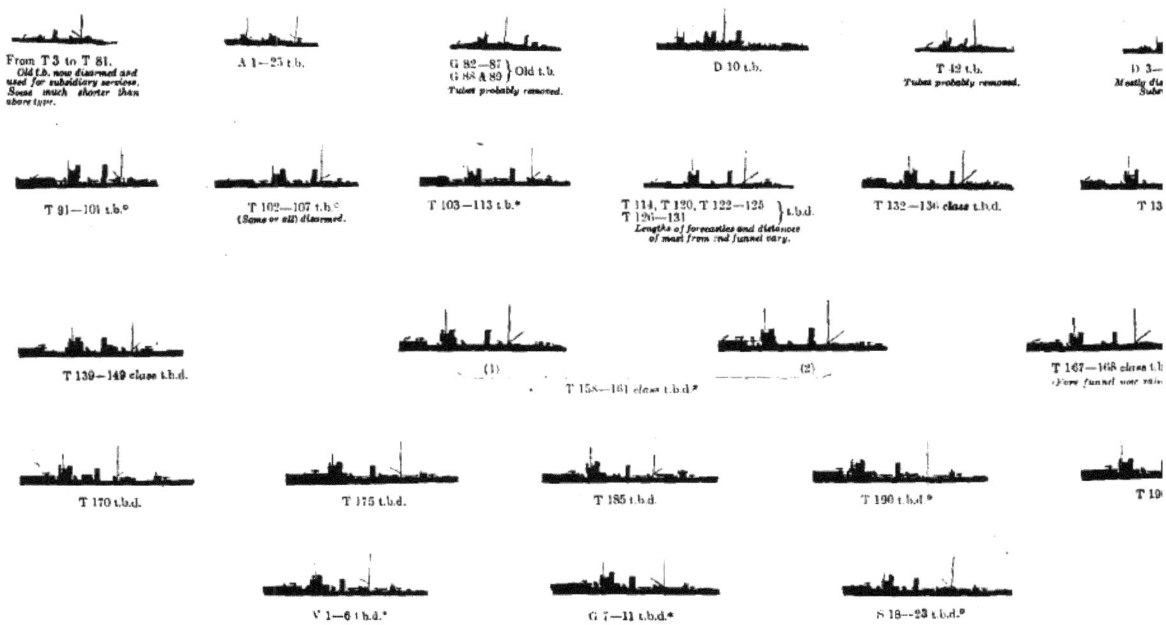

Disposal of Unfinished German Warships, Oct., 1920. All to be converted into Mercantile Vessels.

Type and Name or Number.	*To be dismantled or converted at.*
Battleships :—	
SACHSEN	Kiel
WURTTEMBERG	Hamburg
Battle Cruisers :—	
"Ersatz A"	Wilhelmshaven
"Ersatz Freya"	Hamburg
GRAF SPEE	Kiel
MACKENSEN	
Light Cruisers :—	
"Ersatz Köln"	Bremen
"Ersatz Emden"	Bremen
LEIPZIG	Weserwerft
ROSTOCK	Stettin
WIESBADEN	Stettin
FRAUENLOB	Kiel
MAGDEBURG	Kiel
Destroyers :—	
H 186, H 187	Kiel
H 166—H 169	Kiel
V 162, V 163	Stettin*
V 158—V 161	Stettin
S 152, S 157	Gluckstadt
S 154, S 155	Harburg
S 153, S 156	(Stranded near Elbing, when re-floated, to be converted.)
V 140—V 144	Stettin
B 122—B 124	Hamburg
G 119—G 121	Hamburg
V 117, V 118	Stettin
S 114, S 115	Bremerhaven
Torpedo Boats :—	
A 83—A 85	Kiel
A 67	Stettin

* Conversion under consideration.

So far as can be ascertained, the above List covers all unfinished German Warships now in existence. Various other ships, little advanced in construction, have been broken up on slips. Many Destroyers and Submarines, contracted for in 1918, were never actually laid down.

Battleships.

(DEUTSCHLAND CLASS—4 SHIPS.)

DEUTSCHLAND (Nov., 1904), **HANNOVER** (Sept., 1905), **SCHLESWIG-HO** (Dec., 1906), and **SCHLESIEN** (May, 1906). Displacement, 13,200 tons. Compleme Length (*waterline*), 413 feet. Beam, 72¾ feet. *Mean* draught, 25¼ feet. Length (*over all* *Guns (M. '01): 4—11 inch, 40 cal. (A[4]), 14—6·7 inch, 40 cal., 20—3·4 inch (22 pdr.). ? Torpedo tubes (17·7 inch): 6 *submerged* (bow, stern, and broadside). Armour (Krupp) (amidships), 4" Belt (ends), 3"—1½" Deck, 11" Barbettes, 10" Turrets to these, 8" Lower 6¾" Battery, 12" Conning tower (fore), 6" Conning tower (aft). Machinery: 3 sets 3 cylin triple expansion. 3 screws. Boilers: 12 Schulz Thornycroft, except *Deutschland*: 6 cyli 8 Schulz-T. Designed H.P. 17,000 = 18 kts. Coal: *normal* 800 tons; *maximum* 1750 t 200 tons oil fuel. *Nominal* radius 5500 kts. at 10 kts.

**Note.*—Guns and tubes given above those originally mounted. Present armaments and disposal of ship

Name	*Guns*	*Service*, 1920
Deutschland	Disarmed; guns removed.	Unknown.
Hannover	Disarmed.	
Schleswig-Holstein ..	6—4.1 inch only.	Post-War Ships.
Schlesien	Disarmed.	

Name.	*Builder.*	*Machinery.*	*Laid down.*	*Completed.*	Trials: *22 hrs. at ¾.*	*6 hrs. full p.*	Boilers.
Deutschland	Germania, Kiel	Krupp	1903	1906	11,377 = 17·1	16,935 = 19·53	Schulz to all.
Hannover	Wilhelmsh'vn		1904	1907	12,153 = 16·9	17,768 = 18·7	
Schleswig-Holstein	Schichau, D'zig	Schichau	1905	1908		19,868 = 19	
Schlesien	Germania, Kiel	Krupp	1905	1908		19,465 = 18·5	

Braunschweig, Elsass, Hessen.

1920 Photo.

Preussen (Lothringen same).

(Braunschweig Class—5 Ships.)

BRAUNSCHWEIG (Dec., 1902), **ELSASS** (May, 1903), **LOTHRINGEN** (May, 1904) (Sept., 1903), & **PREUSSEN** (Oct., 1903). Displacement, 13,200 tons. Complen Length (*waterline*), 413½ feet. Beam, 72¼ feet. *Mean* draught, 25¾ feet. Length (*over* *Guns (M. '01): 4—11 inch, 40 cal. (A³), 14—6·7 inch, 40 cal., 18—3·4 inch (22 pdr.) Torpedo tubes (17·7 inch): 5 *submerged* (bow and broadside), 1 *above water* (stern). Arn 8¾" Belt (amidships), 4" Belt (ends), 3" Deck on slopes, 11" Barbettes to these, 11" 5" Lower deck, 6" Battery, 6¾" Small turrets, 12" Conning tower, 6" Conning tower (aft) 3 sets 3 cylinder vertical inverted triple expansion. 3 screws. Boilers: 8 Schulz-Th cylindrical. Designed H.P. 16,000 = 18 kts. Coal: *normal* 700 tons; *maximum* 1670 t tons of oil (in double bottom). *Nominal* radius: 5500 miles at 10 kts.

*Note.—Guns and tubes given, those originally mounted. Guns and disposal of ships, 1919, as foll

Name	*Guns*	*Service.*
Braunschweig	Disarmed; guns removed.	Accommodation Ship. } For Post-War
Elsass	" "	Accommodation Ship. }
Lothringen	" "	Sea-going Parent Ship for Motor Boat ;
Hessen	" "	Accommodation ship. For Post-War
Preussen	" "	Sea-going Parent Ship for Motor Boat

* Each has superstructures for carrying 12 Motor Boat Minesweepers. Both in Reserve.

Name.	Builder.	Machinery.	Laid down	Completed	Full power trials: 24 hrs.	6 hrs.	Bo
Braunschweig	Germania, Kiel		1901	1904	11,382 = 16·3	17,312 = 18·6	
Hessen	Germania, Kiel		1902	1905	11,384 = 16·4	16,990 = 18·23	8 Sc
Preussen	Vulkan, Stettin	By Builders	1902	1905	11,304 = 16·2	18,374 = 18·69	and
Elsass	Schichau, D'zig		1901	1904	11,554 = 16·7	16,812 = 18·7	h
Lothringen	Schichau, D'zig		1902	1906	11,553 = 16·5	16,950 = 18·54	

Obsolete German Battleships.

WITTELSBACH (*Wettin*, *Schwaben*, similar). *Photo. January, 1920, Lieut. Stevens, R.N.*

Note.—Disposal, 1920, cannot be ascertained. Three ships (names unknown) reported sold or for sale; other two believed to be paid off. All are surplus to Post-war Establishment.

WITTELSBACH (1900), **WETTIN** (1901), **ZÄHRINGEN** (1901), **MECKLENBURG** (1901), **SCHWABEN** (1901). Displacement, 11,800 tons. Complement was 683. Dimensions: 410¾ (*w.l.*), 416¾ (*o.a.*) × 68¼ × 28 feet (*maximum* draught). *Guns (M. '99): 4—9.4 inch, 40 cal., 18—5·9 inch, 40 cal., 12—3·4 inch (15 pdr.), 2 M.G. *Torpedo tubes (17·7 inch): 6 *submerged* (bow, stern and broadside). Armour (Krupp): 9" Belt (ends), 3" Deck on slopes, 10" Barbettes, 10" Turrets to them, 5½" Lower deck (redoubt), 5½" Battery, 6" Casemates (4 bow), 6" Small turrets, 10" Conning tower. Machinery: 3 sets vertical triple expansion. 3 screws. Boilers: 6 Schulz-Thornycroft and 6 cylindrical. Designed H.P. 15,000 = 18 kts. Coal: *normal* 653 tons; *maximum*, 1800 tons. Oil: 200 tons.

*Guns and tubes given, those originally mounted. Present armament and disposal of ships as follows:—

Name	*Guns*	*Service*, 1919.
Wittelsbach .. .. .. / *Wettin*	Disarmed.	Parent ships for Minesweepers, Wilhelmshaven.
Zähringen	Turrets removed, 7—5·9 inch.* 4—15 pdr.*	Training Ship, Danzig.
Mecklenburg	Disarmed.	Accommodation Ship, Kiel.
Schwaben	Turrets removed. 6—5·9 inch.* 4—15 pdr.*	Sea-going Parent Ship for Motor Boat Minesweepers, Wilhelmshaven.

*Guns on board, but ships are disarmed.

Old German Light Cruisers.

HAMBURG, BERLIN, as above.

HAMBURG (July, 1903), **BERLIN** (September, 1903). *Normal* displacement, Complement, 303. Dimensions: 362.9 × 43.3 × 17½ feet (*maximum* draught). noted below. Torpedo tubes (17.7 inch): 2 *submerged*, 2 *above water*. Armou 2" Deck, ¾" Deck (at ends), 4" C.T. Machinery: 2 sets 4-cylinder triple expansion Boilers: 10 Schulz-Thornycroft. Designed H.P. 10,000 = 22 kts. Coal: *norm* *maximum*, 860 tons. *Nominal* radius: 5000 at 10 kts. Begun 1902-3, compl Both for Post-War Fleet.

Notes.

Name	*Armament*	*Mines*	*Service*, 1919	:
Hamburg ..	10—4.1 inch, 2 M.G.	108	Paid off.	W
Berlin ..	" "	"	Headquarters S.N.O. Baltic.	K

ARKONA (April, 1902). Displacement, 2700 tons (sheathed). Complement, 281. Dimensions: 342·4 × 40·3 × 17½ feet (*maximum* draught). Guns: 10—4·1 inch, 2 M.G. Torpedo tubes (17·7 inch): 2 *submerged*. Can carry 400 mines. Armour (Krupp): 2" Deck, 4" C.T. Machinery: 2 sets 4-cylinder triple expansion. 2 screws. Designed H.P. 8000 = 21·5 kts. Boilers: 9 Schulz-Thornycroft. Coal: *normal*, 450 tons; *maximum*, 710 tons.

Condition of all Ships named on this page.

Arkona:—Guns on board. Sea-going Parent Ship for Motor Boat Minesweepers (*UZ* boats), North Sea. For Post-War Fleet.

Nymphe:—Disarmed. In Reserve.*

Thetis:—Disarmed. } For Post-War Fleet.

Amazone:—Disarmed. }

Medusa:—Guns on board. Sea-going Parent Ship for North Sea Motor Boat Minesweepers (*UZ* boats). For Post-War Fleet.

Niobe:—Disarmed and paid off. In Reserve.*

* Surplus to Post-War Establishment.

MEDUSA. *Photo, Abrah*

NYMPHE (1899), **THETIS** (1900), **AMAZONE** (1900), & **MEDUSA** (190 displacement, 2650 tons. (Sheathed and Muntz metalled.) Complement, 275. 324·4 × 38·7 × 17½ feet (*maximum* draught). Guns: 6—4·1 inch, 2 M.G. T (17.7 inch): 2 *submerged*. Armour (Krupp): 2" Deck (amidships), ¾" Deck (a Glacis to engine room hatches, 3" C.T., Cofferdam and cellulose amidships. 2 sets 4-cylinder triple expansion. 2 screws. Boilers: 9 Schulz-Thornycroft H.P. 8000 = 21 kts. Coal: *normal*, 380 tons; *maximum*, 580 tons.

As *Medusa*.

NIOBE (1899). *Normal* displacement, 2600 tons. (Sheathed and Muntz metalled.) C 268 Dimensions: 342·5 × 38·7 × 17½ feet (*maximum* draught). Guns: 10 40 cal., 2 M.G. Torpedo tubes (17·7 inch): 2 *submerged*. Armour (Krupp): 2" D ships), ¾" Deck (at ends), 3¼" Glacis to hatches, &c., 3" C.T., Cofferdam and cellulose Machinery: 2 sets 4-cylinder triple expansion. 2 screws. Boilers: 5 Thornycrof Designed H.P. 8000 = 21·5 kts. Coal: *normal*, 300 tons; *maximum*, 600 tons.

55* Destroyers.

*[illegible]

Totals.	Class	Begun	Completed	Displacement	Nominal H.P.	Nominal Speed	T. Tubes	Coal (max)	Complement
3	*S 18*—[illegible]	1912	1914	tons 570	15,000 [illegible]	kts. 32.5	4	tons 133/55	[illegible]
9	*G 11*—[illegible] / *V 6*—[illegible]	1911	1912	564	15,000 [illegible]	32.5	4*	140/80	73
2	*T 197* / *T* [illegible]	1910	1911	648—656	16,000 [illegible]	32.5	4*	165/65	83—93
2	*T 18*[illegible] / *T 175*	1909	1910	636—654	15,000 [illegible]	31.5—32.5	4*	170/65	83
3	*T 17*[illegible] / *T 16*[illegible]—[illegible]	1908	1909	613—636	14,000 [illegible]	30—32	3*	160/50	83
8	*T 16*[illegible]—[illegible]	1907	1908	554	10,250	30	3*	146	83
11	*T 14*[illegible]—[illegible]	1906	1907	550	10,000	30	3*	190	72
1	*T 13*[illegible]	1905	1907	572	10,800 [illegible]	32	3*	170	80
6	*T 130*—[illegible]	1905	1907	487	6500	27	3*	135	60
10	*T 12*[illegible]—[illegible] / *T 12*[illegible]—[illegible] / *T 11*[illegible]	1904 / 1903 / 1902	1906 / 1905 / 1904	485 / 470 / [illegible]	6500 / 6500 / 6500	27 / 27 / 26	3*	116	60—56

*Tubes removed from nearly all these boats.

†An additional deck load of coal can be carried as follows:—+ 10 tons in *T* [illegible], + 15 tons from *T* [illegible] up to *T* [illegible], about 20 tons up to *S* [illegible].

General Notes to Table.

G boats built by Krupp's Germania Yard, *S* boats by Schichau, *V* boats by Vulkan (Stettin). All are officially rated as "*Grosse Torpedoboote*," or "Big Torpedo Boats," and not as Destroyers ("*Torpedoboote-Zerstorer*"). Unless otherwise noted, all boats have Schulz-Thornycroft (or "Marine Type") boilers. The details tabulated are generally simplified to the average design of each group.

"*T*-boats" were built with the usual *G*, *S* and *V* builders' index letters, but were given letter *T* to avoid duplication of numbers, with the later boats numbered from *V 1* up to *S 23*, and the "War Types," *V 25*—*H 187*. In description of *T* boats, *G* refers to Krupp-Germania boats, *S* to Schichau boats, and *V* to Vulcan boats.

Majority of boats with "T" index letter have had armament removed and are fitted for Mine-sweeping.

1912-13.

From the "[illegible]

3 *Schichau*: ***S 18, S 19, S 23.*** Launched 1913. About 570 tons. Dimensions: [illegible] Armament: 2—4.1 inch, 1—3.4 inch (22 pdr.). Torpedo tubes: 4—19.7 inch. Designed S.H.P. [illegible] kts. Melms-Pfenninger turbines. Fuel: 135 tons coal + 55 tons oil. Complement, 75. *S* [illegible] Fleet. *S 1*[illegible] in Reserve.

GERMAN DESTROYERS.

Note.—Nearly all the Destroyers listed on this and succeeding pages have had their tubes removed and are fitted for minesweeping. About a dozen of them have been mined or otherwise injured during the War and not repaired; they are entirely non-effective for fighting purposes. Boilers and engines of many are in a very poor condition, and the *maximum* speeds given must be taken with reserve.

1911-1912 (and two **1913** Replace Boats).

Photo by courtesy of S. Anderson, Esq.

G 7, 9, 8, 10.

Note.—All boats [illegible] [illegible] now.

G 9.

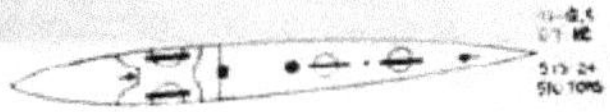

4 *Krupp-Germania*: **G7, G8, G10, G11** (launched 1911-12).

[illegible] *Vulkan*: **V1—V3** (1911), **V5, V6** (1913).

Displacement 564 to 570 tons normal; [illegible] tons *full load* ?. Dimensions: [illegible] × [illegible] × 9¾ feet (mean draught). Armament: 2—[illegible] inch (22 pdr.) 30 cal., 2 machine [illegible]. Torpedo tubes: 4 19·7 inch. Machinery: turbines, Parsons in Germania boats, A.E.G.-Curtis in Vulkan boats. Designed H.P. 15,000 = 32·5 kts. [illegible] made 37 kts. on trials, other boats all exceeded designed speeds. Fuel: 110 tons coal, 60 tons oil. 4 Schulz-Thornycroft boilers (1 oil burning). Complement, 73. [illegible] Post-War Fleet.

Notes.—V 5 and V 6 built to replace original V 5 and V 6 sold to Greece. Disposal of V [illegible] unknown. V 6 in Reserve.

T 186 (sister to *T 190*), at Rosyth, Sept., 1920.

Photo, A. M.

G Boats (*T 196*). Has raised fore funnel now.

1 *Krupp-Germania*, **T 196.** } launched 1910-11.
1 *Vulkan*, **T 190.**

Displacements: *G* boat 648 tons, *V* boat 656 tons normal (750 tons *full load* ?). Dimensions: [illegible] Armament: 2—3·4 inch (22 pdr.) 30 cal., 2 machine guns. Torpedo tubes: 4—19·7 inch. Machi[illegible] turbines in *G* boat, A. E. G.—Curtis in *V* boat. Designed H.P. 16,000 = 32·5 kts. Fuel: [illegible] 55 tons oil. 4 Schulz-Thornycroft boilers (1 oil burning). Complement 83 (93 War Complement).

At present *T 190* and *T 196* are quite useless, but have been selected for Post-War Fleet.

1909-10.

T 1[illegible] (sister to T 1[illegible]), at Rosyth, Sept., 1920. Photo, J. M. Dick, Esq.

T 1[illegible], at Rosyth, Sept., 1920. (T 17[illegible] is similar.) Photo, J. M. Dick, Esq.

1 Vulcan: **T 185**
1 Krupp-Germania: **T 175**
} launched 1909-10.

Displacements: V boat 637 tons, G boat 654 tons *normal* (760 tons *full load*). Dimensions: about 243 × 26 × 10 feet. Armament: 2—3.4 inch (22 pdr.), 30 cal., 4 or 2 machine guns. Torpedo tubes: 4—19.7 inch. Machinery: A. E. G.-Curtis turbines in V boat, Parsons in G boat. Designed H.P. 16,000 = 31.5 to 32.5 kts. Fuel: V boat, 150 tons coal and 40 tons oil; G boat, 190 tons coal and 50 tons oil. Complement, 83. T 1[illegible] Post-War Fleet, T 1[illegible] in Reserve.

1908-9.

[illegible] [illegible] now raised.

1 Krupp-Germania: **T 170.**
2 Schichau: **T 168, T 167.**

All launched 1908-9. Displacements: G boat, 636 tons; S boats, 615 tons (750 tons *full load*). Dimensions: about 243 × 26 × 10 feet. Armament: 2—3.4 inch (22 pdr.), 30 cal., 4 or 2 machine guns. Torpedo tubes: 3—17.7 inch. Machinery: Turbines, Parsons in G boat, Schichau (Melms & Pfenninger) in S boats. Designed H.P. 11,000 [illegible]. Fuel: [illegible] tons coal, [illegible] tons oil. Complement, 83. T 170 in Reserve.

T 1[illegible]-T 1[illegible] as above photo and plan.

No Photo available.

Two funnels between funnels, with ventilator between same. Ventilator in space between 2nd funnel and mainmast, closer to funnel than to mainmast.

T 1[illegible]—T 1[illegible].

8 Vulcan: **T 158-151,** launched 1907-8. Displacement: 5[illegible] tons *normal* 6[illegible] tons *full load*. [illegible] feet (*mean draught*). Guns: [illegible] inch [illegible]. Machinery: T 1[illegible] has A.E.G.-Curtis turbines, the remainder reciprocating engines. Designed H[illegible] Coal: 160 tons; T 1[illegible], 170 tons. Complement, 8[illegible].

OLD GERMAN DESTROYERS (*Grosse Torpedoboote*).

Disposal of all boats on this page unknown. Believed that some are still being used for Mine Clearance Duties.

1906.

T 139. *1920 Photo, Lieut. Stevens, R.D.N.*

11 Schichau: **T149–139,** (launched 1906-7.) Displacement: 530 tons. Dimensions: 231 x 25½ x 8½ feet (mean draught.) Armament: 1—3·4 inch (15 pdr.), 35 cal., 3—1 pdr., 55 cal., 2 machine. Torpedo tubes: 3—17·7 inch. Machinery, reciprocating. Designed H.P. 10000=30 kts. Coal: 190 tons. Complement, 80.

6 Krupp-Germania: **T136–132** (launched 1905-6.)

Displacement: 487 tons. Dimensions: 207 x 23 x 11 feet. Armament: 4—1 pdr., 55 cal., 2 m[illegible] (illustrated above) has special armament of 1—3·4 inch (15 pdr.), 35 cal., 2—4 pdr., 55 cal., 2 mach[illegible] 3—17·7 inch. Machinery, reciprocating. Designed H.P. 6500=27 kts. Coal: 125 tons. Comp[illegible]

Appearance Notes.—Raised forecastle very short, no gun mounted on same. Space between fo[illegible] [illegible] Steam-pipe raised above fore funnel. Funnels spaced wide apart. Mainmast close to derrick on after side. Two guns abaft mainmast.

1905-6.

1 *Krupp-Germania* special boat: **T137** (launched 1907.) Displacement: 565 tons. Dimensions: 234½ x 25 x 9⅞ feet. Armament: 1—3·4 inch (15 pdr.), 35 cal., 3—1 pdr., 55 cal., 2 machine. Torpedo tubes: 3—17·7 inch. Machinery: Parsons turbines. Designed H.P. 10000=32 kts., made 33·9 on trials. Coal: 120 tons. Complement, 80.

1904-2.

T 121 OFF THE MOUTH OF THE EMS.

10 Schichau: **T131, T130,* T128–125** (launched 1903-5.) 485 tons. **T122–120** (launched 1901.) 430 tons. **T114*** (launched 1902.) 420 tons. } *Plans as given in next page*

Armament: 3—4 pdr., 40 cal., 2 machine, 3—17·7 inch tubes. Designed H.P. 6500=27 kts. Coal: 116 tons. Complement, 60.

Appearance Notes.—Low in water. Long forecastle without any gun, long space between forecast[illegible] funnels, with raised steam pipe to fore funnel.

* *T 129, T 124, T 123, 119, 118, 117, 116* and *115* have been sunk. *T 125* used

Torpedo Boats.

Totals.	Name or Number.	Date.	Displacement.	I.H.P.	Max. Speed.	Torpedo Tubes.*	Coal.	Complement.
			tons		kts.		tons	
3	V 105—108	1914-15	320	5200*t*	32	4	..	..
1	D 10	1897-99	355	5500	28	2	80	60
1	D 9	1893-94	380	4040	24½		105	59
2	D 8, D 7	1890-92	350	4000	23		75	49
2	D 6, D 5	1888-90	320	3600	23	3	90	49
2	D 4, D 3	1887-89	300	2500	21		65	49
8	A 1—25	1915	107	1084	18	2	29	..
6	T 113—108	1901	400	5600	27			
6	T 107—102	1900	400	5400	26	3*	92-116	56
11	T 101—91	1899	400	5400	26			
2	T 88 & 89	1898	160	1800	25	2	36	24
6	T 87—82							
7	T 81—74	1889—1897	180—150	1500	25—22	3	30	24—16
5	T 73—69							
6	T 63—55							
2	T 53, T 49	1889—1897	150	1500	25—22	3	30	16
2	T 45, T 42			1350				
22	T 40—T 3	1885-7	85	1000	20	3	30	16

***Note.**—Many have been converted into Mine Sweepers and have no tubes now.

"Divisional Boats."

1 *Thornycroft*: **D 10** (1897). 355 tons. Armament: 5—4 pdr., 2—17·7 inch tubes. Speed (originally) 28 kts. Boilers: Yarrow.

"Divisional Boats"—*Continued.*

D 5, with guns and tubes removed, used for Subsidiary Service. *Photo, Renard.*

Note.—Disposal of all Divisional Boats 1920, not known.

7 *Schichau.* **D 9** (1893), **D 8, D 7** (1891), **D 6, D 5** (1888-89), **D 4, D 3** (1887). For details see Destroyers on preceding page. Armament: 5—4 pdrs., 3—17·7 inch tubes, one *submerged* in bow. Tubes probably removed from all now. Reconstructed and given two funnels 1907-09.

Note.—D 3 had tubes removed and was used for Fisheries Protection before war, being armed with [illegible]. D 5 also had tubes removed and was Tender to Submarine Depot Ships. *D 1* and *D 2* still exist, but useless.

GERMAN FIRST-CLASS TORPEDO BOATS.

1915

A 1—25 TYPE. *Photo, L. N. A.*

8 Fulton "A 1" type boats: **A 1, A 17, A 19, A 21–25.** Built about 1915. Dimensions: 120.1 × 15.1 × 4.1 feet. Guns: 1—4 pdr. (Krupp, 40 cal. "Tpbts"), 1 M.G. Tubes: 2—17.7 inch, singly mounted (removed from Sweepers). Can carry 4 mines each. Oil: 24 tons = 440 miles at 19 kts. Other details as Table.

Disposal 1920 unknown.

Note.—Boats below and in next column previously rated as Destroyers (*Grosse Torpedoboote*) selected as Torpedo Boats for Post-War Fleet.

1901.

6 Krupp-Germania: **T 113—108** (1901–2). 400 tons. Armament: 3—4 pdr., 2 machine, 3—17.7 inch tubes. H.P. 6000 = 26 kts. Coal: about 110 tons. Complement, 56. *T 108—110* and *T 113* for Post-War Fleet.

1900.

No Photo available.

Plan as given above for *T 113—108* boats, but only *one* tube midway between funnels, the third tube being *aloft mainmast.* All for Post-War Fleet.

6 *Schichau*: **T 107—102** (1900–01). As *T 113–108* above, but coal 95 tons and H.P. 5400 = 26 kts. *T 10*[illegible]

1899.

No Photo available.

See Silhouettes. Surplus to Establishment. Disposal, 1920, unknown.

9 *Schichau*: **T 100, T 98—91*** (1899–1900). 400 tons. Armament: 3—4 pdr., 2 M.G. 3- (probably removed from majority). H.P. 5400 = 26 kts. Coal: 95 tons.

No photo available.

Both for Post-War Fleet.

2 *Schichau*: **T 101, T 99.** As above boats, but tubes removed and perhaps guns. Both were before war.

* *T 91* bears name **Sleipner.**

OLD GERMAN TORPEDO BOATS (*Kleine Torpedoboote*).

Above-water tubes probably removed. Used as Mineswepers, Tenders, &c. Disposal 1920 uncertain—probably scrapped.

2 *Krupp-Germania*: ***T 89*** and ***T 88*** (1893-8). 1—4 pdr., 2—17·7 inch tubes.

6 *Schichau*: ***T 87—T 82*** (1898). 1—4 pdr., 3—17·7 inch tubes (1 bow submerged, 2 deck).

5 *Schichau*: ***T 81—74**** (1893-4). Armed as *T 87—T 82* above.
*Last *T 75* War Loss.

T 42 (1887). Appearance after being re-boilered and re-engined.

16 *Schichau*: ***T 73—69, T 63—59, T 55, T 53, T 49, T 45, T 42*** (1889-1897). 1—4 pdr., 3—13·8
(1 bow submerged, 2 deck).

2nd & 3rd Class.

Like *T 81—74* but shorter and with big funnel

22 old boats: ***T 40—33, T 31—27, T 25, T 24, T 21—20, T 16, T 15—13, T 11, T 3*** (1885-188
Speed: 17 kts. (probably much less now). Armament: 1—1 pdr., 3 torpedo tubes (2 deck and 1 submerged). Guns and tubes may have been removed from some. Complement, 16.

Appearance: Similar to *T 81—74* opposite, but much shorter in length.

Notes.—*T 36, 35, 31—27* and *25* converted to Mine sweepers, tubes being removed; some (or all) of the other also have been altered for this service.

Motor Launches (*Schnellboote*).

No photo available.

Total Number uncertain : **LM...** **LM...** 7 tons. Draught, 3.3 feet. Each 3 sets airship 240 B.H.P. motors. 720 B.H.P. = over 30 kts. speed. 1 bow *above water* 17.7 inch tube and 2 M.G.

No photo available.

49 boats : **F...** **F...** (built 1917-18). 19 tons. *About* 55 feet long. 60 B.H.P. motors = 11 kts. 1 M.G.

No photo available.

Total Number uncertain : **UZ 1—50** (built 1916-17). *About* 60 tons. Length, 85 feet. 2—300 B.H.P. motors, 600 B.H.P. Speed : 17 kts. Guns : 1—4 pdr. Endurance : 100 miles.

Note.—Of above 3 types, 111 ordered. *LM* = *Luftschiff-Motoren* (airship-motors). *UZ* = *Unterseeboote-Zerstorer* (Submarine Destroyer). All *UZ* and *F* boats fitted as Anti-Submarine craft and Minesweepers.

"Controlled Torpedoes."

("FL boote = *Fernlenkboote*.)

6 tons. Dimensions : 42.6 × 6.1 × 2 feet. 2—300 H.P. seaplane motors = 30 kts. These craft were used off the Belgian Coast, and were simply a modern development of the old Brennan and Lay torpedoes. The "controlled torpedo" is a small boat with twin petrol motors, partly covered in. The crew leave after it has been started. About 30-50 miles of single core insulated electric cable connect the boat's steering and control gears with the shore directing station. A seaplane, escorted by a strong escort of fighting 'planes, "spots" for the "controlled torpedo," sending instructions by W/T. on course, speed, etc., to the shore controlling station. About 300-500 lbs. of high-explosive are contained in the bows of the "torpedo," being exploded on impact with the target.

Submarines.

Under the Peace Treaty, Germany is absolutely prohibited from building (or maintaining in service) any Submarines, whether of Naval or Mercantile types.

All unfinished German Submarines—of which there were a considerable number at the end of 1918—have been dismantled, broken up, scuttled outside German ports or otherwise disposed of.

Mine Layers (are disarmed).

Albatross after naval action of [illegible] foremast partly shot away.
(*Was Mining School Ship before the war.*)

ALBATROSS (October, 1907). 2200 tons. Complement, 197. Length, 295 feet. Bea[...] *Maximum* draught, *about* 13 feet. Armament : 8—3.4 inch (15 pdr.), 35 cal. Boi[...] Thornycroft. Designed H.P. 6000 = 20 kts. Coal, 450 tons. **Carries 600 mines.**

Note.—Was disabled and driven ashore near Ostergarn Lighthouse, Gothland Island (Baltic) by Russian [...] 1st, 1915. Above illustration shows her being towed to internment at Visby. Released from internment in 1918.

NAUTILUS (1906). 1950 tons. Complement, 198. Length, 305 feet. Beam, 42¾ f[...] 13¾ feet. Armament : 8—3.4 inch, 15 pdr. Boilers : Schulz-Thornycroft. Machinery : [...] Designed H.P. 6000 = 20 kts. Coal, 445 tons. **Carries 200 mines.**

SMALL GERMAN MINE LAYERS AND SWEEPERS.

Disposal 1920 uncertain; some still in service, others sold out for Mercantile and Fishing services.

Appearance unknown.

F.M. 1, 2, 4—36. (Built 1917-18.) 175 tons. *About* 139½ × 19¼ × 4½ feet. Guns: 1—3·4 inch 22 pdr. I.H.P. 800 = 13·5 kts. Coal, 38 tons. Fitted to carry mines.

Notes.—Shallow draught sweepers. Proved of little worth owing to inadequate fuel supply and dangerous instability.

1919 *Illustration.*

M 28—30, M 32—35, M 37, M 38, M 42—46, M 48, M 50—54, M 57—62, M 65, M 66, M 68—82, M 84—90, M 93—94, M 96—122, M 125, M 129, M 137, M 138. (Built 1915-18.) *About* 480 tons. *About* 192 × 24 × 8½ feet. Guns: see next column. I.H.P. 1600 = 17 kts. 2 screws. Coal: 150 tons and 12 tons oil. Complement, 40. Can carry 30 mines.

War Losses: *M 31, M 36, M 39, M 40, M 41, M 47, M 49, M 55, M 56, M 63, M 64, M 67, M 83, M 91, M 92, M 95.*

Not finished:—*M 123, M 124, 126—135, 139—136.*

Similar to boats below.

M 19, M 20, M 21, M 25. (Built 1915.) *About* 400 tons. *About* 164 × 20 × 9 feet ... below. I.H.P. 1200 = 16 kts. Complement, 40. Can carry 30 mines.

War Losses: *M 22, M 23, M 24, M 26, M 27.*

1919 Ill...

M 1—5, M 7, M 8, M 10, M 13, M 17, M 18. (Built 1915.) *About* 350 tons. A... 23 × 9 feet. Guns: see below. I.H.P. 1500 = 16 kts. Complement, 40. Carry 30 min...

War Losses: *M 6, M 9, M 11, M 12, M 14, M 15, M 16.*

Armaments for all "M" Types.

(a) 3—4·1 inch, or (b) 2—4·1 inch, or (c) 1—4·1 inch, or (d) 2—3·4 inch, 22 pdr. Also 2 machine...

Completion as Surveying Ship under consideration, November, 1920.

METEOR (Danzig Yard, 1914 or 1915?). 1150 tons. Dimensions: 219¾ × 33¾ × 10½ feet. Guns: 4—4·1 inch, 2—1 pdr., 2 machine. Machinery: 2 sets triple expansion. 2 screws. Boilers: 4 Schulz-Thornycroft. Designed H.P. 1550 = 14 kts. Coal: tons. Was intended for "surveying service" on East African Coast.

DRACHE (Germania, 1908). 730 tons. Complement 64. Guns: 4—3·4 inch (22 pdr.), 4—1 pdr. Speed: 15 kts. Coal: 150 tons.

HAI (Geestemünde, 1867). 640 tons. Complement 51. Armament: 8—4 pdr. Speed: 12·5 kts. Coal: 75 tons.

DELPHIN (Meyer, Papenburg, 1906). 450 tons. Complement 41. Armament: 4—3·4 inch (22 pdr.), 2—4 pdr., 4 M.G. Speed: 9·5 kts. Coal: 55 tons.

FUCHS (Meyer, Papenburg, 1905). 640 tons. Complement 51. Armament: 2—4·1 inch, 2—3·4 inch (15 pdr.). Speed: 12·5 kts. Coal: 75 tons.

To be converted for Surveying Service, B

PANTHER (April, 1901). Displacement, 1000 ton 210·3 × 31·8 × 10·2 feet. Complement, 130. G 40 cal., 6—1 pdr. 2 Machine. Machinery: T 2 screws. Boilers: 4 Schulz-Thornycroft. De = 14 kts. Coal: 275 tons.

Photo,

BLITZ & PFEIL (1882). 1288 tons. Dimensi 13¾ feet. Complement, 134. Guns: 6—3·4 4 machine. Tubes: 2 *above water* (probably *merged*. Designed H.P. 2700 = 15 kts. Coal: 2

(*Continu*

Gunboats—(continued).

ZIETEN (1876). 1006 tons. Complement, 108. Guns: 6—4 pdr. Tubes (13·8 inch): 2 *submerged* (bow and stern). Speed: under 13 kts. Coal: 140 tons.

No Photo available.

GREIF. No details known.

No Photo available.

PRIMULA. Sea-going Parent Ship, 5th Minesweeping Flotilla.

No Photo available.

SPERBER. No details known.

Also:

Grille (1857). 350 tons. Speed: 13 kts. Guns: 2—3·4 inch.

Condor (Mine Hulk), *Schwalbe* (Target Hulk).

River Gunboats.

(*All on Vistula.*)

No Photo available.

MÖWE. Ex Shallow Draught River Steamer, armed with 1—1 pdr. and 1 M.G.

No Photo available.

LIEBE. Ex Shallow Draught Tug. Guns: 1—3·4 inch (22 pdr.) 1—1 pdr.

No Photo available.

GARDENQO. Ex Shallow Draught Tug. Guns: 1—3·4 inch (22 pdr.), 1 M.G.

Fleet Tender.

NORDSEE (Atlas Werke, Bremen, 1914). 790 ... sions: 175¾ × 20¾ × feet. Armed with 2 ... pdr.) guns. H.P. 1700 = kts. Serves as L... Sea Patrols.

Yachts (*Jachten*).

Not finished.

——— (ex *Ersatz Hohenzollern*, Vulkan, Stettin, ... tons. Complement, 455. Dimensions: 528¼ ... feet. Guns: Not known, but is capable of m... 12—5·9 inch. Machinery: Turbines. 3 scre... 10 Schulz-Thornycroft (2 oil burning). Designe... = 24 kts. Coal: 1000 tons + 500 tons oil.

Surveying Ships (*Vermessungsfahrzeu...*)

No photos available.

TRITON. No details known. Was mercantile vesse... Russians to block entrance to Libau Harbour. Germans later in the War and converted fo... Surveying Ship.

Also four Pilot Vessels in service, 1920, names unk...

Also see *Meteor* and *Panther* on preceding page.

Training Ships, 1920.

Not known.

(ROYAL) NETHERLANDS NAVY.

Revised 1920, by courtesy of the Chief of Staff, Ministry of Marine, The Hague.

(Some notes also extracted from the 1920 Edition of the Official Handbook "Koninklijke Marine.")

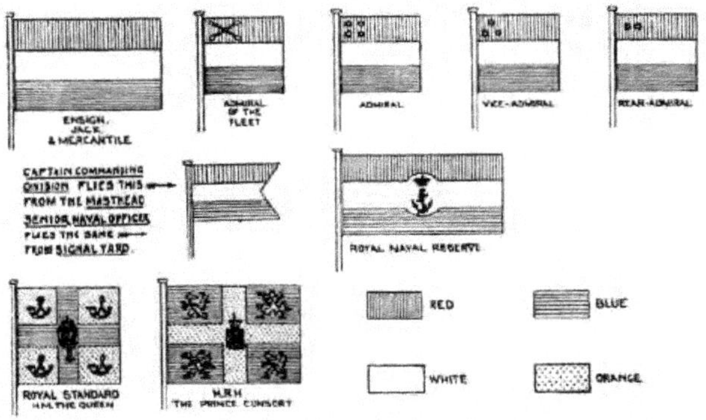

Naval Guns (Krupp Models). All details unofficial.

Position.	Calibre.		Official Mark.	Length.	Weight of Gun.	Weight A.P. shell.	Initial Velocity.	[illegible] penetration with A.P. capped [illegible] K.C.		
	in.	c/m.		cals.	tons.	lbs.	ft. secs.	[illegible] in.	[illegible] in.	[illegible] in.
HEAVY	11	28		42½	31	595	2920	8	13	15½
	9·4	24	"No. 2"	40	24½	375	2788	3¼	7½	10½
	9·4	24	"No. 1"	40	24½	375	2690	4	8½	10½
MEDIUM	5·9	15	"No. 6"	50	5½	101½	3084	—	—	—
	5·9	15		50	5¼	101½	3084	—	—	—
	5·9	15	"No. 5"	40	5	90½	2788	—	—	—
	5·9	15	"No. 4"	40	4½	90½	2444	—	—	—
	5·9	15	"No. 3"*	40	4½	100	2231	—	—	—
	4·7	12	"No. 2" / "No. 1"	40	2	52½	2230	—	—	—
LIGHT	4·1	10·5	"A"	50	—	40	2697	—	—	—

Velocities of 40 calibre guns are not service ones. In addition, there are 40 and 55 calibre [illegible] models of 7·5 c/m. (3 inch, 15 pdr.) guns, 6 9 pdr. (7·2 c/m.) Mark "A" mortar, 5 c/m. 4 pdr., 37 c/m. 1 pdr., and various old types.

*"No. 3" model of the 5·9 inch/15 cm. is the same as the "No. 2" model.

Netherlands Uniforms.

	1	2	3	4	5	6	7	8
	Admiraal	Luitenant-Admiraal	Vice-Admiraal	Schout-by-nacht	Kapitein ter Zee	Kapitein Luitenant ter Zee	Luitenant ter Zee 1e Kl.	Luitenant ter Zee 2e Kl.
Corresponding British or U.S.	*Admiral of the Fleet*	*Admiral*	*Vice-Admiral*	*Rear-Admiral*	*Captain*	*Commander*	*Lieut. Comm'r.*	*Lieut. (under 11 years)*

In relative ranks—

Doctors have insignia 4 to 8 without the curl.
Paymasters „ 5 to 9 „ in silver.
Engineers „ 5 to 8 with the curl with blue velvet between stripes.

The broad stripes in ranks 1—4 are 2 inches wide (5 c/m); 5—8, ½ inch (1 c/m); No. 9, ¼ i
Distance between stripes is ¼ inch (5 m/m).

Naval Construction.

None of the Royal Naval Yards in Holland appear to have built any warships for some
Warship construction is carried out by three firms: (1) Koninklijke Maatschappij "de Sc
Royal Company "de Schelde") at Flushing; (2) Maatschappij "Fijenoord" Rotterd
Nederlandsche Scheepsbouw Maatschappij, Amsterdam. All these firms built warships
except the last named, which does not build submarines. Werf Gusto (A. F. Smulders), Sc
the Conrad Co., Haarlem, have also built special Submarine Depot Ships for foreign Na
Dutch East Indies, some small auxiliaries have been built at Surabaya, Batavia, &c.

Notes:

For Arsenals, Docks, Slips, &c., in Holland and Dutch East Indies, a revised and ex
given on the next page but one.

All displacements are given in *metric* tons. Regarding the two sets of figures gi
plements on later pages, (H) denotes complement when serving in home or European
denotes complement when serving in Dutch East Indies or abroad.

Personnel: About 11,000 all ranks, navy and marine infantry (enlisted men).

Principal Trade Ports: Rotterdam, Amsterdam, Flushing.

Coinage: Guilder (or florin) (100 cents) = 1s. 8d. British, or 40 cents U.S.A.

Oversea Possessions: Dutch East Indies (Sumatra, Java, Borneo, etc.); Dutch Guiana (
Curaçoa.

Minister of Marine: General Pop (*ad interim*). *Naval Attaché (London)*: Captain A.

Colour of Ships: Grey. Submarines have national colours (red, white, blue) painte
stripes over bows.

Mercantile Marine.

From "Bureau Veritas" 1919-20 figures. Ships under 100 tons gross excluded

832 Steam and Motor Ships of 1,714,074 tons gross.

83 Vessels (with Auxiliary Motors) of 20,645 tons gross.

334 Sailing Vessels of 43,783 tons gross.

During 1919, Dutch Shipyards launched 100 vessels of 137,086 tons gross; some
foreign account.

Scale : 1 inch = 160 feet.

NETHERLANDS RECOGNITION SILHOUETTES.

(E. I. denotes ships now serving on Dutch East Indies

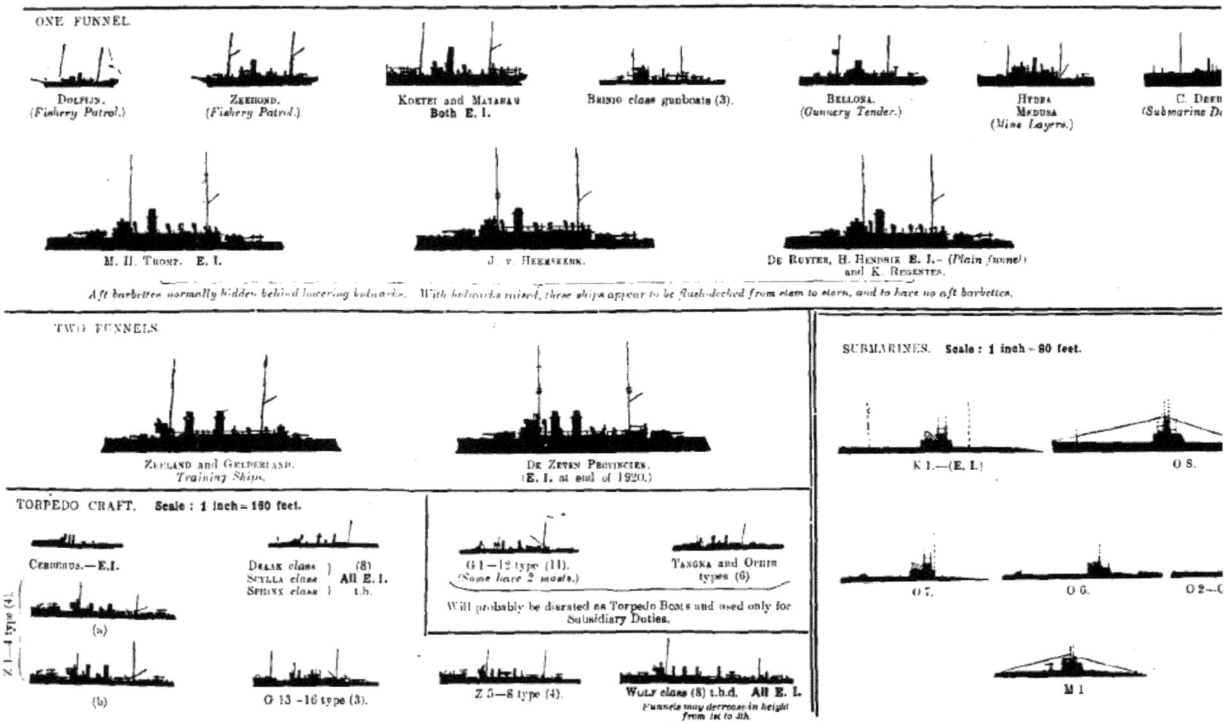

DOCKS.

Note.—All the following particulars have been communicated to "Fighting Ships" by courtesy of Koninklijke Nederlandsche Vereeniging, "Onze Vloot" (Royal Netherlands Association, "Our Fleet") during 1918.

Arsenals.

WILLEMSOORD (Nieuwediep).—No slips. Dry dock No. 1: $360\frac{3}{4} \times 68\frac{3}{4} \times 19\frac{1}{2}$ ft. No. 2: $262\frac{1}{4} \times 65\frac{1}{2} \times 15$ ft. Floating dock: $131\frac{1}{3} \times 23 \times 7\frac{3}{4}$ ft.

HELLEVOETSLUIS.—One slip. Dry dock No. 1: $239\frac{1}{4} \times 51\frac{3}{4} \times 14$ ft. No. 2: $231\frac{1}{3} \times 51\frac{3}{4} \times 7\frac{1}{3}$ ft. Can be used together as one dock, $477\frac{1}{2} \times 53\frac{3}{4} \times 14$ ft.

Private Establishments.

AMSTERDAM:—

Amsterdam Dry Dock Co.—Juliana-dock: Length, 615 ft.; breadth, $118\frac{1}{4}$ ft. For ships with a draught of 20 ft. Carrying capacity, 16,500 tons. Wilhelmina-dock: Length, $424\frac{3}{4}$ ft.; breadth, 96 ft. For ships with a draught of $18\frac{3}{4}$ ft. Carrying capacity, 7500 tons. Koninginne-dock: Length, 400 ft.; breadth, 92 ft. For ships with a draught of 17 ft. Carrying capacity, 4000 tons. Konings-dock: Length, 400 ft; breadth, 72 ft. For ships with a draught of 15 ft. Carrying capacity, 3000 tons.

Floating Dry Dock Co., Ltd.—No. 1: $213\frac{1}{2} \times 62\frac{1}{2} \times 20\frac{1}{2}$ ft. No. 2: $200 \times 62\frac{1}{2} \times 20\frac{1}{2}$ ft. No. 3: $164 \times 62\frac{1}{2} \times 18\frac{1}{2}$ ft.

MIDDELBURG:—

The Royal Co.: "De Schelde" (Koninklijke Maatschappij de Schelde).—Dry dock: $413\frac{1}{2} \times 65\frac{1}{4} \times 13$—15 feet.

ROTTERDAM:—

Municipal Floating Docks.—No. 1: *Length, $157\frac{1}{2}$ ft.; breadth, 69 ft. Carrying capacity, 2000 tons. No. 2: Length, $295\frac{1}{4}$ ft.; breadth, 69 ft. Carrying capacity, 4000 tons. For ships with a draught of $18\frac{3}{4}$ ft. No. 3: Length, 361 ft.; breadth, 69 ft. Carrying capacity, 6000 tons. No. 4: Length, $567\frac{1}{2}$ ft.; breadth, 95 ft. Carrying capacity, 15,000 tons.

* Dock No. 1 can be coupled to dock No. 2.

Ships of $229\frac{1}{2}$ ft. length can be taken into dock No 1, ships of $377\frac{1}{4}$ ft. length into dock No. 2, ships of 459 ft. length into dock No. 3, and ships of 656 ft. length into dock No. 4.

Rotterdam Dry Dock Co.—Prins Hendrikdock No. 1: Length, $439\frac{3}{4}$ ft.; breadth, 95 ft. For ships with a draught of $19\frac{3}{4}$ ft. Carrying capacity, 7500 tons. Prins Hendrikdock No. 2: Length, 300 ft.; breadth, $70\frac{1}{2}$ ft. For ships with a draught of $18\frac{1}{2}$ ft. Carrying capacity, 3000 tons.

ROTTERDAM (*continued*):—

Shipbuilding Co.: "New Waterway."—Dock-harbour: $1312 \times 656 \times 20\frac{1}{2}$ ft. Slips—No 1: Length, $406\frac{1}{4}$ ft.; breadth, $19\frac{3}{4}$ ft. For ships up to 8000 tons. No 2: Length, 400 ft.; breadth, $19\frac{3}{4}$ ft. For ships up to 8000 tons. No. 3: Length, $406\frac{1}{2}$ ft.; breadth, 23 ft. For ships up to 12,000 tons. No. 4: Length, 400 ft.; breadth, 23 ft. For ships up to 12,000 tons. No. 5: Length, 886 ft.; breadth, $39\frac{1}{2}$ ft. For ships up to 18,000 tons. Also 3 floating docks: (*a*) 500 × 105 ft. (10,000 tons), (*b* and *c*) 4,200 tons capacity.

Wilton's Engine Works & Shipbuilding Yard, Ltd.—Docks—No. 1: Length, $426\frac{1}{2}$ ft.: breadth, 70 ft. For ships up to a length of 492 ft. Carrying capacity, 8500 tons. No. 2: Length, $307\frac{1}{2}$ ft.; breadth, 58 ft. For ships up to a length of $426\frac{1}{2}$ ft. Carrying capacity, 4500 tons. Side slip: $347\frac{3}{4}$ ft. long, for ships to 3000 tons. Wilhelmina-dock: Length, 510 ft.; breadth, 81 ft. For ships up to a length of 574 ft. Carrying capacity, 14,000 tons. Also floating dock, 600 × 107 ft. (46,000 tons).

VLISSINGEN (Flushing):—

The Royal Co.: "De Schelde" (K. M. de Schelde.)—Dry dock: $243 \times 52\frac{1}{2} \times 11$ ft.

DUTCH EAST INDIES.

(*Details revised 1920.*)

Arsenals.

SOERABAIA or SURABAYA (Java):—

Three iron floating docks: (*a*), $196\frac{3}{4} \times 57 \times 17\frac{1}{2}$ ft (1400 tons); (*b*), $328 \times 62 \times 22\frac{3}{4}$ ft. (5000 tons); (*c*), $459\frac{3}{4} \times 82 \times 24\frac{1}{2}$ ft. Last named is chartered by Soerabaia Dry Dock Co.

Private Establishments.

BATAVIA (Java):—

Tanjong Priok Dry Dock Co.—Floating dock: 324 × 67 × 22 ft.

SABANG (Sumatra):—

Sabang Bay Harbour & Coal Co.—Floating dock: 296 × 69 ft. (3000 tons), for ships up to 300 ft. long and 16 ft. draft.

SOERABAIA or SURABAYA (Java):—

Soerabaia Dry Dock Co.—Floating dock: $350 \times 58 \times 19\frac{1}{2}$ ft. For chartered dock, *v.* Surabaya Arsenal details above.

1918 *Photo, Fotopersbureau "Holland,*

Will probably proceed to East Indies Station at end of 1920.

DE ZEVEN PROVINCIEN (March, 1909).

Displacement, 6530 tons. Complement, 447 (A), 409 (H).

Length (*o.a.*), 333 feet. Beam, 56·1 feet. *Maximum draught*, 20·2 feet.

Guns (Krupp):
- 2—11 inch, $42\frac{1}{2}$ cal.
- 4—5·9 inch, 40 cal. "No. 5."
- 10—13 pdrs., 55 cal. S.A. "No. 1"
- 1—9 pdr. mortar "A."
- 4—1 pdr.
- 2 machine.

Torpedo tubes:
None.

Armour (Krupp):
- 6" Belt (amidshi
- 4" Belt (ends)
- 2" Deck (on slop
- $9\frac{3}{4}$" Main Barbet
- $9\frac{1}{2}$" Shields ...
- 4" Small Barbet
- $7\frac{1}{4}$" Conning To

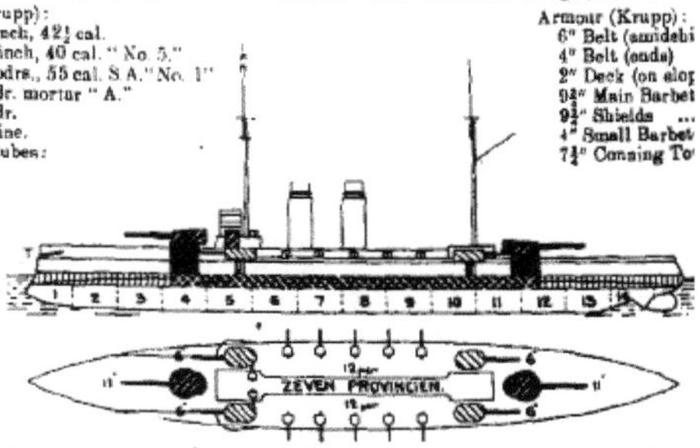

Machinery: 2 sets triple expansion. 2 screws. Boilers: 8 Yarrow. Designed 16 kts. (Trials: 8516 = 16·3 kts.) Coal: *normal*, 700 tons; *maximum*, 1030 tons. Rad 2100 miles at 15·3 kts.; 5100 miles at 8 kts.

Notes.—Main belt is 7 feet wide. Laid down Feb., 1908, at Amsterdam D.Y.

Note.—Aft barbette is hidden behind lowering bulwarks.

1918 *Photo, Fotopersbureau "Holland," Amsterdam.*

JACOB VAN HEEMSKERCK (September, 1906).

Displacement, 5000 tons. Complement, 351.

Length (*o.a.*), 321½ feet. Beam, 49·9 feet; Beam outside sponsons, 55·4 feet. *Max.* draught, 18¾ feet.

Guns (Krupp):
- 2—9·4 in., 45 cal. "No. 2."
- 6—5·9 in., 40 cal. "No. 5."
- 6—13 pdrs. "No. 3."
- 1—9 pdr. mortar. "A."
- 4—1 pdr.
- 2 machine.

Torpedo tubes (18 inch):
- 1 *submerged* (bow).
- 1 *submerged* (stern).

Armour (Krupp):
- 6" Belt (amidships) ...
- 4" Belt (ends) ....
- 2" Deck (reinforcing belt)
- 7¾" Main Barbettes ...
- 4" Hoods to these ...
- 6" Small Barbettes ...
- 7¾" Conning tower (K.C.)

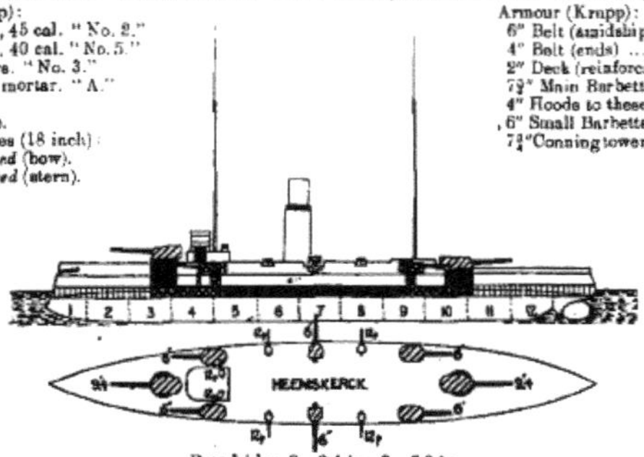

Ahead:
1—9·4 in.
2—5·9 in.

Astern:
1—9·4 in.
2—5·9 in.

Broadside: 2—9·4 in., 3—5·9 in.

Machinery: 2 sets triple expansion. 2 screws. Boilers: 6 Yarrow. Designed H.P. 6400=16·5 kts. Coal: *normal* — tons; *maximum* 610 tons.

Notes.—Main belt is 6 feet wide and about 220 feet long. Begun at Amsterdam D.Y., 1905, and completed, May, 1908. *Trials*: I.H.P. 6306=16·54 kts. Radius of action: 3300 miles at 10 kts.

Note.—Aft barbette is hidden behind lowering bulwarks.

Photo,

On East Indies Station.

MARTEN HARPERTZOON TROMP (1904).

Displacement, 5300 tons. Complement, 349 (H).

Length (*o.a.*), 330·7 feet. Beam, 49·8 feet. *Maximum* draught, 18·7 feet.

Guns—(M. '01):
- 2—9·4 inch, 40 cal. "No. 2."
- 4—5·9 inch, 40 cal. "No. 5."
- 8—13 pdr. "No. 3."
- 1—9 pdr. mortar "A."
- 4—1 pdr.
- 2 machine.

Torpedo tubes (18 in.):
- 3 *above water*.

Armour (Krupp):
- 6" Belt (amidships)
- 4" Belt (ends) ......
- 2" Deck (reinforcing
- 7¾" Barbettes (N.C.)
- 4" Hoods to these
- 3" Small turrets ...
- 7¾" Conning tower

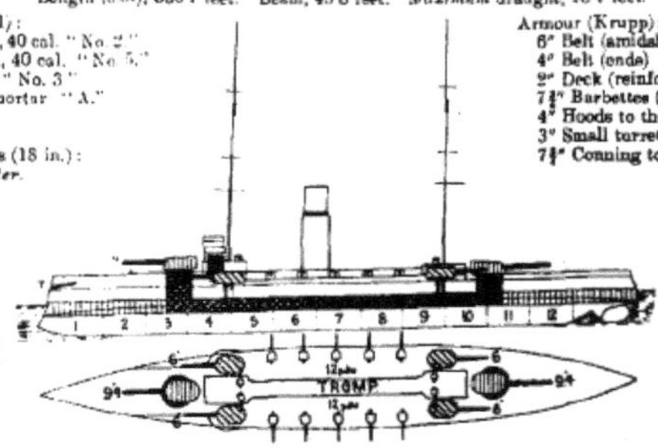

Ahead:
1—9·4 in.
2—5·9 in.

Machinery: 2 sets triple expansion. 2 screws. Boilers: 6 Yarrow. Designed H.P
16 kts. Coal: *normal* 680 tons; *maximum* 830 tons. Radius of action: 1600 miles at 14½ k
miles at 9¼ kts. *Trials* ('06).—6,405 I.H.P.=16·7 kts. Begun at Amsterdam D.Y., 1903, a
pleted 1906.

OLD BATTLESHIPS. (16 knot).—*Pantserschepen.*

191? *Photo, Fotopersbureau, "Holland."*

"H. Hendrik" on East Indies Station.

KONINGIN REGENTES (1900), DE RUYTER (1901), & HERTOG HENDRIK (1902).

Length (o.a.), 310·9 feet. Beam, 49·8 feet. *Maximum* draught, 19 feet.

Displacement, 5080 tons. Complement, 375 (A), 347 (H).

Guns—(M. '99):
- 2—9·4 inch, 40 cal. "No. 1."
- 6—5·9 inch, 40 cal.*
- 4—13 pdr. "No. 3."
- 1—9 pdr. mortar. "A."
- 2—1 pdr.
- 2—1 pdr. (anti-aircraft).
- 2 machine.

Torpedo tubes (18 in.):
- 2 *above water.*†
- 1 *submerged* (bow).†

Armour (Krupp):
- 6" Belt (amidships) ......
- 4" Belt (ends) ....
- 2" Deck (reinforcing belt)
- 10" Barbettes ....
- 4" Hoods to these ....
- 9½" Conning tower (fore)

**Gunnery Notes.*—The 5·9 inch are "No. 2" type in *K. Regentes* and "No. 1" type in other two ships. In 1919-20, 2—5·9 inch were added to each ship and 4—13 pdr. removed from each.

†*Torpedo Notes.*—In *H. Hendrik* all three tubes are *above water.*

Armour Notes.—The belt is [illegible] feet wide and four feet of it is below the waterline.

Machinery: 2 sets triple expansion. 2 screws. Boilers: 5 Yarrow. Designed H.P. 6300 (6400 in *K. R.*) = 16 kts. Coal: *normal* 680 tons; *maximum* 830 tons. Same radius of action as *M. H. Tromp.* *K. Regentes* begun 1898 and *H. Hendrik* 1900, both at Amsterdam D.Y.; *De Ruyter* begun 1900, by Fijenoord Co., Rotterdam.

COLONIAL CRUISERS. *(Verkenningskruisers.)*

JAVA (building) & *SUMATRA (building).*

Normal displacement, 7050 tons. Complement, 474 (b), 482 (a).

Length (o.a.), 509·5 feet. Beam, 52½ feet. *Max.* draught, 18 feet.

Guns (Krupp):
- 10—5·9 inch (50 cal.) "No. 6."
- 4—13 pdr. S.A. (anti-aircraft) "No. 4."
- 4 machine.

Torpedo tubes:

Armour (Krupp
- 3" Belt ......
- " Deck ......
- " Gun shield
- " Funnel bas
- " Conning to

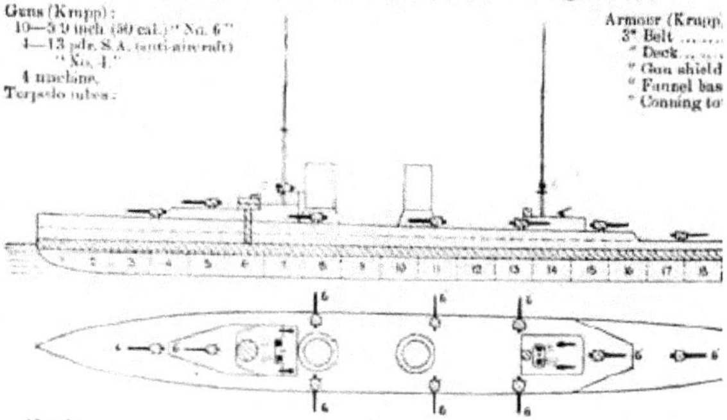

Ahead:
4—5·9 in.

Broadside: 7—5·9 in.

Machinery: 3 sets Krupp-Germania turbines. 3 screws. Boilers: 8 (*type unknown*). H.P., 65,000 = 30 kts. Fuel oil only: *normal*, tons; *maximum*, tons. Radius of ac miles at 12 kts.

Name	*Builder*	*Machinery*	*Laid down*	*Completed*	*Trials.*	*Boilers*
Sumatra	Nederlandsche Scheepsbouw Maatschappij, Amsterdam	Krupp, Germania	July '16	*To be* 1925†		
Java	K. M. de Schelde, Flushing	Krupp, Germania	May '16*	1925†		

*Contracted for, June, 1917. † Building proceeding very slowly. These are only provisional d

General Notes.—These cruisers are being built in Dutch Yards from plans supplied by Krupp. The Ger supplies materials, technical supervision, and a certain amount of skilled labour. The shortage metals and alloys in Germany has resulted in very slow construction of these two ships up to th Date of completion was first intended to be in 1918. Belong to 1915-16 Naval Programme. A t *Celebes*, was provided by 1917-18 Programme, and her construction was contracted for by Fijen Rotterdam. She was to have been of the same design as above two ships, but (*a*) *Celebes* length 519·4 feet; and (*b*) displacement 7205 tons. In 1919 it was decided that work on *Celebes* should be *Celebes* still appears in the official list, but with the foot note "Work stopped (Bouw stopgezet)." S be resumed later on.

Old Protected Cruisers. *(Pantserdekschepen.)*

GELDERLAND. 1918 *Photo, Fotopersbureau "Holland," Amsterdam.*

(HOLLAND CLASS.)

GELDERLAND (1898). Displacement, 4030 tons. Complement, 381 (A), 325 (H). Length, (o.a.) 310·7 feet. Beam, 48¼ feet. *Maximum* draught, 17¾ feet.

also

ZEELAND (1897). Displacement 3900 tons. Complement 381 (A), 325 (H). Length, (o.a.) 306·1 feet. Beam, 48¼ feet. *Maximum* draught, 17¾ feet.

Armour (Harvey and steel): 2" Deck (steel), 4" Engine hatches, 6" Gun shields, 4" Conning tower.

Guns: Re-armed during 1919-20 as follows:—

10—4·7 inch "No. 2" (and/or "No. 1.")
2—13 pdr. S.A. "No. 2."
1—9 pdr. mortar "A."
4—1 pdr.
2 machine.

Torpedo tubes:
1 *submerged* in *Gelderland* only.

Note.

Zeeland has 2—5.9 inch "No. 3" and only 8—4·7 inch "No. 1" at present, but the two 5·9's are to be transferred to *K. Regentes* and 2 more 4·7 inch ("No. 2" or "No. 1") shipped in their place. In *Gelderland* the 4·7's are all "No. 2."

Machinery: 2 sets 3 cylinder triple expansion. 2 screws. Boilers: *Gelderland*, 12 Yarrow; *Zeeland*, 8 Yarrow and 2 cylindrical. Designed H.P. (*n.d.*) 8400 = 18·5 kts. (*f.d.*) about 10,000 = 20 kts. Coal: *normal* 470 tons, *maximum* 930 tons in *Gelderland*; 1000 tons in *Zeeland*. Radius of action: 2100 miles at 16 kts., 4500 miles at 9·7 kts.

Name	*Builder*	*Machinery*	*Laid down*	*Completed*	*First Trials*	*Boilers*	*Best recent speed*
Zeeland	Flushing	De Schelde	'95	'98	9818 = 19·5	8 Yarrow, + 2 cyl.	
Gelderland	Rotterdam	Fijenoord	Feb.'97	'00	9867 = 20	12 Yarrow	

Notes.—Both Training Ships; *Gelderland* for gunnery and especially for fire-control exercises.

8 Destroyers *(Torpedobootjagers).*

No.	Type	Date	Displacement	H.P.	Max. Speed	Coal	Complement	T. tubes	I Dr [illegible]
2 2 4	*Panter* (T) *Lynx* (Y) *Fret* (Y)	'12–'14 '12–'13 '09–'12	tons 510	8000–8500	30	tons 120*	64†	2	[illegible]

Y = Yarrow design. * + 12½ tons oil in 4 later boats. † 81 when on Foreign Service.

*HERMELIJN. 1918 *Photo, Fotopersbureau "Holland," A*[illegible]

(ALL ON EAST INDIES STATION.)

8 *Yarrow type*: **Fret** (1910), **Wolf** (1910), **Jakhals** (1911), **Bulhond** (1911), **Hermelijn** (1913), **Ly**[illegible] all built by K. M. de Schelde, Flushing. **Panter** (1913), **Vos** (1913), both built by Fijenoord Co., R[illegible] 510 tons. Dimensions: 230 (o.a.) × 22 × 6¼ feet *normal* draught, *max.* draught 9¼ feet. Armament: 4—[illegible] cal., semi-auto.) "No. 2," (in *Vos* and *Panter* "No. 1") 1 machine, 2—18 inch tubes. Designed H.P. 8000 [illegible] 30 kts. (30 to 30½ on trials). 4 Yarrow boilers. Radius of action: for first four boats named, 2360 miles and 670 miles at 29 kts. Last four carry an extra 12½ tons oil fuel, and can steam 2700 miles at 8½ kts. and at 30 kts.

*Appearance Note.—In *Bulhond*, the four funnels are of varying heights, the first funnel being highest and [illegible] shortest. Other boats of this class may have the same type of funnels.

37* Torpedo Boats (*Torpedobooten*).

*The 11 *G* Class boats and 6 *Ophir* Class boats will probably be discrated as Torpedo boats in 1920. If this course is taken, then total will be reduced to 20 boats.

No.	Type		Date	Displacement	H.P.	Max. Speed	Fuel	Complement	T. tubes	Max. draught
	25 Seagoing class—			tons			tons			feet
4	Z 8—Z 5		'14–'16	310	5600	27	47	39	4	[illegible]
4	Z 4–Z 1		'15–'19	322	5500(*t*)	27	81	39	4	
3	G 16–G 13		'12–'14	180	2600	25	44	27	3	8
3 / 4 / 4	G 12–G 1		'05–'09 / '05–'08 / '03–'06	140	1900	25	30	25	3	7
3 / 2	*Ophir*		'02–'04 / '00–'02	144	1900	25	40	25	3	7
	8 1st class—									
3 / 3 / 2	*Draak* / *Sphinx* / *Scylla*	All E.I.M.	'05–'07 / '01–'04 / '00–'01	103	1200	24	21	11+6*	3†	6
1	*Cerberus*		'86–'88	81	800	21	12	11+6*	3	7

All Yarrow types. (*t*)=Turbines.

E.I.M.—Belong to East Indies Marine.

*Complements of these boats 11 Europeans 11 East Indians. †3 boats of *Draak* class have only 2 tubes.

1918 *Photo, Fotopersbureau "Holland," Amsterdam.*

Z 8.

4 310 *ton* boats, ***Z 8, Z 7*** (K. M. de Schelde, Flushing, 1915), ***Z 6, Z 5*** (Fijenoord Co., Rotterdam, 1915). 310 tons. Dimensions: 193 × 19¼ × 5½ feet. Guns: 2—13 pdr., "No. 1," 2 machine. Torpedo tubes: 4—17·7 inch (one twin deck mounting + 2 single ditto). Designed H.P. 5700 to 5500=27 kts. (made about 27·4 on trials). Machinery: Reciprocating. Boilers fitted to burn coal or oil. Carry 45 tons coals+7 tons oil. Radius of action: 425 miles at 20 kts. Completed 1916.

1920 *Photo, Lieut. H. Steen Ste*

Z4 (outer boat) with *high* funnels, *short* foremast, *high* mainmast. Z1 (inner boat) with *short* funnels, *short* mainmast.

4 322 *ton* boats, ***Z 4—Z 1*** (Ned. Scheepsb. Mij., Amsterdam, 1916-18.). 322 tons. Dimensions: 2 Guns: 2—13 pdr., "No. 1," 2 machine. Torpedo tubes: 4—17·7 inch (one twin deck mou ditto). Designed S.H.P. 5500=27 kts. Curtis type turbines. Boilers fitted to burn coal or oil coal+50 tons oil. Radius of action: —— miles at — kts.

Note.—The original boats, *Z 4—Z 1*, building by the German Vulkan Co. in 1914, were appropriated Navy, became the German Destroyers *V105—108*. German *V 108* War Loss, while *V105–106* between Poland and Brazil 1921. These four "replace" boats were begun by the Nederlandse Maatschappij, Amsterdam, in 1914, but owing to lack of materials they were not finished until *Z 4—Z 1* boats listed above.

G13.

1919 *Photo, Fotopersbureau "Hollan*

3 180 *ton* boats ('12-'14): ***G 16, G 15*** (Fijenoord Co., Rotterdam, 1911), ***G 13*** (K. M. de Schelde, Flu tons. Dimensions: 162¼ × 17 × 4·6 feet. Armament: 2—13 pdr. "No. 1," 3—17·7 inch deck tubes. 2600=25 kts (26 kts. on trials). Coal: 44 tons.

Note to all Boats in this column.—It was originally intended that all Torpedo boats listed below should be scrapped in 1920. Later, it was proposed that they should be disrated as Torpedo boats and applied to subsidiary duties. No definite decision as to their future disposal had been arrived at up to July, 1920.

G 7 (2 masts). *Photo, Fotopersbureau "Holland," Amsterdam* G6 (1 mast).

3 ('07-'09 boats):—***G 9, 10 & 12.***
4 ('05-'08 boats):—***G 8—5.***
4 ('05-'06 boats):—***G 4—1.***

Armament: 2—1 pdr., 3—17.7 inch deck tubes (or 1 bow submerged, 2 deck tubes). Radius: 1250 miles at 8 kts.

Note.—Names of ***G8—G1*** are *Cornelis Janszen de Haan* (*G 8*), *Jacob Cleydijck* (*G 7*), *Pieter Constant* (*G 6*), *Roemer Vlaeq* (*G 5*), *Willem Willemsze* (*G 3*), *Meijndert Jentjes* (*G 3*), *Jan Danielzoon van de Rijn* (*G 2*), and *Johan van Brakel* (*G 1*). They still bear these names, *G 8—G 1* only being their identification marks. *G 11* rammed and sunk off Vlieland, March 1918. *G 2* mined Feb., 1918, salved in sections and rebuilt.

3 ('02-'04 boats) named:—***Tangka, Smeroe, Wajang.***
3 ('00-'02 boats) named:—***Ophir, Pangrango, Rindjani.***

Armament: 2—1 pdr., 3—17.7 inch deck tubes. Radius: 1650 miles at 8 kts.

All Torpedo Boats listed in this column belong to East Indies Marine.

3 *Draak* class:—***Draak, Krokodil, Zeeslang.*** (1906-7).
3 *Sphinx* class:—***Minotaurus, Python, Sphinx.*** (1902-3).
2 *Hydra* class:—***Hydra, Scylla.*** (1900).

Armament: 2—1 pdr., 3—17.7 inch (3 *Draak* Class, 2—17.7 inch (1 submerged, 1 deck tube.)

(*Nominal* radius of all these boats about 1500 at 8½ kts.)

1 *Ardjoeno* class: ***Cerberus*** (1888), 84 tons. Armament: 2—1 pdr., 2—17.7 inch tubes, viz.—1 deck + bow.

In June, 1920, construction was thus:—

(*a*) **13** boats completed: *O 2—8, K I—V, M I.*
(*b*) 7 boats building or completing: *K VI—X.*
(*c*) 9 boats projected: O 9—14, *K XI—XIII.*

Tabulated Details.

Owing to shortage of materials caused by the war, the completion and commencement of many submarines has been (and will be) much delayed. The following details are not derived from any official source, but are based on very reliable details, specially furnished to "Fighting Ships."

No.	Type	Date	Displacement tons.	H.P.	Max. speed kts.	*Extreme endurance.*	Complement	T. tubes	Max. draug't feet
3	*O 14—O 12*	*Pro.*	...	...	...	...	...	...	...
3	*K XIII—K XI*								
3	*O 11—O 9*	*Pro.*	...	...	...	...	...	...	...
1	*M I* ( )	1915†	168 192	80 175	6 5	900 miles at 7 kts. 30 miles at 2½ kts.	16	6*	9
1	*O 8* (H)	1915†	411 510	480 280	12 10	1250 miles at 12 kts.	29	4	12½
3	*K X—K VIII* (H)	'17-?	520 836	1300 630	16 8	...		6	...
2	*K IV—K III* (H)	'15-'20							
4	*K VII—K V* (HD)	'15-'20	520 800	1200 630 1740 450	14 9 16 9	3500 miles at 11 kts. 25 miles at 8 kts.	...	6	11¼
	K II (HD)	'13-'20							
1	*O 7* (HD)	'13-'16	176 230	360 185	11 10	730 miles at 10 kts. 21 miles at 8½ kts.	12	3	9¼
1	*O 6* (H)	'13-'16	170 208	360 185	11 10	750 miles at 10 kts. 25 miles at 8½ kts.	12	3	9¼
1	*K I* (HW)	'11-'14	330 390	900 650	17 9	2950 miles at 11 kts. 22 miles at 9 kts.	17	3	9¼
4	*O 5—O 2* (HW)	'10-'13	131 150	350 200	11 8	760 miles at 10 kts. 28 miles at [illegible] kts.	10	2	9

* Mine-laying tubes. † Taken over in 1917.

Notes on above Table.

All boats with letter *K* and Roman numerals are for Colonial Service in East Indies. (H) submarines are "Holland" **type, and all boats of this design** are built by the Koninklijke Maatschappij de Schelde, to plans by Electric Boat Co., **U.S.A. (HD) submarines are of the** Hay Denny type, all now being built by the Maatschappij Fijenoord, Rotterdam, **to plans by Marley F. Hay and** Messrs. Denny & Co., Dumbarton. (HW) submarines are of the Hay-Whitehead **type, built by the K. M. de Schelde,** to plans by Marley F. Hay, and Messrs. Whitehead & Co., Fiume; no further **boats of this type are now being** built.

Projected.

3 boats, type not yet settled: *O 14, O 13* and *O 12.* Provided by 1918 Naval Programme, but not yet ordered or begun.

Projected.

3 boats *type not yet settled: K XIII, K XII* and *K XI.* **Provided for by 1918 Naval Programme.**

Projected.

3 boats, *type not yet settled: O 11, O 10* and *O 9.* **Provided for by 1917 Naval Programme. No begun.**

M 1. Appearance generally as above.

1——-type: ***M 1*** (ex-German mine-laying submarine *UC 8*, Vulkan, Hamburg, 1915, strand Island, Nov., 1915, salved, interned and purchased by Dutch Government, 1917). Displa *surface,* 192·5 tons *submerged.* Dimensions: 111½ × 10¼ × 10 feet. Guns: 1 machine. No torpe 12 mines in 6 mine-laying tubes before conning tower. Machinery: 1 4 cyl. Daimler oil engi for short periods with electric motors), on *surface: max. submerged* speed: 5½ kts. (1 hour). E at 5 kts. on *surface;* 50 miles at 2½ kts. *submerged.*

(*Continu*

O 8. *1918 Photo, Fotopersbureau "Holland."*

4 *Holland* type: **O 8** (ex-British submarine *H 6*), built by Canadian Vickers Co., Montreal, 1915, wrecked on Schiermonnikoog Island, Jan., 1916, salved Feb. 1916, interned at Nieuwediep, and subsequently purchased by Dutch Government. Displacement: 440 tons on surface, about 510 tons submerged. Dimensions: 150¼ × 15¾ × 12⅓ feet. Armament: 4—18 inch torpedo tubes. Machinery: 2 sets 6 cyl. Nlseco Diesel engines = 12 kts. on surface. *Max. submerged* speed 10 kts. Endurance: 1350 miles at 12 kts. on surface (16 tons oil carried).

Building.

3 *Holland* type: *K X*, *K IX* and *K VIII*, building by K. M. de Schelde. No details known, but are practically the same as *K IV* and *K III*, described below. Provided for by the 1916 Naval Programme.

K 3 on trials. *1920 Photo, by courtesy of M. F. Hay, Esq.*

2 *Holland* type: **K 4** (1919) and **K 3** (1919). Both by K. M. de Schelde, Flushing. Displacement: about 560 tons *on surface*, 825–836 tons *submerged*. Dimensions: 211¼ × 18¼ × 11½ feet. Armament: 1 gun (calibre unknown, but probably a 12 pdr.) and 8—17·7 inch torpedo tubes. Have Diesel engines. Designed surface speed 17·5 kts., but *K III* on trials only attained 16·6 kts.

Photo wanted.

K II, K V Completed.

K VI, K VII Building or Completing.

4 *Hay-Denny* type: *K VII*, *K VI*, **K V** (launched ——), and **K 2** (launched 1917.) All by M. Fijenoord. Displacements: 550 tons *on surface*, 800 tons *submerged*. Dimensions: 177 × 16¼ × feet. A 13 pdr. (anti-aircraft) and 6—17·7 inch torpedo tubes, of which two are broadside tubes. Ma 8-cyl. Diesel engines = 16 kts. *on surface* in *K VI* and 14·5 kts. in *K V*—*K VII*. *Max. submerged* Endurance: 3500 miles (*normal*), 5500 miles (*extreme*), at 11 kts. *on surface*; 3 hours continuous kts. *submerged*.

K VII, *K VI* and *K V* are an improved *K II* type; in these three boats B.H.P. of Diesel engines and surface speed have been reduced.

1918 Photo, Fotopersbureau "I

1 *Hay-Denny* type: **O 7** (M. Fijenoord, Rotterdam, 1916). Displacements: 176 tons *on surface*, 220 to Dimensions: 105 × 12 × 9½ feet. 3—17·7 inch torpedo tubes (2 bow, 1 stern). Machinery: 1 set engines = 11½ kts. *on surface*. *Max. submerged* speed 10 kts. Endurance: 750 miles at 10 kts 3 hours continuous running *submerged* at 8½ kts.

1918 Photo, Fotopersbureau "H

1 *Holland* type: **O 6** (K. M. de Schelde, Flushing, 1916). Displacements: about 170 tons *on surface* *submerged*. Dimensions: 112¼ × 12 × 9½ feet. 3—17·7 inch torpedo tubes. 1 set of 6-cyl. Diesel e kts. *on surface*; max. *submerged* speed, 10 kts. Endurance as *O 7* above.

(Continued on nex

1918 *Photo by courtesy of Marley F. Hay, Esq.*

1 *Hay-Whitehead type*: **K I** (K. M. de Schelde, Flushing, 1913). Displacements: 336 tons *on surface*, 390 tons *submerged*. Dimensions: 148·3 × 14·2 × 9½ feet. 3—17·7 inch torpedo tubes (2 bow, 1 stern). Machinery: 2 sets 8-cyl. Diesel engines = 17 kts. *on surface*. *Max. submerged* speed 9 kts. Endurance: 1700 miles (*normal*), 2250 miles (*extreme*), at 11 kts. *on surface*; 3½ hours continuous running at 8·9 kts. *submerged*.

4 *Hay-Whitehead type*: **O 2** (1911), **O 3** (1912), **O 4** and **O 5** (both 1913). All built by K. M. de Sch… Displacements: 133 tons *on surface*, 150 tons *submerged*. Dimensions: 88·4 × 9½ × 9 feet. Arm… inch tubes in … torpedo carried for each. Machinery: 1 set 6-cyl. Diesel engines = 11 kts. on … *submerged* speed 8½ kts. Endurance: 700 miles at 10 kts. *on surface*; 3 hours continuous run… *submerged*.

NETHERLANDS "ARMOURED" GUNBOATS. (*Pantserbooten.*)

(*All three photographs by Fotopersbureau "Holland," Amsterdam, 1918.*)

BRINIO. Ventilators used as stanchions to forebridge. After bridge stands abaft mainmast. Small uncapped funnel.

FRISO. Thin and tall funnel.

GRUNO. Capped funnel and ensign gaff to mainmast.

BRINIO (1912), **FRISO** (1912), **GRUNO** (1913). All built by Amsterdam D.Y. 540 tons. Complement 52. Dimensions: 172·2 (*o.a.*) × 27·9 × 9·1 feet *max.* draught. Armament: 4—4·1 inch semi-automatic, 2 machine. Armour: 2" Belt, ¾" Deck, 2" Conning tower. Machinery: 2 sets of Diesel engines (Kromhout, M.A.N., or Werkspoor types). B.H.P. 1200 = 14 kts. Oil, 25 tons. Replacement of Diesel Motors by Steam Engines reported under consideration 1920.

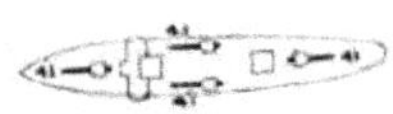

Gunboats (*Kanonneerbooten.*)

(East Indies Marine.)

KOETEI (1898), 790 tons; **MATARAM** (1896), 810 tons. About 173·9 × 31·2 × 12 feet *average*. Armament: 2—4·1 inch, 1—9 pdr. mortar, 4—1 pdr. H.P. 1350—1400=13½ kts. Coal: 140—150 tons. Complement, 100 (55 Europeans, 45 natives). Are officially rated as "Flottielje-vaartuigen" or "Flotilla Vessels."

No photo available.

BRAGA, HEFRING, NJORD, TYR. Are old iron twin-screw Gunboats, built 1876-79. 280 tons (*Hefring* and *Njord* = 270 tons). Complements, 34-30. Dimensions: 91·5 to 91·8 (*w.l.*) × 26·9 × 7·8 to 8·1 feet (*max.* draught). Guns: In *Braga*, *Hefring*, 1—4·7 inch "No. 3" or "No. 1," 3—1 pdr., 1 machine; in *Njord* and *Tyr*, 1—5·9 inch "No. 2," 3—1 pdr. H.P. 100 to 170 = 7½ to 8½ kts. Coal: 30 tons.

Mine Layers (*Mijnenleggers.*)

Building.

VAN MEERLANT, DOUWE AUKES, first begun under 1917 Naval Programme and the second begun under 1918 Naval Programme. Both laid down by Werf Gusto, Schiedam, 1918. 748 tons. Dimensions: 179·8 × 28·5 × 10·4 feet. Guns: 3—13 pdr. semi-auto. "No. 4." Twin screws. No other details known.

Will be of the same type as *Hydra* and *Medusa* (see below), but will be faster and carry more mines.

1918 Photo, Fotopersbureau "Holland."

HYDRA (1911) & **MEDUSA** (1918). Both built at Amsterdam D.Y. 670 tons. Complement 53. Dimensions: 163 × 29·5 × 9 feet. Guns: 3—13 pdr. S.A. "No. 2," 1—1 pdr., 1 M.G. Designed H.P. 800 = 11·5 kts. Coal: 72 tons. Radius of action: 1440 miles at 6 kts. Each carries double laying gears for 80 mines.

Note.—There is also a T.B. named *Hydra*, but she is on the East Indies Station.

Mine Layers—*continued.*

Photo wanted.

(East Indies Marine.)

HERKULES (Surabaya D.Y., 1909). Tug of 254 tons ... ment. Dimensions: 102·8 × 13·4 × 8·4 feet. C... listed. I.H.P. 400 = 10·2 kts. 1 screw. Oil fuel ... 2218 miles at 10 kts.

Photo wanted.

TRITON (N. Shields, 1906). Ex-trawler of 277 tons. ... 34. 120·5 × 21·5 × 11·6 feet. Guns: 2—1 pdr., 1 M... 412 = 10·2 kts. Coal: about 140 tons. Double mine-...

Photo wanted.

VULCANUS (Grimsby, 1902). Ex steam trawler, 410 ... plement, 34. 123·7 × 21·6 × 10·5 feet. Guns: 2—1 ... Double laying gear for mines.

Photo wanted.

(East Indies Marine.)

ASSAHAN* (1900), 800 tons; **SERDANG*** (1897 ... **SIBOGA*** (1898), 790 tons. About 173·2 × 31·2 × 12 ... Armament: 1—4·1 inch, 1—9 pdr. mortar, 4—1 pdr. ... to 1400 = 13 to 14 kts. Coal: 130—150 tons. Comp... (55 Europeans, 45 natives).

* Originally sisters to *Koetei* and *Mataram* (v. first column of this page); on conversion to Mine Layers appearance of above ... considerably altered.

(Continued on ...

Mine Layers—*Continued.*

No photo available.

THOR (1877), **BALDER** (1878), **VIDAR** and **HADDA** (1879), ex-gunboats of 270-280 tons. Complement 34-36. Carry (9 mines each (double mine-laying gear.) Guns: 2—4 pdr. (semi-auto.) in *Balder* and *Hadda*, 2—4 pdr. in *Vidar* and *Thor*; 1 M.G. to all four. Speed: 7 kts. Coal: 30 tons.

No photo available.

DAS (1873), **HAVIK** (1875). Ex-gunboats of 210 tons. Guns: 2—1 pdr., 1 M.G. Speeds: About 6¾-7 kts. Complement, 32. Coal: 18 tons.

Mine Sweepers. (*Mijnenvegers.*)

Photo wanted.

I (v.d. Kuyck en van Ree, Rotterdam), —— tons. **II** (Kaptein, Dordrecht), 300 tons. **III, IV** (v.d. Schuyt, Papendrecht), both 275 tons. Completed and taken over in 1918. Steel. Dimensions: (*I*) 92·5 × 20·6 × 9·6 feet; (*II*) 100·7 × 20·5 × 10·2 feet; (*III & IV*) 90·2 × 20 × 8·2 feet. Guns: 1 machine. Complement, 16. No more details available.

Submarine Depot Ships (*Depotschepen voor Onderzeebooten*).

Building.

(For East Indies Marine.)

ONE SHIP (not yet named) building at Amsterdam. No details available.

Photo wanted.

PELIKAAN (——), —— tons. Dimensions: 207·4 × 42·6 × 12·1 feet. Guns: not known. H.P. 1400 = 12 kts. Oil fuel: 300 tons. Complement, ——.

1918 *Photo, Fotopersbureau, "Holland."*

CORNELIS DREBBEL (1915). 600 tons. Dimensions: 166·8 × 32·8 × 6·8 feet. Guns: 1—1 pdr. B.H.P. 170 = 6 kts. Oil fuel: 71 tons. Complement, 73.

also

Vali (1878). 840 tons. Guns: 1—1 pdr. No engines.

Surveying Ships (*Opnemingsvaartuigen*)

Building or completing.

EILERTS DE HAAN, building by Fijenoord Co., to replace old gunboats *Raaf* and *Geep*. 330 tons. 147·3 × 21·9 × 7 feet. No guns. Designed H.P. Coal: 44 tons. Complement, 18. Begun und[er] Programme.

Photo wanted.

(East Indies Marine.)

TYDEMAN (Soerabaya D.Y., 1916). 1320 tons. Dim[ensions] × 32·8 × 11·9 feet. Guns: none listed. I.H.P. 1 screw. Coal: 194 tons. Complement, —.

Photo wanted.

HYDROGRAAF (1911). 300 tons. Dimensions: 9·6 feet. Guns: none. I.H.P. 300 = 9 kts. Complement, 18.

(*Continued*

Surveying Ships.—*continued.*

No photo available

(**East Indies Marine.**)

VAN DOORN (1901) **VAN GOGH** (1898). Composite and zinced. 700 tons. Complements, 83. H.P. 290 to 370=10 to 9 kts. Coal: 100—110 tons.

No photo.

(**East Indies Marine.**)

LOMBOK, SUMBAWA (1891). 660 tons. Complement, 83. Speed: 12¼ kts. Coal: 90–120 tons.

Old Gunboats *Raaf* (1880) and *Geep* (1875) are employed as Surveying Ships, but are to be replaced by new Surveying Ship, *Eilerts de Haan*, as noted in preceding column. Both 200 tons and 7 kts. speed.

Fishery Patrols.

(Mainmast now removed.)

ZEEHOND (1899). 720 tons. Complement, 72. Armament: 2—1 pdr. Speed: 11 kts. Coal: 78 tons.

Fishery Patrols—*continued.*

DOLFIJN (1889). 290 tons. Complement, 42. Armament: 2—1 pdr. Speed: 10·7 kts. Coal: 29 tons.

Training Ship.

(*For Gunnery and Chief Petty Officers.*)

BELLONA (1892). Gunnery tender. 930 tons. *For Training duty*, carries 1—4·7 inch, 3—13 pdr., 1—9 pdr. mortar, 2—4 pdr., 8—1 pdr., 2 M.G., all of various models. Speed: 8 kts.

Also

Spencer (old Gunboat, 1875). 210 tons. Guns: 1—15 pdr., "No. 2." 3—1 pdr., 1 M.G. Speed: 7 kts.

Non-Effective

Training, Accommodation and Guard

(*Instructie, logement-en Wachtschepen.*)

For Torpedoes and Mining:—

Koningin Emma der Nederlanden (old Gunboat, 1892). 1—

Schorpioen (old Coast Defence Vessel, 1868). 1—1 pdr.

Mercuur. Torpedo Transport.

Pijl. Communication Ship.

Heimdall. Hellevoetsluis Light Station. 1—4 pdr.

For Engineers.—

Bommel.
Bulgia. } Old Gunboats.
Fret.

For Repair Work:—

Van Speyk. Old Gunboat.

Los. Old Gunboat.

For Midshipmen.—

Ever. Old Gunboat.

For Marines and Reservists:—

Atjeh. Old Gunboat.

Nautilus. Old Sloop.

Guardships.—

Neptunus. At Willemsoord. 2 M.G.

Buffel. At Flushing. 2—1 pdr.

Disposition uncertain:—

Schouten, Freyr, and 2 Cutters.

Dutch East Indies Marine.

Consists of the 2 gunboats *Koetei, Mataram*, 3 Mineleyers *Serdang, Sebogo*, 8 torpedo boats of the *Draak* class (torpedo boat *Cerberus*), surveying ships, *Tydeman, van Gogh, Lombok* and *Sumbawa*.

All these ships described on this and preceding pages.

ROYAL BELGIAN NAVY.

(Revised 1920, by courtesy of the Chief of Staff, Ministry of National Defence, Brussels.)

Flags.

The National Flag is also used as the Naval and Mercantile Ensigns.

Royal Standard, as National Flag, but has arms of Belgium in the middle of the centre (yellow) stripe.

Minister of National Defence has no special flag at present.

Uniforms.

No drawings available: see description below.

	Capitaine	Lieut. de Vaisseau de 1e Classe.	Lieut. de Vaisseau de 2e Classe.	Enseigne de Vaisseau de 1e Classe.	Enseigne de Vaisseau de 2e Classe.
Corresponding British or U.S. Ranks	Captain	Lieut.-*Commander*.	*Lieut.*	*Sub-Lieut.*	*Midshipman*, Ensign in U.S. Navy

Indication of the rank as stated hereunder :—

Cloak and Jacket Sleeves:

Captain : Four flat gold stripes, 6 m/m. wide.
First Lieutenant : Three flat gold stripes, 6 m/m. wide.
Second Lieutenant : Three flat stripes 6 m/m. wide, the centre one a silver stripe, the two outer ones gold.
First Sub-Lieutenant : Two flat gold stripes, 6 m/m. wide.
Second Sub-Lieutenant : One flat gold stripe, 6 m/m. wide.

On the Cap:

Captain : Four flat gold stripes, 3 m/m. wide.
First Lieutenant : Three flat gold stripes, 4 m/m. wide.
Second Lieutenant : Three flat stripes, 4 m/m. wide, the centre one silver, the outer ones gold stripes.
First Sub-Lieutenant : Two flat gold stripes, 4 m/m. wide.
Second Sub-Lieutenant : One flat gold stripe, 4 m/m. wide.

The uniform of the personnel of the Torpedo-Boat and Sailors' Detachment is the following :—

Officers (1) : Dark blue cloth jacket with turned down collar and two rows of five uniform buttons, copper gilt, with fouled anchor. Sleeves with a 66 m/m. cuff and a flap with three buttons, 12 c/m. high by 4 c/m. wide.
Dark blue cloth vest with six small uniform buttons.
Dark blue cloth trousers.
Dark blue cloth cap, crown and stand with peak of glazed leather. On the stand a gold embroidered anchor with the Royal Crown above.
White gloves, white collar, black tie.
Dark blue cloth cloak (without any badges or piping), with rank badges and flaps with the rank stripes on the sleeves, movable hood.
Sword and pistol (revolver).
Belt of the Army pattern, in black.
For drilling purposes, dark blue breeches and black leggings of the Army pattern.

Uniforms—*Continued.*

N.C.Os. : Dark blue cloth jacket of the same pattern as the officers and sleeves with the flap, with stripes. Rank badges on the sleeves.
Dark blue cloth vest with six small uniform buttons.
Dark blue cloth trousers.
Cap of the same pattern as the officers, dark blue cloth crown and stand, g peak. On the stand a tricolor enamel cockade of 25 m/m. diameter ; black ri wide, bearing the Belgian Coat of Arms stamped in gold (30 m/m. high by 4
White collar, black tie, jersey for the drill and manœuvres.
Dark blue cloth overcoat, same pattern as the Field Service Infantry Unif regimental badges or piping ; movable hood ; rank badges on the sleeves duties' badges on the left-hand sleeve. For duty ashore, black leggings of pattern.

Indication of the ranks :—

Sleeves of the overcoat and of the jacket.

Boatswain : Of various categories :—
Premier Maitre : Three flat gold stripes, 20 m/m. wide, with a piping, plac the bottom end touching the cuff, red piping between the stripes.
Maitre de Premiere Classe : Two flat gold stripes, 20 m/m. wide, diagonally w between them.
Maitre de Deuxieme Classe : One flat gold stripe, 20 m/m. wide, and a similar stripe, with red piping, placed as above.
Second Maitre : One flat gold stripe, 20 m/m. wide, with red piping, placed dia
Quartier-Maitre : Two red wool stripes, placed diagonally.
The Premier-maitres bear on the stand of their cap a 4 m/m. flat silver stripe place of the Second Sub-Lieutenant.
Sailors : Dark blue jacket with turned down collar of dark blue serge, ornamented stripe of 2 c/m. width, surrounding the collar at a distance of 4 c/m. from it.
In full dress uniform, light blue cloth collar ornamented with a white strip cloth cap with light blue wool tuft and white elastic. On the stand of thei wide ribbon with the Belgian Coat of Arms printed in gold (30 m/m. high by
Dark blue cloth great coat for use at sea.
For drilling purposes ashore, black leggings of the Infantry pattern (Infantry

Minister of National Defence : M. Paul Emile Jansen.

Personnel : officers and men, active list. men, reserve list.

Mercantile Marine.

From "Bureau Veritas" 1919-20 figures.
Ships under 100 tons *gross* excluded.

140 Steam and Motor Ships of 290,992 tons *gross*.
2 Vessels (with Auxiliary Motors) of 476 tons *gross*.
221 Sailing Vessels of 45,368 tons *gross*.

During 1919, Belgian shipyards launched 20 Vessels of 10,499 tons *gross* ; some of ti account.

Colonies : Belgian Congo.

Coinage : Franc (of 100 centimes) = *about* 9½d. English currency at 1914 rate of exch value less.

(Uniform scale. Divided into 2000 yard squares. Soundings in fathoms.)

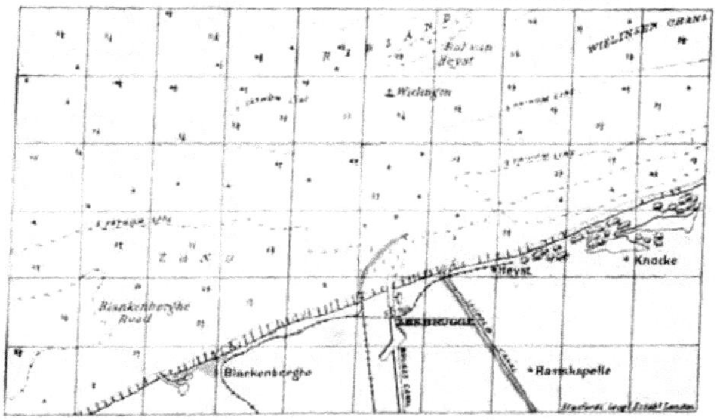

ZEEBRUGGE.—Small port on Belgian coast, occupied and fortified by Germany during the war. Connected to Bruges by canal (6¼ miles) 230 feet wide at water level, 72¾ feet wide at bottom × 26¼ feet deep. 5151 feet of quayage. Lock 840 feet between gates × 65⅔ feet at entrance × 18 feet depth at low tides. Oil tanks situated in Inner Port. Mole 1930 yards long, where the Germans established batteries and depôts for torpedo craft, submarines, and seaplanes. Coast was heavily fortified by Germans, batteries including naval guns of 15 inch calibre and downwards, the largest guns ranging up to 25,000 yards. Port works, &c., wrecked by Germans on evacuation, but salvage and repair work is well advanced. Small wet dock at Blankenberghe (14 feet depth of water). Admiralty Chart No. 120.

OSTEND. Is connected by canal with Bruges. During German occupation coast was heavily fortified. Bomb proof Submarine shelters, also slips for building and repairing small submarines, repair workshops, &c. One or two floating docks existed to take 1 large T.B.D. or 2 small Submarines. The Government have in view the construction of two docks, each 492 × 60 × feet. All port works, &c., were wrecked by the Germans on evacuation, but are now largely re-established. The Tirpitz battery at Hamilton Farm, near Mariakerke consisted of 4—11·2 inch guns and could range up to 30,000 yards *extreme*.

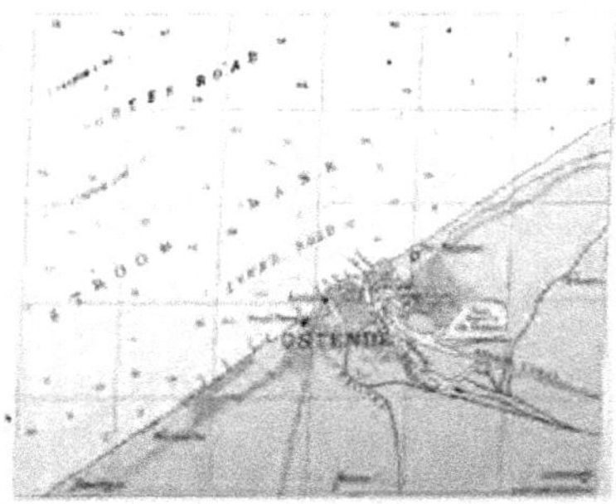

Following Ports within Belgium are connected with sea:—

ANTWERP. Eight Dry Docks. Municipal: No. 1, 506 × 81¼ × 23 feet; No. 2, 248 × … Nos. 4, 5, & 6, 440½ × 49¾ × 16 feet. J. Cockerill: 400 × 41 × 17½ feet. 330 × 42 × 11 and 300 × 36 × 13 feet. About 12 wet docks, coaling basin, and 11 … equipped, with fixed and moveable cranes, and lit by arc lamps.

BRUGES. Connected to Ostend and Zeebrugge by canals. 5 wet docks. Was exten… Germans as a naval base during the war. Various bomb-proof shelters for Submari… here.

GHENT. Two dry docks, No. 1, 446 × 42½ × 18 feet (*available* length 393½ feet). No. 2, … feet (*available* length 220 feet.) 7 wet docks. Carels Works located here.

Shipbuilders.

SOCIETÉ JOHN COCKERILL, Hoboken, near Antwerp.

COMPAGNIE ANGLO-BELGE DE CONSTRUCTIONS NAVALES, Ghent.

RECOGNITION SILHOUETTES.

Scale: 1 inch = 160 feet.

——— (ex-Zinnia.)
(*Sloop*).

A 1—A 9.
(*Torpedo Boats.*)

A 21—A 25.
(*Torpedo Boats.*)

MOTOR LAUNCHES. **Scale: 1 inch = 80 feet.**

V 1—V 17.

Note.—The British Cruisers *Vindictive*, *Intrepid* and *Iphigenia*, which were scuttled at Ostend and Zeebrugge during the War as Blockships, have been presented to Belgium. *Vindictive* has broken in two; *Intrepid* and *Iphigenia* are, of course, badly damaged, and it is not likely that any of these three ships will be put in sea-going order.

Sloop [illegible]

Ex-Zinnia. 1920 Photo, [illegible], [illegible].

——————— (ex-British Fleet Sweeping Vessel *Zinnia*, launched August, 1915, by Messrs. Swan, Hunter & Wigham Richardson, Wallsend. Purchased for Belgian Navy, June, 1920). 1200 tons. Dimensions: 250 (*p.p.*), 262½ (*o.a.*) × 33 × 11 (*mean*) × 11¾ feet (*max.* draught). Guns: Sold without armament. Designed I.H.P. 1400 = 17 kts., but actually requires *about* 2000 I.H.P. for this speed. 1 set, triple expansion engines. Boilers: 2 cylindrical. 1 screw. Coal: 130 tons *normal*; 250 tons *max.* = *about* 2000 miles at 15 kts. Complement (as British ship), 77. Was built under Emergency War Programme as a unit of the *Acacia* group of the "Flower Class" Sloops.

14 Torpedo Boats (*Torpilleurs*).

No.	Class.*	Built.	Displacement tons	I.H.P.†	Speed,† kts.	Max. Oil tons	T.T.	Mines	Complement
5	*A 21*—*A 25* (S)	1915-16	228	3000	25.8	53.7	1§	10	
9	*A 1*—*A 9* (V)	1914-15	107	1125	20	23.8	2§	4	

* (S) = Schichau, Elbing or Danzig. (V) = Vulkan, Hamburg. † Trial data.

Torpedo Boats (*continued*).

5 *Schichau "A 11" type*: **A 21** (ex-German *A [illegible]*), **A 22** (ex-German *A [illegible]*), **A 23** (ex-German *A [illegible]*), **A** [illegible] *A [illegible]*), and **A 25** (ex-German *A [illegible]*). Built at Elbing or Danzig, 1915-16. Dimensions: 164, [illegible] Guns: 2—3.4 inch (22 pdr. Krupp 30 cal. "Ubts. u. Tpbts." model). Torpedo tube: 1—17 [illegible] Navy type).§ Can carry 10 mines each. Machinery and boilers unknown. Oil: *Normal, about* 45 tons = 975 miles at 20 kts. Other details as Table.

Note.—Built for German Navy during War, being a modification of the Schichau design for the Danish [illegible] T.B. Formed part of the German Flanders Flotilla. After the fall of Ostend and Zeebrugge, [illegible] in Holland, and were interned. Transferred to Belgium by the Netherlands Government, 1920.

A 2. 1920 Photo, S.F.B.A.B., furnished by courtesy of the Ministry of National Defe[illegible]

9 *Vulkan "A 1" type*: **A 1** (ex-German *A [illegible]*), **A 2** (ex-German *A [illegible]*), **A 3** (ex-German *A [illegible]*), **A 4** [illegible] **A 5** (ex-German *A [illegible]*), **A 6** (ex-German *A [illegible]*), **A 7** (ex-German *A [illegible]*), **A 8** (ex-German *A [illegible]*) and [illegible] *A [illegible]*). All built at Hamburg, 1914-15. Dimensions: 126.3 × 15.[illegible] × 4.1 feet. Guns: 1—4 p[illegible] "Tpbts." model). Torpedo tubes: 2 single 17.7 inch.§ Can carry 4 mines each. Machinery and [illegible] Oil: 23.8 tons *max.* = 440 miles at 19 kts. For other details *v.* Table.

Note.—The first three boats, *A 1*—*A 3*, abandoned by Germans on evacuation of Flanders and capture[illegible] six boats, *A 4*—*A 9*, took refuge in Holland and were interned. Transferred to Belgium by the Ne[illegible] ment, 1920.

§*General Note to Table and Descriptions.*—The Germans removed the torpedo tubes from the "A" To[illegible] they used as Mine Sweepers. Not certain if one or both tubes have been removed from [illegible] described above.

Submarines (*Sous-marins*).

Two ex-German Submarines were to have been transferred to Belgium, but, up to the summer of 1920, no boats had been delivered to the Belgian Navy. These boats are only to be used for Propaganda purposes, and cannot be incorporated in the Belgian Navy.

Four boats, *UB* 10, *UB* 30, *UB* 59 and *UC* 4 (Minelayer), were scuttled by the Germans on evacuating Flanders during October, 1918. They are believed to be past repair.

Belgian Naval Officers are serving in the French Submarine *Nereide* for instructional purposes.

Sea-going Motor Launches (*Vedettes de mer*).

V 3. *Photo, S. Ph. A. B., furnished by courtesy of the Ministry of National Defence, Brussels.*

17 boats, type unknown: **V 1—V 17.** No further details officially furnished. From above photo, mount 1 machine and 1 Lewis gun. Are probably ex-German Mine-Sweeping Motor Boats, which took refuge in Holland during Autumn of 1918, were interned, and handed over to Belgium by Netherlands Government, 1920.

River Patrol Boats (*Vedettes de Rivière*).

No photo available.

5 boats, *type unknown*, **V 21—V 25.** Are probably ex-German Minesweeping Motor Boats, inte October, 1918, and handed over to Belgium, 1920.

Note.—The Chief of Naval Staff, Ministry of Marine, the Hague, has notified "Fighting Ships" tha German Minelayers *Feronia*, *Nicolaus*, *Theresa*, *Gabo*, *Irma*, *Julie* and *Theodore* have also been Belgium. They do not appear in the list furnished by the Belgian Chief of Staff, and probably hav to mercantile or fishing craft.

Details given below are unofficial:—

Belgian Congo.

Photo wanted.

(*Both on Lake Kivu*).

PAUL RENKIN, Gunboat, launched during the War. No details available.

One Motor Patrol Boat (name unknown), launched during the War. No details known.

No photo available.

(*On Lake Tanganyika*).

NETTA (launched December, 1915). Torpedo Boat. No details known.

There are also believed to be various vessels—mostly stern wheel steamers—on the Cong craft, no modern or reliable details can be obtained.

(ROYAL) SPANISH NAVY.

(Corrected, 1919, from the "Lista Official de los Buques de Guerra y de los Mercantes," issued by the Ministry of Marine. Revised 1920 from information furnished by the courtesy of Teniente de Navio M. Mille).

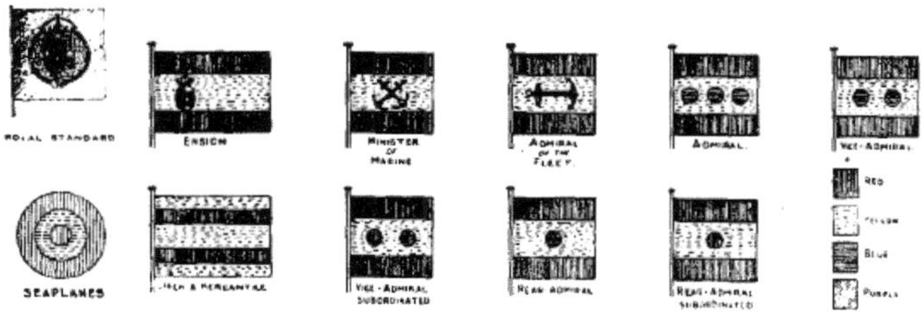

Uniform Insignia.

As Almirante but *four* upper stripes.

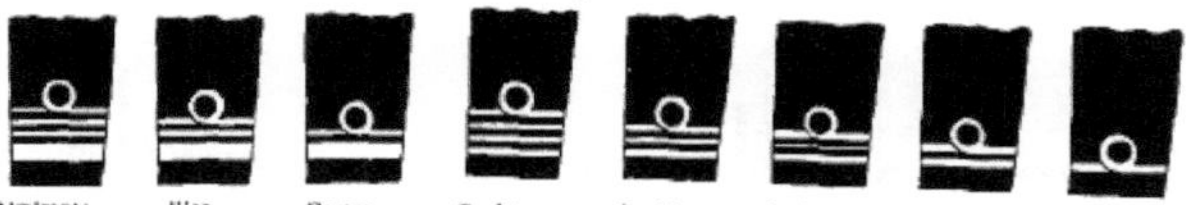

Capitan-General de la Armada. *Admiral of the Fleet* · Almirante. *Admiral.* · Vice-Almirante. *Vice-Admiral* · Contra Almirante. *Rear-Admiral.* · Capitan de Navio. *Captain.* · Capitán de Fragata. *Commander.* · Capitán de Corbeta. *Lieutenant-Commander.* · Teniente de Navio. *Lieutenant.* · Alferez de Navio. *Sub-Lieutenant.*

Other branches, *without the curl*, have distinguishing colours as follows:—

Engineers ...	*green.*	Astronomical ...	*green brown.*
Constructors ...	*blue.*	Pharmaciens ...	*yellow.*
Doctors ...	*red.*	Chaplains ...	*violet.*
Paymasters ...	*white.*		

Personnel: about 11,000 officers and men, 5000 marines—16000 in all. *Minister of Marine*: Señor Dato (*ad interim*). *Chief of Naval Commission, (London)*: Capt. Eliseo Sanchez.

Colour of Ships: Light grey.

Mercantile Marine.

From "Bureau Veritas" 1919-20 figures. Ships under 100 tons gross excluded.

514 Steam and Motor Ships of 791,800 tons gross.

13 Vessels (with Auxiliary Motors) of 4,018 tons gross.

227 Sailing Vessels of 50, 258 tons gross.

During 1919, Spanish Shipyards launched 41 vessels of 52,600 tons gross; some of these for foreign account.

General.

Naval School is at Cadiz on land. At Ferrol, the old frigate *Villa de Bilbao* (1843, 1313 tons), is a school for quartermasters, (*Escuela de Marineria*): there is also the corvette, *Nautilus* (*Escuela de cabos de mar*), built 1868, of 1700 tons, Guns: 1—6 inch, 6—5 pdr. Complement 221. The battleship *Pelayo* is training ship for gunnery. *Carlos V* is Flagship of the Training Squadron, which consists of *Pelayo*, *Nautilus*, *Terror*, and 6 Torpedo Boats, all based at Rios, Vigo Bay. Cruiser *Reina Regente* carries out extended cruises for Training of Cadets.

(All details unofficial).

Principal Naval Guns in Service.

Notation	Nominal Calibre		Model.	Length in calibres.	Weight of Gun.	Weight of A.P. shell.	Initial velocity.	Max. penetration A.P. capped direct impact at K.C. at 5000 yards.	3000 yards.	Danger S... averag... 10,000 yards.	... y...
	c/m.	inches			tons.	lbs.	ft. secs.	inches	inches.		
HEAVY	*30·5*	12	V'09	50	66	850	3010	17	20	...	
	32	12·6	H.	35	47½	1036	2034	8	11	...	
	28	11	H.	35	32½	837	2034	5½	8	...	
	24	9·4	G.	40	19	336	2625	7¼	9½	110	3
MEDIUM	*15*	6	V.	50	7¾	100	3100	...	8¾	...	
	15	5·9	R.	45	4¼	88	2625	...	3	25	2
	14	5·5	H.	35	...	86	2001	...	...	...	
	14	5·5	C.	35	...	86	2034	...	...	...	
LIGHT	*10·5*	4·1	K.	35	...	35·2	...	...	...	...	
	10·2	4	V.	50	2·1	31	3030	...	...	...	

V = Vickers. H = Hontoria. G = Guillen. R = Rueda. C = Canet. K=

Naval Yards, &c.

FERROL. Dry docks. (1) "Victoria Eugenia," 573 × 82 × 37 feet (complete Julian," 437½ × 72½ × 29 feet. Ferrol is the big shipyard. It has a large bay, well fortified. At La Graña (Ferrol Bay) Station for Submarines.

CARTAGENA. One dry dock, 492 × 82 × 31½ feet. One floating, 6000 tons. C and torpedo craft building yard. Station and Repair establishment for Su pedoes to be built here. Very strongly fortified.

LA CARRACA (CADIZ). Dry docks. (1) 238½ × 55½ × 23½ feet, (2) 328 × 70½ × × 57 × 18½ feet, (4) 925 × 85 × 34 feet (in two lengths). No slips. Only Morocco coast gun boats. S.E.C.N. ordnance and munitions factory l Matagorda Shipyard.

MAHON. Very small yard. Station for Torpedo Craft and Submarines.

RIOS (Vigo). }
MARIN. } Both Stations for Torpedo Craft and Submarines.

Private Establishments.

Warship building is carried out in the Naval Yards by the "Sociedad Española de C (S. E. C. N.), an Anglo-Spanish shipbuilding syndicate in which the British firms of Sir Whitworth, Ltd., Vickers, Ltd., and John Brown & Co. have a large interest. In carrying out these three firms supply materials, armour, etc., and skilled supervision of work. The S.E. Ordnance factories for the protection of guns, gun-mountings, shells, &c., at the Arsenal of and at Reinosa, near Santander. S.E.C.N. Shipbuilding Yards have been created at Matagorda, and Sestao (Bilbao). The Italian Ansaldo Co. acquired a shipyard at Valencia in

Private Docks.

BARCELONA. Harbour Board Floating dock, 367 × 68 × 20 feet.
CADIZ. S.E.C.N. dock, 455 × 64 × 22 feet.
BILBAO. Compania Euskalduna dock, 605 × 60 × 21 feet. Two smaller
SANTANDER. Gamazo dock, 445 × 57 × 23½ feet.
GIJON. S.E.C.N. dock, 288 × 46 × 12 feet.

Coinage: Pa. 25·20 = £1.

Colonial Possessions, &c.: Rio de Oro, Adrar, Rio Muni (or Spanish G Fernando Po, Corisco Islands, Great and Little Elobey; all on Atlantic Coast of Islands (Atlantic). There is a Protectorate in the north and west of Morocc Ports at Ceuta, Melilla, Alhucemas, Gomera, and Chafarinas

Scale : 1 inch = 160 inch.

SPANISH RECOGNITION SILHOUETTES. (Re-arranged 1919).

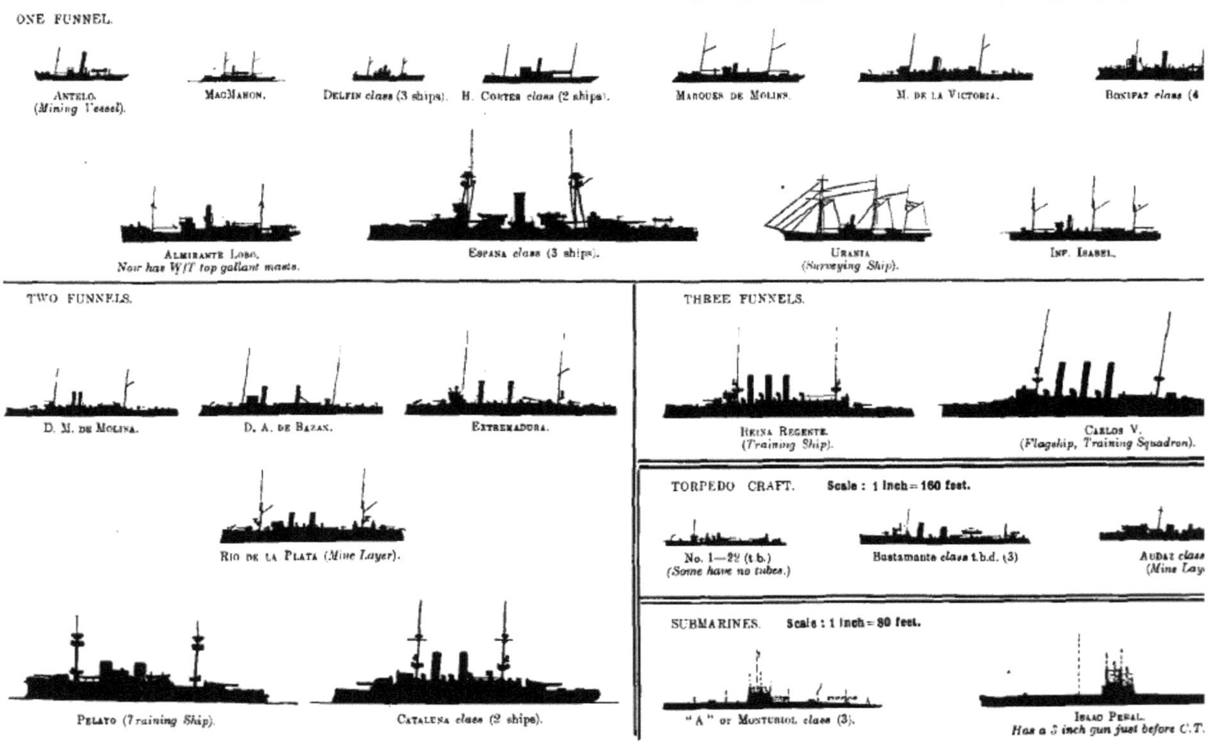

ESPAÑA (Feb., 1912), **ALFONSO TRECE** (May, 1913), & **JAIME PRIMERO** (Sept., 1914).

Normal displacement, 15,452 tons. Complement, 735.

Length (*waterline*), 435 feet. Beam, 78¾ feet. *Max.* draught, 25½ feet. Length (*over all*), 459¼ feet.

Guns (Vickers):
8—12 inch, 50 cal. (**Dir. Con.**)
20—4 inch, 50 cal.
2—3 pdr.
2 Maxims.
(2 landing).
Torpedo tubes:
none.

Armour (Krupp):
8" Belt (amidships)
3" Belt (bow)
4" Belt (aft)
1½" Deck
6" Lower deck side
8" Turrets (N.C.)
10" Turret bases (N.C.)
3" Battery (N.C.)
10" Conning tower
6" After C.T. (Hadfield)
Special protection:
1½" H.T. internal screens over Sections 4—15 on plans.

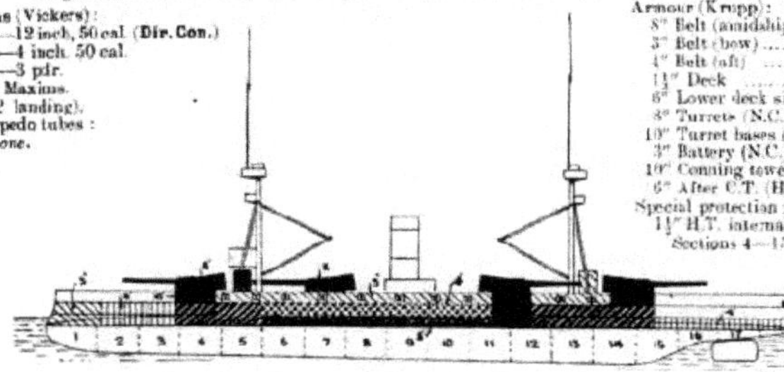

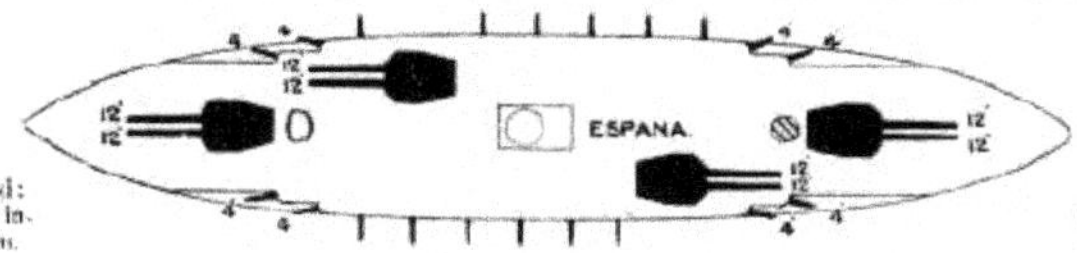

Ahead:
6—12 in.
4—4 in.

Astern:
6—12 in.
4—4 in.

Broadside: 8—12 in., 10—4 in.

Machinery: Parsons turbine. 4 screws. Boilers: 12 Yarrow. Designed H.P. 15,500 = 19·5 kts. Coal: *normal* 900 tons; *maximum* 1850 tons + 20 tons oil fuel = nominal radius of 5000 miles at 10 kts., and 3100 miles at 16¾ kts.

Gunnery Notes.—Vickers type 12 inch gun directors in lower top on each mast. 12 inch manœuvred by hydraulic power; all-round loading at any angle of elevation. Magazine capacity is 80 rounds per 12 inch gun, but there is ample storage for more than this. Arcs of fire: end 12 inch, *about* 270°; echelon 12 inch, 180° on own beam (*nominal*) and about 80° far beam. Heights of guns over normal waterline: 12 inch, 24 feet; 4 inch, 13½ feet. Total weight about 2550 tons. (These notes are not from any official source).

Armour Notes.—Main belt is 6 feet 7 inches deep, 4 feet 7 inches of this being below waterline and 2 feet above. On plans 8 inch belt amidships should extend to bases of end barbettes. Hull without armour = 5600 tons.

Machinery Notes.—Engines = 1320 tons.

Torpedo Notes.—Bullivant net defence. No torpedo tubes installed.

Name	*Builder*	*Machinery*	*Laid down*	*Completed*	*Trials (Full Power)*	*Boilers*	*Best recent speed*
España *Alfonso XIII* *Jaime I*	Ferrol Yard	S. E. C. N.	Dec. '09 Feb. '10 Feb. '12	1914 1915 1919*	21,582 = 20·2 = 20·36 —	Yarrow in all	

General Notes.—Built under Navy Law of 7th January, 1908.

* By April, 1919, only 5—12 inch guns had been delivered for *Jaime I*. She was still short of various other special fittings, but was otherwise complete.

ESPAÑA Photo (1919), T.

ALFONSO XIII. Photo (1919), T

2nd Class Battleship.

(Acorazado de 2ª Clas.)

(Gunnery Training Ship.) 1913 *Photo Casau, Cartagena.*

PELAYO (1887), (1897), (1910.)

Length *(o.a.)* $341\frac{5}{8}$ feet. Beam, $65\frac{1}{4}$ feet. *Maximum* draught, $25\frac{5}{12}$ feet.
Displacement, 9733 tons. Complement, 621.

Guns (Henteria):
- 2—12·6 inch, 35 cal.
- 2—11 inch, 35 cal.
- 9—5·5 inch, 35 cal.
- 12—6 pdr.
- 9—1 pdr.
- 2 landing.

Torpedo tubes: *removed.*

Armour (steel):
- 16" Belt (amidships) ...
- 12" Belt (ends) ....
- $3\frac{1}{2}$" Deck (flat below belt).
- 10" Barbettes............
- 2" Hoods to these......
- 3" Battery (Harvey)...

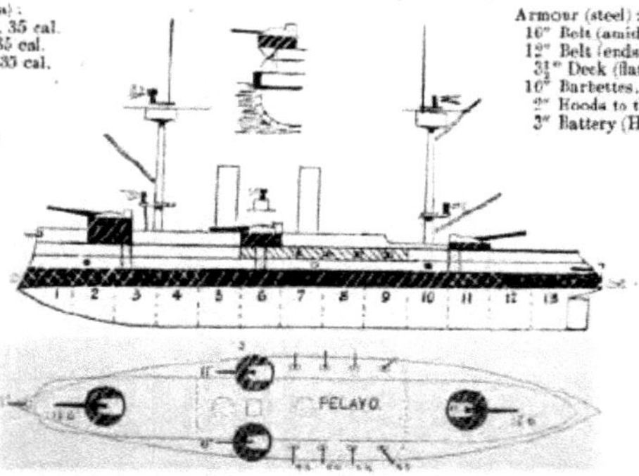

Machinery: 2 sets vertical compound. 2 screws. Boilers: 16 Niclausse (in 4 groups). Designed H.P. 6800 = 15 kts. Coal: *normal* 500 tons; *maximum* 676 tons. Built at La Seyne. Reconstructed there in 1897. Was partially rebuilt in 1910.

(1st Class Cruisers.) *Cruceros de 1ª Clas.*

PRINCESA DE ASTURIAS. (CATALUNA CLASS—2 SHIPS). 1920 *Photo, T. de N.*

PRINCESA DE ASTURIAS (1896) & CATALUNA (1900).

Displacement, *P. de A.* 7427 tons, *C.* 7405 tons. Complement, 546.

Length *(o.a.)*, $347\frac{3}{4}$ feet. Beam, $60\frac{3}{4}$ feet. *Max.* draught, $23\frac{1}{2}$ to $23\frac{3}{4}$ feet.

Guns:
- 2—9·4 inch, 40 cal. Guillen
- 8—5·5 inch, 40 cal. Canet
- 8—6 pdr.
- 10—1 pdr.
- 2 landing.

Torpedo tubes: *Removed.*

Armour (Harvey):
- 12" Belt (amidships)...
- 6" Belt (forward)......
- 8" Belt (aft) ....
- $2\frac{1}{2}$" Deck (flat on belt
- 12" Bulkheads ....
- 8" Barbettes ....
- 4" Hoists to these.....
- $2\frac{1}{2}$" Casemates/shrouds
- 8" Conning tower (fo

Ahead:
1—9·4 in.
2—5·5 in.

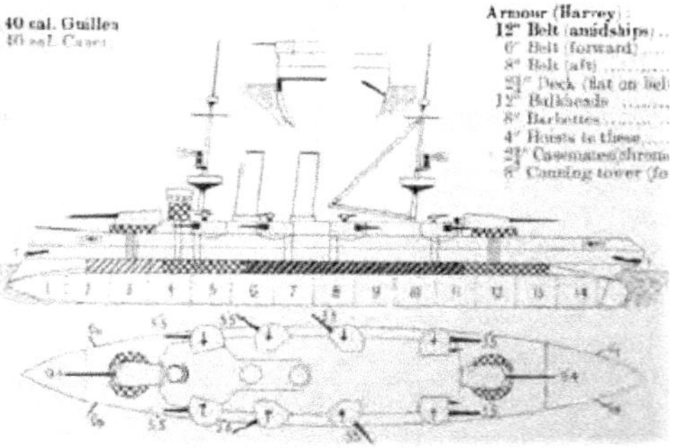

Broadside: 2—9·4 in., 4—5·5 in.

Machinery: 2 sets horizontal triple expansion. 2 screws. Boilers: 6 cylindrical. Designed 10,500 = 18 kts. Coal: *maximum* 1178 tons in *C.* and 1007 in *P. de A.*

(1st Class Cruisers continued on ...

1st Class Cruiser (*Crucero de 1ª Clas*).

(*Flagship, Training Squadron*). 1920 Photo, T. de N. M. Mille.

CARLOS QUINTO (1895). 9903 tons. Comp., 583. Dimensions: 404¾ (*a.a.*) × 67 × 27⅔ feet, *max.* draught; *mean* draught, 25 feet. Armament: 2—11 inch (35 cal. Hontoria), 8—5·5 inch, (35 cal. Hontoria), 4—4·1 inch, (Krupp), 10—6 pdr., 8—1 pdr., 2 Machine. Torpedo tubes (14 inch): 2 *above water*. Armour: 2½" Deck, 10" Barbettes with 4" Hoods and 8" Hoists, 2" over Battery, 12" C.T. Boilers: 12 cyl. Designed H.P. 15,000, made 19 kts. on trial. Coal: *max.* 2008 tons.

2nd Class Cruiser (*Crucero de 2ª Clas*).

(*Cadets' Training Ship*) 1920 Photo, T. de N. M. Mille.

REINA REGENTE (Sept., 1906) 5778 tons. Complement, 452. Dimensions: 337 × 52¾ × 16½ feet. Guns: 10—5·9 inch, (45 cal. Rueda), 12—6 pdr., 8—1 pdr., 2—3 inch Field. Armour: 3¼" Deck and C.T., 3" Gun Shields. Machinery: Reciprocating. 2 screws. Boilers: Belleville. Designed H.P. 11,000 = 19·5 kts. Coal: 1178 tons.

(New) Light Cruisers (*Cruceros de—Clas*).

Building.
Provided by Navy Law of 17th February, 1915.

No. 6 (building) and No. 7 (building).

Normal Displacement: 4725 *metric* tons. Complement: ——.
Length, 439½ feet. Beam, 46 feet. Draught, (*a*) 14¼ feet *mean*. (*b*) 15¼ feet *max.*

Guns:
6—6 inch, 50 cal. **Dir. Con.**
4—9 pdr. anti-aircraft.
6—2 pdr. pom-poms.
(1—12 pdr. landing.)
Torpedo Tubes (21 inch).
4 *above water* (twin mountings).

Armour:
..." Side amidships
..." Side (ends) .....
..." Deck ....
..." C.T. ....
..." Director tower ..

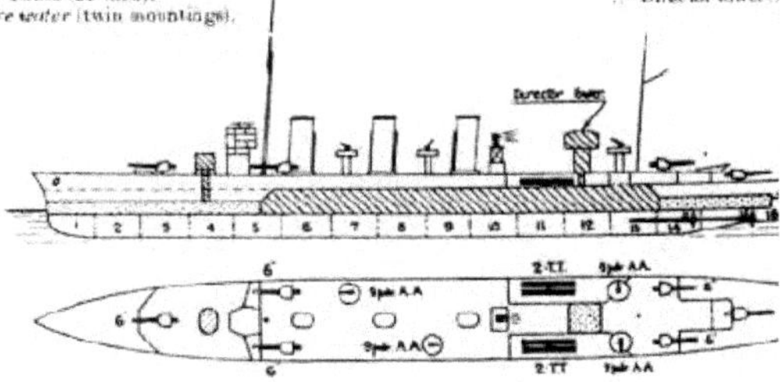

Ahead:
2—6 in.

Broadside: 4—6 inch, 2—21 inch tubes.

Machinery: Parsons (geared) turbines. Boilers: 10 Yarrow; to burn oil fuel only. I 30,000 = 29 knots. Fuel (oil only): *normal*, tons; *maximum*, 840 tons.

Name	Builder	Machinery	Laid down	To be Completed	Trials
No. 6 No. 7	Ferrol D.Y. Ferrol D.Y.	S. E. C. N.	2 April '17 28 Sept. '17	1921 or 1922	

General Notes.—On account of the great shortage in shipbuilding materials construction of these ships very slowly. Two further ships of this class, previously projected, were cancelled in 1919.

(New) Light Cruiser (*Crucero de — Clas*).

Completing.
Built under Navy Law of 30th July, 1914.

REINA VICTORIA EUGENIA (21st April, 1920).

Displacement 5550 tons. Complement 412.

Length (*p.p.*) 440 feet. Beam, $49\frac{1}{4}$ feet. (*Max.* draught, $16\frac{3}{4}$ feet. / *Mean* „ $15\frac{3}{4}$ „) Length *over all*, 462 feet.

Guns (Vickers):
9—6 inch, 50 cal.
4—3 pdr. (AA.?)
1—12 pdr. (Field)
4 Machine
Torpedo tubes: (21 inch).
2 *above water*.

Armour (Nickel and H.T.):
3" Side
" Ends
3" Deck
" Conning tower

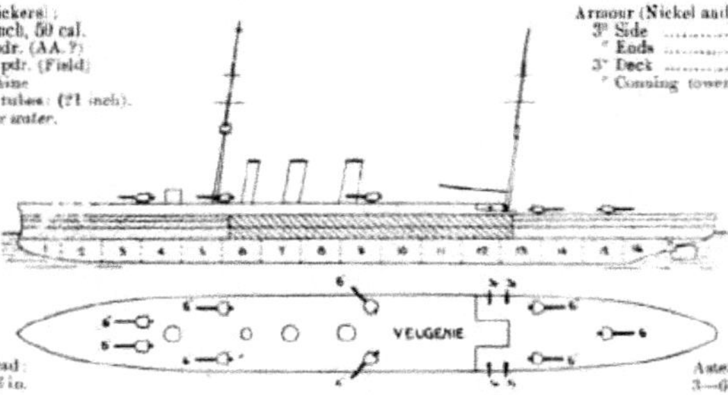

Ahead: 4—6 in.
Astern: 3—6 in.
Broadside: 5—6 in., 1—21 in. tube.

Machinery: Parsons turbines. 2 screws. Boilers: 12 Yarrow (coal and oil burning), in 3 rooms. Designed H.P. 22,500 = 25·5 knots. Coal: *normal* tons; *maximum* 660 tons (inclusive of oil fuel).

Gunnery Notes.—Forecastle 6 inch have 180° arc of training, *i.e.*, 30° inboard and 150° on own beam. One or two anti-aircraft guns will be mounted, either as additional guns or in place of 3 pdr. guns.

Torpedo Notes.—1—21 inch tube on each beam on upper deck between foremast and conning tower.

Name	*Builder*	*Machinery*	*Ordered*	*Laid down*	*Completed*	*Trials*	*Boilers*	*Best recent speed*
R.V. Eugenia	Ferrol D.Y.	S. E. C. N.	Aug. '14	Mar. '15	1920*		Yarrow	

General Notes.—Design is practically identical with British Light Cruisers of *Birmingham* class.

* Construction of this ship has been (and will be) greatly delayed owing to shortage of materials caused by the war.

3rd Class Cruisers (*Cruceros de 3a Clas*).

Photo (1919), *T. de N.*

EXTREMADURA (Cadiz, 1900). 2100 tons. Complement, 266. Dimensions: $288\frac{3}{4} \times 36\frac{1}{4}$ × Guns: 8—4 inch (50 cal. Vickers), 4—6 pdr., 4—1 pdr., no tubes. Armour: $1\frac{1}{4}$" Deck. M 2 screws. Boilers: Thornycroft. H.P. 7000 = 19 kts. Coal: *max.* 425 tons.

(*Is now equipped as a Mine Layer.*) *Photo* (1919), *T. de N.*

RIO DE LA PLATA (Havre, 1898). 1920 tons. Comp., 213. Dimensions: $250\frac{1}{4} \times 35\frac{1}{2}$ × *maximum* draught; *mean* draught, $14\frac{1}{4}$ feet. Guns: 2—5·5 inch, 35 cal. (Canet), 4—4·1 inch 6—9 pdr., 4—1 pdr. No tubes. Armour: $1\frac{1}{2}$" Deck. Machinery: Reciprocating. 2 screws. Normand-Sigaudy. H.P. 7000 = 19 kts. Coal: *max.* 370 tons.

7 + 3 Destroyers (*Destroyers*).

No	Type	Date	Displacement	H.P.	Max. speed	Fuel	Complement	T. tubes	Max draught
			tons		kts.	tons			feet
3	New	*Bldg*	1145	33,000 *t*	34	265 oil	...	4	$10\frac{1}{2}$
3	*Bustamante* (V)	'12-'17	364	6250 (*t*)	28	79	70	2	$5\frac{1}{4}$
1	*Proserpina* (C)	1897	457	7200	30	94	74	2	$9\frac{3}{4}$
2	*Audaz* (C)	1897	457	7700	30	94	74	2	$9\frac{3}{4}$
1	*Terror* (C)	1896	450	6000	28	104	74	2	$9\frac{1}{4}$

C=Clydebank. V=Vickers, etc. built in Spain by S.E.C.N.

3 New Boats.

Building.

Provided by Navy Law of 17th Feb., 1915.

3 *New* Ocean-going Destroyers (names not known) building at Cartagena by S. E. C. N. Displacement: 1145 tons, (*normal*) 1325 tons (*full load*). Dimensions: 275 × 27 × 10½ feet. Designed H.P. 33,000 = 34 kts. Will have Parsons turbines and 4 Yarrow boilers. Guns: 3—4 inch (40 cal.), 2—2 pdr. anti-aircraft. Torpedo tubes: 4—21 inch in 2 twin deck mountings. Fuel 265 tons oil only. To be finished in 1921 (?).

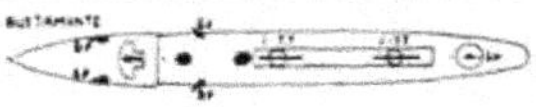

3 Bustamante Class.

BUSTAMANTE. *Photo, Casau, Cartagena.*

3 *Bustamante* class: named **Bustamante** (1913), **Villaamil** (1913), **Cadarso** (ex *Requesens*, 1914). All built at Cartagena D.Y. 364 tons. Dimensions: 221½ × 22 × 5¼ feet. Armament: 5—6 pdr. S.H.P. 6250 = 28 kts. Tubes: 2—18 inch in first two and 2—15 inch in *Cadarso* only. Yarrow or Normand boilers. Parsons turbine, 3 screws. Coal: 79 tons = 900 miles at 15 kts. Complement 70. Built under Navy Law of 7th January, 1908.

4 Audaz Class.

AUDAZ. 1920 Photo, T. de N. M.

Osado (Proserpina, rebuilt 1916, appearance may be changed.) 1920 Photo, T. de N. M.

1st, 2nd, and 3rd boats of the *Audaz* class:—**Audaz, Osado,** and **Proserpina,** Clydebank, 1897, (*Proserp* 1916). 457 tons. 229 × 221 × 9¾ feet. I.H.P. 7200 (7350 in *Osado*) = 29.3 to 31 kts. on trials. Guns: 2—1 pdr., 2—1 pdr. Torpedo tubes: 2—15 inch. Coal: 94 tons = 1020 miles at 15 kts. Complement 74.

1920 Photo T. de N. M

4th boat of *Audaz* Class: **Terror** (Clydebank, 1895). 450 tons. 221 × 22 × 9¼ feet. I.H.P. 6000 = 29.5 kts. (pdr., 2—3 pdr. Torpedo tubes: 2—15 inch. Coal: 104 tons = 1100 miles at 15 kts. Complement 73

(All above 4 boats fitted for Mine Laying.)

22 1st Class Torpedo Boats (*Torpederos de 1a Clase*).

Photo 1911, T. de N. M. Mila.

22 *Vickers-Normand.* ***2*** (1911), ***1, 3, 5, 6*** (1912), ***4*** (1912, rebuilt 1916), ***7, 8, 13*** (1913), ***9, 11, 12, 14*** (1915), ***10, 15, 16*** (1916), ***17, 18, 19, 20*** (1918), ***21, 22*** (1919). 177 tons. Dimensions: 164 × 16¼ × 4⅝ feet (*max.* draught). Armament: 3—3 pdr., 3—18 in. tubes, twin amidships and single aft. (Nos. *1, 4, 5, 17, 18* no tubes; *13* and *16* only 2 tubes.) H.P. 3750 = 26 kts. Parsons turbines and Normand boilers. Coal: 33 tons. 3 screws (Nos. 1—7), 2 screws (Nos. 8—22). Complement 31. 24 boats sanctioned under Law of the 7th January, 1908, but Nos. 23 and 24 were abandoned 1919.

Note. Following two old T.B. still exist: *1—1st Class boat*, **Halcon** (**No. 41**), built 1887, rebuilt 1906. 237 tons. I.H.P. 1650 = 20 kts. Armament: 2—2½ pdr., 2 tubes. Coal: 25 tons. *1—2nd Class boat*; **Habana** (**No. 45**), built 1886. 61 tons. I.H.P. 780 = 19 kts. Armament: 1 M.G. and 2 tubes. Coal: 25 tons.

4 (+6) Submarines (*Submarinos.*)

Note.—All details given on this and next page are *not* based on any official information.

No.	Type	Date	Displacement	H.P.	Speed	Radius of Action	T. tubes	Complement	Max. draught
			tons		kts.				feet
6	"*Improved Peral*" type (H)	'16-'20	610 740	1400 850	16 10½	...	4-6	...	11¼
3	*Monturiol* or *A* class (FL)	'15-'17	260 382	700 500	13 8½	650 miles at 8.5 kts. 85 miles at 4 kts.	2	17	10¼
1	*Peral* (H)	'15-'16	488 750	1200 680	15 10	3100 miles at 11 kts. 100 miles at 5 kts.	4	32	11

(H) = Holland type (FL) = Fiat-Laurenti type.

The 1915 Naval Programme provides for 28 submarines, out of which 4 boats have been completed and 6 are building; the remaining 18 boats may not be proceeded with.

Submarines (*continued*).

Building?

6 *Holland type*: Names and numbers not known. Begun at Cartagena, July, 1916. Are an improved enlarged *Isaac Peral* type. Displacements: 610 tons on *surface*, 740 tons *submerged*. Dimensions about × 18¾ × 11¼ feet. Machinery: on *surface*, 2 sets Diesel engines, totalling 1400 B.H.P. = 16 kts.; *submerged*, electric motors and batteries of 850 H.P. = 10-10½ kts. Guns: 2 (calibre unknown, but one ... types). Torpedo tubes: 4 or 6 (size uncertain). Provided for by Law of 17th February, 1915.

Note.—Will probably be numbered *B1—B6*. Reported unofficially that first two boats will be launched in Ju... and delivered November, 1920; the other four boats to be subsequently delivered at the rate of one eve... months. It appears more likely that the first boat will not get afloat till 1921, and the last may not be de... till 1922 or 1923.

Photo 1919, T. de M. N. M...

MONTURIOL (A1). These boats have a tall telescopic mast for W/T abeam of after periscope on port side of ...

Laurenti-Fiat type: **Narciso Monturiol** (**A1**), **Cosme Garcia** (**A2**) and **A3**, (all built by Fiat San Giorgio Co., Spezia, 1915-17). Displacements: 260 tons on *surface*, 382 tons *submerged*. Dimensions: ... × 13⅝ × 10¼ feet. Machinery: on *surface*, 2 sets of 350 H.P., 6-cylinder, 2-cycle Fiat Diesel engin... 700 H.P.; when *submerged*, 2 electric motors of 250 H.P. + batteries = 500 H.P. Maximum speeds: 13 on *surface* and 8½ kts. *submerged*. Radii of action: on *surface*, 650 miles at full speed and 1600 mil... 8½ kts.; when *submerged*, 18 miles at 8 kts. and 85 miles at 4 kts. Torpedo tubes: 2—18 inch in b... Complement 17. Built under Law of 17th February, 1915.

Note. Each boat is internally sub-divided into 8 w.t.c. Maximum diving depth 130 feet (about 22 fathoms). Detachable lead keel, weighs 9½ tons. Fitted with telephone buoy, Fessenden submarine signalling receiving apparatus, and Marconi W/T. For rescue and salvage work there are 4 lifting rings attach... and 2 escape hatches within hull, for crew. 2 Periscopes about 3½ inches diameter.

(*Continued on next pa...*

(Continued).

Photo (1919), T. de N. M. Mille.

...onning tower; two ventilators on top of C.T. may also ...n in above illustration.

...ry, '16). Displacement: 488 tons on surface, 750 tons ... 11 feet. Machinery: on surface, 2 sets of 500 H.P. (made 15·36 on trials); submerged, electric motors ...tion: 3700 miles on surface, at cruising speed of about ... 3 in. Q.F. (anti-aircraft and disappearing mounting). ...ember, 1915.

(Cañoneros de 1ª Clase).

...ed.

...artagena, Jan., 1920. Displacement, 1500 tons. ...uary, 1915. No other details known.

1st Class Gunboats—(Continued).

RECALDE. Photo (1919), T. de N. M. Mille.

BONIFAZ (1911), **LAURIA** (1912), **LAYA** (1910), **RECALDE** (1910). 787 tons. Complement, 121. Dimensions: 213½ × 30 × 9½ feet (*max*) draught. Guns: 4—14 pdr. 2 machine. Designed H.P. 1100 = 14 kts. Made 14·5 kts. on trial. Yarrow boilers. Coal: 148 tons. Radius, 3000 at 10 kts. Built under 1908 Navy Law.

Photo (1919), T. de N. M. Mille.

DON ALVARO DE BAZAN (1897). 810 tons. 236 × 27 × 11¼ feet. Guns: 6—6 pdr., 2—2½ pdr., 2 M.G. I.H.P. 2500 = 19 kts. Coal: 168 tons. Complement 121.

(Continued on next page).

First Class Gunboats. *Continued.*

Photo (1918), T. de Navio, M. Mille.

MARQUES DE LA VICTORIA (1897, reconstructed 1917-18), 810 tons. 233 × 27 × 10¾ feet. Guns: 8—6 pdr., 2 M.G. I.H.P. 3500 = 19 kts. Coal: 168 tons. Complement, 121.

1916 Photo, T. de Navio, M. Mille.

DOÑA MARIA DE MOLINA (1896). 810 tons. 236 × 27 × 11¼ feet. Guns: 8—6 pdr., 2 M.G. H.P. 3500 = 19 kts. Coal: 183 tons. Complement, 121.

1920 Photo, T. de Navio, M. Mille.

INFANTA ISABEL (1886, reconstructed 1911). 1130 tons. 211½ × 32 × 13 feet. Guns: 4—6 pdr., 16—6 pdr. I.H.P. 1500 = 12 kts. 1 screw. Coal: 194 tons. Complement, 191.

Second Class Gunboats. *(Cañoneros de 2a clas.)*

VASCO NUNEZ DE BALBOA (1895) & **HERNAN CORTES** (1895). 295 tons. 155 × 21½ × 7 feet. Guns: 2—3 pdr., 2—1 pdr. or Maxims. I.H.P. 500 = 10 kts. 1 screw. Coal: 61 tons. Complement, 54.

Since reconstructed, with new funnel. *1920 Photo, T. de N. M. Mille.*

MARQUES DE MOLINS (1895). 560 tons. 177½ × 23 × 10¼ feet. Guns: 4—3 pdr., 3 Machine. Speed, 13 kts. Coal: 134 tons. Complement, 93.

Third Class Gunboat. *(Cañonero de 3a clas.)*

1916 Photo, T. de Navio, M. Mille.

MACMAHON (1887). 114 tons. 91½ × 16 × 5 feet. Guns: 2—2[illegible]

River Gunboats. *(Lanchas-Cañoneros.)*

1919 Photo, T. de Navio, M.

CARTAGENERA (1908). 27 tons. Complement, 12. 53.3 × 3.9 feet. Guns: 1 pom-pom. I.H.P. 120 = 17 kts.

1920 Photo, T. de N. M. Mille.

PERLA (1887). 42 tons. 7 kts. 1 machine gun.

Submarine Salvage Vessel.

(Buque para salvamento de submarinos.)

(Photograph wanted.)

KANGARU (1917). Double hulled type, with interior docking. Built by Werf Conrad, Haarlem, Holland. About 300. Length, 275·4 feet; beam, 65·4 feet; draught (with sub docked), 13 feet. Can dock submarine up to 151 feet long. power: Can raise 650 tons, from 27½ fathoms. H.P. 1000 =

Note: Not delivered. Law suit proceeding with builders. Rotterdam, 1920.

Mining Vessels.

Mine Layers.

Old Cruiser **RIO DE LA PLATA** and old Destroyers of the *Audaz* class have been equipped for this duty. Also some (or all) of the 15 Coastguard and Fisheries Vessels of the *Esmeralda* type (see third column of this page) will be available as Mine Layers when built.

Auxiliaries.

Hercules (Repair Ship), *Bermeo*, *Donostierra*, *Guipozcoa*, *Marisino*, *Primero de Meira*, and three others.

Photo (1915), *T. de N. M. Mille.*

ANTELO (1903). Ocean-going Tug, Mine Layer, Mine Sweeper and Training Ship for Mining Service. 342 tons. 131·2 × 21·6 × 6·8 feet. I.H.P. 650 = 11 kts. Coal: 32 tons. Carries 40 mines.

Transport. (*Transporte.*)

Photo (1919), *T. de N. R. Mille.*

ALMIRANTE LOBO (1909). 2545 tons. Guns: 2—5 pdr. Designed H.P. 4300 = 12 kts. *Nominal* radius, 4540 miles at 10 kts.

Surveying Ship. (*Comisión Hidrograf*

URANIA (1895). 801 tons. 195 × 26 × 10¾ feet. Gu 2 screws. I.H.P. 350 = 10 kts. Coal: 200 tons. N 6400 at 8 kts.

Coastguard and Fisheries Vessels. (G

Projected.

ESMERALDA, *INTREPIDA*, *PRONTA* (Malaga), *LIE MARIA* (Cadiz), *SERPIENTE* (Algeciras), *CUER CEDIDA* (Valencia), *RADIANTE* (Barcelona *GARZA*, *PEZ* (Balearic Islands) and 3 *others*, under new Naval Programme. All to be equip laying and sweeping in war-time. 300 tons. 114·8 × 19·7 × feet. Speed, 10 kts. Machinery engines. Complement, 70.

Note.—Above boats sanctioned under 1915 Navy tracted for by private firms but none begun yet. are not building yards but stations to which these assigned, when completed.

GAVIOTA. 1926 *Photo, T. de Nari*

DELFIN, DORADO, GAVIOTA (1910-11). 15 1—5 pdr. Speed, 11 kts.

PORTUGUESE FLEET.

Officially Revised, 1920, by courtesy of the Ministry of Marine, Lisbon.

Re-illustrated 1919, from photographs furnished by courtesy of Captain Ernest Vasconcellos and A. W. Paterson, Esq., of Lisbon. Also checked by notes made by H.C.L. during visit to Lisbon, 1920.

All displacements in English tons unless otherwise stated.

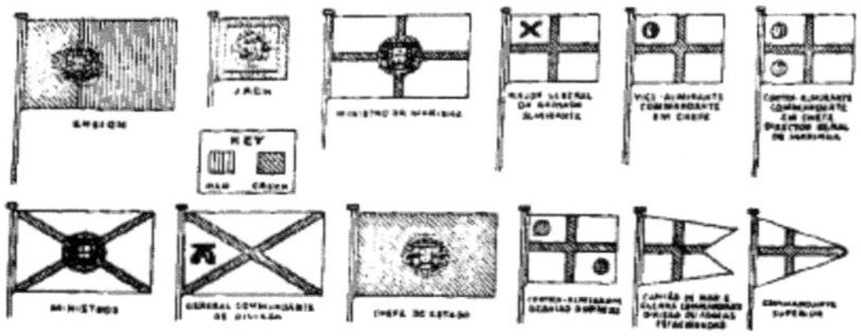

Naval Uniforms.

Almirante.*	Vice-Almirante.†	Contra-Almirante.‡	Capitão de Mar e Guerra.	Capitão de Fragata.	Capitão Tenente.	Primeiro Tenente.	Segundo Tenente.	Guarda Marinha.
Admiral.	*Vice-Admiral.*	*Rear-Admiral.*	*Captain.*	*Commander.*	*Lieut.-Commander.*	*Lieut.*	*Sub-Lieut.*	*Midshipman.*

* Admiral has four *gold* stars. † Vice-Admiral has three *gold* stars. ‡ Rear-Admiral has three *silver* stars.

Staff officers same but *without* executive curl. Colour between stripes: *Surgeons*: red; *Engineers*: violet; *Paymasters*: blue; *Constructors*: purple red. On visor of cap: *Admirals*: 2 oak leaves; *Captains*: 1 oak leaf; *Commanders*: 1 narrow stripe. Uniforms like British Navy. Chin strap of gold cord, but officers of lieutenant's rank and below have black chin straps.

Personnel:

Trade Ports: Lisbon and Oporto.

Oversea Possessions: Cape Verde Islands, Madeira and Azores, Guinea (Senegambia), San Thomé and Prince Islands (Gulf of Guinea), Timor (East Indies), Goa, Damao and Diu (India), Macao (China), Angola (Congo), Portuguese East Africa.

Coinage: Escudo (1000 reis) = 4s. 5d. British, \$1.06 U.S.A.; 4500 reis = £1 English

Colour of Ships: Dark grey.

Minister of Marine: Commander Judice Bicker.

Naval Bases.

LISBON. Some small slips. Government dry dock: 258 × 56¾ × 12 feet. Harbour Board dry docks: No. 1, 540 × 82 × 32½ feet; No. 2, 328 × 49 × 27 feet. Private dock, H. Parry & Sons (Cacilhas): No. 1, 225 × 37 × 9 feet. There are also two smaller docks under construction. A new arsenal is projected on other side of Tagus opposite present arsenal. Will also undertake mercantile building. Estimated time to build new yard, 6 years.

S. PAULO DE LOANDA (ANGOLA). There is a small floating dock here of 1,100 t capable of taking any of the gunboats employed on this station.

Mercantile Marine.

(From "Bureau Veritas" 1919-20 figures. Ships under 100 tons gross exclude

126 steam and motor ships of 229,392 tons gross.

2 vessels (with auxiliary motors) of 476 tons gross.

221 sailing vessels of 45,368 tons gross.

During 1919 Portuguese shipyards launched 20 ships of 10,499 tons gross; some of thes account.

Scale: 1 inch = 160 feet. RECOGNITION SILHOUETTES.

ONE FUNNEL.

For Two Funnel Types, Torpedo Craft, Submarines, see ne

Scale: 1 inch = 160 feet. RECOGNITION SILHOUETTES—*Continued.*

TWO FUNNELS

Patria.

Republica.

C. Afonso.

Vasco da Gama.

Adamastor.
(Dismantled: Bow and Stern Guns removed.)

5 de Outubro.
(Has W/T. gallant mast now.)

Almirante Reis.
(Dismantled at present; has no funnels, guns, topmasts, or boats.)

TORPEDO CRAFT **Scale: 1 inch = 160 feet.**

No. 2, No. 3 (t.b.)

Ex-85F–91F (6)

Tejo t.b.d.
(as reconstructed.)

Douro *class* t.b.d. (2).

SUBMARINES. **Scale: 1 inch = 80 feet.**

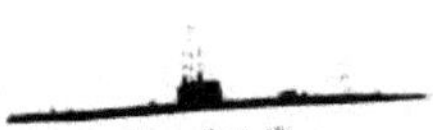

Foca *class* (3)

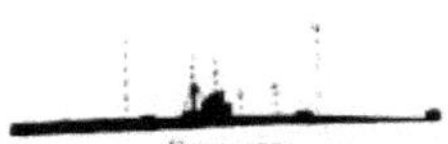

Espadarte.

Old Battleship *(Couraçado)*

VASCO DA GAMA (Blackwall, 1876, reconstructed and lengthened by Orlando 1902). Comp., 259. Dimensions: 232·9 × 40¼ × 18¾ feet. Guns (new in 1902): 2—8 inch, 1—4 inch, 5—3 pdr. No torpedo tubes. Armour (iron): 9"—7" Belt, 7¼" (Terni) Machinery: 2 sets triple expansion. 2 screws. Designed H.P. (new): 6000 = 15·5 kts. Cylindrical. Coal: 300 tons.

Protected Cruisers *(Cruzadores)*

Photo by favour of Major Munsell.

ALMIRANTE REIS (ex *Dom Carlos I*, Elswick, 1896). 4186 tons. Comp., 418. 360·8 × 46¼ × 17·5 feet. Guns (Elswick): 4—6 inch, 45 cal., 8—4·7 inch, 45 cal., 14—3 pdr 3 Machine. Torpedo tubes (14 inch): 2 *above water*. Deck, 4½". Machinery: 2 sets trip 2 screws. Boilers: 12 Yarrow. H.P. 12,000 = 19·5 kts. Coal: 820 tons.

Note.—Dismantled 1918 for re-construction and long refit. Still dismantled, March, 192

Protected Cruisers—(*Continued*).

Photo, Abrahams & Sons.

SAO GABRIEL (F. & C. de la Med., 1898). 1809 tons. Comp., 242. Dimensions: 246 × 35·5 × 14·3 feet. Guns (Canet): 2—5·9 inch, 4—4·7 inch, 8—3 pdr., 2—1 pdr., 3 machine. Torpedo tubes: 1—18 inch (bow) *above water*. 1" Deck, 2½" C.T. Machinery: 2 sets triple expansion. 2 screws. H.P. 3000 = 15 kts. Coal: 289 tons. Boilers: 4 Normand-Sigaudy.

(Dismantled 1919.) *1919 Photo.*

***ADAMASTOR** (Orlando, 1896). 1729 tons. Comp., 206. Dimensions: 242·1 × 35·1 × 15·3 feet. Guns (Krupp): 2—old 5·9 inch (not on board 1920), 4—4·1 inch, 4—9 pdr., 2—1 pdr., 3 machine. Torpedo tubes (14 inch): 3 *above water* (bow and broadside). Armour: 1½" Deck, 2¾" C.T. Machinery: 2 sets triple expansion. 2 screws. Boilers: 8 cylindrical. H.P. *forced* 4000 = 18·19

3 + *2* Destroyers. (*Contra-torpedeiros*).

No.	Type	Date	Displacement	H.P.	Max. speed	Coal	Complement	T. tubes	dr
			tons		kts.	tons			
2	*Vouga* (Y)	'14—?	660	11,000	27*t*	146	73	2	
2	*Douro* (Y)	'11-'13							
1	*Tejo*	'01-'17*	526	7000	27	140	73	2	1

(*t*) = turbine. (Y) = Yarrow. * Rebuilt by this date

GUADIANA. 1919 *Ph*

4 *Yarrow type*. **Vouga** (1920, completing) and *Tamega* (building) both by Lisbon D.Y., **Douro** (1913) and **Gua** (1914). 660 tons. Dimensions: 240 × 23½ × 7¾ feet, *max.* draught. Armament: 1—4 inch, 2—3 inch, 2 tubes. turbines, 3 Yarrow boilers. Coal: 146 tons. 1600 miles at 15 kts. Built by Yarrow and assembled at On trials, *Douro* made 30·5 kts., good for 27 now.

1919 Photo.

1 *Navy type*. **Tejo** (Lisbon D.Y. 1901, *rebuilt* 1917). 526 tons. Dimensions: 225¾ × 23 × 8·7 feet. Armament: 1—1—3 inch, 2—3 pdr. 2—18 inch tubes. Machinery: 2 sets triple expansion. 2 screws. Boilers: White-F H.P. 7000 = 27 kts. (*estimated*). Wrecked in 1910, salved, and reconstructed.

18 **Torpedo Boats.** (*Torpedeiros*).

Photo wanted.
(*For appearance see Silhouettes.*)

Assigned to Portugal for Police Duties only, Sept., 1920.

6 Ex-Austrian Boats. ***Nos. 85F,*[1] *86F,*[1] *88F,*[1] *89F,*[1] *90F,*[1] *91F*[1]** (Ganz-Danubius Co., Porto Re, Fiume, 1913-15). Displacement: 250 tons. Dimensions: 188.5 x 18.7 x 4.7 feet. Guns: Originally armed with 1—11 pdr., 1—11 pdr. AA. and 1 M.G., but only one gun is to be mounted for Police Duties. Torpedo Tubes to be removed. Designed S.H.P. 5000 = 28 kts. Turbine engines and Yarrow boilers. Fuel: —— tons coal + —— tons oil.

[1] Austrian numbers, new names or numbers not reported.

(*For appearance see Silhouettes.*)

2 old **Yarrow** type. ***No. 2*** and ***No. 3*** (Poplar, 1886). 66 metric tons. Dimensions: 119½ x 12½ x 3.4 feet. Armament: 2 M.G. and 2 tubes. Designed H.P. 700 = 19 kts. (much less now). 1 screw. Coal: about 18 tons. Complement, 22.

4 **Submarines.** (*Submersiveis*)

No.	Class	Date	Dis-place-ment	H.P.	Speed	Endurance	T. tubes	Com-ple-ment
			tons		kts.			
3	*Foca class*	'15-'17	260 / 388	550 / 300	14.2 / 8	[illegible] miles at 8½ kts. / 18 = 100 miles.	2	21
1	*Espadarte*	'10-'13	245 / 300	550 / 300	13.8 / 8	1300 miles at 8½ kts. / 21 = 100 miles.	2	21

All above boats are of the Laurenti type. Displacements in metric tons. All have Fessenden submarine sig[illegible]

3 *Laurenti-Fiat* type: ***Foca, Golfinho,*** and ***Hidra*** (all built at Spezia, 1916-1917). Dimensions: 147 x [illegible] Armament: 2—18 inch bow torpedo tubes, 4 torpedoes. Machinery: 2 sets of 275 B.H.P. 6-cylinder Fiat-[illegible] on *surface*; electric motors + batteries *submerged*. Maximum speeds: 14.2 kts. on *surface* and 8½ kts. *subm*[illegible] of action: about 650 miles at full speed, and 1200 miles at 7-8 kts. on *surface*; about 18 miles at 8½ kts. at *submerged*.

1920 Photo, Ministry.

1 *Laurenti-Fiat* type: ***Espadarte*** (Spezia, 1912). Displacement, $\frac{245}{300}$ tons. Dimensions: 148 x 13½ x 9.7 [illegible] tubes: 2—18 inch (bow). 4 torpedoes carried (4000 yds. at — kts.). Machinery: 2 sets 275 B.H.P. 6-[illegible] engines ***on surface***; electric batteries + motors ***submerged***. Maximum speeds, 13 kts. ***above*** and 8 k[illegible] Radius of action: 1300 miles at 8½ kts. *on surface*; 21 miles at 8 kts., and 100 miles at 4 kts. *submerged*.

Fleet-sweeping Vessels (Sloops).

No photo available.

Appearance as British "Flower" Class Sloops with *one* mast.

REPUBLICA (ex-British Fleet-sweeping Vessel *Gladiolus*, launched Oct., 1915, by Messrs. Chas. Connell & Co., Scotstoun, N.B.). 1250 tons. Dimensions: 255¼ (*p.p.*), 267¾ (*o.a.*) × 33½ × 11 (*mean*), 11¾ feet (*max.* draught). Guns: 2—4 inch, 2—3 pdr. *A.A.* Designed I.H.P. 1400 = 17 kts., but actually requires about 2200 I.H.P. for this speed. Machinery: 1 set 4-cyl triple exp. Boilers: 2 cylindrical. 1 screw. Coal: 130 tons *normal*; 260 tons *max.* = about 2000 miles at 15 kts. Complement (79 as British ship). Built under British Emergency War Programme as a unit of the *Azalea* group of "Flower Class" Sloops. Sold to Portugal 1920.

No photo available.

Appearance similar to British "Flower" Class Sloops with *two* masts.

CARVALHAO ARONJO (ex-British Fleet-sweeping Vessel *Jonquil*, launched May, 1915, by Messrs. Chas. Connell & Co., Scotstoun, N.B.). 1200 tons. Dimensions: 250 (*p.p.*), 262¼ (*o.a.*) × 3[illegible] (*mean*), 11¾ feet (*max.* draught). Guns: Not known. Designed I.H.P., speed, machinery, [illegible] &c., as *Republica* opposite, but *max.* coal capacity 250 tons, and complement (as British sh[illegible]). Built under British Emergency War Programme as a unit of the *Acacia* group of "Flower [illegible]" Sloops. Sold to Portugal 1920.

Auxiliary Cruiser (*Cruzador auxiliar*).

1920 Photo, Frenol, Brest.

PEDRO NUNEZ (ex-s.s. *Malange* of Empreza Nacional, Lisbon, 1889). 5574 metric tons. 331 × 38 × feet. Guns: 2—4.7 inch, 2—3 inch, 2—3 pdr. I.H.P. 2300 = about 16 kts. 1 screw. Complement, 209.

"Aviso," or Despatch Vessel.

1919 Photo

CINCO DE OUTUBRO (ex *Amelia*, ex *Banshee*, Ramage & Ferguson, Leith, 1900). 1343 tons. 226·9 × 28·9 × 14 feet. Guns: 2—1 pdr. H.P. 1800 = 15 kts. 2 screws. Cylindrical boilers. Coal: 230 tons. Complement, 225.

Gunboats (*Canhoneiras*).

1 *Ex-Turkish Gunboat.* ———— To be selected from *Aidin Reis* or *Kemal Reis* classes.

Building?

THREE NEW GUNBOATS (names unknown), reported as been seen on stocks at Lisbon D.Y., March, 1920, when [illegible] gress was thus: No. 1 and No. 2 in frame to upper dec[k] and lower deck at stern; No. 3 in frame to main deck at lower deck at stern.

No details known of these Gunboats. The above partic[ulars] purely unofficial.

(*Gunboat types continued on ne[xt page]*)

Gunboats (*Canhoneiras*)—*continued.*

Photo, R. L. Ferrett, Esq.

BENGO (1917), **MANDOVI** (1917), **QUANZA** (1918), **BEIRA** (1910) and **IBO** (1910). All built at Lisbon D.Y. 463 tons. Dimensions: 147 × 27½ × 6·8 feet. Guns: in *Bengo*, 1—3·5 inch, 4—3 pdr.; *Mandovi* and *Ibo*, 1—3 inch, 2—6 pdr., 2—3 pdr.; in *Quanza*, 2—3 inch; in *Beira*, 1—3·5 inch, 1—9 pdr., 2—3 pdr. H.P. 700 = 13 kts. Coal: 85 tons. Radius: 2200 miles at 9 kts. Boilers: Cylindrical. 2 screws. Complement 71.

1919 Photo.

(C) **SAYE** (1908) and (C) **LURIO** (1907). Both built at Lisbon D.Y. 305 tons. Dimensions: 140·4 × 23·6 × 5·9 feet. Guns: 2—3 pdr., 1 machine. H.P. 500 = 12½ kts. Boilers: Cylindrical. 2 screws. Coal: 61 tons. Complement 51.

(C) **PATRIA** (Lisbon D.Y., 1903). 626 tons. Dimensions: 196·8 × 27·5 × 8·4 feet. Guns: 4—3·9 inch, 6—3 pdr., 1 machine. ½" water-line belt. H.P. 1800 = 16·7 kts. Cylindrical boilers. 2 screws. Complement 88.

Gunboats (*continued*).

1919 Photo.

LIMPOPO. (Poplar, 1890). 288 tons. 121 × 21 × 6½ feet. Guns: 3—6 pdr., 1 machine. H.P. 525 = 11·5 kts. Complement, 51.

ZAMBEZE (Lisbon, 1886). Wood. 606 tons. 146 × 25½ × 11 feet. Guns: 2—3 pdr. H.P. 400 = 10 kts. Cylindrical boilers. Coal: 90 tons. Complement, 65.

Photo, Sub-Lieut. Pepeta, R.I.M.

SADO (Birkenhead, 1875). 635 tons. 140½ × 28½ × 10¾ feet. Guns: 2—4·1 inch, 2—9 pdr., 1—1 pdr., 1 machine. H.P. 500 = 11 kts. Cylindrical boilers. Coal: 84 tons. Complement, 30.

(C) On Colonial Service. **Gunboats** (*continued*).

ACOR (ex-s.s. *Gomes VII*, 1874). 330 tons. 128 [illegible] Guns: 1—3 pdr. I.H.P. 300 = 9 kts.; could make 11 [illegible] Complement 53. Has been employed on fisheries a[illegible] duties. Will probably be condemned soon.

For River Gunboats, see next page.

Mine Layers. (*Vapores lança minas*)

VULCANO (1910). Built by Thornycroft. 151 tons [illegible] 4½ feet. 2 dropping gears forward. H.P. 412 = 12 kt[s.] Cylindrical boiler. Complement, 24.

MINING VESSEL, name not known. (1892). 78 [illegible] 58·5 × 13·4 × 7 feet. I.H.P. 150 = 8 kts. 2 screws. Boiler. Coal: 10 tons.

Small River Gunboats. (*Lancha Canhoneira*)

(C) **MACAU** (Yarrow, 1909). 133 tons. 119½ × 19·8 × 2 feet. Guns: 2—6 pdr., 3 machine. H.P. 250 = 11·8 kts. Boilers: Yarrow. Complement, 24.

Note.—On Zambesi River, there is also a Stern Wheel Gunboat *Tete*, built by Yarrows, 1918-19, but she is the property of the Colonial Department. Displaces 70 tons. Armed with 2—1 pdr. (Hotchkiss), and 2 M.G. Speed, 9 kts.

No Photo available.

(C) **FLECHA**, is also given in Official Navy List as **FLEXA** (1909). tons. 68·8 × 13·1 × 2·2 feet. Guns: 1 or 2 small. 1 screw. Speed 10 kts. Complement 7.

Note.—There is also a *Flotilla of 7 launches* for River Service in Portuguese India, with civilian crews. They are under the command of the "Captain of Ports."

1912 Photo.

RIO MINHO (Lisbon D.Y., 1904). 38 tons. Paddle wheel. 80·7 × 13·1 × 1·9 feet. H.P. 64 = 7½ kts. Guns: 2—1 pdr. Complement 68, are nominally borne on her books, the majority of these ranks and ratings being actually at various shore stations on the Portuguese bank of the river Minho, for fisheries, customs duties, &c.

Transports.

1912 Phot

SALVADOR CORREIA (Birkenhead, 1895). 300 metric 140 × 21 × 7 feet. Guns: 1—3 pdr., 2—1 pdr. H.P. 450 = 11 1 screw. Cylindrical boilers. Coal: 80 tons. Complement,

Also

Zaire (1913). 300 metric tons.

Massabi (1886). 296 metric tons.

**Flecha*.

*Above three vessels serve as Transports in the Colonies of Gu and Angola; they have mixed complements of naval officers civilians.

Salvage Vessel. (*Vapor de Salvação*)

Similar to *Berrio* (next page), but hull flush from end to end.

PATRÃO LOPEZ (——). 1100 metric tons. 157·5 (p.p.) × 2 × 14·6 feet. I.H.P. 326 = 10 kts. 1 screw. Coal: 130 to Complement, 53.

(C) = On Colonial Service.

Tugs. (*Rebocadores.*)

1919 Photo.

BERRIO (1898). 198 metric tons. 132·9 × 22·5 × 10·5 feet. H.P. 1050 = 12·27 kts. 2 screws. Complement 14.

1919 Photo.

LIDADOR (1884). 252 metric tons. Iron. 113·8 × 20 × 6·9 feet. Guns: 1—6 pdr. H.P. 400 = 11 kts. 2 screws. Complement, 41.

Auxiliary.

No Photo available.

(C) **SALVADOR.** No details given by the Ministry of Marine.

For Fisheries Protection.

Note.—Gunboat *Acor* and River Gunboat *Rio Minho* may be employed on this service, from time to time.

For Various Duties.

No Photo available.

(C) **DILLY** (is also referred to in Official Navy List as **DILI**). 500 metric tons. 140 (*p.p.*) × 25 × 10·3 feet. I.H.P. 500 = 11 kts. 1 screw. Complement 12.

No photo available.

LINCE (1911). 78 metric tons. 88·5 (*p.p.*) × 14 × 5·5 feet. I.H.P. 300 = kts. 2 screws. Complement (nominally) 53.

Armed Launches.

Appearance as British M.L.

2 boats: Present names or numbers unknown. Are ex-*557* and M.L. *571*, built 1915, purchased for Navy, 1920. Load, displacement = 37 tons. Dimen 88¾ × 12⅓ × 3 feet (*max.* load draught). Guns: B.H.P. 440 = 19 kts. 2 sets Standard petrol motors. gallons petrol. Complement, 9.

Guarda-Marinha Janeiro	No details communica
Tenente Roby	Ministry of M

Also the old wooden sailing frigate *Dom Fernando* (1843), is totally non-effective for fighting purposes. Used Training Ship.

NAVY OF THE SERBS, CROATS AND SLOVENES.

"*Fighting Ships*" is informed by the Military Attaché, Legation of the Serbs, Croats and Slovenes, London, that, up to June, 1920, no decision had been arrived at regarding the future Navy of the Triune Kingdom.

12 Ex-Austrian Torpedo Boats.

(For Coastguard, Preventive and Police Duties only.)

SILHOUETTES. Scale: 1 inch = 160 feet.

Nos. 76T–79T (4). Nos. 87F–97F (4). Nos. 54T–61T, 69F (4).

Photos wanted.

Appearance as Silhouettes above.

4 Ex-Austrian boats: ***76T,* 77T,* 78T, 79T*** [illegible], 1913-15.

4 Ex-Austrian boats: ***87F, 93F,* 96F,* 97F*** [illegible], 1913-15.

Details of above:—Displacement: [illegible] tons. Dimensions: 188½ × 18⅔ × [illegible] feet. Guns: originally armed with [illegible] [illegible] A.A. and 1 M.G. [illegible]. Torpedo tubes to be removed. [illegible] H.P. turbines [illegible] = 28 kts. Yarrow boilers. Fuel: [illegible]

4 Ex-Austrian boats: ***54T,* 60T,* 61T**** (Stab. Tecnico, Trieste, 1906-7), and ***69F**** (Ganz-Danubius Co., Porto Ré, Fiume, 1908-9). Displacement: 200 tons. Dimensions: 189 × 18 × 4½ feet. Guns: 4—4 pdr., 1 M.G. (only one to be retained). Torpedo tubes to be removed. H.P. 3000 = 27 kts. Yarrow boilers. Fuel: 47 tons.

*Are Austrian numbers. These boats will be re-named or re-numbered.

Official Photo, K.N.A.

SERBIA. (Purchased 1916.) Under 100 tons. Speed: 12 kts. Is unarmed. National colours painted on funnel. Mainmast now removed.

ROYAL HELLENIC NAVY.

(GREEK FLEET).

Officially Revised, 1920, by courtesy of The Chief of Naval Staff.

Flags.

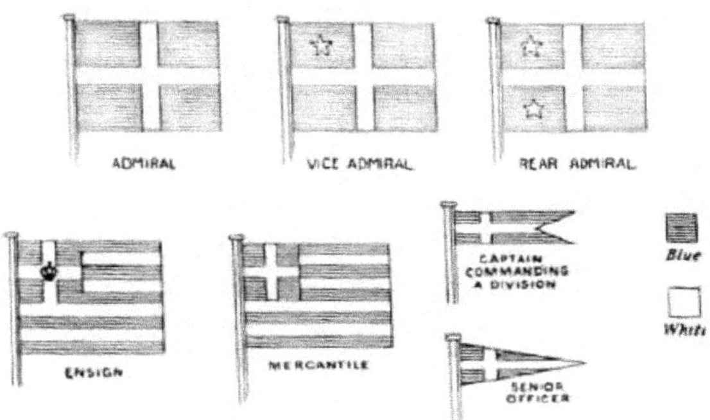

Uniforms.

(Corrected 1920, by courtesy of Rear-Admiral George Chetwode, R.N., Chief of Staff, Naval Mission to Greece).

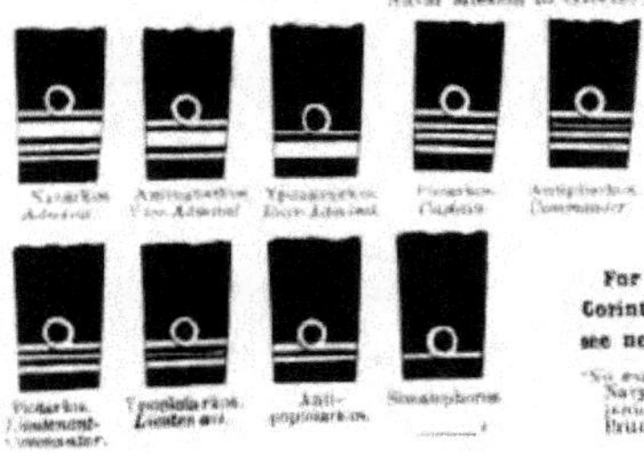

Other branches, without the curl:—

Surgeons: *purple velvet.*
Apothecaries: *green velvet.*
Constructors: *black velvet.*
Engineers: *violet velvet.*
Paymasters: *scarlet cloth.*

For details of Docks, Canal of Corinth, Fleet Anchorages, &c., see next page.

*No exact corresponding rank in British Navy. It is senior to Midshipman but junior to Sub-Lieut. Nearest equivalent British rank is Acting Sub-Lieutenant.

Scale: 1 inch = 160 feet. RECOGNITION SILHOUETTES.

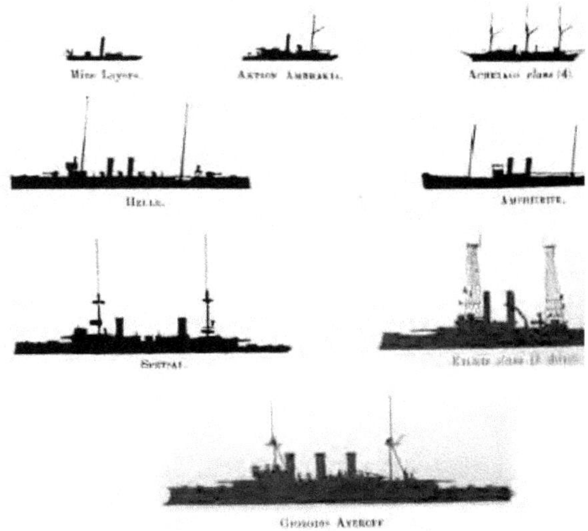

For Torpedo Craft and Submarines, v. next page.

RECOGNITION SILHOUETTES.—*Continued.*

TORPEDO CRAFT. **Scale : 1 inch = 160 feet.**

Ex 98 M—100 M.

Ex 92—95 F.

AIGLI *class* t.b. (6).

NIKI *class* t.b.d. (3).

THYELLA *class* t.b.d. (4).
Ex ULAN t.b.d.

LEON t.b.d.

AETOS *class* t.b.d. (3).

Docks.

SALAMIS (Arsenal). Government floating, 311 × 59 × 23 feet (3,000 tons).

Private Docks: There are two dry docks at Piræus (Harbour Trust), viz., "Vasilefs Giorgios," 460 × 69 × 20½ feet; "Vasilissa Olga," 320 × 51 × 26 feet; and a slipway, "Vassiliadis," 356 × 37 feet. *Max.* hauling capacity, 3500 tons.

Mercantile Marine: ("Bureau Veritas" 1919-20 figures; vessels under 100 tons *gross* excluded). 215 steamers of 349,512 tons *gross*, 582 sailing of 100,234 tons *gross*.

Chief Ports: Piræus, Syra, Volo, Salonika, Patras, Smyrna.

Oversea trade chiefly with England, Russia, and Austria.

Coinage: Drachma (100 lepta) = about 9½d. British, 18 cents U.S.A. (1914 values).

Personnel: About 4000 (conscript 2 years or enlistment).

Minister of Marine: A. Miaoulis. A British Naval Mission has been appointed to Greece.

Naval Attaché (London): Captain Panas, R.H.N.

Canal of Corinth.

Length: 6 kilometres = 3¾ miles *approx.* Breadth: (*a*) at surface 24·6 metres = 80·7 feet, (*b*) at bottom 21 metres = 68·9 feet. Depth: 8 metres = 26¼ feet. Not navigable by Dreadnoughts and Battle Cruisers.

Fleet Anchorages.

During the War, the Allied Fleets made extensive use of Greek Harbours and Anch **MUDROS BAY** (Lemnos) became the chief base of the British Eastern Mediterranean Fle Allied Squadrons during the War, being netted, mined and patrolled. A R.N. Airship Stati established at Mudros for anti-submarine service, with an outlying station at Kassandra. R.A.F. S were also created at Bessonreau, Imbros, Romanos, Syros, Stavros, Skyros, Syra, Thasos, Talikn Thermi-Mitylene (all under C.-in-C., Mediterranean & R.A. Aegean). IMBROS and **SUDA BAY** with R.A.F. Station) were also used as Fleet Anchorages. **ARGOSTOLI** (Cephalonia) was, for the base for the French Fleet, but later **CORFU** became the principal base of the French Squadrons, being netted, mined and patrolled by British R.N. Trawlers from N.E. of Corfu to S Island near Valona. Milo, Salamina and the Gulf of Patras were also netted. Navarin, Skyro other points were used as supply and sub-bases by the French Navy. **SALONIKA** was also mined and defended by shore batteries. The permanent works at these bases and stations have b the property of the Greek Government, and will be used by the Greek Navy in the future.

SMYRNA (Asia Minor) is now an anchorage used by Hellenic Warships.

Note.

This Battle-cruiser has been lying unfinished in German Ports for the past six years. Her present condition is believed to be so ruinous, it is most doubtful if she will ever be finished. The Greek Ministry of Marine has probably terminated the contract. Under the Treaty of Versailles, she cannot be incorporated in the Post-war German Fleet. Therefore she is listed here purely for the sake of reference. With other unfinished German warships she may possibly be converted into a mercantile ship.

(Greek *Salamis*, November, 1914).

Displacement, 19,500 tons.

Length, 570¾ feet. Beam, 82 feet. Draught, 25¼ feet.

Guns (Bethlehem—see *Notes*):
8—14 inch, 45 cal.
12—6 inch, 50 cal.
12—12 pdr.
Torpedo tubes (19·7 inch):
3 *submerged*

Armour (Krupp):
9¾" Belt (amidships) . .
4" Belt (ends) . . . .
6" Battery . . . .
9¾" Barbettes . . . .
11¾" C.T. (fore) . . . .

Launched: hull with armour and part of superstructure in place.

Machinery: A.E.G.-Curtis turbines. 3 screws. Designed S.H.P. 40,000 = 23 kts. Boil . Fuel: 1200 tons coal and also some oil fuel.

Gunnery Notes.—Bethlehem guns never delivered, the 14-inch being bought by British Admiralty for armin *Abercrombie* type Monitors.

Notes.—Begun for Greece by Vulkan, Hamburg, July, 1913. A German Naval Annual says this ship was laun under the name of *Vasilefs Giorgios.* Used as a Hulk from 1914 to 1918, and still unfinished.

Old Battleships.

LEMNOS. 1919 Photo, Captain H. D. Kollerton, R.N.

KILKOS (ex *Mississippi*) (December, 1905), & **LEMNOS*** (ex *Idaho*) (September, 1905). *Normal* displacement, 13,000 tons. *Full load* displacement, 14,465 tons. Complement, 802. Length (*waterline*), 375 feet. Beam, 77 feet. *Max.* load draught, 24¾ feet. Length *over all*, 382 feet. Guns: 4—12 inch, 45 cal., 8—8 inch, 45 cal., 8—7 inch, 45 cal., 12—3 inch, 13 pdr., 4—6 pdr. (saluting), 2—6 pdr. (anti-aircraft), 4—1 pdr., 8 M.G. (1 landing). Torpedo tubes (21 inch): 2 *submerged*. Armour (Midvale): 9" Belt (amidships), 4" Belt (ends), 3" Armour deck, 12"—8" Turrets, 10"—7½" Turret bases (N.C.), 6½"-6" Secondary turrets (N.C.), 7" Lower deck (redoubt), 7" Battery (redoubt), 9" Conning tower. Machinery: 2 sets vertical 4 cylinder triple expansion, 2 screws. Boilers: 8 Babcock. Designed H.P. 10,000=17 kts. (can only make 12 kts. now). Coal: *normal*, 750 tons; *maximum*, 1821 tons=6920 miles at 10 kts. Authorised 1903 for U.S. Navy; laid down at Cramp's, Philadelphia, May, 1904, and completed early in 1908. Sold to Greece early in 1914. *Will probably go on Sale List, 1921.

To go on Sale List, 1921. *Photo, Mons. C. B. Bontiri.*

SPETSAI (*Graville, France*, 1889, rebuilt 1900). Displacement, 4808 tons. Complement, 440. Length (*o.a.*) 344' feet. Beam, 51½ feet. *Mean* draught, 18 feet. Guns (Canet): 2—10·6 inch, 36 cals., 1—10·6 inch, 30 cals., 3—5·9 inch, 45 cals., 1—5·9 inch, 50 cals., 8—9 pdr., 4—3 pdr., 8 to 6—1 pdr. Torpedo tubes: 1—15 inch, 2—14 inch (all *above water*). Armour (Creusot steel): 12"—4¾" Belt (lower), 3" Belt (upper), 2½" Deck (flat), 13¾" Battery redoubt, 12" Barbette (aft), 6" Hood, 6" Hoists (main guns), 12" Conning tower. Machinery: 2 sets vertical triple expansion, 2 screws. Boilers: Belleville (1914-15). H.P. 6700=17 kts. Coal: *normal* 400 tons; *maximum*: others, 690 tons. Built by F. & C. de la Med., 1887-1890.

(1907) Armoured Cruiser.

1919 photo, Comm. R. D. Hal[illegible]

GIORGIOS AVEROFF (March, 1910).

Normal displacement, { 9,960 tons / 10,118 metric tons. } Complement, 550.

Length (*over all*), 462 feet. Beam, 69 feet. *Maximum* draught, 24¾ feet. Length (*p.p.*) [illegible]

Guns (Elswick):
- 4—9·2 inch, 45 cal.
- 8—7·5 inch, 45 cal.
- 16—14 pdr.
- 1—12 pdr. (anti-aircraft)
- 4—3 pdr.
- 2 M.G.

Torpedo tubes (18 inch):
- 2 *submerged* (broadside).
- 1 „ (stern).

Armour (Terni):
- 8" Belt (amidship[illegible]
- 3¼" Belt (ends) [illegible]
- 2" Deck ....
- 7"—6" Upper belt
- 4" Upper belt (a[illegible]
- 8" Main barbettes
- 6½" Turrets (N.C[illegible]
- 7" Citadel ....
- 7" Second'ry turr[illegible]
- 7" Conning towe[illegible]

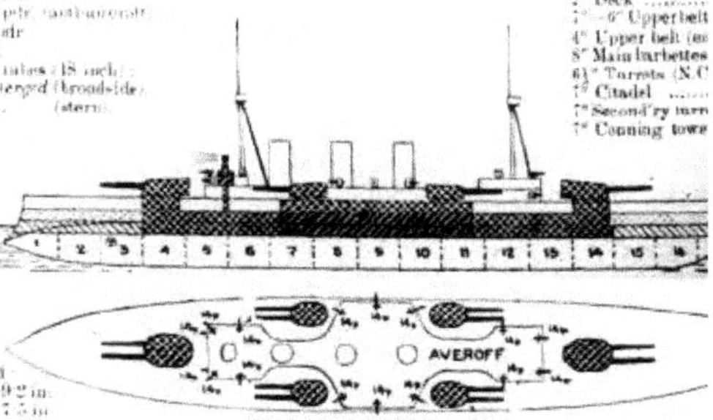

Ahead:
2—9·2 in.
4—7·5 in.

Broadside: 4—9·2 in., 4—7·5 in.

Machinery: 2 sets 4 cylinder triple expansion. 2 screws. Boilers: 22 Belleville. [illegible] 19,000=22·5 kts. Trials: 21,500=23·9. Coal: *normal* 660 tons; *maximum* 1500 tons= [illegible] 10 kts.; 2489 miles at 17½ kts. Built by Orlando, 1905-11.

Gunnery Notes.—All big guns electrically controlled. 12 pdr. anti-aircraft gun mounted on after superstru[illegible]

General Notes.—Cost £956750. Sister to Italian *Pisa* with certain modifications as to armament and rig. [illegible] ship formerly known as "X." Under the will of a deceased Greek millionaire M. Giorgios Averoff, ab[illegible] left for increasing the Navy. This sum was devoted towards the part purchase of the above [illegible] the cost being defrayed by the State. Still has shell holes in funnels from Balkan war. Refitted at Malta, 1919.

HELLE. 1920 Photo, by favour of [illegible], Esq., Smyrna.

HELLE (May, 1913).

Normal displacement, 2600 tons. Complement, 232.
Length (*over all*), 322 feet. Beam, 39 feet. Draught, 14 feet.

Guns (Armstrong):
2—6 inch.
4—4 inch.
2—14 pdr., 3 inch.
1—6 pdr. anti-aircraft.
2—3 pdr.
2—1 pdr.
Torpedo tubes (18 inch):
2 above water.

Armour (steel):
2" Deck on slopes
1" Deck on flat

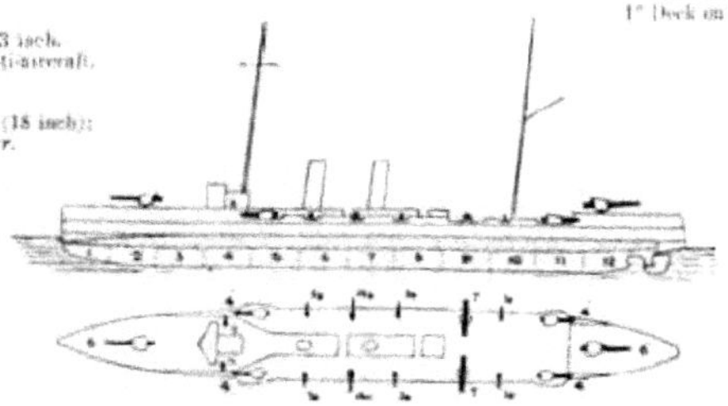

Machinery: Parsons turbine. 3 screws. Boilers: Thornycroft. Designed H.P. 6000 = 20 kts. Coal: *normal*, 400 tons; *maximum*, 600 tons + 103 tons of oil fuel = 5250 miles at 10 kts.; 3320 miles at 17¾ kts.

Name	*Builder*	*Machinery*	*Laid down*	*Completed*	*Trials:* *¼ hours.*	*Full Power.*	*Boilers*	*Best recent speed*
Helle	N.Y. Shipbldg.	N.Y. Shipbldg.	1910	Nov.'13	7500 = 20·3	8650 = 21	Thornycroft	

General Notes.—Built as the *Fei-Hung* for China. Purchased 1914.

12 Destroyers.

No.	Type	Date	Displacement (tons)	H.P.	Max speed (kts.)	Fuel	Complement	T. tubes
4	*Aetos* (L)	'10-'12	980	19,750 (t)	32	For 3000 at 15 kts.	102	4
1	*Ex Ulan* (Y)	'06-'08	400	6000	28	..	65	2
4	*Thyella* (Y)	'05-'07	350	6700	30	From 1140 to 1250 miles at 15 kts.	70	2
3	*Niki* (V)	'06-'07	350	6500	29·5			

(L) = Cammell Laird. (V) = Vulkan Co. (Y) = Yarrow. (t) = Turbines.

* Will go on Sale-List 1921.

T.B.D. LEON, as re-built with *three* funnels. 1920 Photo, by favour of J. [illegible], Esq., Smyrna.

(For Description, see next page.)

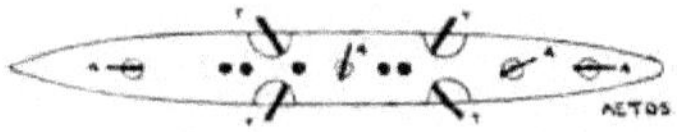

Notes to Plans.—*Aetos, Jerax, & Panther* have 3 funnels, forward pair being set to port and starboard of centre line. *Leon* has *three* funnels. In all boats the 6 pdr. AA. gun mounted on "bandstand" abaft last funnel.

AETOS (JERAX, & PANTHER as above view. For LEON see preceding page). 1920 photo, *F. L. King, Esq.*

4 (Cammell-Laird type: **Aetos, Jerax, Leon, Panther** (all launched 1911). 980 tons normal, 1175 *full load*. Dimensions: 293 × 27½ × 8 feet, *normal* draught; *full load* draught, 6 feet. Armament: 4—4 inch Bethlehem, 1—6 pdr. AA. 4—18 inch tubes. 4 spare torpedoes in carriers on upper deck. 2 searchlights. S.H.P. 19,750 = 32 kts. (30 about max. now). Combined Parsons and Curtis turbines. 5 White-Forster boilers. Coal: 215 tons + 84 tons oil = 3000 miles at 15 miles. These were 4 boats, *San Luis, Santa Fé, Tucuman,* and *Santiago*, built for the Argentine. Purchased by Greece, Oct., 1912.

1 *Ex-Austrian* boat: ———— (ex-*Lion*, Stab. Tecnico, Trieste, 1906). 400 tons. Dimensions: 219.8 × 20.2 × 6 feet. Armament: 4—11 pdr., 1 M.G., 2—17.7 inch tubes (all Austrian marks). Designed I.H.P. 6000 = 28 kts. 4 Yarrow boilers. Coal: 90 tons. Assigned to Greece September, 1920.

1920 Photo, *H. Freund, Brest.*

4 *Thyella* class. Yarrow type: **Thyella, Sphendoni, Navkratousa** and **Lonkhi** (1906-07). 350 tons. Dimensions: 220½ × 20½ × 6 feet, mean draught; max. draught, 9 feet. Armament: 2—14 pdr., 4—6 pdr. 2 tubes (18 inch). Speed: 30 kts. (about 28 max. now). Coal: 80 tons. Endurance: 1140 to 1250 miles at 15 kts.

3 *Niki* class. Vulcan Stettin type: **Aspis** (1906), **Niki** (1905), and **Velos** (1906). 350 tons. Dimensions: … × 6 feet, mean draught; max. draught, 9 feet. H.P. 6700 = 30 kts. (about 28 now). Armament: 2— … pdr. 2 tubes (18 inch). Coal: 90 tons. Complement 70. Endurance: 1140 to 1250 miles at 15 kts. … class was lost during the War.

12 Torpedo Boats.

Photos wanted.

For appearance, see Silhouettes.

For Coastguard and Police Duties only.

3 *Ex-Austrian* boats: **Nos. 98 M*, 99 M*, 100 M*** (built at Monfalcone, Italy, 1913-15). Displace… Dimensions: 188.5 × 18.5 × 4.9 feet. Guns: Originally armed with 1—11 pdr., 1—11 pdr. AA. and … one gun is to be retained for Police Duties. Torpedo tubes to be removed. Designed S.H.P. 3000 = … engines and Yarrow boilers. Fuel: … tons coal + … tons oil. Assigned to Greece, September, …

3 *Ex-Austrian* boats: **92 F*, 94 F*, 95 F*** (built by Ganz-Danubius Co., Porto Ré, near Fiume 191… as 98 M—100 M above.

*Austrian numbers; new names or numbers not reported.

6 Vulcan type: **Aigli, Alkyone, Arethousa, Dafni, Doris, Thetis** (all launched … Dimensions: 147 × 15½ × 4 feet. Armament: 2—3 pdr. Bethlehem, 3—18 inch tubes. S.H.P. 26… *Aigli* 26.2, *Doris* 25.7.

Hospital Ship.

To be sold 1921.

AMPHITRITE (Birkenhead, 1864, 1885). 1950 tons. Dimensions: $298\frac{3}{4} \times 26 \times$ feet. Guns: 4—1 pdr. H.P., 1200 = 14 kts. (Funnels have been lengthened since this photo was taken.)

Mine Layers.

To be sold 1921.

MONEMBASIA, AIGIALIA, NAUPLIA (1881). 290 tons. Dimensions: $75\frac{1}{2} \times 13\frac{1}{4} \times 8\frac{1}{2}$ feet. H.P. 60 = 9 kts. Coal: 60 tons. Carry about 20 mines each.

Old Gunboats.

"Achelaos" to be sold 1921.

ACHELAOS, ALPHIOS (Blackwall, 1884), **PENEIOS, EUROTAS** (Dumbarton, 1884). All rebuilt 1895-7. 404 tons. Dimensions: $131\frac{1}{4}$ (*p.p.*) $\times 21\frac{1}{4} \times 11$ feet. Guns: 2 old 3.7 in. in *Achelaos, Alphios, Eurotas* (+4 M.G. in *Eurotas*.) *Peneios* armed with only 4 M.G. H.P. 400 = 11 kts. Complement, 60.

AKTION — *French official War Photo.*

AKTION & AMBRAKIA (Blackwall 1881, re-built 1910). 433 tons. Dimensions: 128 (*p.p.*) $\times 25 \times 9\frac{1}{4}$ feet. Guns: 1—6 inch (new), 2—11 pdr. (in *Aktion* only), 1—1 pdr. H.P. 380 = 11 kts. Complement, 70.

Note.—Foremast in one or both ships now removed.

Old Gunboats (*continued*).

SALAMINIA (1858). 374 tons. Guns: not known. speed: 10 kts. Coal: 62 tons. Complement, 55.

Used only for transferring crews and officers from the Naval .

Oil Tanker.

Photo wanted.

PROMETHEUS (——). 5000 tons. $320 \times 42 \times$ — fe carry up to 4175 tons of oil fuel.

Salvage Tug.

(*1 funnel, 2 masts, schooner-rigged, swan bow.*)

TENEDOS (——). 100 tons.

Gunnery Training Ship.

(*1 funnel, barque-rigged, ram bow.*)

NAUARKOS MIAULIS (La Seyne, 1879). 1770 tons. various, from 12 pdr. downwards, and including 6 pdr. anti- for training purposes.

BULGARIAN DANUBE FLOTILLA.

Note.—Under the Treaty of Peace, Bulgaria is only allowed to maintain four fast Patrol Boats and six Vedette Boats on the Danube, for Police and Customs Duties.

Note that the red of the flag is crimson.

ENSIGN

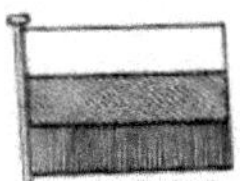

MERCANTILE

Torpedo Gunboat.

Originally belonging to Bulgarian Navy. Present existence uncertain.

NADEZHDA.

Photo, Geiser.

NADEZHDA (Bordeaux, 1898). 715 tons. Dimensions: 220 × 27½ × 10 feet. Guns (Creusot): 2—3·9 in., 2—9 pdr., 2—3 pdr. Torpedo tubes: 2 above water. Armour: Nil. Designed H.P. 2600 = 17 kts. (188? trial). Boilers: Lagrafel d'Allest. Complement 97.

Note.

On the capture of Sevastopol, in 1918, by the Austro-German Armies, *Nadezhda* was sent there to [illegible]. Later, her crew returned and placed their ship at the disposal of the Soviet Authorities. During 1919, she was reported as "Interned at Sevastopol." Nothing further can be traced about her. She has not appeared in the Russian "White" Fleets of General Denikin and General Baron Wrangel, nor can she be identified in lists of warships anchored at Sevastopol, February, 1920. She may have been scuttled during April, 1919, with other Russian (Black Sea) warships then destroyed.

Patrol Boats.

SILHOUETTE.

(Torpedo tubes removed.)

4 Creusot boats: **Smyeli, Khrabry, Deraki, Strogi** (built in sections in France; reassembled at Varna). [illegible] tons. Dimensions: 136½ × 13½ × [illegible] feet. Armament: 1—3 pdr. Torpedo tubes removed. Designed [illegible] = [illegible] kts. 1 screw. [illegible] boilers. Civilian complement, 25.

Vedette Boats.*

SIMEON VELIKI. 600 tons. Guns: 2—12 pdr., 4—3 pdr. 70 H.P. **ASJEN…** Guns: 4—3 pdr. 70 H.P.

Batef, Levski. Small craft of 12 tons, armed with spar torpedoes. Interned at Varna. *Boris, L…*, *Raicos, Rakovski Voivode, Stefan Karaja, Hadji Dimitar*, all small motor Vedette boats.

* Eleven vessels are named. Six (names unknown) are to be used as Vedette Boats, the other five are to be disposed of.

RUMANIAN NAVY.

RECOGNITION SILHOUETTES.

Scale: 1 inch = 160 feet.

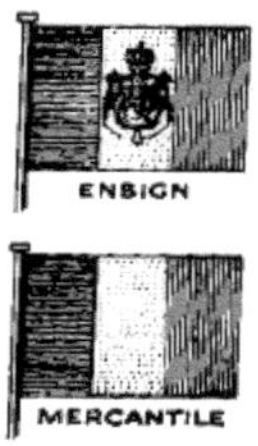

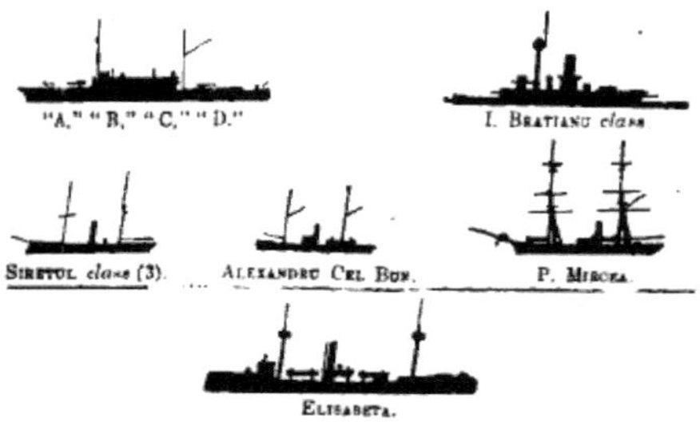

TORPEDO CRAFT.

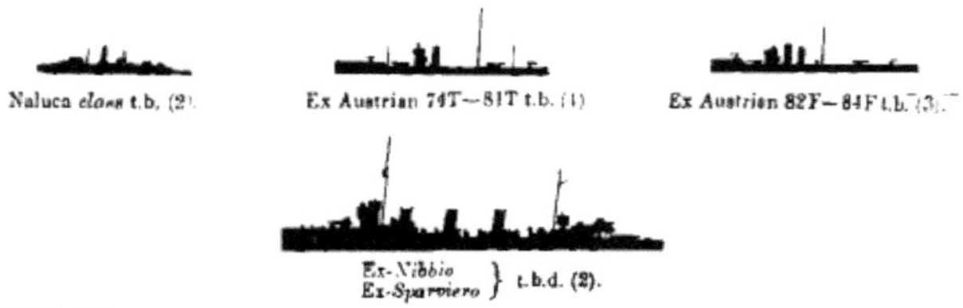

Personnel: ?

Mercantile Marine: (Bureau Veritas 1919-20 totals; ships under 100 tons *gross* excluded): 41 steamers of 73,993 tons *gross*; 1 sailing vessel of 240 tons *gross*.

Docks: At Galatz, on the Danube River, there used to be a floating dock in two sections, 264 × 56 × 16¼ feet (2400 tons). May have been destroyed during war.

Remarks.

Information for the revision of these pages would be much appreciated.

Note.

This space has been reserved, to effect the transfer of British Coast Defence Vessel. **GORGON,** from the British Navy to the Rumanian Navy, in the event of the sale being completed.

GORGON has been offered to the Rumanian Navy by the British Admiralty, but, up to the end of October, 1920, she had not been bought by Rumania. This vessel is accordingly retained in the British Navy Section.

Cruiser.

ELISABETA (Elswick, 1887, rebuilt 1905). 1380 tons. Complement, 150. Length, 239½ feet. Beam, 33¼ feet. *Mean* draught, 12¼ feet. Guns: (4—4·7 inch have been landed for service ashore during war and may not be re-mounted), 4—12 pdr. (modified for anti-aircraft fire), 4 machine, 2 S.L. Torpedo tubes (14 inch): 4 *above water*. Armour (Steel): 3¼" Deck (amidships), 1¾" Deck (ends). Designed H.P. 4700 = 17·3 kts. Coal: 300 tons.

Danube Monitors.

ION BRATIANU (1907), **MIHAIL COGALNICEANU** (1908), **ALEXANDRU LAHOVARI** (1908), and **LASCAR CATARGIU** (1907). Displacement 670 tons. Complement, 110. Dimensions: 208¼ × 33½ × 5¼ feet. Guns: 3—4·7 inch, 2—3 pdr. anti-aircraft, 2 machine, 2—14 inch S.L. Armour: 3" Belt, 3" Deck, 3"—2" Turrets. H.P., 1800 = 13 kts. Coal: 60 tons. Built by Stabilimento Tecnico, Trieste.

Lorient D.Y. **"A"** (ex-French *Chiffone*, launched 1917).
Lorient D.Y. **"B"** (ex-French *Friponne*, launched 1916).
Brest D.Y. **"C"** (ex-French *Impatiente*, launched 1916).
Brest D.Y. **"D"** (ex-French *Mignonne*, launched 1917).
Displacement: 355 tons. Length: 199 ft. 8 in. Beam: 22 ft. 7 in. Draught: 8 ft. 4 in. 3·9 inch. Engines: 2 sets of Sulzer Diesel motors. B.H.P. 900 = 15 kts. Fuel ca[illegible] oil = 3000 miles (10 kts.), 1600 miles (15 kts.) Complement, 50.

Notes.—Purchased from French Navy, 9th January, 1920.

2 Destroyers.

2 *Pattison type.* ———— (ex-Italian *Nibbio*, 1918), and ———— (ex-Italian *Sparviero*, 1919). 139[illegible] tons *full load*. Dimensions: 308·5 (*p.p.*) × 31 × 10⅞ feet. Guns: 3—6 inch, 40 cal., 1—14 pdr., 40 cal. [illegible] Tubes: 4—18 inch in 2 twin deck mountings. Designed S.H.P. 44,000 = 34 kts. Trials: *Sparvie[ro]* 38·04 kts., and has reached 40·94 kts. 2 Tosi turbines. 2 screws. 4 Thornycroft oil burning boilers. 1700 miles (15 kts.), 380 miles (full speed). Complement: 139.

Notes—Belong to class of 4 Destroyers, *Vifor*, *Vartej*, *Vartez*, *Vijelie*, ordered for Rumanian Navy [illegible] requisitioned for Italian Navy during war and re-named *Aquila*, *Falco*, *Nibbio*, *Sparviero*. Last tw[o] Rumania 1920; first two retained by Italy.

*Purchase of further Destroyers from British Navy pending as these pages go to press.

Destroyers—*continued.*

This space reserved to transfer British Destroyers of the "Admiralty M" Class to Rumanian Navy in 1921 Edition. Up to the end of October, 1920, these Destroyers had not been purchased by Rumania.

18 Torpedo Boats.

No.	Type	Date	Dis-place-ment	H.P.	Max. speed	Fuel (Coal)	Com-plement	Tubes or Gears	Max draught
			tons		kts.	tons			feet
4*	*74T—81T* (ST)	1913-15*	250	5000 t	28	—	—	Nil*	5
3*	*82F—84F* (GD)								
8	*Major G. Sontzu* (T)	1906	50	600	18	7	20	2 + 2	[illegible]
3	*Naluca* (F)	1888	60	540	16	7	20	2 + 2	7

(F) = Forges et Chantiers de la Med. (T) = Thames Ironworks.
(ST) = Stab. Tecnico, Trieste.
(GD) = Ganz-Danubius, Fiume.
(t) = Turbines.

*Ex-Austrian boats taken over 1920. To be re-named or re-numbered. Torpedo tubes to be removed from 6 boats (to be used only for Coastguard and Preventive Duties), but can be retained in one boat to be added to Navy.

Photo wanted.
(For appearance, see Silhouettes)

4 Ex-Austrian boats: ***74T,* 75T,* 80T,* 81T**** (Stab. Tecnico, Trieste, 1913-15). Displacement: 250 tons. Dimensions: 188.3 × 18.1 × 13 feet. Guns: Originally armed with 1—11 pdr., 1—11 pdr AA and 1 M.G., but only one gun is to be mounted for Police Duties.† Torpedo tubes to be removed.† Designed S.H.P. 5000 = 28 kts. Turbine engines and Yarrow boilers. Fuel: —— tons coal —— tons oil.

*Austrian numbers; new names or numbers not reported.

†Except one boat which may be added to Navy, retaining all guns and 3—17.7 inch torpedo tubes. (This boat might be selected from *82F—84F* in next column.)

Torpedo Boats (*continued*).

Photo wanted.
(For appearance, see Silhouettes)

3 Ex-Austrian boats: ***82F,* 83F,* 84F**** (Ganz-Danubius Co., Porto Re, Fiume, 1913-15). All details as ref...

*To be re-named or re-numbered.

No photo available.
2 funnels abreast.

8 Major Sontzu class: ***Capt. Nicolae Lascar Bogdan, Capt. Romano Mihail, Capt. V... Maracineau, Lieut. Demetre Calinescu, Major Constantine Ene, Major Dor... Giurescu, Major Nicolae Ivan, Major Giurgo Sontzu*** (Thames Iron Works, 1906). Dimensions: 101 × 12 × [illegible] feet. Armament: 1—3 pdr., 1 machine, 2 sets of dropping gears and 2—14 inch spar torpedoes... H.P. [illegible]. For other details, see Table.

3 *Naluca* class: ***Naluca, Sborul*** (1888, reconstructed 1907). Dimensions: [illegible]. Armament: [illegible] pdr. 2 tubes. 1 spar torpedo. For other details see Table.

Patrol Boats.

No photo available.

ARGESUL, FROTUGUL, GRANICERUL, ILEANA, PANDURUL, PORUMBITSA, POTERAZUL, RANDUNICA, SENTINELA, SILISTRIA, SRUTUL, SVIMULET, TROTUS, TELERMAN, VEDEA, VEGHETORUL.

No details known of above 16 boats.

No photo available.
(All armed with 2 machine guns.)

Opanez, Rahova, Smardan (1882). 45 tons. 1—1 pdr. Speed: 9 kts.
Romania (1882, 1885, 1890). 111 tons. Speed: 9 kts.
Prutul (1893). 21 tons. Speed: 8 kts.

Motor Launches.

6 boats: **A—F.** No details known.

Mine Layer.

(v. Silhouette)

ALEXANDRU CEL BUN (1882). 104 tons. Dimensions: $75\frac{1}{2} \times 15 \times 5\frac{3}{4}$ feet. Guns: 2—1 pdr., 2 machine. H.P. 100 = 9 kts. Coal: 12 tons. Complement, 29.

Coastguard Cruisers.

Bistritsa.

BISTRITSA, OLTUL, SIRETUL (Blackwall, 1888). 95 tons. Dimensions: $100 \times 13\frac{1}{2} \times 5\frac{1}{4}$ feet. Guns: 1—6 pdr., 1—1 pdr. H.P., 380 = 12 kts. Coal: 12 tons. Complement, 30.

1 funnel.

GRIVITZA (1880). 104 tons. Dimensions: $108\frac{1}{2} \times 17 \times 5\frac{1}{2}$ feet. Armament: 2—6 pdr., 2—1 pdr., 2 machine. H.P., 180 = 9 kts. Coal: 18 tons. Complement, 33.

No Photo available.

FULGERUL (1873). 85 tons. Dimensions: $82 \times 15\frac{3}{4} \times 4\frac{1}{2}$ feet. Guns: 1—6 pdr., 1—1 pdr. H.P., 100 = $7\frac{1}{2}$ kts. Coal: 10 tons. Complement, 18.

PRINCEPO MIRCEA (1882). 300 tons. Speed:
and 16 Hospital Barges).

Training Ship.

No photo available.

STEFAN CEL MARE. No details available.

Note.—The Ensign, Jack and Mercantile blue is *sky blue*, not ordinary flag blue. In Ensign, the sun emblem is in the *middle* of centre bar of flag. No yellow border in Jack.

ARGENTINE FLEET.

Revised by courtesy of Ministry of Marine, 1920.

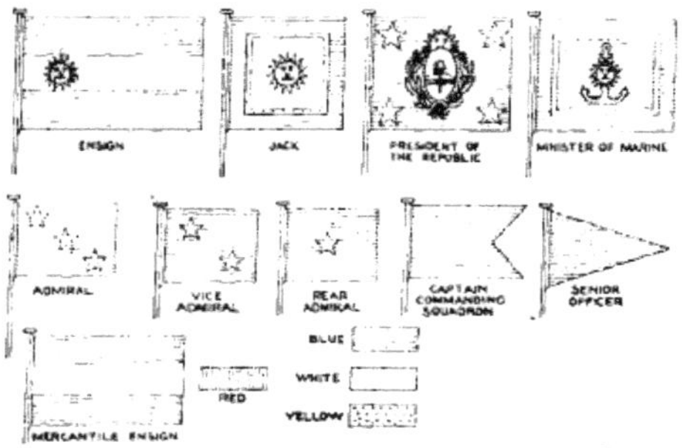

Notation.	Nominal Calibre.		Maker.	Length in Calibres.	Muzzle Velocity.	Weight A.P. Projectile.	Max. penetration against K.C. with capped A.P. at		Danger Space against average warship, at			*Service* Rounds per minute.
							5000 yards.	3000 yards.	10,000 yards.	5000 yards.	3000 yards.	
	inches.	c.m.		calibres.	ft. secs.	lbs.	inches.	inches.				
*A**	12	*30·5*	B	50	2900	870	19	23				2
A	10	*25·4*	E	40	2207	500	7	9½				·6
C	9·4	*24*	K	35	{2300 / 2133}	352	4½	6				·4
A	8	*20·3*	E	45	2060	210	6½	9				1
B	8	*21*	E	40	2280	210	4	6				
E	8	*20*	K	35	2120	200	3	5				
D	6	*15*	B	50	2600	105		8½				
D	6	*15*	C	50	2625	88	4	6	74	240	465	5
*E**	6	*15*		45	2500	100	3	4½	66	210	435	
*F**	6	*15*		40	2200	100	3	3¾	37	150	360.	
*F**	4·7	*12*	C	50	2600	46·3	...	3½				
*F**	4·7	*12*	E	45	2570	45	...	2¾				
*F**	4·7	*12*	E	40	2230	45	...	...				

* = Brass cartridge case used.

In the Maker's column B = Bethlehem E = Elswick ; K = Krupp ; C = Canet ; V = Vickers.

The above details of Argentine Naval Ordnance are not taken from any official source and must be regarded as merely approximate.

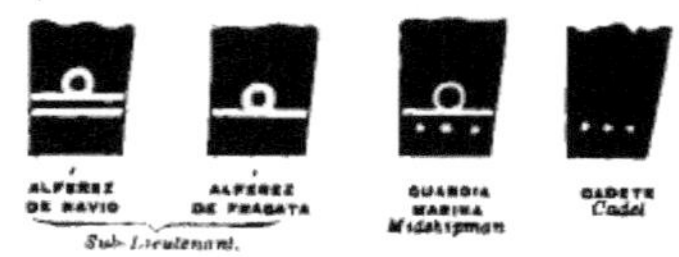

Note :—Coloured velvet cloth be[...] sleeve as follows :—

Engineers : Purple.
Electrical Engineers : Emerald
Surgeons : Red.
Paymasters : Grey-blue.
Constructors : Dark green.

Arsenals.

(1) **BUENOS AIRES.** Dry docks: (Eastern) 590½ × 65½ × 25 feet, and (Western) 492 × 65½ ×
There is also a small private dock at San Fernando, 17 miles above the city.

(2) **PUERTO BELGRANO** (Bahia Blanca). Dry docks: (1) 657 × 84 × 32½ feet, (2) 683 × 11

(3) **RIO SANTIAGO.** New dock, 672 × 114 × 36 feet (*not yet completed*). Two floating
(1) 1500 tons lift ; (2) 300 tons lift.

General Notes.

Personnel : about 9500 all ranks. *Reserve* : about 8000. *Special reserve* : 10,000.

Minister of Marine : Dr. Julio Moreno.

Naval Attaché (*London*) : Capt. Enrique Moreno.

Mercantile Marine.

From "Bureau Veritas" 1919-20 figures. Ships under 100 tons *gross* excluded.

142 steam and motor ships of 163,120 tons *gross*.
2 vessels (with Auxiliary motors) of 361 tons *gross*.
55 sailing vessels of 23,976 tons *gross*.

Scale: 1 inch = 160 feet.

ARGENTINE RECOGNITION* SILHOUETTES. (Revised 1919).

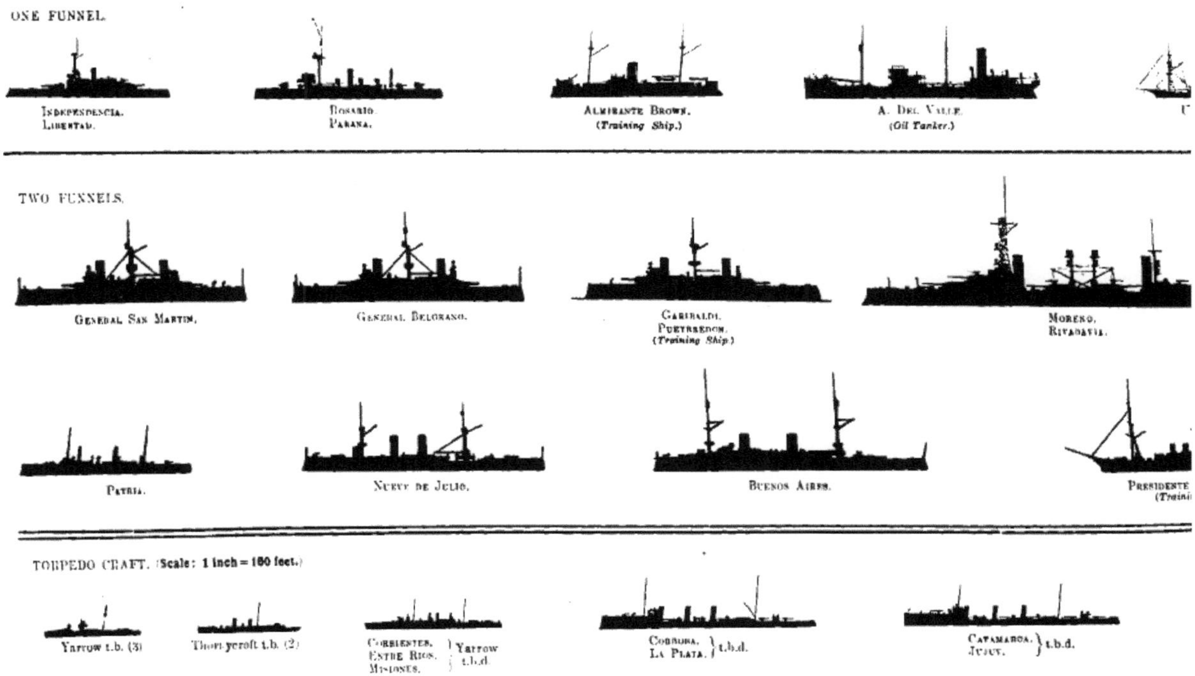

RIVADAVIA (26 Aug., 1911) & MORENO (23 Sept., 1911).

Normal displacement, 27,940 tons. *Full load*, 30,600 tons. Complement, 1215.

Length *w.l.*, 577½ feet. Beam, 95 feet. *Max load* draught, 28 feet. Length *over all*, 585 feet.

Guns (Bethlehem):
- 12—12 inch, 50 cal. **Dir Con ?**
- 12—6 inch, 50 cal.
- 16—4 inch, 50 cal.
- 4—3 pdr.
- 6 M.G.
- (4 landing)

Torpedo tubes (21 inch):
- 2 *submerged* (broadside).

Armour (Krupp):
- 11"—8" Belt (amidships)
- 5" Belt (bow) N.C.
- 4" Belt (stern) N.C.
- 3" Deck (slopes)
- 9"—8" Side above belt
- 12"—9" Big gun turrets
- 6" Secondary battery (N.C.)
- 12" Conning Tower (forward)
- 9" Do. (aft)
- Total weight: 7000 tons

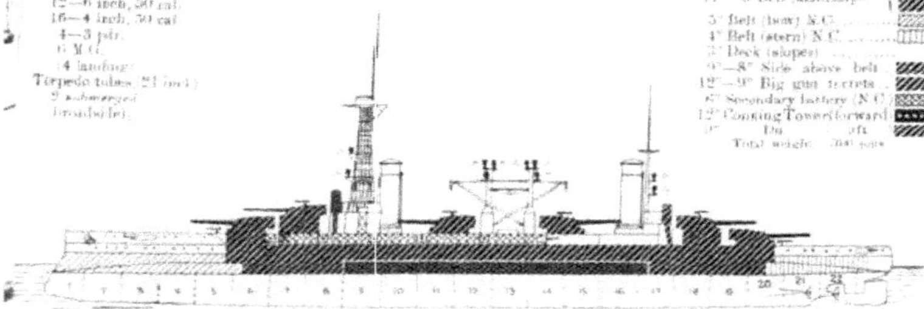

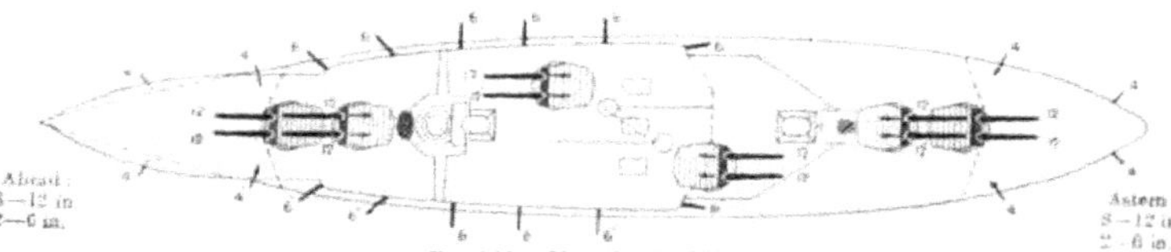

Ahead:
8—12 in.
2—6 in.

Broadside: 12—12 in., 6—6 in.

Astern:
8—12 in.
2—6 in.

Machinery: Curtis Turbines. 3 screws. Designed S.H.P. 39,500 = 22.5 kts. Boilers: 18 Babcock. Coal capacity: *normal*, 1600 tons; *maximum*, 4000 tons. Oil: 660 tons. *Nominal radius*: 3930 miles at full speed, 7740 miles at 10 kts. (see *General Notes*).

Gunnery Notes.—Heights of barbettes over normal draught w.l.: No. 1, 29¼ feet; No. 2, 37¼ feet; Nos. 3 and 4 (echelon), and No. 5, 29 feet; No. 6, 20 feet; 6 inch guns, 10½ feet above w.l. Arcs of fire: end barbettes, Nos. 1 and 6, 270°; Nos. 2 and 5 (super-firing), 300°; Nos. 3 and 4 (echelon) 180° own beam and 100° far beam. R.F. on heads of derrick posts.

Armour Notes.—Main belt is 8 feet deep, 4½ feet above water-line and 3½ feet below same at normal draught. It is 250 feet long, but 11" section is only 2 feet deep from top edge of belt, and then tapers to 5" on lower (under-water) edge. Belt under end barbettes is 10" tapering to 5" as main belt. Upper belt 400 feet long, 9" lower edge, 8" top edge, 6" battery above this. Funnel bases, 1½" nickel steel for 15 feet above deck. Protective decks, 1½" upper, 3" lower. Two Director stations behind upper belt at bases of C.T. communication tubes. Barbettes: bases are 9" where exposed. Shields to these: 12" port-plate, 9" sides, 11" back, 3" roof.

Anti-Torpedo Defence.—3" longitudinal wing bulkheads in way of machinery and magazine spaces. 1" nickel steel flats under magazines, boilers, and engine rooms. Total weight, 680 tons (included in 7,000 tons total weight of armour given above). 12 searchlights.

RIVADAVIA. 1918 *Photo, by courtesy of the Min*

General Notes.—Built under 1908 Programme. Gave considerable trouble on trials, as steam con[illegible] contract rate. Turbines were [illegible] before these ships got through their trials. [illegible] have [illegible] plans are being prepared for conversion to "all oil fuel." On [illegible] Aires—New York—Buenos Aires, in 1918 one of these ships consumed 12,000 tons of coal, or [illegible] the estimated normal cruising consumption with auxiliaries running.

RIVADAVIA. 1918 *Photo, Cress, G. H. H.*

Torpedo Notes.—Bullivant net defence. 12 [illegible] searchlights. Electric installation at [illegible].

Name	*Builder*	*Machinery*	*Laid down*	*Completed*	*Trials*	*Boi[lers]*
Rivadavia	Fore River Co.	Fore River Co.	May '10	Dec. 1914	[illegible]	[illegible]
Moreno	N.Y. Shipbuilding		July '10	Mar. 1915		[illegible]

ARGENTINE ARMOURED SHIPS.

Gen. Belgrano: Searchlight on mast on fore side of 2nd funnel.

Gen. San Martin: This ship is almost of same appearance as Belgrano.

1918 Photo, by courtesy of the Ministry of Marine. (Lieutenant.)

Garibaldi.

1918 Photo, by courtesy of Ministry of Marine.

GENERAL BELGRANO (July, 1897), **GENERAL SAN MARTIN** (1896), both by Orlando, Leghorn, **GARIBALDI** (1895), by Ansaldo, Sestri, Ponente. All 6840 tons. Length (o.a.), 328 ft. Beam, 59¾ feet. *Max.* draught, 23¼ feet. Complements, (*G.B.*) 425, (*G.S.M.*) 431, (*G.*) 414. Armament: *Belgrano* 2—10 inch, 14—6 inch, 4—12 pdr., 4—6 pdr., 4—18 inch tubes *above water*. *San Martin*: 4—8 inch, 10—6 inch, 6—4·7 inch, 2—12 pdr., 5—6 pdr., 4—18 inch tubes *above water*. *Garibaldi*: 2—10 inch, 10—6 inch, 6—4·7 inch, 2—12 pdr., 4—6 pdr. Armour: (Krupp in *G.B.*, Harvey Nickel in *G.S.M.*, Harvey in *G.*), 6" Belt with 2" at ends and 1½"—2" deck, 6" Lower deck side, 6" Battery, 6" Barbettes, 6" Bulkheads, 3" Gunhouses, 6" C.T. Machinery: 2 sets triple expansion. Boilers: 8 cylindrical, single-ended. H.P. 13000=20 kts. Coal: *normal*, about 400 tons; *maximum* (*G.B.*) 1150 tons, (*G.S.M.* & *G.*) 1137 tons. Endurance: about 6000 miles at 10 kts.

Note.—A fourth ship of this type, **PUEYRREDON,** now employed as a Training Ship, will be found on next page.

ARGENTINE COAST DEFENCE BATTLESHIPS.

Independencia.

1918 Photo, by courtesy of the Mini[stry of Marine.]

INDEPENDENCIA (1891), & **LIBERTAD** (ex-*Nueve de Julio*) (1890). Both bu[ilt by Laird,] Birkenhead. Displacement 2300 tons. Complement 196. Length (p.p) 240 feet. [...] *Max.* draught, 13 feet. Guns: 2—9·4 inch, 35 cal. (Krupp), 4—4·7 inch, 40 cal. (Elsw[ick] ...) Armour (compound): 8" Belt (amidships), 2" Deck (flat on belt,) 8" Bulkhead (forward[...]) (aft), 8" Barbettes and bases, 5" Shields to big guns (fronts), 4" Conning tower. [...] Compound vertical. 2 screws. Boilers: 4 double cylindrical. Designed H.P. [...] Coal: *normal* 230 tons; *maximum* 350 tons. Endurance: 3000 miles at 10 kts.

Old Protected Cruisers.

1917 Photo, Abrahams & Son, Devonport.

BUENOS AIRES (Elswick, 1895). Displacement 4780 tons. Sheathed and coppered. Complement 375. Length (o.a.), 403¾ feet. Beam, 47⅛ feet. *Maximum* draught, 19 feet. Guns (Elswick): 2—8 inch, 45 cal. (A), 4—6 inch, 45 cal., 6—4·7 inch, 45 cal., 12—3 pdr. Torpedo tubes: 5 *above water*. Armour (steel): 5" Deck (amidships), 5" Engine hatches, 4½" Gun shields, 3" Hoists to guns, 6" Conning tower (Harvey). Machinery: 2 sets vertical 4-cylinder (2 low-pressure cylinders). 2 screws. Boilers: 4 double-ended and 4 single-ended. Designed H.P. 13,000 = 23 kts.; 17,000 = 24 kts. Coal: *normal* 400 tons; *maximum* 1000 tons. Endurance: 6400 miles at 10 kts.

1920 Photo, by courtesy of the Ministry of Marine.

NUEVE DE JULIO (Elswick 1892). Displacement 3540. Complement 315. Length, (o.a.) 354¾ feet. Beam, 44 feet. *Maximum* draught, 19¼ feet. Guns (Elswick): 4—6 inch, 40 cal., 8—4·7 inch, 40 cal., 10—3 pdr. Torpedo tubes: 3 *above water*. Armour (steel): 4½" Deck, 5" Glacis to engines, Cofferdam amidships. Machinery: 2 sets vertical 4 cylinder (2 low-pressure cylinders). 2 screws. Boilers: 8—single-ended. Designed H.P. (*forced*) 14,500 = 22·5 kts. Coal: *normal* 300 tons; *maximum* 770 tons. Endurance: 5500 miles at 10 kts.

Training Ships.

New fas W/T. aerials from truck of mast to bow and stern, as *Garibaldi* on preceding page

PUEYRREDON (Ansaldo, 1897). Armoured ship. *Normal* displacement, 6840 tons. Co[...] 427. Guns (Elswick): 2—10 inch, 40 cal. (A), 10—6 inch, 40 cal., 6—4·7 inch, 40 cal., 2[...] 4—6 pdr., Torpedo tubes (18 inch): 4 *above water* (behind armour). All other [...] *General Belgrano*, described on a preceding page, except:—(a) Armour (Terni); (b) 16 [...] boilers. (c) *maximum* coal capacity 1000 tons = 4800 miles at 10 kts. Laid down 1896, [...] 1901. Made 19·24 kts. on first trials. Machinery:—Ansaldo-Maudslay.

(*Gunnery Training Ship.*)

ALMIRANTE BROWN (Samuda, Poplar, 1880, reconstructed at La Seyne, 1897). Old Coast [...] Battleship. 4200 tons. Dimensions: 240 (p.p.) × 50 × 20½ feet. Complement, 290. [...] 10—6 inch (50 cal.), 4—4·7 inch (50 cal.), 8—6 pdr. Torpedo tubes: 2—18 inch *above* [...] Armour (compound): 9" Belt with 7" Bulkheads and 1½" Flat Deck, 8" Central battery. D[...] H.P. 4500 = 14 kts. (present speed, 11 kts.) Boilers: Cylindrical. Coal: 622 tons.

[*Continued on next*

Training Ships—*continued.*

PRESIDENTE SARMIENTO (1898). Built at Birkenhead. 2850 tons. Complement 294. Dimensions: $251\frac{1}{2} \times 43\frac{1}{4} \times 23\frac{1}{4}$. Guns: 2—4·7 inch, 50 cal.; 2—4 inch, 2—6 pdr., 2—3 pdr. Torpedo tubes: 3 *above water*. Designed H.P. 2800 = 15 kts. Coal: 330 tons. Boilers: 1 Niclausse, 1 Yarrow, 2 cylindrical single-ended.

Despatch Vessels. (*Avisos*).

No photo available.

PAMPERO. (France, 1900). 240 tons. Dimensions: 130 × 17 × $7\frac{1}{2}$ feet. H.P. 470 = 13.5 kts. Guns: *Not known.*

No photo available.

GAVIOTA. (Howaldt, 1888). 120 tons. Dimensions: $98\frac{1}{2}$ × $17\frac{3}{4}$ × 8 feet. H.P. 290 = 8 kts. Guns: *Unknown.*

No photo available.

GOLONDRINA (1888), **USHUAIA** (ex-*Bahia Blanca*, 1888) both by Howaldt, Germany. 95 tons. Dimensions: $98\frac{1}{2}$ × $17\frac{3}{4}$ × $10\frac{1}{2}$ feet. H.P. 170 = 8 kts. Guns: *Unknown.*

Despatch Vessels—*continued.*

No photo available.

AZOPARDO (Trieste, 1884). 400 tons. Dimensions: $157\frac{1}{2}$ × $26\frac{1}{4}$ × $8\frac{3}{4}$ feet. H.P. 600 = 10 kts. Guns: *Unknown.*

RESGUARDO (1874), **VIGILANTE** (1874). 100 tons. Dimensions: $98\frac{1}{2}$ × 18 × $3\frac{3}{4}$ feet. Guns: *Unknown.* H.P. 168 = 8 kts. Coal: *about* 17 tons.

Gunboats.

PATRIA (Birkenhead, 1893). Displacement, 1070 tons. Dimensions: $251\frac{1}{12}$ × $30\frac{1}{2}$ × $11\frac{1}{4}$ feet. Complement, 136. Guns: 2—4·7 inch, 40 cal., 4—6 pdr., 2—3 pdr. Machinery: 2 sets triple expansion. 2 screws. Boilers: 4 double-ended Yarrow. Designed H.P. 4500 = 20·5 kts. Coal: 260 tons. Endurance: 3200 miles at 10 kts.

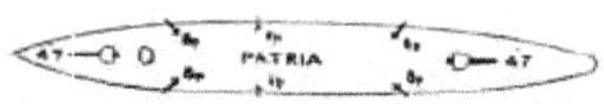

See Silhouettes.

URUGUAY (Birkenhead, 1874). 550 tons (sheathed). Original speed: 11 kts. Coal: 90 tons.

Armoured River Gunboats.

O.

PARANA (April, 1908) & **ROSARIO** (July, 1908) Elswick. Displacement, 1055 tons. Complemen 229¾ feet. Beam, 32¾ feet. Draught, $7\frac{1}{12}$ feet. (Howitzers, 6—12 pdr., 8 Machine, 2 landing. A (amidships), $1\frac{1}{2}$"—1" Deck, 3" Conning tower. kts. 2 Yarrow boilers. Coal: 120 tons. Nomi miles at $10\frac{1}{2}$ kts.

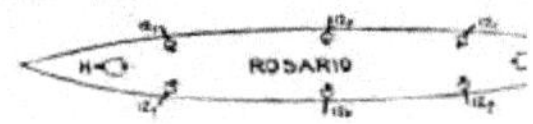

River Police Craft. (*Policia flu*

No photos available.

INACAYAL (1901), **SAYHUEQUE** (1901), Rennie Forrest. 145 tons. Dimensions: 1 feet. H.P. 185 = 9 kts.

NAMUNCURÁ (1901). 120 tons. Dimensions × $2\frac{1}{4}$ feet. H.P. 140 = 9 kts.

TEUCO (1884). 120 tons. Dimensions: 110 Speed: 8 kts.

7 Destroyers. (*Cazatorpederos.*)

No.	Type	Date	Displacement	H.P.	Max. speed	Fuel	Complement	T. tubes	Max. draug't
			tons		kts.	tons			feet
2	*Cordoba* (S)	'10-'12	950	20,000	32	290 coal + 50 oil	99	4	8
2	*Catamarca* (K)	'10-'12	950	18,000	32	250 coal + 100 oil	99	4	$8\frac{1}{2}$
3	*Corrientes* (Y)	'96-'98	340	4000	26	80 coal	66	3	$8\frac{1}{4}$

K = Krupp. S = Schichau. Y = Yarrow.

Photo by courtesy of the Ministry of Marine.

2 (German) type: **Cordoba** (Schichau, Nov., 1910), **La Plata** (Germania, Jan., 1911). 950 tons. 295 × 29¼ × 7½ feet. Armament: 4—4 inch (Beth.), 4—21 inch tubes. Curtis (A.E.G.) turbines. Boilers: 5 Schulz-Thornycroft. Designed S.H.P. 20,000 = 32 kts. Coal: 290 tons. Oil: 50 tons. Endurance: 2700 miles at 15 kts., 715 miles at full speed. Trials (max.): Cordoba 34.7; La Plata, 34.7.

Photo by courtesy of the Ministry of Marine.

2 German type: **Catamarca** (Schichau, Jan., 1911), **Jujuy** (Germania, March, 1911). 950 tons. 288⅔ × 27 × 8½ feet. Armament: 4—4 inch (Beth.), 4—21 inch tubes. Curtis (A.E.G.) turbines. Boilers: Schulz-Thornycroft. Designed S.H.P. 18,000 = 32 kts. Coal: 250 tons. Oil: 110 tons. Endurance: 3000 miles at 15 kts. 800 miles at full speed.

CATAMARCA.

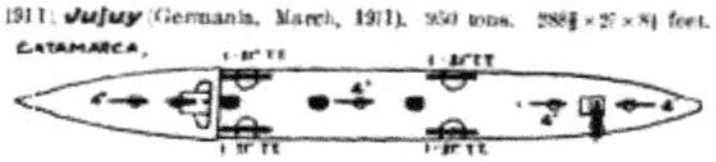

Destroyers—*Continued.*

3 Yarrow type: **Corrientes, Entre Rios, Misiones** (1896). Dimensions: 190 × 19 × 8½ feet. Armament: 1—14 pdr., 3—6 pdr., 2—1 pdr., 3 tubes (18 inch). 6 Yarrow boilers. Endurance: 900 miles at 15 kts. Other details as table.

5 Torpedo Boats. (*Torpederos*).

No.	Type	Date	Displacement	H.P.	Max. speed	Fuel	Complement	T. tubes	Max. draught
			tons		kts.	tons			feet
2	*Comodoro Py* (T)	'90-'91	110	1700	24.5	24	43	3	3½
3	*Bachardo* (Y)	'90-'91	85	1200	23	22	28	2	6

T = Thornycroft. Y = Yarrow.

2 Thornycroft type: **Comodoro Py** (1891) and **Murature** (1890). Dimensions: 150 × 14½ × 5½ feet. Armament: 2—3 pdr., 1 M.G., 3—18 inch tubes. Thornycroft boilers. Endurance: 1200 miles at 15 kts. Other details Table above.

3 Yarrow type: **Bachardo, Jorge, Thorne** (all 1890). Dimensions: 130 × 14 × 6 feet. Armament: 2—3 pdr. 1 M.G., 3—18 inch tubes. Endurance: 1200 miles at 15 kts. For other details, v. Table above.

Minelayer (*Minador*).

Photo wanted.

FULTON (Birkenhead, 1897). 79 tons. Dimensions: 79 × 15 × 6 feet. H.P. 100 = 9 kts. Guns (if any) and number of mines carried: *Not known*.

Transports (*Transportes*).

Photos wanted.

MINISTRO EZCURRA (Greenock & Grangemouth Co., 1914). 2600 tons. Dimensions: 250 × 40 × 18¾ feet. H.P. 1242 = 10½ kts.

BAHIA BLANCA (Reiherstieg, Hamburg, 1911). 9882 tons. Dimensions: 489 × 59 × 27½ feet. H.P. 4200 = 12.5 kts.

VICENTE-FIDEL-LOPEZ (Earles, 1906). 725 tons. Dimensions: 164 × 26 × 11½ feet. H.P. 480 = 9.3 kts.

CHACO (1895), **PAMPA** (1895), both by Connell. 8700 tons. Dimensions: 392½ × 46 × 28¾ feet. H.P. 3500 = 14 kts.

GUARDIA-NACIONAL (Barclay, Curle, 1890). 6500 tons. Dimensions: 385 × 41¾ × 23 feet. H.P. 1850 = 14 kts.

PRIMERO DE MAYO (Howaldt, 1890). 1050 tons. Dimensions: 197 × 29¾ × 14 feet. H.P. 650 = 11 kts.

PATAGONIA (ex-Cruiser, 1885, rebuilt 1919). 2150 tons. Dimensions: 224¾ × 32¾ × 16¾ feet. H.P. 300 = 10 kts. Oil: 263 tons = 4800 miles endurance.

CONSTITUCIÓN (1875). 500 tons. H.P. 320 = 10 kts.

REPUBLICA (1875). 416 tons. H.P. 420 = 9 kts.

PIEDRABUENA (ex-*Parana*, 1874). Ex-Gunboat of 550 tons. H.P. 475 = 10 kts. Guns: 1 machine.

Transports—*continued*.

Rio Negro. 1920 Official Photo.

RIO NEGRO (1864, rebuilt 1920). 1525 tons. Dimensions: 325 × 36 × 19¾ feet. H.P. 2250 = 11 kts. Coal: 720 tons. Endurance: 5000 miles.

CHUBUT (Old ship, rebuilt 1918). — tons. Dimensions: 296 × 47 × 17 feet. H.P. 2000 = 7 kts. Carries 350 tons coal + 150 tons oil. Endurance: 2500 miles.

Oil Tankers.

No photo available.

INGENERIO LUIS A. HUERGO (Fore River, U.S.A., 1917). 7200 tons. Dimensions: 343 × 46 × 22.5 feet. Designed H.P. 1550 = 10 kts.

1918 *Official photo.*

ARISTÓBULO DEL VALLE (Great Lakes, U.S.A., 1917). 5400 tons. Dimensions: 261 × 43 × 21½ feet. Designed H.P. 1400 = 9.5 kts.

Tugs (*Remolcadores*).

Photos wanted.

ONA (1913), **QUERANDI** (1914). Both built by T 615 tons. Dimensions: 137 × 28 × 11 feet. 11 kts. Fitted as Salvage Vessels.

RAWSON (Buenos Ayres, 1906). 95 tons. Dimen 16 × 8½ feet. H.P. 80 = 8 kts.

ALBATROS (Buenos Ayres Navy Yard, 1901). 66 t sions: 60 × 14 × 8 feet. H.P. 130 = 5 kts.

FUEGUINO (1900), **TEHUELCHE** (1900). B Day, Summers. 310 tons. Dimensions: 100 feet. H.P. 520 = 11 kts.

CORMORÁN (1900), **PETREL** (1900), both by I Navy Yard. 44 tons. H.P. 80 = 7½ kts.

DELFIN (1900), **PENGUIN** (1900), both by B Navy Yard. 38 tons. H.P. 80 = 8 kts.

202 B (Liverpool, 1881). 75 tons. H.P. 110 =

Surveying Ships (*Balizadores*)

Photos wanted.

ALFEREZ MACKINLAY (Holland, 1914). 783 520 = 10 kts. Dimensions: 193¾ × 28¼ ×

INGENIERO IRIBAS (France, 1891). 360 tons. 131¼ × 21¼ × 8¼ feet. H.P. 300 = 10 kts.

Miscellaneous Ships (*Buques Varie*

PILCOMAYO (1875). Crane Ship of 410 tons. 9 kts.

EL PLATA, LOS ANDES (1874). Old Monitors H.P. 750 = 9 kts. *nominal*. Guns: 2—8 inch, 1—12 pdr., 4 or 2—6 and 3 pdr.

Training Ships described on earlier pages.

BRAZILIAN FLEET.

Note.—This Navy is being reorganized by a U.S. Naval Mission, 1920.

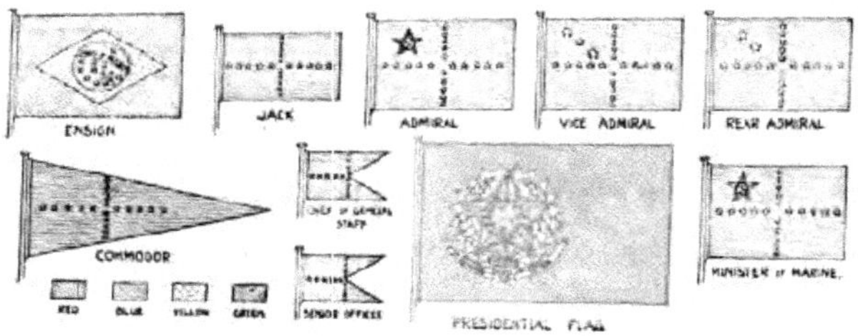

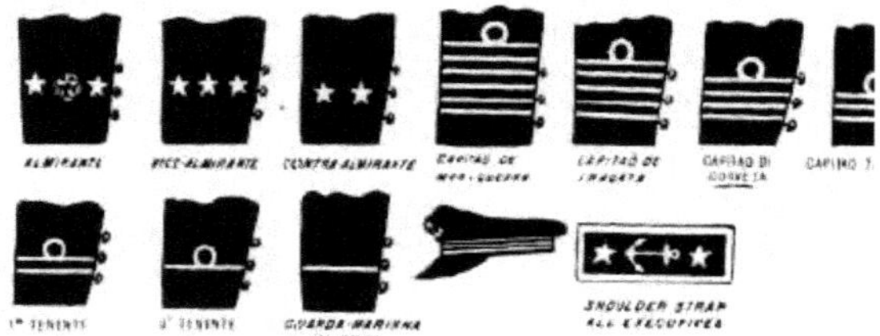

Almirante is only a war appointment: *Vice-Almirante* is the highest place rank.

Capitão di Corveta is equivalent to a British Lieut-Commander. *Capitão tenente* a lieute *2º tenente*—sub-lieutenant. Civilian ranks have the same stripes *without* the curl, and plus col above top stripe and below bottom one as under:

Constructors.—No colour, but a ball above the stripes.
Engineers (Machinista).—Green.
Doctors (Cirurgiao).—Red.
Paymasters (Commissario).—White.

Shoulder straps for engineers carry a screw propeller, for doctors a stethoscope, for paymas two quill pens crossed. These devices are also worn on the collar of tunic.

Caps.—Admirals have a broad gold band with oak leaf device, the other commissioned ra narrow gold bands as on sleeves.

Warrant officers and men carry white distinctive sleeve badges, as follows:—*Torp* torpedoes; *gunnery*, two crossed guns; *navigation*, two crossed anchors; *signal*, two crossed fla *quartermasters*, a wheel; *engine room complements*, a screw propeller.

Marine Infantry.—Scarlet tunics with black insignia: five stripes for first sargento, four second ditto, two for cabo. White helmets as the British. No commissioned officers.

Brazilian sailors enter as boys at about twelve years. They do three years in a school, tl ten months special instruction, then serve twelve years with the fleet. Volunteers three to two years. There are nineteen schools for boys.

Mercantile Marine.

From "Bureau Veritas" 1919-20 figures. Ships under 100 tons *gross* excluded.

291 steam and motor ships, 493,092 tons *gross*
2 vessels (with auxiliary motors), of 301 tons *gross*.
81 sailing vessels of 19,633 tons *gross*.

Personnel.—About 10,000 of all ranks and ratings.

Minister of Marine—Senhor Raul Soares (July, 1919).

Naval Attaché, London.—Commander Oscar Sousa de Spinola.

Guns in Service.*

Notation.	Calibre		Length in calibres	Weight of A.P. shell	Muzzle Velocity.	Max. penetration A.P. capped at K.C. at 5000 yards	3000 yards	Danger Space against average ships at 10,000 yards	5000 yards	3000 yards	Service rounds per minute
	inch.	c.m.	cals.	lbs.	ft. secs.	inch.	inch.				
Heavy	12	30·5	45	850	2800			...	...	...	2
Medium	9·2	23	45	380	2700	9	11½	...	...	...	8
Light	6	15	40	100	2500	4	5	65	200	420	6
	6	15	50	100	2640	4½	5½	72	240	460	6
	4·7	12	40	45	2150	...	...	...	...	...	8
	4·7	12	50	45	2630	...	...	...	...	...	8

* **Note.**—These particulars of the Brazilian Naval Ordnance are not derived from any official source and must be regarded as only approximate.

Arsenals.

RIO DE JANEIRO. 6 Docks. Government are "Guanabara" (Cobras Island), 423 × 70 × feet; "Santa Cruz," 258¾ × 54¾ × 18 feet. Government Floating Dock, "Affonso Penn 550½ × 100 × 31 feet (22,000 tons), requires 51 feet depth of water to dock ship of 30 f draught. Another Floating Dock, 350 (*o.a.*) × 136 feet (to lift 6000 tons), completing Messrs. Vickers Ltd. 1919. This dock was taken over for British Navy during war, l released in 1919.* 4 Private Docks, Mocanguê No. 1, 400 × 60 × 16 feet; Mocanguê No. 370 × 50 × 16 feet. Lage Bros, "Cruzeiro" (Viana Island), 400 × 58 × 18 feet. Lahmey (Toque Toque), 550 × 74 × 30¾ feet, property of Companhia Commercio & Navegação.

Note.—Reported that money is being appropriated for the building of Ordnance and Armo Plate Plant, Torpedo Factory and large Repair Shops at Mai Island (where the Naval Colle and Midshipman's School is now situated).

PARA. 2 Floating Docks, "Affonso Penna" and "Lauro Müller," each 235 × 45 × 12 feet (17 tons).

LADARIO River.

*Large Floating Dock, 680 × 113 × 56 feet (32,000 tons), was used during the war as the British Admiralty T Floating Dock. Not certain if this dock has been towed out to Rio de Janeiro.

Scale : 1 inch = 160 feet.

BRAZILIAN RECOGNITION SILHOUETTES.

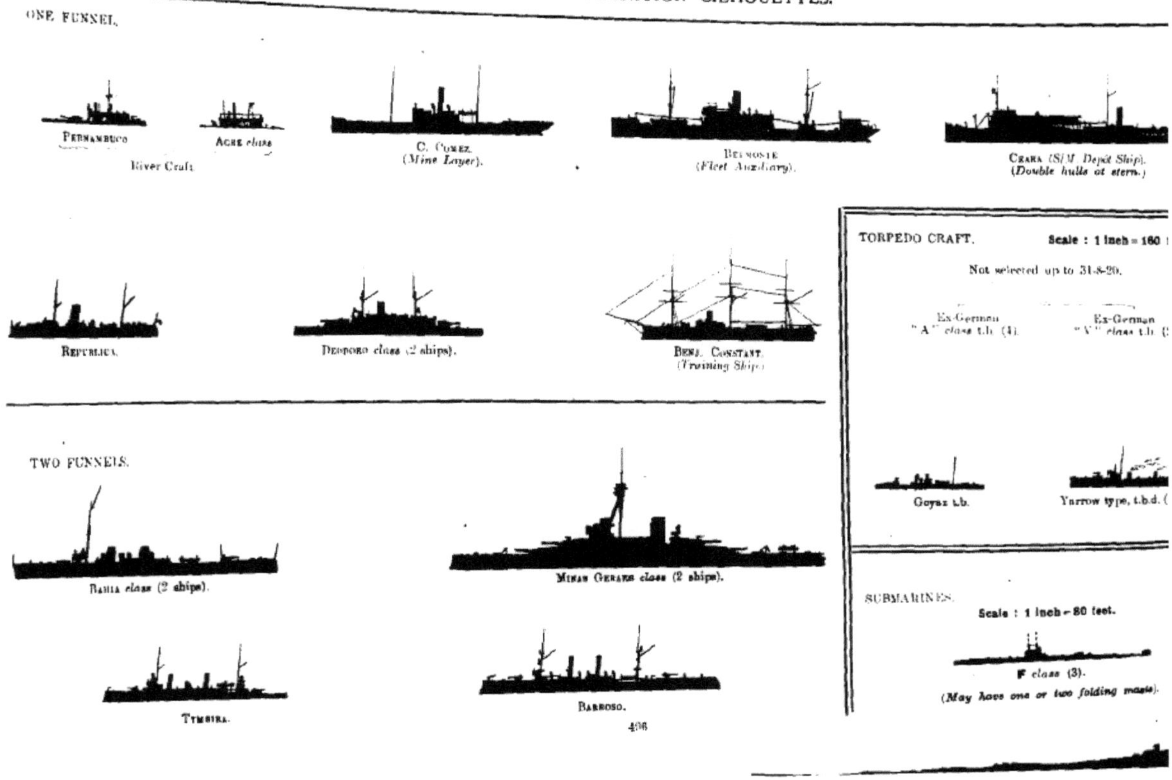

(MINAS GERAES CLASS - 2 SHIPS.)

MINAS GERAES (Sept., 1908), **SAO PAULO** (April, 1909).

Normal displacement 19,200 tons. *Full load* 21,200 tons. Complement 850.

Length (*waterline*) 530 feet; (*p.p.*) 500 feet. Beam, 83 feet. *Max.* load draught 25 feet. Length (*over all*), 543 feet.

Guns (Elswick):
12—12 inch, 45 cal. } Dir. Con.
22—4·7 inch, 50 cal. }
8—3 pdr.
4 M.G.
Torpedo tubes:
Dropping gear for boats and carry torpedoes for this purpose.

Armour (Krupp):
9" Belt ....
6"—4" Belt (bow) N.C. }
6"—4" Belt (aft) N.C.... }
2½" Deck (slopes) .... *
9"—6" Upper belt.........
9" Bulkheads ....
9" Battery (main deck) ...
12"—8" Turrets (K.C.)...
12" Conning tower (fore)
9" Conning tower (aft) ...

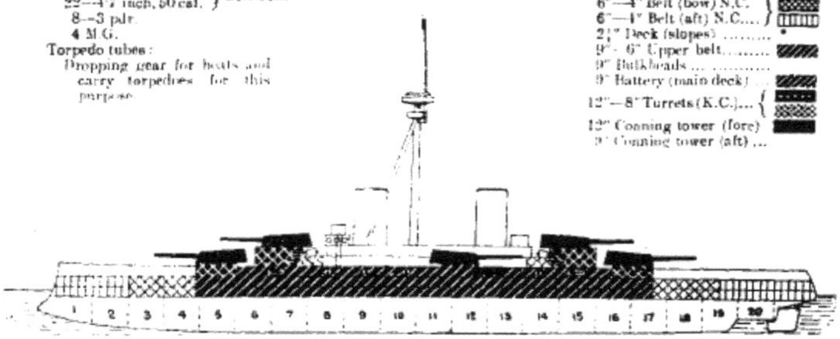

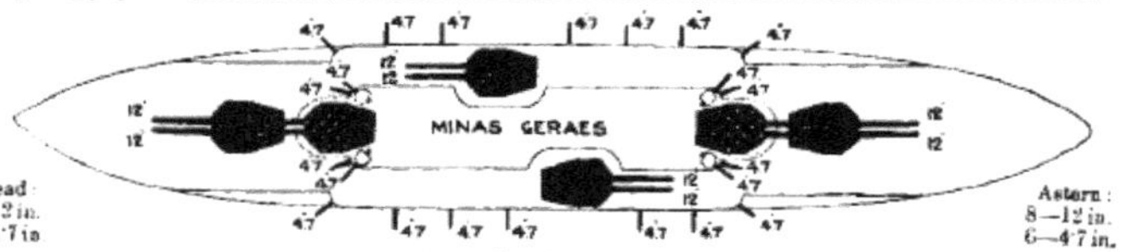

Ahead:
8—12 in.
6—4·7 in.

Astern:
8—12 in.
6—4·7 in.

Broadside: 10—12 in., 11—4·7 in.

Machinery: 2 sets triple expansion. 2 screws. Designed S.H.P. 23,500 = 21 kts. *f.d.* Boilers: 18 Babcock. Coal: *normal* 800 tons; *maximum* 2360 tons + 350 tons oil. Endurance: about 3600 miles at 19 kts., and about 8000 miles at 10 kts.

Gunnery Notes.—Sperry and Ford fire controls added, also large base R.F. and directors, during refits by New York Navy Yard. Arcs of fire: Centre-line 12-inch guns 270–280°; beam 12-inch 180° nominal. Main deck, 4·7 inch, 100°; 4·7 inch in superstructure casemates, 90°. 3 pdr. guns can be transferred to landing mountings.

Engineering Notes.—66 revs. about 10 kts; 126 = 20 kts.; 148 about 21·5 kts. Heating surface, 56,370 sq. ft. Grate area, 1868 sq. ft.

* *Armour Notes.*—Main belt 22½ ft. deep, 5 ft. of this being below waterline. Belt does not run up to stern.

SAO PAULO.* 1920 photo.

* Above photo shows *Sao Paulo* after large refit by New York Navy Yard, 1917-1[illegible] *Minas Geraes* is to be refitted in same way.

Name	Builder.	Machinery.	Laid down	Com-pleted	Trials. 30 hour at ⅗	Trials. Full power	Bo[illegible]
M. Geraes	Elswick*	Vickers	1907	Jan. '10	16,177 = 19·13	25,519 = 21·2	Babc[illegible]
S. Paulo	Vickers*	Vickers*	1907	July '10	16,067 = 19·85	25,517 = 21·2	Babc[illegible]

General Notes.—Both ships were somewhat over normal displacement on trials.

* *Note.*—*Sao Paulo* underwent large refit at New York Navy Yard from May 1917–1919, it bei[illegible] that she might join British Grand Fleet. A large amount of material was sent out fr[illegible] for this purpose. During 1919, she served with U.S. Atlantic Fleet for exercise at Guant[illegible] *Minas Geraes* to commence her refit at New York Navy Yard towards end of 1920.

(1908) Light Cruisers (*Cruzadores*)

Bahia. 1918 *Photo.*

BAHIA (Jan., 1909) & **RIO GRANDE DO SUL** (April, 1909).

Normal displacement 3000 tons. Complement 320.

Length { (p.p.), 380 feet / (o.a.), 401½ .. } Beam, 39 feet. *Max.* draught, 13 feet 5 ins.

Guns (Elswick):
10—4·7 inch, 50 cal.
6—3 pdr.
Torpedo tubes (18 inch):
2 *above water*

Armour:
1½" Deck ....
3" Conning tower ....

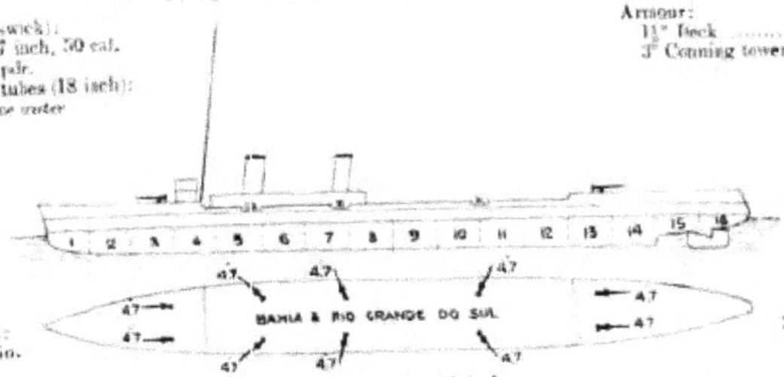

Ahead: 2—4·7 in.

Astern: 2—4·7 in.

Broadside: 5—4·7 inch.

Machinery: Parsons turbine. 3 screws. Boilers: 10 Yarrow. Designed S.H.P. 18,000 = 25 kts. Coal: *normal* 150 tons, *maximum* 650 tons. Endurance: about 1400 miles at 23¾ kts., 3500 miles at 10 kts.

Name	Builder	Machinery	Laid down	Completed	Trials: 6 hours ½	Trials: Full power	Boilers	Best recent speed
Bahia	Elswick	Vickers	1908	1910	=24	20,191 = 27·02	Yarrow	
Rio G. do Sul	Elswick	Vickers	1908	1910	=24	20,900 = 27·41	Yarrow	

General Notes.—Reported to be in poor condition up to 1917 and not capable of raising more than 18 kts. But, during 1918, they may have been repaired and put in effective condition.

(1896) Coast Defence Vessels (*Encouraçados*)

1919 *Photo, Commr. B. D. Bolle*

DEODORO (ex-*Ypiranga*, 1898) & **FLORIANO** (1899). Both built at La Seyne. D 3162 tons. Complement 200. Length (p.p.), 267¾ feet. Beam, 48 feet. *Max.* draugh Guns: 2—9·4 in., 45 cal.; 4—4·7 in., 50 cal.; 4—6 pdr., 2 M.G. (2 field). Armour (Har 13¾" Belt (amidships), 4" Belt (ends), 1½" Deck (reinforcing belt), 8" Turrets, 3" Casemates tower. Machinery: 2 sets triple expansion. 2 screws. Boilers: (1912), Babcock. De 3400 = 15 kts. Coal: 246 tons. Radius of action: 2500 miles at 10 kts. Reported t bad condition.

(1895) Protected Cruiser.

1919 *Photo, Commr. B. D. H*

BARROSO (Elswick, 1896). 3450 tons (sheathed and coppered). Comp. 375. Dim 330 × 43¾ × 16½ feet. Guns (Elswick): 6—6 in., 50 cal., 4—4·7 in., 50 cal.; 10—6 4 Maxims, (2 field.) Armour: 3½" Deck (amidships). Machinery: 2 sets 3-cyl. tr 2 screws. Boilers: cylindrical. Designed H.P. *forced* 7500 = 20·5 kts. Coal: no *maximum* about 850 tons. Endurance: 6500 miles at 10 kts. Refitted 1916-17. T *Abreu* and *Amazonas*, were purchased by the U.S. for the Spanish-American War, U.S. *Albany* and *New Orleans*.

Gunboat.

1915 Photo.

REPUBLICA (Elswick, 1892). 1300 tons. Comp. 152. Dimensions: 210½ (p.p.) × 35 × 15 feet. Guns (Elswick): 6—4.7 inch, 4—6 pdr., 6 M.G. Torpedo tubes (18 inch), 4 above w. Armour: 2¾" deck. Designed H.P. 3300 = 16 kts. Cyl. boilers. Coal: 170 tons.

Torpedo Gunboat.

1916 Photo.

TYMBIRA (ex-*Cananeru*, 1896). 1030 tons. Complement 110. Dimensions: 259 × 30 × 10¾ feet. Armament (Krupp): 2—4.7 inch, or 4.1 inch, 50 cal., 6—6 pdr., 2—1 pdr., 3—14 inch tubes. Armour: 1" Deck. Designed H.P. 7000 = 23 kts. Coal: 200 tons. Built by Krupp.

River Craft.

1916 *Photo, Abrahams.*

PERNAMBUCO (1890-10). River Monitor, built at Rio.) 470 tons. Dimensions: 146 × 24 × 5¼ feet. H.P. 800 = 11 kts. Coal: 15 tons. Guns: 3—4.7 inch, 10 machine. Armour: 6"—4" belt, 1" deck, 3½" conning tower, 6" turret. Took 20 years to build.

ACRE, MISSOES (Yarrow, 1904). 200 tons. Dimensions: 120 × 20 × 2 feet. Armament: 1—3.4 inch (15 pdr.) howitzer, 1—6 pdr., 4 maxims. H.P. 300 = 11 kts. Complement, 30.

No Photo available.

JUTAHY, TEFFÉ (1890-92, refitted 1917). River Gunboats of 33 tons. Dimensions: 90 (p.p.) × 15 × 5 feet. Guns: 1—3 pdr., 2 M.G. H.P. = 11 kts. Coal: 7 tons.

Note:—These two Ships have also—but very doubtfully—been reported as vessels of "180 tons and 9 kts speed, of much the same appearance as *Acre* and *Missoes*." Probably have been confused with Commercial River Steamers bearing same names.

River Craft—*continued.*

No Photo available.

Present existence doubtful.

OYAPOCK (—). Despatch Vessel of 195 tons. 14 k

For Service with Seaplane

The Brazilian Navy possess about ten Seaplanes of C types. It is not known if any Ship has been fitted o carrier or as Tender to Seaplanes.

Mine Layers.

Photo wanted.

(See Silhouettes for appearance.)

CARLOS GOMEZ (ex-*Itaipu*, 1892). 1800 tons ,00. Dimensions: 301 × 39 × 10 feet. Designed kts. (*less now*). Guns: *not known*. Coal: 400 to up to 550 mines *maximum*.

Mine Sweepers.

Tugs *Jaguarayo* and *Mario do Couto* (both listed on probably fitted out for this service.

Armed Mercantile Cruiser and Fleet Collier.

BELMONTE (ex-German S.S. *Valesia*, Rostock, 1912). 5225 tons gross. Dimensions: 364½ × 51 × feet. Guns: 4—4.7 in., 6—6 pdr. H.P. 2500 = 12 kts. Can take about 6500 tons coal as cargo.

Note.—Above ship seized in Brazilian port, 1917, after declaration of war against Germany.

Tugs.

(Officially rated as Despatch Vessels.)

No Photos available of any of the Tugs and Yachts named below.

AMAPÁ (ex-Tug *Uaypes*). **GOYTACAZ.** No details known.

LAURINDO PITTA (1910). 304 tons.

MARECHAL HERMES. Sea-going Tug. No details known.

JAGUARAYO. Sea-going Tug. No details known. Attached to Mining Flotilla; may be Mine-Layer and/or Mine-Sweeper.

MARIA DO CONTO. Sea-going Tug. No details known. Attached to Mining Flotilla; may be Mine-Layer and/or Mine-Sweeper.

Yachts.

JOSÉ BONIFACIO. Ex-U.S. Yacht *Nourmahal* (?), 768 tons; 8½ kts. speed. No other details known.

TENENTE ROSA, TENENTE RIBEIRO (1911). 152 tons. Speed: 12½ kts. One or two small guns.

SILVA-JARDIM. No details known.

Training Ship.

1916 *Photo, Abrahams*

BENJAMIN CONSTANT (La Seyne, 1892, recon... 2750 tons. Comp. 360. Guns (Elswick): 8—4.7... 2—12 pdr., 2—1 pdr. Torpedo tubes (14 inch... Armour: 2" deck. Designed H.P. 4000 *f.d.* = 15 kts. Coal: 260 tons.

Also Brigs *Caravellas*, *Cidade de Manaos* and *Recife*.

10 Destroyers.

10 Yarrow type: ***Amazonas, Matto Grosso, Piauhy, Pará*** (all launched 1908), ***Rio Grande... Parahyba, Alagoas, Santa Catharina*** (all launched 1909), ***Parana, Sergipe*** (both l... Displacement 560 tons. Dimensions: 240 × 23½ × (*Mean* draught) 7⅔ feet. Armament: 2—4 inch... 2—18 in. tubes. Designed H.P. 8000 = 27 kts. Machinery: 2 sets triple expansion reciprocating. 2 Yarrow boilers. Majority of these boats reboilered 1917-18. Coal: 140 tons. Nominal radius: 1600 m... On acceptance trials with 6567 to 8877 H.P. they made 27·1 to 26·7 kts. Hulls of several are said... condition.

Note:—The numbers on after funnel are thus: 1, *Amazonas*; 2, *Pará*; 3, *Piauhy*; 4, *R. G. do Norte*; [illegible] 6, *Alagoas*; 7, *Sergipe*; 8, *Parana*; 9, *S. Catharina*; 10, *Matto Grosso*.

7 Torpedo Boats.

Photos reserved for 1921 Edition

[illegible] German "A" type Torpedo Boats are to be added to the Brazilian Navy, but up to the end of August, 1920, no boats had been selected from those [illegible] by Germany. They will probably be chosen from the German "A" types numbered between A 57 and A [illegible] which the following are [illegible] details:—[illegible] tons. Designed H.P. 4750=[illegible] kts. Fuel: [illegible] tons oil=[illegible] miles at [illegible] kts. Guns: 1—3·4 inch, 30 cal. (Krupp). Torpedo tubes: 1—17·7 inch. [illegible] Built 1916-18.

See Polish Section for photograph

2 ex-German "V" type Torpedo Boats are also to be handed over to the Brazilian Navy, and will probably be selected from the V[illegible]—[illegible] class, whose details are: 310 tons. Dimensions: 200 [illegible] × 20 [illegible] × 6 [illegible] feet. [illegible] H.P. 6000=20 kts. speed [illegible] on trials. Turbine engines. Fuel: 55·7 tons coal + 16·2 tons oil=630 miles at 20 kts. Guns: 3—3·4 inch Krupp, 15 cal. Torpedo tubes: [illegible]—17·7 inch. Can carry 12 mines. Begun by Vulkan Co., for Netherlands Navy, as Z1—Z[illegible]; taken over by German Navy, 1914, and renumbered V[illegible]—[illegible], V[illegible]. War Loss. Third boat of this class assigned to Polish Navy.

*Singly mounted in V[illegible], [illegible], and in one triple mounting in V[illegible].

1 *Yarrow* type: **Goyaz** (1907) 150 tons. Dimensions: 152 × 15½ × 4¾ feet. Guns: 2—3 pdr. Torpedo tubes: 2 single 18-inch. Turbine and reciprocating engines. Designed speed, 25½ kts. Yarrow boilers. Worn out and practically unfit for further service.

3 Submarines.

1917 *Photo, by favour of Ansaldo San G*[illegible]

3 *Laurenti-Fiat* type: **F 5, F 3, F 1** (Spezia, 1913-14). Built by Fiat-San Giorgio Co. Dimensions: 15[illegible] × [illegible] × 12 feet draught. Displacement: 250 tons *on surface*, 305 tons *submerged*. Machinery: 2 sets 6-cylinder, 2-cycle Fiat-Diesel engines *on surface* = [illegible] H.P. 2 sets 250 H.P. electric motors + batteries *submerged*. Speeds: (maximum) 13 kts. *on surface*, 8 to 8½ kts. *submerged*. Radius of action: *on sur*[illegible] 800 miles at full speed, 1600 miles at 8½ kts.; when *submerged*, 18 miles at full speed, or 100 miles [illegible] Torpedo tubes: 2—18 inch in bows. Complement about 20.

Submarine Carrier, Depot, Docking and Salvage Ship

(Also serves as Training Ship for submarine service.)

1917 *Photo, by courtesy of the Fiat San-Gio*[illegible]

CEARA (Spezia, 1915). Length (p.p.) 328 feet. Beam, 52 feet. Draught and displacement (with all stores) va[illegible] (*a*) with dock empty and dock-gate closed 4100 tons at 14 feet draught; (*b*) with dock-gate open a[illegible] flooded to float submarine in, 4130 tons at 17¼ feet; (*c*) with gate closed and submarine docked, 4560 [illegible] feet; (*d*) with gate closed and submarine under hydraulic pressure test in dock, 6460 tons at 20¼ feet; dock empty and gate closed and when raising submarines by double cranes at stern, 1615 tons a[illegible] tons for submarines (sufficient fuel carried to fill tanks of six submarines four times). Radius o[illegible] Machinery: 2 sets 6-cylinder, 2-cycle Fiat-Diesel engines 4400 B.H.P. = 14 kts. Fuel: 400 tons (own [illegible] 4000 miles at 10kts. Guns: 4—4 inch, 2 smaller. Built by Fiat-San Giorgio Co. Completed 1915.

Note.—This ship has been specially designed and completely equipped to serve as a depot ship, salvage [illegible] floating dock for a flotilla of 6 submarines. There is a central, circular caisson dock 216¼ feet long [illegible] submarines up to 198 feet length and 23¼ feet beam) between the double hulls, and two salvage cranes for raising 400 tons deadweight. Equipment for service to submarines includes two 150 Kw. charging [illegible] two electric-driven and one steam-driven 75-150 atmos. air compressors, refrigerating plant, workshops, [illegible] pumps to empty dock in 2 hours, &c.; carries spare batteries, torpedoes, stores, &c., for 6 submarines.

CHILEAN FLEET.

Revised 1920, from information furnished by courtesy of (a) The Ministry of Marine; (b) Rear-Admiral Luis Gomez, Chief of the Chilean Naval Commission; (c) Commander F. Merino, Naval Attache, London; (d) Senor Roberto Kelly G.

Flags.

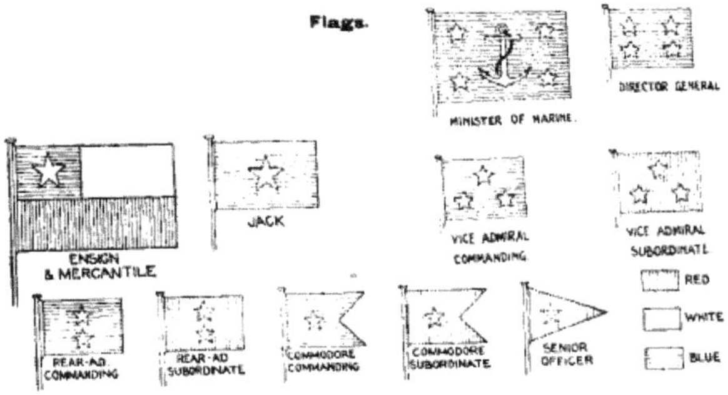

Uniforms.

Corresponding British or U.S.: Vice-Ad. | Rear-Ad. | Captain. | Commander. | Lieut.-Comdr. | Lieut. Senior. | Lieut. Junior. | Sub. Lieut. | Midshipman.

Other Branches the same without the star, and colours as follows: Engineers (blue), Paymasters (white), Doctors (red).

Minister of War and Marine: Senor Ladislao Errazuriz.

Personnel: About 7500, of whom about 1000 are conscripts, the rest volunteers.

Mercantile Marine: 76 steamers, of 71,556 tons gross; 22 sailing vessels, 11,420 tons gross, and 16 vessels with auxiliary motors, 5174 tons gross.

Naval Bases.

TALCAHUANO (Submarines based here.) One dry dock, 614 × 87 × 30½ feet. New dock building, 856 × 117 × 36 feet. One small floating dock, 216 × 42 × 15 feet.

VALPARAISO. One small wooden floating dock privately owned, 300 × 49½ × 19 feet (3500 tons capacity).

There is also a small wooden floating dock at Magallanes.

Naval Ordnance. (All details unofficial.)

	Type*	Calibre. Inches	Calibre. Cm.	Length (cals.)	Weight of Gun (tons.)	Weight of Proj. (lbs.)	Weight of Charge. (lbs.)	M.V. (f.secs.)	M.E. (f.tons.)
HEAVY	E	**14"**	35·6	45	85	1400	324	2700	70,700
MEDIUM	C	**9·4"**	24	36	24	374	198	2230	12,421
	E	**8"**	20·3	45	18·5	210	44	2650	10,226
	E	**8"**	20·3	40	...	...	...	...	...
	E	**6"**	15·2	50	8¾	100	31	3000	6240
	E	**6"**	15·2	40	6½	100	18·3	2500	4334
	—	**5·9"**	15	...	...	...	...	...	...
	—	**4·7"**	12	...	...	...	...	...	...
	E	**4·7"**	12	45	3½ cwt. 53	45	8	2552	2110
	C	**4·7"**	12	45	...	...	...	...	...
LIGHT	...	**4"**	10·2	...	...	...	...	...	...

*E = Elswick. C = Canet.

14 inch, 50 cal., in *Alm. Latorre.*

9.4 inch, 36 cal., in *C. Prat.*

8 inch, 45 cal., in *Chacabuco.*

8 inch, 40 cal., in *O'Higgins, B. Encalada* and *Esmeralda.*

6 inch, 50 cal., in *Alm. Latorre.*

6 inch, 40 cal., in *O'Higgins, Es... B. Encalada, M. Zenteno.*

5.9 inch in *P. Errazuriz.*

4.7 inch in *Alm. Riveros* class (3).

4.7 inch in *Alm. Cochrane, C. Prat, G. Baquedano, Talcahuano Errazuriz.*

4 inch in *Alm. Lynch* class (2) & *Riveros* class (3).

Torpedoes: 21 inch (heater), 18 inch and 14 inch.

Mines: not known.

Naval Programme.

This was sanctioned by the Laws of July 6th, 1910, and October 21st, 1911, and provid... construction of two Dreadnoughts, six Destroyers and some Submarines. The two Dreadn... four of the six Destroyers were requisitioned for the British Navy during the war, but, during... Dreadnought and three Destroyers were reclaimed by purchase. One Dreadnought and one may still be built or purchased, as both are authorized.

Paragraph 3, Law of July 6th, 1910, provides that the sum of £400,000 be set aside an... building a first-class warship, embodying the most recent improvements. But, up to the p... sum has not been expended.

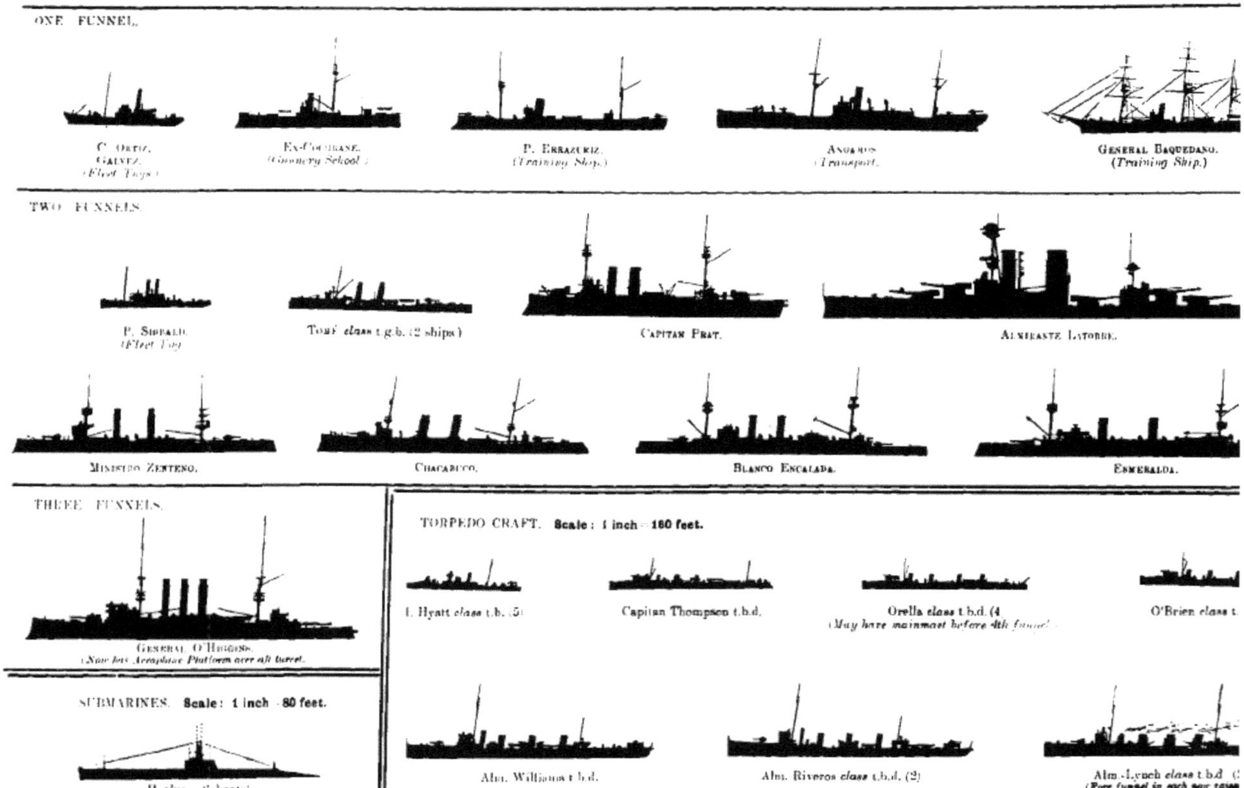
ONE FUNNEL.
C. Ortiz, Galvez. (Fleet Tugs.)
Ex-Cochrane. (Gunnery School.)
P. Errazuriz. (Training Ship.)
Angamos (Transport.)
General Baquedano. (Training Ship.)
TWO FUNNELS.
P. Sibbald. (Fleet Tug.)
Toro class t.g.b. (2 ships.)
Capitan Prat.
Almirante Latorre.
Ministro Zenteno.
Chacabuco.
Blanco Encalada.
Esmeralda.
THREE FUNNELS.
General O'Higgins.
(Now has Aeroplane Platform over aft turret.
SUBMARINES. Scale: 1 inch = 80 feet.
H class (6 boats).
TORPEDO CRAFT. Scale: 1 inch = 160 feet.
I. Hyatt class t.b. (5)
Capitan Thompson t.b.d.
Orella class t.b.d. (4
(May have mainmast before 4th funnel.)
O'Brien class t.
Alm. Williams t.b.d.
Alm. Riveros class t.b.d. (2)
Alm.-Lynch class t.b.d.
(Fore funnel in each now raise

1911 CHILEAN DREADNOUGHT. (*purchased 1920*).

ALMIRANTE LATORRE (ex British *Canada*, ex Chilean *Almirante Latorre*, Nov., 1913.)

Displacement, 28,000 tons (about 32,000 *full load*). Complement, 1176.

Length (*p.p.*), 625 feet. Beam, 92½ feet. { *Mean* draught, 29 feet / *Max.* " 32 " } Length *over all*, 661 feet.

Guns (Elswick):
- 10—14 inch, 45 cal. } **Dir. Con.**
- 14—6 inch 50 cal. }
- 2—3 inch (anti aircraft)
- 4—3 pdr.
- (2 landing)
- 4 M.G.

Torpedo tubes (21 inch):
- 4 *submerged*

Armour:
- 9" Lower belt
- 7" Middle belt
- 4½" Upper belt
- 6"-4" Belt (ends)
- 1½", 4" B'lkh'ds (f. & a.)
- 6" Batteries
- 10" Barbettes
- 10"—7" Gunhouses
- 3" C.T. (base*)
- 11" C.T. (6"—3" hood)
- 6" Fore com. tube
- 6" Torpedo C.T.
- 6" Aft com. tube

*[illegible]

Armour:
- 1" Shelter (over casemates)
- 1" Fo'xle (over battery)
- 1½" Upper (outside battery)
- 1½" Main (aft)
- 1" Protective
- 2" (forward) } lower
- 4" (aft) }

Special Protec.:
- 2"—1½" Internal screens (mags., &c.) Sections 6-8, 15, 19-22 on plans.

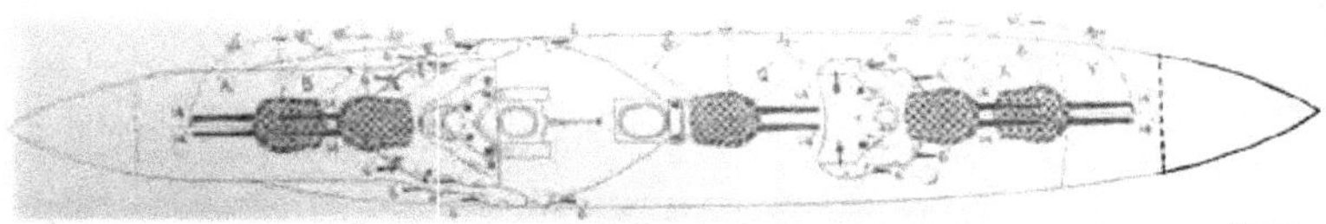

Ahead: 4—14 in. 4 [illegible] in.

Broadside: 10—14 in., 7—6 in., 2—21 in. tubes

Astern: 4—14 in. 3 to 6-6 in.

Machinery: Turbine, 4-shaft: (L.P.) Parsons; (H.P.) Brown-Curtis. Boilers: 21 Yarrow. Designed H.P. 37,000 = 22.75 kts. Coal: *normal*, 1150 tons; *maximum*, 3300 tons. Oil: 520 tons.

Gunnery Notes.—14-inch have a range only limited by max. visibility. Originally had 16—6 inch, but the 2—6 inch on upper deck, abeam of after funnel, were removed and ports plated over. Reason for removal was because guns were only a few feet from muzzle of "Q" turret 14-inch guns on extreme bearing and were damaged by blast.

Armour Notes.—Barbettes, 6" and 4" as they descend behind belts.

Searchlights.—6—24" and 2—20" (signalling) Harrison lamp type burners.

Aircraft Notes.—Flying-off platforms to Turrets B and X. Anti-aircraft guns on after superstructure.

ALMIRANTE LATORRE. 1919 Photo, [illegible]

1919 Official [illegible]

ALMIRANTE LATORRE: Aircraft View.

Engineering Notes.—Reported that the Chilean Ministry of Marine stipulated, at the last [illegible] should be attained without using forced draught. Funnels were accordingly ra[illegible] draught. Reported to have made 23-24 kts. in service.

Torpedo Notes.—Tubes are 21 inch Elswick side-loading, worked by hydraulic power.

Name	*Builder*	*Machinery*	*Laid down*	*Completed*	*Trials*
Alm. Latorre	Elswick	Clydebank	Nov.'11	Sep.'15	[illegible]

General Notes.—Laid down for Chile as the *Valparaiso*, her name being altered afterwa[illegible] Purchased for British Navy on outbreak of War and re-named *Canada*. Addition[illegible] during War is said to have raised her *normal* displacement to over 30,000 tons[illegible] secondary battery of 22—4.7 inch, and 2 stump masts abeam aft. Re-purchase[illegible] Her sister ship, *Almirante Cochrane*, was purchased and taken over for British Navy [illegible] and modified for service as an Aircraft Carrier. (See British Navy Section).

(1888) Battleship.

(?) Copyright photo, G. Allan, Valparaiso.

CAPITAN PRAT (La Seyne, 1890. *Reconstructed* 1909). 6902 tons. Sheathed and coppered. Complement 500. Length (*p.p.*) 328 feet. Beam, 60¾ feet. *Maximum* draught, 22¾ feet. Guns (Canet): 4—9.4 inch, 36 cal., 8—4.7 inch, 45 cal., 10—6 pdr., 4—3 pdr., 16—1 pdr., 4 Maxim. Torpedo tubes (18 inch): 2 *above water*. Armour (Creusot): 12" Belt (amidships), 3" Deck (flat on belt), 10½" Barbettes, 2" Barbette hoods, 4" Redoubt amidships, 2" Small turrets. Machinery: 2 sets horizontal triple expansion. 2 screws. Boilers: new in 1909, 12 Babcock. H.P. 12,000 = 18 kts. (about 16 to 18 kts. now). Coal: *normal* 400 tons; *maximum* 775 tons. Endurance: 1650 miles at 10 kts.

(1895) Armoured Cruiser.

(?) Copyright photo, G. Allan, Valparaiso.

ESMERALDA (Elswick, 1896, re-fitted 1910). 7050 tons. Sheathed and coppered. Complement 700. Length (*p.p.*), 436 feet. Beam, 53 feet, 2 ins. Draught (*mean*), 20¼ feet. Guns (Elswick): 2—8 inch, 40 cal., 12—6 inch, 40 cal., 12—12 pdr. (2 field). Torpedo tubes (18 inch): 2 *above water*. Armour (Harvey): 6" Belt, 2" Deck, 6" Bulkheads, 4½" Shields to guns, 4½" Hoists. Machinery: 2 sets 4 cylinder triple expansion. 2 screws. Boilers: 2-ended cylindrical. H.P. 18,000 = 22.25 kts. (still steams well and can do 21 kts. now). Coal: *normal* 550 tons; *maximum* 1350 tons. Endurance: (*a*) 2400 miles at 21 kts., (*b*) 7680 miles at 10 kts.

(1896) Armoured Cruiser.

Note.—Aeroplane platform over after 8 in. turret. 1919 *Copyright photo, G. Allan, Valpa*

GENERAL O'HIGGINS (Elswick, 1897). 8500 tons. Sheathed and coppered. Complem... Length (*p.p.*), 412 feet. Beam, 62¾ feet. *Max* draught, 22 feet. Guns (Elswick): 4—8 inch, 10—6 inch, 40 cal., 8 or 10—12 pdr., 10—6 pdr., 4 M.G. Torpedo tubes (18 inch): 2 sub. Armour (Harvey-nickel): 7"-5" Belt (amidships), 2" Deck (slopes), 7½" Port plates to 8 in. turrets, 6" Hoists to these, 6" Gun houses to 6in. guns, 6" Casemates (4), 9" Conning ... Machinery: 2 sets triple expansion. 2 screws. Boilers: 30 Belleville (in 3 groups). ... H.P. *natural* 10,000 = 19 kts.; *forced* 16,000 = 21.5 kts. (still steams very well and can do ... now). Coal: *normal* 700 tons, *maximum* 1200 tons. Endurance: (*a*) 2250 miles at 20 ... 6000 miles at 10 kts. Large refit 1919-20.

(1897) Protected Cruiser.

1(?) *Copyright photo, G. Allan, Valpara*

CHACABUCO (Elswick, 1898, purchased 1902). 4500 tons. Complement 400. Length (*p.p.*) ... feet. Beam, 46½ feet. *Maximum* draught, 17 feet. Guns (Elswick): 2—8 inch, 45 cal., 1... inch, 40 cal., 12—12 pdr., 6—3 pdr. Armour (Harvey nickel): 4½" Deck (amidships), 1¾" (ends), 4½" Fronts to 8 in. gun shields, 2½" Sides to 8in. gun shields, 2½" Shields to 4.7 in. 5" Conning tower. Boilers, cylindrical. I.H.P. 15,500 = 24 kts. *forced*. Still steams well, a... do 23 kts. now. 2 screws. Coal: *normal* 300 tons; *maximum* 1028 tons. Electric trainin... elevating gear to 8 inch guns. Laid down at Elswick, 1897, as a speculation. Completed 190... purchased by Chile.

1911 CHILEAN DREADNOUGHT. (*purchased 1920*).

ALMIRANTE LATORRE ex British *Canada*, ex Chilean (*Almirante Latorre*, Nov., 1913.)

Displacement, 28,000 tons (about 32,000 *full load*). Complement, 1176.

Length (*p.p.*), 625 feet. Beam, 92 feet. { *Mean* draught, 29 feet. *Max.* " 32 " } Length *over all*, 661 feet.

Guns (Elswick):
- 10—14 inch, 45 cal. } Dir. Con.
- 14—6 inch, 50 cal. }
- 2—3 inch (anti aircraft)
- 4—3 pdr.
- (2 landing)
- 4 M.G.

Torpedo tubes (21 inch):
- 4 *submerged*.

Armour:
- 9" Lower belt
- 7" Middle belt
- 4½" Upper belt
- 6"—4" Belt (ends)
- 4½", 4" Bulkheads (f. & a.)
- 6" Batteries
- 10" Barbettes
- 10" Gunhouses
- 4" C.T. base*
- 11" C.T. (6"—3" hood)
- 6" Fore com. tube
- 6" Torpedo C.T.
- 6" Aft com. tube

*[illegible]

Armour:
- 1" Shelter (over casemates)
- 1" Fo'xle (over battery)
- 1½" Upper (outside battery)
- 1½" Main (aft)
- 1" Protective
- 2" (forward) } lower
- 4" (aft) }

Special Protec.:
- 2"—1½" Internal screens (mags., sect. 8 c.rooms, 6-8, 15, 19-22 on plans.

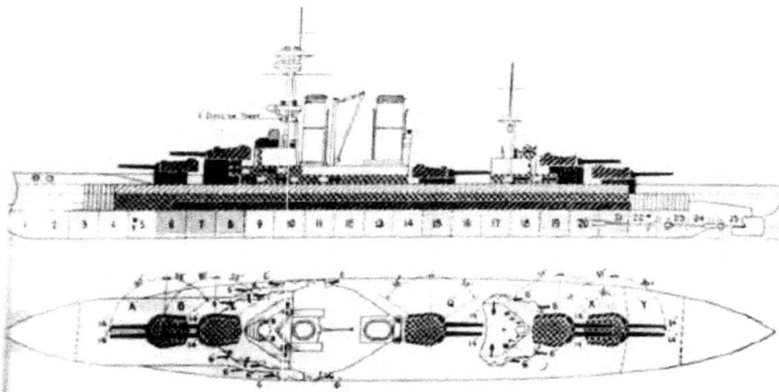

Ahead: 4—14 in. 4 to 5-6 in.

Broadside: 10—14 in., 7—6 in., 2—21 in. tubes.

Astern: 4—14 in., 3 to 6-6 in.

Machinery: Turbine, 4 shaft: (L.P.) Parsons; (H.P.) Brown-Curtis. Boilers: 21 Yarrow. Designed H.P. 37,000 = 22.75 kts. Coal: *normal*, 1150 tons; *maximum*, 3300 tons. Oil: 520 tons.

Gunnery Notes.—14-inch have a range only limited by max. visibility. Originally had 16—6 inch, but the 2—6 inch on upper deck, abeam of after funnel, were removed and ports plated over. Reason for removal was because guns were only a few feet from muzzle of "Q" turret 14-inch guns on extreme bearing and were damaged by blast.

Armour Notes.—Barbettes, 6" and 4" as they descend behind belts.

Searchlights.—6—24" and 2—20" (signalling) Harrison lamp type burners.

Aircraft Notes.—Flying-off platforms on Turrets B and X. Anti-aircraft guns on after superstructure.

ALMIRANTE LATORRE. 1919 *Photo*, [illegible]

ALMIRANTE LATORRE: Aircraft View. 1919 *Official British* [illegible]

Engineering Notes.—Reported that the Chilean Ministry of Marine stipulated, at the last moment, [illegible] should be attained without using forced draught. Funnels were accordingly raised to [illegible] draught. Reported to have made 23-24 kts. in service.

Torpedo Notes.—Tubes are 21 inch Elswick side-loading, worked by hydraulic power.

Name	Builder	Machinery	Laid down	Completed	Trials	Boilers
Alm. Latorre	Elswick	Clydebank	Nov '11	Sep. '15	[illegible]	Yarrow

General Notes.—Laid down for Chile as the *Valparaiso*, her name being altered afterwards to *Almirante* [illegible] Purchased for British Navy on outbreak of War and re-named *Canada*. Additional protection [illegible] during War is said to have raised her normal displacement to over 30,000 tons. First de[illegible] secondary battery of 22—4.7 inch, and 2 stump masts abeam aft. Re-purchased by Chile, [illegible] Her sister-ship, *Almirante Cochrane*, was purchased and taken over for British Navy in 1917, re-na[illegible] and modified for service as an Aircraft Carrier. (See British Navy Section).

(1888) Battleship.

(1909) Copyright photo, G. Allan, Valparaiso.

CAPITAN PRAT (La Seyne, 1890). *Reconstructed* 1909. 6902 tons (sheathed and coppered). Complement 500. Length (*p.p.*) 328 feet. Beam, 60⅔ feet. *Maximum* draught, 22½ feet. Guns (Canet): 4—9·4 inch, 36 cal., 8—4·7 inch, 45 cal., 10—6 pdr., 4—3 pdr., 10—1 pdr., 4 Maxim. Torpedo tubes (18 inch): 2 *above water*. Armour (Creusot): 12" Belt (amidships), 3" Deck (flat on belt), 10½" Barbettes, 2" Barbette hoods, 4" Redoubt (amidships), 2" Small turrets. **Machinery**: 2 sets horizontal triple expansion. 2 screws. Boilers (new in 1909), 12 Babcock. H.P. 12,000 = 18 kts. (about 16 to 18 kts. now). Coal: *normal* 400 tons; *maximum* 775 tons = circa 1650 miles at 10 kts.

(1895) Armoured Cruiser.

(1919) Copyright photo, G. Allan, Valparaiso.

ESMERALDA (Elswick, 1896, re-fitted 1910). 7050 tons. Sheathed and coppered. Complement 700. Length (*p.p.*), 436 feet. Beam, 53 feet, 2 ins. Draught (*mean*), 20¼ feet. Guns (Elswick): 2—8 inch, 40 cal., 12—6 inch, 40 cal., 12—12 pdr. (2 field). Torpedo tubes (18 inch): 2 *above water*. Armour (Harvey): 6" Belt, 2" Deck, 6" Bulkheads, 4½" Shields to guns, 4½" Hoists. Machinery: 2 sets 4 cylinder triple expansion. 2 screws. Boilers: 2-ended cylindrical. H.P. 18,000 = 22·25 kts. (still steams well and can do 21 kts. now). Coal: *normal* 550 tons; *maximum* 1350 tons. Endurance: (*a*) 2400 miles at 21 kts., (*b*) 7080 miles at 10 kts.

(1896) Armoured Cruiser.

Now has Aeroplane platform over after 8-in. turret. 1919 Copyright photo, G. Allan, Val[illegible]

GENERAL O'HIGGINS (Elswick, 1897). 8500 tons. Sheathed and coppered. Complement [illegible] Length (*p.p.*), 412 feet. Beam, 62¾ feet. *Max.* draught, 22 feet. Guns (Elswick): 4—8 inch [illegible] 10—6 inch, 40 cal., 8 or 10—12 pdr., 10—6 pdr., 4 M.G. Torpedo tubes (18 inch): 2 [illegible] Armour (Harvey-nickel): 7"—5" Belt (amidships), 2" Deck (slopes), 7½" Fort plates to 8 [illegible] turrets, 6" Hoists to these, 6" Gun houses to 6in. guns, 6" Casemates (6), 9" Conning [illegible] Machinery: 2 sets triple expansion. 2 screws. Boilers: 30 Belleville (in 3 groups). H.P. *natural* 10,000 = 19 kts.; *forced* 16,000 = 21·5 kts. (still steams very well and can [illegible] now). Coal: *normal* 700 tons; *maximum* 1200 tons. Endurance: (*a*) 2250 miles at 20 [illegible] 6900 miles at 10 kts. Large refit 1919—20.

(1897) Protected Cruiser.

1919 Copyright photo, G. Allan, Valpar[illegible]

CHACABUCO (Elswick, 1898, purchased 1902). 4500 tons. Complement 400. Length (*p.p.*) [illegible] feet. Beam, 46½ feet. *Maximum* draught, 17 feet. Guns (Elswick): 2—8 inch, 45 cal. [illegible] inch, 40 cal., 12—12 pdr., 6—3 pdr. Armour (Harvey nickel): 4½" Deck (amidships), 1[illegible] (ends), 4½" Fronts to 8 in. gun shields, 2½" Sides to 8in. gun shields, 2½" Shields to 4.7 in[illegible] 5" Conning tower. Boilers, cylindrical. I.H.P. 15,500 = 24 kts. *forced*. Still steams well, [illegible] do 23 kts. now. 2 screws. Coal: *normal* 300 tons; *maximum* 1026 tons. Electric train[illegible] elevating gear to 8 inch guns. Laid down at Elswick, 1897, as a speculation. Completed 19[illegible] purchased by Chile.

1896. Protected Cruisers (continued).

(Copyright photo, G. Allan, Valparaiso.)

MINISTRO ZENTENO (Elswick, 1896). 3420 tons (sheathed and coppered). Complement, 300. Length, (p.p.), 330 feet. Beam, 43 feet. Max. draught, 16 feet. Armament: 8—6 in. (40 cal.), 10—6 pdr., 4—1 pdr., 4 Maxims. Armour: 3" Deck, 4" Conning tower. Designed H.P. 7500 = 20 kts. (16 to 18 kts. now). Cylindrical boilers. Coal: maximum 850 tons. Endurance: 6450 miles at 10 kts.

1893.

(Copyright photo, G. Allan, Valparaiso.)

BLANCO ENCALADA (Elswick, 1893). 4420 tons (sheathed and coppered). Complement 427. Length (p.p.), 370 feet. Beam, 46½ feet. Maximum draught, 19½ feet. Armament: 2—8 inch, 40 cal., 10—6 inch, 40 cal., [illegible] 4.7 inch, [illegible]—12 pdr., 8—3 pdr., 4—1 pdr. Armour: 4" Deck, 6" Shields to 8" guns, 6" Conning tower. Boilers: cylindrical (retubed 1908-9). Designed H.P. forced 14,500 = 22.75 kts. (about 21.5 kts. now.) Coal: maximum 850 tons. Endurance: 7000 miles at 10 kts. To be refitted, 1920, at Talcahuano.

1890. Protected Cruiser (and Training Ship).

(Copyright photo, G. Allan, [illegible])

PRESIDENTE ERRAZURIZ (La Seyne, 1890, refitted 1908). 2080 tons (sheathed and [illegible]). Complement, 170. Dimensions: 268¼ (p.p.) × 35¼ × 14½ feet. Guns (new in 1908): 4—5[illegible] 4.7 inch, 4—6 pdr., 4—3 pdr. Armour: 2½" Deck, 3" shields to guns. Machinery: 2 sets triple expansion, 2 screws. Boilers: Belleville (1908). Designed H.P. 5400 = 19 kts. (12.5 [illegible]) Coal: 280 tons. Endurance: 2980 miles at 10 kts.

Sloop (and Training Ship).

(Copyright photo, G. Allan, [illegible])

GENERAL BAQUEDANO (1898). 2500 tons, sheathed and coppered. Armed with 4—[illegible] guns, 2—12 pdr., and 2—6 pdr. I.H.P. 1500 = 13.75 kts. Belleville boilers. Coal: 300 tons. [illegible] by Elswick. Best recent speed: 12 kts.

Torpedo Gunboats.

Mainmast removed [illegible] 1920. *Photo, Allan, Valparaiso.*

TOMÉ (ex *Almirante Lynch* 1890) & **TALCAHUANO** (ex *Almirante Condell*, 1890). Displacement, 720 tons. Length, 240 feet (o.a.). Beam, 27½ feet. Maximum draught, 11 feet. Complement, 87. Guns: *Tomé* 3—12 pdr., 2 M.G.; *Talcahuano*, 1—4.7 inch (aft), 2—12 pdr., 2 M.G. Torpedo tubes (*in both*): 5—14 inch, (1 stern, 1 bow, 4 broadside), all *above water*. Armour: Double skin (amidships), 1" Bulkheads. Machinery: 2 sets triple expansion. 2 screws. Boilers: Belleville, *without economisers*. Designed I.H.P. 4500 = 20 kts. (exceeded on *trial*). Coal: 220 tons. Endurance: 5100 miles.

Laid down at Birkenhead, 1889. Completed 1891. Reconstructed 1898. Attached to Torpedo and Engineering Schools.

12 Destroyers.

No.	Type	Date	Displacement	H.P.	Max. speed	Fuel	Complement	T. tubes
			tons		kts.	tons		
3	*Alm. Riveros* (W)	'11-'15	1730†	30,000	31-5†	403 coal, 85 oil	205	4
2	*Alm. Lynch* (W)	'11-'14	1430	30,000	31†	127 coal, 80 oil	160	3
2	*Capitan O'Brien* (L)	'00-'02	311	6000	30	80 coal	65	2
1	*Capitan Thompson* (E)	'98-'00	350	6500	30	120 coal	65	2
4	*Capitan Orella* (L)	1896	300	6000	30	90 coal	65	3

(E) = Elswick. (L) = Laird. (W) = White. († = Turbines.)

*When in British Navy 1914-20, these three boats were rated as Flotilla Leaders. The two *Alm. Lynch* [illegible] much the same design, can also be rated as Leaders.
[illegible]

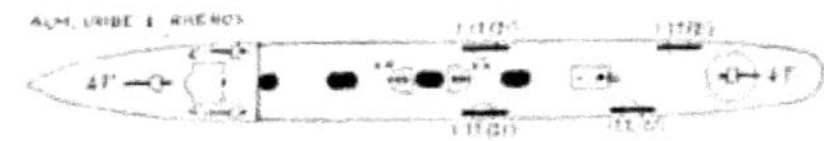

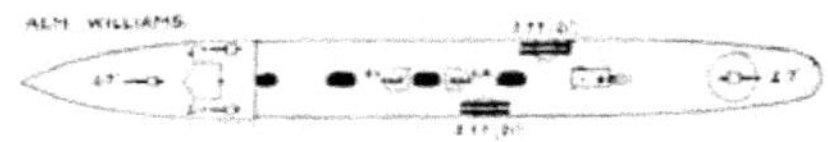

ALMIRANTE WILLIAMS. (**For description v. next page.**)

ALMIRANTE RIVEROS.

3 White: **Almirante Williams** (ex British *Botha*; ex Chilean *Almirante Williams Rebolledo*, 1914), 1742 tons, **Almirante Uribe** (ex British *Broke*; ex Chilean *Almirante Goñi*, 1913), 1704 tons; **Almirante Riveros** (ex British *Faulknor*; ex Chilean *Almirante Simpson*, 1913), 1694 tons. Dimensions: 331 × 32½ × 11 ft. 7 in. Armament: 2—4·7 inch, 2—4 inch, 2—2 pdr. pom-poms, 4—21 inch tubes.* Designed H.P. 30,000 = 31 kts. in *Almirante Williams*, 32 kts. in *Almirante Uribe* and *Almirante Riveros*. Machinery: Turbines. Boilers: White-Forster, 3 screws. Fuel: [illegible] 403 coal + [illegible] oil. Complement 205.

Note.—Another of the class, British *Tipperary*, ex Chilean *Almirante Riveros*, sunk in Battle of Jutland. The above three boats purchased August, 1914, on outbreak of war, from Chile. Refitted by British Navy 1918-19, and re-purchased by Chile, April, 1920.

*Tubes mounted in pairs in *Almirante Williams*; singly mounted in *Almirante Uribe* and *Almirante Riveros*.

ALM. CONDELL (centre), AL. LYNCH (behind). (Fore funnels now raised.) *Copyright photo, G. Allan, Valparaiso.*

2 White: **Almirante Lynch** 1912, **Al. Condell** (1913). Dimensions: 320 × 32½ × 11 feet. Normal displacement 1430 tons. Full load: 1850 tons. Armament: 6—4 inch, 4 M.G., 4—18 inch tubes. Designed H.P. 30,000 = 31 kts. Turbines: Parsons. Boilers: White-Forster. 3 screws. Fuel: 127 tons coal + 80 tons oil = 2750 miles at 15 kts. Trials: *Al. Lynch*, 31·8 kts. (6 hours). *Al. Condell*, 32·4 kts. Complement 160.

CAPITAN O'BRIEN. Both boats now have mainmasts. *Copyright photo, G. Allen.*

2 Laird: **Capitan O'Brien** & **Capitan Merino Jarpa** (both 1901). Armament: 1—12 pdr., 5 [illegible] inch tubes, one amidships, one aft. Boilers: Laird-Normand. Other details as Laird boat below.

Now has mainmast. *Photo*

1 Elswick: **Capitan Thompson** 1896. 300 tons. Dimensions: 215 p.p. × 21 × 5¼ feet. Armament: 4—6 pdr., 2—18 inch tubes. Designed H.P. 6500 = 30 kts. Coal: 120 tons = 1150 miles at 15 kts.

No photographs available.

Appearance as O'Brien class above, except that the funnels are shorter. May have mainmast stepped before fourth funnel.

4 Laird: **Capitan Orella, Capitan Munoz Gamero, Teniente Serrano, & Guardia Riquelme** (all 1896). 311 tons. Dimensions: 213 p.p. × 21½ × 8½ feet. Armament: 1—12 pdr., 5—6 pdr., 2—18 inch tubes. Boilers: Normand. Designed H.P. 6500 = 30 kts., but some can only do 25 now. Nominal radius 900 miles at 15 kts. All reboilered and refitted 1909. Complement 65.

5 Torpedo Boats.

[illegible] type: ***Injeniero Hyatt, Cirujano Videla*** (Yarrow, Poplar, 189[illegible]), ***Injeniero Mutilla, Guardia Marina Contreras*** and ***Teniente Rodriguez*** (Laird[illegible], 1896-190[illegible]). [illegible] tons. Dimensions: 152 × 15¼ × [illegible] feet. Armament: [illegible]—6 pdr., 3—18 inch torpedo tubes. Designed H.P. [illegible] = 27.5 kts. Yarrow boilers. Coal: [illegible] tons. Complement 28 and 32.

6 Submarines.

112 — Copyright photo, G. Afr[illegible], Val[illegible]

[illegible]: ***H1, H2, H3, H4, H5, H6*** Fore River Co., U.S.A. (1915-17). Displacement [illegible] tons. Dimensions: [illegible] × 15 × 12½ feet. H.P. [illegible] = [illegible] kts. Machinery: *for surface* 2 sets [illegible] H.P. Nlseco Diesel. When *submerged* electric motors and batteries can develop [illegible] H.P., [illegible] kts for one hour only. Radius [illegible] miles at 11 kts *on surface*, [illegible] miles at [illegible] kts *submerged*. Oil: 17½ tons. Torpedo tubes: [illegible]—18 in. Complement 22.

Note.—These boats were originally *H 13* and *H* [illegible] of the British *H 11*—[illegible] group of submarines, built by Fore River Co., U.S.A., during 1915. The Admiralty intended to take them over and equip them with torpedo [illegible] Canadian Vickers Co., Ltd. at Montreal—provided that these boats could legally be delivered in an unfinished [illegible]. The U.S. Government decided the submarines could not leave any U.S. port, so long as the United States remained neutral, and all the boats were interned at Boston. They were released on the U.S. declaration [illegible] 1917. With the approval of the U.S. authorities, the above six boats were ceded to Chile by Great Britain [illegible] part payment for the Chilean warships building in British yards in August, 1914, and appropriated [illegible] British Navy.

Transports.

Note.—The following three ships are registered under the ownership of the Chilean Government, and do not belong to the Regular Navy. They are, however, largely used as Naval Transports, being directed to that service by the Chilean Ministry of Shipping as emergencies arise.

Photo wanted.

MAIPO (ex-*Manitola*, Thompson, Sunderland, 1901). 4687 tons *gross* (about 6600 tons displacement). Dimensions: 390 × [illegible] feet. I.H.P. [illegible] = 12 kts. speed. Machinery: [illegible]. No. of screws: [illegible]. Boilers: [illegible]. Coal capacity: [illegible] tons.

Transports—*continued*

Photo wanted.

(One funnel and four masts?)

RANCAGUA (ex-*Idaho*, Swan Hunter, Wallsend, 1898). 6210 tons *gross* (about 8600 tons displacement). Dimensions: 460 × 50 × 27½ feet. I.H.P. 6250 = 15 kts. Machinery: Triple expansion. No. of screws: 2. Boilers: Cylindrical. *Max.* coal capacity: 1073 tons.

Transports—*continued.*

Photo wanted.

(See Silhouette for appearance.)

ANGAMOS (ex-*Citta di Vincenzo*, Swan Hunter, Wall[illegible] rebuilt in Chile at an unknown date). Displace[illegible] tons. Dimensions: 377 × 44¼ × 23¼ feet. I.H.P. [illegible] kts. Machinery: Triple expansion. No. of [illegible] Boilers: Cylindrical. *Max.* coal capacity: 800 [illegible]

The following Steamers of the Compania Sud Americana are also subventioned for use as Transports in e[illegible]: *Aysen, Huasco, Cachapoal, Castro, Imperial, Palena, Renaico.*

Sea-going Fleet Tugs. *(Remolcadores.)*

P. SIBBALD. 1929 Photo, by courtesy of the Chilean [illegible].

PILOTO SIBBALD (Hose, McLachlan, Paisley, March, 1915). 885 tons. Dimensions: $141\frac{3}{4}$ *(p.p.)* × 29 × 16 feet. I.H.P. 1200 = $11\frac{1}{2}$ kts. Triple expansion engines. 2 screws. 2 cyl. boilers. Coal: 120 tons. Complement, 18.

Notes.—Begun under above name for Chilean Navy; taken over for British Navy during War and re-named *Slave*. Re-purchased by Chile 1920, when she resumed her original name. Can carry 240 tons fresh water. Fitted as Fire Float. Also has submersible electric and petrol-driven centrifugal salvage pumps.

No photo available.

ARTILLERO (1916). 95 tons. Speed: 5 kts.

C. ORTIZ (GALVEZ similar). 1926 Photo, Robertson, Gourock.

CONTRAMAESTRE ORTIZ (1909). 690 tons. Speed: 9 kts.

GALVEZ (1897). 475 tons. Speed: 9 kts.

GALVARINO (1885). 330 tons. Speed: 8 kts.

Coastguard Vessels *(Escampavias).*

Note.—The following vessels are normally employed as Surveying Ships and Lighthouse Tenders, but in the event of war they would be used as Mine Layers and Patrol Vessels.

Photo wanted.

COLOCOLO (1919) and **LEUCOTON** (1919), both built by Sandvikens Skeppsdocka Co., Helsingfors; **ELICURA** (1919) and **OROMPELLO** (1919), both built by Maskin—Och—Brobyggnads Co., Helsingfors. All four completed 1919. 520 tons. Length (o.a.), 172·6 feet. Beam, 24·6 feet. *Maximum* draught, 11 feet. Guns: (*not known*). I.H.P. 1400 = $14\frac{1}{2}$ kts. speed. 2 sets triple expansion engines. 2 screws. Boilers: 3 cylindrical. Coal capacity: *Colocolo* and *Leucoton*, 50 tons; *Elicura* and *Orompello*, 65 tons. Complement, 42. Refitted by Messrs. J. Samuel White & Co., Cowes, 1920.

Photos wanted.

AGUILA (1906). 820 tons. Speed: 10 kts.
PORVENIR (1906). 450 tons. Speed: 8 kts.
YELCHO (1906). 280 tons. Speed: 8 kts.

No photographs available of these ships.

YANEZ (1889). 320 tons. Speed: 8 kts.
CÓNDOR } (1889). 145 tons. Speed: 8 kts.
HUÉMUL }
PISAGUA (1874). 300 tons. Speed: 8 kts.

Training Ships.

1913 Photo, [illegible]

Ex-COCHRANE (1874, reconstructed 1890). Ship. 3500 tons. Dimensions: 210 (p.p.) [illegible] Guns: 4—4·7 inch, 2—12 pdr., 4—3 pdr., [illegible] (iron): 9"—4½" Belt. 8" Battery. H.P. 43[illegible] over 8 kts. now). Cylindrical boilers. C[illegible] Complement, 306.

PRESIDENTE ERRAZURIZ (previously [illegible] serves as a Gunnery Training Ship.

TOMÉ, TALCAHUANO (previously describ[illegible] to Torpedo and Engineering Schools.

Torpedo Boats of the *Ingeniero Hyatt* class are Tr[illegible] Schools.

Corvette *Abtao*, Navigation School Ship; Frig[illegible] Boys.

GENERAL BAQUEDANO (previously descr[illegible] as a Training Ship.

Hulk *Huascar* is Parent Ship for Submarines at T[illegible]

COLOMBIA.

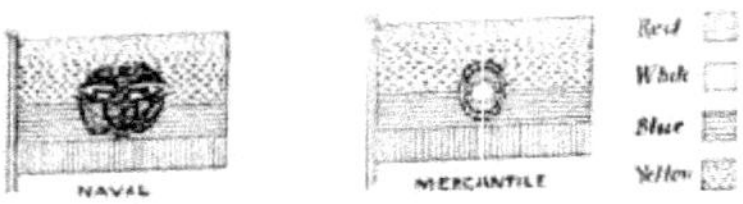

Minister of War: Señor Jorge Roa.

Gunboat.

Photo wanted.

CHERCINTO (1896). 643 tons. Dimensions: 185 (p.p.) × 31 × 12 feet. Guns: not known. Designed H.P. 100 = about 12 kts.

River Gunboats.

Photo wanted.

ESPERANZA (1897), **GENERAL NERINO** (1895), both built at Perth Amboy, U.S.A. Stern-wheelers of 400 tons. Dimensions: 140 (*p.p.*) × 9 × 3 feet. Guns: 3 M.G. Designed H.P. 430 = 15 kts. No other details known.

M. L. (*Motor Launches*).

No photo available.

Cauca (1913). 50 tons. Dimensions: 110 × 17 × 4 feet. Guns: none at present ?* Speed, 12 kts.

Guarda Costas 1, 2, 3, 4 (Yarrow, 1913). 26 tons. Dimensions: 80 × 12¼ × 3½ feet. Guns: *No. 2* has 1—1 pdr., no guns in others yet?* H.P. 100 = 12 kts. Fuel: 1800 galls. petrol = 2400 miles at 10 kts.

* Guns were detained in Great Britain during the war; may be delivered and mounted now.

Old Transport *Bogota* is still in service.

CUBA.

Revised 1920, by courtesy of Captain Julio Morales Coello (Director of the Naval

Mercantile Marine: 35 steamers of 23,256 tons gross, 140 sailing vessels of 15,152
also 6 German steamers of 19,164 tons, seized on declaration of war, 1917.

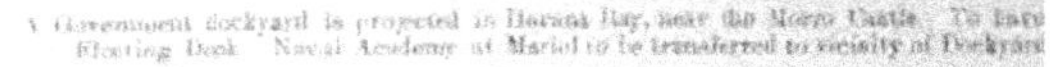

Floating Dock. Naval Academy at Mariel to be transferred to vicinity of Dockyard

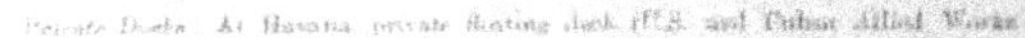

Corp.) 200 × 46 × 17½ feet (2600 tons).

Minister of War: Señor Jose R. Villalon.

RECOGNITION SILHOUETTES.

Scale: 1 inch = 80 feet.

Scale: 1 inch = 160 feet.

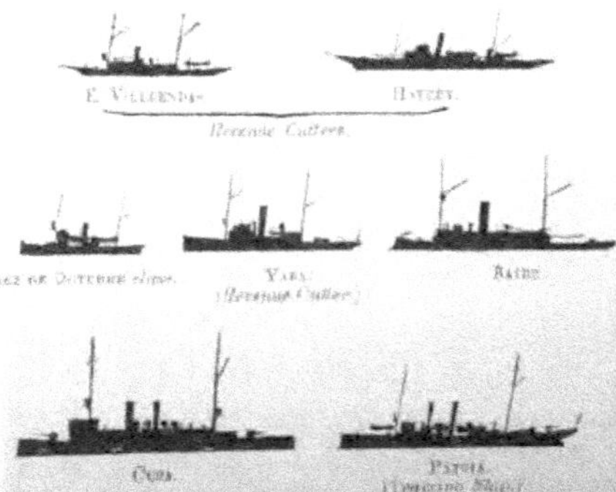

Light Cruisers.

CUBA (Cramp, 1911). 2055 tons. Dimensions: (p.p.) 260 ×
Armament: 2—4 inch, 4—6 pdr., 4—3 pdr., 4—1 pdr.
H.P. 6000 = 18 kts. Babcock boilers. Coal, 250 tons.

(*Training Ship.*)

PATRIA (Cramp, Philadelphia, 1911). 1200 tons. Dimens
(*p.p.*) 176 × 13 feet. Guns: 2—6 pdr., 4—3 pdr. 4
machine. I.H.P. 4000 = 16 kts. Babcock boilers. Coa

Gunboats.

Photo received too late for publication.

HABANA (1912) and **PINAR DEL RIO** (1912). Both wooden vessels of 80 tons, built at Havana. Dimensions: 100 × 18 × 6 feet. Guns: 1—1 pdr. I.H.P. 200 = 12 kts. 1 screw. Coal: 20 tons.

DIEZ DE OCTUBRE (J. Samuel White, Cowes, 1911). **VEINTE Y CUATRO DE FEBRERO** (J. Samuel White, Cowes, 1911). 218 tons. Dimensions: 110 × 20 × 8 feet. Guns: 1—3 pdr., 2—1 pdr. Speed, 12 kts. Coal, 30 tons.

Photo by courtesy of the Ministry of War.

BAIRE (Schichau, Danzig, 1906). 500 tons. Dimensions: 182 × 23 × 6 feet. Guns: 4—3 inch, 2—3 pdr., 1 machine. H.P. 1200 = 14 kts. Babcock boilers. Coal, 120 tons.

Note.—Refitted in a U.S. Navy Yard, 1917, may now have a new armament.

Transport.

KYDONIA (Neptunwerft, Rostock, 1906). 2500 tons *gross*. Dimensions: 291.4 × 41.2 × 16.1 feet. No other details known.

Submarine Chasers.

As Illustration for "SC boats," U.S. Navy Section.

No. 1 (ex U.S. *SC 274*, Mare Island, N.Y.). **Nos. 2—4** (ex U.S. *SC 302, 311, 312*, Puget Sound, N.Y.). All built 1917-18. About 77 tons. Dimensions: 105 (*pp.*) × 14¾ × 5½ feet. Guns: 1—3 inch (23 cal.), 2 Colt M.G. B.H.P. 660 = 16½ to 18 kts. Fuel: 2400 gallons petrol. Engines: 3 sets Standard motors. Wooden hulls.

Revenue Cutters.

MATANZAS (1912), **VILLAS** (1912). Details as *Habana* and *Pinar del Rio*.

HATUEY (ex U.S. *Delaware*, Bath, U.S.A., 1903, bought 1907). 258 tons. 217 × 27 × 15 feet. Guns: 2—6 pdr., 2—3 pdr., 2—1 pdr. I.H.P. 1250 = 12 kts. Coal, 100 tons.

Photo by courtesy of the Ministry of War.

ENRIQUE VILLUENDAS (Chester, U.S.A., 1899). 178 tons. 132 × 20 × 9 feet. Guns: 2—3 pdr., 1 M.G. I.H.P. 600 = 15 kts. Coal, 55 tons.

Revenue Cutters.—*Continued*

Photo received too late for publication.

GUAIMARO. (Lewis Nixon, New York, 1900).

MARTI (Seabury, New York, 1898). 50 tons. Speed 15 kts. Coal: 15 tons.

ALFREDO (Seabury, New York, 1898). 39 tons. Speed, 11 kts.

CESPEDES (ex-Spanish gunboat, 1896). 40 tons. Speed, 12 kts. Coal: 10 tons.

MACEO (ex-Spanish gunboat, 1896). 35 tons. Speed, 10 kts. Coal, 8 tons.

AGRAMONTE (ex-Spanish gunboat, 1896). 15 tons. pdr. Speed, 11 kts. Coal, 10 tons.

Photo by courtesy of Capt.

YARA (Middlesborough, 1895). 449 tons. 155 × 26 2—6 pdr., 2—1 pdr. I.H.P. 600 = 12 kts. Coal

VEINTE DE MAYO (Glasgow, 1895). 203 tons. feet. Guns: 2—3 pdr., 2—1 pdr. I.H.P. 500 50 tons.

ECUADOR.

(Revised in Ecuador, 1920.)

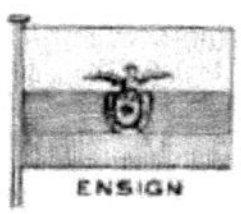

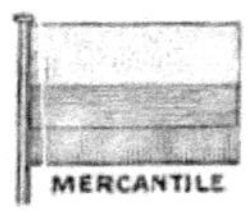

Flag Notes.—To distinguish from Colombian flag, note that centre bar is *sky-blue*. Mercantile flag also has no badge, compared with Colombian.

Minister of War and Marine: Sr. R. Parry Rosa.

Mercantile Marine: 7 steamers of 1912 tons gross; 4 sailing vessels of 1882 tons gross.

Silhouette Scale: 1 inch = 160 feet.

Torpedo-Gunboat.

1918 *Photo, G. Chambers, Esq.*

LIBERTADOR BOLIVAR (ex *A. Simpson*, 1896, purchased from Chile, 1907). Displacement 750 tons. Length, 240 feet. Beam, 27½ feet. *Maximum* draught, 10½ feet. Guns (Elswick): 4—3 pdr., 2 Maxims. Torpedo tubes (18 inch): 3 *above water*. Armour (Harvey): 1" Belt, 4½" Gun shields, 1" Hood to steering gear, 1" Bulkheads. Machinery: 2 sets triple expansion. 2 screws. Boilers: 4 Normand. Designed I.H.P. *forced* 4500 = 21·5 kts. Coal: 100 tons. Built by Laird.

Gunboat.

1918 *Photo, G. Chambers, Esq.*

COTOPAXI (1884). 300 tons. 135 (*p.p.*) × 21 × 9 feet. Guns: 2 small Q.F. I.H.P. 175. *Max.* speed, 10½ kts.

Also one small Tug, *Tarqui*.

(HAYTI.)

According to information received from Port-au-Prince, March, 1919, all Haytian Warships were sold out of service during 1915.

This Republic now only maintains two unarmed Auxiliary Schooners for Coastguard Duties.

MEXICAN FLEET

Minister of Marine:

Mercantile Marine: "Bureau Veritas" 1919-20 figures (ships under 100 tons gross excluded): 35 steamers of 32,349 tons gross; sailing vessels of 2630 tons gross.

Docks.

At Vera Cruz (Atlantic Coast), Government floating dock, 262 feet (1600 tons).

At Salina Cruz (Pacific Coast), Government dry dock, 604 × feet (leased to Tehuantepec National Railway Co.)

RECOGNITION SILHOUETTES.

Silhouette Scale: 1 inch = 160 feet.

Vicente Guerrero.

Tampico and Vera Cruz.

Blasquez and Bravo.

Torpedo Gunboats.

Photo, J. Romanez, Esq.

BLANQUET (ex *Morelos*) and **BRAVO** (Odero, 1903). 1210 tons. Comp. 130. Dimensions: 240 (*p.p.*) × $32\frac{5}{8}$ × 10 (*mean*) feet. Armament (Bethlehem): 2—4 inch, 4—6 pdr., 1—14 inch tube (bow) *above water*. Designed H.P. 2500 = 17 kts. Blechynden boilers. 2 screws. Coal: 280 tons.

TAMPICO (1902) **VERA CRUZ** (1902). Both built at Elizabeth Port, N.J. 984 tons. Comp. 164 (*Vera Cruz*), 110 (*Tampico*). Dimensions: 200 × 33 × 11 feet. Armament: in *Tampico*, 2—4 inch, 4—6 pdr.; in *Vera Cruz*, 4—4 inch, 6—6 pdr. In both ships 1—14 inch tube (bow) *above water*. Designed H.P. 2400 = 15 kts. Mosher boilers. 2 screws. Coal: 200 tons.

Gunboats.

Photo wanted.

GENERAL JESUS CARRANZA.* No details known.

*Probably re-named, 1920, after fall of Carranzist Government. New name (if any) not known.

Transports.

VICENTE GUERRERO (Vickers, 1908). 1850 tons. Dimensions: 245 × 35 × 17 feet. Guns: 6—4 inch, 2—3 pdr. Designed H.P. 1200 = 12 kts.

No Photos available.

PLAN DE GUADALUPE (ex *Dolphin*, Wilmington, 1892). 821 tons. Dimensions: 213 (*o.a.*) × 32 × 15 feet. Guns: not known. Designed H.P. = 14 kts.

Also *Sonora*, subventioned merchantman.

Patrol Boats.

No photo available.

——————, ——————, and ——

Names not known; are ex-Canadian Drifters, pu[t on] duty as Pacific Coast Patrols. Built about 191[8] Lauzon, Canada. 150 tons. Dimensions: 95 × I.H.P. 225 = 9 kts. Coal: 31 tons. 1 screw. 1

Also two Revenue Cutters, names unknown.

Armed Tugs.

No photo available.

PACIFICO. No details known.

NICARAGUA.

Mercantile Marine: 2 steamers of 2262 tons *gross*.

Gunboat.

Photo wanted.

MOMOTOMBO (——) 400 tons. Guns: 2—12 pdr., 1—6 pdr. No other details known.

Auxiliaries.

No photo available

OMOTEPE. No details known.

Illustrated in "Fighting Ships" 1912-17.

MAXIMO JERAZ (ex s.s. *Venus*). No details known.

PARAGUAY.

Extracted 1920 from Official List published by the Ministry of War.

Minister of War: Col. Adolfo Chirife.

Gunboat.

Photo wanted.

ADOLFO RIQUELME. No details available.

Despatch Vessels.

Photo wanted.

ITAPURU. No details available.

Despatch Vessels—(*continued*).

Photo wanted.

TENIENTE HERREROS. No details available.

Photo wanted.

TRIUMFO. No details available.

Transports.

Photo wanted.

GENERAL DIAZ. No details available.

Photo wanted.

INDEPENDENCIA. No details available.

PERUVIAN FLEET.

Note.—Appointment of U.S. Naval Mission to Peru approved by Act of Congress, June, 1920.

PERU ENSIGN

Mercantile flag the same but without centre device.

PERUVIAN VICE-ADMIRAL

Rear Admiral has one sun

PERUVIAN CAPTAIN

Colour of Ships: Brown grey. *Minister of Marine*: Commander Ontaneda.

Mercantile Marine: "Bureau Veritas" 1919-20 figures (ships under 100 tons gross excluded): 41 steamers of 45,353 tons *gross*; 40 sailing vessels of 22,805 tons *gross*.

Docks.

At Callao, Floating Dock "El Peruano," 385×70×22 feet (7,000 tons), property of La Compania Peruana de Vapores y Dique de Callao. Is only dock in Peru.

Uniforms.

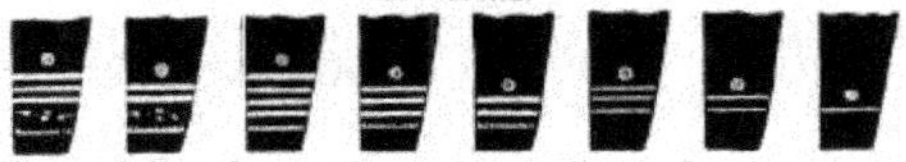

Vice-Almirante. Contra-Almirante. Capitan de Navio. Capitan de Fragata. Capitan de Corbeta. Teniente 1°. Teniente 2°. Alferez de Fragata.

Captain. Commander. *Lieut. Comm'r. Lieut. Junior. Sub. Lieutenant.*

Engineers the same without the sun.
Other branches have the same stripes without the sun, and in addition colours as follows:—
Torpedo, *light blue*; Constructor, *green*; Electricians, *light green*; Doctors, *red*; Paymasters, *blue*.

PERUVIAN RECOGNITION SILHOUETTES.

Scale: 1 inch = 160 feet.

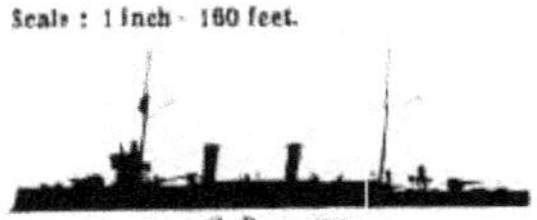

C. Bolognesi.

A. Grau.

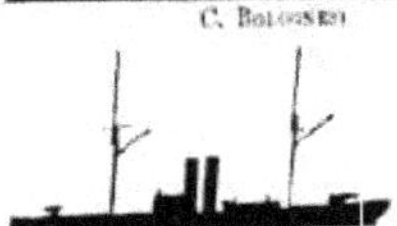

Lima.

T. Rodriguez t.b.d.

Submarines:—Scale 1 inch = 80 feet.

Ferre and Palacios.

Light Cruisers.

Coronel Bolognesi. The *A. Grau* is identical *except* that she has a poop.
("*A. Grau*" *Submarine Depot Ship*.)

ALMIRANTE GRAU (March, 1906) & CORONEL BOLOGNESI (Nov.,

Displacement: 3,200 tons. Complement, 315.

Length (*p.p.*): 370 feet. Beam, $40\frac{1}{2}$ feet. *Maximum* draught, $14\frac{1}{4}$ feet.

Guns (Vickers):
2—6 inch, 50 cal.
8—14 pdr.
8—1 pdr.
Torpedo tubes (18 inch):
2 *submerged*.

Armour:
$1\frac{1}{2}$" Deck (amids
" Coaming (ove
" Gun Shield

Ahead:
1—6 in.

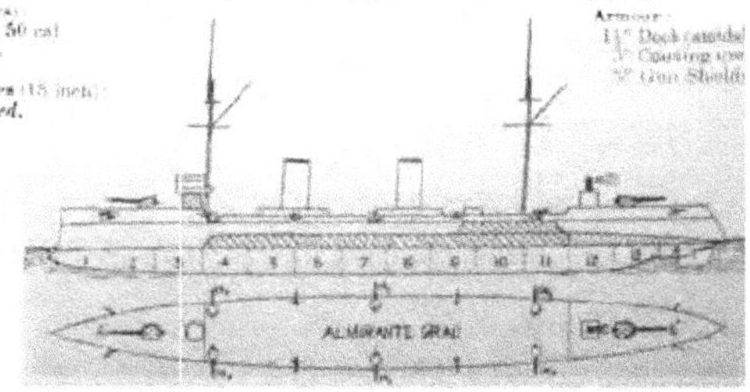

Broadside: 2—6 in.

Machinery: 4 cylinder vertical triple expansion. 2 screws. Boilers: 10 Yarrow. De
14,000 = 24 kts. (not above 20 now, unless re-boilered). Coal: *normal*, tons; *maximu*
Endurance: 3,700 miles at 10 kts.

General Notes.—Built and engined by Messrs. Yarrow, Ltd., about 1905-7.

Destroyer

Photo by courtesy of Messrs. Schneider et Cie.

1 *Schneider-Creusot*: **Teniente Rodriguez** (1909, ex-*Actée*). 430 tons. Dimensions: 212 × 20.3 × 10½ feet. H.P. 7000 = 28 kts. (much less now). Schneider-Zoelly turbines; du Temple boilers. Coal: 100 tons = 1200 miles at 10 kts. Armament: 6—9 pdr., 3—18 inch tubes (one deck amidships and 1 bow tube above water). Complement 69. Is sister boat to the French *Voltigeur*, *Carabinier*, *Tirailleur* type Destroyers.

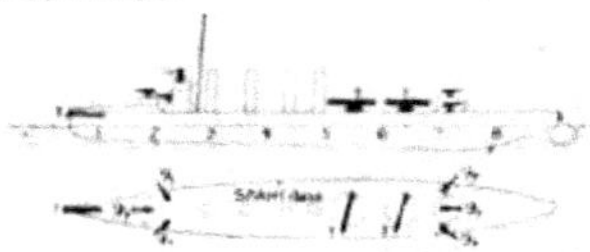

Submarines.

2 *Laubeuf* type: **Ferre** (1912), and **Palacios** (1913). Built by Schneider-Creusot. $\frac{400}{500}$ tons. Complement 19. Schneider-Carels Diesel motors. H.P. $\frac{400}{200}$ = $\frac{14}{9}$ kts. (present speed about 12·2 kts. *on surface*, 7·5 kts. *submerged*). Endurance: 2000 miles at 10 kts. 1 bow torpedo tube, 4 external dropping gears for torpedoes.

Gunboat.

Photo wanted.

AMERICA (1904). 100 tons. Guns: 2—3 pdr. No other details known.

Transports.

LIMA (Ex *S. S. Socrates*, Kiel, 1880). 1790 tons. 2000 = 10 kts. Coal: 300 tons. Guns: 4—8—3 pdr. Complement, 131.

Chalaco is still chartered out to a private company.

Vedette Boats.

Cuzco, *Puno*. No photo or details available.

River Launches.

Cahuapanas, *Francisco Pizarro*, *Iquitos*, *Orellana*, *Requena*. No photo or details available.

URUGUAYAN FLEET.

Revised, 1920, by courtesy of Captain Taylor.

RECOGNITION SILHOUETTES.

Scale: 1 inch = 160 feet.

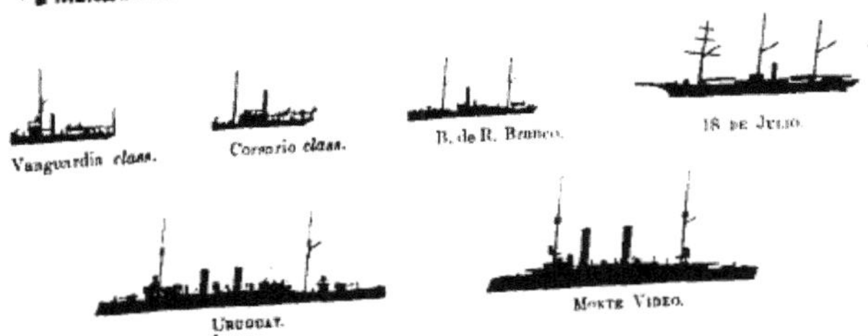

Minister of War and Marine:—

Personnel:—1306.

Coinage: Gold Peso of 100 Centesimos = 4s. 3d.

Mercantile Marine.

From "Bureau Veritas" 1919-20 figures. Ships under 100 tons *gross* excluded.

22 Steam Ships of 24,038 tons *gross*.

43 Sailing Vessels of 18,168 tons *gross*.

Ports, Yards, &c.

MONTE VIDEO.—There is a fort here, but only for returning salutes. The National Dock (Dique Nacional) is operated by the Naval Authorities. Length, 459 feet. Breadth, 55 feet. Maximum depth, 17 feet. Steam and electric pumps can empty dock in eight hours. Dique Mauá (south of city) is property of Monte Video Gas Co. Dimensions: 272 × 50 × 12 feet 4 inches to 15 feet 2 inches. Has 2 steam pumps and 25 ton crane, also fixed and moving cranes. The firm Varadero del Cerro have a yard, well equipped for repairs, with a small slipway. The National Port Administration bought the Varadero Lussich, a repairing yard with a small slipway during 1919.

Old Cruiser.

1918 *Photo, A. J*

MONTE VIDEO (ex-*24 de Agosto*, ex-*Dogali*, ex-*Salamina*, Elswick, 1890, purchased from
2050 tons. Complement, 250. Dimensions: 250 (*p.p.*) × 37 × 16¼ feet (*max.* draugh
6—6 inch, 9—6 pdr., 6 M.G. Torpedo tubes: 4 *above water*. Armour: 2"—1" dec
shields, 2" conning tower. Machinery: 2 sets horizontal and 4 double-ended boilers. E
17 kts. Coal: 480 tons.

(*Continued on*

Torpedo Gunboat & Training Ship.

URUGUAY (Vulkan, Stettin, 1910). 1150 tons. Complement, 125. **Dimensions**: $278\frac{3}{4}$ (*p.p.*) × $30\frac{3}{4}$ × 12 feet (*max.* draught). Guns (Skoda): 2—4·7 inch, 45 cal., 4—12 pdr., 6—1 pdr. (Vickers), 4 M.G. **Torpedo tubes**: 2—18 inch *above water*. Armour: $\frac{3}{4}$" nickel steel over boilers and engines. 4 Normand boilers. H.P. 8,000 = 23 kts. **Coal**: 210 tons = 3,000 miles at 10 kts.

Note.—Is fitted for service as Training Ship for mid-shipmen. Old Gunboat, *Gen. Suarez*, is a Harbour Training Ship.

Training Ship.

Now rigged as Silhouette.

DIEZ Y OCHO DE JULIO. 678 tons. Complement, 85. 4 small guns and 2 machine guns. Speed, 12 kts. Re-fitted and re-boilered 1920.

Gunboat.

1918 *Photo*, *A*

BARON DE RIO BRANCO (*ex-Maldonado*). Old gunboat. 300 tons. Carries 4 small Q.F. and 2 M.G.

Tenders.

VANGUARDIA class. 1918 *Photo, A. J. Carbone.*

CORSARIO class. 1918 *Photo, A. J. Carbone.*

INGENERIO, LA VALLEJA, ORIENTAL, OYARVIDE. These are four armed tugs, built about 1908, displace about 60 tons each. Guns: 3 pdr. or M.G. Speed, 13 kts. Three other tugs, *Chapicuy*, *Corsario*, *Yaguary*, also exist, and are classified as tenders. No information has been supplied to assist in determining which of the seven ships named above belong to *Vanguardia* class or which to *Corsario* class.

VENEZUELAN FLEET.

ENSIGN

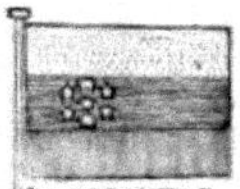

MERCANTILE

Personnel :

Minister of War and Marine: General C. Jimenez Rebolledo.

Warship Port: Puerto Cabello. Steel floating dock here 282 x 90 x 22 feet (2400 tons capacity) built in five sections on self-docking system, and worked by electricity. Also a wooden floating dock, 180 x 60 x 19 feet (1200 tons capacity). Both these docks are Government property.

Mercantile Marine: "Bureau Veritas 1919-20 figures. Ships under 100 tons gross excluded": 2 steamers of 4,902 tons gross, 9 sailing vessels of 1,108 tons gross.

RECOGNITION SILHOUETTES.

Scale 1 inch = 160 feet.

MIRANDA.

J. F. RIBAS.

GENERAL SALOM.

M. SUCRE.

Revised from the 1920 official Report of the Ministry of War and Marine.

Gunboats.

[illegible]

MARISCAL SUCRE ex Spanish *Isla de Cuba*, 1886, captured by U.S., 1898, and sold to Venezuela, 1912. 1125 tons. Dimensions: 192 x 30 x 12½ feet (*Mean* draught) 14 feet (*Max.* draught). Guns: 2—4 inch, 2—6 pdr., 6—3 pdr., 2—1 pdr. H.P. 2900 = 13 kts., *max.* continuous speed, 10 kts. Coal: 200 tons. Endurance: 2250 miles at 9.5 kts. Complement 100.

[illegible]

[illegible]

GENERAL SALOM (ex *Restaurador*, purchased 1900; built as U.S. private yacht *Atlanta*, 1884). 750 tons. Dimensions: 246 x 26 x 13 feet (*max.* draught). Guns: 1—12 pdr., 4—6 pdr., 1 machine. H.P. 1900 = 13 kts., *max.* continuous speed 10 kts. Coal: 200 tons. Endurance: 2000 miles at 10 kts. Complement 63.

Gunboats—*Continued.*

MIRANDA (Clydebank, 1895; purchased from Sp[illegible] tons. Dimensions: 135 x 19 x 8 feet. G[illegible] H.P. 315 = 10 kts. Coal: 30 tons. Endurance [illegible] 8 kts. Complement 46.

Armed Tug.

[illegible]

JOSÉ FELIX RIBAS (ex-*Zumbador*; built 18[illegible] Dimensions: 127 x 23 x 12 feet. Guns: 2—6 pdr [illegible] Coal: 100 tons. Endurance: 1440 miles at 10 kts.

Transports.

Photo wanted.

SALIAS (ex-S.S. *Julia*, transferred from Ministry of [illegible] Navy, 1914). 80 tons *net*. Dimensions: 120 x 1[illegible] draught). H.P. 150 = 10 kts., *max* continuous spe[illegible] —— tons. Endurance: 1000 miles at 7 kts. [illegible] Employed as Naval Collier, 1920. Not known [illegible] mounted.

Brigantine ***Antonio Dias***, used as Naval Coal trans[illegible] Dimensions: 109 x 27.5 x 11 feet. Complement [illegible] some extent.

CHINESE FLEET.

Officially Revised 1920 by courtesy of Rear-Admiral Chen Ngen Tao, Chief of Bureau of Operations, Board of the Navy, Peking.

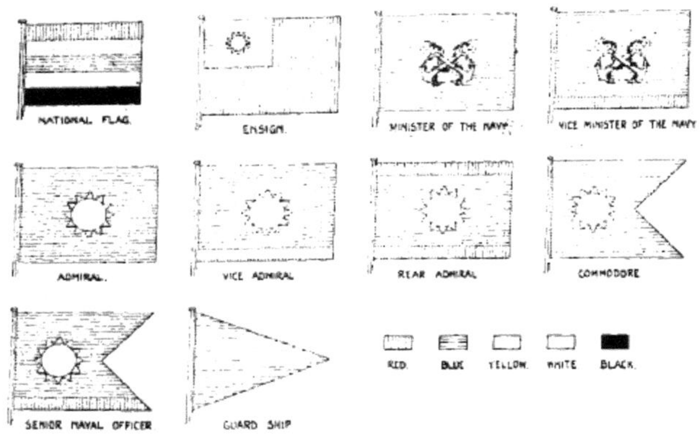

Uniforms.

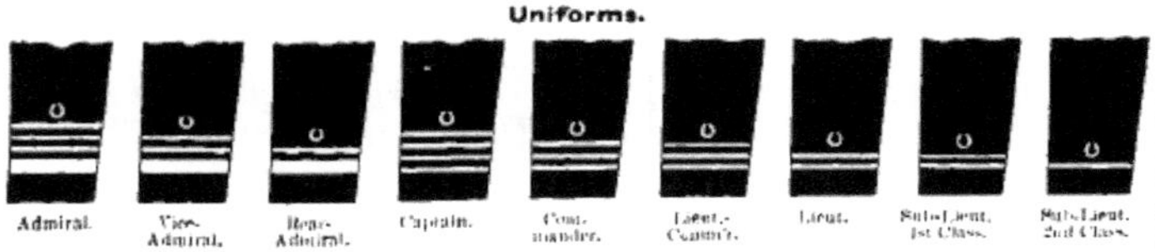

Colour between stripes:—*Engineers* (none); *Surgeons*, red; *Paymasters*, white; *Engine Constructor*, purple; *Ship Constructor*, drab grey or ash colour; *Navigating Officers*, light indigo blue. All these branches *without* the curl or emblem over top stripe.

Personnel: *Minister of Marine*—Admiral C. P. Sah.

Naval Attaché, London: Commander S. K. Chen.

Mercantile Marine: (From "Bureau Veritas" 1919-20 figures. Ships under 100 tons *gross* excluded). 90 Steam and Motor Ships of 134,214 tons *gross*; 7 Sailing Vessels of 865 tons *gross*. During 1919, Chinese Shipyards launched 9 vessels of 12,307 tons *gross*; some of these for foreign account.

Docks.

FU-CHAU. Government Dock, 480 × 90 × 23½ feet.

TAKU. *Arsenal.* Government and Private Docks:—

A:—320 × 92 × 20 feet. C:—300 × 83 × 16 feet. E:—315 × 55 ×
B:—295 × 80 × 17 feet. D:—220 × 44 × 17 feet. F:—400 × 115

SHANGHAI. *Arsenal.* Government Dock, 390 × 79 × 19 feet. Other Private Doc

(1) SHANGHAI DOCK & ENGINEERING CO., LTD. (*a*) Cosmopolitan Dock, 560 (*b*) International Dock, 540 × 77½ × 23 feet. (*c*) New Dock, 475 × 74 × 21½ feet. Dock, 362 × 67 × 16 feet. (*e*) Old Dock, 400 × 55 × 16 feet.

(2) NEW ENGINEERING & SHIPBUILDING WORKS. Yangtsepu Dock, 570 × 7

(3) KIANGNAN DOCK & ENGINEERING WORKS. Kiangnan Dock, 556 × 70 ×

WHAMPOA. *Arsenal.* No Government Docks, but 2 Private: Cooper, 430 × 65 × 16 450 × 75 × 18 feet.

AMOY. Dock, 360 × 60 × 16½ feet. Property of the Government of the Province of Fukie

TONGKU. Private Docks: (*a*) 540 × 52 × 14 feet. (*b*) 260 × 50 × 9 feet.

Private Yards.

Small gunboats are built by the Kiangnan Dock & Engineering Co., Shanghai (12 buil the Yangtse Works, Hankow.

Secessionist Warships.

During 1919-20, the following Warships were at Canton, under the Local Gover declared its independence of the Central Government at Peking:—

Light Cruisers: **CHAO HO, HAI CHEN, HAI CHI.**

Destroyers: ***Tung An, Yu Chang.***

Gunboats: **YUNG PENG, YUNG HSIANG, FEI YING.**

River Gunboat: **CHU YU.**

Despatch Vessel: **WU FENG.**

Transport: **FU AN.**

Reported during summer 1920, that differences between the two Governments had be Notes—as given above by the Chinese Bureau of Operations, (during May, 1920)—n hold good, if the Cantonese Government has now submitted to the Peking Central

Note on Spelling of Chinese Warship Names.

The transliteration of Chinese names depends first upon the dialect in which t happen to be spelt, and secondly, on the system used by the foreigner to represent The official dialect is Pekingese and a widely adopted system of representing it is t "Wade's Spelling." In the 1919 Edition of "Fighting Ships," it was thought advis some standard form of spelling Chinese Warship Names. Accordingly, "Wade's S been introduced, all aspirates, accents and tonic marks being omitted.

Scale: 1 inch = 160 feet.

RIVER GUNBOATS.

CHIEN CHUNG *class* (3).

LI CHIEH.

CHIANG HSI *class* (2)

LI SUI.

LIEN CHING.

CHIANG HENG *class* (4 ships)

CHU TAI *class* (6 ships).

YING JUI *class* (2 ships).

CHIEN WEI *class* (2 ships).

HAI YUNG *class* (3 ships).

HAI CHI.

TORPEDO CRAFT.

HU PENG *class*, or Nos. 7—10 (t.b.)

FEI YING.

CHAO HO.

YING JUI (July, 1911) & OHAO HO (Oct., 1911).

Normal displacement { *Ying Jui*: 2,750 tons. *Chao Ho*: 2,600 tons. } Complement, 279.

Ying Jui: Length (*p.p.*) 330 feet. Beam, 42 feet. *Max.* load draught, 14 feet 11 inc
Chao Ho: „ 320 „ „ 39 „ „ „ 14 „

Guns (*see Notes*):
2—6 inch, 50 cal.
4—4 inch, 50 cal.
2—14 pdr., 50 cal.
6—3 pdr.
2—1 pdr.
Torpedo tubes (18 inch):
2 *above water*.

Armour:
3" Deck (amid
1½" Deck (en
3" Conning t
" Gun shiel

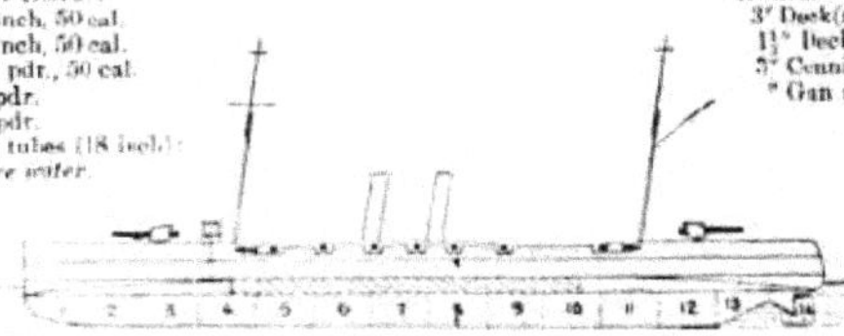

Ahead:
1—6" in.
2—4 in.

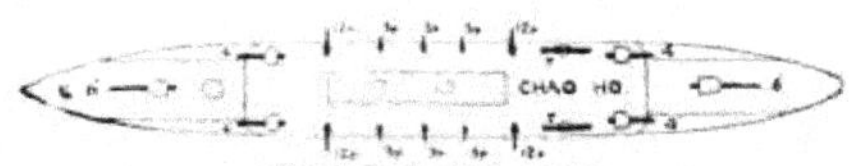

Broadside: 2—6 in., 2—4 in.

Machinery: Parsons turbine. Boilers: see notes. 4 screws. Designed H.P. (6000 S.H.) 8000 = 22 kts. Coal: *normal*: 220 tons; *maximum*: 550 tons. Oil: 50 tons in *Ying Jui*, *Chao Ho*. Endurance: about 4500—5000 miles at 10 kts., and about 2900 miles at 18 kts.

Gunnery Notes.—Vickers models in *Ying Jui*; Elswick in *Chao Ho*.

Name	*Builder*	*Machinery*	*Laid down*	*Completed*	*Trials* ⅘ths	Full power	I
Ying Jui	Vickers	Vickers	1910	1912	5391 = 20·7	8793 = 22·23	Whit
Chao Ho	Elswick	Hawthorn	1910	1912		8622 = 22·12	6

Hai Yung. 1920 Photo

HAI YUNG (1897), **HAI CHOU** (1897) and **HAI CHEN** (1898). Displacement, 2950 metric tons. Complement 295. Length (*p.p.*) 314 feet. Beam, 40¾ feet. *Maximum* draught, 19 feet. Guns (Krupp): 3—5·9 inch, 40 cal., 8—4·1 inch, 40 cal., 6—1 pdr., and 6 machine. Torpedo tubes (14 inch): 1 bow (*submerged*). Armour: 2¾" Deck (amidships), 1½" deck (ends), 2" gun shields, 1½" conning tower. Designed H.P. 7,500 = 19·5 kts. 8 cylindrical boilers. Coal: *normal* 200-220 tons; *maximum* 500-580 tons. All three ships built by Vulkan Co., Stettin.

Photo, Symonds & Co.

HAI CHI (Elswick, 1898). Displacement, 4,300 tons. Complement, 431. Length (*p.p.*), 424 feet. Beam, 46¾ feet. *Mean* draught, 16¾ feet; *max.* 19 feet. Guns (Elswick): 2—8 inch, 45 cal., 10—4·7 inch, 45 cal., 12—3 pdr., 4—1 pdr., 6 machine. Torpedo tubes (18 inch): 5 *above water*. Armour (Harvey): 5" Deck (amidships), 1½" deck (ends), 4½" gun shields, 4" ammunition hoists, 6" conning tower. Designed H.P.: 17,000 (*f.d.*) = 24 kts. Boilers: 12 cylindrical. Coal: *normal* 400 tons, *maximum* 990 tons.

Destroyers.

Photo wanted.

3 *Schichau* type: ***Chien Kang, Tung An, Yu Chang*** (—1912), 390 tons. Dimensions: 198 ×
Armament: 2—12 pdr., 2—3 pdr., 2—18 inch tubes. Designed H.P. 6000 = 30 kts. Coal, 80 tons. Com

First called *Fu Po*, *Fei Hung* and *Cheng Feng* but names were altered by Yuan Shih Kai.

Torpedo Boats.

Photograph wanted of Kawasaki-Normand boats.

(*See Silhouettes.*)

4 *Kawasaki-Normand* type: ***Hu Peng*** or ***No. 7, Hu E*** or ***No. 8, Hu Ying*** or ***No. 9, Hu Chun*** (Kobe, Japan, 1907-8), 97 tons. Dimensions: 135 × 15½ × 7½ feet. Armament: 2—3 pdr., 3 M.G., tubes (1 bow *above water* and 2 deck). Designed H.P. 1200 = 23 kts. Coal, 28 tons *normal*. Complem

No Photo available.

2 *Vulcan* type: ***Su**** or ***No. 4, Chen**** or ***No. 2*** (1895). 90 tons. 144 × 17 × 7 feet. Armament: 4 M.G tubes. H.P. 700 = 18 kts. Complement 38.

No Photo available.

2 *Schichau* type: ***Lieh**** or ***No. 3, Chang**** or ***No. 1*** (1895). 62 tons. 130 × 15 × 5 feet. Armament torpedo tubes. H.P. 600 = 16 kts. Complement 30.

* These four T.B. sometimes have suffix *Ting* (Boat) or *Tzu* (meaningless) added to their nam

Torpedo Gunboats.

CHIEN WEI (1902) and **CHIEN AN** (1900). Displacement 850 tons. Complement, 133. Guns: 1—4·1 inch, 3—9 pdr., 6—1 pdr. Torpedo tubes: 2 above water. Armour: 1" belt and deck (amidships). Designed H.P. 6500 = 23 kts. (about 18 now). Coal: 180 tons. Built at Fu-Chau.

FEI YING (July, 1895). Displacement, 850 tons. Complement, 145. Guns (Krupp): 2—4·1 inch, 6—3 pdr., 4 M.G., 3 tubes (above water). Designed H.P. 5500 = 22 kts. Boilers: 8 Yarrow. Coal: 170 tons. Built at Stettin.

Gunboats.

Photo wanted.

FEI-HU, FEI-LUNG (built or building by Kiangnan Dock Co., at [illegible] about 1911-?). Dimensions: 220 × 29 × [illegible] feet. [illegible] tons. Speed: 14 kts. (designed).

Note.—These apparently do not appear in the Official Chinese Navy List, but they evidently do exist. It is possible they are not finished yet.

Gunboats—*Continued.*

Photo wanted.

YUNG PENG, YUNG HSIANG (Kawasaki Co., Kobé, Japan, 1912—13). 830 tons. Dimensions: 205 (*p.p.*) × 29½ × 8 feet. Guns: 1—4·1 inch, 1—12 pdr., 4—3 pdr., 2—1 pdr. H.P. 1350 = 13·5 kts. Coal: 190 tons. Complement, 105.

River Gunboats.

Photo wanted.

HAI BUNG* (1920), **HAI OU** (1916-17), both built by Taku Dockyard. Shallow-draught Gunboats of 227 tons. Dimensions: 108 × 18 × 9·8 feet. Guns: 4—1 pdr., 2 machine. H.P. = kts. Fuel: tons. Complement,

* Chinese characters not being available, this name cannot be spelt in accordance with "Wade's System."

Photo wanted.

HAI KU (1919), **HAI HUNG** (1916-17), both built by Fu-Chau Dockyard. Shallow-draught Gunboats of 190 (?)* tons. Dimensions: 65 × 12 × 4·7 feet. Guns: 2—2 pdr., 2 machine. H.P. = kts. speed. Fuel: tons. Complement,

* Displacement as officially listed: probably a mistake. *Hai Yen* of same dimensions, displaces 56 tons.

River Gunboats—*Continued.*

Photo wanted.

HAI HAO, HAI FU, both shallow-draught Gunbo[ats] Kiangnan Dock Co., Shanghai, 1916-17. 106 tons. 105 × 17 × 9 feet. Guns: 2—1 pdr., 2 machine. kts. speed. Fuel: tons. Complement,

Photo wanted.

HAI YEN. Shallow-draught Gunboat, built by Taku [Dockyard] 56 tons. Dimensions: 65 × 12 × 4·7 feet. Guns: machine. H.P. = kts. speed. Fuel: tons.

Photo wanted.

YUNG CHIEN, YUNG CHI. Begun 1915, by K[iangnan] Co., Shanghai. 900 tons. Dimensions: 215½ × 2[illegible] Guns: 1—4 inch, 1—12 pdr., 4—3 pdr., 2—1 pdr. [H.P.] 1400 = 13·5 kts. Coal: 150 tons.

CHIEN CHUNG (6/15). **YUNG AN** (191[illegible])
KUNG CHEN (1915).

All built by Yangtse Works, Hankow. 90 tons. Dim[ensions] [illegible] × 18½ × 3 feet. H.P. 450 = 11 kts., but *Chien-C[hung]* [illegible] kts. on trials in 1915. Guns: 1—3·4 inch howi[tzer] [illegible] Complement, 42.

(Contin[ued])

River Gunboats—*Continued.*

CHIANG HSI (Krupp, 1911). **CHIANG KUN** (Vulkan, 1912). Dimensions: 146 (*p.p.*) × 24 × 3 feet. Displacement: 140 tons. H.P. 900 = 12 kts. Boilers: 2 Schulz. Coal: 30 tons. Guns: 1—3·4 (22 pdr.) howitzer, 1 machine. Complement, 42.

Note.—The three River Gunboats, *Chiang Ku*, *Chiang Kung* and *Chiang Ta*, supposed to have been built at Hankow about 1908, are now reported as never having been commenced.

LI SUI (ex-German *Otter*, Tecklenborg, 1909. Interned in China, 1914, and taken over for Chinese Navy about 1917). 256 tons. Dimensions: 173·9 × 28·5 × 2·6 feet. Guns: 2—6 pdr., 3 M.G. Armour: 2" on waterline, 2" C.T. I.H.P. 1300 = 14 kts. Boilers: 4 Schulz. Coal: 79 tons. Complement (was 47).

LI CHIEH (ex-German *Vaterland*, Elbing, 1903. Interned at Nankin, August, 1914, and taken over for Chinese Navy, 1917). 217 tons. Dimensions: 157·5 × 26·2 × 2·6 feet. Guns: 1—3·4 inch (22 pdr.) howitzer, 1—4 pdr., 2 M.G. I.H.P. 1400 = 13 kts. Boilers: 2 Schulz. Coal: 79 tons. Complement (was 47).

River Gunboats—*Continued.*

CHU CHIEN. CHU KUAN. CHU TAI. CHU TUNG. CHU YU (1).* CHU YU (2).

All built by Kawasaki Yard, Kobe, 1906-07. 745 tons. Complement, 91. Dimensions: 200 × 29½ × 8 feet. Armament: 2—4·7 inch, 2—12 pdr., 4 machine. Machinery: 2 sets vertical triple expansion. 2 screws. Boilers: Water-tube. Designed H.P. 1350 = 13 kts. Coal: 150 tons.

* Also known as *Chu Yien*, but *Chu Yu* is correct transliteration by "Wade's System."

CHIANG CHEN. CHIANG HENG. CHIANG LI. CHIANG YUAN.

All built by Kawasaki Yard, Kobe, 1906-07. 550 tons. Complement, 91. Dimensions: 180 × 28 × 7 feet. Armament: 1—4·7 inch (bow), 1—12 pdr. (aft), 4—3 pdr., 4 Maxims. Machinery: 2 sets vertical triple expansion. 2 screws. Boilers: water-tube. Designed H.P. 950 = 13 kts. Coal: 113 tons.

Despatch Vessels.

Photo wanted.

WU FENG (Tsingtao, 1912). 200 tons. Dimen... × 7 feet. H.P. 300 = 10 kts. Guns: 4—3 pdr.

LIEN CHING (Shanghai, 1910). Dimensions: ... 9½ feet (mean). 500 tons. I.H.P. 900 = 13·5 ... boilers fitted with Howden's f.d. Coal: 95 ... 4—6 pdr. Complement, 61.

Transport.

No photo available.

FU AN. No details known.

Training Ship.

No photo available.

(Barque rigged: one funnel.)

TUNG CHI (1895). Steel. 1700 tons. Guns: 2— ... inch, 5—3 pdr., 2 M.G. Speed: 11·2 kts. ... details known.

Note.—These Ships do not belong to the Chinese Navy, but to the Water Police of the Provinces named. They might, however, be used for National Defence in the event

Gunboats.

(All belong to Water Police of Kwang Tung Province.)

No photo available.

KUANG YU (1891). Wooden vessel of 600 tons displacement. Dimensions: 156 × 24 × 10 feet. Guns: 9 machine. Designed H.P. 400 = 11 kts. (never attained). Coal: 110 tons. Complement 72.

No photo available.

KUANG OHEN. **KUANG HENG.** **KUANG LI.** **KUANG YUAN.**

Built 1886. Steel. 300 tons. Dimensions: 110 × 18 × 7 feet. Guns: one old Krupp, 4½ tons (5·9 inch), 1 old 3·5 inch. I.H.P. 78 = 7 kts.

No photo available.

KUANG CHIN (China, 1885—7). Wooden vessel of 560 tons. Dimensions: 250 × 31 × 10 feet. Guns: 2—4·1 inch, 1—12 pdr., 2—2 pdr. H.P. 312 = 9 kts. 1 screw. Coal: 110 tons. Complement 71.

Gunboats (Water Police of Kwang Tung Provence.)—*Continued.*

No photo available.

KUANG KENG (China, 1885). Composite vessel of 560 tons. Guns: 1—4·7 inch, 1—12 pdr. H.P. 400 = 11 kts.

River Gunboats.

(Both belonging to Water Police of Chechiang Province.)

CHE CHING *class* (2).
(Scale: 1 inch = 160 feet).

No photo available.

CHE CHING, CHE TING (Kiangnan Dock Co., Shanghai, 1914). Dimensions: 146 × 24 × 2 feet. 141 tons. H.P. 750 (*f.d.*) = 13 kts. Boilers: Yarrow. Guns: 1—3·4 inch howitzer, 2—3 pdr., 8 machine. Complement, 41.

Old Torpedo Boats.

(All belong to Water Police of Kwang Tung Pr

No photo available.

5 *Schichau type*:

Lei Chung	***Lei Kan*** (1)	***Le***
Lei Li	***Lei Kan*** (2)	

All built in Germany, 1884-5. 26 tons. Armament: 1 torpedo tube. H.P. 420 = 19 kts. (much less now).

No photo available.

4 type unknown:

Lei Chen	***Lei Ken***
Lei Kun	***Lei Sun***

All built 1884-5. No details known.

ROYAL SIAMESE NAVY.

(Revised in Siam, 1920.)

Flags (1919).

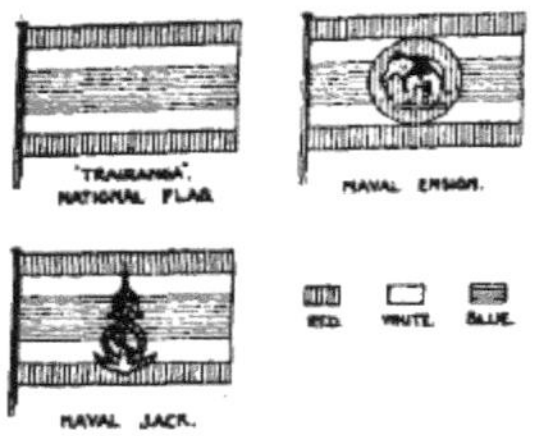

Minister of War:

Mercantile Marine: 17 Steamers of 17,019 tons *gross*; no return for Sailing Vessels.

RECOGNITION SILHOUETTES.

Scale: 1 inch = 160 feet.

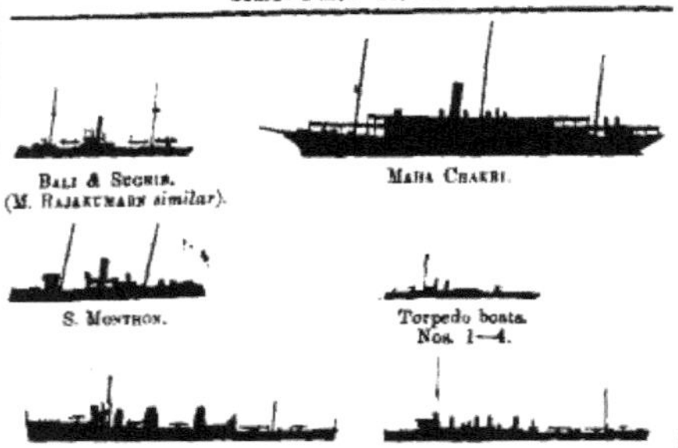

Arsenal.

BANGKOK.—Dry dock: 321½ × 48 × 14½ feet H.W. highest springs). Sheer legs with lift of 40 tons and crane with lift of 10 tons. Workshops and plant are capable of effecting repairs for the whole fleet.

Private Docks. (Bangkok.)

Dry Dock Co., No. 1; 300 × 45 × 11½ feet. Samsen Dry Doc × 13 feet.

Personnel: 5000. Reserve: 20,000. Royal Naval Academy gation and Engineering at Phra Rajwangderm, with acco for 120 cadets. There are about six Provincial Establishments, so that men recruited locally can re liminary training before being drafted to ships of the R

Naval W/T Stations: At Bangkok and Singora, with rad miles by day and 600 miles at night.

3 Destroyers.

As British "Thornycroft R" Class Destroyers.

Phra Ruang (ex-British *Radiant*, launched by Thornycrofts, Nov., 1917. Purchased by Siam, July, 1920). 1035 tons. Dimensions: 274 (*o.a.*) 265 (*p.p.*) × 27½ × 11 feet (*max.* draught) and 8½ feet (*mean*). Guns: 3—4 inch, 1—2 pdr. pom-pom, 1 M.G. Tubes: 4—21 inch, in two twin deck mountings. Machinery: Brown-Curtis (all geared) turbines. Designed S.H.P. 29,000 = 35 kts. (39·67 trials). 3 Yarrow boilers. Oil fuel: about 285 tons max. Complement (was 82).

Notes.—Built for British Navy under Emergency War Programme. Begun December, 1915, completed February, 1917. For the purchase of this boat, about two million ticals was raised by voluntary subscription in Siam.

Sua-Tayan-Chol. *1919 Photo, J. Bailey.*

Sua-Gamron-Sindhu (1912), 385 tons, and ***Sua-Tayan-Chol*** (1908), 375 tons. Both built by Kaw Kobe, Japan. Dimensions: 227 (*p.p.*) × 21½ × 6 feet. H.P. 6000 = 27 kts. Coal: 100 tons = 1000 miles Armament: 1—12 pdr., 5—6 pdr., 2 M.G.; 2—18 inch tubes. Complement: 75.

Notes.—Same type as the Japanese "*Arare*" class Destroyers.

TORPEDO BOATS.

Note.

This space was reserved for the inclusion of details of ex-German "A" Type Torpedo Boats, which, according to report, were to be assigned to the Royal Siamese Navy.

Up to the end of August, 1928, no surrendered German Torpedo Boats had been allocated to Siam.

No. 2 — 1914 *Photo, J. Bailey*

4 Kawasaki boats:—Nos. **1, 2** and **3** (1908), and No. **4** (1913). 150 tons. 1916 × 16·2 × 26 feet. H.P. 1900

Armament: 1—6 pdr., 1—3 pdr., 2—18 inch tubes. Complement: 29.

There is also a small "Minor Torpedo Boat" built at Bangkok about 1913-14. Has 1—14 inch tube and paraffin motors. Believed that this boat has been presented to the "Sea Scouts." The "Sea Scouts" own two Motor Launches, each armed with one 3 or 1 pdr. gun and fitted with "Caster" motors.

MISCELLANEOUS SHIPS.

Gunboats.

BALI (1901), **SUQRIB** (1901). 580 tons. 162 × 23 × 10 feet. Guns: 1—4·7 inch, 5—6 pdr., 4—1 pdr. Designed H.P. 500=11·4 kts. Complement, 83.

See next column.

1 funnel, 2 masts, with one military top.

MURATHA (1898). 530 tons. 145 × 25 × 10 feet. Guns: 1—4·7 inch, 4—6 pdr., 3—1 pdr. Designed H.P. 500=11·4 kts.

Gunboats—*continued.*

Not certain if above photo is of *Muratha* or *M. Rajakumarn.*

MONGKUT RAJAKUMARN (ex-*Filipinas*, Hongkong and Whampoa Dock Co., 1887; purchased 1891). 700 tons. 175 × 23½ × 11 feet. Guns: 2—4·7 inch, 2—6 pdr., 2—3 pdr. Speed 11 kts. 2 screws. Complement about 100.

Note.—Originally built for Philippines Government, but having failed to reach designed speed on trials, delivery was refused. Sold by her builders to Siam.

Despatch Vessels.

Photo wanted.

SATHIT RAJAKARN (1907). 144 tons. H.P. 112. Complement, 25. No other details known.

Photo wanted.

KECHON (1907). 110 tons. H.P. 120=9 kts. Complement No other details known.

(*Continued on next page*).

Despatch Vessels.—*continued.*

No photo available.

PRAP PARAPAKS (1900). 200 tons. H.P. 225 = 10 kts. Complement, 25. No other details known.

No photo available.

TEWA SURARAM (1898). 145 tons. Guns: 1—6 pdr., 2 machine. H.P. 145 = 10 kts. Complement, 25.

Transports.

No photo available of any of these ships.

VIDES, KICHKAR (ex. *Buk*, ex *Lygidas*, Ritson & Co., Maryport 1902). 850 tons. 176 × 27½ × 10½ feet. H.P. 780 = 9 to 10 kts. Rigged with 3 masts and has one funnel just before mizzen mast.

UTHAI RAJAKITR (1897). 134 tons. H.P. 135 = 8 kts.

BANCHU (ex. S.S. *Beagle*, Blackwood and Gordon, Glasgow 1892). 250 tons. 110⅔ × 20⅘ × 6⅕ feet. H.P. 120 = 8 kts. Has one mast stepped at break of short, high forecastle, and 1 funnel right aft.

Royal Yacht.

1919 Photo, J. Bowling, Esq.

MAHA CHAKRI (Kawasaki Co., Kobe, Japan, 1918). About 2,400 tons *gross*. Complement, Length, 298 feet (*w.l.*), 335 feet (o.a.). Beam, 40 feet. Draught, feet. Guns: Not known. Machinery (*see Notes*): 2 sets triple exp. Boilers: 4 (type unknown) coal and oil burning. I.H.P. about 2000 = 15 kts. Oil fuel: 200 tons = 2000 miles endurance.

Notes.—The hull of the old Royal Yacht *Maha Chakri* was sold to the Kawasaki Co. in 1917, but the engines and other fittings were removed, overhauled and renovated for installation in the new *Maha Chakri* described above.

Coastguard Vessel.

SRIYA MONTHON (Thornycroft, 1908). Displacement 225 tons. Dimensions: (*p.p.*) × × (*mean* draught) feet. Guns: 1—6 pdr., 4 or 6 smaller. I.H.P. 700 = 14·5 knots speed. Coal: tons. No. of Screws: . Boilers: . Complement, . Used for Revenue and Customs Duties.

Also 3 ex N.D.L. steam lighters, re-named *Chen Thale*, *Han Thale*, *Loe Thale*, built by Hongkong and Whampoa Dock Co. About 160 × 27 × feet, 440 tons *gross*. Nom. H.P. 60 = kts. speed. Seized during the war and now used for various purposes, including occasional surveying duties.

There are various small craft used for River Duties of which no precise details are available.

PERSIA.

Minister of Marine:—Vossukh-es-Sultaneh (July, 1920).

Mercantile Marine:—2 steamers of 1327 tons *gross*.

Docks:—None can be traced.

ENSIGN

MERCANTILE

Silhouette Scale: 1 inch = 160 fee

PERSEPOLIS.

Gunboat.

Photo, Lieut. Preece

PERSEPOLIS (1885). Dimensions: 207 × 29½ × 19½ feet. 12 H.P. 450 = 10 kts. Guns: 4 old 2·7 in., 2 machine.

Note: Gunboat *Muzaffir* served with and in H.M. Navy during It is not known what has become of this ship.

Not known if Persia has any Armed Vessels in the Caspi or not.

Royal Yacht *Selika* (bought 1892) of 400 tons may exist sti

LIBERIA.

RECOGNITION SILHOUETTE.

Scale: 1 inch = 80 feet.

MESURADO.

Revenue Cutter.

By courtesy of "Motor Ship."

MESURADO (Blandy Bros., Las Palmas, C.I., 1914). 24 tons. Dimensions: 85 × 12 × 4½ feet. Guns: 1 Nordenfelt, 1 M.G. H.P. 120 (2—60 H.P. Parsons paraffin engines) = 12–14 kts. Fuel: 500 galls. paraffin. 1 small searchlight.

Note.—The report that the above vessel had been sold to Honduras has been proved to be incorrect. *Mesurado*, during July, 1920, was still lying at Las Palmas, and will not be delivered until the Liberian Government has paid for her.

LATE ADDENDA: ALLOCATION OF AUSTRO-HUNGARIAN DANUBE FLOTILLA AND TURKISH WARSHIPS.

AUSTRO-HUNGARIAN DANUBE FLOTILLA.

ASSIGNED TO.	MONITORS.	PATROL BOATS.
AUSTRIA	Nil.	***Fogas*** ***Kompo*** ***Neretva*** (ex-*Barsch*) ***Poszony*** (ex-*Stor*)
HUNGARY	Nil.	***Bregalnica*** (ex-*Vels*) ***Csuka*** ***Komaron*** (ex-*Lachs*) ***Viza***
SERBS, CROATS AND SLOVENES	**DRAVA** (ex-*Enns*) **MORAVA** (ex-*Körös*) **SAVA** (ex-*Bodrog*) **VARDAR** (ex-*Bosnia*)	Nil.
RUMANIA	**DRINA** (ex-*Temes*) **INN** **BOCA** (ex-*Sava*)	Nil.
INTER-ALLIED DANUBE COMMISSION	**LEITHA** **MAROS** **SZAMOS**	Nil.

Remarks.—Conditions under which above River Craft have been assigned are not known at present, but it is probable that they have been allocated as Police, Preventive and Coastguard Vessels, each retaining only one small gun for such duties.

TURKISH NAVY.

ASSIGNED TO.	TYPE. SHIPS.
GREAT BRITAIN* ..	Battle Cruiser :—**SULTAN SELIM** Obsolete Battleships :—**MUIN-I-ZAFFER, IDJLALIEH** Cruisers :—**HAMIDIEH, MEDJIDIEH** Torpedo Gunboats :—**BERC-I-SATVET,† PEIK-I-SHEVKET†** Destroyers :—***Muvanet-i-Milet, Noumoune-i-Hamiyet, Basara,§ Sameoun,§ Tasho Berc-Efsan*** Sloop :—**ZUHAF** Gunboat :—**MALATIA** Depot Ship :—**NEDJMI-SHEVKET** Mine Layer :—**NUSRAT**
JAPAN*	Battleship :—**TOORGOOD REIS**

Remarks.—*All ships allocated to Great Britain and Japan will probably be broken as Target Ships for Experimental purposes.

†One of these two Torpedo Gunboats will probably be assigned to Portug date.

§May be returned to Turkey later, to replace three worst Torpedo Boats o and *Younnous* types.

RETAINED BY TURKEY (For Police, Preventive and Coastguard Duties only):—

Gunboats :—**AIDAN REIS, BURACK REIS, SAKIZ, PREVESAH, REIS, ISSAH REIS, KEMAL REIS, BARIK-I-ZAFFER.**

Torpedo Boats :—***Sivri Hissar, Sultan Hissar, Drach, Moussoul, Ac. Younnous.***

AUSTRO-HUNGARIAN DANUBE FLOTILLA; RUSSIAN AND TURKISH NAVIES.

(For Partition of Austro-Hungarian Danube Flotilla and Turkish Navy, v. Addenda Page 530a).

Notes.

UNDISPOSED VESSELS OF THE LATE AUSTRIAN NAVY.

Disposal of the following vessels not traced up to Nov., 1920:—

Torpedo Boats: **Nos. 3, 7, 19—25, 27, 29, 30, 31—40.**
Mining Vessel: **CARNIOL.**
Mine Depot Ships: **AURORA, DELTA, GASTEIN.**
Submarine Depot Ships: **GAA, PELIKAN.**

The Torpedo Boats Nos. 3 and 7 were captured by French Navy at the end of the War, but their present disposal is unknown. The other T. B. are obsolete, probably worn out and useless. **CARNIOL, GAA, GASTEIN, PELIKAN** are possibly suitable for conversion to Mercantile vessels. **AURORA** and **DELTA** are only Hulks. Old Battleships *Kaiser Max* and *Rudolpha* have been converted to pontoons. The two old Battleships, **KRONPRINZ ERZHERZOG RUDOLF** and **MARS**, were removed from the Austrian Navy in 1916. It is not deemed worth while to list and illustrate these vessels again as the Austrian Navy has ceased to exist.

Late AUSTRO-HUNGARIAN DANUBE FLOTILLA.

Note.

Except 5 or 6 vessels, all ships interned at Novisad, (Ujvidek, Neusatz), a Serbian port about 80 miles above Belgrade.

Disposal of ships not yet decided on.

During the regime of the Bela Kun Government at Buda-Pest, a Danube Monitor was re-named **POSZONY.** Identity of this ship not known.

RECOGNITION SILHOUETTES.

Scale: 1 inch = 160 feet.

Patrol Boats.

Szamos.

Maros *class monitors.*

Koros.

(As re-built.)

Almos & Enns type.

(Approximate.)

Bodrog.

Monitors.

For *Temes*, see below. Has been so largely reconstructed, she might be regarded as a new Monitor, launched in 1915.

Photo wanted.

ALMOS (1915), **SAVA** (1915), **BOSNIA** (1915), **RAGUSA** (1916). All completed 1916. Reported to be identical with *Enns* and *Inn* below. *Bosnia* and *Sava* built under 1914-15 Austro-Hungarian Naval Programme. All surrendered. *Bosnia* and *Sava* taken over by British and Serbs.

Photo wanted.

ENNS* (1913), **INN*** (1913). 528 tons. Dimensions: 190 × 33¾ × 4¼ feet. Designed H.P. 1500 = 13 kts. Boilers: Yarrow. Armament: 2—4.7 inch + 3—4.7 inch howitzers, 2—11 pdr., 6 M.G. Armour: 1½" Belt and Bulkheads, 1" Deck, 2" C.T. and Turrets (2). Complement, 86. Fuel: Oil only. Belong to 1912 Austro-Hungarian Naval Programme.

* *Inn* mined and sunk in the Danube River, Sept. 22nd, 1917. Subsequently salved, repaired and returned to active service. *Enns* taken over by British and Serbs.

Photo wanted.

TEMES. Originally sister to *Bodrog*, and launched 1904. Sunk, October, 1914; raised, June, 1916. Has been so largely rebuilt, she is now practically a new Monitor. Present armament said to be: 2—4.7 inch, 4—14 pdr., 6 M.G. Other details not known.

Monitors—*Continued.*

BODROG (1904). 433 tons. Dimensions: 18
Designed H.P. 1400 = 13 kts. Boilers: Yar
2—4·7 inch (45 cal.) + 1—4·7 inch howitzer,
machine. Armour: 1½" Belt and Bulkhead
Turrets and Conning tower. Complement, 79.

(Contin

Monitors—*continued*.

***KÖRÖS** (1892), ***SZAMOS** (1892). 440 tons. Dimensions: 177½ × 29½ × 4 feet. Designed H.P. 1200 = 10 kts. Boilers: Yarrow. Guns: 2—4.7 inch (35 cal.), 2—9 pdr., 2 M.G. Armour: 2" Belt, ¾" Deck, 3" Turret, 2" C.T. Complement, 79-80. Coal: 80 tons.

*Körös sunk, April 1915, but subsequently raised, repaired, and returned to active service.

***LEITHA** (1871), ***MAROS** (1871). Both reconstructed, 1894. 305 tons. Dimensions: 164 × 20¼ × 3¼ feet. H.P. 700 = 8 kts. Guns: 1—4.7 inch (35 cal.), 2 or 3—11pdr., 3 M.G. Armour: 1¼" Belt, 1" Deck, 2" Turret. Complement, 57. Coal: 20 tons.

*Taken over by British and Serbs.

Torpedo Vessel.

Photo wanted.

VEKSEL. (——) No details known, but said to be armed with 1—3.9 inch, 1 M.G. and 2 torpedo tubes.

Mining Craft.

No Photo available.

MINELAYER (NAME UNKNOWN). No exact details available, but said to be armed with 2—11 pdr. and 2—3 pdr. guns and to carry 30 mines.

No Photo available.

TULIN. Barge, propelled by motor engines and carrying 140 mines.

Scouts.

Barsch,* Barsto, Fogos,* Iauka,* Kompo, Vels,* Viza or 2—3 pdr.

* Taken over by British and Serbs.

Patrol Boats.

2 boats (numbers not known), **g, h** (1909), 18 tons, 22 kts. **f** (1907), 12 tons, 22 kts., 1 M.G.; **b, d** (1906-9), 14 kts. Guns: 1 or 2 small. Complement, 13.

Also several motor launches.

Miscellaneous.

Klina, *Samarsa*, *Val* 594 tons *gross* — Armed Steamers, each having 2—

Hebe. Armed Yacht.

Bayo, *Boski* — Armed Trawlers.

RUSSIAN FLEET.

FLAGS.—Not known. A new National Ensign is reported to have been adopted with the Russian words, "The Soviets of the Federal Republics," on a red ground.

UNIFORMS.—Shoulder markings abolished, and rank is now indicated by stripes on cuff, as in British Navy. *All* ranks now have gold buttons. Officers' caps have an anchor badge, on a red ground, and encircled by a wreath, much the same as in the British Navy. Ranks are distinguished thus:—

Admiral: ¾-inch top stripe with curl and three stars round curl; two 1-inch lower stripes.

Vice-Admiral: Same as Admiral but only *two* stars.

Kontre-Admiral: Same as Admiral but only *one* star, above curl.

Kapitan (*I Ranga*): Same as Admiral but no stars.

Kapitan (*II Ranga*): Two 1-inch gold stripes, top with curl.

Starchi Leitenant: One 1-inch or ¾-inch top stripe with curl, and *three* ½-inch gold stripes below.

Lietenant: Same as above but only *two* ½-inch gold stripes below.

Mitchman: Same as above but only *one* ½-inch gold stripe below.

Pra'porchi: This is a temporary rank, about equivalent to Sub-Lieut., R.N.R. in British Navy. One ½-inch gold stripe *without* curl.

Note.—The above particulars relate to changes introduced by the Kerensky Government. Under the Bolshevist Administration it is said that all marks of rank in the Fleet and Army have been abolished.

SEVASTOPOL to	miles.
Batum	400
Bizerta	1300
Constantinople	298
Gibraltar	2000
Kertch	175
Malta	1000
Nikolaieff	200
Novorossisk	200
Odessa	165
Piraeus	660
Salonika	643
Toulon	1654

(LIBAU)* to	miles.
Brest	1430
Cherbourg	1220
Copenhagen	465
Danzig	260
Gibraltar	2330
Kiel	550
Memel	230
Reval	30
Rosyth	1010
Vladivostok	12,405
Windau	50

KRONSTADT to	miles.
Abo	250
Bornholm	627
Brest	1680
Cherbourg	1470
Copenhagen	715
Danzig	558
Gibraltar	2580
Helsingfors	150
Kiel	800
Libau	250
Memel	480
Reval	150
Rosyth	1345
Petrograd (By sea canal)	21
Vladivostok	12,655
Windau	200

VLADIVOS[TOK] to	
Hong Kong	
Kiao-chau	
Maizuru	
Manila	
Matsumai	
Nagasaki	
Port Arthur	
Sassebo	
Wei-hai-wei	

Note.

The present bo… Russia are so inde… has been deemed … suspend the Gener…

Modern Naval Guns (Obukhoff).

Situation.	Designation.	Length in Calibres.	Weight of Guns.	Weight of A.P. Shot.	Initial Velocity (approximate).	Maximum Penetration with A.P. Capped Shell against K.C. Armour: 6000 yds.	5000 yds.	3000 yds.	Danger Space against average warship at: 10000 yards.	5000 yards.	3000 yards.	Usual Rounds per minute.
	inch.	calibres.	tons.	lbs.	f.s.	in.	in.	in.	yards.	yards.	yards.	
A^7	14	45	82·6	1486	2525	...	...	22¼	...	...	...	...
A^6	12	50	48·2	714	3000	...	14½	20	...	...	...	...
A^6	12	40	43·5	714	2600	6¼	14	18	150	400	750	0·30
A A	12	35	56	714	2200	4¼	8	11	120	380	580	0·25
A^3	10(V)	50	35	497	3000	6	12¾	15¼	...	...		
A^3	10	45	32	400	2500	5	10½	13	130	450	700	0·45
A	8(V)	50	17	216	3300		7¼	10¾	...	...	...	...
A	8	45	15¼	188	2800		6	9½	165	430	625	1
D*	6	45	7¼	89	2900		4¼	6	60	250	485	3
F*	6	45	7	80	2500	...	3¼	5	50	200	430	3
F*	5·1	55	...	...	...	...	...	...	...	...	...	...
F*	4·7	45	2¼	46	2600	...	...	...	...	...	...	about 4
F*	3	60	14 cwt	13½	2700	...	...	...	...	...	...	...
F*	3	35	12	13½	2600	...	...	...	...	...	...	...

* = Brass cartridge cases. V = Vickers. *Maximum* rates of 10 inch and 8 inch are 3 and 6 rounds per minute. These guns are in the *Rurik*.

Guns.

Baltic Ships.

$14''_{45}$ in *Kinburn* class (4 ships) not yet completed.
$12''_{50}$ in *Gangoot* class (4 ships).
$12''_{40}$ in *Respublika* class (2 ships), *Slava*.
$12''_{35}$ in *Tchesma*.
$10''_{50}$ in *Rurik*.

$8''_{50}$ in *Respublika* class (2 ships), *Petr Veliki* and *Rurik*.
$8''_{45}$ in *Bayan* class (2 ships), *Gromoboi* and *Rossiya*.
$6''_{45}$ in *Slava*, *Grajdanin*, *Petr Veliki*, *Bayan*, and all cruisers.
$5{\cdot}1''_{55}$ in *Kinburn* class (4 ships) and *Admiral Butakoff* class (4 ships).

Black Sea Ships.

$12''_{50}$ in *Gen. Alexeieff*.
$12''_{40}$ in *Borets za Svobodu* and *Evs*… ships).
$10''_{45}$ in *Rostislav*.

$8''_{50}$ in *Evstafi* class (2 ships).
$6''_{45}$ in *Borets za Svobodu*, *Ro*… *Sviatitelia* and *Evstafi* class (2…
$5{\cdot}1''_{55}$ in *Admiral Nakhimov* class … *Gen. Kornilaff*.

Note.—A considerable number … been dismounted from Russian … service on various land fronts.

RUSSIAN BALTIC AND PACIFIC DOCKYARDS, Etc.

Divided into squares of 2000 yards. Uniform scale. Soundings in fathoms. Heights in feet.

Mercantile Marine.

From "Bureau Veritas" 1919–20 figures.
Ships under 100 tons *gross* excluded.

639 Steam and Motor Ships of 665,628 tons *gross*.
7 Vessels with Auxiliary Motors of 1377 tons *gross*.
1728 Sailing Vessels of 289,754 tons *gross*.

Chief Trade Ports.

Petrograd, Vladivostok.

COINAGE:

Colour of Ships.

Big Ships: Light grey. *Destroyers & Submarines*: Not known.

Personnel.

Before the War exceeded 60,000 (including Reserves ?) During the War, *active* personnel, excluding mobilised Reserves, is said to have been 51,670 officers and men (inclusive of Coastguards and Marines). Owing to numerous desertions present strength is entirely unknown.

Overseas Possessions: None.

Naval Arsenals, Etc.

KRONSTADT. Fitting out and repairing yard for Baltic Fleet. No slips. Four large dry docks; (*a*) $594 \times 85\frac{1}{2} \times 28\frac{1}{2}$ feet; (*b*) $487 \times 71 \times 27$ feet; (*c*) $1058 \times 62\frac{3}{4} \times 19\frac{1}{2}$ feet; (*d*) $522\frac{3}{4} \times 52 \times 17\frac{1}{2}$ feet.

Kronstadt is fed from the shipbuilding yards at Petrograd, the New Admiralty (2 slips), Galernii Ostrov (3 slips), Baltic works (3 slips), and torpedo-boat yards at Izhora (on the Neva). The channel above Kronstadt is now about 35 feet deep.

Admiralty Chart No. 2215.

(White Sea).

Murmansk, on the Kola Inlet, reported to have been used as a subsidiary base for torpedo craft, mine-sweepers and ice-breakers.

Commercial harbours at Kronstadt and Petrograd.

Private yards at Petrograd: Putilov Works, Nevski Shipbuilding and Mechanical Works.

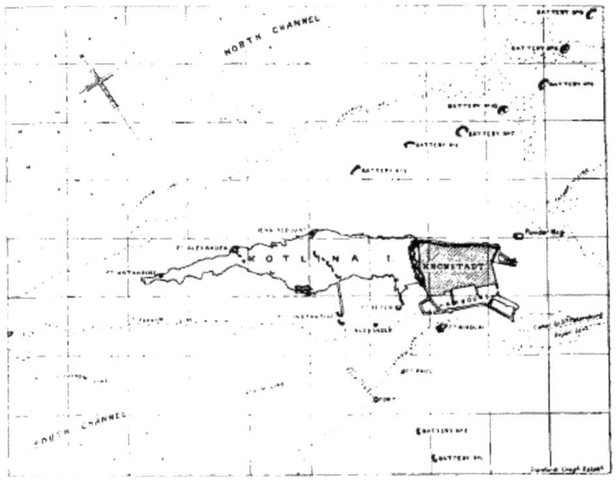

KRONSTADT. (Fort Peter was blown up in Dec., 1917.)

(In the Pacific).

Occupied by Japanese Forces.

VLADIVOSTOK. Small dockyard. Government wharf 1500 feet long, 27 feet depth of water alongside. Two docks, each $618\frac{1}{2} \times 90 \times 34\frac{1}{2}$ feet, and a third $572 \times 90 \times 30$. One small floating dock 300 feet long (8000 tons). Two small basins. Some repairing shops. Three floating cranes, 50, 75 and 100 tons. Water shallow. Good *tideless* protected harbour. Harbour is frozen in January and February, but passage is cleared by ice-breakers. Strongly fortified (high-site batteries with 10 in. disappearing guns). Admiralty Chart No. 1011. Allied troops landed here during 1918.

Commercial harbours: Vladivostok, Nikolaievsk.

VLADIVOSTOK (occupied by Japane

Summer, 191

C-in-C. (Army, Navy, &c.):
Baltic Naval Forces: Raskolnikoff.
Black Sea:
Volga and Caspian: Kozanov.

RUSSIAN NAVY.

BALTIC SECTION

(INCLUSIVE OF SHIPS IN THE WHITE SEA, AT VLADIVOSTOK AND ON AMUR RIVER).

Reported that the Baltic Fleet is to he completely re-organised, but owing to desertion of *personnel*, executions of naval officers, looting and mishandling of ships, very few ships can now be rendered effective. In September, 1920, the Bolshevist Active Fleet was reported to consist of **PETROPAVLOVSK, SEVASTOPOL, ANDREI PERVOSVANNI,** five large Destroyers ***Azard, Samsun, Garibaldi, Kapitan 2r Isyimettiev, Pobieditel,*** five smaller Destroyers ***Sibirski-Strelok, Vsadnik, Gaidamak, Amurets, Ussuriets,*** four old Torpedo Boat Destroyers ***Lovki, Likoi, Metki, Krepki,*** and three or seven Submarines.

As regards ships listed as building and completing: all work ceased in June, 1917, and practically nothing further has been done to advance construction during the past three years.

It is not worth while dividing the Russian Navy pages into Bolshevist and Anti-Bolshevist sections. On the next page a list is given of ships in the White Sea and at Vladivostok. Vessels named in heavy type are listed on following pages; those in small type cannot be described for lack of details. Ships not included in the White Sea, Vladivostok or Amur River lists may be taken as under Bolshevist control.

Various ex-Russian Baltic Warships have been transferred to the Esthonian and Finnish Navy Sections (*q.v.*).

Warning must be added that it is extremely difficult to secure accurate information regarding the Russian Navy, and the authenticity of all information given in these pages cannot be vouched for with confidence.

RUSSIAN WARSHIPS, FLEET AUXILIARIES, &c. (NOT IN BALTIC OR BLACK SEA).

WHITE SEA.

Old Battleship :—

TCHESMA

Destroyers :—

Leit. Sergieff
Kapt. Yurasesvski
Bestrashni
Beshumni

Submarines :—

Sviatoi Georg
Delfin

Trawlers :—

Total, 36.

Yachts :—

Alvina (ex-*Sokolitsa*)
Josephine (ex-*Gorislava*)
Yaroslavna

Auxiliaries :—

TAIMYR
VAIGATCH
BAKAN
Ludmilla
Kolguieff

Ice Breakers :—

Kniaz Pojarski
Solovei Budimirovitch
Alexander Sibiriakoff

Sailing Steamers :—

Bellaventure
Beothic
Bonaventure
Bruce

Miscellaneous :—

PAHKTUSOV
XENIA
Advance
Antony
Cheviot
Inei
Kingfisher
Koupava
Olen
Orlik
Sniedzhinka
Strepet

Tugs and Ice Breakers :—

About 10.

Motor Patrol Boats :—

About 8 or 10.

AT BRITISH PORTS.

Cruiser :—

ASKOLD

Destroyers :—

Grosovoi
Vlastni

Ice Breakers :—

SVIATOGOR*
ALEXANDER*
Kozma Minin (at Liverpool)

*Under White Ensign as R.F.A. Paid off at Rosyth. Are described in British Navy Section.

AT FRENCH PORTS.

Ice Breaker :—

Mikula Salianinovich (at Cherbourg)

VLADIVOSTOCK.

Note.

The Japanese took over all ships of Russian Naval Department, April, 1920. Present disposal not known.

Destroyers :—

Bespooshtchadni
Bravi
Boiki
Bodri
Grosni
Leit. Maleieff
Insh. Mech. Anastassoff
Tverdi
Totchni
Trevoshni
Smyeli
Serditi
Statni
Skori

Gunboat :—

MANDJUR (Minelayer)

Mine Layers :—

DIOMID
PATROKL
ULIS
MONGUGAI

Miscellaneous :—

Aleut
Kamchadal
Okhtosk
Orel

SHILKA
Takut

AMUR RIVER FLOTI

GROSA
SHKWAL
SHTORM
SMERTCH
TYFUN
URAGAN
VIKHR
VYUGA
OROKANIN
MONGOL
BURIAT
VOGUL
SIBIRIAK
KOREL
KIRGIZ
KALMUIK
SIRIANIN
VOTYAK

(OROKANIN to VOTYAK: Believed to … of Japan…)

Kopya
Palash
Pika
Pistolet
Pulya
Shtik

ONE AND TWO FUNNELS. **(For Two Funnel Battleship types, see next page)**

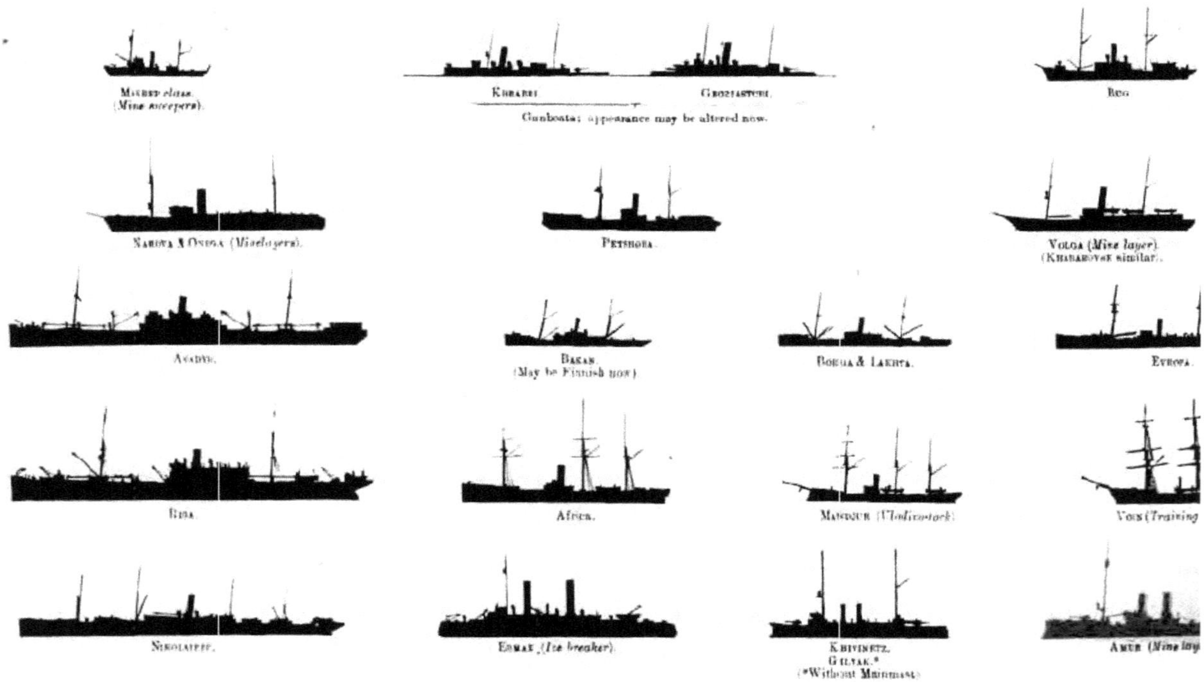

Scale: 1 inch = 160 feet.

RUSSIAN WARSHIPS RECOGNITION SILHOUETTES.

TWO FUNNELS (*continued*).

Petr Veliki.

Tchesma (*White Sea*).

Grajdanin.

Gangut *class* (4 ships).

Respublika *class* (2 ships).

Ex-Shtandart and Kretchet. (*Yachts*).

THREE FUNNELS.

Bogatyr.

Avrora *class* (2 ships or 1 ?).

Okean (*Training Ship*).

Rurik.

FOUR FUNNELS.

Rossia & Gromoboi.*

*Reported Sunk.

Bayan *class* (2 ships).

FIVE FUNNELS.

Askold *Laid up in the Clyde.*

Scale: 1 inch = 160 feet.

RECOGNITION SILHOUETTES—RUSSIAN TORPEDO CRAFT.

(Revis

Notes.

Destroyers of classes marked * taken over by Vladivostok or last reported in White Sea.

Destroyers and Submarines of classes marked Volga and Caspian Flotilla.

The Grosovoi class t.b.d. marked § in British Po

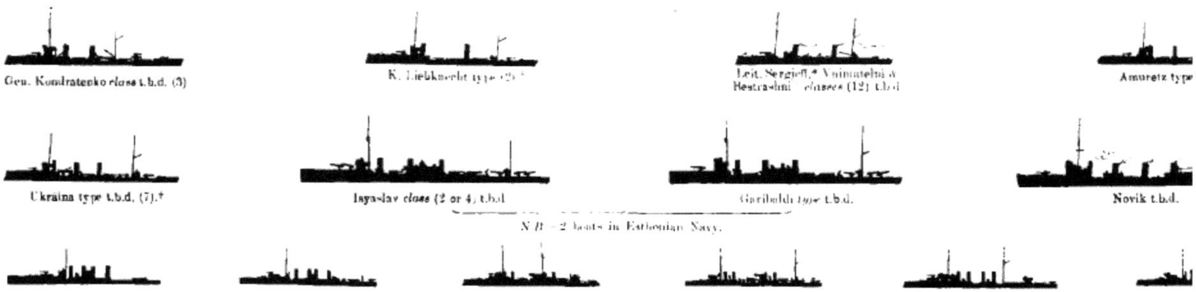

Tverdi class t.b.d. (5)* appearance uncertain; probably as Smyeli or Dyelni classes.

Scale: 1 inch = 80 feet.

RECOGNITION SILHOUETTES—RUSSIAN SUBMARINES.

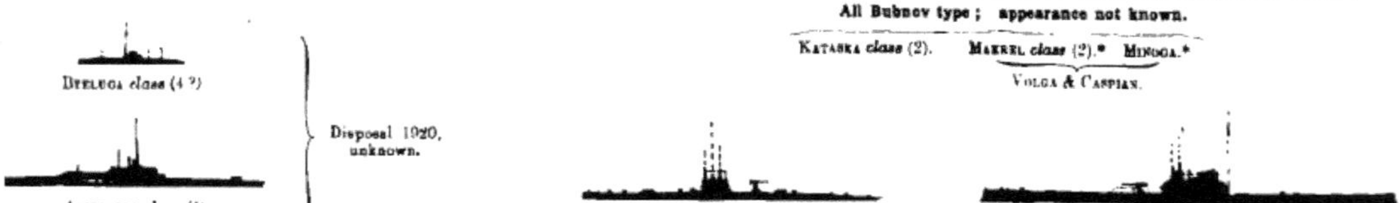

1912 RUSSIAN BATTLE-CRUISERS. Nos. 8, 7, 6, 5

BORODINO (1917?) **ISMAIL** (June, 1915). **KINBURN** (Oct., 1915), **NAVARIN** (1917?)

Normal displacement, *about* 32,500 tons. Complement,

Length, (o.a.) 728¼ feet. Beam, 98 feet. Draught, 28⅞ feet.

Guns:
12—14 inch, 45 cal. (A?)
24—5·1 inch, 55 cal.
4—4 inch (anti-aircraft)
4 machine
Torpedo tubes:
6 *submerged.*

Armour:
12" Belt ....
" Belt (bow) ....
" Belt (aft) ....
" Deck ....
" Turrets ....
" Battery ....

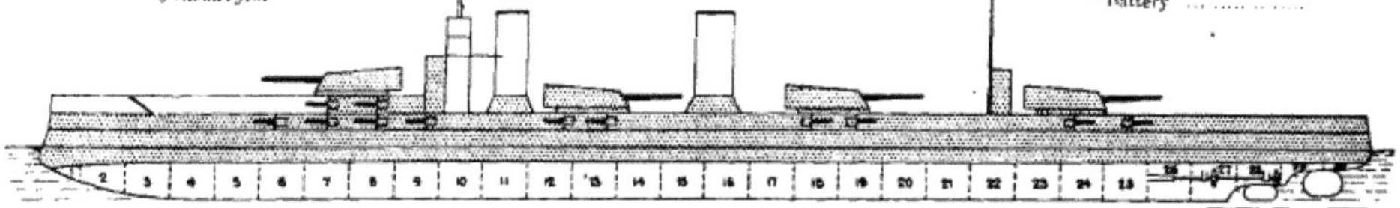

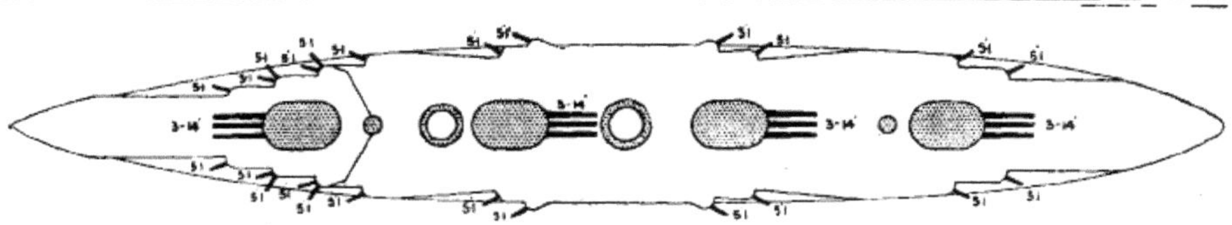

Notes.

Owing to lack of materials and lat was stopped on these ships in 1917. doubtful if they will ever be comple

Plans are based on information an furnished by the Naval Attache of th Embassy, 1917.

Machinery: Parsons turbines. screws. Boilers: 25 Yarrow. Designed S.H.P. 68,000 = 26½ kts.
Coal: *normal* tons; *maximum* 1,950 tons + 1575 tons oil fuel.

Name	*Builder*	*Machinery*	*Laid down*	*Completed*	*Trials*	*Turbines*	*Boilers*	*Best recent speed*
Borodino	Galernii	All by Franco-Russian Co.	Dec.'12	?			All Yarrow	
Ismail	Baltic Works		Dec.'12					
Kinburn	Baltic Works		Dec.'12					
Navarin	Galernii		Dec.'12					

GANGŪT (Oct., 1911), **POLTAVA** (July, 1911), **PETROPAVLOVSK** (Sept., 1911), **SEVASTOPOL** (June, 1911.)

Normal displacement, 23,370 metric tons. Complement, 1125.

Length *over all*, 594 feet. Beam, 87 feet. *Mean* draught, 27½ feet.

Guns:
12—12 inch, 50 cal. (*A*⁶).
16—4·7 inch, 50 cal.
2—9 pdr. (anti-aircraft).*
1—3 pdr.
8 Machine.
Torpedo tubes (18 inch):
4 *submerged*.
*2—11 pdr. AA. in Poltava

Armour (Krupp):
8¾" Belt (amidships) ...
5"-2" Belt (ends)
3"-4" Internal belt (see notes)
3" Deck
12"-10" Turrets
8" Turret bases
4" Battery
10" Conning tower

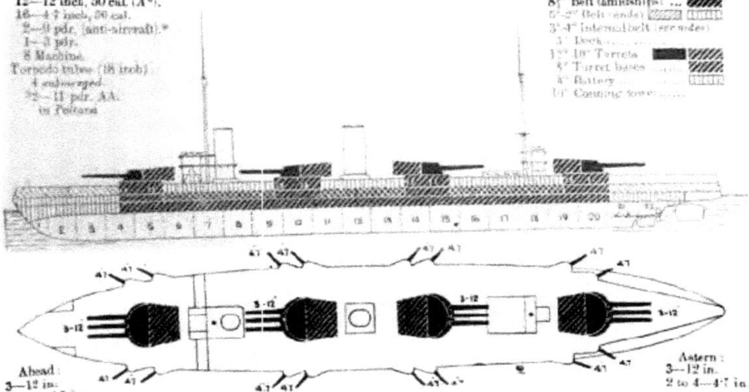

Broadside, 12—12 in., 8—4·7 in.

Machinery: Parsons turbine. 4 screws. Boilers: 25 Yarrow. Designed H.P. 42,000 = 23 kts. Coal: *normal*, 1000 tons; *maximum*, 3000 tons. Also 1170 tons oil.

Gunnery Notes.—The port plate above each gun is in the form of a hinged flap, allowing each 12-inch gun to elevate to 35—40° *maximum*. Arcs of fire: End triple 12-inch barbettes, 310°; central barbettes, 130° *each beam*; aft group of 4—4·7 inch, 90°; other 4·7 inch, 80°.

Armour Notes.—Belt is about 15 feet wide, 5 feet of it below water, uniform thickness. There is a *secondary* 3"-4" internal belt 11 feet inboard above protective deck, extending between the end barbettes. The space between main belt and internal belts is divided up into w.t. compartments.

Name	*Builder*	*Machinery*	*Laid down*	*Completed*	*Trials: Full power*		*Boilers*	*Best recent speed*
Gangut	Galernii I.	Franco-Rus. Wks.	June,'09	Jan. '14	About 50,000	=23	Yarrow	
Poltava	Galernii I.	Franco-Rus. Wks.	June,'09	Jan. '15		=23	Yarrow	
Petropavlovsk	Baltic Works	Baltic Works	June,'09	Jan. '14		=23	Yarrow	
Sevastopol	Baltic Works	Baltic Works	June,'09	Jan. '15		=23¼	Yarrow	

General Notes.—The late Gen. Vittorio Cuniberti prepared the original designs for this class—compare general resemblance to his Italian *Dante Alighieri*. The Ministry of Marine afterwards altered the plans to include Russian ideas of armouring, ice-breaking bows and other special features. Said to be most unhealthy, insanitary and badly ventilated vessels.

POLTAVA.

Note.—**PETROPAVLOVSK, after being torpedoed by British C. M. B., s water and on an even keel at Kronstadt. She has been raised, repa service, 1920. POLTAVA aground in Neva. GANGŪT laid up SEVASTOPOL in service 1920.**

GANGŪT.

(RESPUBLIKA CLASS—2 SHIPS).

RESPUBLIKA (*ex-Imperator Pavel Pervi*, Sept., 1907) & **ANDREI PERVOSVANNI*** (Oct., 1906).

Normal displacement, 17,680 metric tons. Complement, 933.

Length (*waterline*), 454 feet. Beam, 80 feet. *Mean* draught, 27 feet. Length *over all*, 460 feet.

Guns:
- 4—12 inch, 40 cal. (A*).
- 14—8 inch, 50 cal. (A).
- 12—4·7 inch, 50 cal.
- 2—3 pdr.
- 8 Machine.

Torpedo tubes (18 inch):
- 2 *submerged* (broadside).

***Docked, Aug. 1919, after being torpedoed by British C.M.B. Repaired and in Service, 1920.**

Armour (Krupp):
- 8½" Belt (amidships)
- 5" Belt (bow)
- 4" Belt (aft)
- 3" Deck
- 2" Splinter deck
- 10" Big turrets (N.C.)
- 5" Turret bases (*see notes*)
- 7" Small turrets
- 4" Bases (*see notes*)
- 5"—3" Main deck side
- 5" Battery (N.C.)
- 3" Q.F. Battery (amidships)
- 8" Conning tower (N.C.)
- 6" After C.T. (N.C.)

(Total weight 4400 tons.)

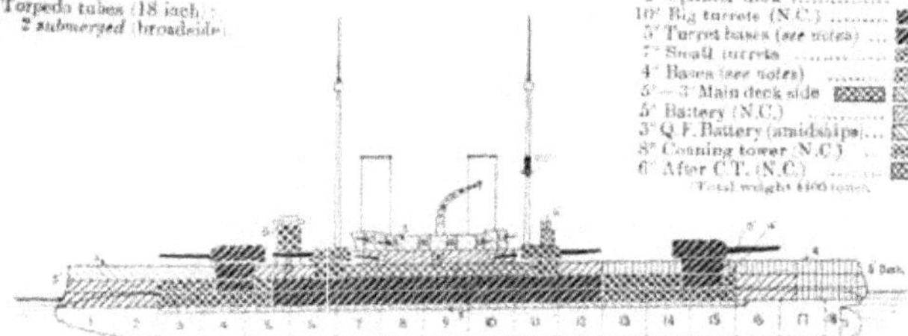

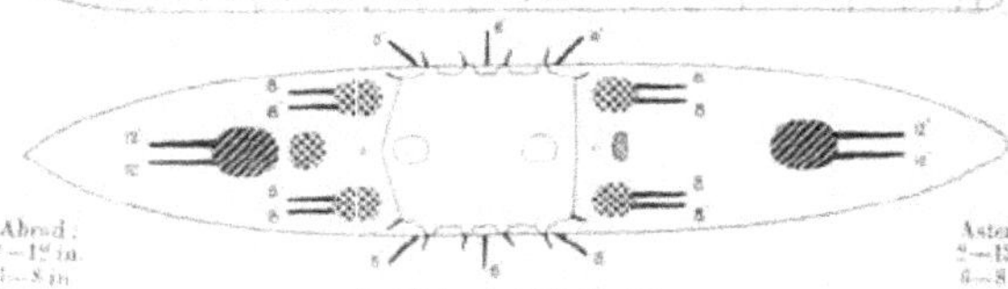

Ahead: 2—12 in. 6—8 in.

Astern: 2—12 in. 6—8 in.

Broadside: 4—12 in., 7—8 in.

Machinery: 2 sets triple expansion. 2 screws. Boilers: 25 Belleville. Designed H.P. 17,600=18 kts. (*e.f.* trials). Coal: *normal* — tons; *maximum* *Respublika* 1,325 tons, *A. P.* 1,500 tons.

Armour Notes.—Belt is 7½ feet wide, 5 feet of it below waterline; lower edge is 3" thick. Turret bases 5"—4" according to the side armour forming additional protection to them. 3" floor to batteries. 2" lateral "anti-torpedo" bulkheads.

Gunnery Notes.—Loading positions, big guns: all round. Hoists: electric. Big guns manœuvred electrically; secondary turrets, electrically. Arcs of fire: Big guns, 270°; secondary guns, 150°.

Torpedo Notes.—6 searchlights.

Name	*Builder*	*Machinery*	*Laid down*	*Completed*	*Last refit*	*Trials*	*Boilers*	*Best recent speed*
A. Pervosvanni	Galernii Island	Baltic Works	Jan., '03	Aug. '10		18,396=17	Belleville	14*
Respublika	Baltic Works	Baltic Works	Sept., '03	Sept. '10		17,775=17·7	Belleville	

*Full speed, 1918.

General Notes.—Design altered during construction in order to embody Japanese war lessons: armour thinned and extended to cover hull, Q.F. battery increased, etc., etc. Hulls unpierced; all ventilation through decks. These ships have been described as "veritable infernos" to live in, on account of their bad ventilation and lighting.

A. PERVOSVANNI. (Note derricks.) *Photo Seward.*

Note.—Both ships now rigged as Silhouette.

RESPUBLIKA. (Note cranes.)

GRAJDANIN (*ex-Tsesarevitch*) (Feb., 1901). *Photo: Abrahams, Devonport.*
Normal displacement, 13,210 metric tons. Complement, 803.
Length (*over all*), 388¾ feet. Beam, 76 feet. *Maximum* draught, 28¼ feet.

Guns:
4—12 inch, 40 cal. (A⁴)
12—6 inch, 45 cal.
10—11 pdr.
4—6 pdr.
2—1 pdr., AA.
6 Machine
Torpedo tubes (18 inch) removed.

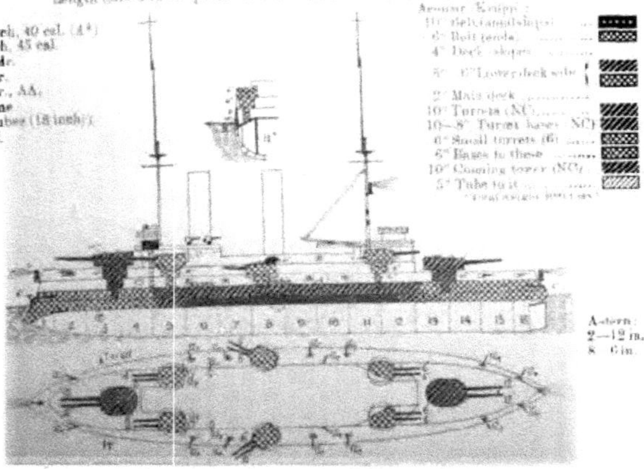

Ahead:
2—12 in.
8—6 in.

Broadside: —4—12 in., 6—6 in.

Machinery: 2 sets triple expansion. 2 screws. Boilers: 20 Belleville, without economisers. Designed H.P. 16,300 = 18 kts. Coal: *normal* 800 tons; *maximum* 1350 tons. Laid down at La Seyne, Toulon, 1899, completed 1903.

(*In White Sea.*)

TCHESMA (*ex-Tango, ex-Poltava*) (Nov., 1894). Displacement 10,960 tons. Co Length, 369 feet. Beam, 72¼ feet. *Maximum* draught, 27 feet. Guns (Obukhoff): 4—12 incl 12—6 inch, 45 cal., and 8—12 pdr. Armour (compound): Belt 15" (partial). T 10", small turrets 5". Designed H.P. 11,000 = 17 kts. Boilers: 16 Miyabara. Coal: *maxim* Sunk at Port Arthur in the Russo-Japanese War 1904; salved in 1905 by the Japanese *Tango.* Retroceded to Russia by Japan, March, 1916.

PETR VELIKI (1872, re-constructed 1905). Displacement, 9665 tons. Armament (4—8 inch (50 cal.), 12—6 inch (45 cal.), 12—12 pdr., 4—6 pdr., 8—3 pdr., 2 M.G = 13 kts. Coal: 1200 tons. Present speed: 11 kts.

Has F.C. top on foremast now. *Photo, Abrahams, Devonport.*

RURIK (Nov., 1906).

Normal displacement, 15,190 tons. Complement, 899.

Length *over all*, 529 feet. Beam, 75 feet. *Mean* draught, 26 feet. Length (*p.p.*), 490 feet.

Guns (Vickers):
- 4—10 inch, 50 cal. (*A**)
- 8—8 inch, 50 cal. (*A*)
- 20—4·7 inch, 50 cal.
- 2—3 pdr.
- 2—1 pdr., AA
- 2 machine.

Torpedo tubes (18 inch):
- 2 *submerged* (broadside).

Armour (Krupp):
- 6" Belt (amid.) (K.C.)
- 4" Belt (bow) (N.C.)
- 3" Belt (stern) (N.C.)
- 1½" Armour deck
- 8" Turrets (K.C.)
- 6" Turret bases
- 7" Secondary turrets
- 3" Redoubt (N.C.)
- 3" Battery (N.C.)
- 8" Conning tower (N.C.)

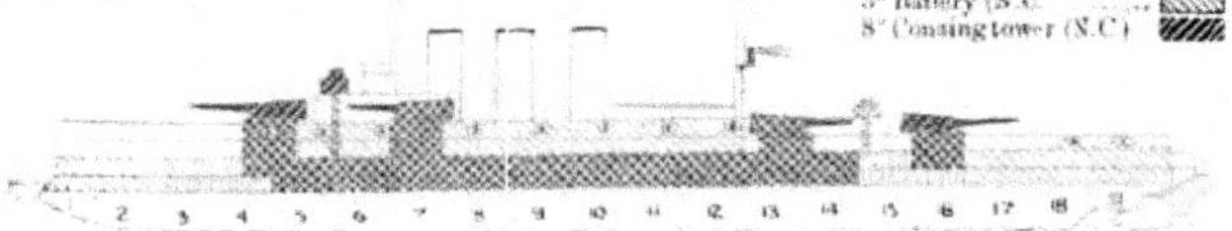

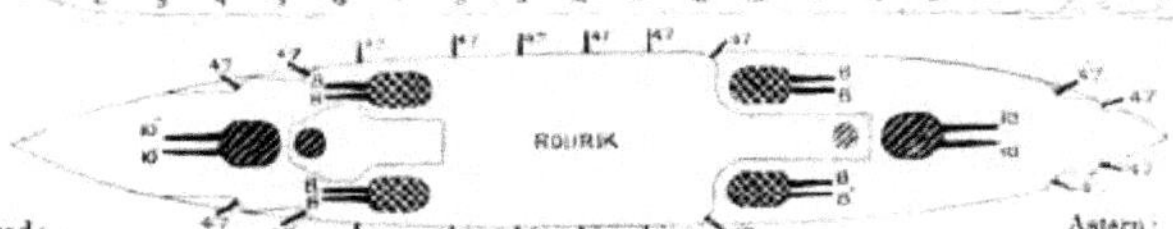

Ahead: 2—10 in. 4—8 in.

Broadside: 4—10 in., 4—8 in.

Astern: 2—10 in. 4—8 in.

Machinery: 2 sets 4 cylinder quadruple expansion. 2 screws. Boilers: 28 Belleville. Designed H.P., with ⅞ power, 19,700 = 21 kts., full power not specified. Coal: *normal* 1200 tons; *maximum* 1916 tons.

General Notes.—Built and engined by Vickers, Ltd. Laid down September, 1905, completed September, 1908. ***On trials:*** Mean for ten hours was 20,675 H.P. = 21·5 kts.; has made 22·4 kts. Big guns 35° elevation. Armoured magazines and special anti-torpedo protection.

(Officially rated as "Cruisers.") *Photo, Abrahams, De*

ADMIRAL MAKAROV (May 1906, re-fitted 1910). BAYAN (August, 1907)

Displacement, 7900 metric tons. Complement, 568—593.

Adm. Makarov:—Length (*o.a.*), 449⅔ feet, Beam, 57·7 feet, *Mean* draught, 21·4 feet.
Bayan:— " " 450 " " 57·5 " " " " 21·3 "

Guns:
- 3—8 inch, 45 cal. (*A*).
- 12—6 inch, 45 cal.
- 2—11 pdr. (*A.M.*).
- 1—1 pdr., AA. (*Bayan*.)
- 2—3 pdr.

Torpedo tubes (18 inch):
- 2 *submerged*.

Armour (K
- 7" Belt (
- 4" Belt
- 2" Armo
- 2½" deck
- 6" bulkh
- 3¼"-2½"
- 6" Turre
- " Turre
- 2½" Batt
- 5½" Conn

Note.

Re-armed during War. Extra 8 inch mounted without shield between 4th funnel and mainmast. 4—6 inch added—positions unknown.

Machinery: 2 sets vertical triple expansion. 2 screws. Boilers: 26 Belleville. Designed H.P.
21 kts. Coal: *normal* 750 tons; *maximum*, *Bayan* 950 tons, *A.M.* 1050 tons.

Notes.—*Bayan* laid down at New Admiralty Yard, Petrograd, 1905, and completed 1910. Machinery by Russian Co. *Admiral Makarov* begun by La Seyne, Co., Toulon, 1905, completed 1908. Machinery by
Pallada sunk by *U 26*, in Finnish Gulf, October, 1914.

1898-3 Armoured Cruisers.

Photo, Abraham & Sons, Devonport.

Reported scuttled by crew, Oct., 1920.

Громобой. **GROMOBOI** (May, 1899). Displacement, 13,430 metric tons. Sheathed and coppered. Complement, 868. Length (*waterline*), 472½ feet. Beam, 68½ feet. *Mean* draught, 27½ feet. Guns (as re-armed): 6—8 inch, 45 cal. (*A*), 20—6 inch, 45 cal., 2—9 pdr. AA., 2—3 pdr. Torpedo tubes: 2 *submerged* (broadside). Armour (Harvey nickel): 6" Belt, 2" Deck, 6" Bulkheads, 4" Lower deck (amidships), 6" Casemates (14), 2" Hoists, 10" Conning tower (Total *about* 2100 tons). Machinery: 3 sets triple expansion. 3 screws. Boilers: ————. Designed H.P. 14,500 = 19 kts. Coal: *normal* 800 tons; *maximum* 2300 tons. Laid down at Baltic Works, 1898; completed 1901; re-fitted 1907, when 2 *above water* tubes were removed, and upper deck guns casemated in. Made 20.1 kts. on first *trials*, with 15,499 H.P., but best recent speed known of is 18.5 kts.

(For particulars see next column.)

Rossiya.

Россія **ROSSIYA** (May, 1896). Displacement, 12,195 tons (sheathed and coppered ... ment, 833. Length (*over all*), 480 feet. Beam, 68½ feet. *Mean* ... feet. Guns (as re-armed): 4—8 inch, 45 cal. (*A*), 14—6 inch, 45 cal., 2—3 pdr., 2 M. ... (Harvey): 8"—5" Belt, 4" Belt (aft), 2½" Deck (Deck flat on belt), 6" Bulkheads, 4" ... (amidships), 6" Battery bulkheads, 2" Casemates to 8", 2" Screens in battery. Ma... sets vertical triple expansion. 3 screws. Boilers (fitted 1907): 32 Belleville ... Designed H.P. 14,500 = 19 kts. Coal: *normal* 1000 tons; *maximum* 2500 ... down at Baltic Works, 1893; completed 1898; re-fitted 1907, when all torpedo ... removed. Made 20.25 kts. on first *trials*, with 18,446 H.P.

1913 Light Cruisers.

Note.
No information available regarding present condition of these ships.

ADMIRAL BUTAKOFF (*not launched*), **ADMIRAL GREIG** (Dec. 1916), *ADMIRAL* ... (*not launched*), and **SVIETLANA** (June, 1915).

Displacement, 6800 metric tons. Complement,
Length, 507¾ feet. Beam, 50½ feet. Draught, 18¼ feet.

Guns:
15—5.1 inch, 55 cal.
4—4 in. (antiaircraft).
4—3 in.
4 machine.
Torpedo Tubes:
2 *submerged*.
Carry 100 mines

Armour:
3" Belt ...
1" Deck ...
3" Gun sh...
" Conning...

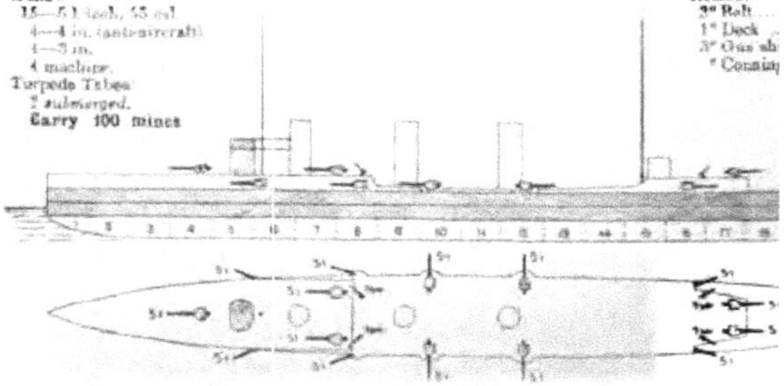

Machinery: Brown Curtis or Parsons turbines. Boilers: 25 Yarrow. H.P. 50,000 = 29.5 ...
normal, tons; *maximum*, 1170 (with oil fuel).

Name	*Builder and Machinery*	*Laid down*	*Com-pleted*	*Trials*	*Turbine...*
Ad. Butakoff	Putilov Yard, Petrograd	Nov. '13	?		
Ad. Greig	Baltic Works, Reval*	Dec. '13			
Ad. Spiridoff	Putilov Yard, Petrograd	Nov. '13			
Svietlana	Baltic Works, Reval*	Dec. '13			

General Notes.—Built under 1912 Naval Programme. Are equipped for carrying seaplanes. *Ad. B...* *Spiridoff* were originally intended for Siberian Fleet in Pacific. Four sister ships of the *Admiral* ... building in Black Sea.

* Towed to Petrograd, 1918.

Laid up in the Gareloch, Firth of Clyde, 1919.

ASKOLD (April, 1900). 6000 metric tons. Complement, 600. Dimensions: 430¼ (*o.a.*) × 55¼ × 20¾ feet (*mean draught*). Armament: 12—6 inch, 45 cal., 10—11 pdr., 2—3 pdr. AA., 2 M.G. Torpedo tubes (18 inch): 2 *submerged* (broadside), 2 *above water*. Fitted for Minelaying. Armour: 3" Deck, 4" Funnel bases, 6" C.T. Machinery: 3 sets triple expansion. 3 screws. Boilers: 18 Schulz-Thornycroft. Designed H.P. 19,500 = 23 kts. Coal: *normal*, 720 tons; *maximum*, 1050 tons. Was in Far East, 1914, and served with Allied Fleets during the War.

BOGATYR (January, 1901). 6650 tons. Complement, 565. Dimensions: 433 (*w.l.*) × 54½ × 20¾ feet (*max.* draught). Length (*over all*), 440 feet. Armament: 16—5.1 inch, 55 cal., 4 M.G. Torpedo tubes (18 inch): 2 *submerged* (broadside). Armour: 3" Deck, 3" Turrets, 3" Casemates (4). Machinery: 2 screws. Boilers: Normand-Sigaudy. Designed H.P. 19,500 = 23 kts. Coal: *normal*, 900 tons; *maximum*, 1270 tons. Fitted for Minelaying. *Oleg* sunk by British C.M.B. *No. 4*, June 17th, 1919.

(Діана: Аврора.) **DIANA** (October, 1899) & **AVRORA*** (May, 1900). 6830 metr Complement, 573. Dimensions: 410 (*w.l.*) × 55 × 21¼ feet (*mean d* Armament: *Avrora*, 10—6 inch, 45 cal., 5—6 pdr. AA., 2 M.G.; *Diana*, 10—ß 4—6 pdr., 4—6 pdr. AA. Carries 125 mines. Torpedo tubes (18 inch), no tubes in 2 *submerged* (*broadside*) in *Avrora*. She is also fitted for Minelaying. Machinery horizontal 3-cylinder. Designed H.P. 11,600 = 20 kts. 3 screws. Boilers: 24 B Coal: *normal*, 960 tons; *maximum*, 970 tons. Looted by Bolshevist crews, 19 abandoned in the Neva, 1918.

* Reported mined and sunk in Baltic, Aug., 1920.

Armoured Gunboat.

KHRABRI (1895). 1740 tons. Complement, 197. Guns: 1—6 inch, 45 cal.; 5—5.1 in cal.; 2—1 pdr., 1—1 pdr. AA., 2 M.G. Armour: 5" Harvey, waterline belt ami 3" ditto aft. Designed H.P. 2000 = 14 kts. Nielausse boilers. Coal: *maximum*, 16 Built by Burmeister & Wain, Copenhagen.

Note.—Re-fitted and re-armed, 1916. Appearance may be altered.

(*For Gunboats v. next page.*)

For Amur River Monitors of the *Groza* type, and the Caspian Sea gunboats *Ardahan* and *Kars*, see later pages.

Gunboats.

Note.—Mainmast has been removed.

(Abandoned at a Finnish Port, 1918.)

GILYAK (1907). 875 tons. Dimensions: 218¼ (*w.l.*) × 36 × 8 feet (*max.* draught). Guns: 2—4.7 inch, 4—11 pdr. Used to carry about 50-60 mines. I.H.P. 900 = 12 kts. Coal: 135 tons.

(Disarmed, April, 1919.)

KHIVINETZ (1905). Displacement, 1340 tons. Complement, 161. *Guns*: 4—4.7 inch, 2—3 pdr. AA. Machinery: Triple expansion. 2 screws. Boilers: 8 Belleville. Designed H.P. 1400 = 13½ kts. Coal: *normal*, 100 tons; *maximum*, 190 tons.

Gunboats (Mine Layers).

Photo, Lieut.-Col. Alan Burgoyne, M.P.

(Disarmed.)

GROZIASTCHI (1890). 1640 tons. Complement, 178. Guns: 4—6 inch, 2—11 pdr. Armour: 5" Belt amidships, 3" Aft, with 3¼" Bulkhead forward. Designed H.P. 2000 = 14 kts. Coal: 150 tons. Belleville boilers. Carries 20-30 mines.

Note.—Re-fitted and re-armed, 1916. May now be of different appearance.

(At Vladivostok.)

MANDJUR (1888). 1440 tons. Guns (old): 2—8 inch, 1—6 inch, 4—9 pdr. Torpedo tubes: 1—14 inch (in bow). Designed speed: 13 kts. Coal: 250 tons. Complement, 160.

Fitted as Minelayer. Carries about 50 mines.

River Gunboats—Location unknown.

8 *Gunboats*: (Names or Numbers unknown, built 1915-17). —— tons. Dimensions: 66 × 19 × 2½ feet. Guns: 2—3 inch (mountain type), 2 M.G. Speed: 12½ kts.

Despatch Vessels.

(Both abandoned in Finnish Ports, 1918.)

POSADNIK (1892) and **VOEVODA** (1893), 400 tons. 2 or 3—11 pdr., 2 M.G. Designed H.P. 3500 = 21 (now). Coal: 80 tons.

No Photos.

Lastotchka (Yarrow, 1904). 120 tons. Speed 18 kts. Tu

Razvyedchik and ***Dozorni*** (both 1904). 100 tons. Guns: 1—1 pdr., 1 machine. Speed, 16 kts. Both re-fitte

(May be in Finnish Navy.)

BAKAN (1896). 883 tons. Guns: 2—11 pdr., 4- 800 = 11 kts. Coal: 130 tons. Complemen

Note.—Re-fitted and re-armed, 1916. Appearance may be d

Also:—

Porashayustchi (ex-Destroyer of *Pritki* type.

Petrol T. B. No. 1.

Abrek, Berkut, Harpun, Kondor, Kontoir, Nanet, Periskop, Bogatvola, Spatnik, Yakor.

RUSSIAN DESTROYERS (*Eskadrenny Minonossetz* if over 400 tons, *Minonossetz* if under).

21 Building?
97* Completed. } = **118** **Destroyers.**

* *Konstantin, Svoboda* reported sunk; if so totals are 95 completed and 116 total T.B.D.

No.	Type	Date	Displacement	H.P.	Max. speed	Coal or Oil	Complement	Tubes Mines	Max. Draught
4	"Hogland type"	1916–?	1550	34,000	*32t*	tons *500*	..	12/80	11
3	*Zabiaka*	1912–15	1610	32,000	*35t*	*400*	110	12/80	9½
4	*Izyaslav*	1912–?	1350	32,700	*35t*	*150*	110	9/80	9¾
27	*Gavryil*	1912–?	1260	30,000	*35t*	*150*	110	9/80	9¾
1	*Novik*	1910–12	1200	33,000	*32t*	400	140	8/80	10¼
3	*Kondratenko* class (K)	1904–06	625	7300	25	190	98	3/50	9½
6	*E. Bukharski* class (K)	1904–07	580	6500	25	150–205	94	3/25	8½
7	*Ukraina* class	1903–07	500	6200	25	130	85	*3/16*	8
2	*Leit. Sergieff* (S)	1903–05	350	6000	29	95	64	3	7½
10	*Vnimatelni* (S) / *Bestrashni* (S)	1904–07 / 1899–01	355	6000	27	125	64	3/16	6
8	*Lovki* class (N)	1904–05	330	5700	27	60	65	2/15	7
2	*Grosovoi* class (N)*	1899–01	310	5700	26	80	59	2/6	8
10	*Dyelni* class (Y)	1904–06	350	5700	26	100	64	2/14	6
4	*Bravi* class (Y)	1902–06							
5	*Tverdi* class (Y)*	1905–06	300	3800	25	60	59	2	6
4	*Smyeli* class (Y)*	1899–03	380	3800	26½	70	52	1/18	6½
8	*Pritki* class (Y)*	1895–01	220	3800	27	60	52	1/18	7½

* These are officially rated as torpedo boats.

Types:—(K) = Krupp. (S) = Schichau. (N) = Normand. (Y) = Yarrow.

4 Hogland type (Mine Layers).

Condition unknown.

4 boats (not named), begun 1916, by Petrograd Metal Works. 1550 tons. Dimensions: 324¾ × 31¼ × 11 5—4 inch, 2—9 pdr. AA., 4 M.G. Torpedo tubes (18 inch): 12 in 4 triples. To carry 80 mines c 34,000 = 32 kts. Turbines. Oil: *about* 500 tons. Belong to 1916 Programme.

3 Zabiaka type (Mine Layers).

Photo wanted.

3 *Petrograd Metal Works*: **Orphei, Pobieditel*, Zabiaka** (all 1914). 1610 tons. Dimensions: × 9½ feet. Guns: 4—4 inch, 1—2 pdr. AA. pom-pom, 2 M.G. Torpedo tubes (18 inch): 12 in 4 tri 80 mines. Designed S.H.P. 32,000 = 35 kts. Oil: 400 tons. Complement, 110.

* In service, 1920.

2 (+ 2) Isyaslav type (Mine Layers).

1919 *Photo.*

4 *Reval Shipbuilding Co.:* **Bristshislav, Feodor Stratilat, Isyaslav, Priamislav,** 1350 tons. Dimensions: 344½ × 31½ × 9¾ feet. Designed H.P. 32,700 = 35 kts. Parsons turbines and Normand boilers. Armament: 5—4 inch, 1—9 pdr. AA., 2 M.G. Torpedo tubes (18 inch): 9, in 3 triple deck mountings. Oil: 150 tons. Builders are sometimes referred to as Lange & Becker, Böcker & Co., or Chantiers de Nord-Ouest.

Note.—*Avtroil* captured by British, now Esthonian *Lennuk*. *Feodor-Stratilat* towed to Petrograd 1918, not yet finished *Bristshislav* not finished.

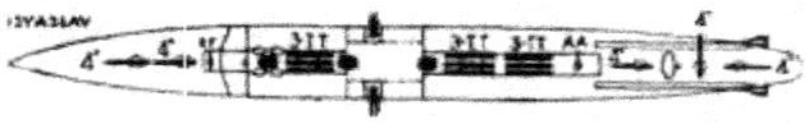

Note for "Isyaslav type" above, and "Gavryil type" below:—Were designed to mount 2—4 inch and 12 tubes in 4 triples, but during completion 2 or 3—4 inch added and one triple T.T. mounting omitted.

12 (+ 15 ?) Gavryil type (Mine Layers).

V. Wambola, Esthonian Navy Section.

Konstantin, Svoboda, reported sunk.

4 *Petrograd Metal Works:* **Azard,* Desna, Lieutun, Samsun.***

7 *Putilov Works, Petrograd:* **Garibaldi** (ex-*Leit. Ilyin*) ¶ **Kapitan 2r. Isylmettiev,* Kapitan 2r. Korn,* Kapitan 2r. Konon Zotov,* Kapitan 2r. Kroun,* Kapitan-Leit. Belli,* Leit. Dubassov*.**

7 *Russo-Baltic S. B. Co. (Reval):* **Gavryil, Konstantin†, Mietchislav, Mikhail,* Sokol,*† Svoboda†, Vladimir.*†**

6 *Ziese Yard, Riga:* **Hogland,*†; Grengamn,*†; Khios,*†; Kulm,*†; Patras,*†; Rymnik,*†; Smolensk,*†; Stirsuden,*†; Tenedos.*†;** (Condition of these boats unknown.)

* Not finished, 1918. Towed from Reval (?) or Riga (?) to Petrograd, February, 1918. †Reported sunk 1920, but not deleted from totals. ¶ In service, 1920.

Displacement: 1260 tons. Dimensions: 314½ × 30½ × 9¼ feet. Designed H.P. 30,000 = 35 kts. All have turbines and oil fuel only. Armament: 4—4 inch, 1—2 pdr. AA., 2 M.G. Torpedo tubes (18 inch): 9, in 3 triple deck mountings. Carry 80 mines. Complement, 110. Oil: 150 tons. 1912 Programme.

Note.—Bolshevist *Spartak*, ex Kerensky *Mikula Maklei*, ex Czarist *Kapitan Kinsbergen*, captured by British and now Esthonian *Wambola*.

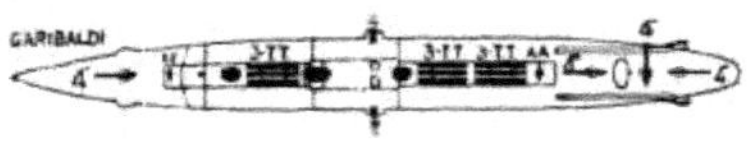

1911 Special Boat (Mine Layer).

1 *Putilov-Vulkan boat:* **Novik** (Putilov Yard, Petrograd, 1911). 1200 tons. Dimensions: 336 × Designed H.P. 33,000 = 32 kts. A. E. G.-Curtis turbines and 400 tons oil fuel only. Compleme *Trials:* 42,800 = 36.3 kts. Armament: 4—4 inch, 1—2 pdr. AA., 4 M.G., 8—18 inch tubes in 4 twi Carries 80 mines.

Note.—Built at the expense of the Putilov Co., as an experiment, and taken over for the Russian N

3 (Gen. Kondratenko type (Mine Layers.)

3 "Kondratenko" class, named **Gen. Kondratenko, Sibirski-Strelok*** (Helsingfors, 1906), (Crichton Yard, Abö, 1905). 625 tons. Dimensions: 246.7 × 26.9 × 8.5 feet. Designed H Coal: 215 tons. Complement, 101-98. All armed with 3—4 inch, 2—3 pdr., 3 M.G., and 3- one twin and one single mounting. Carry 50 mines.

Note.—Much the same build as *Amurets* class, (next page), but have a raised poop. *Okhotnik* lost

*In service, 1920.

6 Emir Bukharski type (Mine Layers).

4 Later boats, named ***Gaidamak,**** ***Vsad'nik.**** (Krupp-Germania Yard, Kiel, 1905), ***Amuretz,**** (Riga, 1905), ***Ussurietz**** (Helsingfors, 1907), the last two being built from materials supplied by Krupps. 570 tons. Dimensions: *Gaidamak* and *Vsadnik*, 233 × 24.2 × 7.5 feet. *Amuretz* and *Ussurietz*, 233 × 23.6 × 7.8 feet. Designed H.P. 6500 in first two, 6200 in second two = 25 kts. Coal: 205 tons. Complement, 97–94. All armed with 2—4 inch, 1—3 pdr., 1—1 pdr. AA., 4 M.G., and 3—18 inch deck tubes. Carry 25 mines.

*In service, 1920.

(With Bolshevist Volga Flotilla.)

2 Earlier boats, named ***Yakob Sverdlov*** (ex *Emir Bukharski*), ***Karl Liebknecht*** (ex *Finn*), (Helsingfors, 1905-4). 580 tons. Dimensions: 237.8 × 26.9 × 7.7 feet. Designed H.P. 6200 = 25 kts. Machinery of *Y. Sverdlov* by Schichau. Coal: 150 tons. Complement: 97–94. Armament: 2—4 inch, 2 or 1—1 pdr., 4 M.G., and 3 deck torpedo tubes. Carry 25 mines. *Moskvityanin* sunk in Caspian Sea, 1919.

7 Ukräina type (Mine Layers).

Fore funnel now raised.

7 "Ukräina" class, named ***Donskoi-Kasack*** (1907), ***Storeguchi Strashni*** (all 1905), ***Tur*** *Turkmenets-Stavropolski*),* ***Ukraina, Voiskovoi*** (all 1904), ***Zabaikaletz*** (1907.) All built Yard, Riga. 500 tons. Dimensions: 240 × 23¼ × 7½ feet. Designed H.P. 6200 = 25 kts. Boilers: Coal: 50 tons *normal*, 135 tons *full load*. Complement, 88–85. Armament: 3—4 inch, 1—1 p and 2 torpedo tubes (18"). Carry 16 mines.

*Bolshevist Volga Flotilla.

2 Leit. Sergieff type.
10 Vnimatelni type (Mine Layers).

2 *Schichau, Danzig.* ***Leit. Sergieff,**** ***Kapitan Yurassovski**** (both launched 1905). 350 tons. Di 202½ × 22½ × 7½ feet. Designed H.P. 6000 = 29 kts. 4 Thornycroft boilers. Coal: 95 tons. Compl All armed with 2—11 pdr., 5—3 pdr., and 3 tubes.

3 *Schichau, Elbing.* ***Bestrashni,**** ***Bezposhtchadni,***† ***Beshumni**** (all launched 1899).

Also:—

7 *Schichau, Danzig.* ***Boyevoi, Burni, Insh. Mech. Svereff, Insh. Mech. Dmitrieff, Vnin Vnushitelni, Vulnositvi.*** (All built by Schichau, Danzig, and launched 1905–6). 355 tons. Dir 208½ × 23 × 6 feet. Designed H.P. 6000 = 27 kts. 4 Schichau Boilers. Coal: 125 tons. Complement armed with 2—11 pdr., 6 M.G., and 2—18" tubes. Carry 16 mines.

*In White Sea. †At Vladivostok.

8 Lovki type (Mine Layers).

8 "*Lovki*" class, named **Lovki,* Likoi,*** (Normand Le Havre, 1905), **Logki, Iskusni, Krepki*** (La Seyne, 1905), **Metki,* Molodetski, Mostchni** (F. & C. de la Med. Havre, 1905). 330 tons. Dimensions: 185¾ × 21 × 5½ feet. Designed H.P. 5700=27 kts. Boilers: 4 Normand. Coal: 100—110 tons. Complement, 65. All armed with 2—11 pdr., 2 M.G., and 2 deck tubes (18"). Carry 15 mines.

*In Service, 1920.

2 Grosovoi type (1 boat Mine Layer).

(Are officially rated as Torpedo Boats. Both in British Ports.)

2 "*Grosovoi*" class, named **Grosovoi, Vlastni.** (F. et C. de la Med., Le Havre, 1900). 310 tons. Dimensions: 185¾ × 19½ × 6½ feet. Designed H.P. 5700=26 kts. Boilers: 4 Normand. Coal: 60 tons. Complement, 59. Both armed with 2—11 pdr., 1 M.G., and 2 deck tubes (18"). *Vlastni* carries 4 mines.

31 Yarrow or Modified Yarrow type (1907-1894.)

No Photos available.
v. Silhouettes for appearance.

10 "*Dyelni*" class, named **Dyelni,* Dostoini,* Dieyatelni,* Rastoropni,* Razyas[illegible] Stroini,* Storoshevoi,*** (launched 1906-5). **Gremyastchi, Vidni** (1905-6). Re-armed [illegible] 2 M.G. Tubes: 2—18". Carry 14 mines.

*Bolshevist Volga Flotilla.

4 "*Bravi*" class, named **Bravi,† Boiki, Bodri,† Grosni†** (launched 1903-2). No mainmast.

†At Vladivostok, disarmed.

Above T.B.D. all built by Nevski Works, Petrograd, and displace 350 tons. Dimensions: 210 × 21 × 6½ fe[illegible] H.P. 5,700=26 kts. Boilers: 4 Normand in *Dyelni* class, 4 Yarrow in *Bravi* class. Coal: 80 tons. [illegible] 64. *Bravi* class armed with 1—11 pdr., 5—3 pdr., and 2 deck torpedo tubes, but *Bravi, Boiki, Bo*[illegible] tube.

PRITKI, PODVIJNSI, POSLUSHNI, as above Photo: other boats of *Pritki* Class same but with strai[illegible] *Smyeli* Class, straight stems, no mainmast.

Following Boats officially rated as Torpedo Boats:

5 "*Tverdi*" class, named **Leit. Maleieff,† Insh. Mech. Anastassoff,† Tverdi,† Totchni,†** (Okhta Works, Petrograd, 1906; re-erected at Vladivostok).

4 "*Smyeli*" class, named **Smyeli,† Serditi,† Statni,† Skori,†** (Nevski Works, Petrograd, 1901, Port Arthur).

8 "*Pritki*" class, named **Pritki*** (ex *Sokol*, Yarrow, Poplar, 1895), **Podvishni,§ Poslushni,§ Protchni*** (Ishora Works, Petrograd, 1900-1899). **Retivi,* Ryani,§ Ryezvi§** (Nevski Wo[illegible] 1901).

Above T.B.D. displace 180-220 tons (see Summary Table). Dimensions: about 190 × 18½ × 5½-6 feet, 27-25 kts. (4000=28 kts. in *Pritki*). Boilers: 4 locomotive in *Tverdi* class, 4 Yarrow in others. [illegible] Complement, 58-62. All armed with 2—11 pdr., 2 M.G., 2—18" tubes in *Tverdi* class; others 1-[illegible] 18 mines.

*Bolshevist Volga Flotilla. †At Vladivostok, disarmed.
§May be in Finnish Navy now.

RUSSIAN (BALTIC) TORPEDO BOATS.

No.	Type	Date	Displacement	H.P.	Max speed	Fuel	Complement	T. Tubes	Max. draug't
5	*Nos. 220—214* (N)	1902-04	150	5700	29	30	25	2	5
1	*No. 213* (N)	1902	100	3800	24	60	26	2	5
3	*Nos. 142, 129, 128* (N)	1894–98	120	2000	21	20	20	3	6½
1	*Petrol T.B. No. 1*	1904	..	..	..	..	..	..	..

(N) = Normand type.

Nos. 214—220, 222.

8 *Normand* type. **Nos. 214–220, 222** (1902-4). Guns: 2—3 pdr. and 2—15 in. tubes (double mounted). Some of these boats may now be in the Finnish Navy.

1 Normand type: **No. 213** (1902). 2—1 pdr. 2—15 inch tubes.

3 Normand type: **Nos. 142, 129, 128** (1896-7). 2—1 pdr. 3—15 inch tubes.

2 pole masts.
No funnel.

1 ——— type. **Petrol T.B. No. 1.** 3 small guns. 1 tube. Used now as a despatch vessel.

29 { 20 building? / **15** modern / **14** old } **Submarines** (*Podvodnija Lodki.*)

(Inclusive of Baltic and Siberian flotillas.)

No.	Type	Date	Displacement	H.P.	Max. speed kts.	Fuel tons	Tubes and Gear	C...
10	*G 1—10* (H)	?	960/1280	4200/700	16/9	60	16*	..
10	*B 1—10* (B)	?	970/1285	3150/700	17/9	58	16*	..
1	*Sviatoi Georg* (LA)	1916–17	255/376	700/500	13/8.5	For 1200 miles / For 2080 miles	2	..
2	*Ford* (B)	1912–17	630/784	840/500	11/9	40	4*	..
2	*Edinorog* (B)	1912–17	Do.	2640/900	16/9	40	12	..
2	*Volk* (B)	1912–16	Do.	500/445	10/9	40	12	5
7	*Tigr* (B)	1912–17	Do.	500/900	10/9	40	8	..
4	*Alligator* (L)	1908	150/155	160/400	8/7	15	4	2[illegible]
1	*Minoga* (B)	1908	117/147	480/140	11/5.5	5	2	2[illegible]
2	*Makrel* (B)	1907	150/200	120/100	7.5/5	2¼	4	1[illegible]
4	*Byeluga* (H)	1904	110/120	160/60	8.5/5	2	1	16
2	*Kataska* (B)	1904	150/200	120/100	8/7	3	4	20
1	*Delfin* (B)	1903	115	..	10/8	..	2	12

(B) = Bubnov type. (LA) = Laurenti-Ansaldo type. (L) = Lake type. (H) = Holland type.

* Mine Layers.

10? Holland Type (Mine-layers).

Doubtful if ever begun.

10 *Holland* boats: *G 1–10*, ordered from Nobel & Lessner 1916. 262 × 22½ × 14 feet. Guns: 2–12 or 14 pdr., 2 M.G. Tubes: 16. Dropping gears: 6. Mines: 40. Other details as Table.

10? Bubnov Type (Mine-layers).

Doubtful if ever begun.

10 *Bubnov* boats: *B 1–4* (Baltic Works, Petrograd), *B 5–10* (Russo-Baltic Works, Reval). 264 × 23 × 14 feet. Guns, tubes, dropping gears, mines, as *G 1–10* above. Other details as table.

1 Laurenti Boat.

1919 *Photo by courtesy of Messrs. Ansaldo-San Giorgio.*
(*In White Sea.*)

1 *Laurenti-Ansaldo* type: **Sviatoi Georgi** (Ansaldo-San Giorgio, Spezia, 1916 or 1917). **Dimensions: about** 150 × 14 × 10 **feet. Machinery: 2 sets of 350 H.P., 6-cyl. 2-cycle Fiat Diesel engines on *surface*. Radius of action: about 1500 miles at cruising speed on *surface*, 70 or 80 miles at a low speed *submerged*. Armament: 1 small Q.F. and 2 bow 18-inch torpedo tubes. Is generally similar to Spanish *Monturiol* and Brazilian *F* classes. Arrived at Archangel, Sept., 1917.**

13 "Improved Bubnov Type (1 boat Mine-layer).

Photo Wanted.

2 Special boats: **Forel**, Baltic Works, Petrograd, 1917–1?; **Ersh** Nobel & Lessner, Reval, 1917–(?). [illegible] Guns: 1–6 pdr., 1–1 pdr. A.A., 1 M.G., 4 [illegible] 40–50 Mines.

Photo Wanted.

2 *Edinorog* type: **Smaya, Yasy**, (Nobel & Lessner, Reval, 1912–1?). Guns: 2 or 1–6 pdr., 1–1 pdr. Tubes: 4. Dropping Gears: 8.

Photo Wanted.

2 *Volk* type: **Vyepr, Volk** (Baltic Works, Petrograd, 1912–16). Guns: 2–6 pdr., 1–1 pdr. A.A. Dropping Gears: 8. Have Diesel engines removed from Amur River Gunboats.

Photo Wanted.

7 *Tigr* type: **Kuguar, Leopard, Pantera, Ryss, Tigr, Tur, Yaguar** (Nobel & Lessner, R[illegible] Guns: 2–6 pdr., 1–1 pdr. A.A., 1 M.G. Tubes: 4. Dropping Gears: 8.

General Notes: All launched 1915-16. Dimensions: 22? × 11½ × 12½ feet. Other details as Table. *Bars, G*[illegible] *Ugor* and *Edinorog* war losses.

4 Lake Type.

Photo, Lieut.-Col. Alan H. Burgoyne, M.P.

Present disposal uncertain.

4 *Lake* type: ***Alligator, Drakon, Kaiman, Krokodil*** (1908). Dimensions: 131 × 14 × 16 feet. Naptha (or light oil) motors for surface. Armament: 1—3 pdr., 1 machine, 4—18 inch tubes, 2 dropping gears for 18 inch torpedoes.

1 Bubnov Type.

No photo available.

Bolshevist Volga Flotilla.

1 *Bubnov* type: ***Minoga*** (1908, re-built 1912). Dimensions: 107 × 18⅔ × 8 feet. Diesel engines. 1—1 pdr. gun. Two 18 inch tubes.

2 Bubnov Type.

No photo Available.

Bolshevist Volga Flotilla.

2 early *Bubnov* type: ***Makrel, Okun*** (1907). Displacement, 150/200 tons. 4—18 inch dropping gears.

4 Old Holland Type.

No photo available.

Present disposal uncertain.

4 early *Holland* type: ***Byeluga, Piskar, Sterliad, Tchuka*** (1904). 1—1 pdr. gun. 1—15 incl

2 Old Bubnov Type.

No photo available.

Bolshevist Volga Flotilla.

2 early *Bubnov* type: ***Ugor*** (ex *Graf Sheremetier*), ***Kasatka,*** (Built about 1904 re-built 1914-15), and *Okun*, but have 1—1 pdr. and 1 M.G.

1 Old Bubnov Type.

Illustrated in
"Fighting Ships" up to
1915-16.

In White Sea.

1 old *Bubnov* boat ***Delfin*** (1903). Used to be armed with 2—15 inch dropping gears. Present value almc

Motor Patrol Boats.

Total number unknown. Some in the Baltic; others in the White and Black Seas. Builders: chiefly U.S. firms, 1915-16. —— tons. Dimensions: 60 × 10 × 2½ feet, but others may be 80-85 feet long. Machinery: Twin or triple sets of petrol motors (12-cyl., Standard, Dusenberg, Van Blerck, &c.). H.P. 500-600 = 25-26 kts. designed speed. Some made 30.1 kts. on trials in smooth water, but best sea speed is only about 20 kts. for short periods. Fuel: 4000 galls. petrol = 600 miles at cruising speed. Guns: One (11 pdr., 6 or 3 pdr. ?), and perhaps M.G. Depth charges may also be carried. Crew, about 8-10.

Coast Patrol Boats:—

Barsuk	*Moscn*	*Strel*
Dela	*Pulash*	*Tumba*
Gornostai	*Pistolet*	*Tsapfa*
Griff	*Pulee*	*Vladimir*
*Kobchik**	*Ruslan*	*Voron*
Korsun	*Shaska*	*Vuidra*
Kunitza	*Sobol*	*Yastreb*

* *Abandoned in Finnish Port, 1918.*

Destroyer Depot Ships.

Repair Ships, **ANGARA** and **KAMA** attached to T.B.D. Flotillas.

Submarine Depot Ships.

Cruiser **DIANA**: for details, see preceding pages with other cruisers.

No photo available.
(Resembles German "Vulkan.")

VOLKHOV (Putilov, 1913). 2400 tons. Dimensions: 315 × 69 × 11¼ feet. Diesel engines. H.P. 1200 = 10 kts. Radius 3600 miles. Can raise 1000 tons.

Note.—Carries all essential stores for submarines (oil fuel, reserve accumulators, &c.), and has compressed air, distilling and charging plant, workshops, &c.

KHABAROVSK (1895). 2800 tons. H.P. 1800 = 12·5 kts. Guns: 2—11 pdr. 2—3 pdr. 2 M.G. Coal: 390 tons. Complement, 132.

EVROPA (1878). 3200 tons. H.P. 12,000 = 13 kts. Guns: 1—3 pdr.; tubes: 2 *above water*, 1 *submerged*. Coal: 390 tons.

Seaplane Carrier.

No photo available.

ORLITZA (——). 3000 tons. 10 kts. speed. G pdr., 2 M.G. Carries 4 seaplanes.

Mine Layers.

No photo available.

(At Vladivostok.)

DIOMID, PATROKL, ULIS (1913-14). 600 tons. 11 Dimensions: 160 × 26¼ × feet. Guns: Not —— = 12 kts. speed.

AMUR (1907). 3608 tons. Complement, 318. Dimen 46 × 14½ feet. Armament: 9—4·7 inch, 4 M.G. 17 kts. Boilers: 12 Belleville. Coal: 670 tons. mines. Yenesei of this type lost in the war.

(Continued

Mine Layer *(continued)*.

VOLGA (1905). 1711 tons. Complement, 266. H.P. 1600 = 13 kts. Babcock boilers. Armament: 4—3 pdr. Carries 236 mines. Coal: 160 tons.

No photo available.

USSURI (1901). 3200 tons. Complement, 210. Dimensions: 236 × 36 × 15 feet. Guns: 3—11 pdr., 4—3 pdr., 2 M.G. H.P. 1056 = 10 kts. Carries 500 mines.

No photo available.

(Transport, at Vladivostok.)

SHILKA (Elswick, 1896). 3500 tons. Complement, 205. Dimensions: 278 × 39.4 × 17 feet. Guns: 4—4.7 inch, 8—11 pdr., 4 M.G. H.P. 1650 = 11 kts.

No photo available.

(At Vladivostok.)

MONGUGAI (Flensburg, 1891). 2500 tons. Complement, 193. Dimensions: 202 × 30 × 18 feet. Guns: 7—3 pdr. H.P. 6000 = 9 kts.

Mine Layer *(continued)*.

ONEGA (ex *Gertzog Edinburgski*, 1875). 4840 tons; and **NAROVA** (ex *General Admiral*, 1872). 5030 tons. Speed: 12-13 kts. Guns: 4—11 pdr., 4 M.G. Complete 6" iron belt. Carry 600 mines.

Also:—

Ilmen, Lena, Lovat, Luga, Msta,* Svir,* Ural.*

*Abandoned in Finnish Ports, 1918.

Mine Sweeper.

No photo available.

ISKRA, PLAMYA, PATRON (Middlesborough, 1913-14). 500 tons. Dimensions: 146 × 24½ × 10 feet. Guns: 2—11 pdr. H.P. 650 = 11 kts.

MINREP. (Now has a main mast.)

FUGAS, MINREP,* PROVODNIK, SAPAL,* VZRUIV (1911). 150 tons. Guns: 2—3 pdr. Speed: 10 kts.

*Abandoned in Finnish Ports, 1918.

Also about 14 other small craft and about 5 Torpedo Boats fitted as Sweepers.

Repair Ships.

No photo available.

OKA (1914), **KAMA** (1911). 2000 tons. Speed

PETSHORA (1910). 2000 tons. Speed: 10 kt tons.

No photo available.

(In White Sea.)

XENIA (ex *Cayo Soto*, 1900). 5900 tons. Sp Guns. 2—11 pdr., 2—3 pdr., 2 M.G. Coal:

Repair Ships—(*continued*).

ANADYR (1896). 12,000 tons. Speed: 13 kts.

Salvage Ships.

Assistance *Hero* *Karin* *Meteor* *Nordmir* *Protektor* *Silach* *Solid*

Transports.

32 ships laid up in Baltic. Auxiliary Cruiser *Orel* and 3 or 4 Transports at Vladivostok.

Transports—(*continued*).

BORGA (1882). 4700 tons, and **LAKHTA** (1889), 5000 tons. Speed: 10 kts.

SAMOYED (1895). 1300 tons. Speed: 12 kts.

PAKHTUSOV (——). 1100 tons. Speed: 9 kts In White Sea.

TAIMYR (1908), **VAIGATCH** (1908). In White Sea. 1290 tons. Speed: 10 kts. (*Surveying Ships.*)

MURMAN (1912, for White Sea). 500 tons.
(Also other ships, chartered or requisitioned for period of war.)

Hospital Ships.

14 laid up in Baltic.

Oil Tankers.

Aramis *Elena* *Nina* *Olga* *Spinoza* *Tamara* *Tatiana*

Ice Breakers.

ALEXANDER SIBIRIAKOFF
KNIAZ POJARSKI
SOLOVEI BUDIMIROVITCH

Above 3 ships in White Sea. Not certain they are Nav *Kanada* returned to private owners, 1920. Disp Canadian *J. D. Hazen* and *Minto* (sold to Russia ment, 1914-15) not known.

No photo available

TSAR MIKHAIL FEODOROVITCH (Vulkan, Stet 1391 tons *gross*. Dimensions: 244 × 57 × 27½ f = 10 kts. speed.

(*In the Baltic.*)

ERMAK (Elswick, 1898). 8000 tons. Dimensions × 25 feet. 1¼" Steel belt. H.P. 10,000 = 16 3000 tons.

Gorod Revel also reported to be in the Baltic.

Armed Yachts.

Ex-SHTANDART (1895). 5557 tons. Guns: 8—3 pdr. H.P. 10,600 = 21 kts. Belleville boilers.

(*Flagship, Bolshevist C.-in-C. Baltic.*)

KRETCHET (ex-*Polarnaya-Sviedza*, 1890). 3640 tons. Guns: 6—12 pdr. H.P. 5600 = 18·8 kts. Cylindrical boilers.

Ex-ALEXANDRIA (1903). 500 tons. 14·5 kts. Paddle-wheel Yacht.

There are also 9 small Yachts of from 800—180 tons.

Gunnery School Ship.

(*And Transport.*) *Photo, Seward, Weymouth.*

RIGA (ex *Belgravia*), (Blohm & Voss, 1899). 14,500 tons. Complement, 273. Guns: 2—6 inch, 8—4·7 inch, 8—12 pdr., 4—6 pdr., 4—3 pdr., 18 machine. H.P. 4000 = 12 kts. An Enemy Naval Annual refers to a sister ship, *Argun*, of which nothing is known in this country. Also *Vyemi*.

Diving School.

Old *Afrika* (1877), 2600 tons.

Engineers' Training Ships.

(*And Transport.*)

OKEAN (Howaldt, 1902). 11,900 tons. H.P. 11,000 = 18 kts. Coal: 1600 tons. Boilers: 6 Belleville, 6 Niclausse, 3 Yarrow, 2 Thornycroft. Complement, 700.

Also Despatch vessels *Lastotchka*, *Petrol T.B. No. 1*, Auxiliaries *Nicolai Pogolski*, *Bukeir*, *Byelomor*, *Kotka*, *Samoyed*.

Mine and Torpedo School

NIKOLAIEFF (ex-*Palatia*) (1894). 13,500 tons. H.P. 4900 = 13 kts. Guns: 2—11 pdr., 4—4 above water.

Also *Angara*, *Onesbodtelna*, *Khoper*,* *Terek*,* Motor

*Mine-layers.

Old Training Ship.

VOIN (1893). 1250 tons. Guns: 4—12 pdr. Sp

AMUR RIVER FLOTILLA.

NO FUNNELS.

Scale: 1 inch = 80 feet.

Korti class (6)
(Patrol Boats).

Scale: 1 inch = 160 feet.

Grosa class (8 ships).

ONE FUNNEL.

Scale: 1 inch = 160 feet.

Buriat class (10 ships).

River Monitors & Gunboats.

Photo, "Motor Ship & Motor Boat."

GROSA, SHKWAL, SHTORM, SMERTCH, TYFUN, URAGAN, VIKHR, VYUGA (——1910). Displacement, 950 tons. Complement, . Length (waterline), 233 feet. Beam, 42½ feet. Draught, 4½ feet. Guns: 2—6 inch (one forward, one aft), 4—4·7 inch in pairs in midship turrets, 4 machine. Armour: 1½" turrets. Machinery: Nobel-Lessner Diesel motors. 1000 h.p. = 11 kts.

Note.—The machinery has been removed from all of the above ships to provide engines for new Submarines built during the War. Their 6-inch guns have been removed for shore service.

River Monitors & Gunboats (continued).

No photo available.
Believed to be in hands of Japanese, 1920.

OROKANIN, MONGOL, BURIAT, VOGUL, SIBIRIAK, KOREL and **KIRGIZ, KALMUIK, SIRIANIN, VOTYAK** (1909). 190 tons. Complement, 40. Dimensions: 164 × 27 × 2 feet. Guns: 2—11 pdr.,* 2 machine,* but last four have 1—4·7 inch* and 1 machine gun.* Designed H.P., 500 = 11 kts. Coal: 145 tons.

* The Bolshevists removed all guns from these ships.

No photo available.

Kopyo, Palash, Pika, Pistolet, Pulya, Shtik.

These are motor patrol-boats (1908) of 24 tons, 16 kts. speed, each armed with 1—3 inch mountain gun and a M.G.

Note.—*Kinjal, Rapira, Sabiya, Shashka*, transported to Black Sea and Danube during the War.

FLOTILLAS on

(a) **Pripet River,**

(b) **Lake Ladoga,**

(c) **Lake Onega,**

(d) **Lake Peipus,**

(e) **Dvina River?**

Engagements by Flotillas as located above, r[illegible] Bolshevik Military Reports, 1919-20. Nothi[illegible] the ships constituting these Flotillas, but they [illegible] to be mostly River Steamers, Barges, small Mo[illegible] Armed with field and machine guns. In some ca[illegible] crude "armouring" composed of wooden b[illegible] rails.

BLACK SEA SECTION

(INCLUSIVE OF CASPIAN SEA AND VOLGA RIVER FLOTILLAS.)

1919–20 NOTES.

(Continued from 1919 "Fighting Ships," p. 576).

By the capture of Odessa and Nicolaieff, early in 1920, the Bolshevik Forces secured possession of various Warships. These vessels are the Dreadnought **DEMOKRATIYA,** the four Light Cruisers of the *Admiral Nakhimoff* Class, Destroyers ***Levkos*** and ***Korfu,*** Submarines ***AG 23, 24, 25, 26, Nerpa,*** the Submarine Depot Ship *Elborovess*, and six more small Auxiliary Craft. None of these ships are completed, but **ADMIRAL NAKHIMOFF** and Submarine ***AG 23*** are nearly ready. It having been reported that the Bolsheviks intended to commission ***AG 23*** at the end of 1920, the British Government addressed a Note to the Moscow Government, giving warning that any Soviet Submarines found at sea, will be attacked on sight by H.M. Warships.

The Bolsheviks have established a Flotilla of some description in the Sea of Azov; its constitution is uncertain, but it is believed to comprise some Armed "K" Barges.

All other Black Sea Warships are attached to General Baron Wrangel's Forces. The various vessels, taken over by the Allied Governments during 1919, were returned to the Russian (White) Forces during 1920. Many of the ships of the "Wrangel Navy" require refits of one to seven months duration before they can become effective fighting units.

As regards deletions effected:—Destroyer ***Zante*** has been completely wrecked on Cape Fintanki. Submarines ***Lebed, Pelikan*** re scuttled at Odessa, February, 1920, on the approach of the Soviet Army, the hulls of both boats being entirely destroyed.

Developments ensuing on the defeat of General Baron Wrangel's Army and the fall of Sevastopol, November, 1920, not vn. It is very probable that more ships were scuttled or disabled before the port was abandoned by the White Forces. The hevik Commissary of the Navy, Raskolnikoff, says four Battleships were captured, with about "ten steamers," but no ships n[illegible] are given. General Baron Wrangel still retains the Cruiser **GENERAL KORNILOFF:** disposal of other ships of the "ngel Navy" not yet ascertained.

NO FUNNELS.

Scale: 1 inch = 80 feet.

K Barges.
(Bolshevik—Silhouette approximate.)

Kinjal type
(Patrol Boats).

ONE FUNNEL. **Scale: 1 inch = 160 feet.**

Kitoboi.

Terets class (3 ships).

Albatros class (2 ships).

Kazbek.
(Disarmed).

Dunai & Beshtau.

Czarevitch ...

TWO FUNNELS. **Scale: 1 inch = 160 feet.**

Machinery of Ships below destroyed:—

Sinop, (? Poluedrovanni.)

Tri Sviatitelia.
(Has been reconstructed. Present appearance probably differs from above.)

Rostislav.

Berezan.

General Alexieff.

Almaz.

Admiral Nakhimoff.
(Bolshevik; almost completed). Silhouette is approximate only.

THREE FUNNELS.

Evstafi class (2 ships).
Machinery destroyed.

Borets Za Svobodu
Machinery destroy

Gen. Korniloff.
P. Merkuriya (Ma...ry destroyed.)

Scale: 1 inch = 1 foot.

DESTROYERS. (All Gen. Baron Wrangel's Forces.)

KAP. SAKEN.

BESPOKOINI.
DERSKI.
BISTRI (engines destroyed).
POSPYESHNI.
PULKI.
GNIEVNI.

ZAVIETNI.
ZAVETNI.
JEEVOI.
JARKI.
ZVONKI.
ZORKI.

SVIREPI.
STROGI.

Scale: 1" = 80 feet. Submarines.

AG 22 (Wrangel).
AG 23—26 (Bolshevik).

SEVASTOPOL. Two slips. Two large docks (487 × 85 × 32 and 445 × 85 × 28). One basin. Admiralty Chart No. 963.

NIKOLAIEFF. Strongly fortified approaches. Channel only navigable for ships of 30 feet *maximum* draught.

NOVORISSISK. Used as Base by Bolshevik Warships.

Other fortified Black Sea bases without yards or docks; Batûm, Kinburn, Otchakoff, Azov, Kertch, Yenikale, Poti (all well fortified).

Commercial harbours at Odessa, Batûm.

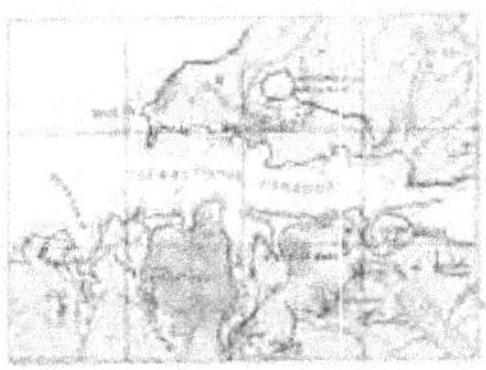

SEVASTOPOL.

Shipbuilders.

NICOLAIEFF SHIPBUILDING & ENGINEERING CO., NICOLAIEFF. (Sometimes, but incorrectly referred to as "Franco-Belgian Co.") 1 concrete Dreadnought Slip, 650 × 100 feet. Five other slips, 400-425 feet. Floating dock, 558 × 136 × 30 feet (32,000 tons). Also a patent slipway. Employees, 3000 (1914).

RUSSIAN SHIPBUILDING CO., NICOLAIEFF. Old Admiralty Yard, taken over 1910. Three big concrete slips, (a) 1000 × 100, feet (b) 950 × 100 feet, (c) 650 × 75 feet. Two other slips and one Martin's Patent Slipway, 1000 tons haul by electric power. Employees, 3800 (1914).

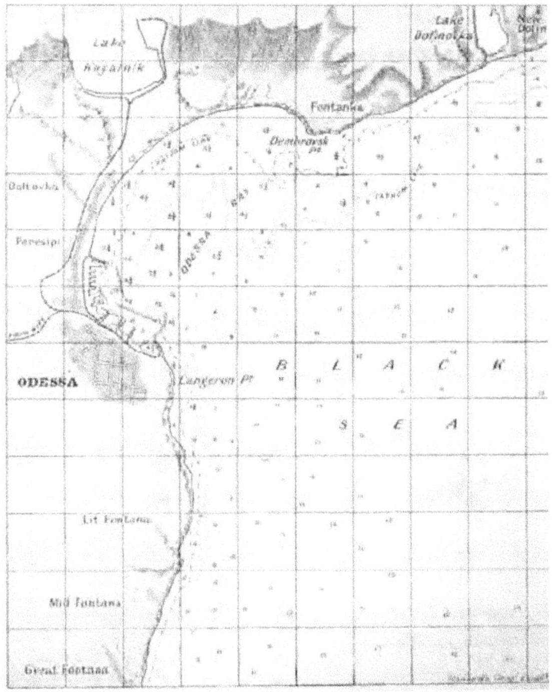

ODESSA. One Private Floating Dock (Russian Steam Navigation and Trading Co.) 381 × (clear between sides) × 19½ feet (on blocks). Patent slip for vessels up to 1000 t
Admiralty Chart No. 603.

GENERAL ALEXEIEFF (ex-*Volya*, April, 1914), and **DEMOKRATIYA** (18th Oct., 1916).

	Length. (*o.a.*)	Beam.	Draught.	Displacement.	
G. Alexeieff	551 feet 2 inches ×	89 feet 8 inches ×	27 feet 4 inches (*max.*)	= 22,600 tons	Complement.—1252.
Demokratiya	597 feet 1 inch ×	95 feet 1 inch ×	29 feet 5 inches	= 27,300 tons	

Guns: (*G. Alexeieff*)
12—12 inch, 50 cal. (A[4])
18—5·1 inch, 55 cal.*
8—11 pdr.*
4—9 pdr. (anti-aircraft)*
4—3 pdr.*
4 machine*
Torpedo tubes (18 inch):
4 submerged.
*In *Demokratiya*:
20—5·1 inch.
4—4 inch AA.
4—M.G.

Corrections to Plans: Fore funnel should be close to foremast, mainmast should be much further aft with C.T. abaft same. Also have overhanging or "yacht" stems. Two foremost 5·1 inch removed from *Volya*.

Armour:
12"–8" Belt (amidships)
5"–4" Belt (bow)
4" Belt (stern)
3" Deck
9"–8" Upper Belt
12" Turrets and Bases
5" Battery
12" Conning tower (fore)
8" Conning tower (aft)

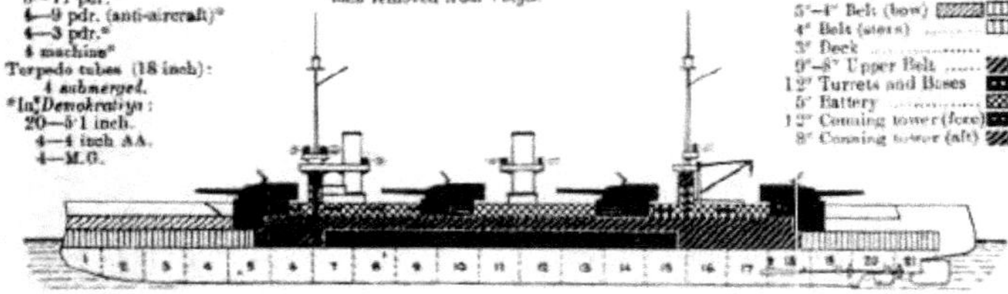

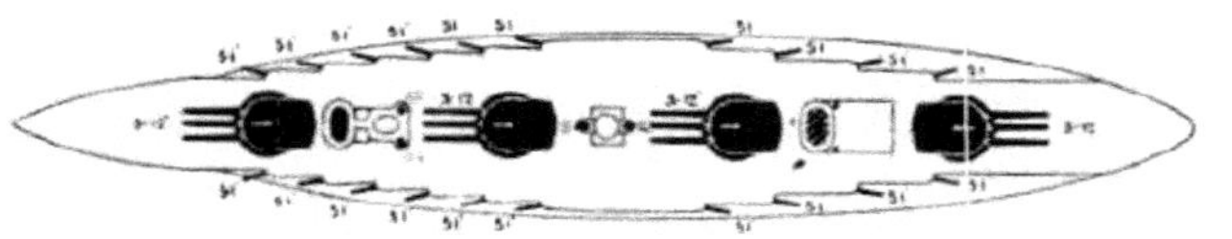

Machinery: Parsons or Brown-Curtis. 4 screws. Boilers: 20 Yarrow. Designed H.P., *Gen. Alexeieff*, 26,500; H.P. *Demokratiya* 27,900. Speed of both, 21 kts. *G. A.*, 2,300 tons maximum. Oil 720 tons. (Fuel for *Demokratiya* unknown).

Name.	Builder.	Machinery.	Laid down.	Completed.	Trials.	Turbines.	Boilers.	Best recent speed.
Gen. Alexeieff	Nikolaieff (B)		Aug. '11	June, '17	2900 = 21		Yarrow	
Demokratiya	Nikolaieff (B)		July '14	**No**			Yarrow	

B=Russo & Ivanoff—Brown (Clydebank).

General Notes.—Design generally resembles *Gangut* class. Original names were *Imperator Alexander Treti* (now *Gen. Alexeieff*) and *Imperator Nikolai Pervi* (now *Demokratiya*). Completion of first two ships *Imp. Maria* and *Imp. Ekaterina Velikaya* (latter re-named *Svobodnaya Rossia*), was accelerated after the beginning of the war. *Imp. Maria*, the first ship of this class, destroyed by explosion at Sevastopol during the war.

Gen. Alexeieff. — *1920 Photo, Mr. H. G. Peck, Midn.*

Details of forward barbette difficult to distinguish in above photograph, but possibl
guns now mounted there, reducing main armament to 11—12 inch guns. Photo
vertically by rigging of H.M.S. *Marlborough*.

Gen. Alexeieff.

Notes.

DEMOKRATIYA not finished. All work stopped on her two years ago; resu
Stopped Jan., 1920. Captured by Bolsheviks, Feb., 1920.

Imperatritsa Maria salved 1919. Funnels, masts, barbettes, C.T. &c., removed.
docked *upside down* at Sevastopol, and is beyond repair as a warship.

IOANN ZLATOUST (AND EVSTAFI). *Photo by favour of Kapitan Kelinsky.*

BORETZ ZA SVOBODU.

EVSTAFI CLASS—(2 SHIPS).

EVSTAFI (Nov., 1906) & **IOANN ZLATOUST** (May, 1906). Displacement, 13,000 metr Complement, 879. Length (*over all*), 385¾ feet. Beam, 74 feet. *Maximum* draught, 27 feet. 4—12 in., 40 cal. (*A*⁴); 4—8 in., 50 cal. (*A*); 12—6 in., 45 cal.; torpedo tubes (18 in *submerged* (broadside). Armour (Krupp): 9"—7" Belt (amidships), 2"—Belt (bow), 7" Bul 2½" Armour deck (behind belt), 3" Deck (forward and aft), 10" Turrets (N.C.) and bases, 6' deck side, 5" Battery (N.C.), 9" Conning tower (5" tube to it.) **Machinery: Destroyed** Coal: *normal* 670 tons, *maximum* 1000 tons. *Evstafi* built at Nikolaieff, *I. Zlatoust* at Sev Both laid down 1903. Completed 1910. Offered for sale by Wrangel's Government, but captured by Bolsheviks on fall of Sevastopol, Nov., 1920.

BORETZ ZA SVOBODU (ex *Pantelimon*, ex *Kniaz. P. Tavritcheski*) (Oct., 1900). Displa 12,840 metric tons. Complement, 741. Length (*over all*), 378½ feet. Beam, 73 feet. Dra feet *maximum*. Guns: 4—12 in., 40 cal. (*A*⁴); 12—6 in., 45 cal.; (other guns and tubes during war). Armour (Krupp): as *Evstafi* class. Machinery: **Destroyed** 25/4/19. Coal: 670 tons, *maximum* 1100 tons. Laid down at Nikolaieff 1898, completed 1903.

РОСТИСЛАВЪ **ROSTISLAV** (Sept., 1896). Displacement, 9920 metric tons. Compleme Length (*waterline*), 341 feet. Beam, 69 feet. *Mean* draught, 25 feet. Length *over* feet. Guns (Obukhoff): 4—10 inch, 45 cal. (*A*³); 8—6 inch, 45 cal.; 18—11 pdr. pdr. AA., 3 M.G. Armour (Harvey): 15"—8" Belt (amidships), 3" Deck (flat on b Fore turret, 9" After turret, 6" Turret bases, 6" Lower deck redoubt, 6" Small turrets, turret bases, 10" C.T. Machinery: **Destroyed** April 25th, 1919. Coal: *normal* 50 *maximum* 800 tons. Liquid fuel also carried. Laid down at Nikolaieff, 1895, and co for sea, 1898. Cost about £850,000.

(*Continued on ne*

Old Battleships—*Continued.*

(Before reconstruction). *Photo by favour of H.I.H. Grand Duke Alexander.*

TRI SVIATITELIA. (Nov., 1893, reconstructed, 1912). 13,530 tons. Complement, 744. Guns: 4—12 inch (40 cal.), 14—6 inch; 2—3 pdr. AA., 2—1 pdr., 3 M. G. Tubes: 2—18 inch *submerged.* Armour (Creusot special): 16" Belt (incomplete) with 12" bulkheads and 3" deck flat on top edge of belt. 16" lower deck redoubt and 5" battery redoubt over. 16" turrets with 12" bases. 12" (fore) C.T. Machinery: **Destroyed** 25/4/19. Coal: *normal* 750 tons; *maximum* 900 tons.

Note.—Another old battleship, *Georgi Pobiedonostz*, is in service still. Only a Port Guardship, and is disarmed.

SINOP (1887, *re-constructed*, 1910). 11,230 tons. Complement, 650. Dimensions: 339 × 69 × 29 feet. Armament: 4—8 inch, 12—6 inch, 2—3 pdr., 4 M.G. Armour (compound): 16"—10" Belt, 12" Redoubt. Machinery: **Destroyed** 24/5/19. Coal: *max.* 850 tons.

Note.—Another old Battleship, *George Pobiedonosetz*, is in service still. Only a Port Guardship, and is disarmed.

Light Cruisers.

Note.—All in Bolshevik hands. For condition, 1920, v. Table.

ADMIRAL NAKHIMOFF (Oct., 1915),* **ADMIRAL LAZAREFF** (June 19…

ADMIRAL KORNILOFF (not launched). *ADMIRAL ISTOMIN* (not launche…

* Almost finished, 1920. Towed to Odessa, Feb., 1920.

Displacement, 6750 metric tons. Complement .

Length (*waterline*), 505 feet. Beam, 50½ feet. Draught, 18½ feet. Length (*over all*), 5…

Guns:
- 15—5.1 inch (55 cal.)
- 4—9 pdr. (anti-aero)
- 4 machine

Torpedo tubes (18"):
- 2 *submerged*

Carry 100 mines.

Plans as "Admiral Butakoff" class, p. 546.

Armour:
- 3" Belt ……
- 1" Deck ……
- 3" Gun Shields …

Machinery: Curtis turbine. Boilers: 13 Yarrow. Designed H.P. 55,000 = 29.75 kts. C… *only*): *normal* tons; *maximum* tons.

Name	*Builders*	*Machinery*	*Laid down*	*Condition, 1920*	B…
Ad. Nakhimoff	Nikolaieff†		Nov. '13	Nearly finished.	Y
Ad. Lazareff	Nikolaieff†		Nov. '13	[illegible]% advanced.	Y
Ad. Korniloff	Nikolaieff*		Sept. '14	Both [illegible]% advanced.	
Ad. Istomin	Nikolaieff*		Sept. '14		

General Notes.—Sister ships to *Admiral Butakoff* class, described on a previous page in Balti…
… Are equipped for carrying seaplanes.

†Begun by Russian S.B. Co., Nicolaieff, to be completed by Nicolaieff S.B. & Engineering Co. *A. Nakhim…* Odessa, February, 1920.

GENERAL KORNILOFF (October, 1902). **PAMIYAT MERKURIYA** (ex-*Kag…* 1905). 6750 tons. Complement, 573. Dimensions: 436 × 54 × 20½ feet. Ar… *Gen. Korniloff*, 16—5.1 inch; *P. Merkuriya* had 16—6 inch and 2—11 pdrs., all … dismounted now. Tubes (18 inch): 2 torpedo tubes *submerged.* Armour: 3" Deck, … and Casemates (4). Machinery: 2 sets 4 cylinder triple expansion. 2 screws. 16 Normand. Designed H.P. 19,500 = 23 kts.* Coal: *normal* 700 tons; *m…* 1100 tons.

* Machinery of *P. Merkuriya* destroyed 25/4/19.

16 (+ 2) Black Sea Destroyers.

No.	Type	Date	Displacement	H.P.	Speed	Fuel	Complement	T. tubes Mines	Max. draught
			tons		kts.	tons			
3	*Cerigo*	'11-'17	1325	29,000	33(?)	390 oil only	121	12	10½
3	*Derski*	'11-'14	1088	25,500	34(?)	270 ,,	131	12	9
3	*Bistri*	'11-'14	1110	23,000	34	350 ,,	110	10/80	9¾
1	*Kap.-Saken*	'05-'09	800	6500	25	260 coal	91	3/80	9
6	*Zavidni* (L.)	'03-'06	350	5700	26	90	71	2/18	7½
2	*Strogi* (Y)	'00-'02	376	3800	26½	70	56	2	6¼

(L) = Laird type. (Y) = Yarrow type. (t) Turbines.

Completing at Nicolaiev

3 boats: ***Cerigo,* Korfu,† Levkos*** † 1325 tons. Dimensions: 303 × 29¾ × 9 feet. Armament: 4—4 inch, 1—2 pdr., (anti-aircraft), 12 tubes in 4 triple deck mountings. Turbine engines. Oil fuel only: 390 tons. *Zante* wrecked 1920.

*Gen. Wrangel's Army. †Bolshevik; at Nicolaiev, 70% complete.

No photo available

3 boats: ***Bespokoini,* Derski,* Gnievni**** (1914). 1088 tons. Dimensions: 307½ × 29½ × 9 feet. Armed as *Bistri* type below, but carry no mines, although *Derski* has laying rails. S.H.P. 25,500 = 34 kts. Oil: 255-280 tons.

*Wrangel's Army.

No photo available—see Silhouettes.
2 masts, 3 funnels.

3 boats (1913-14), named: ***Bistri,* Pospyeshni,† Pulki,†*** 1110 tons. Dimensions: 321 × 30½ × 9¼ feet. Armament: 4—4 inch, 2—3 pdr.(anti-aircraft), 5-twin 18 inch tubes. Carry 80 mines. H.P. 25,500=34 kts. Parsons turbine. 5 boilers. Oil fuel only. 351 tons.

* Engines destroyed, April 27th, 1919.

†Attached to Gen. Baron Wrangel's Army.

1 boat (1907): ***Kapitan Saken**** (ex *Pushkin*). 800 tons. Dimensions: 242½ × 27½ × 8 feet [illegible] kts. Armament: 2—4.7 inch (?? cal.), 2—4 pdr. A.A., 2 M.G., 3—18 inch tubes. Carries [illegible] Boilers. Coal: 260 tons.

*Wrangel's Army.

6 boats (*Laird type*): ***Zavidni,† Zavetni,† Jooiki,† Jarki,† Zvonki,† Zorki*** [illegible] Dimensions: 210 × 21 × 6½ feet (normal draught). *Maximum draught*: 7½ feet. H.P. 5700 = [illegible] 4 Yarrow. Carry 90 tons coal or coal + oil. Armament: 2—11 pdr., 2 M.G., 2—18 inch [illegible] *Jooni* sunk by Bolshevik Flotilla, Sea of Azov, September, 1920.

†Wrangel's Army.

2 boats (early *Yarrow type*): ***Strogi,* Svirepi**** (1901). 376 tons. Dimensions: 196 × 18½ × [illegible] = 26 kts. Boilers: Yarrow. Carry 70 tons coal. Armament: 2—11 pdr., 2 M.G., 2—18 inch [illegible] *Maximum draught*: 6¼ feet.

*Wrangel's Army.

9 Submarines (*Podvodniya Lodki.*)

No.	Type	Date	Displacement	H.P.	Max. speed	Fuel	Complement	Tubes and Gear	Max. draught
			tons		kts.	tons			
5	*AG 21–26* (H)	'18–?	[illegible]	[illegible]	[illegible]	30	...	4	15¼
2	*Burevestnik* (B)	'15–'18	[illegible]	[illegible]	[illegible]	30	32	8	12½
	Utka, only. (B)	'15–'16		[illegible]	[illegible]	40	32	12	12'
1	*Tyulen.* (B)	'11–'15	[illegible]	[illegible]	[illegible]	21	32	12	12½

(H) = Holland type. (B) = Bubnov type.

AG 22 attached to General Baron Wrangel's Fleet.

AG 23–26 being completed by Bolsheviks. Appearance as Chilean "*H*" Class.

5 *Holland* type: *AG 21–26* (were still in packing cases, at Nicolaieff, March, 1918), **AG 22** (completed), **AG 23** (launched June, 1920). Dimensions: 150¼ × 15¾ × 12½ feet. 4—18 inch bow tubes. Nlseco Diesel engines. Same type as Chilean *H* Submarines. Other details as Table.

No photo available.

2 *Improved Bubnov* type: **Burevestnik,* Utka,*** Dimensions: 223 × 14¼ × 12½ feet. Guns: *Burevestnik*, *Lebed*, 1—11 pdr.; *Utka*, 2—11 pdr., 1—1 pdr., 1 M.G.; *Pelikan*, 1—11 pdr., 1—1 pdr., 1 M.G. Have 4 tubes and 4 dropping gears (8 gears in *Utka*). B.H.P. and speeds vary, see Table. Have Diesel engines taken from Amur River Gunboats. *Lebed*, *Pelikan*, scuttled at Odessa, February, 1920. Other details as Table.

*Gen. Baron Wrangel's Forces.

No photo available.

2 *Improved Bubnov* type (*Volk* class): **Nerpa,* Tyulen†** (1913), built at Nikolaieff. Dimensions: 220 × feet. Guns: 1—11 pdr., 1—3 pdr. Tubes: 4. Dropping gears: 8. Have Diesel engines taken fr River Gunboats. *Are faster submerged than on surface.* Other details as Table.

**Nerpa*, hull only at Nicolaieff, in hands of Bolsheviks.
†*Tyulen*, General Baron Wrangel's Fleet

Convoy Vessels ?

No photo available

Information received is very contradictory, and can be summarised thus:—

Report A.—20 "Convoy Vessels" of 1200 tons each, laid down by Russian S. B. Co., at Nic 1914. Six completed 1914-17; other 14 lying unfinished on stocks.

Report B.—The six completed vessels given in Report A are probably the "Scouts" *A* *Luculic, Okotnik, Razvyedtchik, Sevastopol, Vesta,* which were at Sevastopol, January,

Report C.—"Convoy Vessels" are only Trawlers, each armed with 2—5.1 inch guns, and ger known as "the *Elpidofori* Type." Nothing known of the six ships named in Report B. are probably only small Auxiliaries.

Gunboats.

GROZNI. Re-constructed from Minelayer. Probably ex-Armed Barge *K* 15. Has 2—4.7 and 2—3 inch guns. Speed: 12 kts.

Fore and mizzen masts now removed.

TERETZ (1887), **DONETZ** (1887), **KUBANETZ** (1887). 1280-1295 tons. Complement, 135. Guns (new in 1904-06): 2—6inch (45 cal.); 1—4.7 inch (45 cal.); (2—11 pdr. in *Donetz*), 4—3 pdr., 2—1 pdr. H.P. *Donetz* 2000 = 12 kts., *Teretz* 1500 = 11 kts., *Kubanetz* 1500 = 13.5 kts. Boilers (new in 1904-06): Belleville. Coal: 220-237 tons.

Note.—All three attached to Gen. Baron Wrangel's Army

Patrol Boats.

ALBATROSS, BAKLAN (1909). 100 tons. Guns: 6—1 pdr. Speed: 9½ kts. Identical with *Minrep* class in Baltic Section.

Chaika (1910). 25 tons. 1—1 pdr.

Also: *Ledorez, Lidia, Pantikopela.*

Anti-Submarine Patrols.

16 Motor Launches.

Danube Motor Patrol Boats.

Kinjal, Shaska, Sabiya, Rapira (1910). 24 tons. 1—3 inch mountain gun, 1 M.G. B.H.P. 200 = 16 kts. Transported from Amur River during War.

Armed Barges.

Similar to British "K" type Motor Lighters used in Gallipoli Campaign.

K 12, (K 15 ?). Guns: 2—6 inch.
K 1, 3, 5, 7, 10, 11, 17. Guns: 1—6 inch.
K 2, 4, 6, 8, 14, 16, 19. Guns: 2—3 inch.
K 9. Guns: 1—3 inch.

Note.—In Sea of Azov and Don River. Believed captured by Bolsheviks, 1920.

Submarine Depot Ships.

Not finished.

ELBOROVESS. Was 30% advanced at Nicolaieff when port was abandoned to Bolsheviks, February, 1920

BERESAN (1870). 3050 tons. Guns: 6—3 pdr. or 2—1 pdr. Machinery destroyed, April 28th, 1919.

Also: *Dneprovets* and *Voz* (*Vozrozhdinia ?*).

Seaplane Carrie

No photo available.

NICOLAI. Small Ice Breaker Tug, used details known.

Another Aircraft Carrier, *Donskoi Geria*, has service, but information is not regarded

Mine Layers.

BESHTAU, DUNAI (1891). 1620 tons. Guns: 2—6 pdr., 4—3 pdr., 2 M.G. Spee 130 tons. Carry 350 mines.

Mine Sweepers.

Various Small Craft fitted out for this servic

All Ships on this page attached to General Baron Wrangel's Forces.

Auxiliary Cruiser.

v. Silhouettes

CZAREVITCH GEORG. Armed with 2—3 inch guns. In 1919 had mixed Army, Navy and female complement.

Transports.

Anton	*Margarita*
Dobycha	*Maryosh*
Dnepai	*Rion*
Julia	*Rize*
	Shilka

Ice Breakers.

No photo available.

ILYA MOUROMETZ. No details.

No photo available.

***MIKULA SALIANINOVITCH** (Vickers, 1916). tons. Dimensions: 292 × 57½ × 19¼ feet. I.H.P. 8000 = 15½ kts. Coal: 1300 tons. 2 triple expansion engines. 4 S.E. and 2 D.E. boilers. Fitted with Towing Gear.

*At Cherbourg, under management, on behalf of Wrangel's Government. Will probably proceed to Black Sea.

Ice Breakers—*Continued.*

No photo available.

***KOZMA MINIM** (Swan Hunter, 1916). tons. Dimensions: 248 × 57 × 19 feet. I.H.P. 6400 = 14½ kts. Coal 840 tons. 6 S.E. boilers. 3 sets triple expansion engines. Fitted for towing.

*At British Port, but controlled by Russian Committee on behalf of Gen. Baron Wrangel's Government.

For *Nicolai*, see preceding page.

Oil Tanker.

Bakou.

Salvage Ship.

Chersonese.

Yacht.

ALMAZ (1903). 3300 tons. Complement, 294. Dimensions: 363 × 43½ × 17¾ feet. Guns: 7—4.7 inch. Armour: 3" Deck over machinery. Designed H.P. 7500 = 19 kts. Belleville boilers. Coal: 850 tons. Carries 4 seaplanes.

Despatch Vessels.

KITOBOY. *About* 430 tons. Complement, 30. Guns: 2—11 pdr. Speed: *about* 12 kts. Taken over for Navy in Baltic, 1914. Captured by "White" officers, 1918. Engaged in Anti-Bolshevik operations in Baltic, 1918-19. Proceeded to Black Sea and joined General Baron Wrangel's Naval Forces, 1920.

(Disarmed Hulk at Sevastopol.)

KASARSKI (1889, ex Torpedo Gunboat). 400 tons. Complement, 64. Armament: 2—11 pdr., 2 M.G. Designed H.P. 3500 = 21 kts. Coal: 85 tons.

Trawlers, &c.

Albatross	*Graf Ignatieff*
Ambrosia	*Jason*
Bakkan	*Skiff*
Horaan	*Volga*
	No. 411
	No. 412

These ships have access to the Caspian Sea, provided that there is free passage to the delta of the Volga. The Bolshevist Naval Base at Stalo Terrchenya on the Caspian Sea, was destroyed during British naval operations in the summer of 1919.

Destroyers :—

Karl Liebknecht
Yakob Sverdlov
Turkman
Dostoini
Rostoropni
Dyatelni
Dyaini
Storoshevoi
Prytki
Proschni
Retyvi

Submarines :—

Yaguar
Minoga
Makrel
Okun
Katasko

Bolshevik Army Report, April 10th, 1920, mentioned an engagement fought on Caspian between the Red T.B.D. *K. Liebknecht* and "the White Cruisers *Milutin* and *Opyt*." Nothing known of these latter two Vessels; probably Armed Motor Vessels now returned to Merchant Service.

Scale : 1 inch = 160 feet.

Kars & Ardagan.

Gunboats.

Photo: "Motor Ship & Motor Boat."

ARDAGAN (1909) & **KARS** (1909). 630 tons. Armament: 1—4.7 inch, 2—4 inch, 1—3 inch. Machinery: Nobel-Lessner Diesel motors. 1000 H.P. = 14 kts.

Despatch Vessels.

Photo wanted.

ASTRABAD (1900). 325 tons. Guns: 5—3 pdr. Speed: 11 kts.

Despatch Vessels—*continued.*

Photo wanted.

GEOK TEPE (1883). 1010 tons. Guns: 4—4 pdr., 4 M.G. Speed: 11 kts.

Surveying Ship.

Photo wanted.

ARAKS (1900). 745 tons. Guns: 4—3 pdr. Designed H.P. 800 = 12½ kts.

Hospital Ship.

Skobelev.

Miscellaneous.

All Non-Bolshevist Vessels.

Asia	*Emile Nobel*	*Slava*
Australia	*Leila*	*Venture*
Caspia	*Kruger*	*Windsor Castle*
Chaseval	*Sergie*	*Zoroaster*

(Late) TURKISH FLEET. (INTERNED).

Note:

Under the Treaty of Sevres, 1920, the Turkish Fleet is to be surrendered to, and disposed of by, the Allied Governments. Hereafter, Turkey may only maintain a few small and lightly-armed vessels for Police and Coastguard Duties. No warships may be built or bought other than those required to replace vessels in permitted Establishment.

Up to November, 1920, no steps had been taken by the Allied Governments to dispose of Turkish warships beyond assigning one gunboat to Portugal. All ships are laid up at the Golden Horn, in the Bosphorus, and Gulf of Ismid under Allied supervision.

Uniforms.

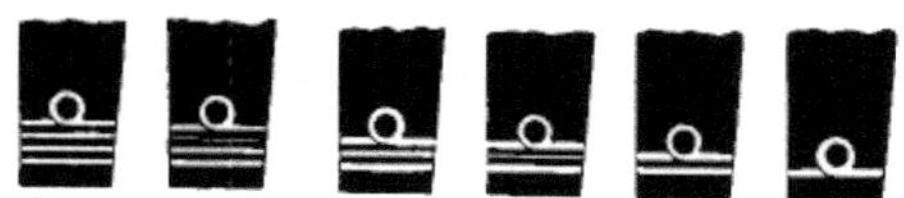

British or U.S.A.		Turkey.
Captain	=	Galion Capitani.
Commander	=	Fregate Capitani.
"	=	Corvette Capitani.
Lieut. Commander	=	Birindji Yuzbachi.
Lieutenant	=	Yuzbachi.
Sub-Lieut.	=	Mulazim.
Midshipman	=	Mehendis.

Mercantile Marine.

From "Bureau Veritas" 1919-20 figures.
Ships under 100 tons *gross* excluded.

93 Steam and Motor Ships of 72,662 tons *gross*.
2 Vessels (with Auxiliary Motors) of 248 tons *gross*.
949 Sailing Vessels of 209,624 tons *gross*.

Colour of ships: Khaki.

Principal Ports. Trebizond, Sinope, Samsoun, Djeddah.

Coinage: Lira (Medjidié) of 100 piastres = 18s. British, D4.28 U.S.A. One piastre (4 cents = *about* 2d. British. These are *nominal* values. Rate of exchange has depreciated o the War.

Tables of Distances.

(A) CONSTANTINOPLE
to
(Mediterranean Ports.)

	miles.
Alexandria	722
Alexandretta	810
Beyrout	810
Brindisi	786
Cattaro	880
Corfu	682
Gibraltar	1800
Malta	830
Mudros Bay	190
Piraeus	344
Pola (a)	996
" (b)	1098
Port Said	786
Salonika	531
Smyrna	273
Toulon	1313
Triest (a)	1050
" (b)	1152
Wilhelmshaven (c)	3355
" (d)	3893
Zeebrugge (c)	3089
" (d)	3905

(a) By Canal of Corinth.
(b) By Cape Matapan.
(c) By English Channel.
(d) By north of Scotland & west of Ireland.

(B) CONSTANTINOPLE
to
(Black Sea Ports.)

Batoum	...
Burgas	...
Nicolaieff	...
Odessa	...
Sulina (mouth of Danube River)	...
Trebizond	...
Varna	...

SMYRNA
to

Alexandretta	...
Beyrout	...
Corfu	...
Cattaro	...
Gibraltar	...
Malta	...
Mudros Bay	...
Piraeus	...
Pola (a)	...
" (b)	...
Salonika	...

(a) By Canal of Corinth.
(b) By Cape Matapan.

Scale: 1 inch = 160 feet.

RECOGNITION SILHOUETTES: LATE TURKISH NAVY.

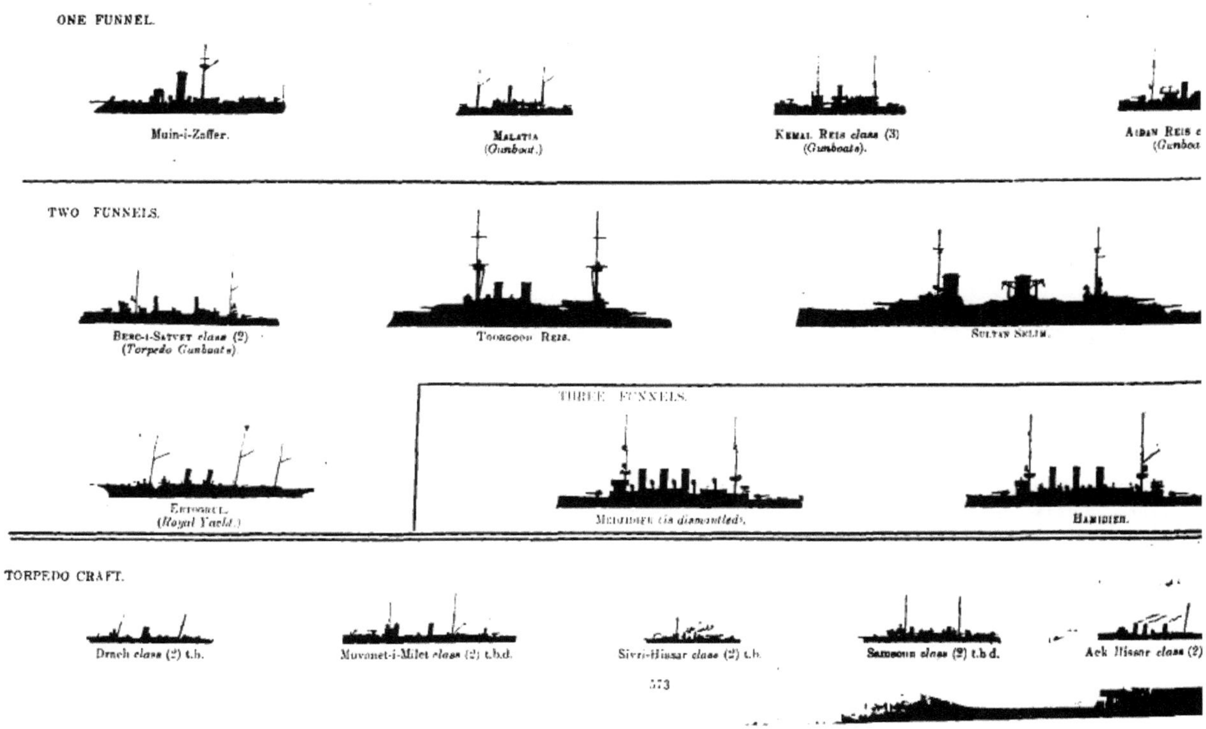

SULTAN SELIM (ex German *Goeben*, March, 1911).
Displacement, 22,640 tons. Complement, 1013.
Length (*w.l.*), 610¼ feet. Beam, 96 ft. 10 in. Draught (*max. load*), 26 ft. 11 in.

Guns (*see Notes*):
10—11 in. 50 cal. } Dir Con
12—5·9 in. 45 cal. }
12—3·4 in. 22 pdr.
2 M.G.
(1 landing)
Torpedo tubes (19·7 in.):
1 bow } all
2 broadside } *submerged*
1 stern }

Anti-Torp. Pro.:—
2"—1" deep H.T. Steel B.H., between extreme barbettes; Minute internal subdivision.

Armour (Krupp):
11" Belt (amidships) .....
5"—4" Belt (bow) .....
4" Belt (stern) .....
9"—8" Barbettes .....
10"—8" Gunhouses .....
5" Battery .....
10" Fore C.T. .....
8"—3" Double Com. Tube
8" After C.T. .....
6"—3" Com. Tube .....
"— " R.F. Towers .....
3½"—1" Decks .....
" Roofs to Gunhouses.....
3" Roof, Fore C.T. .....

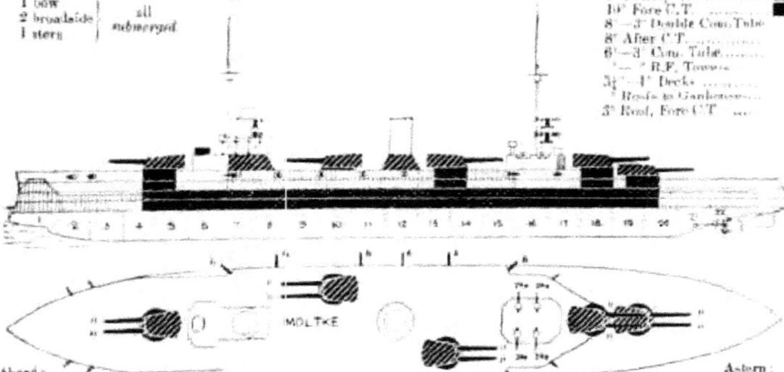

Ahead:
6—11 in.
2—5·9 in.
1—19·7 in. T.T.

Astern:
6—11 in.
2—5·9 in.
1—19·7 in. T.T.

Broadside: 10—11 in., 6—5·9 in., 2—19·7 in. T.T. (angling 30° before to 20° abaft beam).

1919

Note.—In above illustration, the fourth 5·9 inch gun from bows does not appear to be have been destroyed when this ship was mined in 1914, and never replaced. Figu on plan opposite for 5·9 inch guns are those originally mounted.

Machinery: Parsons turbine, 4 shaft, direct drive. Boilers: 24 Schulz-Thornycroft — German "Marine Type." Designed H.P. nominally 52,000 for 25 kts., but when new she could develop 70,000 H.P. = 27 kts. Coal: *normal* about 1000 tons; *maximum* 3050 tons. Oil fuel: 200 tons.

Gunnery Notes.—All guns German Navy Models, with German Navy type of fire-controls, R.F., &c. 11-inch have large degree of elevation, and range is only limited by maximum visibility. 5,9 inch range up to 16,500 yards. Two R.F. towers are sunk into the deck, near amidships (echelon) barbettes.

Torpedo Notes.—Torpedoes and torpedo tubes, German Navy type. Torpedoes probably German G VII*** (23-ft. steam heater) type with 430 lbs. charge and *max.* ranges of (*a*) 11,700 yards at 28 kts., (*b*) 5,500 yards at 35 kts. Stern *submerged* torpedo tube is on starboard quarter. 8—50" controlled S.L. Torpedo nets removed during War.

Armour and Protection Notes.—Like all other German Battle Cruisers, this ship is heavily armoured and minutely subdivided. Main belt between extreme barbettes is about 350 feet long, 11" at w.l. tapering to about 8" on upper edge, and 4" on lower edge. Barbette bases only 1" where covered by belt.

Engineering Notes.—Endurance is (*a*) 5350 miles at 10 kts., (*b*) 2,370 miles at a continuous *max.* sea-going speed of 23 kts.

Name	*Builder*	*Machinery*	*Ordered Begun*	*Began Trials*	*Trials (ma H.P. k*
Sultan Selim	Blohm & Voss, Hamburg	Blohm & Voss, Hamburg	Dec. 7, '08 April, 1909	May, 1912	71,275 = 2

*Stated to have developed 85,661 H.P. = 27.9 kts. on trials, but this must be *max.* H.P. and best ru
†Extra trials run up to Oct., 1912.

General Notes.—Built under 1909 German Navy Programme as "H." as a sister to German *Molt* Mediterranean about 1912, with Cruiser *Breslau*, it being intended that these two ships sh Austro-Hungarian naval bases. With *Breslau*, this Battle Cruiser escaped from Messina to Con first days of War, and was nominally transferred to the Turkish Navy as the *Sultan Selim*, the *I* was also becoming the pseudo-Turkish *Midilli*. On November 18th, 1914, *Goeben* fought an inde the three Russian pre-Dreadnoughts of the *Evstafi* type. During this action, or shortly after heavily injured by a double mine explosion. The first mine struck the starboard bows over t rooms, shattering the hull from the belt to the docking keel; the second mine exploded on the po bunker, disabling the third barbette. It took three months to effect repairs at Constantinople by me *Goeben* appears to have been again mined on another date during 1915 or 1916. During 1917 she by bombs dropped from a British Handley-Page aeroplane during a raid on Constantinople. *Goeben* and *Breslau* sank the British Monitors *Raglan* and *M 28* at Imbros, but both ships were sank, and *Goeben* was beached near Nagara Point, where she was repeatedly attacked by bom unsuccessful attempt was also made by submarines against her. She was towed off, and aft Sevastopol she was taken there to be docked, re-fitted, overhauled, and to have her boilers re-tu probably had more narrow escapes from destruction than any other Dreadnought or Battle Cru she certainly is a remarkable testimonial to the thorough under-water protection of German D Battle Cruisers. She is also known to the Turkish Navy by the unofficial name of *Yavuz*. In Ismid with other Turkish warships at end of War.

(*May be Depot or Special Service Ships now.*)

TOORGOOD REIS (1891, rebuilt 1903). 9961 tons. Complement, 579. Dimensions: 379¾ × 64 × 24¾ feet (*maximum* draught). Guns: 4—11 inch, 40 cal. + 2—11 inch, 35 cal. (all probably removed), 8—4.1 inch, 35 cal.; 8—3.4 inch (15 pdr.), 2 landing, 2 AA. guns, 4 M.G. Torpedo tubes (18 inch): 2 *submerged* (broadside). Armour (compound): 15¾" Belt amidships, 12" Belt at ends, 2½" Deck, flat on belt, 11¾" Barbettes, with 3" shields, 3" Battery, 12" C.T. Machinery: 2 sets vertical triple expansion. Boilers: 12 cylindrical return flame (new in 1909). Designed H.P. 9000 = 17 kts.; present speed much less. Coal: *normal*, 600 tons; *maximum*, 812 tons + 98 tons oil. Endurance: 5300 miles at 10 kts., 2600 miles at 15¼ kts. Originally was German battleship *Weissenburg*. Purchased 1910. Reported to be practically worn out.

(*Hulk—No guns.*)

MUIN-I-ZAFFER (1869). Rebuilt by Ansaldo, 1904-7. 2362 tons. Complement, 226. Guns None. Torpedo tubes (14 inch): 4. Armour (old, iron): 6"—3" Battery, 6" Battery, 5" Bulkheads. Machinery: Horizontal compound. 1 screw. Boilers (new in 1906): cylindrical. Designed H.P. 6500 = 15.5 kts. Coal (*max.*): 220 tons.

Note.—Old *Idjlalieh* still exists but is utterly useless.

HAMIDIEH (ex *Abdul Hamid*, Elswick, Sept., 1903). 3830 tons. Complement sions: 368 × 47½ × 16 feet (*mean* draught). Guns: 2—5.9 inch, 45 cal.; als on board. Torpedo tubes (18 inch): 2 *above water*. Armour: 4" Deck. Ma 4 cylinder triple expansion. 2 screws. Boilers: Cylindrical. Designed H.P kts. (*forced draught*). Coal: *normal*, 275 tons; *maximum*, 750 tons. End miles at 10 kts.

(*Dismantled; only 1 mast and 1 funnel now. Badly strained by mining. Hull and deck side bulged up.*)

MEDJIDIEH (ex Russian *Prut*, ex Turkish *Medjidieh*, launched 1903). Mined Black Sea, 1915. Salved, repaired and re-fitted at Nicolaieff for Russian Nav Seized by Austro-German Armies at Sevastopol, 1918, and returned to Turkish tons. Complement, 365. Dimensions: 225 × 40 × 17½ feet. Guns (as r Russian Navy): 4—5.1 inch (Vickers 1914),* 4 M.G. Deck 1". H.P. 12,0 Niclausse boilers, but may have another type now. Coal: 600 tons. Endur miles at 10 kts. Built by Cramps, Philadelphia.

* And [illegible]

Old Sloop.

No photo available.

ZUHAF (Germania, 1894). 632 tons. Dimensions: 180 × 24 × 11¾ feet. Guns: 4—6 pdr., 2—1 pdr. Torpedo tubes (14 inch): 2 above water. Designed H.P. 650 = 11 kts. Coal: 120 tons. Complement, 82.

Note.—Reported to have been fitted out as a Mine Layer.

Gunboats (*Gambot*)

Note.

One Gunboat, name not yet reported, [illegible] to Portuguese Navy.

Turkey may be allowed to keep a few of these vessels for Police and Coastguard Duties.

4 *St. Nazaire* type: **AIDAN REIS** (June, 1912), **BURACK REIS** (May, 1912), **SAKIZ** (Feb., 1912), and **PREVESAH** (Jan., 1912). 502 tons. Dimensions: 178½ × 27¼ × 8 feet. H.P. 1025 = 14 kts. Guns: 2—3.9 inch (50 cal.), 2—6 pdr., 4 M.G.

Gunboats—*Continued.*

3 *La Seyne* type: **KEMAL REIS** (ex *Burack Reis*, Feb., 1912), **HIZIR REIS** (Feb., 1912) and **ISSA REIS** (Dec., 1911). 413 tons. Dimensions: 154¼ × 25¾ × 4½ feet. Guns: 3—3 inch (12 pdr.), 2—3 pdr., 2 M.G. H.P. 850 = 14 kts.

1 *Crescent* type: **MALATIA** (1897). 240 tons. Dimensions: 154¼ × 19¾ × 6¼ feet. Guns: 2—4 pdr. Torpedo tubes, 1 above water. H.P. 350 = 12½ kts. Coal: 11 tons. Complement, 30.

Note.—This class originally consisted of eight vessels, viz., *Aintab*, *Baffra*, *Malatia*, *Ordu*, *Gendedim*, *Refahie*, *Tashkeupru*, *Vechihir*. Five of the class (including *Refahie*) were sunk in the Turco-Italian War, 1912. The others were destroyed in the late war.

No photo available.

BARIK-I-ZAFFER (Ansaldo [illegible] 1893). 1[illegible] tons. Dimensions: 121½ × 18 × [illegible] feet. Guns: 1—12 pdr., 2 M.G. Torpedo tubes (14 inch): 2 above water.

Note.—Various sister ships of *Barik-i-Zaffer* believed to have been destroyed during the War, but their names cannot be ascertained exactly.

Gunboats—Existence Doubtful

Yozgad and *Timsah* reported to have been seen at Co[illegible] in June, 1919. *Timsah* mined and injured be[illegible] *Yozgad*—if she exists—is utterly useless.

Torpedo Gunboats (*Torpedo Muhrib*

BERC-I-SATVET (Dec., 1906) and **PEIK-I-SHEV[illegible]** (1906). Both built at Krupps' Germania [illegible] 1014 tons. Dimensions: 262½ × 27¼ × 9½ [illegible] 2—3.0 inch, 4—6 pdr., 2—1 pdr., 2 M.G. Torpe[illegible] Designed H.P. 5100 = 22 kts. Coal: 240 tons.

Note.—*Berc-i-Satvet* torpedoed by British Submarin[illegible] *Peik-i-Shevket* by *E 14*, both during 1915. E[illegible] both boats) beached and salved. They have no[illegible] as was believed during the War.

Best available scan

6 **Destroyers** *(Torpedo Muhriby).*

No.	Type	Date	Displacement	H.P.	Max speed	Fuel Coal/Oil.	Complement	T. tubes	Max draught
			tons		kts.	tons			feet
2	*Muvanet-i-Milet* (S)	'09-'10	687	10,250*t*	33	145/30	87	3	10½
3	*Samsoun*	'07-'10	290	6800	29	70	67	2	10
1	*Berk-Efaan* (K)	1894	270	3400	25	75	..	2	9

(S) = Schichau. (…) = Chantiers de la Loire. (K) = Krupp.

2 *Schichau boats*: **Muvanet-i-Milet** and **Noumoune-i-Hamiyet.** Both 1909. 687 tons. Dimensions: 233½ × 25¼ × 10½ feet. Armament: 2—3·4 in. (22 pdr.), 2 machine, 3—18 in. tubes. Schulz-Thornycroft boilers and Schichau (Melms and Pfenninger) turbines. On trials from 36·2 to 34·5 kts., but can only make about 27 now. Endurance: 1200 miles at 15 kts. Were originally built for the German Navy as *S 165—S 168*, and sold to Turkey in 1910. Another boat (called *Mahabet-i-Milet* or *Gairet-i-Vataniye*), both sunk during the War.

Note.—Hull of a third boat, *Yadighiar-i-Milet* (mined during War), lies at Pera, but is beyond repair.

Destroyers—*(continued).*

3 *French type boats.* **Samsoun, Bassra,** both launched at Bordeaux (1907), and **Tashoz** launched by … (1907). 290 tons. Dimensions: 185 × 21 × 9½ feet. Armament: 1—12 pdr., 6—3 pdr., 2—18 inch tubes … 975 miles at 15 kts. Are practically the same as French *Durandal* class.

(No photo available.)

(A view of this boat's model was published in "*Fighting Ships*" up to 1910.)

1 *Krupp boat*: **Berk-Efsan** (1894). 270 tons. Dimensions: 197 × 21½ × 8 feet. Armament: 6—6 pdr., 2—…

Best available scan

6 Torpedo Boats.

No.	Type	Date	Displacement	H.P.	Max. speed	Fuel	Complement	T tubes	Max. draught
			tons		kts.	tons			feet
2	*Sivri-Hissar* (C)	'06-'07	98	1900	26	15 coal	18	3	5
2	*Drach* (A)	'04-'07	160	2700	24	60 "	20	2	7
2	*Younous* (A)	1901	145	2400	27	60 "	20	2	6

(C) = Creusot. (A) = Ansaldo.

2 Creusot boats: ***Sultan-Hissar, Sivri-Hissar*** (1906). Dimensions: 129 × 14 × 4½ feet. Armament: 2—1 pdr., 3—18 inch tubes. (1 bow, 2 deck). For other details v. Table.

2 Ansaldo boats: ***Drach*** (1905) and ***Moussoul*** (1905). Dimensions: 167 × 18 × 4½ feet. Armament: 2—1 pdr. 2—14 in. tubes. For other details v. Table.

Torpedo Boats—(*continued*).

2 Ansaldo boats: ***Ack-Hissar*** (1904) and ***Younous*** (1902). Dimensions: 165½ × 18½ × 5¼ fe[et]. 2—1 pdr., 2—18 inch tubes. Endurance: 2000 miles at 14 kts. For other details v. Table.

Turkey is absolutely forbidden, under the Treaty of Peace, from buying, building, or otherwi[se] Submarines of any description—naval or mercantile.

9 Motor Launches.

No photo available.

4 boats of 15 kts. speed, built about 1912-13. No other details known.

Photo, Messrs. J. I. Thornycroft & Co.

5 boats: *Names or numbers not known.* Are Thornycroft type motor patrol boats, built 1911, for customs, police and coast-guard duties. Dimensions: × × 2½ feet. Guns: 2 M.G. or 1 small Q.F. 140 H.P. Thornycroft petrol motor = 11 kts. Bullet-proof steel over conning position and engines.

Note.—Originally 20 of these boats were built. Two were captured during the Turco-Italian War. One captured during the war on the Tigris river and served for a time as H.M.S. *Flycatcher*, but was transferred later to British Army.

Depot Ships.

All have been (or will be) reduced to commercial service. The same course is to be followed with Fleet Auxiliaries, and all vessels given in next two columns may now be on Mercantile Duties.

Mine-Layers.

Photo Wanted.

NUSRAT (Germania, 1912). 364 tons. Dimensions: 131¼ × 24½ × 8¼ feet. Guns: 2 M.G. H.P. 1200 = 15 kts. Carries 25 mines.

KIRESUND (1877). 3056 tons *gross*. 10 kts.

INTIBAH. No details known.

Mine Sweepers.

No photo available

CASTOR, POLLUX. No details known.

Yachts.

(Royal Yacht.)

ERTOGRUL (Elswick, 1903). 900 tons. Dimensions: 260 (*p.p.*) × 27½ × 11¼ feet. Guns: 8—3 pdr. H.P. 2500 = 21 kts. Machinery: Hawthorn Leslie. Boilers: Cylindrical.

(There is also another small Royal Yacht, **Senguediu** built at Elswick in 1903, of 180 tons, and 14 kts. speed.)

Yachts—*continued.*

IZZEDIN (——, 1865). 1075 tons. Guns: H.P. 350 = 12 kts.

STAMBOUL (——, 1865, 1888). 910 ton inch, 2 M.G. H.P. 350 = 17 kts. *nominal.* less now.

GALATA }
SUQUTLI } No details known.

Another small yacht, *Izziah*, exists.

www.ingramcontent.com/pod-product-compliance
Ingram Content Group UK Ltd.
Pitfield, Milton Keynes, MK11 3LW, UK
UKHW050953290926
15470UKWH00006B/1793

9 781608 882342